SUPERHYDROPHOBIC SURFACES

SUPERHYDROPHOBIC SURFACES

RUSSELL J. CRAWFORD
School of Science, Swinburne University of Technology, Hawthorn, Victoria, Australia

ELENA P. IVANOVA
School of Science, Swinburne University of Technology, Hawthorn, Victoria, Australia

ELSEVIER

AMSTERDAM • BOSTON • HEIDELBERG • LONDON
NEW YORK • OXFORD • PARIS • SAN DIEGO
SAN FRANCISCO • SINGAPORE • SYDNEY • TOKYO

Elsevier
Radarweg 29, PO Box 211, 1000 AE Amsterdam, Netherlands
The Boulevard, Langford Lane, Kidlington, Oxford OX5 1GB, UK
225 Wyman Street, Waltham, MA 02451, USA

Notices

Knowledge and best practice in this field are constantly changing. As new research and experience broaden our understanding, changes in research methods, professional practices, or medical treatment may become necessary.

Practitioners and researchers must always rely on their own experience and knowledge in evaluating and using any information, methods, compounds, or experiments described herein. In using such information or methods they should be mindful of their own safety and the safety of others, including parties for whom they have a professional responsibility.

To the fullest extent of the law, neither the Publisher nor the authors, contributors, or editors, assume any liability for any injury and/or damage to persons or property as a matter of products liability, negligence or otherwise, or from any use or operation of any methods, products, instructions, or ideas contained in the material herein.

ISBN: 978-0-12-801109-6

British Library Cataloguing in Publication Data
A catalogue record for this book is available from the British Library

Library of Congress Cataloging-in-Publication Data
A catalog record for this book is available from the Library of Congress

CONTENTS

CONTRIBUTORS

Kateryna Bazaka
Chapter 5, College of Science, Technology and Engineering, James Cook University, Townsville QLD, Australia

Olga Bazaka
Chapter 5, College of Science, Technology and Engineering, James Cook University, Townsville QLD, Australia

Chris M. Bhadra
Chapter 3, School of Science, Swinburne University of Technology, Hawthorn, Victoria, Australia

Boris N. Chichkov
Chapter 6, Institut für Quantenoptik, Leibniz Universität Hannover and Laser Zentrum Hannover e.V.

Russell J. Crawford
Chapters 2, 3 and 8, School of Science, Swinburne University of Technology, Hawthorn, Victoria, Australia

Elena Fadeeva
Chapter 6, Institut für Quantenoptik, Leibniz Universität Hannover and Laser Zentrum Hannover e.V.

Gediminas Gervinskas
Chapter 7, Australian Synchrotron, 800 Blackburn Rd, Clayton, Australia

Elena P. Ivanova
Chapters 2, 3 and 8, School of Science, Swinburne University of Technology, Hawthorn, Victoria, Australia

Tim S. Jakubov
Chapter 1, School of Science, Swinburne University of Technology, Hawthorn, Victoria, Australia

Saulius Juodkazis
Chapter 7, School of Science, Swinburne University of Technology, Hawthorn, Victoria, Australia

Vi Truong Khanh
Chapter 4, School of Science, Swinburne University of Technology, Hawthorn, Victoria, Australia

Jürgen Koch
Chapter 6, Institut für Quantenoptik, Leibniz Universität Hannover and Laser Zentrum Hannover e.V.

Sivashankar Krishnamoorthy
Chapter 7, Nanomaterials Unit, Science et Analyses des Materiaux, Centre Recherche Public, Gabriel Lippmann

David E. Mainwaring
Chapters 1 and 4, School of Science, Swinburne University of Technology, Hawthorn, Victoria, Australia

Pandiyan Murugaraj
Chapter 4, School of Science, Swinburne University of Technology, Hawthorn, Victoria, Australia

Song Ha Nguyen
Chapters 2 and 8, School of Science, Swinburne University of Technology, Hawthorn, Victoria, Australia

Vy T. H. Pham
Chapter 3, School of Science, Swinburne University of Technology, Hawthorn, Victoria, Australia

Hayden K. Webb
Chapters 2, 3 and 8, School of Science, Swinburne University of Technology, Hawthorn, Victoria, Australia

EDITORS BIOGRAPHIES

Russell J. Crawford is a Professor of Chemistry at Swinburne University of Technology. He obtained a Master of Science from Swinburne and a PhD from The University of Melbourne. He has held leadership positions in the university, including Dean of Science and Dean, Faculty of Life & Social Sciences. His research is in surface and colloid science, with early work focusing on mineral flotation and the removal of heavy metals from aqueous environments. His recent research focuses on understanding the ways in which biological organisms interact with solid substrate surfaces, such as those used in the construction of medical implants.

Elena P. Ivanova is a Professor at Swinburne University of Technology. She received a PhD from the Institute of Microbiology and Virology, Ukraine; a ScD from the Pacific Institute of Bio-organic Chemistry, Russian Federation; and a JD from The University of Melbourne. Before joining Swinburne University of Technology, hold postdoctoral positions at the New Energy and Industrial Technology Development Organization, Osaka, Japan and at the Center of Marine Biotechnology, University of Maryland, USA. Professional interests are concentrated on the fundamental and applied aspects of Nano/Biotechnology including planar microdevices, biomaterials, immobilization of biomolecules and microorganisms in micro/nano/ environments, bacterial interactions with micro/nanostructured surfaces.

PREFACE

The degree to which water can spread on a surface, commonly referred to as "wetting," is of great importance in many scientific, industrial, and medical processes. The degree of wettability of a surface determines the extent of interfacial contact area that can be established between a solid surface and water with which it comes into contact. This, in turn, determines the chemical or physical interactions that may occur between the two phases.

A thorough understanding of the way in which water interacts with solid surfaces provides us with the ability to control the many equilibrium processes that may take place between the phases. This occurs principally through the ability to "tune" the surface.

The term "superhydrophobicity" refers to a condition of extreme water repellency of a solid surface. A water droplet in contact with such a surface retains an almost spherical shape, with no spreading taking place, together with a minimal extent of contact between the two phases. Superhydrophobicity can lead to other interesting phenomena, most notably the ability for a surface to undergo "self-cleaning." Here, a water droplet coming into contact with such a surface can easily roll across the surface on which it rests, collecting dust or other contaminating particles through adsorption or absorption as it moves, eventually rolling off the surface and taking with it the contaminating particles. Several naturally occurring self-cleaning surfaces such as that found on the upper surface of the lotus leaf are known to exist, and considerable efforts have been made in the last 20 years to reproduce their self-cleaning properties.

This book will discuss the underlying mechanisms responsible for the condition of superhydrophobicity and the major theories that ultimately determine the wettability of a surface. Following this, a summary of the naturally existing superhydrophobic surfaces, both biotic and abiotic, will be presented. A substantial part of this book has been dedicated to describing the methods currently being used to fabricate synthetic superhydrophobic surfaces. These will be discussed, together with an in-depth description of the various synthetic superhydrophobic materials that have been produced. Finally, the implications of superhydrophobicity and the role that this condition has played in biological systems will be explored.

Russell J. Crawford and Elena P. Ivanova
Melbourne, 2015

ACKNOWLEDGEMENT

We would like to extend our appreciation and sincere thanks to those colleagues, postgraduate students, friends, and family who assisted us during the process of writing this book. Most importantly, we thank our team of expert and outstanding coauthors; without their highly valued scholarly contribution, this book could not have been written.

BC and EF kindly acknowledge financial support of interdisciplinary research consortium of Hannover Biofabrication and The German Research Foundation (DFG SFB599 "Sustaintable Bioresorbing and Permanent Implants of Metallic and Ceramic Materials" and DFG Project "Electrode optimization for neuroprostheses").

Russell J. Crawford and Elena P. Ivanova

CHAPTER ONE

Superhydrophobicity – An Introductory Review

GLOSSARY

TPL three-phase contact line
S-L-V solid-liquid-vapour phases
CMC critical micelle concentration

Superhydrophobicity as a phenomenon has become an increasing focus of research and technological activity, where its fundamental aspects span surface chemistry, chemical physics, and cellular biology. Additionally, its significance to the behavior of natural systems, interfacial fluid dynamics, and biotechnology represents an area rapidly gaining potential importance. Detailed reviews have progressively explored superhydrophobicity from a number of viewpoints (e.g., Ma and Hill, 2006; Quéré, 2002; Shirtcliffe et al., 2010). Here, aspects underlying this wetting behavior are illustrated. It has long been recognized that surface roughness has a profound effect on wetting behavior, in particular through apparent contact angles and subsequent contact angle hysteresis (Bico et al., 2001; Quéré, 2008). Quéré (2008) points out that both chemical and structural surface heterogeneity can cause pinning of the three-phase contact line (TPL) of an advancing wetting front, whereby the difference in the advancing and receding contact angles produces a Laplace pressure and hence a force resisting further liquid advancement. Movement of the wetting front (advancing and receding) can be viewed as a kinetic process in response to changing forces at the TPL that characteristically produce jumps in the movement of this line. Rough and microstructured surfaces inherently increase hydrophobicity of hydrophobic surfaces through two very different mechanisms: a purely geometrical increase in the actual surface area with respect to its projected area generally termed the Wenzel state (Wenzel, 1936) and a composite interfacial effect arising from an air–water interface when air is trapped between microstructural features of the surface ahead of the advancing wetting front forming a Cassie–Baxter state (Cassie and Baxter, 1944), as illustrated in Figure 1. As such, these conditions represent homogeneous and heterogeneous surface wetting systems, respectively, and in both cases are derived from the result of variations in the interfacial energy of the substrate phase(s) solid or solid-vapor.

Superhydrophobic Surfaces
http://dx.doi.org/10.1016/B978-0-12-801109-6.00001-X

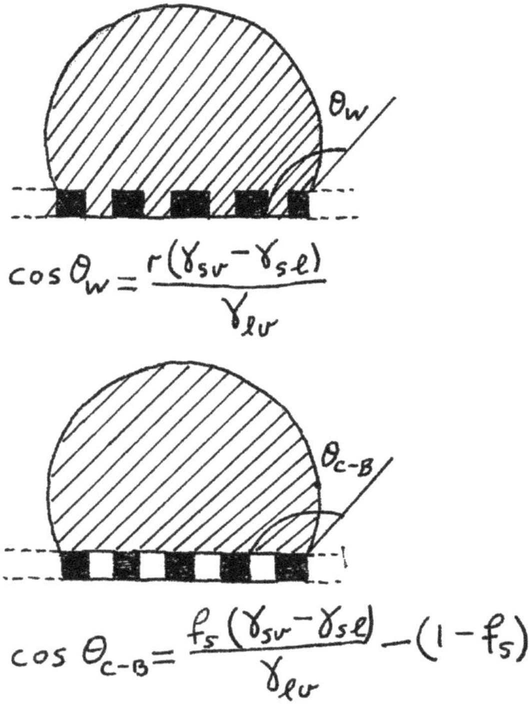

Figure 1 Contact angles on structured surfaces in the classic Wenzel (W) and Cassie–Baxter (C–B) states, respectively, where *r* is the physical amplification of surface area due to roughness and f_S is the fractional area in contact with air. *Adapted from Shirtcliffe et al. (2010).*

Marmur characterized superhydrophocity according to two criteria (Marmur, 2004), a very high contact angle and very low drop roll-off angle, as well as addressing the transition between these states in terms of their metastability and its impact on superhydrophobicity (Marmur, 2003). Since the free energy increases with increasing contact angle, the state that is most stable is represented by that with lowest contact angle. Gao and Yan (2009) and Quéré (2008) point out that due to local energy minima, a water drop cannot only assume multiple energy states but also coexist on a particular surface, and from the energy barrier for the transition Cassie–Baxter to the Wenzel state, surface geometry influences the interfacial energy requirement to reach equilibrium whether homogeneous or heterogeneous wetting is involved.

Pinning of the advancing TPL is also dependent on the topology (feature size, spacing, and shape). Shirtcliffe et al. (2010) and McHale (2007) note that this also determines the observed contact angle and consistency with the

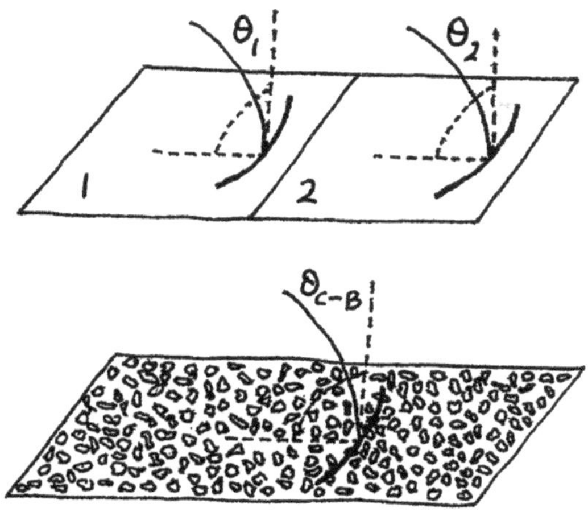

Figure 2 Advancing wetting front across homogeneous zones 1 and 2 with respective contact angles $\theta_{1,2}$ and across a microscopically random surface (1,2) with contact angle θ_{C-B}.

Wenzel and Cassie–Baxter models since the incremental advancing area is assumed to characterize the surface overall. And, as a result, features on either side of the TPL do not influence the contact angle over the increment but must be of sufficiently small size to ensure that an average of the overall topology is sampled by the TPL at any one time, as shown by McHale (2007) and Shirtcliffe et al. (2010) for a random surface, illustrated in Figure 2.

Interestingly, these studies also illustrate that a systematic variation in topological pattern within the bounds of a TPL yields a driving force for liquid transport not reliant on capillary suction (Figure 3), which may have implications for microfluidic device design.

Transition between a Cassie–Baxter suspended state and complete surface immersional Wenzel wetting readily occurs through the imposition of external forces whether mechanical, hydrostatic, or compositional or through the size of the drop and its internal Laplace pressure (Ishino et al., 2004; Lafuma and Quéré, 2003). The enhancement of superhydrophocity, through a hierarchical surface roughness where a nanoscale topology is superimposed on surface microscale roughness, has now been widely studied. In many cases, this has been related to fractal behavior (Bottiglione and Carbone, 2013). Nosonovsky (2007), in investigating the instability of the Cassie–Baxter wetting state and its transformation to Wenzel wetting, established hierarchical roughness as a criterion enhancing the composite wetting state, which was then shown to apply to both natural and synthetic

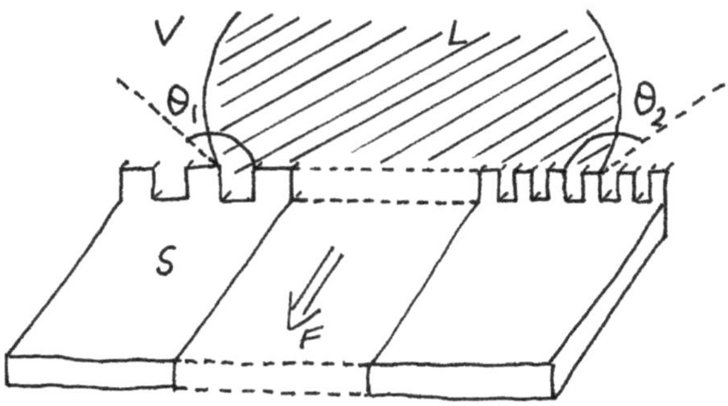

Figure 3 Influence of a variation in surface topology producing an additional wetting force and fluid flow. *Adapted from Shirtcliffe et al. (2010).*

surfaces and suggested as a general requirement when other destabilizing scale-dependent effects are present such as in microprocesses. Noting the transition to a stable Cassie state under imposed vibration on the Lotus leaf surface (Boreyko and Chen, 2009), Boreyko and Collier (2013) then examined the obverse situation where drops in the Wenzel condition can assume the Cassie–Baxter state through a dewetting transition. Here, they show that pinned Wenzel drops cannot be totally stretched during a transition to reduce their contact angle below a critical curvature since pinch-off of the drop surface occurs, resulting in a partial liquid Wenzel droplet remaining on the surface. They thereby provide a condition for a Wenzel dewetting where the receding contact angle needs to be moderately large (~90°).

Contact angle and movement at the TPL (wetting or dewetting) have been noted as a response to forces at the intersection of interfaces between the three principal phases (S-L-V). Within this line of contact, where the feature size of surface topology exerts influence, an additional line tension (Rusanov et al., 2004) exists akin to a "one dimensional surface tension" that may influence the contact angle and hence the Wenzel and Cassie–Baxter models and their respective transitions, if the droplet and feature size are small. Amirfazli and Neumann (2004) point out, in a detailed analysis of the status of line tension, that line tension influences contact angle through a modified Young equation (Figures 1 and 4) where κ represents the geodesic curvature of the TPL.

Amirfazli and Neumann also point out the significance of line tension in microbial and cell–surface systems where contact is of the order of 100 nm and where line tension can ultimately influence contact and interactions (Ellwood and Pethica, 1980). Line tension was shown to have a significant impact on the wetting behavior of nanoporous materials by Raspal et al. (2012)

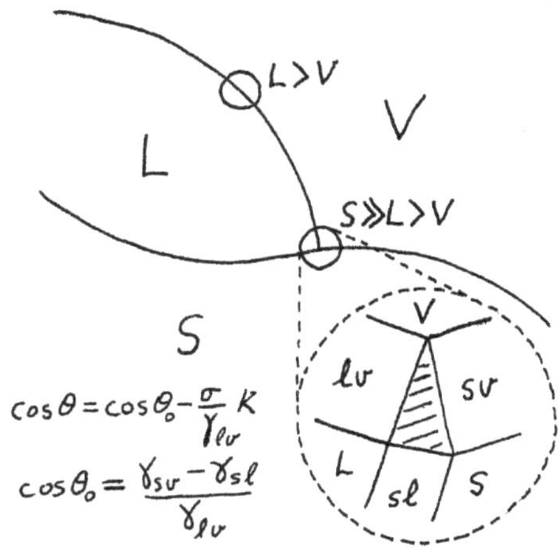

Figure 4 Interfaces (*sl, lv, sv*) existing at a point on the TPL of a droplet wetting front and the relative magnitude of the intermolecular forces at the interfaces, together with the contribution of line tension σ along the TPL of curvature κ. Shaded area represents an S-L-V interphase zone. *Adapted from Amirfazli and Neumann (2004).*

due the relatively high contribution of the total TPL within the nanopores compared with the droplet wetting perimeter, which was evaluated by the Wenzel (full pore penetration) and Cassie–Baxter (no penetration) models as well as a modified Laplace (partial penetration) model. Microhydrodynamic studies confirm that structured superhydrophobic surfaces can, when maintained in the Cassie–Baxter state, reduce pressure gradients for flow up to 40% corresponding to slip-lengths where the usual parabolic liquid velocity profile is zero over up to ~10 μm from the surface rather than the usual "no slip condition" zero at the surface (Ou and Rothstein, 2005; Quéré, 2008; Roach et al., 2007). Finally, the wetting behavior of superhydrophobic surfaces in the presence of surfactants above and below their critical micelle concentration (CMC), nonaqueous liquids, and immiscible phases (Ferrari and Ravera, 2010) is beginning to open further regimes for study and application. Here, the contact angles, hysteresis, and transition between wetting states are shown to be highly dependent on the nature of the adsorbed state and its interactions at the molecular level.

Hydrophobic behavior has a rich history stretching from the myth of Jason returning from Colchis with the Golden Fleece, said to be placed in streams to gather fine alluvial gold in its oily fibers, to Benjamin Franklin spreading oils on Clapham Common ponds in 1765 to form monolayers capable of reducing surface turbulence. With the advent of nanoscale materials and technology, surface architectures at the microscale and nanoscale have extended its influence to superhydrophobicity. In the chapters of this monograph, the breadth of this behavior is surveyed where it spans—natural surfaces and biological interfaces, their design and fabrication, specific metallic and polymer superhydrophobic surfaces, and its impact on soils and their productivity—and set in the context of the very many research articles and detailed reviews available on the topic.

REFERENCES

Amirfazli, A., Neumann, A.W., 2004. Status of the three-phase line tension: a review. Adv. Colloid Interface Sci. 110, 121.
Bico, J., Tordeux, C., Quéré, D., 2001. Rough wetting. Europhys. Lett. 55, 214.
Boreyko, J.B., Chen, C.H., 2009. Restoring superhydrophobicity of lotus leaves with vibration-induced dewetting. Phys. Rev. Lett. 103, 174502.
Boreyko, J.B., Collier, C.P., 2013. Dewetting transitions on superhydrophobic surfaces: when are wenzel drops reversible? J. Phys. Chem. C 117, 18084.
Bottiglione, F., Carbone, G., 2013. Role of statistical properties of randomly rough surfaces in controlling superhydrophobicity. Langmuir 29, 599.
Cassie, A.B.D., Baxter, S., 1944. Wettability of porous surfaces. Trans. Faraday Soc. 40, 546.
Ellwood, D.C., Pethica, B.A., 1980. Microbial Adhesion to Surfaces. John Wiley & Sons, Canada.
Ferrari, M., Ravera, F., 2010. Surfactants and wetting at superhydrophobic surfaces: water solutions and non aqueous liquids. Adv. Colloid Interface Sci. 161, 22.
Gao, N., Yan, Y., 2009. Modeling superhydrophobic contact angles and wetting transition. J. Bionic Eng. 6, 335.
Ishino, C., Okumura, K., Quéré, D., 2004. Wetting transitions on rough surfaces. Europhys. Lett. 68, 419.
Lafuma, A., Quéré, D., 2003. Superhydrophobic states. Nat. Mater. 2, 457.
Ma, M., Hill, R.M., 2006. Superhydrophobic surfaces. Curr. Opin. Colloid Interface Sci. 11, 193.
Marmur, A., 2003. Wetting on hydrophobic rough surfaces: to be heterogeneous or not to be? Langmuir 19, 8343.
Marmur, A., 2004. The lotus effect: superhydrophobicity and metastability. Langmuir 20, 3517.
McHale, G., 2007. Cassie and wenzel: were they really so wrong? Langmuir 23, 8200.
Nosonovsky, M., 2007. Multiscale roughness and stability of superhydrophobic biomimetic interfaces. Langmuir 23, 3157.
Ou, J., Rothstein, J.P., 2005. Direct velocity measurements of the flow past drag-reducing ultrahydrophobic surfaces. Phys. Fluids 17, 103606.
Quéré, D., 2002. Surface chemistry: fakir droplets. Nat. Mater. 1, 14.
Quéré, D., 2008. Wetting and roughness. In: Clarke, D.R., Ruehle, M., Tomsia, A.P. (Eds.), Annual Review of Materials Research, p. 71.
Raspal, V., Awitor, K.O., Massard, C., Feschet-Chassot, E., Bokalawela, R.S.P., Johnson, M.B., 2012. Nanoporous surface wetting behavior: the line tension influence. Langmuir 28, 11064.
Roach, P., McHale, G., Evans, C.R., Shirtcliffe, N.J., Newton, M.I., 2007. Decoupling of the liquid response of a superhydrophobic quartz crystal microbalance. Langmuir 23, 9823.
Rusanov, A.I., Shchekin, A.K., Tatyanenko, D.V., 2004. The line tension and the generalized young equation: the choice of dividing surface. Colloids Surf. A Physicochem. Eng. Asp. 250, 263.
Shirtcliffe, N.J., McHale, G., Atherton, S., Newton, M.I., 2010. An introduction to superhydrophobicity. Adv. Colloid Interface Sci. 161, 124.
Wenzel, R.N., 1936. Resistance of solid surface to wetting by water. Ind. Eng. Chem. 28, 988.

CHAPTER TWO

Natural Superhydrophobic Surfaces

GLOSSARY

CA Contact angle
CAH Contact angle hysteresis
RO Radial outward
Superoleophobicity The state of high repellency of oils
WCA Water contact angle

INTRODUCTION

The term "waterproofing" was first developed in the 1900s, with waterproofing agents usually being applied to lightweight woven or knit fabrics (Marangoni, 1871; Bartell and Bartell, 1934; Wenzel, 1936; Cassie and Baxter, 1944). Wenzel and Cassie and Baxter were the first to develop the wettability theories that are currently being widely applied (Wenzel, 1936; Cassie and Baxter, 1944). While Wenzel's theory explains the relationship between surface hydrophobicity and surface roughness (Wenzel, 1936), it is the theory developed by Cassie and Baxter (1944) that is most often applicable for superhydrophobic surfaces. In the same year that Cassie and Baxter published their wettability model, the observation of a natural surface that possessed water contact angles (WCAs) above 150°, the *Triticum* plant, was first reported (Fogg, 1944). It was only after Barthlott and Neinhuis introduced the archetype "lotus effect" in 1997 that naturally occurring surfaces possessing superhydrophobic properties began to draw attention from both scientists and industry (Barthlott and Neinhuis, 1997).

Superhydrophobicity arises from the combination of hierarchical surface structures and low surface energy materials. When superhydrophobicity is combined with the ability to self-clean, we witness surfaces that can perform multiple functions (Jung and Bhushan, 2006; Bhushan et al., 2009; Yan et al., 2011). To achieve these properties, the presence of multiscale or hierarchical structures on naturally hydrophobic materials is required (Nosonovsky and Bhushan, 2005; Verho et al., 2012; Papadopoulos et al., 2013). Nosonovsky and Bhushan determined that a composite interface

Superhydrophobic Surfaces
http://dx.doi.org/10.1016/B978-0-12-801109-6.00002-1

containing a large proportion of trapped air needs to be present in order to maintain superhydrophobicity, which correlates very closely with the requirements of the Cassie–Baxter theory for the condition of superhydrophobicity (Neinhuis and Barthlott, 1997; Nosonovsky and Bhushan, 2007a,b).

SELF-CLEANING PROPERTIES ARISING FROM HIERARCHICAL STRUCTURES

Insects

Insect wings have attracted a great deal of attention by researchers due to the sophisticated structures that exist on their surfaces. These serve to afford the wings unique properties, such as anti-fogging and anti-wetting behavior. These surfaces have also been the source of a great deal of valuable information regarding the mechanisms responsible for superhydrophobicity and self-cleaning behavior. In the past few years, insects such as the butterfly, cicada, water strider, dragonfly, and damselfly have been recognized as insects that possess these two unique properties. Their outermost layer, the cuticle, is the barrier that directly interacts with environment. This layer has been optimized through millions of years of evolution to protect the insects from ever-changing environments. The cuticle is secreted by a single layer of epidermal cells, forming a lipophilic structure (Lockey, 1980; Lockey, 1985; Nelson and Blomquist, 1995; Buschhaus et al., 2007; Jetter and Kunst, 2008; Samuels et al., 2008; Buckner, 2010; Moussian, 2010). This structure consists of two major components, which can be distinguished by their solubility in organic solvents (Buschhaus and Jetter, 2011). The outermost layer of the cuticle, the epicuticle, is composed of a mixture of aliphatic hydrocarbons and their derivatives; these compounds contain one or more oxygen functional groups, including esters, ketones, alcohols, aldehydes, and fatty acids (Koch and Ensikat, 2008; Samuels et al., 2008). The intracuticular layer, located beneath the epicuticle, is a mixture of chitin (poly-*N*-acetylglucosamine) and protein (Lockey, 1980, 1985, 1988). This mixture of organic components is self-organized in the epicuticular layer of the cuticle, producing either a smooth two-dimensional (2D) wax film or a three-dimensional (3D) wax crystal structure. These structures have afforded the organisms the ability to adapt to their environmental living conditions.

The surface of the *Hemianax papuensis* dragonfly wing is covered by an array of nanopillars that form a fractal structure (Figure 1). This structure enables the surface to retain pockets of air when it comes into contact

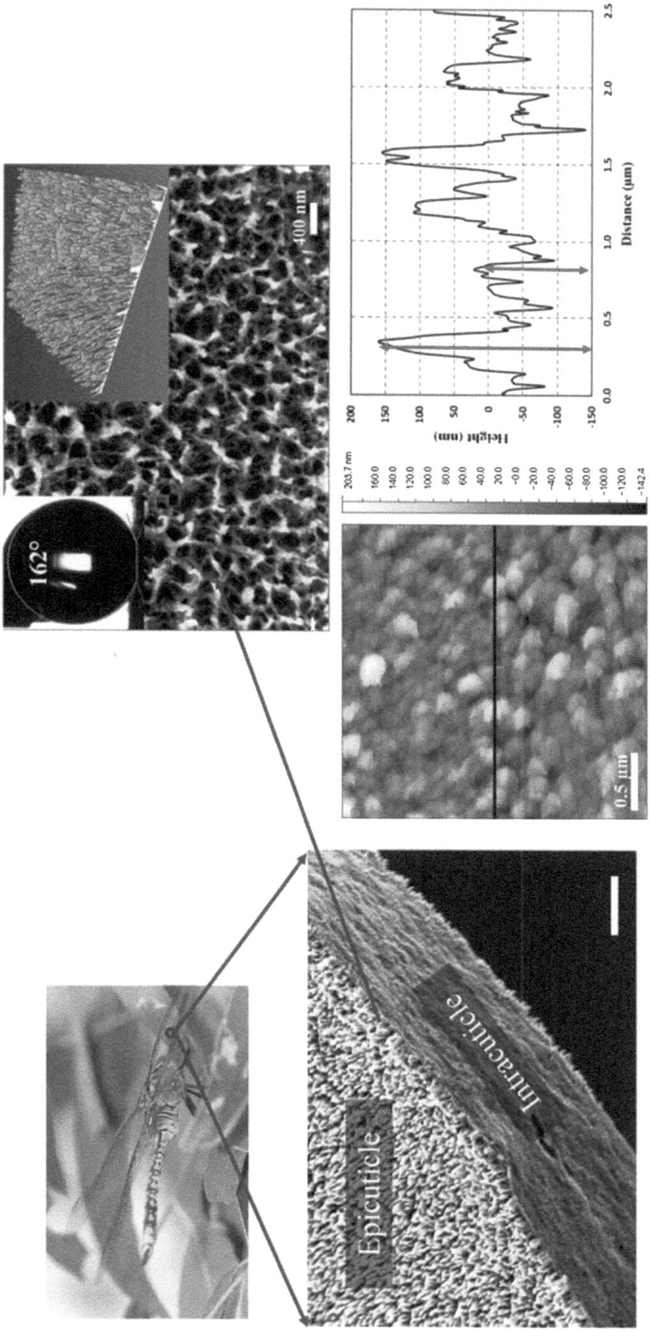

Figure 1 The surface morphologies of the *Hemianax papuensis* dragonfly wing.

with water, producing a high WCA. In the case of the *H. papuensis* wing surface, contact angles (CAs) can be as high as 161°. Moreover, low contact angle hysteresis (CAH), and water droplet sliding angles less than 10° (approximately 6°), indicating the ability to self-clean, are observed (Ivanova et al., 2013b). These self-cleaning properties can be further accelerated as the result of the wings being moved in turbulent conditions during flight (Nishimoto and Bhushan, 2013).

To date, there is no universally adopted system for the description of the surface morphologies of insect wings. Byun et al. used the terms "layered cuticle," "setae," "denticles," and "fractal" to describe the morphological features present on the surfaces of the insect wings in their study, and a majority of the insect surfaces characterized to date could be readily classified according to this system. The term "layered cuticle" generally refers to a surface that contains scale-type structures that overlap, such as those typically found on butterfly wings. Surfaces with "setae" contain high aspect ratio needles or hairs. "Denticle" structures refer to tooth-like projections, and these can vary in their morphology, ranging from small hemispheres to taller pillars. "Fractal" structures are composed of an irregular array of fine nanoscale protrusions (Byun et al., 2009). Among these structural types, the presence of layered cuticles, denticles, and fractal structures appears to result in the production of the most superhydrophobic surfaces, whilst the presence of setae alone on a surface does not induce superhydrophobic properties (Table 1).

The surface topologies of the wings of several insects are presented in Figure 2. The wings of the *Nasutitermes* sp., which belongs to the order Isoptera, exhibit superhydrophobic properties (WCA = 180°, Figure 2(A)) (Watson et al., 2010). The surface of their wings is covered by microscale setae together with a star-shaped nanoscale structure (micraster), which produces a combined hierarchical structure that exhibits superhydrophobic properties. Similarly, the surface of the *Acrida cinerea* grasshopper wing (WCA = 151°) is covered by microdenticles that are enclosed by nanoscale wax crystals (Figure 2(B)). High-resolution scanning electron microscopic images indicate the presence of a binary structure (containing micro and nano features), which results in the wings possessing superhydrophobic properties. The *Meimuna opalifera* (Walker) cicada wings also exhibit superhydrophobicity (WCA = 165°) despite not having a hierarchical surface structure (Figure 2(C)) (Byun et al., 2009). The surface is covered with a layer of dense nanopillars that result in a high degree of air entrapment on the surface. It is noteworthy that not all cicada wings surfaces exhibit the

Table 1 Microscale and Nanoscale Wax Crystal Morphologies on the Epidermal Cells of insect Wing Surfaces and Their Water Contact Angles[a]

Order	Species	Structural Morphology	Water Contact Angles (°)
Coleoptera	*Allomyrina dichotoma*	Setae	54
Coleoptera	*Chrysolina virgata*	Setae	71
Isoptera	*Schedorhinotermes* sp.	Setae	71
Coleoptera	*Amphizoa sinica*	Setae	109
Hymenoptera	*Vespa simillima xanthoptera*	Setae	121
Hymenoptera	*Vespa dybowskii*	Setae	126
Hemiptera	*Meimuna microdon*	Denticle	140
Orthoptera	*Atractomorpha lata*	Denticle	148
Orthoptera	*Acrida cinerea cinerea*	Denticle	151
Odonata	*Hemicordulia tau*	Fractal	157
Odonata	*Hemianax papuensis*	Fractal	161
Lepidoptera	*Artogeia canidia*	Layered cuticle	162
Lepidoptera	*Papilio xuthus*	Layered cuticle	168

[a]This table was modified and updated from Byun et al. (2009). Terminology regarding structural morphology was also adopted from Byun et al. (2009).

same properties; *Hemiptera Meimuna microdon* cicada wings are also covered by a layer of denticle structures but exhibit a WCA of only 140° (Table 1). In this case, the density, size, shape, and composition of the denticle clearly significantly contributes to the extent of surface hydrophobicity. It is clear that the presence of a surface topographical hierarchy is an important factor in imparting superhydrophobicity to a surface, but it is also clear that other factors play important roles. According to the Cassie–Baxter theory, the two important components to determine the hydrophobicity of a surface are the surface chemistry of the materials composing the surface and the ability of the surface to entrap air, this ability being a function of the roughness and topology of the surface. The presence of a hierarchical surface clearly assists in the entrapment of air; however, this is also a function of the density, size, and shape of the hierarchical roughness features, as seen for the cicada wing surface shown in Figure 2(C).

The superhydrophobicity of an insect wing surface and its ability to self-clean are very important factors that contribute to an insect's ability to survive. The nanoarray structures present on the surfaces of some insect wings such as those of the cicada and dragonfly afford the insect antireflective properties, which can assist in protecting them from attack from predators (Watson et al., 2008). The superhydrophobic and self-cleaning properties can assist to keep their surfaces clean and free from contaminants that may

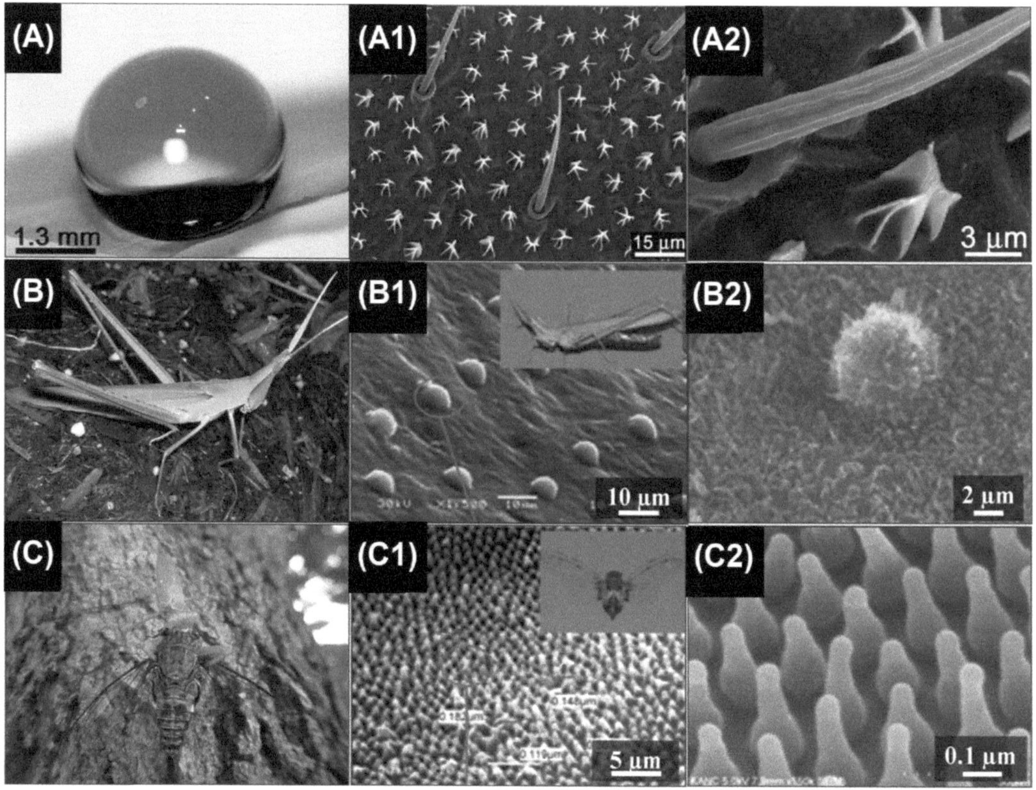

Figure 2 Insect wings and their corresponding high-resolution scanning electron microscopic images: (A-A2) Isoptera *Nasutitermes* sp.; (B-B2) Orthoptera *Acrida cinerea cinerea*; (C-C2) Hemiptera *Meimuna opalifera*. *Watson et al. (2010), Byun et al. (2009).*

also adversely impact their antireflective properties. Moreover, dragonfly wings were recently reported to possess antibactericidal properties (Ivanova et al., 2012, 2013a). The mechanism by which this occurs is unclear; however, it was postulated that this bactericidal behavior was a function of the wettability of the surface (Pogodin et al., 2013).

Plants

Since Barthlott and Neinhuis first reported the "lotus effect," the lotus has become the archetype surface for superhydrophobicity and self-cleaning abilities (Barthlott and Neinhuis, 1997). For centuries the lotus has been known as symbol of purity in many Asian religions. For over a decade, this purity has become a phenomenon within the scientific community, with many studies having been performed with a view to determining the mechanism responsible for their self-cleaning ability.

As is the case with insect surfaces, plant surfaces are cover by a cuticular surface layer. Depending on the plant species, this layer is secreted by the epidermal cells either throughout the lifetime of the plant, at certain stages of their development, or on a single occasion only. The cuticular layer consists of two compartments: a layer of epicuticlar wax crystals and a "cutin" layer. These two layers can be considered to be analogous to the epicuticle and intracuticle present on the surfaces of insects. The composition of the plant epicuticular waxes is very similar to that found on insects, while "cutin" is a composed of a polymer of predominantly ω- and mid-chain hydroxyl and epoxy C_{16} and C_{18} fatty acids in addition to glycerol (Samuels et al., 2008). Some examples of plants with superhydrophobic surfaces are presented in Figure 3. The upper surface of the lotus leaf is covered by a layer of microscale papillae, which is, in turn, covered by an array of nanoscale asperities (Figure 3(A)). These nanoscale asperities are composed of epicuticular hydrophobic wax tubules (Nosonovsky and Bhushan, 2007b). Water droplets that land on the top surface of a lotus leaf remains almost spherical and can roll off the surface readily with a very low tilting angle; their WCA and CAH are 164° and 3°, respectively (Koch et al., 2009).

Other plant leaf surfaces that exhibit superhydrophobic and self-cleaning properties include the India canna (Figure 3(B)), Taro (Figure 3(C)), and perfoliate knotweed (Figure 3(D)). India canna leaves are covered by many wax platelets, distributed randomly on a series of rod-like structures. This increases the proportion of air that can be trapped within the surface, producing WCAs in excess of the 150° CA condition for superhydrophobicity (i.e., 165°). The surface of Taro leaves (Figure 3(C)) is composed of a layer

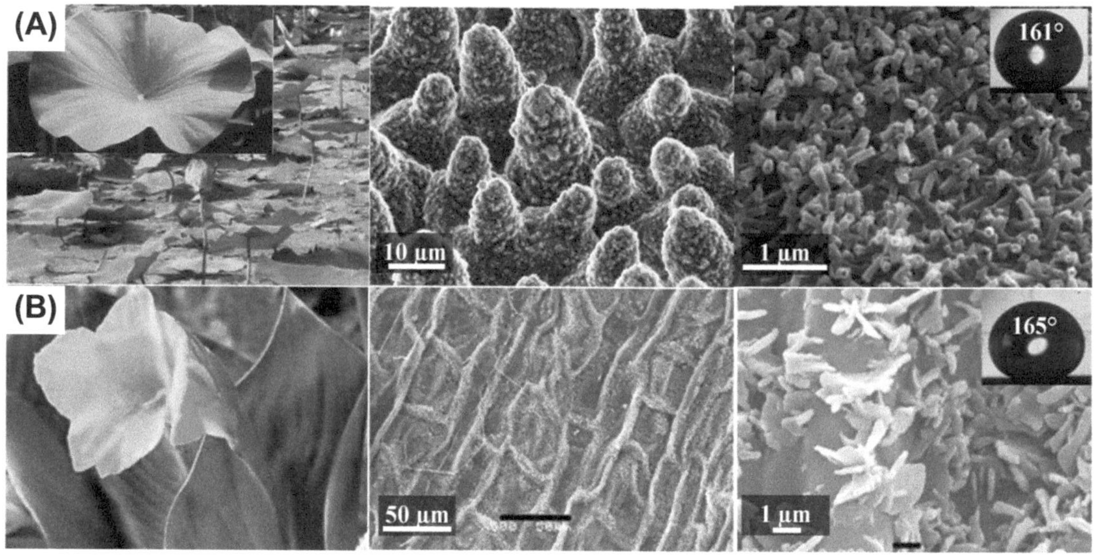

Figure 3 Images of some superhydrophobic plant surfaces and their corresponding high-resolution scanning electron microscopic images: (A) lotus leaves; (B) India canna leaves; (C) Taro leaves; (D) frontal face of perfoliate knotweed leaves. *Guo and Liu (2007).*

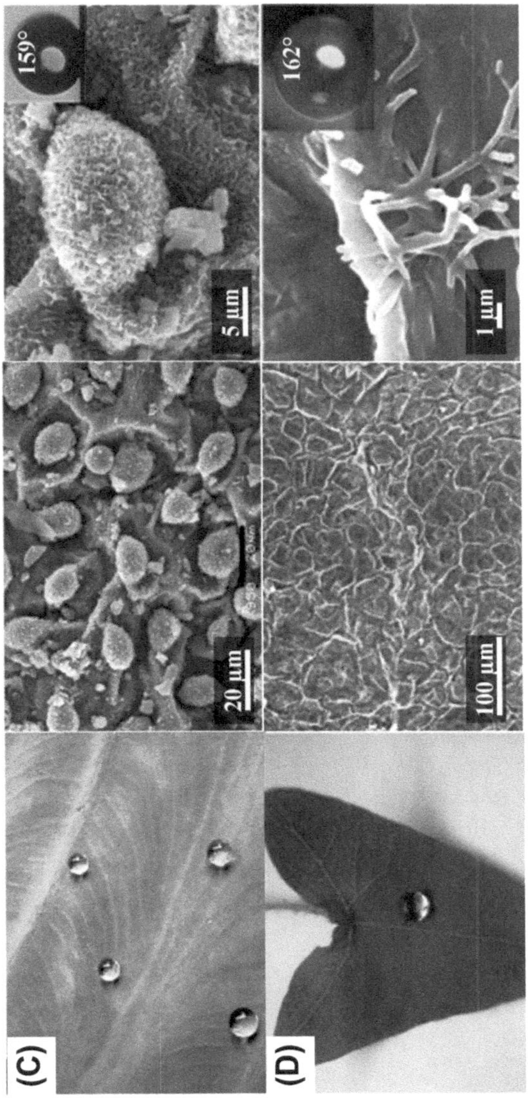

Figure 3 Cont'd

of elliptic protrusions, which are uniformly arranged on microscale cave-like structures, resulting in a WCA of 159° and CAH of 3°. Similar surface hydrophobicity characteristics were observed for perfoliate knotweed leaves (Guo and Liu, 2007).

The examples presented in Figure 3 are just a few of the plants that have been identified as possessing superhydrophobic surfaces. Neinhuis and Barthlott reported the static WCA of 200 water-repellent plant species (Neinhuis and Barthlott, 1997). Most of these plants were classified as having superhydrophobic surfaces, as they exhibited a WCA greater than 150°. The common feature shared by these surfaces is that each of them possesses a very dense layer of 3D cuticular wax crystals arranged randomly or uniformly on their corresponding microscale surface features (papillae).

Certain wax types found on plant surfaces, together with their orientation, are species specific. Therefore, their characteristic composition and orientation have been used in taxonomic classification (Barthlott et al., 1998, 2003; Koch et al., 2008). The morphology of the waxes can range from being irregularly placed to being arranged in highly organized structures on the surface, prompting the question as to what the factors are responsible for determining the three dimensional structure of the surface wax crystals. There is a lack of information available on these factors, but it has been postulated that the cutin network may play a role in controlling the orientation of the wax crystals (Jeffree, 2006).

Unique Superhydrophobic Surfaces

Superhydrophobic Surfaces with a High Adhesive Force

The hierarchical surface structure of the petals of many flowers imparts superhydrophobic properties to their surface, resulting in high CAH (greater than 10°) values. These surfaces possess sufficient microstructure and nanostructure to induce superhydrophobic properties; however, these surfaces also display a high adhesive force toward water. Water droplets in contact with these petal surfaces remain spherical, but these water droplets do not roll off the surface, even when the petal is turned upside down (Figure 4(A)). This is referred to as the "petal effect" (Feng et al., 2008). The microstructure and nanostructure of the petals of, for example, the rose, Chinese Kafir lily, sunflower, have a smaller pitch value compared with that of the lotus leaf (i.e., the distance between the micropapillae is approximately 20 μm in petals compared with 40–50 μm of lotus leaf). As a result, water droplets coming in contact with the surface can impregnate the microstructure and partially penetrate the nanostructure, inducing the

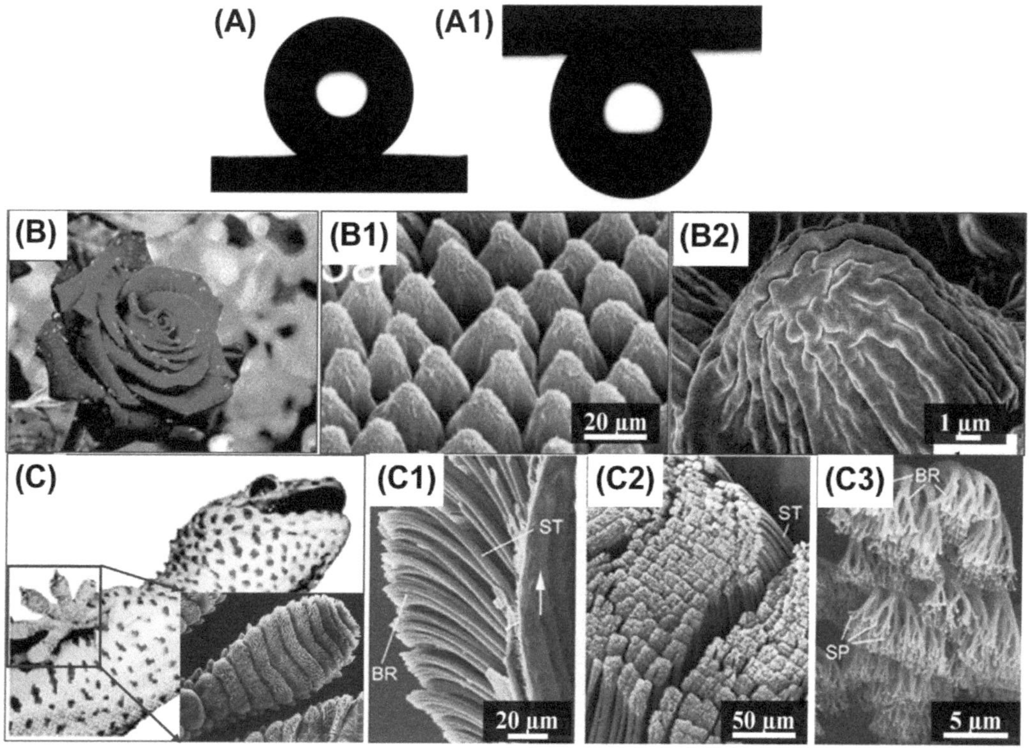

Figure 4 Superhydrophobic surfaces with high water adhesion. (A) Surfaces that exhibit the "petal effect" exhibit high water contact angles (WCAs) and high adhesion to water. (B) Rose petals, for which the effect is named, possess a similar microsctructure to that of lotus leaves; however, the nanostructure is composed of cuticular folds rather than nanocrystals. (C) Gecko feet also exhibit the petal effect. The soles of their feet possess highly hierarchical surface structures that exhibit high WCAs, but the water adheres strongly to the surface. These structures also enable the geckoes to adhere to solid surfaces, enabling them to climb walls. ST, setae; SP, spatula; BR, branched. *Bhushan (2012),Feng et al. (2008), Gao et al. (2005).*

"Cassie impregnating wetting state." The wettability of this type of surface is less than that of a surface exhibiting the Wenzel regime but greater than that of a surface exhibiting the Cassie–Baxter wetting regime. This also implies that the CAH of such surfaces will increase with increasing wetting area, which is governed by the hierarchical structure of the surface.

Bhushan and Her identified the surface structural properties that distinguish the "petal effect" from the "lotus effect" (Bhushan and Her, 2010). It was demonstrated that if a microstructure possesses a large pitch value, small peak to base height, and low density of nanostructure, water droplets coming in contact with the surface can impregnate the microstructure and partially wet the nanostructure. This will result in high CAH and the development of high adhesive forces that have minimal effect on the magnitude of the high static CA (Figure 4(B)–(B2)). On the other hand, a microstructure possessing a small pitch and high density of nanostructure will create a surface that can exhibit an increased propensity to form air pockets, leading to low adhesion with water droplets coming into contact with the surface. Like lotus leaves, rose petals, presented in Figure 4(B)–(B2), are covered with micropapillae; however, instead of having nanocrystals on the surface, the papillae on rose petals have cuticular wrinkles or folds that increase the degree of surface roughness. This structure enables the surface to trap air-pockets, inducing superhydrophobicity; however, the structure is not self-cleaning.

The foot attachment pads of several animals are capable of repeatedly attaching and detaching from a variety of surfaces. This is useful for their locomotion on vertical walls or on inverted surfaces (Hu et al., 2003; Liu et al., 2012). This dynamic attachment mechanism is referred to as reversible or "smart" adhesion and has resulted from their biological evolution over a long period of time (Bhushan, 2012). There are two kinds of foot pads, being either relatively smooth or with a hair-like coating. The former can be found in some amphibians, such as tree frogs, cockroaches, and grasshoppers (Federle et al., 2006). The latter can be found in many insects, such as beetles, spiders, and lizards (Bhushan, 2012). Gecko feet have a hair-like surface and are considered an animal equivalent to the plant surfaces that exhibit the petal effect (Figure 4(C)–(C3)). These surfaces exhibit static WCA larger than 150°, while their adhesive force to water is approximately 66 μN (Liu et al., 2012). The explanation for these properties originates from the morphology of the surface of the skin on their toes (Gao et al., 2005) (Figure 4(C)), which are covered by hundreds of thousands of setae. Each seta contains hundreds of spatula, and their hierarchical structure ranges from macroscale (mm) to nanoscale. This hierarchical structure allows the geckos to adjust the shear adhesion of their feet when in contact with a range of surfaces under different environmental conditions (Stark et al., 2013).

Anisotropic Superhydrophobic Surfaces

Surfaces that possess directionally dependent or anisotropic superhydrophobicity have been the subject of a number of studies due to their potential application in microfluidic devices, their ability to form evaporation-driven surface patterns, and their potential to form coatings that are readily cleaned (Higgins and Jones, 2000; Liu et al., 2006, 2010). Anisotropic superhydrophobicity has been observed on the surfaces of a number of plants and insects, including rice leaves and butterfly wings (Figure 5). These surfaces were found to exhibit properties associated with anisotropic hydrophobicity, these being low water drag, superhydrophobicity, and the ability to self-clean (Bixler and Bhushan, 2012, 2013), properties that arise from the unique hierarchical structures present on the surface (Bixler and Bhushan, 2013).

Rice leaves possess a hierarchical structure similar to that of the lotus leaf; however, the arrangement of the microstructure and nanostructure on the rice leaves is anisotropic, whereas the arrangement is homogenous on the upper surface of lotus leaves (Figure 5(A)–(A2)). The surfaces are patterned with sinusoidal grooves covered by micropapillae that are arranged in a quasi–one-dimensional order parallel to the edge of the leaf. These micropapillae also contain wax nanobumps on their surface (Feng et al., 2002). As a result of this hierarchical structure, water droplets making contact with the surface roll off the leaf along the parallel grooves rather than in a direction perpendicular to these grooves, enhancing the ability of the leaf to self-clean. The tilting angles associated with these two directions of movement have been reported to be 4° and 12°, respectively (Nishimoto and Bhushan, 2013).

Butterfly wing surfaces have been shown to demonstrate anisotropic superhydrophobity and are known for their multifunctional properties such as structural color, chemical sensing capacity, and ability to fluoresce (Gu et al., 2003; Vukusic and Sambles, 2003; Vukusic and Hooper, 2005). The directional adhesion of water droplets on the surface of the wings, both along and against the radial outward (RO) direction away from the body center, can be observed (Figure 4(B)–(B2)). When the wing is tilted along the RO direction, the water droplets initiated the process of "roll off" at an

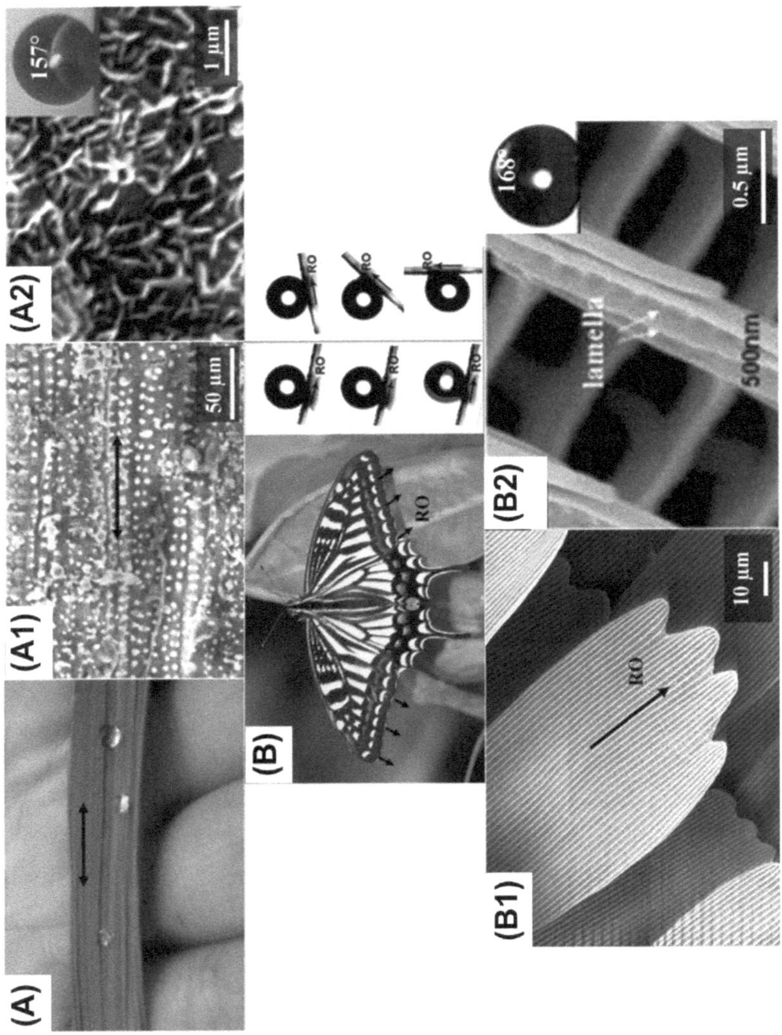

Figure 5 The anisotropic surface structures of (A) rice leaves and (B) butterfly wings. The superhydrophobic structures present on both surfaces are directionally ordered, resulting in a variable wettability that is dependent on the direction of movement. *Guo and Liu (2007), Peng et al. (2011).*

angle of 9°, whereas when tilted perpendicular to the RO direction, the water droplets ere tightly pinned to the wing surface. This unique property of butterfly wings arises from the alignment of the anisotropic shingle-like microscales, together with the radially aligned nanogrooves.

HIERARCHICAL STRUCTURE OF SURFACES ON AQUATIC SPECIES

Not all surfaces that possess hierarchical structure are superhydrophobic. Several aquatic species exhibit superoleophobicity rather than superhydrophobicity, exhibiting oil CA greater than 150° when submerged in water. These organisms possess hierarchical surface structures that are self-cleaning, are anti-fouling, and promote low drag when moving through water (Bixler and Bhushan, 2013). The skin of a shark is one such example; it is covered by minute scales known as dermal denticles, which are shaped like small ribs (or "riblets"). They are positioned such that they align with the direction of fluid flow as the shark swims through the water. It is believed that sharks have the ability to remain clean due to their flexible, low-drag riblet microstructure together with a mucous layer on the surface (Bushnell and Moore, 1991; Bechert et al., 1997; Dean and Bhushan, 2010). The design of the riblets, together with the mucous layer, enables the shark to move through water with reduced drag. This surface structure also assists in the protection of the sharks from abrasion, which in turn minimizes the extent to which microscopic organisms can adhere (Bhushan, 2012).

The scales of other fish species are another example of self-cleaning surfaces in aquatic environments. Their living conditions can be subjected to pollution arising from oil leaks or garbage originating from branched rivers (Hay, 1996). The body of the fish body is, however, well protected from the attachment of microorganisms and hence remains clean. The scales of fish display a similarity to the shark skin in that they are covered by a hierarchical structure, consisting of sector-like scales (diameter of 4–5 mm) covered by papillae (100–300 μm in length and 30–40 μm in width), and exhibit a particularly high oil CA in water (163°).

The surfaces of snail shells also possess the ability to remain clean, despite their dwelling environment and their appearance on rainy days. The surface of snail shells is comprised of a rough structure consisting of line grooves (pitch of 0.5 mm), smaller grooves crossing the line groove (pitch of 0.1 mm), and microgrooves between the line grooves (pitch of 10 μm). As a result, snail shells have a rough but regular hierarchical structure. There is one major difference, however, between these structures and those found on superhydrophobic surfaces that entrap air within their hierarchical structure—the snail shells trap water molecules within their rough surface. This macroscopic surface roughness, in fact, facilitates the entrapment of water, ensuring that the shells remain in a wet condition. This is a key factor that contributes to their ability to self-clean, in that their usually wetted surface is rarely able to be contaminated (Nishimoto and Bhushan, 2013).

SUMMARY

The diversity of unique wetting properties that can be found on natural surface reflects the power of evolution. In all of the examples given throughout this chapter, the surface structures on each organism have evolved as a result of the environmental conditions and needs of the plant or insect, and the structure has evolved to perform very specific functions. Insects and plants commonly inhabit humid environments, and thus their surfaces have often have evolved to become superhydrobic, and in many cases have developed the ability to self-clean and minimize the extent of any surface contamination. Similarly, aquatic organisms have evolved to possess superoleophobic rather than superhydrophobic structures. These structures afford these organisms the ability to cope with any contaminants that may be found in their environment. In each case, it is clear that environmental conditions have played a major role in determining the direction in which these natural surfaces have evolved.

REFERENCES

Bartell, F.E., Bartell, L.S., 1934. Quantitative correlation of interfacial free surface energies. J. Am. Chem. Soc. 56, 2205.

Barthlott, W., Neinhuis, C., 1997. Purity of the sacred lotus, or escape from contamination in biological surfaces. Planta 202, 1.

Barthlott, W., Neinhuis, C., Cutler, D., Ditsch, F., Meusel, I., Theisen, I., Wilhelmi, H., 1998. Classification and terminology of plant epicuticular waxes. Bot. J. Linn. Soc. 126, 237.

Barthlott, W., Theisen, I., Borsch, T., Neinhuis, C., 2003. Epicuticular waxes and vascular plant systematics: integrating micromorphological and chemical data. In: Stuessy, T.F., Mayer, V., Horandl, E. (Eds.), Deep Morphology Toward a Renaissance of Morphology in Plant Systematics. Gantner Verlag, Ruggell, p. 189.

Bechert, D.W., Bruse, M., Hage, W., Van Der Hoeven, J.G.T., Hoppe, G., 1997. Experiments on drag-reducing surfaces and their optimization with an adjustable geometry. J. Fluid Mech. 338, 59.

Bhushan, B., 2012. Bioinspired structured surfaces. Langmuir 28, 1698.

Bhushan, B., Her, E.K., 2010. Fabrication of superhydrophobic surfaces with high and low adhesion inspired from rose petal. Langmuir 26, 8207.

Bhushan, B., Jung, Y.C., Koch, K., 2009. Self-cleaning efficiency of artificial superhydrophobic surfaces. Langmuir 25, 3240.
Bixler, G.D., Bhushan, B., 2012. Bioinspired rice leaf and butterfly wing surface structures combining shark skin and lotus effects. Soft Matter 8, 11271.
Bixler, G.D., Bhushan, B., 2013. Fluid drag reduction and efficient self-cleaning with rice leaf and butterfly wing bioinspired surfaces. Nanoscale 5, 7685.
Buckner, J.S., 2010. Oxygenated derivatives of hydrocarbons. In: Blomquist, G.J., Bagnères, A.-G. (Eds.), Insect Hydrocarbons: Biology, Biochemistry, and Chemical Ecology. Cambridge University Press, Cambridge, p. 187.
Buschhaus, C., Herz, H., Jetter, R., 2007. Chemical composition of the epicuticular and intracuticular wax layers on adaxial sides of *Rosa canina* leaves. Ann. Bot. 100, 1557.
Buschhaus, C., Jetter, R., 2011. Composition differences between epicuticular and intracuticular wax substructures: how do plants seal their epidermal surfaces? J. Exp. Bot. 62, 841.
Bushnell, D.M., Moore, K.J., 1991. Drag reduction in nature. Annu. Rev. Fluid Mech. 23, 65.
Byun, D., Hong, J., Saputra, Ko J.H., Lee, Y.J., Park, H.C., Byun, B.K., Lukes, J.R., 2009. Wetting characteristics of insect wing surfaces. J. Bionic Eng. 6, 63.
Cassie, A.B.D., Baxter, S., 1944. Wettability of porous surfaces. Trans. Faraday Soc. 40, 546.
Dean, B., Bhushan, B., 2010. Shark-skin surfaces for fluid-drag reduction in turbulent flow: a review. Philos. Trans. A Math. Phys. Eng. Sci. 368, 4775.
Federle, W., Barnes, W.J.P., Baumgartner, W., Drechsler, P., Smith, J.M., 2006. Wet but not slippery: boundary friction in tree frog adhesive toe pads. J. R. Soc. Interface 3, 689.
Feng, L., Li, S., Li, Y., Li, H., Zhang, L., Zhai, J., Song, Y., Liu, B., Jiang, L., Zhu, D., 2002. Super-hydrophobic surfaces: from natural to artificial. Adv. Mater. 14, 1857.
Feng, L., Zhang, Y., Xi, J., Zhu, Y., Wang, N., Xia, F., Jiang, L., 2008. Petal effect: a superhydrophobic state with high adhesive force. Langmuir 24, 4114.
Fogg, G.E., 1944. Diurnal fluctuation in a physical property of leaf cuticle. Nature 154, 515.
Gao, H., Wang, X., Yao, H., Gorb, S., Arzt, E., 2005. Mechanics of hierarchical adhesion structures of geckos. Mech. Mater. 37, 275.
Gu, Z.Z., Uetsuka, H., Takahashi, K., Nakajima, R., Onishi, H., Fujishima, A., Sato, O., 2003. Structural color and the lotus effect. Angew. Chem. – Int. Edition 42, 894.
Guo, Z., Liu, W., 2007. Biomimic from the superhydrophobic plant leaves in nature: binary structure and unitary structure. Plant Sci. 172, 1103.
Hay, M.E., 1996. Marine chemical ecology: what's known and what's next? J. Exp. Mar. Biol. Ecol. 200, 103.
Higgins, A.M., Jones, R.A.L., 2000. Anisotropic spinodal dewetting as a mute to self-assembly of patterned surfaces. Nature 404, 476.
Hu, D.L., Chan, B., Bush, J.W.M., 2003. The hydrodynamics of water strider locomotion. Nature 424, 663.
Ivanova, E.P., Hasan, J., Webb, H.K., Truong, V.K., Watson, G.S., Watson, J.A., Baulin, V.A., Pogodin, S., Wang, J.Y., Tobin, M.J., Löbbe, C., Crawford, R.J., 2012. Natural bactericidal surfaces: mechanical rupture of *Pseudomonas aeruginosa* cells by cicada wings. Small 8, 2489.
Ivanova, E.P., Hasan, J., Webb, K.W., Gervinskas, G., Juodkazis, S., Truong, V.K., Wu, A.H.F., Lamb, R.N., Baulin, V.A., Watson, G.S., Watson, J.A., Mainwaring, D.E., Crawford, R.J., 2013a. Bactericidal activity of black silicon. Nat. Commun 4, e.2838.
Ivanova, E.P., Nguyen, S.H., Webb, H.K., Hasan, J., Truong, V.K., Lamb, R.N., Duan, X., Tobin, M.J., Mahon, P.J., Crawford, R.J., 2013b. Molecular organization of the nanoscale surface structures of the dragonfly *Hemianax papuensis* wing epicuticle. PLoS One 8, e67893.
Jeffree, C.E., 2006. In: Riederer, M., Müller, C. (Eds.), Biology of the Plant Cuticle. Blackwell Pub., Oxford, UK, p. 11.

Jetter, R., Kunst, L., 2008. Plant surface lipid biosynthetic pathways and their utility for metabolic engineering of waxes and hydrocarbon biofuels. Plant J. 54, 670.
Jung, Y.C., Bhushan, B., 2006. Contact angle, adhesion and friction properties of micro- and nanopatterned polymers for superhydrophobicity. Nanotechnology 17, 4970.
Koch, K., Bhushan, B., Barthlott, W., 2008. Diversity of structure, morphology and wetting of plant surfaces. Soft Matter 4, 1943.
Koch, K., Bhushan, B., Jung, Y.C., Barthlott, W., 2009. Fabrication of artificial lotus leaves and significance of hierarchical structure for superhydrophobicity and low adhesion. Soft Matter 5, 1386.
Koch, K., Ensikat, H.J., 2008. The hydrophobic coatings of plant surfaces: epicuticular wax crystals and their morphologies, crystallinity and molecular self-assembly. Micron 39, 759.
Liu, H., Zhai, J., Jiang, L., 2006. Wetting and anti-wetting on aligned carbon nanotube films. Soft Matter 2, 811.
Liu, K., Du, J., Wu, J., Jiang, L., 2012. Superhydrophobic gecko feet with high adhesive forces towards water and their bio-inspired materials. Nanoscale 4, 768.
Liu, K., Yao, X., Jiang, L., 2010. Recent developments in bio-inspired special wettability. Chem. Soc. Rev. 39, 3240.
Lockey, K.H., 1980. Insect cuticular hydrocarbons. Comp. Biochem. Physiol. B, Biochem. Mol. Biol. 65, 457.
Lockey, K.H., 1985. Insect cuticular lipids. Comp. Biochem. Physiol. B, Biochem. Mol. Biol. 81, 263.
Lockey, K.H., 1988. Lipids of the insect cuticle: origin, composition and function. Comp. Biochem. Physiol. B, Biochem. Mol. Biol. 89, 595.
Marangoni, C., 1871. Ueber die Ausbreitung der Tropfen einer Flüssigkeit auf der Oberfläche einer anderen. Annu. Rev. Phys. Chem. 143, 337.
Moussian, B., 2010. Recent advances in understanding mechanisms of insect cuticle differentiation. Insect Biochem. Mol. Biol. 40, 363.
Neinhuis, C., Barthlott, W., 1997. Characterization and distribution of water-repellent, self-cleaning plant surfaces. Ann. Bot. 79, 667.
Nelson, D.R., Blomquist, L.G., 1995. Insect waxes. In: Hamilton, R.J., Christie, W.W. (Eds.), Waxes: Chemistry, Molecular Biology and Functions. The Oily Press Ltd Hamilton RJ, Dundee, p. 1.
Nishimoto, S., Bhushan, B., 2013. Bioinspired self-cleaning surfaces with superhydrophobicity, superoleophobicity, and superhydrophilicity. RSC Adv. 3, 671.
Nosonovsky, M., Bhushan, B., 2005. Roughness optimization for biomimetic superhydrophobic surfaces. Microsyst. Technol. 11, 535.
Nosonovsky, M., Bhushan, B., 2007a. Hierarchical roughness makes superhydrophobic states stable. Microelectron. Eng. 84, 382.
Nosonovsky, M., Bhushan, B., 2007b. Multiscale friction mechanisms and hierarchical surfaces in nano- and bio-tribology. Mater. Sci. Eng. R. Rep. 58, 162.
Papadopoulos, P., Mammen, L., Deng, X., Vollmer, D., Butt, H.J., 2013. How superhydrophobicity breaks down. Proc. Natl. Acad. Sci. USA 110, 3254.
Peng, W., Hu, X., Zhang, D., 2011. Bioinspired fabrication of magneto-optic hierarchical architecture by hydrothermal process from butterfly wing. J. Magn. Magn. Mater. 323 (15), 2064.
Pogodin, S., Hasan, J., Baulin, V.A., Webb, H.K., Truong, V.K., Nguyen, T.H.P., Boshkovikj, V., Fluke, C.J., Watson, G.S., Watson, J.A., Crawford, R.J., Ivanova, E.P., 2013. Biophysical model of bacterial cell interactions with nanopatterned cicada wing surfaces. Biophys. J. 104, 835.
Samuels, L., Kunst, L., Jetter, R., 2008. Sealing plant surfaces: cuticular wax formation by epidermal cells. Annu. Rev. Plant Biol. 59, 683.

Stark, A.Y., Badge, I., Wucinich, N.A., Sullivan, T.W., Niewiarowski, P.H., Dhinojwala, A., 2013. Surface wettability plays a significant role in gecko adhesion underwater. Proc. Natl. Acad. Sci. USA 110, 6340.
Verho, T., Korhonen, J.T., Sainiemi, L., Jokinen, V., Bower, C., Franze, K., Franssila, S., Andrew, P., Ikkala, O., Rasa, R.H.A., 2012. Reversible switching between superhydrophobic states on a hierarchically structured surface. Proc. Natl. Acad. Sci. USA 109, 10210.
Vukusic, P., Hooper, I., 2005. Biomaterials: directionally controlled flourescence emission in butterflies. Science 310, 1151.
Vukusic, P., Sambles, J.R., 2003. Photonic structures in biology. Nature 424, 852.
Watson, G.S., Cribb, B.W., Watson, J.A., 2010. How micro/nanoarchitecture facilitates anti-wetting: an elegant hierarchical design on the termite wing. ACS Nano 4, 129.
Watson, G.S., Myhra, S., Cribb, B.W., Watson, J.A., 2008. Putative functions and functional efficiency of ordered cuticular nanoarrays on insect wings. Biophys. J. 94, 3352.
Wenzel, R.N., 1936. Resistance of solid surfaces to wetting by water. Ind. Eng. Chem. 28, 988.
Yan, Y.Y., Gao, N., Barthlott, W., 2011. Mimicking natural superhydrophobic surfaces and grasping the wetting process: a review on recent progress in preparing superhydrophobic surfaces. Adv. Colloid Interface Sci. 169, 80.

CHAPTER THREE

The Design of Superhydrophobic Surfaces

GLOSSARY

CNT Carbon nanotubes
CVD Chemical vapor deposition
DRIE Deep reactive ion etching
EBL Electron beam lithography
LbL Layer-by-layer self-assembly
PANI Polyaniline
PDMS Polydimethylsiloxane
PS Polystyrene
PVA poly(vinyl alcohol)
RIE Reactive ion etching
WCA Water contact angle

METHODS TO PREPARE SUPERHYDROPHOBIC SURFACES

Various methods have been developed for the widespread fabrication of superhydrophobic surfaces. Depending on the type of procedure that was used to fabricate the surfaces, these methods can be categorized as "top-down" or "bottom-up" approaches. A top-down approach refers to the generation of superhydrophobic nanoscale structures/functional devices with the preferred shapes and characteristics, starting on a larger scale, where the surface nanotopography required for the condition of superhydrophobicity was imparted through the removal of material, using, for example, etching techniques. Conversely, bottom-up approaches tend to employ construction approaches whereby the topographical conditions for superhydrophobicity are achieved through the nanoscale assembly of structures based on intricate techniques and systems (Biswas et al., 2012).

Top-down Fabrication Techniques

Templating

Templating techniques replicate a pattern or shape, allowing the inverse of the original pattern to be produced, for example, by either two- or three-dimensional printing, pressing or growing a structure against the voids of

http://dx.doi.org/10.1016/B978-0-12-801109-6.00003-3

the original surface. Similar techniques are then applied to the inverse mask, resulting in the formation of a replica of the original surface. Templating the surfaces is cost- and time-efficient, reproducible, and therefore is widely used for the preparation of polymeric surfaces (Peng et al., 2013).

Replicas of plant leaves containing superhydrophobic properties have been widely fabricated. For example, Singh et al. (2007) cast a polydimethylsiloxane (PDMS) negative template of lotus and *Colocasia* leaves (Figure 1(A)) thus allowing the construction of synthetic positive surfaces that accurately mimicked the surface structure of the natural substrates. Sun et al. (2007) also produced a PDMS replica of a Lotus leaf through a casting technique. The formation of microprotrusions and the nano-architecture created by the original Lotus leaves resulted in a water contact angle (WCA) of 160° (Sun et al., 2007).

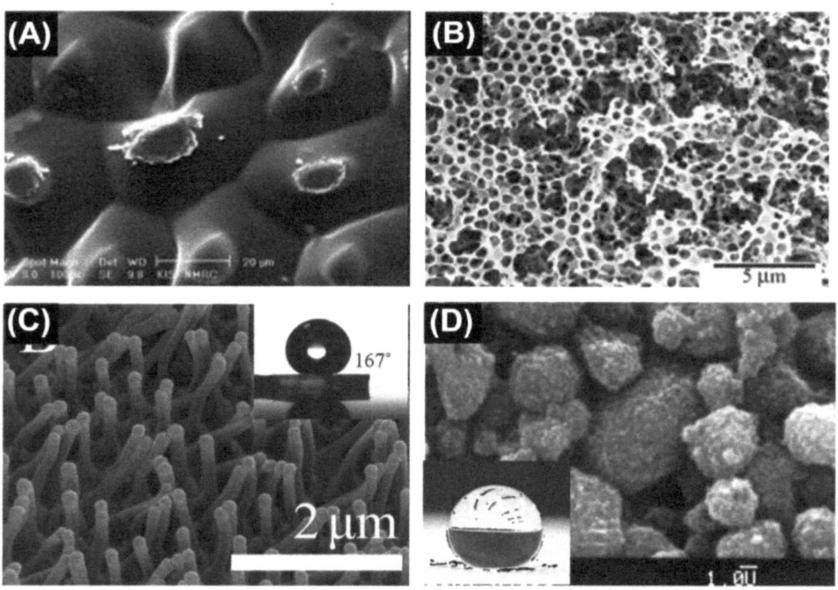

Figure 1 Typical superhydrophobic surface structure fabricated by templating method (A) replica of *Colocasia*-like leaf surface on PDMS template (Singh et al., 2007), (B) a microporous pattern of polyvinylidene fluoride film (Li et al., 2006), (C) superhydrophobic hair shaped polymer surface grown through an AAO template (Zhang et al., 2006) and (D) water droplet resting on a polymer hot-press transferred pattern (Bormashenko et al., 2006). PDMS, polydimethylsiloxane; AAO, aluminum oxide. *Images reprinted with permission from (A, B) Elsevier, Copyright 2007 and 2006 respectively, (C, D) American Chemical Society, Copyright 2006.*

Bico et al. (1999) produced materials via photolithography that had features such as spikes, shallow cavities, and stripes at the microscale, which were then replicated using an elastomeric mould to cast silica features onto a silicon wafer. Following the formation of a hydrophobic self-assembled monolayer coating, the spikes afforded the surface an increase in WCA from 118° of the original surface to 170° on the modified surface. Zhu et al. (2006b) later described the use of a template-based extrusion method in which an anodic aluminum oxide (AAO) membrane was used as a template for the production of a poly(vinyl alcohol) (PVA) aligned nanofiber surface. A similar method was applied for the construction of perfluoropolyether-based nanopillars that possessed superhydrophobic characteristics (Roach et al., 2008).

In another study developed by Love et al. (2002), metals were deposited onto a colloidal array made from silica to produce superhydrophobic surfaces by generating submicron half-shell patterns. After the template was removed, superhydrophobicity was imparted by treating the nanosized metal cup surface with alkanethiols. A similar approach is shown in Figure 1(B), where a macroporous solid pattern template on a fluoropolymeric film around a silica-sphere array is represented (Li et al., 2006). Sun et al. (2007) described two methods by which a colloidal array could be templated: the first involved coating an array of microspheres with nanospheres, and the second involved producing a crystalline solid of nanospheres by templating an array of microspheres in PDMS. Using similar methods, a WCA of 160° was obtained by pressing an aggregated polymer-beaded surface against a crimped stamp (Figure 1(D)) (Bormashenko et al., 2006).

Lithography

Lithography is one of the best known processes for preparing superhydrophobic surfaces. The various versions of this technique allow easy control over the topological features of the surface, i.e., the structure and precise morphology. Soft lithography is a common technique, which is used to develop superhydrophobic patterns on polymers or organic substrata using relief patterns on a PDMS stamp. The structures generated using this method have been reported to be highly adhesive and superhydrophobic (Miyamura et al., 2011). This method, however, is not considered a robust technique, as the patterns generated are often found to be thermally unstable and also have a tendency to disintegrate upon exposure to organic solvents.

Colloidal lithography is another technique of interest, as the process incorporates the use of soluble particles as removable templates. Recent studies have reported that nanostructured surfaces fabricated using an organic template such as polystyrene (PS) can produce surfaces that exhibit WCAs as high as 150° (Sun et al., 2011). The optical properties of the structured surfaces developed using colloidal lithography often suffer from loss of reflectance (Li et al., 2010; Chen et al., 2012). This results in surfaces not being considered suitable in laser applications and for the fabrication of sensors. This problem can be overcome through the combination of colloidal lithography with reactive ion etching (RIE) procedures, resulting in reflective surfaces that retain their superhydrophobicity. Such surfaces can therefore be used as dual purpose near optical, superhydrophobic materials (Ellinas et al., 2011).

Electron beam lithography (EBL) is a commonly used lithographic technique for the fabrication of surfaces with high WCAs. EBL offers significant advantages toward the development of hierarchical structures that display high WCAs due its flexibility and high precision. The hierarchical structures fabricated by EBL are particularly useful for gaining a better understanding of the concept of superhydrophobic surfaces, as they offer the ability to fabricate well-defined primary microstructures with secondary nanostructures (Feng and Rothstein, 2010). A successful attempt to biomimic such hierarchical structures using EBL has been reported (Liu et al., 2011b). SU-8 is a very commonly used epoxy-based chemically-amplified i-line negative-tone resist. The key constituents of SU-8 comprise of a base resin with eight epoxy rings and the photo acid generator (PAG). Acid is produced by the PAG upon UV exposure that attacks the epoxy rings during the postexposure bake. The cross-linkage of the polymer then proceeds through linkages between the opened epoxy rings. In the final stage of the cross-linkage reaction, acid is generated again from the base resin and therefore acts as the catalyst for subsequent cross-linkage reactions (Akihiko et al., 2014). SU-8 is also commonly used in EBL, as SU-8 is highly sensitive and has the inherent ability to reduce the exposure dosage and exposure time of the underlying substratum. The hierarchical surfaces fabricated using dual electron beam exposure contain a uniform coverage of micrometer-sized patterns covered with nanometer-sized patterned features (Shieh et al., 2010). These surfaces show a WCA near 120°, which is not a function of the varied spacing between the microfeatures. EBL has proven to be a truly versatile technique, as not only can the surface hierarchy be accurately controlled, but also the average spacing between the features. The technique is also relatively fast to perform (Feng et al., 2011).

A wide range of size and patterns of etched silicon were fabricated by Öner and McCarthy (2000), ranging from square posts of 20–140 μm in height and 2–128 μm in side lengths (Figure 2(A)) to star-shaped patterns (Figure 2(B)). A similar pattern was achieved by Zhu et al. (2006a) with square pillars in the range of 10–85 μm (Figure 2(D)). Circular pillars, 2–40 μm in diameters and approximately 80 μm in height, were produced by Shirtcliffe et al. (2005) on the surface of an SU-8 photoresist. This method was followed by Tsuruta et al. (2006) who further improved the surface hydrophobicity by coating the surface with hexamethyldisiloxanes. Wagterveld et al. (2006) combined photochemical laser ablation with the SU-8 process, which resulted in the formation of a superhydrophobic surface with a WCA of 165° when coated with a hexafluoropropene layer.

Yoshimitsu et al. (2002) demonstrated a mechanical technique that utilized a dicing saw to fabricate microstructured silicon with a square shape of 50 μm size, as shown in Figure 2(C). Choi and Kim (2006) combined deep RIE and holographic lithography to produce nanosized pillar structures on silicon wafer with the tips of the resulting nanopillars being sharpened by thermal oxidation and subsequent removal of the oxide. One year later, Cao et al. (2007) used a combination of photolithographic patterning and various etching processes to create pillar patterns on silicon substrata. The resulting nanowire pattern generated by chemical vapor deposition (CVD) also afforded these surfaces, overhanging structures and hierarchical roughness (Figure 2(E)) (Cao et al., 2007).

Chemical and Plasma Etching

A number of etching techniques have been applied to various substrata, such as polymers, for precise fabrication of engineered surfaces (Guo et al., 2011). Plasma etching of polymeric substrates is an effective method for the fabrication of superhydrophobic surfaces, as it involves the utilization of a highly reactive plasma species to modify the surface functionality of the substrate (Gupta et al., 2013). Conventional plasma etching processes involve the application uses of gases such as CF_4, C_4H_8, SF_6, Ar, and O_2 to positively or negatively etch the surface depending on the mode of the exposure of the surface to the etchants (Jafari et al., 2013). The scientific advantage of using this technique lies in the fact that the plasma etching method can be used in conjunction with other fabrication techniques such as lithography or templating to create the desired surface. Plasma processes, such as dry etching, allow for easy and versatile texturing of surfaces, even those of large area polymeric surfaces (Palumbo et al., 2011). Plasma

Figure 2 Lithographic surface modification (A) photolithographic towers and (B) indented square posts (Öner and McCarthy, 2000), (C) diced silicon wafer (Yoshimitsu et al., 2002), (D) photolithographic towers (Zhu et al., 2006a) and (E) silicon nanowires grown on those silicon islands (Cao et al., 2007). *Images reprinted with permission from (A, B, C, and E) American Chemical Society, Copyright 2000, 2002 and 2007 respectively, (D) Elsevier, Copyright 2006. Adapted with permission of The Royal Society of Chemistry.*

reactive processes, such as the deep reactive ion etching (DRIE), use chemically reactive ionized gases generated from an electromagnetic field to fabricate surfaces that exhibit superhydrophobicity. This technique results in the uppermost surface of substratum being etched, giving rise to the formation of hierarchical defined structures such as nanopillars or nanocones (Kondrashov and Ruhe, 2014). The surface features significantly change upon variation of the etching parameters, such as the etching gas ratio, reaction exposure time, and applied power. Considerable differences in the hydrophobicity of the surface can result. Plasma etching has been extensively used to fabricate superhydrophobic surfaces on metals, polymers, and elastomer-based materials (Fernández-Blázquez et al., 2011; Zahner et al., 2011). Superhydrophobic polymers are fabricated by either plasma or ion etching or laser ablation if the superhydrophobicity of metals is usually afforded by wet etching (Roach et al., 2008). A WCA as high as 158° has been reported for a surface etched using a dual combination of Ar and O_2 gases (Barshilia and Gupta, 2014). Plasma ablation treatment has also been used to construct micro/nano-fluidic channels within substrata, which are able to sustain high-pressure flow. This has enabled the conversion of the no-slip conditions along the walls of the fluidic channels to those that allow slip to take place (Papageorgiou et al., 2013).

Chemical etching assists in the modification of rough substrata of low surface energy materials, allowing simultaneous modulation of the geometrical microstructure and chemical composition of the surfaces (Hu et al., 2010). Hierarchical surfaces fabricated using such chemical etching processes exhibit a high degree of hydrophobicity, with resultant WCAs nearing 150° (Celia et al., 2013; Zhang et al., 2013). Chemical etching has been used to produce superhydrophobic surfaces on substrata such as nickel, silica, stainless steel, and titanium dioxide, producing nanostructured surfaces that display WCA in the range of 130°–150° (Kang et al., 2012; Zhou et al., 2011).

Wet chemical etching has been widely used to produce superhydrophobicity for a wide range of metal surfaces, as illustrated in Figure 3. The substrata are generally etched using a dislocation etchant by dissolving the dislocation sites in the metal grains to give a rough surface (Qian and Shen, 2005). For example, wet etching was applied to copper to improve the superhydrophobicity of the surface, which also resulted in the formation of multiscale roughness (Shirtcliffe et al., 2005). Guo et al. (2006) reported an application of wet etching of an aluminum alloy, which increased the superhydrophobicity of the surface by forming microprotrusions and nanoparticles (Figure 3).

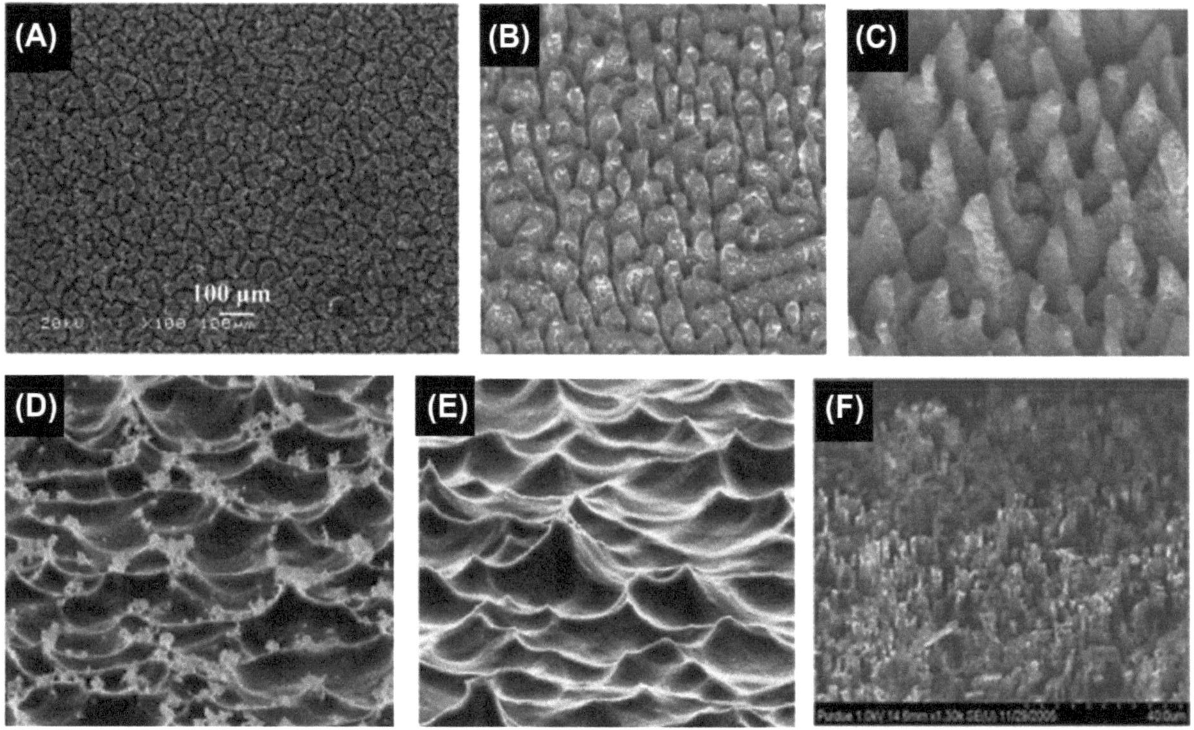

Figure 3 Illustration of (A) etched aluminum alloy (Guo et al., 2006), (B) laser-etched silicon wafer with using SF_6 3.2 kJ/m^2 and (C) using 9 kJ/m^2 (Baldacchini et al., 2006), (D) silicon wafer/photoresist layer overetched by SF_6 plasma before sonication, (E) after cleaning of residual photoresist and (F) micropillar structured silicon after buffered oxide etching. *Images reprinted with permission from (A, D, and E) Elsevier, Copyright 2006, 2005, and (B, C, and F) American Society, Copyright 2006, 2007. Adapted with permission of The Royal Society of Chemistry.*

Anodic Oxidation

Anodization methods have been applied to metallic substrata to generate porous nanoscale hierarchical structures, on substrates such as aluminum (Wu et al., 2011; Chung et al., 2013), titanium (Chen et al., 2011; Tupala et al., 2014), niobium (Yoo and Choi, 2010; Yoo et al., 2013) and copper (Allam and Grimes, 2011; Li et al., 2012a). Porous anodic aluminum oxide films have been widely used in the fabrication of nanomaterials. Films with high uniformity and pore density can be readily obtained from aluminum metal by adjusting the preparation parameters. The process is moderately inexpensive and appropriate for the large-scale fabrication of nanoscale porous structures (Wu et al., 2012). The process can be easily fine-tuned to achieve a desirable surface structure by modifying parameters such as reaction time, applied voltage, and the composition of the anodizing electrolyte. It has also been observed that superhydrophobicity can be induced by the fabrication of an array of dead-end nanopores on the metallic surface. These dead-end nanopores alternatively act as "air-pockets" and help to strike a balance between the adhesion properties and the hydrophobicity of the surface (Park et al., 2010). Superhydrophobic coatings based on either randomly organized surface "unevenness" or ordered arrays have also been anodically oxidized onto the surface nanostructures (Mozalev et al., 2012). Titanium and its alloys exhibit high resistance to corrosion and fracture due to the natural formation of a compact and chemically stable oxide film that results from the metal surface being exposed to air, however, it has been observed that the anticorrosive nature of titanium needs to be increased in special circumstances (Hongxi et al., 2012). This problem can be solved by applying a superhydrophobic coating on the titanium metal layer. The application of such a coating on the base metal substratum not only increases the adhesion value but also significantly increases the hydrophobicity of the surface (Zhang et al., 2011). A uniform coating of 1H, 1H, 2H, 2H-perfluorooctyl-triethoxysilane (PTES) applied on nanotubular titania films has the ability to transform the surface from being moderately hydrophilic to hydrophobic. This coating has also proven helpful toward reducing bacterial adhesion on the resultant nanotubular titania surface (Tang et al., 2011). At present, various facets of the nanotube geometry such as the tube length, pore diameter, wall thickness, intertube spacing, and the wall roughness, can be modified to the desired level by monitoring the anodization settings (Sun et al., 2014a). Additionally, surfaces produced using this technique show significant resistance against corrosion.

Bottom-up Fabrication Techniques

Hydrothermal Reaction

Hydrothermal reaction is recognized as an efficient method for the production of superhydrophobic materials with a variety of patterns and morphologies. Wang et al. (2007a) successfully produced nanolamellar structures on titanium surfaces using an in situ hydrothermal synthesis technique. The superhydrophobicity was obtained by spin coating the surface with vinyl-terminated PDMS after the physical modification (Wang et al., 2007a).

Hydrothermal methods are a simple way for generating hydrophobic surfaces on a large scale and with low cost (Sun et al., 2014b). Hydrothermal treatment has been used on a variety of substrata due to the fact that it is fast, cheap, and affords good tenability through temperature and pressure control of the reaction (Ming et al., 2012). This process results in a high level of surface hydrophobicity, and WCAs of more than 150° have been observed on specific substrata such as silicon, titanium, and zinc (Guo et al., 2011). This process treats the sample surface, mostly polycrystalline metals, in a highly acidic or highly alkaline solution in a controlled temperature and pressure environment. Presently, a template-free and rapid hydrothermal method has also been used to fabricate a superhydrophobic coating on a magnesium alloy (Gao et al., 2014). The roughness of the surface was increased considerably after the hydrothermal process, which takes 3 h. Application of a fluoroalkylsilane (FAS) coating resulted in a further reduction in the surface energy of the Mg alloy, and the measured WCA was as high as ~160° as a result (Gao et al., 2013). Hydrothermal reactions are relatively simple to design, since the additional problems associated with creating a specific template can be avoided. In addition, the anticorrosion properties of titanium and magnesium alloys can be easily enhanced through the use of hydrothermal processes.

Sol–Gel Methods

Sol–gel methods have been frequently used to fabricate superhydrophobic structures on surfaces that exhibit a WCA as high as 160° (Xiang et al., 2011). This particular technique has been used on a variety of substrata such as wood (Wang et al., 2011b), silica (Mahadik et al., 2010; Latthe et al., 2010), textiles, metals, glass, alumina, and stainless steel (Yang et al., 2010). The substratum is often coated with methyltriethoxysilane (MTES) or tetraethoxysilane (TEOS) to enhance the hydrophobicity of the surface. Using TEOS as the precursor adds certain benefits to the procedure, such as the convenience of being able to control the size distribution of the surface features and the ability to afford the surface, greater compatibility with other organic additives. A MTES precursor was used to prepare hydrophobic coatings on copper substrata, which not only improved the adhesion between fiber surfaces and the substratum but also reduced the active permeability of corrosive substrata (Rao et al., 2011). This is a prerequisite step toward the coating of metals such as copper as they are frequently prone to corrosion and hence the active surface area for subsequent sol–gel processing is reduced. The surface additive not only assists in increasing the surface hydrophobicity, but also tends to decrease the roll-off angle to as low as ~2°. In a conventional sol–gel process, the starting material is modified into a glass-like substance through repeated cycles of hydrolysis and polycondensation reactions. It has also been shown that the dipping of the sample in saturated fatty acids also generates a surface that possesses a high WCA. The surface roughness that results from using the sol–gel process can be easily fine-tuned by varying the working protocol and the reacting mixture composition. This process is rapidly replacing other processes, since it allows ready functionalization of the particles, ensuring that they adhere to the substratum. In addition, this process is suitable for low-temperature processing and generates a high degree of surface homogeneity (Liu et al., 2011a).

The sol–gel method has been known to be highly applicable to a variety of solid materials including glass, polymers, silicon wafer, metals, and textiles (Guo et al., 2011). An example is presented in Figure 4(A), which shows a hexagonal-ordered structure of a colloidal crystal film with an advancing tailored superhydrophobicity (Figure 4(A)) (Wang et al., 2007b). Figure 4(B) depicts the work of Hoefnagels et al. (2007), who hydrophobized a cotton textile by growing silica microparticles on the hydrophilic substrate. Superhydrophobicity was imparted to a copper alloy using hexamethylenetetramine and ethylene glycol, which are the chelating agents to Cu^{2+} and Fe^{2+} ions, reported by Huang et al. (2007) (Figure 4(C)). Another interesting study on indium oxide films showed an ordered porous pattern with superhydrophobic and lipophobic characteristics after sol–gel processing (Figure 4(D)) (Li et al., 2007a). The nanosized gaps forming between the pores during the annealing processing were found to be essential to the development of superhydrophobicity (Li et al., 2007a).

Electrospinning

Inspired by the intriguing hierarchical structures formed on the lotus leaf, electrospinning is a highly versatile technique that allows the continuous fabrication of synthetic and natural polymers, nanoparticles, metals, and

Figure 4 (A) Hexagonal packed ordered structured of colloidal films made from latex spheres assembled at 30°C (Wang et al., 2007b), (B) superhydrophobic cotton textile modified with PDMS (Hoefnagels et al., 2007), (C) nanorod structure of copper alloy film formed by sol–gel process (Huang et al., 2007), and (D) porous array of indium oxide film generated by sol-dipping method (Li et al., 2007a). PDMS, polydimethylsiloxane. *Images reproduced by permission of (A) Wiley-VCH, Copyright 2007, (B) and (C) the American Chemical Society, Copyright 2007, (D) Elsevier, Copyright 2007. Reprinted with permission from Elsevier, Copyright 2011.*

ceramics in nano- to microscale (Wang et al., 2011c). The basic apparatus for electrospinning consists of three main components: (1) the feed system, (2) a high voltage power supply, and (3) the ground base collector for the electrospun needle/fiber. The fibers so formed are collected as nonwoven mats and have high surface-to-volume ratio and porosity, small pore size, and a high degree of flexibility (Sas et al., 2012). More recently, electrospinning has been shown to possess several advantages for the fabrication of polymers or composite nanofibrous mats with high specific surface area and high surface roughness. The advantage of electrospinning as a technique is the ease with which the apparatus can be used with no specialized expertise being required (Lin et al., 2011). In addition, this technique has a large number of machine- and object-related parameters that can be easily adjusted to reproduce and mimic natural superhydrophobic surfaces. Surfaces similar to the Lotus leaf have been produced by the fabrication of composite films.

A controlled concentration of PS solution during electrospinning resulted in the fabrication of porous microspheres, which were superhydrophobic (Lee et al., 2013b). A composite film made using polyaniline (PANI)/PS resembling the surface of a Lotus leaf has also been successfully produced (Zhang and Rutledge, 2012). This particular composite has the potential to exhibit extreme superhydrophobicity. Electrospun fibers have been shown to form surfaces with variable WCA values. However, the highest degree of superhydrophobicity has been recorded on the formation of bead-containing fibers. The high WCA (greater than 150°), result from the small diameters of the fibers combined with the presence of hydrophobic polymers (Asmatulu et al., 2010).

Superhydrophobic properties have been extensively reported in electrospun fabricated surfaces (Guo et al., 2011). For instance, trifluoroethoxy polyphosphazene is known as one of the most hydrophobic polymers, which is also repellent to the action of oils and high-radiation activity (Guo et al., 2011; Wang et al., 2014). In 2005, Ma et al. employed the electrospinning technique to increase the hydrophobicity of copolymer poly(styrene-block dimethylsiloxane), as shown in Figure 5(A). It was found that the resulting superhydrophobicity was not only a result of the fibrous mat structure but also due to the specific phase formation of the fibers (Ma et al., 2005). Fluorinated polymer was electrospun with different parameters to generate advanced hydrophobic surfaces with different textures and patterns, including cylinders, globules, and nanofibers (Figure 5(B)) (Agarwal et al., 2006). Electrospun PS also appeared to have similar structures and superhydrophobicity with larger connecting microparticles (Figure 5(D)) (Zheng et al., 2006).

Solvent control has been shown to efficiently increase the WCA of electrospun fibers by creating nano- and microporous structures, such as the study reported by Ma et al. (2007), as demonstrated in Figure 5(C). Other work involves additional agents during the electrospinning process in order to produce specific morphologies or to create distinct properties to the electrospun fibers. For examples, Figure 5(E) shows Fe_3O_4-based carbon nanofibers with the size of 20–30 nm diameter with ball-like structured surfaces, possessing WCA of 157° compared to 27° of original surfaces with a microfiber diameter of 180 nm (Figure 5(E)) (Zhu et al., 2006b).

Layer-by-Layer Assembly

Self-assembly and layer-by-layer (LbL) assembly are convenient and economical methods to produce superhydrophobic surfaces. These methods

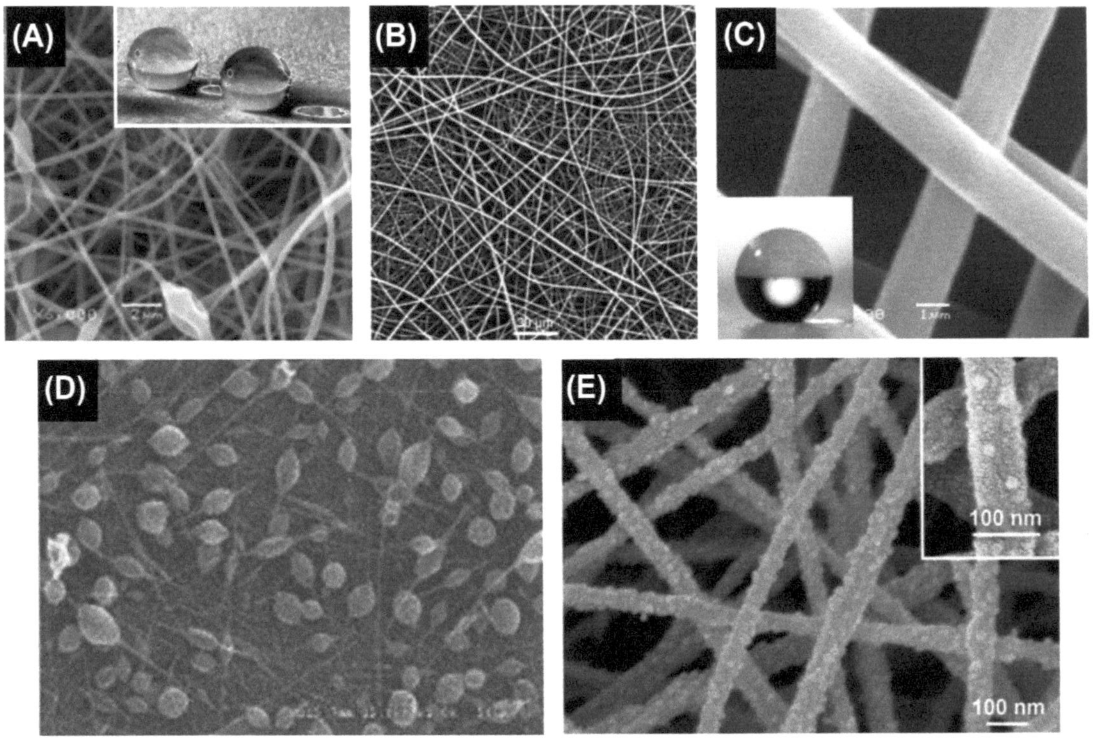

Figure 5 (A) Superhydrophobicity of electrospun fiber mat PS-PDMS/PS (Ma et al., 2005), (B) electrospun fluorinated polymer (Agarwal et al., 2006), (C) fibrous electrospun fluorinated polymer (Ma et al., 2007), (D) electrospun polystyrene with microparticle connecting fibers (Zheng et al., 2006) and (E) Fe_3O_4-based carbon nanofibers (Zhu et al., 2006b). PDMS, polydimethylsiloxane; PS, polystyrene. *Images reprinted with permission from (A) American Chemical Society, Copyright 2005, (B, C, and E) Wiley-VCH Verlag GmbH & Co. KGaA, Copyright 2006, 2007 and 2006 respectively (D) Elsevier, Copyright 2007. Adapted with permission of The Royal Society of Chemistry.*

also allow the accurate control of the finely tailored film growth that occurs as part of the formation process (Li et al., 2012b). This is a conventional process in which oppositely charged thin films are deposited in layers. The precise parameters of the process can be easily controlled to generate hierarchical micro/nanostructures. A multilayer film can be fabricated by alternately dipping a charged substratum into aqueous solutions of oppositely charged materials with intermediate steps of rinsing in water and drying under nitrogen. Polyelectrolyte multilayer (PEM) films can be designed by successively spraying aqueous solutions of oppositely charged polyelectrolytes onto substratum surfaces (Schlenoff et al., 2000). Customized dipping LbL assembly is commonly used to formulate ultrathin polymeric films that are less than 100 nm in thickness; each individual polymeric layer usually has a thickness of several nanometers and the deposition of one layer of polymer takes several tens of minutes (Decher, 1997). Tailoring of chemical composition, physicochemical properties, surfaces topography, and 3D arrangements with adjacent layers can be obtained by controlling the assembling parameters including concentration, temperature, pH, and ionic strength of the dipping solutions (Such et al., 2011). The fabrication of superhydrophobic surfaces by this LbL technique is dependent on the dense electrostatic interaction between the oppositely charged polycations and polyanions. This protocol allows the control of the thickness of the resulting deposition layer with molecular precision (Ariga et al., 2011; Almodóvar et al., 2011). The LbL assembly has distinct advantages in its ability to adjust the size/ratio of the hierarchical dual nanostructures without the need to incorporate the use of any organic solvent, hence lowering the amount of reaction time to nearly half of an hour. Accurate tailoring of the number of depository layers results in generating a highly superhydrophobic surface with a WCA as high as 150° and a minimum slide-off value ~1° (Checco et al., 2014). This optimum value of superhydrophobicity is achieved when the maximum number of deposited layers is from 1 to 3 (Latthe et al., 2012). Commonly used raw materials for LbL assembly include silica (Mahadik et al., 2010; Wang et al., 2012), cotton (Liu et al., 2011b), polymers such as polyethyleneimine, polyacrylic acid, and poly(allylamine hydrochloride) (PAH).

Figure 6(A) is an illustration of a superhydrophobic, self-assembled silica microsphere film reported by Sun et al. (2007). Figure 6(B) is a 3D atomic force microscopy image of hierarchically ordered microporous polymer formed by spontaneous assembly of water vapor condensation, so-called breath figure method (Zander et al., 2007). Tsai and Lee (2007) reported a raspberry-like structure of superhydrophobic films by assembling one

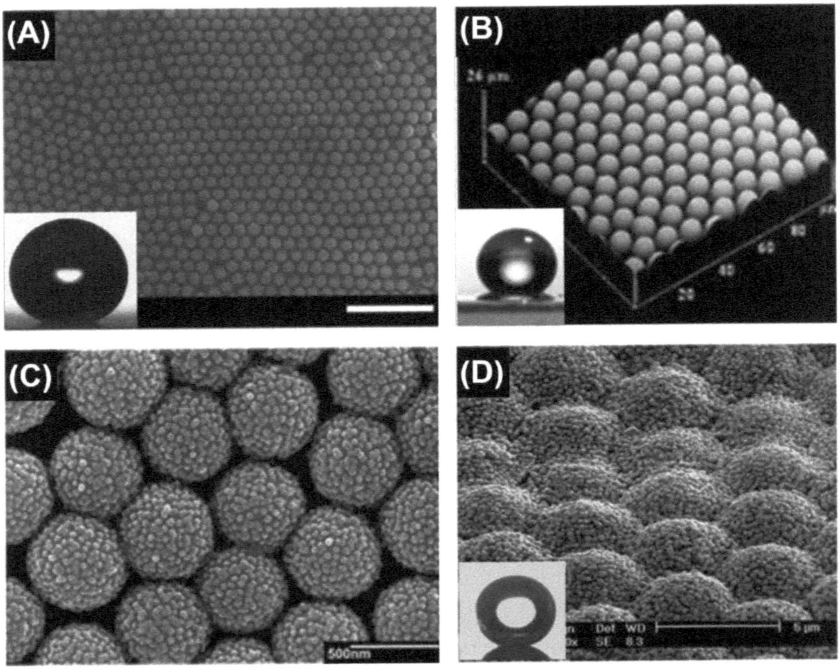

Figure 6 (A) Hierarchical ordered assembled silica microspheres, WCA of approximately 161° (Sun et al., 2007), (B) 3D AFM image of micropillars formed on breath figure (Zander et al., 2007), (C) raspberry-like particulate film fabricated using LB deposition method (Tsai and Lee, 2007) and (D) superhydrophobic coating fabricated by depositing silver nanoparticles using layer-by-layer method, resulted in a WCA of 168° (Li et al., 2007b). *Images reproduced by permission of (A) Elsevier, Copyright 2007, (B, C, and D) American Chemical Society, Copyright 2007. Reprinted with permission from Elsevier, Copyright 2011.*

layer of silica nanoparticles on large silica particulate film prepared using Langmuir–Blodgett (LB) deposition (Figure 6(C)). Similar structure was achieved by depositing silver nanoparticles on a monolayer array of PS microspheres, generating a superhydrophobic coating on the resulted nanostructured surfaces (Figure 6(D)) (Li et al., 2007b).

Electrochemical Deposition

In an effort to mimic naturally occurring superhydrophobic and super "sticky" surfaces, electrochemical methods that incorporate the use of common chemical reactions are frequently used to fabricate superhydrophobic nanostructures. Cathodic electrodeposition offers an efficient way of fabricating nanostructures through a process that is low in cost, suitable for large scales, environmental friendly, easy to use, and suitable for low-temperature processing. Distinct electrodeposition parameters, such as growth temperature, time period of deposition, proper concentration, and composition of the working electrolyte, and the working voltage (Wang and Guo, 2013), are found to significantly affect the morphology of metallic films fabricated through this method (Hsieh et al., 2010). By modifying the monomer concentration and the electrodeposition conditions such as electrolyte deposition method and the charge, the surface chemistry of the substratum can also be easily modified. The electrochemical deposition of zinc oxide (ZnO) films has resulted in WCAs close to 167° (He and Wang, 2011). Electrochemical deposition has also been employed to generate biomimetic superhydrophobic surfaces that contain hierarchical microscale and nanostructures (Vourdas et al., 2008), and regular porous structures grown in acid electrolyte with pore diameters ranging from five to several hundred nanometers. The porous structure generated through the electrochemical deposition exhibits a high hydrophobicity, with the WCA being as high as 152° (Shirtcliffe et al., 2004; Wang et al., 2011a).

Chemical Vapor Deposition

CVD is a deposition process whereby the initiating chemical precursors are transformed into a vapor phase and subsequently allowed to decompose or react on a preheated substratum to form a thin film (Camilli et al., 2011). This process can either be applied to induce superhydrophobicity on the reactant surface or to cast a thin film on the surface, which in turn makes the base substratum superhydrophobic. The electrochemical deposition method is sometimes used to subsequently deposit a film or powder on the base surface. The CVD method has several advantages, allowing the resulting structures to be controlled using various growth parameters (Choi et al., 2000). Carbon nanotubes (CNTs) fabricated using microwave plasma chemical vapor deposition have been used as effective superhydrophobic coatings. It has been reported that the WCA decreases with the simultaneous increase in the working pressure for the reaction (Lee et al., 2013a).

CONCLUSIONS AND OUTLOOK

There is a continuing need for the development of multifunctional materials. The ability to develop a material that contains a user-defined blend of two or more functions into an exclusive composite is desirable in the assembly of novel, multifunctional self-cleaning surfaces. Naturally occurring superhydrophobic surfaces have inspired researchers to devise fabrication techniques for biomimetic synthetic surfaces that generate an exact replica of the natural template surface. An understanding of the fabrication techniques discussed in this chapter for the accurate design of superhydrophobic surfaces allows us to delve deeper into the mechanisms by which these natural surfaces function, and as a result, these properties can be incorporated into synthetic biomimetic surfaces. The resulting multifunctional smart surfaces will be useful in various applications, ranging from fluid handling and transport, optical sensing, and medicine, and as self-cleaning and antifouling materials that are suitable for use in extreme environmental conditions. Biological solutions continue to represent a source of inspiration for the development of multifunctional smart surfaces. Hence, further interdisciplinary collaboration is essential in the design of multifunctional, smart, self-cleaning surfaces in science and engineering.

REFERENCES

Agarwal, S., Horst, S., Bognitzki, M., 2006. Electrospinning of fluorinated polymers: formation of superhydrophobic surfaces. Macromol. Mater. Eng. 291, 592.

Akihiko, K., Yu, A., Takeshi, M., Masashi, Y., Yousuke, G., Seiji, T., Takashi, N., Hideo, H., 2014. High removal rate of cross-linked su-8 resist using hydrogen radicals generated by tungsten hot-wire catalyzer. Thin Solid Films 562, 632.

Allam, N.K., Grimes, C.A., 2011. Electrochemical fabrication of complex copper oxide nanoarchitectures *via* copper anodization in aqueous and non-aqueous electrolytes. Mater. Lett. 65, 1949.

Almodóvar, J., Place, L.W., Gogolski, J., Erickson, K., Kipper, M.J., 2011. Layer-by-layer assembly of polysaccharide-based polyelectrolyte multilayers: a spectroscopic study of hydrophilicity, composition, and ion pairing. Biomacromolecules 12, 2755.

Ariga, K., Lvov, Y.M., Kawakami, K., Ji, Q., Hill, J.P., 2011. Layer-by-layer self-assembled shells for drug delivery. Adv. Drug Deliv. Rev. 63, 762.

Asmatulu, R., Ceylan, M., Nuraje, N., 2010. Study of superhydrophobic electrospun nanocomposite fibers for energy systems. Langmuir 27, 504.

Baldacchini, T., Carey, J.E., Zhou, M., Mazur, E., May 23 2006. Superhydrophobic surfaces prepared by microstructuring of silicon using a femtosecond laser. Langmuir 22(11), 4917–4919. http://www.ncbi.nlm.nih.gov/pubmed/16700574.

Barshilia, H.C., Gupta, N., 2014. Superhydrophobic polytetrafluoroethylene surfaces with leaf-like micro-protrusions through ArO_2 plasma etching process. Vacuum 99, 42.

Bico, J., Marzolin, C., Quere, D., 1999. Pearl drops. Europhys. Lett. 47, 220.

Biswas, A., Bayer, I.S., Biris, A.S., Wang, T., Dervishi, E., Faupel, F., 2012. Advances in top–down and bottom–up surface nanofabrication: techniques, applications & future prospects. Adv. Colloid Interface Sci. 170, 2.

Bormashenko, E., Stein, T., Whyman, G., Bormashenko, Y., Pogreb, R., 2006. Wetting properties of the multiscaled nanostructured polymer and metallic superhydrophobic surfaces. Langmuir 22, 9982.

Camilli, L., Scarselli, M., Del Gobbo, S., Castrucci, P., Nanni, F., Gautron, E., Lefrant, S., De Crescenzi, M., 2011. The synthesis and characterization of carbon nanotubes grown by chemical vapor deposition using a stainless steel catalyst. Carbon 49, 3307.

Cao, L., Hu, H.A., Gao, D., 2007. Design and fabrication of micro-textures for inducing a superhydrophobic behaviour on hydrophilic materials. Langmuir 23, 4310.

Celia, E., Darmanin, T., Taffin de Givenchy, E., Amigoni, S., Guittard, F., 2013. Recent advances in designing superhydrophobic surfaces. J. Colloid Interface Sci. 402, 1.

Checco, A., Rahman, A., Black, C.T., 2014. Robust superhydrophobicity in large-area nanostructured surfaces defined by block-copolymer self-assembly. Adv. Mater. 26, 886.

Chen, X., Wang, J.S., Li, H.Y., Huang, K.L., Sun, G.S., 2011. Characterization of TiO_2 nanotube arrays prepared via anodization of titanium films deposited by dc magnetron sputtering. Res. Chem. Intermed. 37, 441.

Chen, J.K., Qui, J.Q., Fan, S.K., Kuo, S.W., Ko, F.H., Chu, C.W., Chang, F.C., 2012. Using colloid lithography to fabricate silicon nanopillar arrays on silicon substrates. J. Colloid Interface Sci. 367, 40.

Choi, Y.C., Shin, Y.M., Lee, Y.H., Lee, B.S., Park, G.-S., Choi, W.B., Lee, N.S., Kim, J.M., 2000. Controlling the diameter, growth rate, and density of vertically aligned carbon nanotubes synthesized by microwave plasma-enhanced chemical vapor deposition. Appl. Phys. Lett. 76, 2367.

Choi, C.H., Kim, C.J., 2006. Fabrication of a dense array of tall nanostructures over a large sample area with sidewall profile and tip sharpness control. Nanotechnology 17, 5326.

Chung, C.K., Liao, M.W., Chang, H.C., Chang, W.T., Liu, T.Y., 2013. On characteristics of pore size distribution in hybrid pulse anodized high-aspect-ratio aluminum oxide with Taguchi method. Microsyst. Technol. 19, 387.

Decher, G., 1997. Fuzzy nanoassemblies: toward layered polymeric multicomposites. Science 277, 1232.

Ellinas, K., Smyrnakis, A., Malainou, A., Tserepi, A., Gogolides, E., 2011. "Mesh-assisted" colloidal lithography and plasma etching: a route to large-area, uniform, ordered nano-pillar and nanopost fabrication on versatile substrates. Microelectron. Eng. 88, 2547.

Feng, J., Rothstein, J., 2010. Effects of the secondary length scale on hierarchical superhydrophobic surfaces fabricated by double-layer electron beam lithography. In: APS Division of Fluid Dynamics Meeting Abstracts 2010.

Feng, J., Tuomien, M.T., Rothstein, J.P., 2011. Hierarchical superhydrophobic surfaces fabricated by dual-scale electron-beam-lithography with well-ordered secondary nanostructures. Adv. Funct. Mater. 21, 3715.

Fernández-Blázquez, J.P., Fell, D., Bonaccurso, E., de Campo, A., 2011. Superhydrophilic and superhydrophobic nanostructured surfaces via plasma treatment. J. Colloid Interface Sci. 357, 234.

Gao, R., Wang, J., Zhang, X., Yan, H., Yang, W., Liu, Q., Zhang, M., Liu, L., Takahashi, K., 2013. Fabrication of superhydrophobic magnesium alloy through the oxidation of hydrogen peroxide. Colloids Surf A physicochem Eng Asp 436, 906.

Gao, R., Liu, Q., Wang, J., Zhang, X., Yang, W., Liu, J., Liu, L., 2014. Fabrication of fibrous szaibelyite with hierarchical structure superhydrophobic coating on az31 magnesium alloy for corrosion protection. Chem. Eng. J. 241, 352.

Guo, Z., Zhou, F., Hao, J., Liu, W., 2006. Effects of system parameters on making aluminum alloy lotus. J. Colloid Interface Sci. 303, 298.

Guo, Z., Liu, W., Su, B.-L., 2011. Superhydrophobic surfaces: from natural to biomimetic to functional. J. Colloid Interface Sci. 353, 335.

Gupta, N., Kavya, M.V., Singh, Y.R.G., Jyothi, J., Barshilia, H.C., 2013. Superhydrophobicity on transparent fluorinated ethylene propylene films with nano-protrusion morphology by ArO_2 plasma etching: study of the degradation in hydrophobicity after exposure to the environment. J. Appl. Phys. 114, 164307.

He, G., Wang, K., 2011. The superhydrophobicity of ZnO nanorods fabricated by electrochemical deposition method. Appl. Surf. Sci. 257, 6590.

Hoefnagels, H.F., Wu, D., De With, G., Ming, W., 2007. Biomimetic superhydrophobic and highly oleophobic cotton textiles. Langmuir 23, 13158.

Hongxi, L., Qian, X., Xiaowei, Z., Chuanqi, W., Baoyin, T., 2012. Wear and corrosion behaviors of Ti_6Al_4V alloy biomedical materials by silver plasma immersion ion implantation process. Thin Solid Films 521, 89.

Hsieh, C.T., Yang, S.Y., Lin, J.Y., 2010. Electrochemical deposition and superhydrophobic behavior of ZnO nanorod arrays. Thin Solid Films 518, 4884.
Hu, Y.W., Liu, S., Huang, S.Y., Pan, W., 2010. Fabrication of superhydrophobic surfaces of titanium dioxide and nickel through electrochemical deposition on stainless steel substrate. Key Eng. Mater. 434, 496.
Huang, Z., Zhu, Y., Zhang, J., Yin, G., 2007. Stable biomimetic superhydrophobicity and magnetization film with Cu-Ferrite nanorods. J. Phys. Chem. C 111, 6821.
Jafari, R., Asadollahi, S., Farzaneh, M., 2013. Applications of plasma technology in development of superhydrophobic surfaces. Plasma Chem. Plasma P 33, 177.
Kang, C., Lu, H., Yuan, S., Hong, D., Yan, K., Liang, B., 2012. Superhydrophilicity/superhydrophobicity of nickel micro-arrays fabricated by electroless deposition on an etched porous aluminum template. Chem. Eng. J. 203, 1.
Kondrashov, V., Ruhe, J., 2014. Microcones and nanograss: towards mechanically robust superhydrophobic surfaces. Langmuir 15, 30.
Latthe, S.S., Imai, H., Ganesan, V., Rao, A.V., 2010. Porous superhydrophobic silica films by sol–gel process. Microporous Mesoporous Mater. 130, 115.
Latthe, S.S., Gurav, A.B., Maruti, C.S., Vhatkar, R.S., 2012. Recent progress in preparation of superhydrophobic surfaces: a review. JSEMAT 2, 76.
Lee, J.-H., Lee, S.H., Kim, D., Park, Y.S., 2013a. The structural and surface properties of carbon nanotube synthesized by microwave plasma chemical vapor deposition method for superhydrophobic coating. Thin Solid Films 546, 94.
Lee, M.W., An, S., Latthe, S.S., Lee, C., Hong, S.J., Yoon, S.S., 2013b. Electrospun polystyrene nanofiber membrane with superhydrophobicity and superoleophilicity for selective separation of water and low viscous oil. ACS Appl. Mater. Interfaces 5, 10597.
Li, J., Fu, J., Cong, Y., Wu, Y., Xue, L., Han, Y., 2006. Macroporous fluoropolymeric films templated by silica colloidal assembly: a possible route to super-hydrophobic surfaces. Appl. Surf. Sci. 252, 2229.
Li, Y., Duan, G., Cai, W., 2007a. Controllable superhydrophobic and lipophobic properties of ordered pore indium oxide array films. J. Colloid Interface Sci. 314, 615.
Li, Y., Lee, E.J., Cho, S.O., 2007b. Superhydrophobic coatings on curved surfaces featuring remarkable supporting force. J. Phys. Chem. C 111, 14813.
Li, Y., Zhang, J., Zhu, S., Dong, H., Jia, F., Wang, Z., Tang, Y., Zhang, L., Zhang, S., Yang, B., 2010. Bioinspired silica surfaces with near-infrared improved transmittance and superhydrophobicity by colloidal lithography. Langmuir 26, 9842.
Li, Y., Chang, S., Liu, X., Huang, J., Yin, J., Wang, G., Cao, D., 2012a. Nanostructured CuO directly grown on copper foam and their supercapacitance performance. Electrochim. Acta 85, 393.
Li, Y., Wang, X., Sun, J., 2012b. Layer-by-layer assembly for rapid fabrication of thick polymeric films. Chem. Soc. Rev. 41, 5998.
Lin, J., Cai, Y., Wang, X., Ding, B., Yu, J., Wang, M., 2011. Fabrication of biomimetic superhydrophobic surfaces inspired by lotus leaf and silver ragwort leaf. Nanoscale 3, 1258.
Liu, J., Huang, W., Xing, Y., Li, R., Dai, J., 2011a. Preparation of durable superhydrophobic surface by sol–gel method with water glass and citric acid. J. Solgel Sci. Technol. 58, 18.
Liu, Z., Bucknall, D.G., Allen, M.G., 2011b. Inclined nanoimprinting lithography-based 3D nanofabrication. J. Micromech. Microeng. 21, 065036.
Love, J.C., Gates, B.D., Wolfe, D.B., Paul, K.E., Whitesides, G.M., 2002. Fabrication and wetting properties of metallic half-shells with submicron diameters. Nano Lett. 2, 891.
Ma, M., Hill, R.M., Lowery, J.L., Fridrikh, S.V., Rutledge, G.C., 2005. Electrospun poly(styrene-block-dimethylsiloxane) block copolymer fibers exhibiting superhydrophobicity. Langmuir 21, 5549.
Ma, M., Gupta, M., Li, Z., Zhai, L., Gleason, K.K., Cohen, R.E., Rubner, M.F., Rutledge, G.C., 2007. Decorated electrospun fibers exhibiting superhydrophobicity. Adv. Mater. 19, 255.

Mahadik, S.A., Kavale, M.S., Mukherjee, S., Rao, A.V., 2010. Transparent superhydrophobic silica coatings on glass by sol–gel method. Appl. Surf. Sci. 257, 333.

Ming, B., Li, J., Kang, F., Pang, G., Zhang, Y., Chen, L., Xu, J., Wang, X., 2012. Microwave–hydrothermal synthesis of birnessite-type MnO_2 nanospheres as supercapacitor electrode materials. J. Power Sources 198, 428.

Miyamura, Y., Park, C., Kinbara, K., Leibfarth, F.A., Hawker, C.J., Aida, T., 2011. Controlling volume shrinkage in soft lithography through heat-induced cross-linking of patterned nanofibers. J. Am. Chem. Soc. 133, 2840.

Mozalev, A., Habazaki, H., Hubálek, J., 2012. The superhydrophobic properties of self-organized microstructured surfaces derived from anodically oxidized Al/Nb and Al/Ta metal layers. Electrochim. Acta 82, 90.

Öner, D., McCarthy, T.J., 2000. Ultrahydrophobic surfaces. Effects of topography length scales on wettability. Langmuir 16, 7777.

Palumbo, F., Muo, R.D., Cappelluti, D., d'Agostino, R., 2011. Superhydrophobic and superhydrophilic polycarbonate by tailoring chemistry and nano–texture with plasma processing. Plasma Processes Polym. 8, 118.

Papageorgiou, D.P., Tsougeni, K., Tserepi, A., Gogolides, E., 2013. Superhydrophobic, hierarchical, plasma-nanotextured polymeric microchannels sustaining high-pressure flows. Microfluid. Nanofluid. 14, 247.

Park, B.G., Lee, W.H., Kim, J.S., Lee, K.B., 2010. Superhydrophobic fabrication of anodic aluminum oxide with durable and pitch-controlled nanostructure. Colloids Surf. A 370, 15.

Peng, P., Ke, Q., Zhou, G., Tang, T., 2013. Fabrication of microcavity-array superhydrophobic surfaces using an improved template method. J. Colloid Interface Sci. 395, 326.

Qian, B., Shen, Z., 2005. Fabrication of superhydrophobic surfaces by dislocation-selective chemical etching on aluminum, copper, and zinc substrates. Langmuir 21, 9007.

Rao, A.V., Latthe, S.S., Mahadik, S.A., Kappenstein, C., 2011. Mechanically stable and corrosion resistant superhydrophobic sol–gel coatings on copper substrate. Appl. Surf. Sci. 257, 5772.

Roach, P., Shirtcliffe, N.J., Newton, M.I., 2008. Progress in superhydrophobic surface development. Soft Matter 4, 224.

Sas, I., Gorga, R.E., Joines, J.A., Thoney, K.A., 2012. Literature review on superhydrophobic self-cleaning surfaces produced by electrospinning. J. Polym. Sci. B Polym. Phys. 50, 824.

Schlenoff, J.B., Dubas, S.T., Farhat, T., 2000. Sprayed polyelectrolyte multilayers. Langmuir 16, 9968.

Shieh, J., Hou, F.J., Chen, Y.C., Chen, H.M., Yang, S.P., Cheng, C.C., Chen, H.L., 2010. Robust air like superhydrophobic surfaces. Adv. Mater. 22, 597.

Shirtcliffe, N.J., McHale, G., Newton, M.I., Chabrol, G., Perry, C., 2004b. Dual-scale roughness produces unusually water-repellent surfaces. Adv. Mater. 16, 1929.

Shirtcliffe, N.J., Mchale, G., Newton, M.I., Perry, C.C., 2005. Wetting and wetting transitions on copper-based super-hydrophobic surfaces. Langmuir 21, 937.

Singh, R.A., Yoon, E.S., Kim, H.J., Kim, J., Jeong, H.E., Suh, K.Y., 2007. Replication of surfaces of natural leaves for enhanced micro-scale tribological property. Mater. Sci. Eng. C 27, 875.

Such, G.K., Johnston, A.P.R., Caruso, F., 2011. Engineered hydrogen-bonded polymer multilayers: from assembly to biomedical applications. Chem. Soc. Rev. 40, 19.

Sun, C., Ge, L.Q., Gu, Z.Z., 2007. Fabrication of super-hydrophobic film with dual-size roughness by silica sphere assembly. Thin Solid Films 515, 4686.

Sun, X.L., Fan, Z.P., Zhang, L.D., Wang, L., Wei, Z.J., Wang, X.Q., Liu, W.L., 2011. Superhydrophobicity of silica nanoparticles modified with polystyrene. Appl. Surf. Sci. 257, 2308.

Sun, L., Wang, X., Li, M., Zhang, S., Wang, Q., 2014a. Anodic titania nanotubes grown on titanium tubular electrodes. Langmuir 30, 2835.

Sun, Q.F., Lu, Y., Li, J., Cao, J., 2014b. Self-assembly of a superhydrophobic ZnO nanorod arrays film on wood surface using a hydrothermal method. Key Eng. Mater. 609, 468.

Tang, P., Zhang, W., Wang, Y., Zhang, B., Wang, H., Lin, C.T., Zhang, L., 2011. Effect of superhydrophobic surface of titanium on *Staphylococcus aureus* adhesion. J. Nanomater. 2011, 2.

Tsai, H.J., Lee, Y.L., 2007. Facile method to fabricate raspberry-like particulate films for superhydrophobic surfaces. Langmuir 23, 12687.

Tsuruta, S., Morimoto, K., Hirotsu, T., Suzuki, H., 2006. Superhydrophobic phenomena on three-dimensional surface structures coated with plasma polymer. *Jpn. J. Appl. Phys. Part 1 Regular Papers and Short Notes and Review Papers* 45, 8502.

Tupala, J., Kemell, M., Hyvönen, H., Ritala, M., Leskelä, M., 2014. Voltage-dependent properties of titanium dioxide nanotubes anodized in solutions containing EDTA. J. Electrochem Soc. 161, E61.

Vourdas, N., Tserepi, A., Boudouvis, A.G., Gogolides, E., 2008. Plasma processing for polymeric microfluidics fabrication and surface modification: effect of super-hydrophobic walls on electroosmotic flow. Microelectron. Eng. 85, 1124.

Wagterveld, R.M., Berendsen, C.W.J., Bouaidat, S., Jonsmann, J., 2006. Ultralow hysteresis superhydrophobic surfaces by excimer laser modification of SU-8. Langmuir 22, 10904.

Wang, D., Guo, Z., Chen, Y., Hao, J., Liu, W., 2007a. In situ hydrothermal synthesis of nanolamellate $CaTiO_3$ with controllable structures and wettability. Inorg. Chem. 46, 7707.

Wang, J., Wen, Y., Hu, J., Song, Y., Jiang, L., 2007b. Fine control of the wettability transition temperature of colloidal-crystal films: from superhydrophilic to superhydrophobic. Adv. Funct. Mater. 17, 219.

Wang, J., Chen, H., Chen, D., 2011a. Synthesis of biomimetic superhydrophobic surface through electrochemical deposition on porous alumina. J. Bionic Eng. 8, 122.

Wang, S., Liu, C., Liu, G., Zhang, M., Li, J., Wang, C., 2011b. Fabrication of superhydrophobic wood surface by a sol–gel process. Appl. Surf. Sci. 258, 806.

Wang, X., Ding, B., Yu, J., Wang, M., 2011c. Engineering biomimetic superhydrophobic surfaces of electrospun nanomaterials. Nano Today 6, 510.

Wang, S., Guo, X., Xie, Y., Liu, L., Yang, H., Zhu, R., Gong, J., Peng, L., Ding, W., 2012. Preparation of superhydrophobic silica film on Mg–Nd–Zn–Zr magnesium alloy with enhanced corrosion resistance by combining micro-arc oxidation and sol-gel method. Surf. Coat. Technol. 213, 192.

Wang, A., Guo, Z., 2013. Superhydrophobic copper mesh films with rapid oil/water separation properties by electrochemical deposition inspired from butterfly wing. Appl. Phys. Lett. 103, 063704.

Wang, Y., Shi, L., Zhang, W., Jiang, Z., Mu, J., 2014. A comparative structure-property study of polyphosphazene micro-nano spheres. Polym. Bull. 71, 275.

Wu, Y., Cui, J.W., Wang, Y., Xu, G.Q., Zhang, X.Y., 2011. Effects of pretreatment and the first anodization time on morphology of anodic aluminium oxide templates. Adv. Mater. Res. 181, 707.

Wu, R.M., Liang, S.Q., Chen, H., Pan, A.Q., Jiang, H.Y., Deng, J., Yu, Y., Yuan, Z.Q., Liu, Q.L., 2012. Fabrication of a super-hydrophobic micro-nanoporous aluminum surface by anodic oxidation. Appl. Mech. Mater. 200, 190.

Xiang, H., Zhang, L., Wang, Z., Yu, X., Long, Y., Zhang, X., Zhao, N., Xu, J., 2011. Multifunctional polymethylsilsesquioxane (PMSQ) surfaces prepared by electrospinning at the sol–gel transition: superhydrophobicity, excellent solvent resistance, thermal stability and enhanced sound absorption property. J. Colloid Interf. Sci. 359, 296.

Yang, H., Pi, P., Cai, Z.Q., Wen, X., Wang, X., Cheng, J., Yang, Z., 2010. Facile preparation of super-hydrophobic and super-oleophilic silica film on stainless steel mesh via sol-gel process. Appl. Surf. Sci. 256, 4095.

Yoo, J.E., Choi, J., 2010. Surfactant-assisted growth of anodic nanoporous niobium oxide with a grained surface. Electrochim. Acta 55, 5142.

Yoo, J.E., Park, J., Cha, G., Choi, J., 2013. Micro-length anodic porous niobium oxide for lithium-ion thin film battery applications. Thin Solid Films 531, 583.

Yoshimitsu, Z., Nakajima, A., Watanabe, T., Hashimoto, K., 2002. Effects of surface structure on the hydrophobicity and sliding behaviour of water droplets. Langmuir 18, 5815.

Zander, N.E., Orlicki, J.A., Karikari, A.S., Long, T.E., Rawlett, A.M., 2007. Super-hydrophobic surfaces via micrometer-scale templated pillars. Chem. Mater. 19, 6145.

Zahner, D., Abagat, J., Svec, F., Fréchet, J.M.J., Levkin, P.A., 2011. A facile approach to superhydrophilic–superhydrophobic patterns in porous polymer films. Adv. Mater. 23, 3030.

Zhang, F., Chen, S., Dong, L., Lei, Y., Liu, T., Yin, Y., 2011. Preparation of superhydrophobic films on titanium as effective corrosion barriers. Appl. Surf. Sci. 257, 2587.

Zhang, Y., Rutledge, G.C., 2012. Electrical conductivity of electrospun polyaniline and polyaniline-blend fibers and mats. Macromolecules 45, 4238.

Zhang, H.F., Liu, X.W., Zhao, M., Wang, D.B., Shi, C.Z., 2013. Fabrication of super-hydrophobic surface on stainless steel using chemical etching method. Key Eng. Mater. 562, 33.

Zheng, J., He, A., Li, J., Xu, J., Han, C.C., 2006. Studies on the controlled morphology and wettability of polystyrene surfaces by electrospinning or electrospraying. Polymer 47, 7095.

Zhou, Y., He, B., Yang, Y., Wang, F., Liu, W., Wang, P., Zhang, W., Bello, I., Lee, S.-T., 2011. Construct hierarchical superhydrophobic silicon surfaces by chemical etching. J. Nanosci. Nanotechnol. 11, 2292.

Zhu, L., Feng, Y., Ye, X., Zhou, Z., 2006a. Tuning wettability and getting superhydrophobic surface by controlling surface roughness with well-designed microstructures. Sens. Actuators A Phys. 595, 130.

Zhu, Y., Jing, C.Z., Zhai, J., Yong, M.Z., Feng, L., Jiang, L., 2006b. Multifunctional carbon nanofibers with conductive, magnetic and superhydrophobic properties. ChemPhysChem 7, 336.

CHAPTER FOUR

Hydrophobicity of Nonwetting Soils

GLOSSARY

AFM Atomic force microscopy
MED Molarity of an ethanol droplet
PA Palmitic acid
SOM Soil organic matters
ST Stigmasterol
WDPT Water drop penetration time

NONWETTING SOIL AND ITS IMPACT ON WATER TRANSPORT

Dekker et al. point out that water repellency appears to be a rule rather than an exception in the field soils of many countries (Dekker et al., 2005), and hence soil water repellency (SWR) as a phenomenon has been an active topic of study for soil scientists and hydrologists, particularly during the past two decades (Dekker et al., 2005, 2010). SWR exists over a wide range of sandy, loamy, and clay soil types (Doerr et al., 2000; Diehl, 2013), where their wetting behavior has become an increased research focus due to its impact on (1) reduced infiltration rates and corresponding increases in surface runoff and erosion; (2) greater spatial variability in infiltration (fingering and preferential flow) and uneven distribution of soil moisture retention; and (3) increased susceptibility to wind erosion caused by drier soils with reduced cohesion of soil particles. Diehl (2013) showed that the infiltration capacity is lowered by about 6-fold compared with a similar hydrophilic equivalent. Water repellency is also a relative property whose severity depends on the initial soil moisture and particle size, as well as the type of organic matter coating the soil particle surfaces. Such a water-repellent surface layer causes rainwater to pond and, if rainfall is sufficient and surface retention is exceeded, runoff is induced. The poor distribution and dynamics of soil moisture within the root zone of plants also result in limited agricultural yields (Abadi Ghadim, 2000; Harpergilkes, 1994). It is estimated, for example, that about 5.75 million hectares of Australian

Superhydrophobic Surfaces
http://dx.doi.org/10.1016/B978-0-12-801109-6.00004-5

agricultural soils are nonwetting sands (Abadi Ghadim, 2000), principally limiting grain production with an annual average loss of 40% in these farming systems. The importance of understanding such nonwetting and poorly wetting soils more thoroughly has gained further priority due to the increased global demand for global fresh water and food production.

SUPERHYDROPHOBICITY OF SOIL SURFACES

The wetting behavior of an ideal flat surface is determined by its chemical composition and the corresponding molecular properties of wetting liquid. Soils are, in general, characterized by extreme large surface-to-volume ratios and surfaces formed by soil are far from flat. Basic interfacial properties of mineral soils, such as the surface charge density, polarity, and specific surface area, may therefore significantly influence swelling and shrinkage, water adsorption and permeability, or adsorption of soluble and colloidal species from soil solutions within bulk soil (McHale et al., 2005, 2007; Bachmann and McHale, 2009).

Correlations between superhydrophobicity in materials science and extreme water repellency in soil science have only occasionally been mentioned in the literature (McHale et al., 2005, 2007; Bachmann and McHale, 2009), where it is generally termed "severe repellency". The requirements for superhydrophobicity are 2-fold: (1) the small scale surface topography must be suitable to provide a composite surface of solid–air phases with gaps or pores sufficiently small for surface tension forces to dominate and (2) the surface composition of the solid must be hydrophobic to reduce capillary penetration and be imparted on the solid without clogging interparticle pores. It has been suggested that the size range of grains and their packing density are appropriate to contribute superhydrophobicity (McHale et al., 2005, 2007; Bachmann and McHale, 2009; Hamlett et al., 2013). In-field soil specific features also contribute to their superhydrophobic behavior as shown by soil exhibiting extreme water repellency usually within the upper part of the soil profile, suggesting a strong surface effect; enhancement by soil drying consistent with water droplets sitting on a composite air–solid surface; and, finally, forest fires or intense heating of soil causes significant water repellency by volatilized (hydrophobic) waxes from organic matter that subsequently condense on to soil particles.

These characteristics of nonwetting soils are intimately related to the formation of soil organic matter (SOM) coatings, where they have been shown to play a crucial role in soil packing density and overall hydrophobicity (Oades, 1984; Sollins et al., 1996; Piccolo and Mbagwu, 1999).

ROLE OF SOIL ORGANIC MATTER ON WATER REPELLENCY

Most mineral compounds that constitute soil are hydrophilic; however, the modification of organic matter coating the soil particles is instrumental for the transformation of these hydrophilic surfaces to hydrophobic soils (Doerr et al., 2000; Mainwaring et al., 2013). SOM is considered to be a byproduct of plant and animal residues at various stages of decomposition, root exudates, cells and tissues of soil organisms, and substances produced by fungal or soil microbial activity (Llewellyn et al., 2004). Attempts to find correlations between the level of SOM and SWR have to date yielded inconsistent results.

Considerable research progress has been made in the identification of compounds causing this SWR. In particular, most studies have used Soxhlet methods to extract organic matter from nonwetting soils using polar and nonpolar solvents followed by identification of these extracted organic compounds by combined gas chromatography–mass spectroscopy (Llewellyn et al., 2004; Doerr et al., 2005; Atanassova and Doerr, 2010; Cheng et al., 2010; Mainwaring et al., 2013). Polar and amphiphilic solvents were demonstrated to be more effective at removing water repellency from nonwetting soils. Solvents such as chloroform, dichloromethane, and *iso*-propanol were shown not to be efficient in removing organic coatings from soils when assessed for water repellency. Soils treated with non–hydrogen-bonding organic solvents resulted in increased hydrophobicity, which was attributed to solvent-induced conformational changes in surface-exposed SOM.

The effective extraction of SOM by organic solvents indicates that the bulk of the hydrophobic compounds in these soils are not covalently linked to the soil surfaces and that, when redeposited on acid-washed hydrophilic sands, hydrophobicity is restored to a certain degree but not to a sufficient extent that the original nonwetting characteristic were restored (Cheng et al., 2010). This suggesting that the organic matter components remaining on the extracted soil may provide adsorption sites for the other nonpolar material that was removed in the extracts. In general, this highlights the question of whether the hydrophobization of surfaces occurs from selective adsorption of organic matter or the interaction between solvent systems and the organic matter on the extracted soil.

The chemical composition of SOM was found to be more influential on water repellency than the level of SOM on the surfaces. Amphiphilic compounds (mainly long-chain fatty acids), alkanes, amides, aldehydes/ketones, and sterols are the major groups of hydrophobic compounds found on the soil surfaces in addition to insoluble Ca and Mg fatty acid salts, waxes, phytanes, and phytones (Piccolo and Mbagwu, 1999; Doerr et al., 2000; Graber et al., 2009; Atanassova and Doerr, 2010; Mainwaring et al., 2013). Such extraction studies indicate that the outer most organic layer is largely responsible for the particle hydrophobicity. Suggested mechanisms of formation of hydrophobic coatings include conformational changes in the arrangement of molecules and functional groups, hydration, or the spatial rearrangement of organic molecules on the surfaces of SOM coatings resulting from contact with water.

The microstructure of the SOM films formed on the surface was suggested (Mainwaring et al., 2013) to be monolayers of closely packed and steeply orientated species resulting in expanded films, which remain coherent but occupy a much larger area than that in highly condensed films in which the molecules are separate and independently mobile (Figure 1).

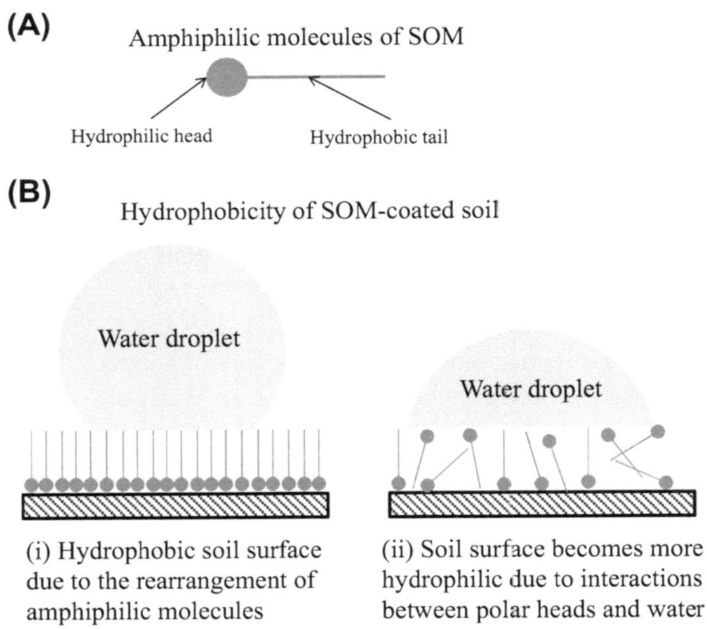

Figure 1 Schematic diagram of (A) an amphiphilic molecules of SOM and (B) changes in SOM molecular orientation leading to the change in hydrophobicity of soil surfaces.

The formation of such films are favored by (1) bulky head groups, (2) more than one polar group, (3) more than one hydrocarbon chain orientated in different directions from the polar part of the molecule (e.g., esters), (4) bent hydrocarbon chains such as *cis*-unsaturated acids, and (5) branched hydrocarbon chains.

MICROSTRUCTURE OF SOIL ORGANIC MATTER COATINGS

Atomic force microscopy (AFM) topology and phase angle measurements together with adhesion force measurements have been used to probe the nanoscale properties of the natural soil particles that control the physiochemical interaction of material within a soil column as well the nature of soil after extraction of the hydrophobic coating (Cheng et al., 2009, 2010). While the topology provided the general morphology and the roughness of the coatings, phase angle mapping broadly differentiated the SOM compounds, and AFM measurements provided information on the degree of hydrophobicity of SOM-coated soil surfaces, including those with model coatings such as humic acids. The broad adhesion force distributions of natural particles illustrated the complexity of surface structure and nature of organic materials on soil surfaces (Cheng et al., 2009, 2010).

Recent studies, as shown in Figure 2, indicate the microstructural features of two key SOM films deposited on flat quartz substrate: (1) palmitic acid (a long-chain fatty acid, as a potential SOM component responsible for SWR) forms discontinuous films consisting of discrete molecular aggregates (or hemimicelles) of diameter 200–400 nm and of height 50–100 nm. Such a microstructure is consistent with the observation of Ravi et al. (2006), who reported clear difference in moisture adsorption behavior of mineral soil surfaces before and after palmitic acid treatment. Environmental scanning electron microscopy revealed the formation of a uniform coating of adsorbed water on surfaces prior to coating, while moisture adsorbed onto the 0.1% palmitic acid–treated soil had an irregular distribution of water droplets condensed on grain surfaces, which is indicative of discontinuous SOM film formation. (2) Stigmasterol (potential SOM component) similarly showed discrete 30–60 nm molecular aggregates of 2–4 nm height without the formation of a continuous film. In contrast, an equimolar mixture of palmitic acid and stigmasterol film formed large aggregates 100–150 nm together with a continuous film coating that itself had a topology consisting of undulations of 3–5 nm, suggestive of a gel-like

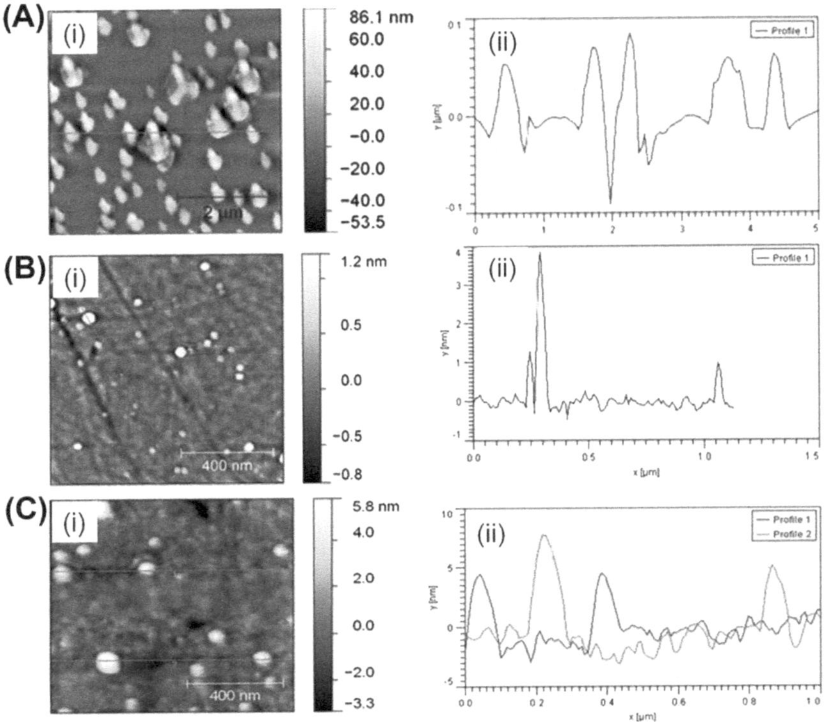

Figure 2 AFM in tapping mode using silicon tips showing the topography of (A) palmitic acid (0.22 µg/cm^2), (B) stigmasterol (0.35 µg/cm^2), and (C) equimolar mixture of palmitic acid and stigmasterol mixture (0.11 µg PA + 0.18 µg ST/cm^2). Line analyses indicated the dimensions of the nanostructural features formed on the respective surfaces. SOM concentration used is equivalent to the quantities assayed in the natural nonwetting soils.

disordered phase separated from the more structured aggregates. Such phase formation behavior is consistent with the binary solubility reported for these compounds where the mixing of palmitic acid and cholesterol was demonstrated to produce phase separation with 80% forming crystalline phase together with remaining 20% existing as liquid or gel like phase in equilibrium (Brief et al., 2009).

Depending on the nature of SOM coatings, the soil surfaces can span the characteristics of being completely wetting, partially wetting, and severely hydrophobic. Palmitic acid and stigmasterol coatings produced water contact angles of 63.1° ± 1.6° and 78.8° ± 4.0°, respectively, consistent with their partial wetting characteristics while equimolar blend of palmitic acid and stigmasterol produced a contact angle of $89.1° \pm 1.6°$, indicative of severe water repellency due to the formation of a continuous hydrophobic film.

ASSESSMENT OF WATER REPELLENCY OF SOIL SURFACES

A number of well-established methods are used to identify nonwetting soils, e.g., sessile drop, Wilhelmy plate and capillary rise techniques. However, the chemical heterogeneity and microscopic and macroscopic surface roughness of soil material makes the direct application of most of these methods for determining soil wetting repellency difficult. (1) Sessile drop methods, including water drop penetration time (WDPT) and molarity of an ethanol droplet (MED) tests are simple methods used widely to diagnose the degree of soil hydrophobicity. In WDPT test, a drop of water is placed on a flattened soil surface, with the time needed for the drop to penetrate the soil reflecting the degree of hydrophobicity. WDPT tests are, however, only considered to measure persistence of repellency rather than actual wettability. In MED tests, drops with decreasing ethanol molarity (increasing surface tension) are applied to the soil until one drop resists instant infiltration. In contrast to the WDPT test, the MED test assesses the actual repellency, i.e., soil surface energies. (2) Also, application of the Wilhemy plate method (WPM) requires significant adaption, since soil particles need to be adhered to a rectangular glass slide. The so obtained advancing and receding contact angle are influenced by surface roughness and chemistry. Here the surface is assumed to approximate close-packed solid hemispheres in the Wenzel state, however WPM contact angles of soil samples are only comparable to each other for samples with comparable grain size and grain form and measured with the same immersion depth and speed.

Moisture adsorption and infiltration in porous media that has been formed from particulates have been the object of much concentrated attention. In particular, the complexity of systems such as soil arises from the diversity of system parameters, such as their irregular shape, size, packing arrangement of particles, particle roughness and size distribution, as well as wettability of solid surface of particles themselves. Among the proposed models for the predicting the wettability behavior of such porous media, three models have been recognised as most realistic. Network models and their more recently refined versions (Lago and Araujo, 2001; Camassel et al., 2005; Lindquist, 2006) use irregular three-dimensional network and simulation processes to explain wetting behavior (Bakke and Øren,

1997). The flow and hydraulic conductivity of one-component liquids in unsaturated porous media and stability and the dynamics of wetting and drainage fronts have been investigated in series of studies (Or and Tuller, 2000, 2003; Shahraeeni and Or, 2010). Additionally, fractal-based models are being increasingly used, at times invoking the percolation theory in self-similar systems of Katz and Thompson (1985). Fragmentation, as a pervasive phenomenon, leads to greater precision in particle size and mass distribution inputs (Brown et al., 1983). The fragmentation model of Turcotte (1986) has added this greater precision to fractal models since it provided a mechanism by which scale-independent fracture properties could form a fractal distribution of particles. Rieu and Sposito (1991) then developed a model of fractal pore space linked to such a fractal particle model (Hunt and Ewing, 2009).

Experimentally, wetting and spreading phenomena in packed beds are investigated by measuring time-dependent capillary imbibition, which provides an indirect method for determining the contact angle of the advancing wetting front. The original Washburn expression for capillary wetting has been extended to porous media to determine the dynamic liquid contact angles of particulate solids. Here, neglecting the effects of slip, gravity, and inertial forces and in the absence of external pressure gradients, the Washburn relation describes the flow in a packed bed modeled as an effective single capillary:

$$\frac{\mathrm{d}h}{\mathrm{d}t} = \frac{r\,\gamma_{lv}\cos\theta}{4\eta h}$$

where h is the height of liquid front at any given time t, r is the effective capillary (pore) radius, γ_{lv} is surface tension, and η is viscosity of liquid, where the plot h^2/t is linear at sufficiently short times t. Performing the experiment with a liquid that completely wets the particulate packing i.e., has a zero contact angle and therefore $\cos\theta = 1$, a value of r is obtained. Further measurements using other liquids for the same packing densities, values of $\cos\theta$ and hence effective wettability of the latter liquids can be determined (Siebold et al., 1997; Michel et al., 2001; Dang-Vu and Hupka, 2005).

The capillary imbibition behavior of packed bed columns with a range of model hydrophobic soil types prepared by using two routes to hydrophobicity is shown: (1) trimethylchlorosilane and (2) model hydrophobic SOM consisting of either individual or equimolar mixture of palmitic acid (PA) and stigmasterol (ST) at quantities equivalent to that in natural nonwetting soils. With increasing proportion of hydrophobic particulate content in the packed bed column, the

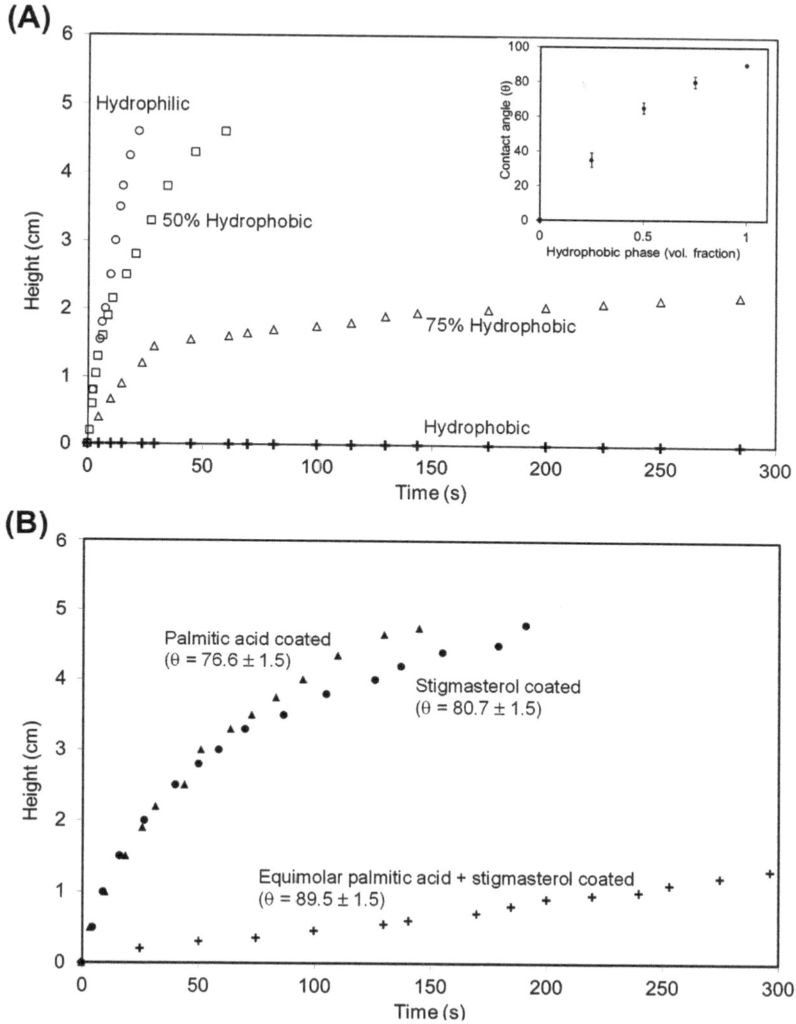

Figure 3 Capillary imbibition in model hydrophobic particulate beds using (A) silanized quartz (inset, water contact angle from initial slope method) and (B) SOM components (water contact angle from initial slope is provided in brackets).

imbibition rate decreased (Figure 3(A)) together with increasing contact angles calculated from the initial slope of the kinetic curve (Figure 3(A), inset) which shows asymptotic behavior approaching 90° and increasing insensitivity.

The capillary imbibition characteristics of single SOM components, PA- or ST-coated particulate beds (Figure 3(B)), were similar to that of the partially hydrophobic model hydrophobic compositions (Figure 3(A)), and the contact angles evaluated from the initial slope were consistent with such wetting behavior, which confirms the presence of partially wetting pore interfaces consistent with the microstructural features on the equivalent planar surfaces as seen by AFM (Figure 2). The hydrophobic particulate bed prepared with equimolar PA- + ST-coated particles exhibited severe hydrophobicity (Figure 3(B)), confirming the presence of a continuous nonwetting film, as seen by AFM.

INFLUENCE OF SURFACTANTS ON NONWETTING SOILS

Application of surfactants is one of the indirect remediation strategies to address the problem of water repellency in soil (Müller and Deurer, 2011). Surfactants enhance the distribution and dynamics of soil moisture within water-repellent soil profiles, improving water transport over the short term. Nonionic surfactants are typically used to improve the wettability of water-repellent soils, which include synthetic ethyleneoxide-polypropyleneoxide block copolymer-based surfactants and glycoside-based surfactants, which are being evaluated for turf, horticulture, and broad acre agricultural soils. While there are many reports in the literature on the impact of surfactant application on the crop yield (Cisar et al., 2000; Feng et al., 2002; Oostindie et al., 2008, 2012), a basic understanding of the detailed mechanism of surfactant interaction with the water-repellent surfaces and its influence on wetting and water transport is limited.

The influence of surfactants on the wetting of planar SOM-coated hydrophobic surfaces in Figure 2 is shown in Figure 4. Here, nonionic surfactant solutions of different concentrations provided a range of surface

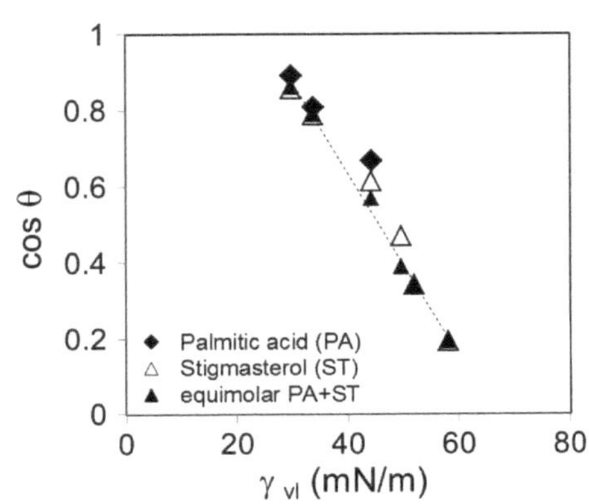

Figure 4 Advancing contact angle versus surface tension of surfactant solutions (Zisman type plot) for SOM-coated planar quartz surfaces.

tension between 58 and 28 mN/m. The wetting behavior of a surfactant solution on these model hydrophobic surfaces as expressed by the cosine of contact angle showed a linear relationship with surface tension. Zisman plots for different SOM-coated surfaces had similar slopes and yielded a critical surface tension that correlated with the surface energy of these SOM films: palmitic acid 22.5 mN/m, stigmasterol 24 mN/m, and the equimolar mixture of palmitic acid and stigmasterol 24 mN/m.

The influence of surfactants on the water imbibition in a packed bed, whether comprising model or natural soils, introduces complexities due to the multicomponent nature of the solution and its surface adsorption behavior, which may continuously vary as imbibition proceeds. Among the models for surfactant wetting of hydrophobic surfaces, two appear to be most appropriate for application to nonwetting soils. Starov and colleagues (Starov et al., 2000, 2004, 2007, 2010; Starov, 2004a, 2004b) focused on the adsorption of surfactant molecules onto a bare hydrophobic substrate in front of the moving three-phase contact line of a meniscus or droplet, where this molecular transfer provides an energy barrier and kinetic constraint to water spreading. A different view on spreading mechanisms was introduced by von Bahr et al. (2001). The key feature of this approach is that during the initial spreading phase, surfactant molecules adsorb at the expanding solid–liquid interface behind the moving liquid–vapor interface. Little work has been reported on the detailed mechanisms underlying the infiltration, permeability, and hydraulic conductivity in poorly wetting and superhydrophobic soils in the presence of such surfactants.

The influence of surfactant concentration and the accompanying surface tensions are shown in Figure 5 for a nonionic surfactant solution being applied to severely hydrophobic packed beds. With decreasing surface tension, the imbibition rate progressively increased. A Zisman plot of the contact angle data obtained from the Washburn initial slope method showed a linear relationship with a similar slope to the planar hydrophobic surface described above, although systematically shifted to lower surface tensions.

Water infiltration and hydraulic conduction in the packed bed occurs, in general, via two major processes: capillary imbibition and infiltration under hydraulic pressure. At initial depths, where the hydraulic pressure is negligible, capillary imbibition is the dominating process influencing the water transport, and with the progressing wetting front, a hydraulic pressure component increases and becomes a major contribution to the infiltration process and hence influences overall hydraulic conductivity. Surfactants play a crucial role in modulating the water transport in the initial stages where the

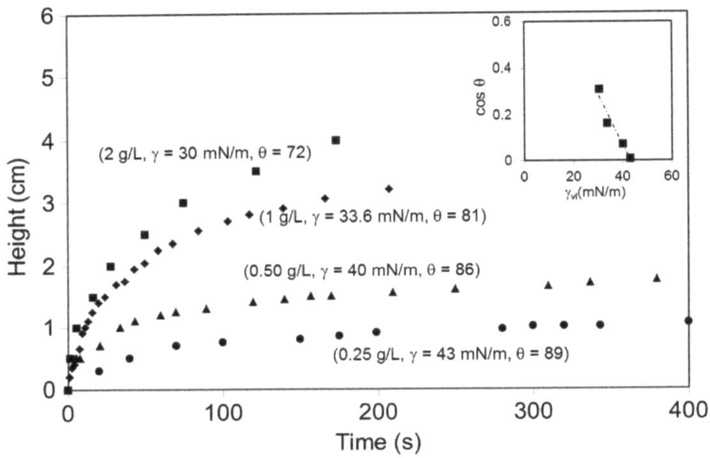

Figure 5 The influence of the surfactant concentration on the capillary imbibition in 100% silane-coated hydrophobic quartz particulate beds (surfactant concentration, corresponding surface tension, γ, and contact angle, θ, evaluated from initial slope are provided in brackets).

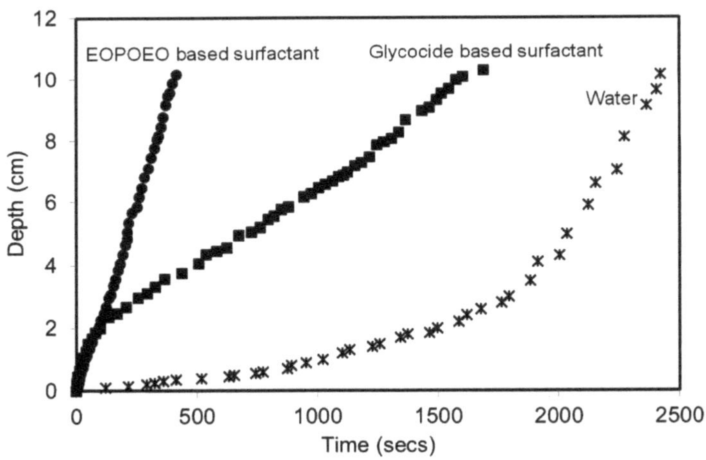

Figure 6 Infiltration in packed bed columns of severely hydrophobic soil at constant pressure head of 0.5 cm (surfactant concentration = 4 g/L).

capillary imbibition dominates this infiltration. Figure 6 demonstrates this, where, due to the high initial capillary entry pressure experienced without surfactant, the infiltration at initial stages is very slow (~40 mm/h) even with constant pressure head of 5 mm. With an advancing wetting front, the effective hydraulic pressure increases, resulting in an additional rate of infiltration due the increased effective hydraulic head. In contrast, surfactants reduce the surface tension of water, which reduces the respective initial pore entry pressures and improves the initial infiltration. The effectiveness of the surfactants is further influenced by differences in surfactant adsorption and depletion, which is seen in the subsequent relative rates of infiltration.

REFERENCES

Abadi Ghadim, A.K., 2000. Water repellency: a whole-farm bio-economic perspective. J. Hydrol. 231–232, 396.

Atanassova, I., Doerr, S., 2010. Organic compounds of different extractability in total solvent extracts from soils of contrasting water repellency. Eur. J. Soil Sci. 61, 298.

Bachmann, J., McHale, G., 2009. Superhydrophobic surfaces: a model approach to predict contact angle and surface energy of soil particles. Eur. J. Soil Sci. 60, 420.

Bakke, S., Øren, P.E., 1997. 3-d pore-scale modelling of sandstones and flow simulations in the pore networks. SPE J. 2, 136.

Brief, E., Kwak, S., Cheng, J.T.J., Kitson, N., Thewalt, J., Lafleur, M., 2009. Phase behavior of an equimolar mixture of n-palmitoyl-d-erythro-sphingosine, cholesterol, and palmitic acid, a mixture with optimized hydrophobic matching. Langmuir 25, 7523.

Brown, W.K., Karpp, R.R., Grady, D.E., 1983. Fragmentation of the universe. Astrophys. Space Sci. 94, 401.

Camassel, B., Sghaier, N., Prat, M., Nasrallah, S.B., 2005. Evaporation in a capillary tube of square cross-section: application to ion transport. Chem. Eng. Sci. 60, 815.

Cheng, S., Bryant, R., Doerr, S.H., Wright, C.J., Williams, P.R., 2009. Investigation of surface properties of soil particles and model materials with contrasting hydrophobicity using atomic force microscopy. Environ. Sci. Technol. 43, 6500.

Cheng, S., Doerr, S.H., Bryant, R., Wright, C.J., 2010. Effects of isopropanol/ammonia extraction on soil water repellency as determined by atomic force microscopy. Soil Sci. Soc. Am. J. 74, 1541.

Cisar, J.L., Williams, K.E., Vivas, H.E., Haydu, J.J., 2000. The occurrence and alleviation by surfactants of soil-water repellency on sand-based turfgrass systems. J. Hydrol. 231–232, 352.

Dang-Vu, T., Hupka, J., 2005. Characterization of porous materials by capillary rise method. Physicochem. Probl. Miner. Process. 39, 47.

Dekker, L.W., Oostindie, K., Ritsema, C.J., 2005. Exponential increase of publications related to soil water repellency. Aust. J. Soil Res. 43, 403.

Dekker, L.W., Ritsema, C.J., Oostindie, K., Moore, D., Wesseling, J.G., 2010. Methods for determining soil water repellency on field-moist samples. Water Resour. Res. 46, W00D33.

Diehl, D., 2013. Soil water repellency: dynamics of heterogeneous surfaces. Colloids Surf. A Physicochem. Eng. Asp. 432, 8.

Doerr, S.H., Llewellyn, C.T., Douglas, P., Morley, C.P., Mainwaring, K.A., Haskins, C., Johnsey, L., Ritsema, C.J., Stagnitti, F., Allinson, G., Ferreira, A.J.D., Keizer, J.J., Ziogas, A.K., Diamantis, J., 2005. Extraction of compounds associated with water repellency in sandy soils of different origin. Aust. J. Soil Res. 43, 225.

Doerr, S.H., Shakesby, R.A., Walsh, R.P.D., 2000. Soil water repellency: its causes, characteristics and hydro-geomorphological significance. Earth Sci. Rev. 51, 33.

Feng, G.L., Letey, J., Wu, L., 2002. The influence of two surfactants on infiltration into a water-repellent soil. Soil Sci. Soc. Am. J. 66, 361.

Graber, E.R., Tagger, S., Wallach, R., 2009. Role of divalent fatty acid salts in soil water repellency. Soil Sci. Soc. Am. J. 73, 541.

Hamlett, C.A.E., Atherton, S., Shirtcliffe, N.J., McHale, G., Ahn, S., Doerr, S.H., Bryant, R., Newton, M.I., 2013. Transitions of water-drop impact behaviour on hydrophobic and hydrophilic particles. Eur. J. Soil Sci. 64, 324.
Harpergilkes, R.J., 1994. Soil attributes related to water repellency and the utility of soil survey for predicting its occurrence. Aust. J. Soil Res. 32, 1109.
Hunt, A., Ewing, R., 2009. Percolation Theory for Flow in Porous Media. Springer, Berlin Heidelberg.
Katz, A.J., Thompson, A.H., 1985. Fractal sandstone pores: implications for conductivity and pore formation. Phys. Rev. Lett. 54, 1325.
Lago, M., Araujo, M., 2001. Threshold pressure in capillaries with polygonal cross section. J. Colloid Interface Sci. 243, 219.
Lindquist, W.B., 2006. The geometry of primary drainage. J. Colloid Interface Sci. 296, 655.
Llewellyn, C.T., Doerr, S.H., Douglas, P., Morley, C.P., Mainwaring, K.A., 2004. Soxhlet extraction of organic compounds associated with soil water repellency. Environ. Chem. Lett. 2, 41.
Mainwaring, K., Hallin, I.L., Douglas, P., Doerr, S.H., Morley, C.P., 2013. The role of naturally occurring organic compounds in causing soil water repellency. Eur. J. Soil Sci. 64, 667.
McHale, G., Newton, M.I., Shirtcliffe, N.J., 2005. Water-repellent soil and its relationship to granularity, surface roughness and hydrophobicity: a materials science view. Eur. J. Soil Sci. 56, 445.
McHale, G., Shirtcliffe, N.J., Newton, M.I., Pyatt, F.B., 2007. Implications of ideas on super-hydrophobicity for water repellent soil. Hydrol. Process. 21, 2229.
Michel, J.C., Rivière, L.M., Bellon-Fontaine, M.N., 2001. Measurement of the wettability of organic materials in relation to water content by the capillary rise method. Eur. J. Soil Sci. 52, 459.
Müller, K., Deurer, M., 2011. Review of the remediation strategies for soil water repellency. Agric. Ecosyst. Environ. 144, 208.
Oades, J.M., 1984. Soil organic matter and structural stability: mechanisms and implications for management. Plant Soil 76, 319.
Oostindie, K., Dekker, L.W., Wesseling, J.G., Ritsema, C.J., 2008. Soil surfactant stops water repellency and preferential flow paths. Soil Use Manage. 24, 409.
Oostindie, K., Dekker, L.W., Wesseling, J.G., Ritsema, C.J., Moore, D., 2012. Influence of a single soil surfactant application on potato ridge moisture dynamics and crop yield in a water repellent sandy soil. Acta Hortic. 938, 341.
Or, D., Tuller, M., 2000. Flow in unsaturated fractured porous media: hydraulic conductivity of rough surfaces. Water Resour. Res. 36, 1165.
Or, D., Tuller, M., 2003. Hydraulic conductivity of partially saturated fractured porous media: flow in a cross-section. Adv. Water Resour. 26, 883.
Piccolo, A., Mbagwu, J.S.C., 1999. Role of hydrophobic components of soil organic matter in soil aggregate stability. Soil Sci. Soc. Am. J. 63, 1801.
Ravi, S., D'Odorico, P., Herbert, B., Zobeck, T., Over, T.M., 2006. Enhancement of wind erosion by fire-induced water repellency. Water Resour. Res. 42.
Rieu, M., Sposito, G., 1991. Fractal fragmentation, soil porosity, and soil water properties: I. Theory. Soil Sci. Soc. Am. J. 55, 1231.
Shahraeeni, E., Or, D., 2010. Pore-scale analysis of evaporation and condensation dynamics in porous media. Langmuir 26, 13924.
Siebold, A., Walliser, A., Nardin, M., Oppliger, M., Schultz, J., 1997. Capillary rise for thermodynamic characterization of solid particle surface. J. Colloid Interface Sci. 186, 60.
Sollins, P., Homann, P., Caldwell, B.A., 1996. Stabilization and destabilization of soil organic matter: mechanisms and controls. Geoderma 74, 65.
Starov, V., Ivanova, N., Rubio, R.G., 2010. Why do aqueous surfactant solutions spread over hydrophobic substrates? Adv. Colloid Interface Sci. 161, 153.

Starov, V., Velarde, M., Radke, C., 2007. Wetting and Spreading Dynamics. CRC Press, Taylor and Francis Group, London, UK.
Starov, V.M., 2004a. Spontaneous rise of surfactant solutions into vertical hydrophobic capillaries. J. Colloid Interface Sci. 270, 180.
Starov, V.M., 2004b. Surfactant solutions and porous substrates: spreading and imbibition. Adv. Colloid Interface Sci. 111, 3.
Starov, V.M., Kosvintsev, S.R., Velarde, M.G., 2000. Spreading of surfactant solutions over hydrophobic substrates. J. Colloid Interface Sci. 227, 185.
Starov, V.M., Zhdanov, S.A., Velarde, M.G., 2004. Capillary imbibition of surfactant solutions in porous media and thin capillaries: partial wetting case. J. Colloid Interface Sci. 273, 589.
Turcotte, D.L., 1986. Fractals and fragmentation. J. Geophys. Res. 91, 1921.
von Bahr, M., Tiberg, F., Yaminsky, V., 2001. Spreading dynamics of liquids and surfactant solutions on partially wettable hydrophobic substrates. Colloids Surf. A Physicochem. Eng. Asp. 193, 85.

CHAPTER FIVE

Superhydrophobic Polymers

GLOSSARY

FAS Perfluoroalkysilane
LDPE Low density polyethylene
PDMS Poly(dimethyl siloxane)
PET Poly(ethylene terephthalate)
PTFE Poly(tetrafluoro ethylene)
PU Polyurethane
PVC Polyvinyl chloride
UV−NIL Ultraviolet nanoimprint lithography

INTRODUCTION

Significant progress has been made in the development of techniques that can support fabrication of superhydrophobic materials, yet, many of them require fairly expensive and complex processing. Superhydrophobic polymer materials are being actively researched as they offer a number of advantages in the production of such materials, including their low-cost and abundance, and ability to be processed under mild conditions, such as ambient temperature and under solution conditions. They can be easily chemically modified and combined with a variety of additives. These polymers can be used on their own; however, they are often considered as coatings for materials that are more expensive or difficult to process or where processing might undermine the mechanical integrity of the structure of the material.

In this chapter, a number of the top-down and the bottom-up approaches for the fabrication of superhydrophobic polymers will be presented. A brief discussion on the mechanisms of surface wettability will also be provided.

DESIGN OF SUPERHYDROPHOBIC POLYMERS

In the production of superhydrophobic polymers, to ensure that Cassie–Baxter is the dominant state of the surface, geometric parameters should be chosen so as to ensure this state has lower energy than the Wenzel state (Patankar, 2004). Surface properties that affect the stability of the Cassie–Baxter state include the dimensions and surface organization of nanoscale features

Superhydrophobic Surfaces
http://dx.doi.org/10.1016/B978-0-12-801109-6.00005-7

and the intrinsic contact angle. Rough surfaces with features at the microscale, nanoscale, and subnanoscale are most conducive to the stability of the Cassie–Baxter state (Tuvshindorj et al., 2014) and are displayed by a multitude of naturally occurring superhydrophobic surfaces, the lotus leaf being one of the best known examples. Others have developed surfaces that favor the Wenzel state, where both superhydrophobic properties and good adhesion to water are desired, the so-called rose petal effect (Barthlott and Neinhuis, 1997; Byun et al., 2009; Gao and Jiang, 2004; Feng et al., 2008; Sun et al., 2009). Many of these natural surfaces have structural anisotropism at the microscale, submicroscale, and nanoscale to assist with the self-cleaning process.

Interestingly, surfaces in the Wenzel state can display an ultra-low sliding angle if they are infused with a low surface energy liquid. This is the strategy used by the Nepenthes pitcher plant on its upper rim (Wong et al., 2011; Zhang et al., 2014). Since it is the trapped liquid that repels other liquids, it is potentially more robust to wear and can regenerate its water-repellent property after sustaining damage. Furthermore, the infusing liquid can be selected so as to repel a variety of liquids, such as oils and organic solvents, thus preventing contamination that might undermine the superhydrophobic properties of the surface. The key challenge here is finding surface–infusing liquid combinations, as the infusing liquid should have stronger wetting preference for the substrate and not for the liquid it is repelling. Furthermore, there should be no intermixing between the infusing liquid and the one being repelled.

Hydrophobic Polymers

For any polymer material, molecules at the surface have a higher potential energy than those in the bulk. A molecule in the bulk has attractive interactions with other molecules, whereas on the surface, these attractive interactions are unbalanced, leading to excess energy (i.e., surface or interface free energy). Materials with low surface energy are especially suited for the fabrication of superhydrophobic surfaces. Among the most widely used polymers are those containing fluorine, such as polytetrafluorethylene (PTFE). Other polymers are often treated to introduce organic groups of $-CF_3$, $-CH_3$, and $-CH_2-$ and thus render the polymer more hydrophobic. Polyethylene, polystyrene, polyvinyl chloride, polydimethylsiloxane (PDMS), and certain types of polyesters and polyurethanes are also hydrophobic and are used extensively in the fabrication of superhydrophobic constructs.

If used, hydrophilic polymers will need to be either subjected to surface modification to impart the relevant hydrophobic functionality or coated with a thin layer of a hydrophobic polymer.

Surface Topography and Roughness

A desirable surface roughness and morphology in polymers can be attained by using a variety of methods that can be broadly divided into top–down and bottom–up approaches. In the top–down approach, the existing material is first fabricated in bulk and then modified to exhibit the desired surface properties (e.g., by plasma etching or deposition of thin hydrophobic layer). In the bottom–up approach, the material is gradually built to possess the desired morphology and surface chemistry (e.g., the fabrication of an aerogel). Chemical vapor deposition is most often used for the fabrication of top thin films from a hydrophobic precursor, typically containing fluorine and silane functionalities. Plasma-assisted deposition can also be used for the application of the top hydrophobic layer, whereas plasma-assisted etching is used for masked and maskless surface etching. Lithography is widely used for template fabrication and for nanomodification of polymer substrates. Particle deposition, self-assembly, and self-organization are used to impart microscale and nanoscale features on the surface, which can then be coated with a polymer layer to create a superhydrophobic surface. Catalytic growth and plasma-assisted fabrication are used to grow nanoparticles, such as, nanotubes, nanoparticles, and nanofibers, on the temperature-resistant surfaces to form the desired nanotopography. There are, however, many techniques based on a mixed approach, such as etching of the bulk material to achieve the desired morphology followed by chemical vapor deposition of the hydrophobic chemistry onto the surface. Conversely, in this study, the fabrication of superhydrophobic microchannels involved first molding a composite material of PDMS/PTFE particles, followed by isotropic plasma etching (Tropmann et al., 2012).

Processing Considerations

The choice of processing method is highly dependent on the properties of the polymer (e.g., its thermal and physicochemical stability) and the resolution of the pattern to be achieved. From an industrial perspective, other considerations include scalability, simplicity and cost of processing, safety and environmental impact, ease of integration into existing manufacturing workflow, availability and cost of input and auxiliary materials. For example, superhydrophobic materials have great potential for used in the construction of microfluidic devices, where their low-adhesive properties can enable efficient liquid transport. Processing of these devices is, however, challenging due to their multimaterial composition and complex three-dimensional structure, which is rich in closed difficult-to-access channels. Moreover, for many real-life applications, the resistance of the superhydrophobic surface to oils and/or other solvents, such as hexadecane, dodecane, octane, and ethylene glycol, is highly desirable. Indeed, a superhydrophobic coating contaminated by an oily substance can suffer a significant loss of its water-repelling properties.

Poor physical stability of the surface structure and/or chemistry can also greatly reduce the lifetime of the superhydrophobic surface. The loss of nanoscale features is particularly damaging. To protect the fragile fine-scale surface topographies against wear and thus providing enhance mechanical durability of the superhydrophobic effect, the use of larger-scale sacrificial micropillars has been proposed (Huovinen et al., 2014). The wetting properties of superhydrophobic polypropylene surfaces enriched with protective structures (Figure 1) were found to remain in mechanical compression up to 20 MPa and in abrasive wear tests up to 120 kPa at the optimal surface density of the protective pillars of ~15%.

FABRICATION TECHNIQUES

Template-Based Fabrication and Lithography

The surfaces of polymers with suitable hydrophobic chemistry can be shaped to display superhydrophobic features by using molds. A common method of polymer shaping is hot embossing, where a mold is pressed against the polymer substrate at temperatures just above their melting temperature. Good resolution of features can be obtained using this approach. For large scale, high-speed industrial fabrication, melt-flow casting is more attractive. It also uses template replica molding and can be easily integrated into existing manufacturing processes. This process is challenged by difficulties in fabricating rollers with complex micrometer- and nanometer-scale two-dimensional ordered holes and incomplete penetration of the melt into these microstructures and nanostructures (Feng et al., 2010). Processing under vacuum condition can promote polymer filling of the mold, resulting in a more hydrophobic surface compared with that fabricated via thermal pressing alone (Feng et al., 2009).

The molds themselves can be fabricated to replicate a naturally occurring superhydrophobic surface, such as that of lotus leaf, lizard skin, and butterfly wing (Barthlott and Neinhuis, 1997; Byun et al., 2009; Feng et al., 2008; Gao and Jiang, 2004; Sun et al., 2009) or designed de novo. Biomimicking of superhydrophobic surfaces is fairly straightforward and typically involves the fabrication of a polymer-based high-resolution negative template, which is then used to produce nanostructured polymer films. PDMS is often the material of choice for the mold, as good resolution of the

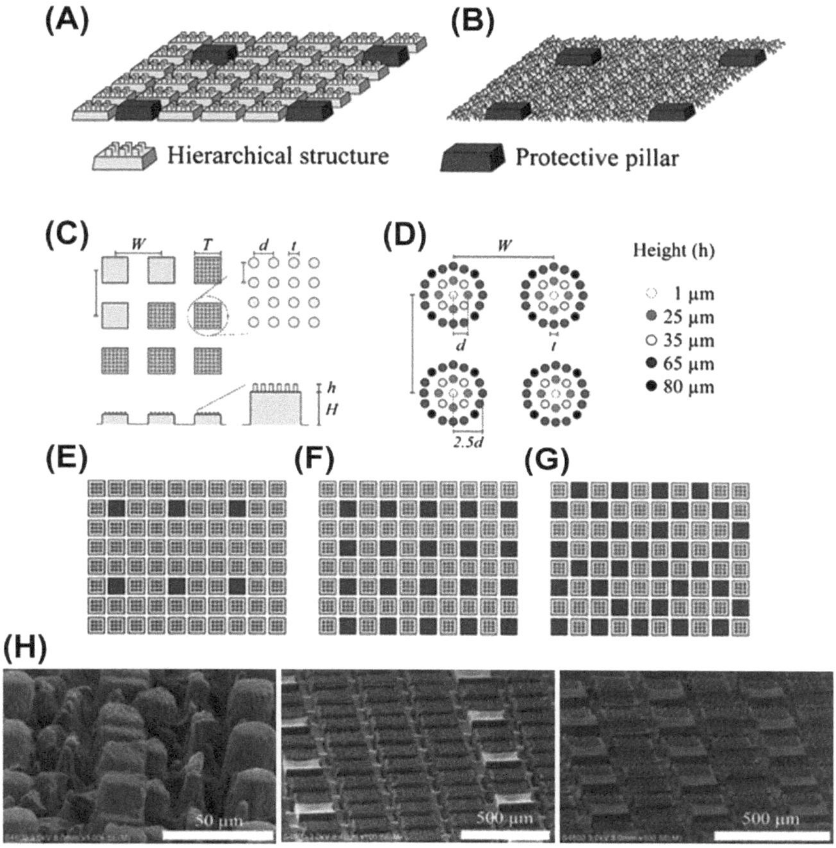

Figure 1 A design schematic of fine-scale surface topographies protected with micropillars (A) micromicrostructure, (B) multiscale structure, (C) fabrication parameters for hierarchical structures, (D) fabrication parameters for multiscale structures, and positioning of the replacement grids (E) p5, (F) p15, and (G) p25 (5%, 15%, and 25% of the total surface area, respectively). (H) SEM images of the examples of fabricated structures on the polypropylene surfaces. *Reproduced from Huovinen et al. (2014) by permission of the American Chemical Society.*

replica can be obtained under mild processing conditions. The quality of the pattern can be affected when the mold is used above complex, multifaceted shapes. For example, one of the methods to fabricate microchannels with superhydrophobic bottom and sidewall surfaces involves the transfer of molding-generated micropillars on PDMS films to the channel surfaces using a hot embossing process (Xinchuan and Cheng, 2010). While the middle sections of the wall and bottom surfaces had a similar nanotopography to the original design, the compressing and stretching of the PDMS films over the bottom and top corners of the channel, respectively, altered the spacing of the micropillars. For higher temperature and/or vacuum enabled processing, loss of fidelity can take place due to the deformation of the pattern, which can be at times beneficial, such as when vacuum was used to promote fiber elongation on the surfaces of superhydrophobic polyethylene (Feng et al., 2009).

PDMS molds are often produced from patterns designed and etched onto synthetic surfaces, such as micropatterned silicon wafers. The fabrication of patterned silicon wafers is far more challenging and costly and is typically done using photolithography. The mold itself is produced by liquid precursor being first cast onto the patterned wafer, cured and then peeled off. The process of making PDMS mold is simple; however, during peeling, the molds can get physically damaged and/or contaminated with the photoresist that can be lifted off the wafer. Furthermore, once the photoresist pattern is damaged, the defects will be replicated in the PDMS mold. In addition to photoresist-patterned silicon, masters can be produced using microstructured metal (e.g., stainless steel, brass, aluminum, and silicon surfaces, metal meshes, and polymer membranes) (Huang et al., 2011). Metal-based masters are much more robust and can withstand harsher processing conditions. The microstructures and nanostructures on their surfaces are typically produced using chemical etching (Feng et al., 2010). Transmission electron microscopy grids can be used as photomasks to create micropatterned nanostructures directly on PDMS solid surfaces without the need for a photoresist (Xue et al., 2011).

In addition to PDMS, molds can be produced via electroforming from nickel, with natural (e.g., lotus leaf) and synthetic materials (e.g., PDMS negative replica) used as a master (Seung-Mo and Tai Hun, 2006). The molds produced directly from the original nanostructured surface have better resolution, with positive replicas produced using such mold having contact angles only 2°–5° below those of the original leaf. The mold can then be used for polymer casting or UV–nanoimprint lithography (NIL). Thermoplastic and UV–NIL are also commonly used to prepare nanostructured molds. In the case of the thermoplastic NIL, a thin film of thermoplastic polymer (resist) is spin coated onto the substrate, then pressed against a patterned surface, and heated until the polymer softens. In UV–NIL, a photocurable polymer is applied to the substrate and then light is used to transfer the pattern from a high-resolution mask onto the surface, by etching the areas that are illuminated. After the pattern-bearing surface is removed and the polymer is fully cured, the substrate is etched, with the areas that are not covered by the polymer having some of the substrate material removed. In soft lithography, the pattern is applied with the use of the elastic stamp, frequently PDMS. The pattern can also be assembled using microparticles and nanoparticles, such as microspheres or nanorods in colloidal lithography (Lai et al., 2011; Li et al., 2010), and molecular imprinting (Shi et al., 2007). The polymer resist layer can also be patterned using electron beam lithography and X-ray lithography.

Optical lithography can be used to transfer the patterns from the mold onto the substrate. In one example, a sequential application of polymer molds engraved with microscale and nanoscale features with a UV-curable resin of polyurethane acrylate resulted in a surface with two-tier roughness morphology. A subsequent treatment with trichloro(1H,1H,2H,2H-perfluorooctyl) silane rendered the surface superhydrophobic and displayed both the Cassie–Baxter and the Wenzel state (Jeong et al., 2009).

Chemical Etching

Chemical etching can impart the desired roughness and porosity onto a variety of polymer surfaces, where surface topography is controlled by the concentration of solvent and the exposure time. In one step, acetone treatment can impart a hierarchically textured topography onto the surface of polycarbonate, transforming the moderately hydrophobic material into superhydrophobic surface (Cui et al., 2012). In a separate study, acetone was deposited as a thin layer onto the surface of polycarbonate and allowed to react under ambient conditions (Zhao et al., 2006). The superhydrophobic surface topography was obtained by either depositing a small amount of coagulator, such as water or methanol, on the acetone-swollen surface or immersing the swollen polymer into the coagulator bath for several seconds. This provided an inexpensive and quick method for surface modification; however, acetone treatment also rendered the polymer opaque, which may not be suitable for many applications.

Plasma Modification

Plasma-assisted processing is well-suited for etching, surface functionalization, and deposition of thin films from organic, inorganic, and mixed precursors on a variety of substrates, including temperature-sensitive polymers, thin film, and fine-structured assemblies. Highly reactive species and field effects that occur in a plasma environment allow the reaction to take place at much lower temperatures. Plasma processing conditions will determine whether the process is that of etching, activation, functionalization, or deposition. For etching, no mask is required; instead, the topography is controlled by controlling the processing conditions, such as the type of gas used, energy supplied to the treatment rector, the duration of the treatment, etc. (Fresnais et al., 2006). Unlike chemical etching, plasma-assisted etching supports anisotropic removal of material, which may be used to create deep and narrow surface features and vertically aligned nanofeatures and sub-nanofeatures. Gases commonly used for etching include air, nitrogen, argon, oxygen, nitrous oxide, helium, tetrafluoromethane, water vapor, carbon dioxide, methane, and ammonia. Exposure to these plasmas can result in the alterations to the surface chemistry, which can be beneficial or detrimental to achieving hydrophobicity. For highly controlled surface functionalization, pulsed plasma deposition is often used, as it is favors the retention of functionalities. This is achieved by minimizing the UV-, ion-, or electron-induced damage to the growing film, while maintain good film stability and covalent attachment to the substrate (Wood et al., 2012). By controlling the input power (Wells et al., 2013), pulsing frequency, and duration of the pulse, it is possible to tune chemical functionality, surface morphology, and density of desired chemical functionality at the polymer surface (Denis et al., 2010; Kwok et al., 2010).

Exposure of low-density polyethylene (LDPE) to oxygen plasma for 5 min resulted in the formation of nanofibril arrays on its surface (Li et al., 2012). Interestingly, in its as-modified state, the surface was superhydrophilic, with a contact angle ~0°. Subsequent heating at 90°C for 24 h resulted in a hydrophobic overrecovery, with the surface displaying water contact angles of greater than 150°. Hydrophobic recovery is a well-documented behavior in polymers and is typically attributed to reorientation of polar groups from the surface and into the bulk, reorientation of nonpolar groups from the bulk and toward the surface, and the diffusion of low-molecular-weight polymers from within the bulk toward the surface. The inherent flexibility of the polymer chains allows for these processes to occur. While this is a significant issue when modifying the surface to be more hydrophilic, in this example, it is beneficially exploited. Similarly, high water contact angles were obtained by oxygen plasma treatment of a PTFE hydrophobic surface (Vandencasteele et al., 2010). Here, surface etching relied on the simultaneous presence of atomic oxygen and negatively charged species (electrons) at the PTFE surface. Again using oxygen plasma, bulk and film poly(methyl methacrylate) surfaces were simultaneously roughened at nanoscales and microscales (Figure 2) (Kontziampasis et al., 2014). Subsequent vacuum deposition of a perfluorinated self-assembled monolayer yielded superhydrophobic and superamphiphobic surfaces.

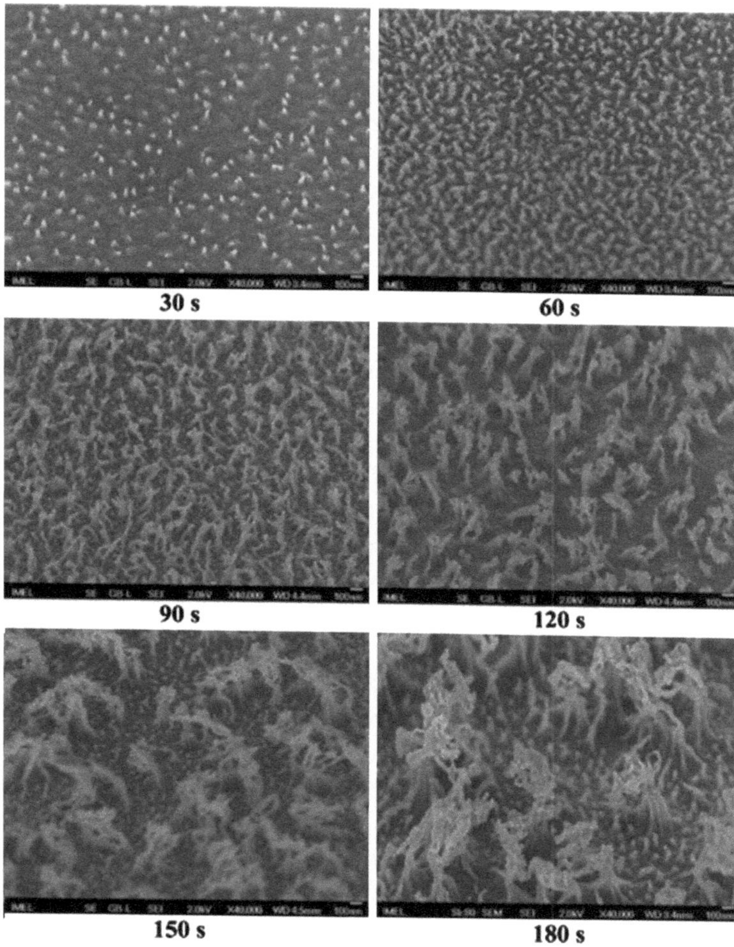

Figure 2 Evolution of morphology on PMMA plates etched and nanotextured in oxygen plasma as a function of etching time (etching conditions: power 1900 W, bias power 250 W (bias voltage −100 V), electrode temperature 15°C, gas pressure 0.75 Pa, oxygen gas flow 100 standard cubic cm). SEM images are shown for 30, 60, 90, 120, 150, and 180 s (magnification 40,000×, tilt 45°). *Reproduced from Kontziampasis et al. (2014) by permission of Elsevier.*

Plasma-assisted modification can be applied to solid, porous, and fibrous substrates. Nonwoven fabrics of polyethyleneterephthalate nanotextured with oxygen plasma and coated with a low-surface-energy nanofilm were highly effective in suppressing vapor condensation and repelling condensed droplets. The plasma-nanostructured surface retained its hydrophobicity after being subjected to water–vapor condensation and oil wetting, such stability attributed to the presence of long and thin nanohairs on the fibers (Shin et al., 2012). In another example, inherent microstructure of cellulose paper was enhanced by plasma-assisted nanostructuring to create a surface that once coated with a thin fluorocarbon film was superhydrophobic (Balu et al., 2008). The nanotopography was varied with plasma treatment, yielding a surface in either the Wenzel or Cassie–Baxter states.

Polymer Brushes

Surface modification using polymer brushes is particularly useful where dynamic, stimulus-responsive change in wettability is required. For example, polyelectrolyte brushes with quaternary ammonium groups undergo reversible switching between superhydrophobicity and superhydrophilicity as a result of direct counterion exchange (Lim et al., 2008). Poly(*N*-isopropylacrylamide) and poly(dimethylamino)ethyl methacrylate brushes grafted onto initiator-modified micro/nanostructured anodized alumina substrates reversible switching between the Wenzel and Cassie–Baxter states when temperature, pH, and electrolytes are changed (Liu et al., 2010, 2011).

Attachment of polymer brushes to the surface with nanoscale and microscale features may provide additional level of control over the assembly of polymer brushes. For example, a two-layer structure was built from a rough polymer layer with needle-like structures of micrometer size and a layer formed by the nanoscopic self-assembled domains of a demixed polymer brush (carboxyl-terminated poly(styrene-co-2,3,4,5,6-pentafluorostyrene) and carboxyl-terminated poly(2-vinylpyridine)) irreversibly grafted onto the needles (Minko et al., 2003). The wettability of the surface can be altered by exposing the surface to solvents that are selective to one of the components of the brush. The microscale features amplify the response, allowing the surface enrichment by the favored component while the rest of the brush is collapsed and hidden within the microstructures of the film.

Grafting to a variety of surfaces can be enabled by the use of dopamine derivatives, which mimic the adhesive substances produced by a variety of marine organisms, such as mussels, when attaching to surfaces. Polydopamine layers can attach to virtually any substance and can provide a mechanically flexible and versatile platform for subsequent brush attachment (Lee et al., 2011; Wei et al., 2014). Plasma modification can be used to activate or functionalize the surface, providing suitable chemistry for subsequent covalent attachment of molecules, such as polymer brushes. The nature, density, and organization of the desired chemical functionalities can be controlled by the treatment gas and the energy. The process is highly versatile in terms of resultant surface chemistry, with pyridine, anhydride, amine, ester, hydroxyl, sulfonic acid, carboxylic acid, cyano, epoxide, halide, thiol, furan, and other functionalized surfaces reported. Surfaces decorated by more than one desirable functional group can be manufactured. For example, plasma treatment of PTFE with oxygen followed by ammonia plasmas resulted in the formation of a superhydrophobic surface decorated with amino functional groups (Minko et al., 2003).

Solution Processing

The roughness and porosity of solution-processed polymer films can be controlled by changing the solvent composition, in particularly by substituting part of the solvent with a nonsolvent. For example, the surface morphology of polyvinyl chloride (PVC) was controlled by using a combination of the solvent (i.e., tetrahydrofuran) and either ethanol or methanol. As a result of the modification, the smooth and moderately hydrophobic (water contact angle of 80°) surface of PVC was rendered superhydrophobic, with contact angle increasing up to 150° (for an ethanol concentration of ~35% v/v) (Loo et al., 2012). Films fabricated with ethanol were more hydrophobic, displaying an increasing number of micron-sized particulates and porous structures with higher concentrations of ethanol. The modified surfaces effectively delayed colonization by pathogenic *Pseudomonas aeruginosa* bacteria by 18–24 h. In a separate study, the superhydrophobicity of ethanol-modified PVC films were found to remain stable under a range of conditions, such as in the pH range from 1 to 13, temperatures up to 50°C, and outdoor exposure (Chen et al., 2009).

Porous polypropylene gels fabricated by dissolving polypropylene in a good solvent, such as *o*-xylene or decalin, and a nonsolvent, such as acetone or dimethyl formamide, have the structure of spongy aggregates with diameters between 10 and 50 μm (Erbil et al., 2003). Thin films of porous polypropylene gel were obtained by evaporating the solvent mixture, which allowed it to be deposited onto a variety of substrates, so long as they are not solvophilic for the solvent/nonsolvent combination. The roughness was controlled by selection of the appropriate solvent combination and the temperature of deposition.

Perfluorinated dopamine derivative formed a structurally rough film with contact angles greater than 150° and low hysteresis (Hong et al., 2014). The highly attractive property of dopamine-based polymers is their ability to attach to a wide range of substrates, with the ability to undergo oxidative polymerization under wet and dry conditions. Dopamine layers can also be solution-processed onto superhydrophobic surfaces using a capillary micromolding, soft-lithographic technique to obtain a superhydrophobic micropatterned surface exhibiting high water adhesion (Kang et al., 2010). Surfaces in the Wenzel state and those with Wenzel-like behavior are well suited for applications that require capture and collection of water droplets.

Aerogels

The microscale and nanoscale surface morphology of aerogels can be used to produce surfaces that are both superhydrophobic and superoleophobic. Unlike solid materials, where morphological features and patterns are restricted to the surface, the topographical patterns of aerogels are present across the bulk of the sample. Considering that mechanical wear is one of the most significant factors that affect the durability of superhydrophobic materials, particularly the stability of the Cassie–Baxter state of the surface, aerogels possess the ability to maintain air trapping and low surface adhesion even after the material is subjected to mechanical abrasion. Furthermore, such wear should not significantly change the chemistry of the surface, which is the case for solid surfaces coated/decorated with hydrophobic chemistries. Finally, the morphology and curvature of the aerogel nanoscale features are conducive to the fabrication of oil-repellent materials.

Aerogels can be fabricated from natural and synthetic materials and modified postdeposition to enhance their nonwetting properties. Aerogels fabricated from unmodified and carboxymethylated, negatively charged cellulose nanofibers displayed some oil-repellent properties and were lightweight and mechanically robust (Aulin et al., 2010). Fluorosilane-treated nanocellulose aerogels displayed high-adhesive superhydrophobic and high-adhesive superoleophobic properties (Jin et al., 2011). The mechanical properties of nanocellulose aerogels arise from the strong hydrogen bonding of ordered cellulose polysaccharide chains within its structure. These super-strong, superhydrophobic structures may provide a new alternative to silk-based biomaterials and nanodevices. The renewable, low-cost nature of the source and ease of fabrication are also highly attractive. Indeed, cellulose nanofibers can be produced by mechanical grinding of cellulose in aqueous conditions, with the resulting aqueous hydrogel then able to be freeze-dried into a highly porous aerogel structure. The freezing conditions (e.g., cryogenic versus vacuum), speed of cooling, and solid faction and properties control the resultant aerogel nanostructure and surface area.

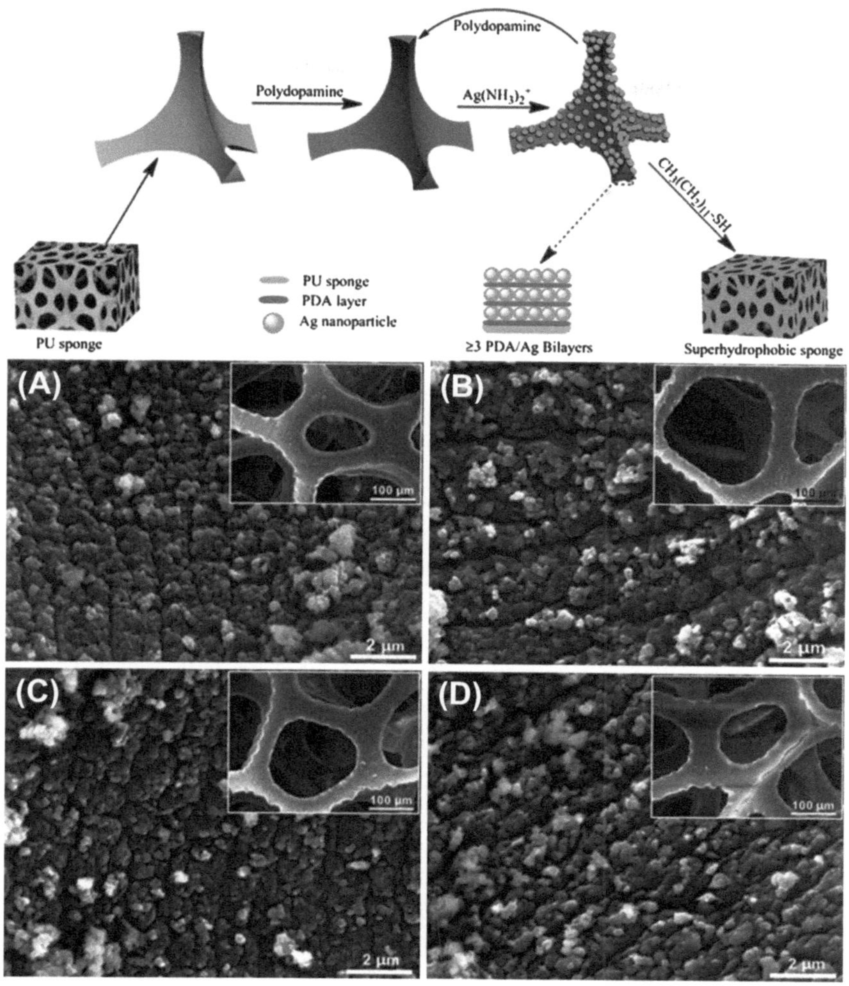

Figure 3 (A) Illustration of the fabrication of a highly compressible and stretchable superhydrophobic coating on polyurethane sponge through layer-by-layer deposition. (B) SEM images of the superhydrophobic coating after compression for (a) 1000 cycles, (b) 2000 cycles, (c) 3000 cycles, and (d) 4000 cycles. *Reproduced from Liu et al. (2014) by permission of The Royal Society of Chemistry.*

and oleophilic coatings (Asthana et al., 2014). The superhydrophobicity was dependent on the concentration and the type of nanoparticles. The carbon black/fluoropolymer coating most effectively resisted wetting by a lower surface tension liquid (5 vol% isopropyl alcohol water mixture), whereas nanocomposites containing a mixture of carbon back and graphene nanoplatelets were effective in separating mineral oil and water.

Hydrophobic poly(alkyl siloxane) solutions with high concentrations of hydrophilic 25, 35, and 150 nm alumina (Al_2O_3) nanoparticles deposited via spraying onto glass surfaces displayed a contact angle and hysteresis of 160° and 5°, respectively (Karapanagiotis et al., 2012). The superhydrophobicity was found to be dependent on particle concentration but not on particle size. The latter directly affected the minimum critical particle concentration required to induce superhydrophobicity, which was found to decrease exponentially with specific surface area.

CONCLUSIONS

The reports described in the chapter constitute only a very small portion of the available literature (and by no means are an exhaustive representation) of the techniques being used to fabricate novel superhydrophobic polymers or render existing surfaces superhydrophobic. As more reports are published, it becomes increasingly difficult to separate them by the type of technique used, as two or more approaches are often combined to attain the most desirable outcome. Clearly, there has been significant shift toward fabricating surfaces that have two or more tiers of features, often organized, as these yielded the most promising self-cleaning, nonwettable surfaces. There is also an increasing appreciation of how the adhesive state of superhydrophobicity can be used to create self-cleaning surfaces.

Nanofabrication and nanomodification are expected to play an increasingly important role in the processing of superhydrophobic surfaces. Such nanoscale precision is often accompanied by increasingly complex and expensive processing methodology, therefore, there is a need to develop facile techniques that can enable nanoscale ordering. Obtaining a better understanding of the mechanisms that govern superhydrophobicity will enable the development and tailoring of superhydrophobic surfaces for specific applications.

REFERENCES

Asthana, A., Maitra, T., Büchel, R., Tiwari, M.K., Poulikakos, D., 2014. Multifunctional superhydrophobic polymer/carbon nanocomposites: graphene, carbon nanotubes, or carbon black? ACS Appl. Mater. Interfaces 6.

Aulin, C., Netrval, J., Wagberg, L., Lindstrom, T., 2010. Aerogels from nanofibrillated cellulose with tunable oleophobicity. Soft Matter 6.

Balu, B., Breedveld, V., Hess, D.W., 2008. Fabrication of "roll-off" and "sticky" superhydrophobic cellulose surfaces via plasma processing. Langmuir 24.

Barthlott, W., Neinhuis, C., 1997. Purity of the sacred lotus, or escape from contamination in biological surfaces. Planta 202.

Byun, D., Hong, J., Saputra, Ko, J.H., Lee, Y.J., Park, H.C., Byun, B.-K., Lukes, J.R., 2009. Wetting characteristics of insect wing surfaces. J. Bionic Eng. 6.
Chen, H., Yuan, Z., Zhang, J., Liu, Y., Li, K., Zhao, D., Li, S., Shi, P., Tang, J., 2009. Preparation, characterization and wettability of porous superhydrophobic poly (vinyl chloride) surface. J. Porous Mater. 16.
Choi, G.-R., Park, J., Ha, J.-W., Kim, W.-D., Lim, H., 2010. Superamphiphobic web of ptfema fibers via simple electrospinning without functionalization. Macromol. Mater. Eng. 295.
Cui, Y., Paxson, A.T., Smyth, K.M., Varanasi, K.K., 2012. Hierarchical polymeric textures via solvent-induced phase transformation: a single-step production of large-area superhydrophobic surfaces. Colloids Surf., A 394.
Denis, L., Marsal, P., Olivier, Y., Godfroid, T., Lazzaroni, R., Hecq, M., Cornil, J., Snyders, R., 2010. Deposition of functional organic thin films by pulsed plasma polymerization: a joint theoretical and experimental study. Plasma Processes Polym. 7.
Erbil, H.Y., Demirel, A.L., Avc, Y., Mert, O., 2003. Transformation of a simple plastic into a superhydrophobic surface. Science 299.
Feng, J., Huang, M., Qian, X., 2009. Fabrication of polyethylene superhydrophobic surfaces by stretching-controlled micromolding. Macromol. Mater. Eng. 294.
Feng, J., Lin, F., Zhong, M., 2010. Stretching-controlled micromolding process with etched metal surfaces as templates towards mass-producing superhydrophobic polymer films. Macromol. Mater. Eng. 295.
Feng, L., Zhang, Y., Xi, J., Zhu, Y., Wang, N., Xia, F., Jiang, L., 2008. Petal effect: A superhydrophobic state with high adhesive force. Langmuir 24.
Fresnais, J., Benyahia, L., Poncin-Epaillard, F., 2006. Dynamic (de)wetting properties of superhydrophobic plasma-treated polyethylene surfaces. Surf. Interface Anal. 38.
Gao, X., Jiang, L., 2004. Biophysics: water-repellent legs of water striders. Nature 432.
Gong, G., Wu, J., Liu, J., Sun, N., Zhao, Y., Jiang, L., 2012. Bio-inspired adhesive superhydrophobic polyimide mat with high thermal stability. J. Mater. Chem. 22.
Han, D., Steckl, A.J., 2009. Superhydrophobic and oleophobic fibers by coaxial electrospinning. Langmuir 25.
Hong, D., Bae, K., Hong, S.-P., Park, J.H., Choi, I.S., Cho, W.K., 2014. Mussel-inspired, perfluorinated polydopamine for self-cleaning coating on various substrates. Chem. Comm. 50.
Huang, Y.-H., Wu, J.-T., Yang, S.-Y., 2011. Direct fabricating patterns using stamping transfer process with pdms mold of hydrophobic nanostructures on surface of micro-cavity. Microelectron. Eng. 88.
Huovinen, E., Takkunen, L., Korpela, T., Suvanto, M., Pakkanen, T.T., Pakkanen, T.A., 2014. Mechanically robust superhydrophobic polymer surfaces based on protective micropillars. Langmuir 30.
Jeong, H.E., Kwak, M.K., Park, C.I., Suh, K.Y., 2009. Wettability of nanoengineered dual-roughness surfaces fabricated by uv-assisted capillary force lithography. J. Colloid Interface Sci. 339.
Jin, H., Kettunen, M., Laiho, A., Pynnönen, H., Paltakari, J., Marmur, A., Ikkala, O., Ras, R.H.A., 2011. Superhydrophobic and superoleophobic nanocellulose aerogel membranes as bioinspired cargo carriers on water and oil. Langmuir 27.
Kang, M., Jung, R., Kim, H.-S., Jin, H.-J., 2008. Preparation of superhydrophobic polystyrene membranes by electrospinning. Colloids Surf. A 313–314.
Kang, S.M., You, I., Cho, W.K., Shon, H.K., Lee, T.G., Choi, I.S., Karp, J.M., Lee, H., 2010. One-step modification of superhydrophobic surfaces by a mussel-inspired polymer coating. Angew. Chem. Int. Ed. 49.
Karapanagiotis, I., Manoudis, P.N., Savva, A., Panayiotou, C., 2012. Superhydrophobic polymer-particle composite films produced using various particle sizes. Surf. Interface Anal. 44.

Kontziampasis, D., Boulousis, G., Smyrnakis, A., Ellinas, K., Tserepi, A., Gogolides, E., 2014. Biomimetic, antireflective, superhydrophobic and oleophobic pmma and pmma-coated glass surfaces fabricated by plasma processing. Microelectron. Eng. 121.
Kwok, D.T.K., Tong, L., Yeung, C.Y., Remedios, C.G.d., Chu, P.K., 2010. Hybrid plasma surface modification and ion implantation of biopolymers. Surf. Coat. Technol. 204.
Lai, K.Y., Lin, Y.-R., Wang, H.-P., He, J.-H., 2011. Synthesis of anti-reflective and hydrophobic si nanorod arrays by colloidal lithography and reactive ion etching. Cryst. Eng. Comm. 13.
Lee, B.P., Messersmith, P.B., Israelachvili, J.N., Waite, J.H., 2011. Mussel-inspired adhesives and coatings. Ann. Rev. Mater. Res. 41.
Li, Y., Zhang, J., Zhu, S., Dong, H., Jia, F., Wang, Z., Tang, Y., Zhang, L., Zhang, S., Yang, B., 2010. Bioinspired silica surfaces with near-infrared improved transmittance and superhydrophobicity by colloidal lithography. Langmuir 26.
Li, Y.P., Li, S.Y., Shi, W., Lei, M.K., 2012. Hydrophobic over-recovery during aging of polyethylene modified by oxygen capacitively coupled radio frequency plasma: a new approach for stable superhydrophobic surface with high water adhesion. Surf. Coat. Technol. 206.
Lim, H.S., Lee, S.G., Lee, D.H., Lee, D.Y., Lee, S., Cho, K., 2008. Superhydrophobic to superhydrophilic wetting transition with programmable ion-pairing interaction. Adv. Mater. 20.
Lin, J., Cai, Y., Wang, X., Ding, B., Yu, J., Wang, M., 2011. Fabrication of biomimetic superhydrophobic surfaces inspired by lotus leaf and silver ragwort leaf. Nanoscale 3.
Liu, F., Sun, F., Pan, Q., 2014. Highly compressible and stretchable superhydrophobic coating inspired by bio-adhesion of marine mussels. J. Mater. Chem. A 2.
Liu, X., Ye, Q., Song, X., Zhu, Y., Cao, X., Liang, Y., Zhou, F., 2011. Responsive wetting transition on superhydrophobic surfaces with sparsely grafted polymer brushes. Soft Matter 7.
Liu, X., Ye, Q., Yu, B., Liang, Y., Liu, W., Zhou, F., 2010. Switching water droplet adhesion using responsive polymer brushes. Langmuir 26.
Loo, C.-Y., Young, P.M., Lee, W.-H., Cavaliere, R., Whitchurch, C.B., Rohanizadeh, R., 2012. Superhydrophobic, nanotextured polyvinyl chloride films for delaying pseudomonas aeruginosa attachment to intubation tubes and medical plastics. Acta Biomater. 8.
Ma, M., Hill, R.M., Lowery, J.L., Fridrikh, S.V., Rutledge, G.C., 2005. Electrospun poly(styrene-block-dimethylsiloxane) block copolymer fibers exhibiting superhydrophobicity. Langmuir 21.
Minko, S., Müller, M., Motornov, M., Nitschke, M., Grundke, K., Stamm, M., 2003. Two-level structured self-adaptive surfaces with reversibly tunable properties. J. Am. Chem. Soc. 125.
Muthiah, P., Hsu, S.-H., Sigmund, W., 2010. Coaxially electrospun pvdf–teflon af and teflon af–pvdf core–sheath nanofiber mats with superhydrophobic properties. Langmuir 26.
Patankar, N.A., 2004. Transition between superhydrophobic states on rough surfaces. Langmuir 20.
Seung-Mo, L., Tai Hun, K., 2006. Mass-producible replication of highly hydrophobic surfaces from plant leaves. Nanotechnology 17.
Shi, F., Liu, Z., Wu, G.L., Zhang, M., Chen, H., Wang, Z.Q., Zhang, X., Willner, I., 2007. Surface imprinting in layer-by-layer nanostructured films. Adv. Funct. Mater. 17.
Shin, B., Lee, K.-R., Moon, M.-W., Kim, H.-Y., 2012. Extreme water repellency of nanostructured low-surface-energy non-woven fabrics. Soft Matter 8.
Singh, A., Steely, L., Allcock, H.R., 2005. Poly[bis(2,2,2-trifluoroethoxy)phosphazene] superhydrophobic nanofibers. Langmuir 21.
Sun, G., Fang, Y., Cong, Q., Ren, L.-q., 2009. Anisotropism of the non-smooth surface of butterfly wing. J. Bionic Eng. 6.
Tropmann, A., Tanguy, L., Koltay, P., Zengerle, R., Riegger, L., 2012. Completely superhydrophobic pdms surfaces for microfluidics. Langmuir 28.

Tuvshindorj, U., Yildirim, A., Ozturk, F.E., Bayindir, M., 2014. Robust Cassie state of wetting in transparent superhydrophobic coatings. ACS Appl. Mater. Interf. 6.

Vandencasteele, N., Broze, B.r.r., Collette, S.p., De Vos, C., Viville, P., Lazzaroni, R., Reniers, F.o., 2010. Evidence of the synergetic role of charged species and atomic oxygen in the molecular etching of ptfe surfaces for hydrophobic surface synthesis. Langmuir 26.

Verma, R., Tomar, N., Creager, S.E., Smith Jr., D.W., 2012. Statically non-wetting electrospun perfluorocyclobutyl (pfcb) aryl ether polymer doped with room temperature ionic liquid (rtil). Polymer 53.

Wei, Q., Yu, B., Wang, X., Zhou, F., 2014. Stratified polymer brushes from microcontact printing of polydopamine initiator on polymer brush surfaces. Macromol. Rapid Commun. 35.

Wells, G.P., Estrada-Raygoza, I.C., Thamban, P.L.S., Nelson, C.T., Chung, C.-W., Overzet, L.J., Goeckner, M.J., 2013. Understanding the synthesis of ethylene glycol pulsed plasma discharges. Plasma Processes Polym. 10.

Wong, T.-S., Kang, S.H., Tang, S.K.Y., Smythe, E.J., Hatton, B.D., Grinthal, A., Aizenberg, J., 2011. Bioinspired self-repairing slippery surfaces with pressure-stable omniphobicity. Nature 477.

Wood, T.J., Hurst, G.A., Schofield, W.C.E., Thompson, R.L., Oswald, G., Evans, J.S.O., Sharples, G.J., Pearson, C., Petty, M.C., Badyal, J.P.S., 2012. Electroless deposition of multi-functional zinc oxide surfaces displaying photoconductive, superhydrophobic, photowetting, and antibacterial properties. J. Mater. Chem. 22.

Xinchuan, L., Cheng, L., 2010. Fabrication of super-hydrophobic channels. J. Micromech. Microeng. 20.

Xu, Q.F., Wang, J.N., Sanderson, K.D., 2010. A general approach for superhydrophobic coating with strong adhesion strength. J. Mater. Chem. 20.

Xue, C.-Y., Zhang, W., Choo, W.H.S., Yang, K.-L., 2011. Simplest method for creating micropatterned nanostructures on pdms with uv light. Langmuir 27.

Zhang, J., Wang, A., Seeger, S., 2014. Nepenthes pitcher inspired anti-wetting silicone nanofilaments coatings: Preparation, unique anti-wetting and self-cleaning behaviors. Adv. Funct. Mater. 24.

Zhang, X., Shi, F., Yu, X., Liu, H., Fu, Y., Wang, Z., Jiang, L., Li, X., 2004. Polyelectrolyte multilayer as matrix for electrochemical deposition of gold clusters:Toward superhydrophobic surface. J. Am. Chem. Soc. 126.

Zhao, N., Weng, L., Zhang, X., Xie, Q., Zhang, X., Xu, J., 2006. A lotus-leaf-like superhydrophobic surface prepared by solvent-induced crystallization. Chem. Phys. Chem. 7.

CHAPTER SIX

Metallic Superhydrophobic Surfaces

GLOSSARY

Ablation the removal of material from solid samples using laser irradiation.

Advancing contact angle (Θ_A) the maximum value of contact angle by the fixed contact line measured when the volume of a drop is being increased and therefore advancing over the surface.

Cassie-Baxter-equation is suitable for applications where heterogeneous wetting is taking place (composite solid/liquid-air interface): $\cos\theta_{struc} = f_{sl}(\cos\theta_{flat} + 1) - 1$.

Contact angle hysteresis ($\Delta\Theta$) the difference between the advancing and receding contact angles.

f_{sl} the ratio of the solid surface in contact with liquid.

Laser fluence (F) the laser pulse energy (E_p) delivered per unit area, for a Gaussian beam: $F = \frac{2E_p}{\pi w^2}$, where $w(1/e^2)$ is the effective focal spot radius.

Quasi-periodic surface structures are generated under laser irradiation exposure and they exhibit regularity in their pattern. Their geometrical sizes (e.g., height, periodicity) are much smaller than the diameter of the focused laser beam.

r the structure parameter, defined as the ratio between the structured enlarged surface area and the same area projected on a flat surface.

Receding contact angle (Θ_R) the minimum value of contact angle by the fixed contact line measured when the volume of a drop is being decreased and therefore receded from the previously contacted surface.

Restitution coefficient $\varepsilon = |V'/V|$, where V' is the velocity after impact and V is the velocity before impact.

Retentive force $F_g = \varrho g V \sin\alpha$, can be determined from the critical value of gravitational force at which liquid drop begins to move, where ρ is liquid density, g is the acceleration due to gravity, V is the drop volume, and α is the tilt angle.

Small-angle X-ray scattering (SAXS) the elastic scattering of x-rays with wavelengths in the range of 0.1–0.2 nm; this technique enables the detection of sample inhomogeneities in the nm range.

Specific heat capacity the quantity of heat energy (in J) per unit weight (in kg) required to change the temperature of an object by 1 K.

Spikes quasi-periodical structures that are conical in shape.

Θ_{flat} the contact angle of liquid on a flat surface.

Θ_{struc} the contact angle of liquid on a structured surface.

Tilt angle (α) the angle at which the a test liquid drop begins to move when a solid surface is tilted.

Superhydrophobic Surfaces
http://dx.doi.org/10.1016/B978-0-12-801109-6.00006-9

Wenzel-equation is suitable for situations where homogeneous wetting is taking place (uniform solid/liquid interface): $\cos\,\theta_{struc} = r\cos\,\theta_{flat}$.

XPS analysis (X-ray photoelectron spectrometry) is a surface chemical analysis technique with a detection range of up to 10 nm in depth.

INTRODUCTION

The fabrication of robust metallic surfaces that can repel liquids has a wide range of applications, including optimization of the performance of construction materials, marine crafts, aircrafts, automobiles, refrigeration machines, wind power stations, and medical devices among many others. It was shown that surface structure is one of the key factors required to enable a surface to obtain superhydrophobic properties (Barthlott and Neinhuis, 1997). Based on a bionic approach, particularly focusing on the surface properties of lotus leaves, specially structured, man-made, water repellent surfaces, reliant on the formation of a stable air–liquid interface, have been developed. Generally speaking, by the interaction of liquids with a structured surface, different wetting scenarios can be realized. (1) Liquid wets the structured surface completely, forming a homogeneous solid–liquid interface. In this case the relation between the contact angle of liquid on a flat surface Θ_{flat}, the contact angle of liquid on the structured surface Θ_{struc}, and the structure parameter r, defined as the ratio of the structured enlarged surface area to its flat projected area, is given by the Wenzel equation (Wenzel, 1936):

$$\cos\,\theta_{struc} = r\cos\,\theta_{flat}.$$

Due to this effect, the interface and therefore adhesion between the solid and liquid are increased compared with that observed on the flat surface of the same material. (2) If the liquid wets the structured surface incompletely, a composite solid/liquid–air interface is formed. In this case the relation between the contact angle of liquid on the flat surface Θ_{flat}, the contact angle of liquid on the structured surface Θ_{stuc}, and f_{sl} the ratio of the solid surface in contact with liquid is given by Cassie–Baxter equation (Cassie and Baxter, 1944):

$$\cos\,\theta_{struc} = f_{sl}\left(\cos\,\theta_{flat} + 1\right) - 1.$$

As a result, the extent of the interface that forms between liquid droplets and solid can be strongly reduced, enabling the droplet to roll off the surface easily. Most of the state-of-the-art man-made superhydrophobic materials are based on the formation of a stable composite solid–air interface. It has been demonstrated that for structured surfaces, there is a certain value of *r* from which the stable solid/liquid–air interface is formed (de Gennes et al., 2003). Later, the shape and geometrical sizes of surface structure were investigated, on both a theoretical and an experimental basis (Extrand, 2004; Marmur, 2003, 2004; Patankar, 2003) with the aim to characterize and to mimic stable superhydrophobic surfaces. Upon investigation of water-repellent plants, it was found that these plants often exhibit similarity in the way in which their leaves are assembled: leaf surfaces exhibit a concave structure with a diameter in the range of several tens of micrometers. These micrometer structures are further covered with nanostructures (Koch et al., 2008). Moreover, it has been pointed out that the dual-scale (nano and micro) roughness and fractal roughness are favorable structures for the fabrication of robust water-repellent surfaces (Bhushan et al., 2009). For the fabrication of superhydrophobic metal surfaces, different structuring techniques can be used (e.g., electrochemical deposition, lithography, oxidation, ablation by laser irradiation). In this chapter, we focus mainly on metal structuring by laser irradiation due to its universal applicability to all metals. The book chapter is organized as follows: first, the physical processes occurring when metal surfaces interact with laser irradiation are discussed. A description of the structures generated is included. Second, the influence of laser irradiation on the chemical composition of metals is referred. Third, the characterization of the wettability of superhydrophobic metals is discussed. Finally, the performance of superhydrophobic metal surfaces created through a combination of laser structuring with coatings and lubricants is reviewed.

STRUCTURING OF METAL SURFACES BY ULTRA-SHORT PULSED LASER IRRADIATION

Surface structuring with ultra-short pulse laser radiation is a versatile technology that has many advantages: (1) it is applicable to almost all solid materials, including all metals; (2) it enables flexible and controllable structuring of samples with complex shapes; and (3) it is particularly suitable for small-scale manufacturing and can be done under normal ambient conditions, without a need of a clean room environment. Thereby, material processing is done by ablation (i.e., material removal from the samples by laser irradiation). Nowadays, surface structuring by laser ablation has been thoroughly investigated as a result of its ability to generate different quasi-periodic and geometrically defined topographies with a resolution down to a submicrometer range (Vorobyev, 2013). Laser-generated structures can be used for a wide range of applications, including the tuning of the wetting properties of metals. Physical processes, which occur due to the interaction of ultra-short laser pulses with matter, have been intensively studied in the past (Hohlfeld et al., 2000; von der Linde and Sokolowski-Tinten, 2000; Rethfeld et al., 2004). Irradiation of metals with ultra-short laser pulses, which have energy higher than the ablation threshold, results in the following processes:

- First, the laser pulse energy is absorbed by free electrons due to the inverse bremsstrahlung within a skin layer. The thickness of the skin layer is determined by the laser light wavelength and metal properties and is in the range of several tens of nanometers. Thermalization of the excited electrons is determined by the electron–electron interaction time and is very fast.
- Following the absorption of the laser pulse energy, the electron and lattice subsystems can be characterized by two temperatures: T_e (the temperature of the hot electrons) and T_l (the lattice temperature). The energy transfer into the metal can be described by two-temperature diffusion model (Anisimov et al., 1974). The equilibrium between these two subsystems will be reached within a few picoseconds, due to electron–phonon interactions and electron diffusion out of the irradiated region.
- If the lattice temperature exceeds the material specific threshold values, melting and ablation will begin. Melting occurs between a few and hundreds of picoseconds. Ablation occurs on a time scale from a few 10 s of picoseconds to several nanoseconds.
- Finally, the ablation crater cools at a rate of about 10^{13}–10^{15} K/s and the melt then resolidifies, resulting in the formation of a unique surface topography.

The morphology of the surface topography depends on the laser processing parameters in addition to the material properties. It is noteworthy that all structures are formed on the surfaces long after the laser pulse absorption has taken place.

For ultra-short pulsed laser irradiation of metals, with pulse duration smaller than 1 ps, two different ablation regimes, termed "gentle" and "strong" ablation, are observed (Nolte et al., 1997). At laser fluences (laser pulse energy per unit area) just above the ablation threshold, the average ablation rate is determined by the optical penetration depth (for example, 13 nm for copper) of laser radiation in metal. The average ablation rate is low. The processed area is relatively smooth with nanometer-sized roughness being formed. For pulses longer than 1 ps, the "gentle" ablation regime is no longer observed.

At higher laser fluences (starting from 0.7 J/cm^2 for copper), the "strong" ablation regime is observed. This regime is characterized by a higher penetration depth (80 nm for copper), an increased average ablation rate, and a processed area surface that contains greater roughness. The size of the resultant surface features under these conditions can be up to several tens of micrometers.

The study of ablation craters generated in different ablation regimes shows that, apart from laser machined structures with defined geometry (e.g., hole, grooves), there are many substructures having geometrical sizes much smaller than that of the laser beam focus. Since these structures exhibit regularity in their pattern, they are also called "quasi-periodic surface structures." The typical quasi-periodic surface structures, which are observed in the "gentle" ablation regime of metals, are laser-induced periodic surface structures or ripples (Figure 1, left) and nanoroughness. The typical quasi-periodic surface structures for "strong" ablation of metals are spikes (Figure 1, right).

Surfaces containing spike structures show promise for the fabrication of superhydrophobic metallic surfaces due to their particular shape (concave structure with high aspect ratio, covered with nanoroughness). Note that the spikes that have been generated on semiconductor surfaces have been intensively studied in the past, inter alia, because of their superhydrophobic properties (Baldacchini et al., 2006; Zorba et al., 2006, 2008).

To understand the processes underlying the formation of spikes on metal surfaces, it is helpful to view the experimental results of metal ablation experiments carried out without moving the samples (stationary conditions). The photographs given in Figure 2 show the development of the spike topography

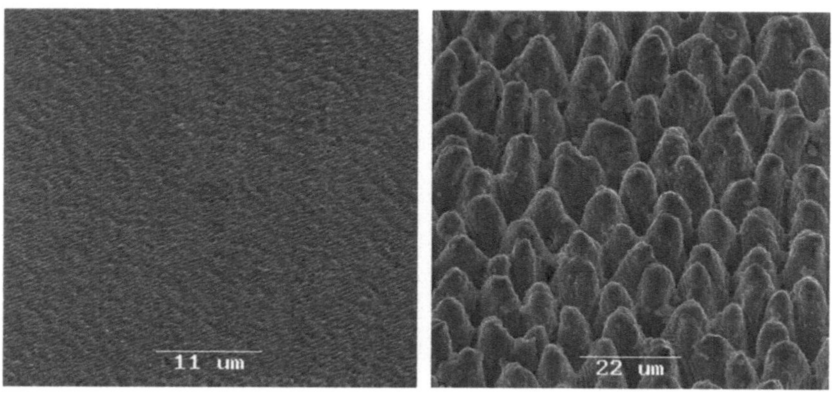

Figure 1 Ripples (left) and spikes (right) on a titanium surface.

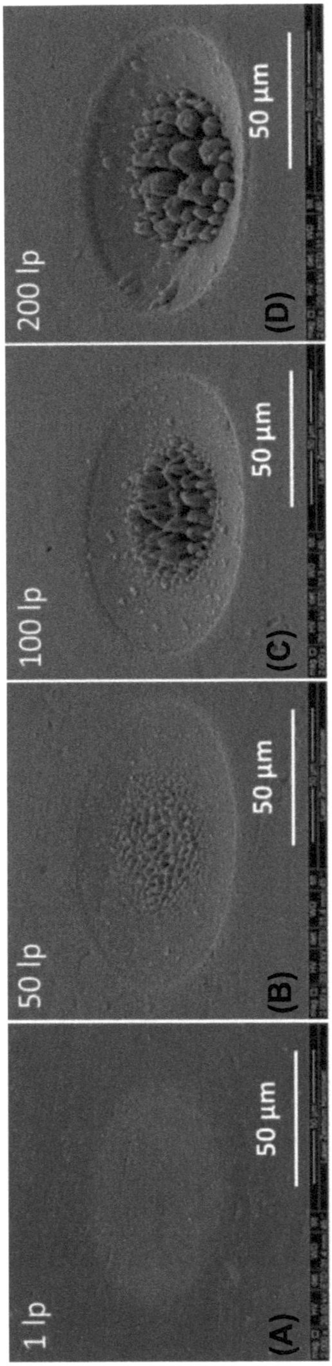

Figure 2 Titanium surfaces after irradiation at 5.09 J/cm^2 with 1, 50, 100, and 200 laser pulses (stationary conditions).

on titanium surfaces after irradiation at laser fluence of 5.09 J/cm^2, with an increasing number of laser pulses. It is evident that the diameter of the ablation crater increases with the number of laser pulses due to an accumulation effect. Thereby, after irradiation with the first laser pulse, nano-sized surface features are generated (Figure 2(A)). After irradiation by approximately 50 laser pulses, a quasi-periodic wavy surface structure with an average wavelength of 1.5–2.5 μm can be observed (Figure 2(B)). After irradiation with 100 pulses, spikes are generated. Thereby, spike tips are located above the original titanium surface level (Figure 2(C)). After irradiation with 200 pulses, the ablation crater becomes deeper, the diameters of single spikes increase, and the spike becomes rounded in shape. In this case, the spike tips deepen into the crater and are located below the original titanium surface.

As can been seen in Figure 2, the crater generated by laser pulses, with the energy higher then threshold for the onset of "strong" ablation regime, has two characteristic regions. The "strong" ablation region located in the crater center, where the Gaussian laser beam has higher intensity, and the "gentle" ablation region is located in the outer area of the crater. Figure 3 shows the magnified views of both regions. Thereby, the central area exhibits a columnar structure or spikes, whose tips protrude above the original sample surface. It was found that the spike height reached its maximum value after irradiation with approximately 200 laser pulses (Oliveira et al., 2009). The outer area on the titanium exhibits ripples with a periodicity of approximately 700 nm and nanostructures that consist of redeposited nano-sized ablation particles. The thresholds for these two ablation regimes are

Figure 3 The magnified view of the center of the ablation crater and the region of "strong" ablation (left) and the outer region of the ablation crater and the region of "gentle" ablation (right) on titanium surface generated at stationary conditions.

dependent on the number of applied laser pulses, and can be determined experimentally (Mannion et al., 2004).

To fabricate a larger spike-structured surface area, the sample has to be moved with respect to the laser focus (nonstationary conditions). The sample translation velocity has to be chosen depending on laser focus diameter, so that the average laser pulse numbers per surface unit area will be in the range of 100–200 pulses. The translation step between two raster lines has to be smaller than the diameter of the "strong" ablation area determined under stationary conditions. The structured areas with homogeneously distributed spikes can be fabricated using a pulse overlap that is greater than 50%. When the laser focus is moved with respect to the samples, regions of "strong" and "gentle" ablation become superimposed, leading to the superimposition of micro- and nanostructures. Thereby, a particular hierarchical two-tier surface structure is generated (Figure 4).

The first tier is composed of spikes. The second tier, generated on the spike surface, contains nanostructures that are typical for the gentle ablation regime and redeposited ablation nanoparticles. The hierarchical surface structures of this type have been shown to be one of the key factors for a surface to exhibit superhydrophobic properties (Bhushan et al., 2009).

The influence of processing parameters on the resultant shape of the spikes has been investigated in detail in studies involving the ablation of semiconductors (Crouch et al., 2004; Carey, 2004; Her et al., 1998, 2000). The main important parameters are summarized here and can be applied

Figure 4 Particular hierarchical two-tier surface structure generated under nonstationary conditions.

for the generation of metal spikes. Briefly, the spike shape can be controlled by the number of absorbed laser pulses, fluence of the absorbed laser radiation, laser polarization, wavelength, pulse duration, the process gases used, and focusing optics.

Surface spikes such as these cannot be generated on all metals. Metals where the spikes can be easily generated and where spikes have been already thoroughly investigated are titanium and steel (Fadeeva et al., 2011; Mannion et al., 2004; Nayak and Gupta, 2010; Oliveira et al., 2009). Spikes have not been observed on copper and the noble metals (gold, platinum) using this laser ablation technique (Fadeeva et al., 2009; Nayak and Gupta, 2010; Vorobyev and Guo, 2006). Differences in surface structures generated in the "strong" ablation regime on steel (where spikes can be easily generated) and copper (where spikes are not observed) are shown in Figure 5. In these images, the ablation craters on steel and copper, generated by 100 ultra-short laser pulses at laser fluence of 10.18 J/cm^2, are presented.

In the craters shown in Figure 5, more material is removed from the copper sample compared with the steel sample, and therefore, ablation from copper seems to be more effective. On both materials, resolidified molten areas exist in the center of the ablation craters. The height of solidified melt features is in the range of 20–30 μm on steel and 2–3 μm on copper. A similar morphology of the ablation crater found on steel can be also observed on titanium. The morphology of the craters formed on copper is similar to that produced using gold and platinum substrates. Presumably, the observed differences in the melt dynamics for these two groups of metals can be explained by the differences in the melt properties (viscosity, surface

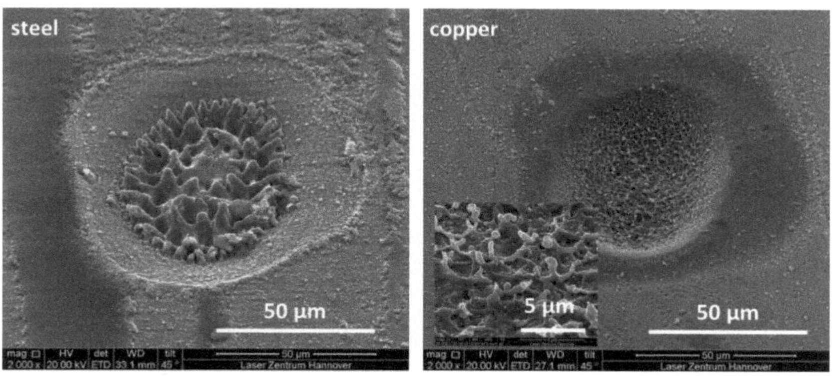

Figure 5 Craters on steel (left) and copper (right) after irradiation with 100 laser pulses of 30 fs at 800 nm laser wavelength and fluence of 10.18 J/cm^2.

tension, etc.). In Mannion et al. (2004) the spike generation was explained by the occurrence of phase explosion in the layer of molten material due to superheating. In the course of this superheating, the spikes in the center of ablation craters can be compared with the rapid expulsion of a liquid, which then quickly cooled, resolidifying.

Apart from the spikes, the irregular composition of the microroughness and nanoroughness generated by ultra-short pulsed laser ablation can be used for the fabrication of superhydrophobic metallic surfaces. The combination of microroughness and nanoroughness can be generated on metals, where the spike formation is not observed (e.g., on copper, platinum, gold). Fadeeva et al. (2009) reported an investigation of the surface structures that are generated on platinum by exposure to ultra-short pulsed laser irradiation, together with the wettability of the surface. By varying the irradiation fluence from 0.084 J/cm^2 to 7.7 J/cm^2 and the number of laser pulses, different nanoroughness is obtained, together with different combinations of irregular nanoroughness and microroughness. The structures with the irregular combination of nanoroughness and microroughness generated at the laser fluence of 7.7 J/cm^2 by approximately 50 laser pulses resulted in the formation of a surface wih superhydrophobic properties (see "wetting characterisation"). An example of this structure is shown in Figure 6.

It was also reported that a similar structure could be produced on copper surface results in superhydrophobic properties (Kietzig et al., 2011).

In summary, different surface structures can be generated using laser irradiation. These surfaces, containing spikes and irregular levels of microroughness and nanoroughness, have been found to exhibit superhydrophobic

Figure 6 Composition of irregular microroughness and nanoroughness on platinum.

properties. It is important to also consider whether the chemical composition of the metal surface is changed as a result of the laser irradiation.

INFLUENCE OF LASER IRRADIATION ON THE CHEMICAL COMPOSITION OF METAL SURFACES

Pure metals, over which nearly all liquids spread, have high surface energies (de Gennes et al., 2003), and, being in contact with air, all base metals oxidize. This process is accelerated by heating. Therefore, the degree to which metals oxidize increases uponinteracting with laser radiation. The resultant oxidized metal surfaces typically demonstrate hydrophilic behavior, however, the high surface energy of pure metals and metal oxides can also induce their contamination with e.g., inorganic carbon from the atmosphere (Weiss, 1971).

In the work by Fadeeva et al. (2011) and Fadeeva (2014), X-ray photoelectron spectrometry (XPS) analysis was performed to compare the chemical composition of a base metal (titanium) and noble metal (platinum) before and after laser structuring (Table 1). These XPS measurements were performed approximately 2 months after the laser structuring had taken place. The untreated, as well as laser structured samples, were stored under normal ambient conditions.

The XPS analysis did not reveal any significant difference in the surface chemistry of untreated and structured samples. Therefore, both base and noble metals are characterized by a slight increase of oxygen after laser irradiation. This can be attributed to the increased oxidization of metals due to heating. It is also noticeable that both metals have a high amount of carbon in their surface composition.

Table 1 Atomic Fractions[a] of Elements[b] in Percent Detected by XPS on Titanium (Ti) and Platinum (Pt) before and after Laser Structuring

Element	Ti before Laser Structuring (At. %)	Ti ~60 days after Laser Structuring (At. %)	Pt before Laser Structuring (At. %)	Pt ~60 days after Laser Structuring (At. %)
Ti $2p_{3/2}$	9.6	13.5	–	–
Pt $4f_{7/2}$	–	–	27.4	21.9
O 1s	35.5	38.1	17.4	20.2
C 1s	50.5	46.9	45.8	44.3

[a]It is conventional to list surface compositions obtained from XPS analyses as atomic fractions (or concentrations), rather than mass fractions.

[b]The uncertainty in the elemental atomic fractions in each value is at best 5%, so only the first decimal place is retained.

Table 2 Atomic Fractions[a] of Elements[b] in Percent Detected by XPS on Stainless Steel (AISI 304L) before and after the Laser Structuring

Element	Before Laser Structuring (At. %)	5 h after Laser Structuring (At. %)	~52 days after Laser Structuring (At. %)
C 1s	26.56	41.52	54.69
O 1s	56.86	48.95	38.73
Fe $2p_{3/2}$	16.57	9.52	6.59

[a]It is conventional to list surface compositions obtained from XPS analyses as atomic fractions (or concentrations), rather than mass fractions.
[b]The elemental atomic fractions are cited from Keitzig et al. (2011).

In Keitzig et al. (2011), XPS analysis was performed on stainless steel (AISI 304L) before, 5 h after, and ~52 days after the laser structuring had taken place. Also in this case, the untreated and laser structured samples were stored under normal ambient conditions. It was found that the main elements of the surface were carbon, oxygen, and iron. Ii is interesting to note that the chemical composition of this substrate changed over time, where a nonpolar carbonaceous layer accumulated on the sample surface (Table 2). The amount of carbon present on the surface was found to increase during the first 2 months after laser structuring, after which time carbon content remained constant. It is important to note that the carbonaceous layers demonstrate hydrophobic behavior.

Quasi-periodic spike structures on metals (Fadeeva et al., 2011; Kietzig et al., 2009, 2011) as well as irregular combination of microroughness and nanoroughness (Fadeeva et al., 2009; Kietzig et al., 2011) have been found to make the surfaces of different types of steel, titanium, aluminum, copper, and platinum superhydrophobic (see "Wetting Characterization"). In Kietzig et al. (2009, 2011), the influence of the accumulated nonpolar carbonaceous layer on the wetting characteristics of the laser structured metal samples was investigated. The authors pointed out that directly after the laser structuring, metals show superhydrophilic behavior, and after another ~2 months, the structured surfaces become superhydrophobic. This change in the wetting behavior has been explained in terms of growing carbon contamination over time. In our own experiments, we also observed that directly after the laser structuring, the spike-structured metal surfaces were hydrophilic; however, after cleaning the structured samples using acetone in an ultrasonic bath followed by vacuum desiccation for 2 h, the spike structures became superhydrophobic (Fadeeva, 2014).

WETTING CHARACTERIZATION

In this section, we focus on the quantitative characterization of the wetting properties of superhydrophobic metal surfaces. The main

Table 3 Quantitative Characterization of Superhydrophobic Metal Surfaces Structured by Ultra-Short Pulsed Laser Irradiation

Sample	Structure Type	$\Theta_{str,A}$	$\Theta_{str,R}$	$\Delta\Theta_{str}$	α	References
Titanium	Spikes	161.4°	160.2°	1.2°	–	Kietzig et al. (2011)[a]
	Spikes	163 ± 4°	–	10 ± 4.5°	5 ± 1°	Fadeeva et al. (2011)[a]
Steel	Spikes	162 ± 4°	–	6°	4 ± 1°	Fadeeva (2014)
Iron	Spikes	160.0°	155.9°	4.1°	–	Kietzig et al. (2011)[a]
Aluminum	Spike	161.5°	158.7°	2.8°	–	Kietzig et al. (2011)[a]
	Spikes	166 ± 4°	–	9 ± 6°	5 ± 1°	Fadeeva (2014)
Copper	ir m- and n-r[b]	158.3°	155.5°	2.8°	–	Kietzig et al. (2011)[a]
Platinum	ir m- and n-r[b]	158°	–	–	–	Fadeeva et al. (2009)
		158 ± 5°	–	8 ± 4°	10 ± 4°	Fadeeva (2014)

[a]Data have been obtained on samples irradiated with ultra-short laser pulses at the laser fluence of 5.16 J/cm^2 and stored under normal ambient conditions for approximately 1 month.
[b]ir m- and n-r = irregular microroughness and nanoroughness.

characteristics of superhydrophobic behavior of solid surfaces were summarized by Extrand (2006) as follows: (1) force that is needed to dislodge a water droplet from the surface is very low; (2) small water droplets with a radius of up to several millimeters have spherical shape with minimal contact to the surface; (3) both apparent advancing (Θ_A) and receding (Θ_R) water contact angles are greater than those on the chemically equivalent smooth surface; (4) water contact angle hysteresis ($\Delta\Theta = \Theta_A - \Theta_R$) is minimal (as a rule ≤10°); and (5) liquid drops may bounce on the surface.

To quantify the superhydrophobicity of laser structured metallic surfaces, the apparent advancing (Θ_A) and receding (Θ_R) water contact angles, water contact angle hysteresis ($\Delta\Theta$) and the tilt angle (α) need to be measured. The tilt angle is defined as the angle at which the solid surface is tilted to allow the test water droplet to move. This parameter can be used for the estimation of force that is needed to dislodge a water droplet from the surface. Furthermore, the water bouncing experiments need to be performed in order to evaluate the superhydrophobic performance of the surface.

The results of the quantitative characterization of superhydrophobic metal surfaces structured by ultra-short pulsed laser irradiation are summarized in Table 3. The water contact angles measured prior to the laser irradiation were 65°–85° on copper, titanium, iron, and aluminum

Figure 7 A dust particle (marked with the red arrow gray in print version) is removed from the superhydrophobic surface by the water droplet moving in the direction of the particle.

(Kietzig et al., 2011), 73 ± 3° on titan (Fadeeva et al., 2011), and 81° ± 3° on platinum (Fadeeva et al., 2009).

As can be seen from the data presented in Table 3, metal surfaces structured with quasi periodic spikes, as well as irregular combination of micro- and nanoroughness, meet the criteria of superhydrophobic surfaces.

The dynamic behavior (tilting plate experiments) of a water droplet on laser structured superhydrophobic metals was found to be similar. Figure 7 shows an example of the dynamic behavior of an 8 μL water droplet on a superhydrophobic stainless steel surface structured with spikes. The solid surface was tilted at an angle of 4°. At this angle, the water droplet started to move. The balance of two forces (retentive and gravitational) determined the droplet behavior. If the gravitational force exceeds the effect of the retentive force, the droplet began to move as a whole. The determination of this critical value of gravitational force enables a simple estimation of the retentive force, which can be calculated as.

$$F_g = \rho g V \sin \alpha. \tag{1}$$

where ρ is liquid density, g is the acceleration due to gravity, V is the drop volume, and α is the tilt angle. The retentive force, otherwise considered as adhesion of the droplet to the superhydrophobic surface for the case depicted in Figure 7, is estimated to be 5.4×10^{-6} N. As can be seen, a moving drop has the potential to clean the surface, as any dust particles present could be retained by the droplet and removed from the surface. A dust particle (marked with the red arrow) has a stronger adhesion affinity to the water droplet compared with the superhydrophobic surface and will be removed together with the moving drop (Figure 7).

Moreover, on superhydrophobic structured metallic surfaces, water droplets can be observed to bounce when dropped onto the surface. The kinetic energy of the falling water droplet is transferred to the surface energy without the droplet spreading, then bounces numerous times before coming to rest. The time lapse pictures presented in Figure 8 show the first two

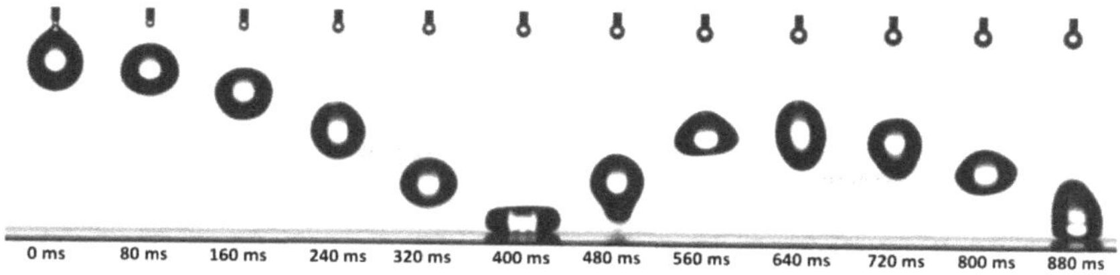

Figure 8 A water droplet bouncing on the superhydrophobic stainless steel surface (initial impact velocity is 35 cm/s).

bounces of the water droplet with a radius of 1.3 mm falling from the height of 6.3 mm onto superhydrophobic stainless steel surfaces. The velocity of impact is 35 cm/s. The droplet trajectory between two impacts is parabolic as anticipated in the gravity field. After two bounces, seven further bounces were observed. Finally, the drop comes to rest on the surface, having a large contact angle in conformity with the Cassie–Baxter equation. Note that such behavior is not observed on flat metal surfaces.

The pictures contained in Figure 8 show that the water contact angle remains high during the impact of water droplet with the superhydrophobic surface. The drop expands, deforms and takes a pancake shape; therefore its kinetic energy is stored in the surface energy. Then, due to the superhydrophobicity of the surface and drop surface tension, the droplet retracts and bounces off the surface. The restitution coefficient $\varepsilon = |V'/V|$, where V' is the velocity after and V is the velocity before the impact, can be estimated from video images and is 0.8. The elasticity of the impact on superhydrophobic surfaces was shown to be limited due to transfer of a part of kinetic energy into droplet vibrations even under other optimal conditions (moderate impact velocity and small contact angle hysteresis) (Richard and Quere, 2000). It can be seen that the drop oscillates after the jump.

Generally speaking, the droplet behavior after falling onto the superhydrophobic surface can be classified into three different regimes: (1) nonbouncing due to low impact velocity; (2) bouncing; and (3) droplet penetration into the structure with a loss of superhydrophobic properties of the structured surface at high impact velocities above a certain critical value. In Bartolo et al. (2006) a critical value of impact velocity for the penetration regime on superhydrophobic micropatterned polydimethylsiloxane was found to be in the range of 0.8–0.9 m/s. Resistance of superhydrophobic surfaces to wetting can be considered as "competition" between the entrapment of air inside the structure (heterogeneous wetting scenario) and complete liquid wetting of the structure (homogeneous wetting scenario). Transition between the heterogeneous (superhydrophobic) wetting regime and homogeneous wetting was shown to be induced also, for example, by vibrations (Bormashenko et al., 2008) or by external pressure (Lafuma and Quere, 2003). Therefore, the material loses its superhydrophobic properties.

Such transition between superhydrophobic wetting and homogeneous wetting has been studied for titanium samples structured with quasi-periodical spikes while immersed in water (Fadeeva et al., 2011). These authors observed that most of the air trapped on the superhydrophobic surface, directly after immersion in water, was gradually replaced by water over time (Figure 9).

The images presented in Figure 9 were taken using a CCD camera coupled with 10× microscope objective. This method enables the qualitative observation of the dynamic wetting of superhydrophobic surfaces while they are immersed in water. The quantitative estimation of the percentage of air trapped in the superhydrophobic titanium surface obtained from these images is, however, inaccurate due to insufficient resolution of the equipment. Using this technique, microscale pockets of air in the structure can be clearly observed, whereas nanoscale bubbles cannot be resolved. Therefore, in Truong et al. (2012), transmission small-angle X-ray scattering (SAXS) based on the technique described previously (Scardino et al., 2009; Zhang et al., 2007) was used to quantify the amount of air that was on superhydrophobic titanium samples structured with quasi-periodic spikes after their immersion in water. Custom-built fluid cells allowed remote fluid injection without changing the sample position relative to the incident X-ray beam. As the scan position and surface structure was not altered between the scans, any change in scattering over time could only be due to a change in the nature of interface between the structured metal surface and the liquid. SAXS measurements were made on dry samples, every 3 min up to 60 min while being wetting by water, and on fully wetted samples. These measurements demonstrated that the immersed superhydrophobic titanium surfaces exhibited a certain degree of resistance to being wetted by the liquid. A rapid decrease in the amount of the dry surface remaining occurred within the first 69 min of immersion (Truong et al., 2012). After 50 min, ~45% of the surface of the superhydrophobic titanium surface remained dry.

It is noteworthy that the superhydrophobic properties of laser structured metal surfaces with quasi-periodic spikes could be restored by cleaning the samples in acetone using an ultrasonic bath, followed by drying in a desiccator.

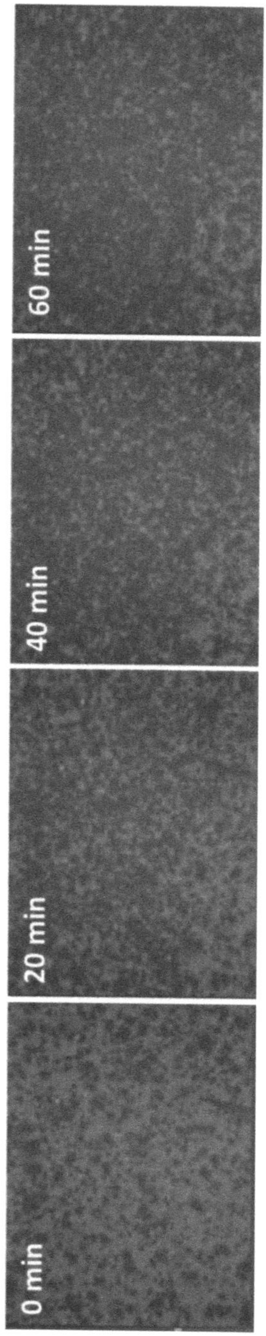

Figure 9 The surface of a superhydrophobic titanium sample structured with quasi-periodic spikes after immersion in water; bright areas of the image correspond to air trapped in the structure, dark areas correspond to the wetted surface.

COMBINATION OF LASER STRUCTURING WITH COATINGS AND LUBRICANTS

Experiments performed using falling drops (Bartolo et al., 2006), vibration (Bormashenko et al., 2008), external pressure (Lafuma and Quere, 2003), and immersion in liquid (Fadeeva et al., 2011; Truong et al., 2012) highlighted the limitations of superhydrophobic surfaces: if the structure is filled with water, the surface loses its water-repellent properties. The resistance of superhydrophobic metallic surfaces to the factors that lead to the loss of water-repellency can be improved by using low-surface-energy coatings. The interplay between coating and the surface structure is illustrated in Figure 10, which shows the relationship between the contact angles measured on flat surfaces cos Θ_{flat} and those measured on the structured surfaces cos Θ_{str} for the fixed geometry of the surface structure, fixed test liquid, and altered chemistry of the solid surface (Lafuma and Quere, 2003). As can be seen, both the heterogeneous and homogeneous wetting regimes can exist on the same surface structure for the same test liquid, depending on the contact angle measured on the flat surface, which is in turn determined by surface chemistry. Thereby, a threshold value Θ_c exists, which separates the regimes of homogeneous and heterogeneous (superhydrophobic) wetting.

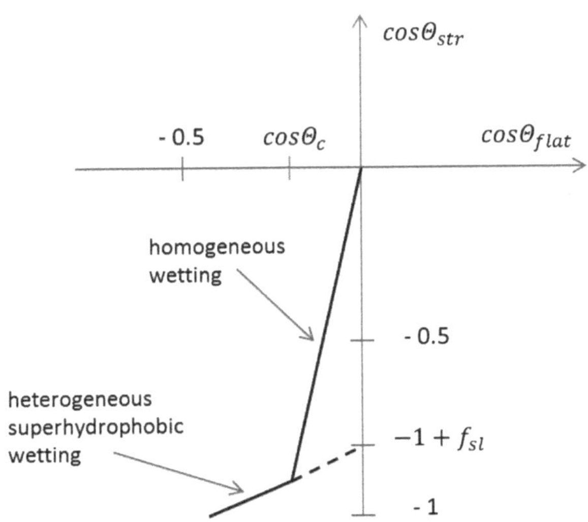

Figure 10 Relation between the contact angles measured on flat surfaces cos Θ_{flat} and structured surfaces cos Θ_{str} for the same material; graph is modified following Lafuma and Quere (2003).

The threshold value Θ_c is determined by the geometrical parameters of surface structure. Note that it has been often reported that the heterogeneous wetting can be observed for $\Theta_{flat} < \Theta_c$. In this case, the heterogeneous regime is metastable (Lafuma and Quere, 2003). This situation is represented by the dotted line in Figure 10. In other words, by applying different coatings onto the spike structures, the superhydrophobic and homogeneous wetting regimes, allowing liquid penetration into the structure, can be realized.

Special polymeric coatings can be used to reduce the surface energy of metals. As can be seen from the data presented in Figure 10, the lower the surface energy of the applied coating, the higher the resulting water contact angle on the structured surface and in turn, the more stable is the heterogeneous wetting regime obtained. The different coatings applied onto laser generated structures are summarized in Table 4.

Stable superhydrophobic surfaces obtained after applying the listed coatings onto different metals and semiconductors are presented in the references given in Table 4.

An alternative method that can be used for the fabrication of stable water-repellent surfaces on metals is to infuse the surface structures with lubricants. Wong et al. (2011) reported the fabrication of slippery lubricant infused structured surfaces based on the following criteria:

1. The lubricating liquid must wick into wet and stably adhere within the structured solid surface;
2. The solid must be preferentially wetted by the lubricating liquid rather than by the liquid needing to be repelled; and
3. The lubricating and repellent test liquids must be immiscible.

Table 4 Hydrophobic Low-Surface-Energy Coatings Applied to the Surfaces Structured by Laser Irradiation

Coating	θ_{flat} [a]	References
(Heptadecafluoro-1,1,2,2-tetrahydrodecyl) trichlorosilane ($CF_3(CF_2)_7CH_2CH_2SiCl_3$)	115°	Baldacchini et al. (2006)
Dimethyldichlorosilane $(CH_3)2SiCl_2$	104°	Zorba et al. (2008)
SAM based on perfluorinated octyl trichlorosilane	120 ± 3°	Jagdheesh et al. (2011)
Silanized using silane reagent (trichloro(1H,1H, 2H, 2Hperfluorooctyl) silane, 97%)	113.00 ± 0.98°	Wu et al. (2009)
Fluoraliphatic polymer antispread E 2/50 FE 60 (Dr Tillwich GmbH)	110°	Fadeeva (2014)

[a] Water contact angle measured on flat surfaces after coating.

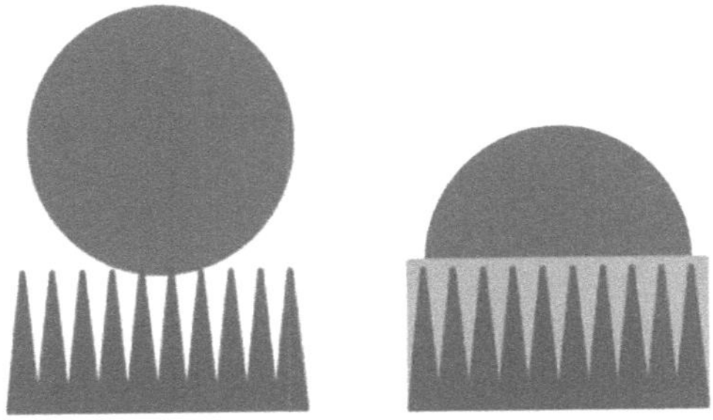

Figure 11 (Left) A water droplet sitting on a superhydrophobic composite air–solid surface: the contact surface between water droplet and solid is strongly reduced, and the water droplet is located mainly on air. As a result, contact angle hysteresis and adhesion between the liquid and solid are greatly reduced. (Right) A water droplet sitting on a slippery lubricant infused structured surface. The lubricating liquid infiltrates the structured surface, forming a physically smooth and chemically homogeneous lubricating film on the solid surface. As a result, contact angle hysteresis and adhesion between the liquid and solid are greatly reduced.

In this way, a low-surface-energy, chemically inert lubricant forms a physically smooth and chemically homogeneous lubricating film on the structured surface, which leads to low contact angle hysteresis and a strongly reduced adhesion of the test liquids (e.g., water) to be repelled. The schematic diagram presented in Figure 11 illustrates the physical action principle of slippery liquid infused structured surfaces in comparison to superhydrophobic surfaces based on composite solid–air interfaces.

As lubricants, low-surface-tension perfluorinated liquids (e.g., DuPont Krytox oils) that are nonvolatile and are immiscible with both aqueous and hydrocarbon liquids can be used. Kim et al. (2012) reported the development of lubricant infused structured surfaces fabricated on aluminum substrates. Polypyrrole was initially electrodeposited on the surface of aluminum samples, which was then followed by the subsequent fluorination of the structured coating and infiltration with lubricant. The resulting water contact angle measurements showed that these surfaces were characterized by low contact angle hysteresis ($\Delta\Theta = \Theta_{str,A} - \Theta_{str,R} = 117° - 115° = 2°$) and reduced ice adhesion strength (Kim et al., 2012). Moreover, the slippery liquid infused structured surfaces were shown to be capable of repelling water and liquid hydrocarbons at, and while transitioning to, a pressure of ~676 atm (equivalent to the hydrostatic pressure at a depth of ~7 km (Wong et al., 2011)), and to maintain a robust liquid repellent properties even for low-surface-tension liquids at temperatures up to at least 200°C (Daniel et al., 2013).

Table 5 Comparison of the Performance of Superhydrophobic Metal Surfaces without and with Additional Coatings (Fadeeva, 2014)

Structure Type	$\Theta_{str,A}$	$\Delta\Theta_{str}$	α	Bouncing
Spikes on aluminum	$166 \pm 4°$	$9 \pm 6°$	$5 \pm 1°$	Yes
Spikes on aluminum + antispread coating	$167 \pm 1°$	$2 \pm 1°$	2°	Yes
Spikes on aluminum + antispread coating + infusing with lubricant	$110 \pm 1°$	$2 \pm 1°$	3°	Yes[a]
ir m- and n-r on platinum	$158 \pm 3°$	10°	10°	Yes
ir m- and n-r on platinum + antispread coating	$164 \pm 1°$	$4 \pm 1°$	3°	Yes
ir m- and n-r on platinum + antispread coating + infusing with lubricant	$112 \pm 1°$	$2 \pm 1°$	2°	Yes

[a]The elasticity of the impact was decreased compared with that obtained using superhydrophobic air-contained surfaces.

We have tested the performance of superhydrophobic metal surfaces:

1. directly after laser irradiation, cleaning in acetone using an ultrasonic bath, and drying in an desiccator;
2. being coated with Antispread E 2/50 FE 60 (Dr Tillwich GmbH);
3. being coated with Antispread E 2/50 FE 60 (Dr Tillwich GmbH) and infused with lubricant GPL 100 (DuPont™ Krytox®)

The experimental results obtained are summarized in Table 5.

As can be seen from the data presented in Table 5, an additional coating of the structured metal surfaces with low-surface-energy Antispread reduces the contact angle hysteresis, and decreases the tilt angle for both types of structures—spikes and irregular microroughness and nanoroughness. As already discussed, the superhydrophobic properties of surface structured metals are dependent on the amount of carbon that is absorbed onto the metal surface. The absorption of carbon from the ambient air is not perfectly uniform in relation to the surface area, leading to an inhomogeneity in the chemical composition of the surface. This in turn results in a relatively high (for superhydrophobic surfaces) value of the contact angle hysteresis. Coatings of the low-surface-energy polymer Antispread reduces the surface energy of the metal, making the surface heterogeneous in chemical composition. In the course of this change, the values of the contact angle hysteresis and tilting angle become reduced to ~2°. These are the best achievable superhydrophobic characteristics obtained on air-trapped metal surfaces generated by present laser technologies, with subsequent surface coating.

After the additional infusion of the superhydrophobic structures with lubricant, a smooth and chemically homogeneous lubricant film was formed on the structured surface. As a result, the contact angle hysteresis becomes smaller and water droplets can be easily removed by tilting the surface at an angle of 2°. Bouncing of water droplets on lubricant infused surfaces has also been observed, however, the elasticity of the bouncing was reduced on the spike structured infused surfaces. On these surfaces, less bounces were observed, together with a lower height of redound.

These experiments show that the performance of superhydrophobic surfaces on metals can be improved by an additional application of a low-energy polymer coating and infusion with lubricants. The lubricant infusion method is particularly interesting for the maintenance of superhydrophobic properties under special ambient conditions (e.g., high pressure or high temperature).

SUMMARY

In this chapter, a laser structuring technique that can be applied for the fabrication of superhydrophobic surfaces on all metals was introduced. This technique enables the flexible and controllable structuring of samples with complex shapes. This technique is particularly suitable for small-scale manufacturing, and can be applied under normal ambient conditions, without the need for a clean room environment. Depending on the properties of the metal substrate being used, different types of structures, for example spikes and irregular micro and nano-roughness, can be generated. Metal surfaces modified to contain both types of structures results in the surface exhibiting superhydrophobic characteristics. It was also found that superhydrophobic properties of metals are dependent on the amount of carbon absorbed on the metal surface. Therefore, metal surfaces become more superhydrophobic (exhibiting lower contact angle hysteresis and tilting angle) after approximately one month of storage under normal ambient conditions. Combination of laser structuring with an additional low-surface-energy polymer coating results in the best achievable superhydrophobic properties for laser structured metals. In these cases, the contact angle hysteresis and the tilting angle were found to be in the range of 2°–3°. The fact that these superhydrophobic metal surfaces trap air within their structure, resulting in strongly reduced liquid–solid area, means that these solid/air–liquid interfaces are limited in their applicability. In instances where surfaces are exposed to vibration, external pressure, or immersion in liquid, a transition between the superhydrophobic (heterogeneous) and homogeneous wetting regimes can be induced, which leads to loss of superhydrophobic properties. Infusion of these superhydrophobic structures with a lubricant, which must be immiscible with the liquid to be repelled, results in the formation of a stable, defect-free and inert interface, maintaining low contact angle hysteresis under high pressures and high temperature conditions.

REFERENCES

Anisimov, S.I., Kapelovich, B.L., Perel'man, T.L., 1974. Electron emission from metal surfaces exposed to ultra-short laser pulses. Sov. Phys. JETP 375, 375.

Baldacchini, T., Carey, J.E., Zhou, M., Mazur, E., 2006. Superhydrophobic surfaces prepared by microstructuring of silicon using a femtosecond laser. Langmuir 22, 4917.

Barthlott, W., Neinhuis, C., 1997. Purity of the sacred lotus, or escape from contamination in biological surfaces. Planta 202, 1.

Bartolo, D., Bouamrirene, F., Verneuil, Ã., Buguin, A., Silberzan, P., Moulinet, S., 2006. Bouncing or sticky droplets: impalement transitions on superhydrophobic micropatterned surfaces. Europhys. Lett. 74, 299.

Bhushan, B., Jung, Y.C., Koch, K., 2009. Micro-, nano- and hierarchical structures for superhydrophobicity, self-cleaning and low adhesion. Phil. Trans. R. Soc. A 367, 1631.

Bormashenko, E., Pogreb, R., Stein, T., Whyman, G., Erlich, M., Musin, A., Machavariani, V., Aurbachc, D., 2008. Characterization of rough surfaces with vibrated drops. Phys. Chem. Chem. Phys. 10, 4056.

Carey, J.E., 2004. Femtosecond-Laser Microstructuring of Silicon for Novel Optoelectronic Devices (Ph.D. thesis). Harvard University.

Cassie, A., Baxter, S., 1944. Wettability of porous surfaces. Trans. Faraday Soc. 40, 546.

Crouch, C.H., Carey, J.E., Warrender, J.M., Aziz, M.J., Mazur, E., Genin, F.Y., 2004. Comparison of structure and properties of femtosecond and nanosecond laser-structured silicon. Appl. Phys. Lett. 84, 1850.

Daniel, D., Mankin, M.N., Belisle, R.A., Wong, T.-S., Aizenberg, J., 2013. Lubricant-infused micro/nano-structured surfaces with tunable dynamic omniphobicity at high temperatures. Appl. Phys. Lett. 102, 231603.

von der Linde, D., Sokolowski-Tinten, K., 2000. The physical mechanisms of short-pulse laser ablation. Appl. Surf. Sci. 154, 1.

Extrand, C.W., 2006. Encyclopedia of Surface and Colloid Science. Taylor, Francis, Chapter Super Repellency, p. 5846.

Extrand, C.W., 2004. Criteria for ultralyophobic surfaces. Langmuir 20, 5013.

Fadeeva, E., 2014. (Ph.D. thesis). Leibniz Universität Hannover.

Fadeeva, E., Schlie, S., Koch, J., Chichkov, B.N., Vorobyev, A.V., Guo, C., 2009. Contact angle, wettability and adhesion. VSP, Chapter Femtosecond Laser-induced Surface Structures on Platinum and Their Effects on Surface Wettability and Fibroblast Cell Proliferation, vol. 6. p.163.

Fadeeva, E., Truong, V.K., Stiesch, M., Chichkov, B.N., Crawford, R.J., Wang, J., Ivanova, E.P., 2011. Bacterial retention on superhydrophobic titanium surfaces fabricated by femtosecond laser ablation. Langmuir 27, 3012.

de Gennes, P.-G., Brochard-Wyart, F., Quere, D., 2003. Capillarity and Wetting Phenomena. Springer.

Her, T.H., Finlay, R.J., Wu, C., Mazur, E., 2000. Femtosecond laser-induced formation of spikes on silicon. Appl. Phys. A: Mater. Sci. Process. 70, 383.

Her, T.-H., Finlay, R.J., Wu, C., Deliwala, S., Mazur, E., 1998. Microstructuring of silicon with femtosecond laser pulses. Appl. Phys. Lett. 73, 1673.

Hohlfeld, J., Wellershoff, S.-S., Güdde, J., Conrad, U., Jähnke, V., Matthias, E., 2000. Electron and lattice dynamics following optical excitation of metals. Chem. Phys. 251, 237.
Jagdheesh, R., Pathiraj, B., Karatay, E., Römer, G.R.B.E., Huis in't Veld, A.J., 2011. Laser-induced nanoscale superhydrophobic structures on metal surfaces. Langmuir 27, 8464.
Kietzig, A.-M., Hatzikiriakos, S.G., Englezos, P., 2009. Patterned superhydrophobic metallic surfaces. Langmuir 25, 4821.
Kietzig, A.-M., Mirvakilia, M.N., Kamalb, S., Englezosa, P., Hatzikiriakosa, S.G., 2011. Nanopatterned metallic surfaces: their wettability and impact on ice friction. J. Adhes. Sci. Technol. 25, 1293.
Kim, P., Wong, T.-S., Alvarenga, J., Kreder, M.J., Adorno-Martinez, W.E., Aizenberg, J., 2012. Liquid-infused nanostructured surfaces with extreme anti-ice and anti-frost performance. ACS Nano 6, 6569.
Koch, K., Bhushan, B., Barthlotta, W., 2008. Diversity of structure, morphology and wetting of plant surfaces. Soft Matter 4, 1943.
Lafuma, A., Quere, D., 2003. Superhydrophobic states. Nat. Mater. 2, 457.
Mannion, P., Magee, J., Coyne, E., O'Connor, G., Glynn, T., 2004. The effect of damage accumulation behaviour on ablation thresholds and damage morphology in ultrafast laser micro-machining of common metals in air. Appl. Surf. Sci. 233, 275.
Marmur, A., 2004. The Lotus effect: superhydrophobicity and metastability. Langmuir 20, 3517.
Marmur, A., 2003. Wetting on hydrophobic rough surfaces: to be heterogeneous or not to be? Langmuir 19, 8343.
Nayak, B.K., Gupta, M.C., 2010. Self-organized micro/nano structures in metal surfaces by ultrafast laser irradiation. Opt. Lasers Eng. 48, 940.
Nolte, S., Momma, C., Jacobs, H., Tünnermann, A., Chichkov, B.N., Wellegehausen, B., Welling, H., 1997. Ablation of metals by ultra-short laser pulses. J. Opt. Soc. Am. B 14, 2716.
Oliveira, V., Ausset, S., Vilar, R., 2009. Surface micro/nanostructuring of titanium under stationary and non-stationary femtosecond laser irradiation. Appl. Surf. Sci. 255, 7556.
Patankar, N.A., 2003. On the modeling of hydrophobic contact angles on rough surfaces. Langmuir 19, 1249.
Rethfeld, B., Sokolowski-Tinten, K., von der Linde, D., Anisimov, S., 2004. Timescales in the response of materials to femtosecond laser excitation. Appl. Phys. A 79, 767.
Richard, D., Quere, D., 2000. Bouncing water drops. Europhys. Lett. 50, 769.
Scardino, A.J., Zhang, H., Cookson, D.J., Lamb, R.N., Nys, R., 2009. The role of nano-roughness in antifouling. Biofouling 25, 757.
Truong, V., Webb, H., Fadeeva, E., Chichkov, B., Wu, A., Lamb, R., Wang, J., Crawford, R., Ivanova, E., 2012. Air-directed attachment of coccoid bacteria to the surface of superhydrophobic lotus-like titanium. Biofouling 28, 539.
Vorobyev, A.Y., Guo, C., 2006. Femtosecond laser nanostructuring of metals. Opt. Express 14, 2164.
Vorobyev, A.Y., Guo, C., 2013. Direct femtosecond laser surface nano/microstructuring and its applications. Laser & Photonics Reviews 7, 385.
Weiss, P., 1971. Clean surfaces: their preparation and characterization for interfacial studies. J. Polym. Sci., Part B: Polym. Lett. 9, 701.
Wenzel, R.N., 1936. Resistance of solid surfaces to wetting by water. Ind. Eng. Chem. 28, 988.
Wong, T.-S., Kang, S.H., Tang, S.K.Y., Smythe, E.J., Hatton, B.D., Grinthal, A., Aizenberg, G., 2011. Bioinspired self-repairing slippery surfaces with pressure-stable omniphobicity. Nature 477, 443.
Wu, B., Zhou, M., Li, J., Ye, X., Li, G., Cai, L., 2009. Superhydrophobic surfaces fabricated by microstructuring of stainless steel using a femtosecond laser. Appl. Surf. Sci. 256, 61.

Zhang, H., Lamb, R.N., Cookson, D.J., 2007. Nanowetting of rough superhydrophobic surfaces. Appl. Phys. Lett. 91, 254106.
Zorba, V., Persano, L., Pisignano, D., Athanassiou, A., Stratakis, E., Cingolani, R., Tzanetakis, P., Fotakis, C., 2006. Making silicon hydrophobic: wettability control by two-lengthscale simultaneous patterning with femtosecond laser irradiation. Nanotechnol. 17, 3234.
Zorba, V., Stratakis, E., Barberoglou, M., Spanakis, E., Tzanetakis, P., Fotakis, C., 2008. Tailoring the wetting response of silicon surfaces via fs laser structuring. Appl. Phys. A: Mater. Sci. Process. 93, 819.

CHAPTER SEVEN

Applications of Nanotextured Surfaces: Three-dimensional Aspects of Nanofabrication

INTRODUCTION

This chapter provides an overview of the latest achievements in research fields that are relevant to the fast expanding field of nanotextured surfaces and discusses possible applications of these surfaces. This chapter provides a better understanding of the mechanisms and mode of operation of nanotextured surfaces in various applications rather than thoroughly reviewing the expanding array of research in this area.

The fast developing fields of microfluidics/optofluidics are expected to repeat the trend that was seen in microelectronics when it started ~70 years ago in terms of miniaturization and the integration of different functionalities; from optical to electrical and chemical control of processes in nano- and micro-volumes, where different analytes and complex flows can be managed and characterized and new materials synthesized. Chiral nanostructures and microstructures on the surface are promising to be the next generation of optical vortices—with light beams carrying mechanical angular momentum via their spin and/or orbital momentum. Such beams provide optical hand function, which can hold and twist nanomaterials on nanoscale dimensions and with great precision. This mode of interfacing at the nanoscale is in high demand. A separate field of sensing, in which light on larger area surfaces can be concentrated into hot spots, is breaking new sensitivity levels and increasing control over the surface morphology and functionality.

NANOSCALE STRUCTURES AND THEIR FUNCTIONS

Optofluidics

Extensive reviews on the development of optofluidic technology (Levy and Shamai, 2008; Monat et al., 2007; Psaltis et al., 2006; Whitesides, 2006) show different examples in a few categories, including fluid-solid, purely fluidic

Superhydrophobic Surfaces
http://dx.doi.org/10.1016/B978-0-12-801109-6.00007-0

interfaces, and colloidal suspensions (Psaltis et al., 2006). It is always difficult to classify optofluidic devices and applications due to the various functions they incorporate and effects they exploit. Some of these optofluidic devices are under further development and continue to evolve into more functional platforms. Integration and reconfigurability are two major advantages of with optofluidic devices. The main integration area discussed by Monat et al. (2007) is the creation of fluidically tunable lasers and channel-incorporated optic elements (fibers, waveguides, lenses, gratings) to control light inside an optofluidic chip. A similar topic—tunable optical element integration into a microfluidic channel—is discussed widely by Levy and Shamai (2008); the authors focus on a reconfigurable optical lens (using the effects of electrowetting, liquid immiscibility, and a soft plastic shell filled with a liquid), to apply laser radiation in fluidic channels filled with rhodamine 6G (Song et al., 2009) or rhodamine 101, changing the index via altering the temperature and the mixing of various liquids.

In another review on optofluidics by Whitesides (2006), the origins of the optofluidics field are discussed as having been conceived through the intersection of (i) molecular analysis (which emerged from the microanalytical methods of gas-phase, high-pressure liquid chromatography and capillary electrophoresis), (ii) biodefence (the high level of support that was received from the U.S. government in the 1990s had a huge impact in this area), (iii) molecular biology (the explosion of research in the area of genomics in the 1980s required the development of devices that could accommodate a greater throughput, higher sensitivity, and greater resolution than microfluidics could offer at the time), and (iv) microelectronics (it was intended to clone techniques used in photolithography for creating microfluidic channels, but silicon and glass substrates appeared to be inconvenient in many applications). Most of the microfluidic/optofluidic devices are tunable via an external pressure source, which is not desirable for applications when miniaturization and integration are required. Integration of tunable optical devices with on-chip microvalve networks and on-chip micropumps (Laser and Santiago, 2004) would be an important step forward. Some researchers (Monat et al., 2007; Psaltis et al., 2006; Whitesides, 2006) make the observation that optofluidics is a field still in its infancy; it is full of promise, and many opportunities exist for creating inventive devices and novel new elegant solutions.

A review published by (Monat et al., 2008) describes the integration of optical elements used in microfluidic chips. Microphotonics has been driven by the idea to integrate different kinds of optical elements onto the one chip, combining microphotonics with microfluidics (Figure 1). Some microfluidic proof-of-principle devices, which incorporate single-mode fibers and can

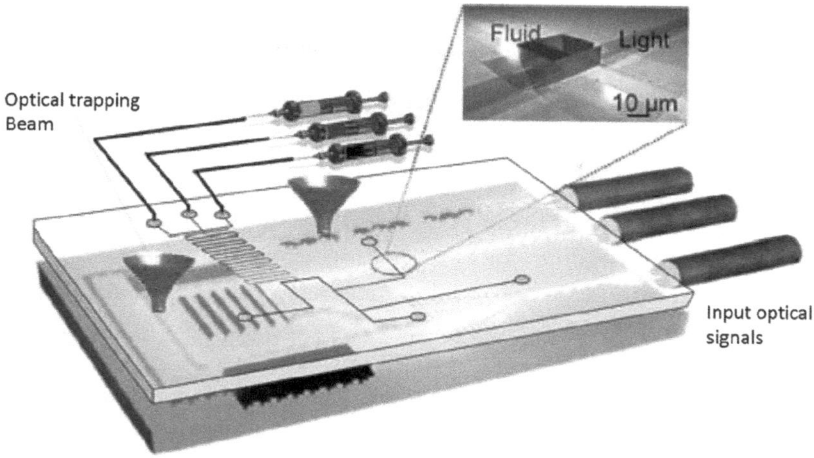

Figure 1 Illustration (Monat et al., 2008) of a multifunctional optofluidic chip. The chip includes (i) a photonic chip (bottom) that is coupled to optical fibers, (ii) a microfluidic circuit (on top) that is connected to fluid sources with different refractive indexes, and (iii) optical trapping beams to manipulate micro-mechanical components. The schematic diagram on the top right-hand side shows a close-up image of the chip, where the optical signal intersects with a fluidic channel on the micrometer scale.

optical trap up to 20 μm diameter spheres, as well as tapered optical fibers and planar photonic crystals, are examined in more detail. The authors conclude that due to the dual applications of sensing and reconfigurable devices offered by optofluidics, the development of this field is expected to have a broad impact, ranging from photonics to biological sciences (Monat et al., 2008)

Some other recent trends in optofluidics include pneumatically tunable dye lasers (Song and Psaltis, 2010); an ultrafast nonlinear device with a checkerboard pattern of liquid strands embedded in photonic crystal fibers acting as nonlinear waveguides (Vieweg et al., 2010); a simple method for creating monodisperse emulsions within the microfluidic device using a hand-operated syringe (Abate and Weitz, 2011); a modulator, based on peristaltic nematogen microflows embedded in a polydimethylsiloxane (PDMS) chip (Cuennet et al., 2011); and surface plasmon polariton nanohole array excitation sensing applications obtained via plasmonic extraordinary transmission. It uses electrokinetic forces to induce nanoflow in the nanochannel of a suspended membrane for nanoparticles (NPs) to trigger the self-alignment of the particles onto the underlying pattern of the nanochannel (Pang et al., 2011); light switching was achieved using a tunable air gap mirror that was actuated by compressed air. A reconfigurable light circuit for dual-channel absorption spectroscopy measurement was demonstrated (Song and Psaltis, 2011), and a 2×3 optofluidic switch, based on a pneumatic control of a waveguide, was able to achieve a switching time that took only 10s of milliseconds (Lim et al., 2011).

Ion-beam Lithography: Three-dimensional Nanostructuring Tool

Tripathi et al. (2012) presents a method that simplifies and improves the ability to form nanopatterns on the surface of a diamond. This direct-write technique implants gallium ions into the top 15 nm of the diamond surface using a focused ion beam (FIB). The implanted areas become hard masked, therefore protecting the substrate material during reactive ion etching. The resulting patterns have a high aspect ratio of 6:1, high resolution down to 18 nm, well-controlled structures (Figure 2) that form millimeter-size

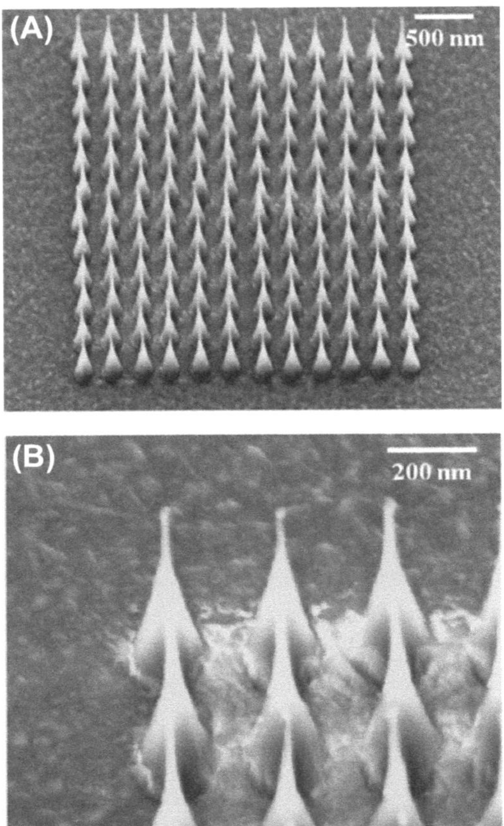

Figure 2 SEM images (Tripathi et al., 2012) demonstrating high-resolution, high aspect ratio diamond nanopillar matrix. The pillar tips are 18 nm in diameter and total height is 110 nm suggesting an aspect ratio of greater than 6.

patterns up to 1.5×1.5 square millimeters on natural, high pressure, high temperature, and chemical vapor deposition–grown diamond films. The authors predicted that this technology may be used in various prototyping and mastering applications ranging from micro-electromechanical systems (MEMS) to semiconductor or LED photonics.

Garoli et al. (2011) present a technique that combines FIB milling and soft molding (Figure 3). Fabrication of high aspect ratio (up to 10) nanorocket-tip arrays using FIB was demonstrated on silicon substrates, then the pattern was transferred into PDMS and finally into UV curable thiolene optical adhesive to obtain a positive mold, which was in turn coated with a 40 nm film of silver. The originally fabricated nanorod structures had a radius of ~19 nm, while the replicated structures had a tip radius of ~40 nm. These authors demonstrated simulation results that highlighted how metal-coated structures can focus surface plasmons, resulting in the enhancement of light intensity at their tips. They foresee the use of nano-tipped structures in high-sensitivity molecular detection research, cell analysis, and DNA hybridization studies.

There is a constant search for new materials that would provide higher sensitivity and efficiency. Ruffato et al. (2012) investigated nanoporous gold made of 75% Ag and 25% Au that provided a greatly enhanced surface-to-volume ratio and demonstrated potential for better reaction efficiency. They presented work on the design, fabrication, and characterization of nanoporous gold (NPG) slit arrays and performed experiments that exploited the phenomenon of extraordinary transmission in the near-IR spectrum. The authors introduced a novel, protected FIB lithography process that used a protective bilayer of 150 nm of PMMA and 120 nm of Ag that shielded the surface of porous structures during direct ion milling (Figure 4).

When protective coating was applied, the original porosity was better preserved and the material maintained its original morphology. The gold slit arrays were functionalized with a dodecanethiol coating, and their transmission response was compared with that of bulk gold. In conclusion, a more effective surface was demonstrated and metallic behavior beyond the NIR spectral range was achieved. Ruffato et al. (2012) stated that NPG demonstrated increased detection sensitivity and reaction efficiency, thus representing a promising material for plasmonic platforms for sensing.

The FIB milling process used to fabricate the nanochannels described by Menard and Ramsey (2010) shows the effect of ion beam shape transfer into a substrate when milling of narrow lines is performed. A two-layered architecture consisting of a 300 nm thick chromium film deposited onto a quartz substrate was used as shown in Figure 5. A 5 nm thick nanochannel was

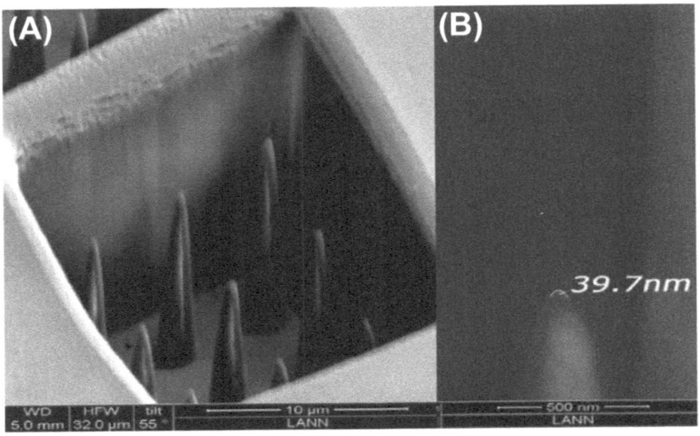

Figure 3 SEM micrographs (Garoli et al., 2011) of the replicated and metallized array of nanorocket tips. (A) Tips array. (B) Detail of the tip radius.

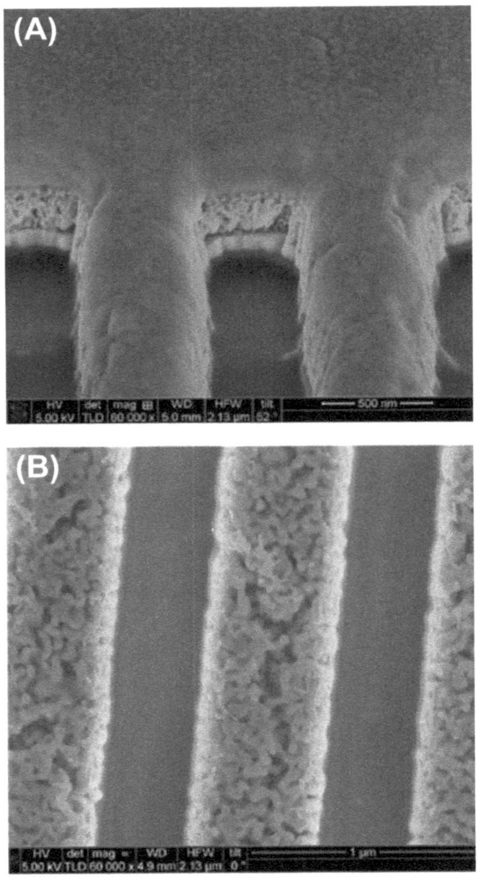

Figure 4 SEM micrographs (Ruffato et al., 2012) of the sample prepared by ion milling on Ag/PMMA/NPG trilayer. (A) After the milling; (B) after removing the PMMA/Ag.

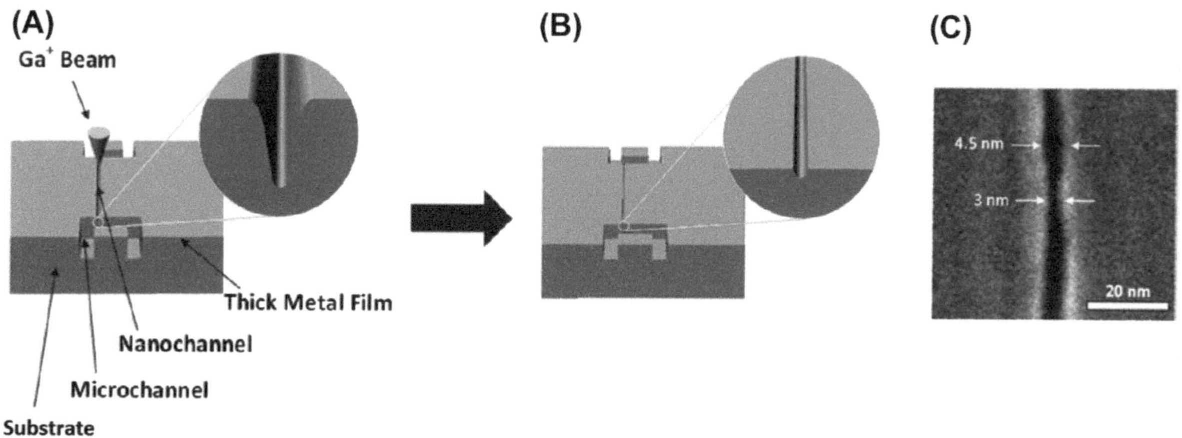

Figure 5 Schematic diagram (Menard and Ramsey, 2010) of the proposed milling technique and SEM of an actual channel. (A) Milling a nanochannel through the thick metal film. (B) Removal of the metal film using an etching solution. (C) Top-view SEM image of a sub 5 nm wide nanochannel milled into a quartz substrate through a 300 nm Cr layer.

milled through a chromium layer until the bottom of a groove just reached the set depth. The metal was removed after the ion milling step using commercially available standard wet chrome etchant. The fabricated nanochannels were sealed with a cover glass slide using oxygen plasma activation of both surfaces, and a permanent bond was formed by annealing the channel in the oven at 900 C for 10 h. The authors tested their method using quartz, single-crystal silicon, and PDMS substrates to demonstrate the versatility of the technique for hard and soft materials. These authors suggest that in future the resulting devices will used in single-molecule DNA transport studies.

Neon (Winston et al., 2011) or helium (Li et al., 2012) ion lithographies rather than gallium are used when milling sub 10 nm features. Li et al. (2012) suggest the use of a combination of high-resolution helium ion beam patterning and nanoimprint lithography when channels of this size are required. These researchers used a Carl Zeiss Orion Plus scanning Helium Imaging Microscope operated at 35 kV and equipped with a Raith ELPHY pattern generator to fabricate a series of line patterns into a 12 nm thick hydrogen silsesquioxane (HSQ) resist and demonstrated that a half-pitch of the structures down to 5 nm was possible. They claim to resolve the smallest half-pitch of line patterns to be 4 nm but the image is very faint (Figure 6). The patterns formed in HSQ were successfully used as a nanoimprint template, and structures were transferred into a UV-curable nanoimprint resist. Winston et al. (2011) demonstrated the production of channels with a half-pitch of 7 nm using neon ion lithography in 16 nm thick HSQ.

Chiral Plasmonics: Three-dimensional Light–Matter Interaction

Recently, chiral NPs have been most often formed using electron beam lithography (EBL), sputtering metals on rotating substrates at oblique angles or by using self-assembled NPs for meta-material applications. When trying to exploit their possible plasmonic response, NPs are usually fabricated using silver or gold coatings. Planar and three-dimensional (3D) structures are prepared, with the latter being more popular.

Dietrich et al. (2012) demonstrated EBL fabrication on nonplanar substrates. They introduced a technique called on-edge lithography that can fabricate 3D and/or chiral nanomaterials using a single lithographic step that is able to produce nanometer-scale features. These authors demonstrated the fabrication of chiral "L"-shaped nanostructures (Figure 7) made of 30 nm thick gold on a prepatterned fused silica substrate. The measurement of the optical response from chiral structures showed a notable and broadband circular dichroism response in the near-IR region. The authors suggested possible applications for their structures as plasmonic polarization filters and in biosensing applications.

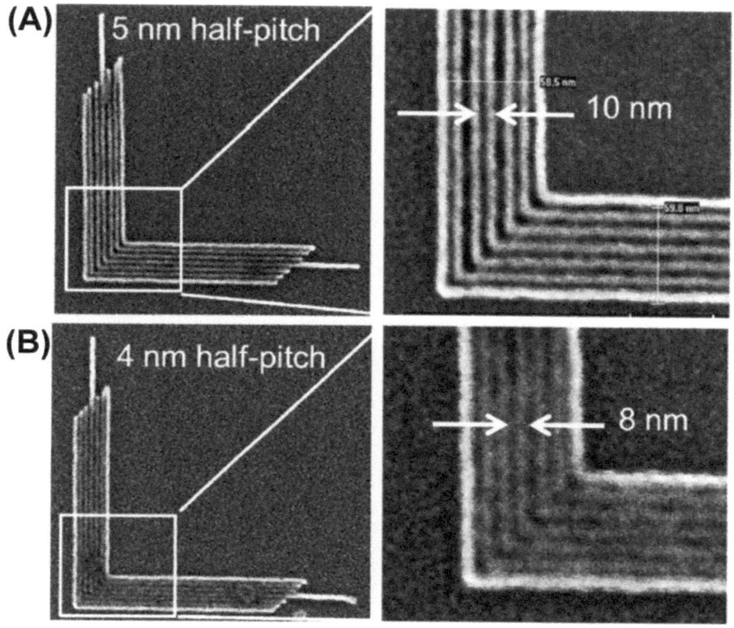

Figure 6 SEM images (Li et al., 2012) of (A) 5 nm and (B) 4 nm half-pitch nested Ls formed by helium ion beam lithography in an HSQ layer that was subsequently developed to remove any unexposed resist. Five nanometer half-pitch patterns were easily resolved. The authors claimed that 4 nm half pitch patterns can also be clearly resolved.

Another EBL fabrication on a prepatterned surface is reported by Staude et al. (2013a, 2013b). In this case, these researchers used a laser to directly write the pattern onto the substrate. Sub-100 nm features made of 20 nm thick gold were fabricated on top of laser polymerized elongated spheres, as shown in Figure 8.

Abdulrahman et al. (2012) demonstrated a new technique for the fabrication of chiral plasmonic nanomaterials. These researchers were able to produce combined structures on a quartz substrate, these being 55 to 130 nm thick achiral gold crosses with a thin layer of flavin mononucleotide, a biomolecule that displays a strong chiroptical signal, on top. They stated that chirality can be obtained from a surrounding chiral molecular material to a plasmonic resonance of an achiral metallic nanostructure. These authors demonstrated that plasmonic chirality can be induced by far-field radiative coupling between a localized plasmon of an achiral metallic nanostructure and a surrounding chiral molecular layer. The experimental results were compared with a simple electromagnetic model, where good agreement was demonstrated.

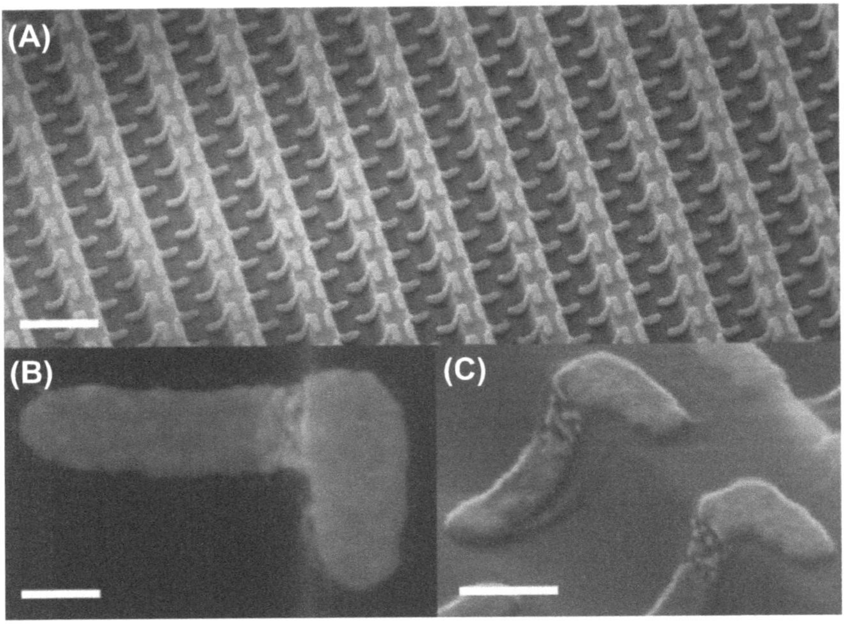

Figure 7 SEM images (Dietrich et al., 2012) of (A) tilted view of the fabricated chiral nanoparticle array. Magnified view of a single 3D-L particle under (B) top and (C) oblique view. The scale bars are (A) 1000 nm, (B) 80 nm, and (C) 100 nm.

Kuzyk et al. (2012) demonstrated another popular bottom-up approach of DNA- and peptide-directed chiral NP self-assembly (Figure 9). These researchers presented the fabrication of nanometer-scale helices formed using 10 and 16 nm diameter spherical gold NPs. The experimental results from the circular dichroism (CD) effect measurements in the visible part of the spectrum (~500 nm) demonstrated that the chiral morphology of organic molecules used with the plasmonic properties of gold NPs is a working combination. These researchers demonstrated that using their proposed method, a high overall yield of 77% and 86% of perfect assemblies of right- and left-handed helices, respectively, with nine NPs could be achieved. Long-term (up to 3 months) storage of a sample and fine-tuning of the spectral NP response by forming a 3 nm thick silver shell were also presented. The authors compared their experimental results with theoretical calculations based on classic electrodynamics and obtained a good agreement for their 16 nm diameter gold spheres. They concluded that DNA-origami could be a valuable tool in designing new materials.

The oblique angle deposition technique, together with a rotating substrate holder was used by Esfandiar et al. (2013) to make graded slanted chiral silver metal films. These researchers used scanning electron microscopy (SEM)

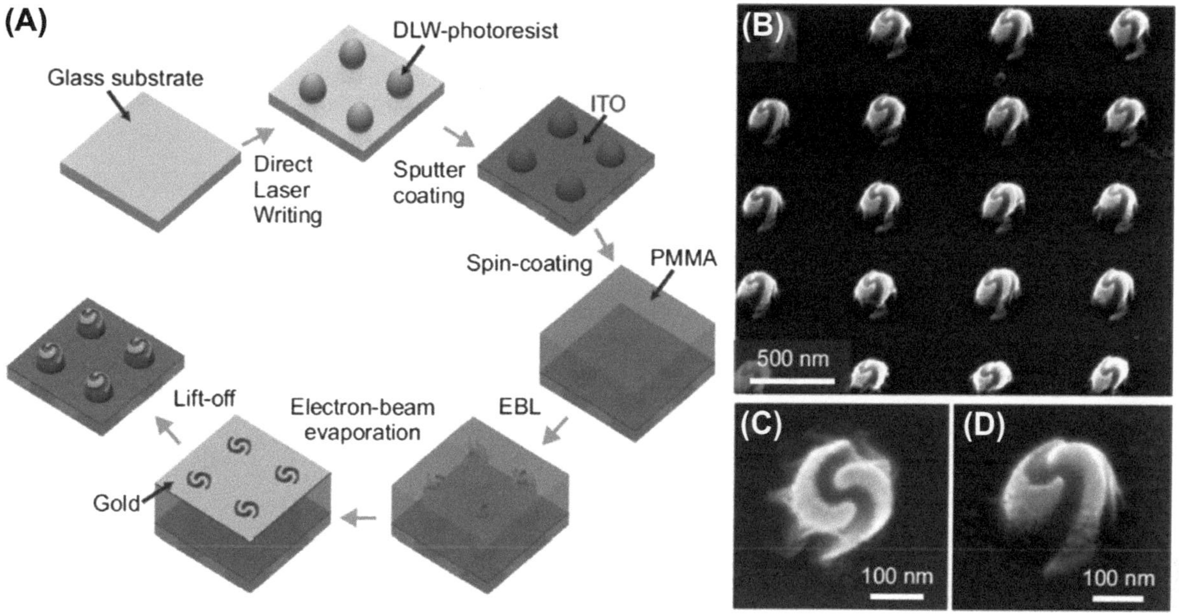

Figure 8 (A) Schematic diagram (Staude et al., 2013a) of the hybrid fabrication approach. A glass substrate is prepatterned via direct laser writing and coated with a thin layer of ITO in order to make it conductive. Next, the electron-beam resist PMMA is spin-coated onto the prepatterned substrate and patterned by EBL. PMMA is developed, the gold film is evaporated, and a standard lift-off procedure is finally performed. (B) SEM micrograph of the fabricated arrays of tapered double-helices on a slanted view. (C,D) Magnified top and slanted views of a single double-helix.

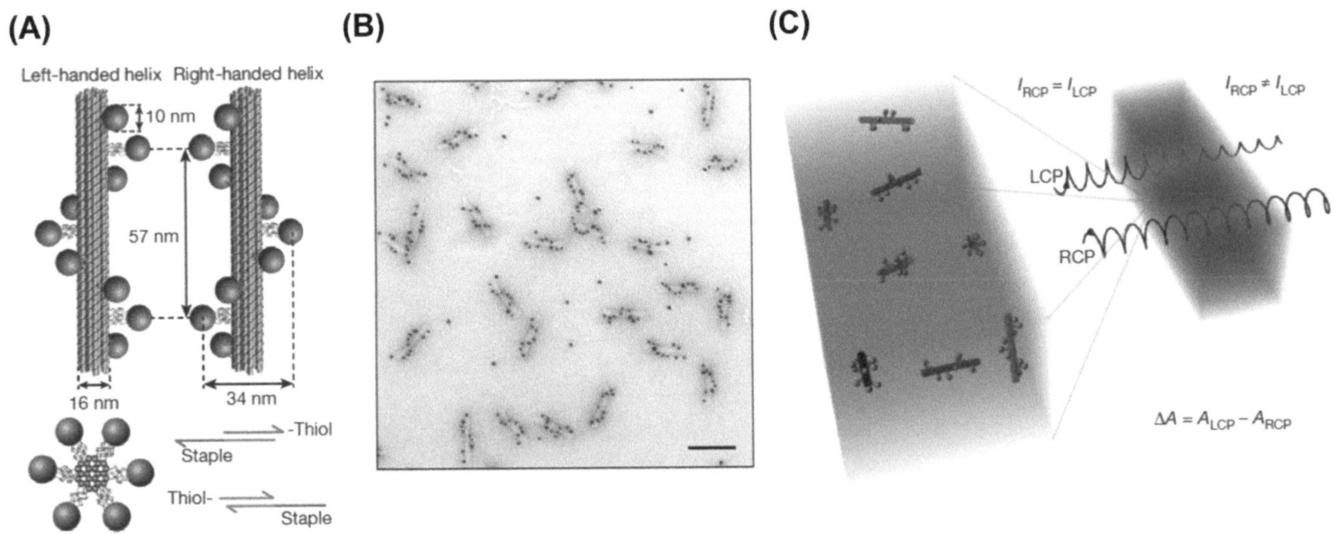

Figure 9 Assembly of DNA origami gold nanoparticle helices and the principle of circular dichroism (Kuzyk et al., 2012). (A) Left- and right-handed nanohelices are formed by nine spherical gold nanoparticles that are attached to the surface of DNA origami 24-helix bundles. (B) Transmission Electron Microscopy (TEM) image of DNA-assembled left-handed gold nanohelices (scale bar, 100 nm). (C) Circular dichroism is measured as the difference in absorbance $\Delta A = A_{LPC} - ARPC$ of left-hand circularly polarized (LCP) and right-hand circularly polarized (RCP) light as a function of wavelength.

and atomic force microscopy (AFM) to characterize the fabricated films. The results of a structural analysis showed the presence of a thickness gradient that increased from 2.05 to 2.16 μm with a distance increase from 10 to 30 mm from the edge of the shadowing block that was placed in the center of a rotating stage. These authors also noted a decrease in the volume of void fraction and an increase in the average grain size from 215 to 245 nm when moving away from the center. The angle at which nanospirals were growing (see Figure 10) also changed toward the normal to the substrate surface when moving away from the center.

An optical analysis of the chiral nanospiral samples revealed a dependence between the reflectance and sample illumination angle as well as the distance

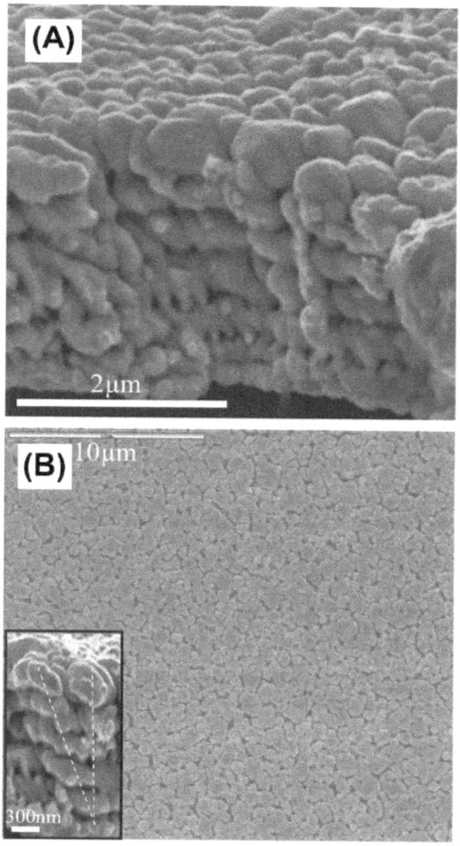

Figure 10 SEM images (Esfandiar et al., 2013) of graded chiral slanted silver nanosculptured thin films grown at 10 mm distance from the edge of the shadowing block placed in the center of a sample holder: (A) side view, (B) top, and (inset) cross-sectional views.

from the center. These authors suggested that the main factor responsible for this change was the grain size of the deposited silver thin film. The photoluminescence (PL) spectra were also measured on a fabricated sample and showed an intensity decrease with increasing distance from the center. The authors concluded that easily tunable nanostructures have the potential to be used in many biosensing and cancer therapy applications.

Han et al. (2013) described a shadowing vapor deposition (SVD) technique for the fabrication of chiral metamaterials. The structures are formed by coating 100 to 140 nm wide column nanoarrays with 40 nm of silver (see Figure 11). Chirality control is achieved by carefully adjusting the deposition angle with respect to the substrate. The deposition angle is varied from 0° to 60°, which is large enough to cast shadows on the desired structures, but small enough not to introduce oblique angle deposition distortions to the deposited film roughness. The measured optical response from the resulting fabricated structures demonstrated a very large circular dichroism in the 450 to 850 nm range for circularly polarized light. The authors see the SVD method as a promising technique for the mass production of large areas of chiral metamaterials that could open new possibilities for the study of negative refractive index materials based on their chiral properties.

Black Silicon

In this section, applications and the methods for fabricating black silicon are presented. Black silicon can be produced by means of dry (Jansen et al., 2009; Pezoldt et al., 2011; Steglich et al., 2014) and wet etching (Srivastava et al., 2010; Toor et al., 2011; Wang et al., 2013) as well as laser fabrication (Lu et al., 2013), to the formation of prepatterned structures using UV or EBL resists and NPs as masks. A great variety of applications for black silicon structures have been presented; many of their uses are as optimized antireflection coatings and light absorbers for solar cell applications, with some other applications exploiting the surface features of black silicon to trap live cells or interlock two surfaces.

In the review by Jansen et al. (2009), the authors presented a detailed comparison between Bosch and cryostat deep reactive ion etching (DRIE) for manufacturing black silicon. The authors discussed the use of different gasses and masking materials to optimize the silicon etch rates and to determine the resulting etch profiles. Lowering the etching temperature down to −120 °C prolongs the life of the mask being used, in turn providing better selectivity, with the only inconvenience being the need to use liquid nitrogen. Some design improvements for DRIE tools were presented; these increase the etch rate and improve the profile of the resulting etched shape. When a procedure to

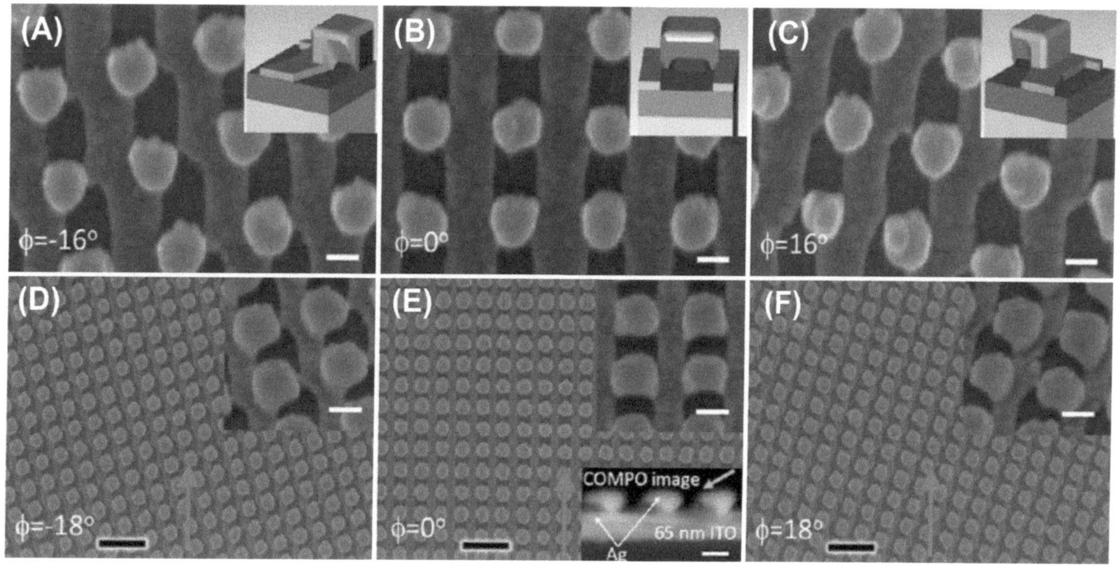

Figure 11 SEM images (Han et al., 2013) of Ag/PMMA arrays (lattice spacing 225 nm) using a deposition angle $\theta=60°$. The width of columns of the PMMA arrays are (A–C) 100 nm and (D–F) 125 nm. The insets in (A–C) are 3D schematic diagrams of the silver-coated PMMA columns. Gold-red color (Black in print version) is for Ag, dark green (Dark grey in print version) for PMMA, blue (Black in print version) for ITO, and green (Grey in print version) for glass. The top insets in (D–F) are SEM images taken at a ~40° tilt. The bottom inset of (E) is a COMPO (back scattering) SEM cross-sectional image showing the Ag and the ITO layers. The black and white scale bars are 500 and 100 nm, respectively. The red arrows shown in (D–F) indicate the deposition directions for all images.

form vertical wall etching is investigated, a "micrograss" is often formed the bottom of the etched structures, called "black silicon". The name comes from the fact that if needle-shaped structures with a height in the order of hundreds of nanometers in height are formed on silicon surface, they tend to trap light, causing the reflection to reduce to below 99% in the visible spectrum range, making the surface of the sample appear black. Researchers use this effect to optimize the etching recipes, suggesting that the oxygen content should be lowered in order to fabricate perfectly vertical profile structures. Black silicon (Figure 12) is obtained in two ways: (i) via the spontaneous self-alignment of nanoneedles from hard particles formed from an etching process waste falling on the surface and creating micromasks and (ii) by patterning the surface using a mask and performing the Bosch etching process that uses passivation and etching steps repeatedly until the desired depth is reached. Substrate temperature management is much more important as the etch rate (up to 32 μm/min) increases, and helium backside cooling becomes the bottleneck in heat transfer.

The cathodoluminescence response of black silicon was investigated by Pezoldt et al. (2011) in the blue-green and red regions of the visible spectra. The authors suggested that the nanosized tips and their coating with fluorine-doped nonstoichiometric silicon dioxide are responsible for this effect. The dry etching SF_6 and O_2 gas mixture recipe (Jansen et al., 2009) was used to make the black silicon structures. These researchers also tried cleaning the surface with argon (Ar) plasma and reflection measurements (below 1% in visible spectrum range). The argon plasma made the silicon nanospikes less sharp, and the overall spike density on the surface decreased as smaller spikes were consequently removed. The authors concluded that the cathodoluminescence spectra could be related to quantum confinement effects and recombination mechanisms related to Si-O bonds or the Si-SiO_2 interface in the vicinity of the needle tip that was measured to have a 15 nm radius.

Steglich et al. (2014) presented a broadband absorber that was on highly boron-doped ($10^{-3}\,\Omega m$) nanostructured (black) silicon that was fabricated using the dry etching conditions proposed by Jansen et al. (2009). Needle-like structures that trap visible light and make the silicon surface appear black were prepared to be 1600 ± 150 nm in height. The entire sample surface was also covered with a 30 nm thick Al_2O_3 coating, deposited by a plasma-enhanced atomic layer deposition technique (PE-ALD), which increased the absorbance of the surface in the near-IR spectrum range. Final samples were demonstrated to reach an absorbance above 99.5%.

Figure 12 SEM images (Jansen et al., 2009) of (A,B) back silicon obtained during cryogenic temperature mixed-mode DRIE and (C,D) room temperature pulsed-mode (Bosch) DRIE where structures were found to be scalloped but straight.

Another DRIE process was developed by Mehran et al. (2011) for the fabrication of hybrid black silicon structures (Figure 13) containing high aspect ratio spikes of height ranging from 40 to 50 nm. Using this technique, one can selectively select areas on the sample where black silicon is grown and where they are free of micrograss. The authors suggest a novel, hydrogen assisted, reactive ion etching method that originally used only oxygen and sulfur-hexafluoride (SF_6) gasses to perform the silicon etching. The wetting and trapping of kidney cells on the needle-covered areas were investigated. The wetting by water of the bare and processed silicon surface changed dramatically, with water contact angles reducing from 40° to 6°, respectively, and for oil, a similar effect was observed, with the contact angle changing from 23° on the polished silicon to close to 0° on the nanoroughened surface. The authors suggested that the nanoneedle patterned silicon surface represents a good substrate for cell adhesion and might be useful in biochips and in cell study applications.

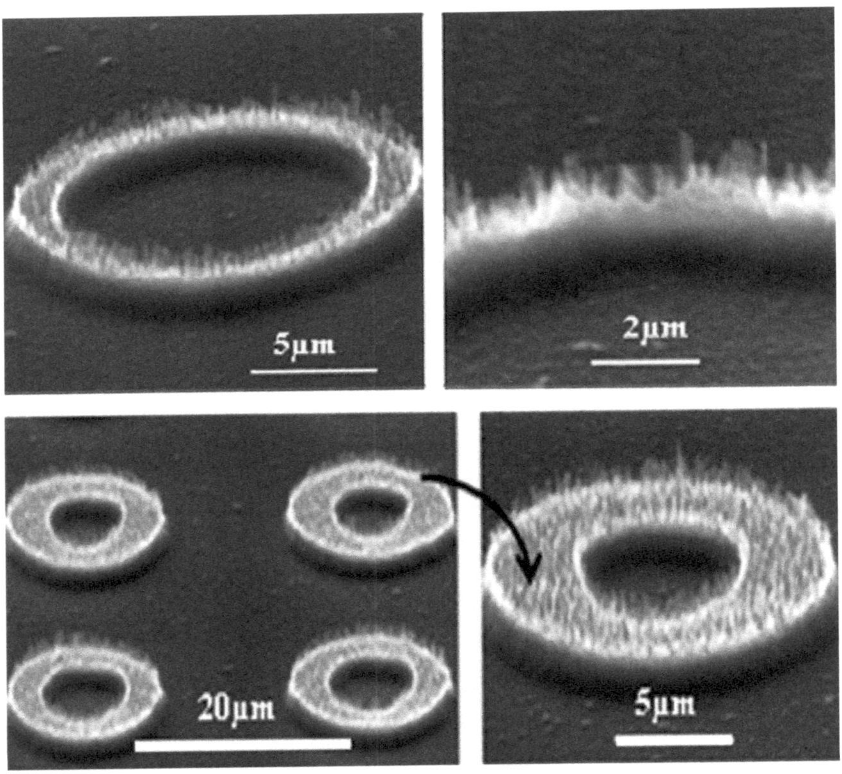

Figure 13 SEM images (Mehran et al., 2011) of selected regions that are fully covered by silicon grass; the surface of the silicon substrate on other regions of the surface remain clean and grass-free.

Murias et al. (2012) described using the DRIE etching process for the formation of black silicon (pyramid-like structures in this case) using an $SF_6/O_2/CH_4$ gas mixture. The authors reported that the reflectance of silicon was reduced from ~30% to ~1%, and suggested that this technique be used in low-cost solar cell manufacture.

Lilienthal et al. (2010) discussed the generation of needle or tube-like structures being formed on the surface after the application of DRIE, obtained from the nanomasking spots created from residues in the plasma. The researchers obtained structures with heights ranging from 3 to 25 μm and from 150 to 500 nm in diameter and demonstrated their possible application in the fields of microfluidics with catalysts, micro-optical, or mechanical mountings or carrier wafer bonding in microelectronics. The development of novel dry etching processes for fused silica was described. Characterisation of the resulting samples suggested a great similarity to black silicon (Figure 14), and it was suggested

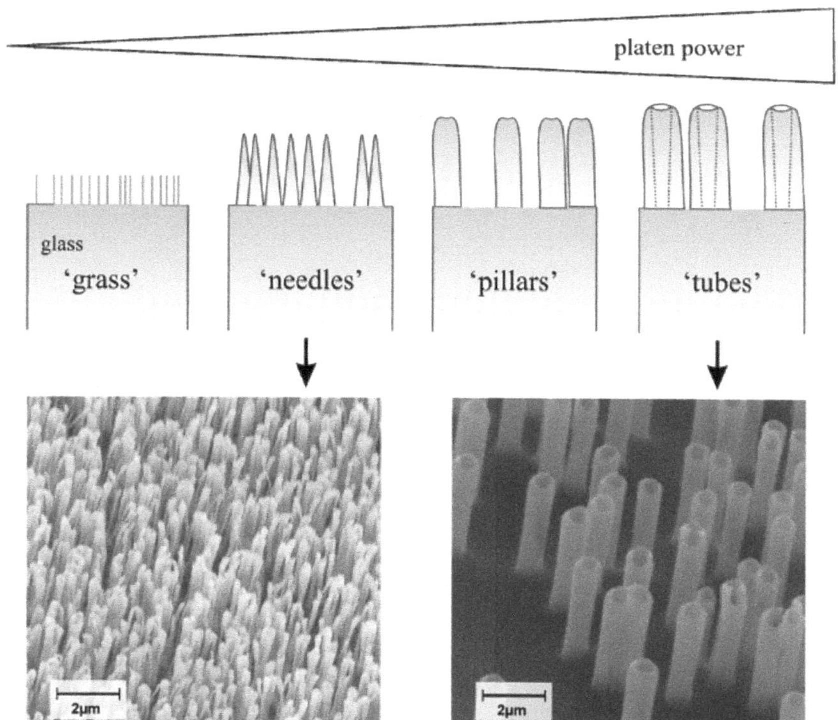

Figure 14 Drawings and SEM images (Lilienthal et al., 2010) of different fabricated needle shapes in fused silica showing dependence of platen power use in DRIE process. Grassy structures, needles, pillars, and tubes can be formed on a surface. Two SEM images are showing examples for dense needle-like nanostructures (left) and tubes (right).

that the technologies used for the formation of such black silicon samples could also be applied to fused silica substrates to make similar structures. One of the examples presented is the production of a nanostructured surface as a Velcro®-like interlocking structure that may allow polymer-to-silicon/glass-bonding.

A wet chemical etching process to form black silicon was presented by Srivastava et al. (2010). Pieces of p-type silicon were submerged into an hydrofluoric (HF) solution at room temperature together with $AgNO_3$ for different times to produce arrays of vertically aligned silicon nanowires (VA-SiNWA) as seen in Figure 15. While the diameter of nanowires remained constant in the 50 to 300 nm range, the height of the spikes was found to be increasing linearly with time at the rate of ~250 nm/min. The

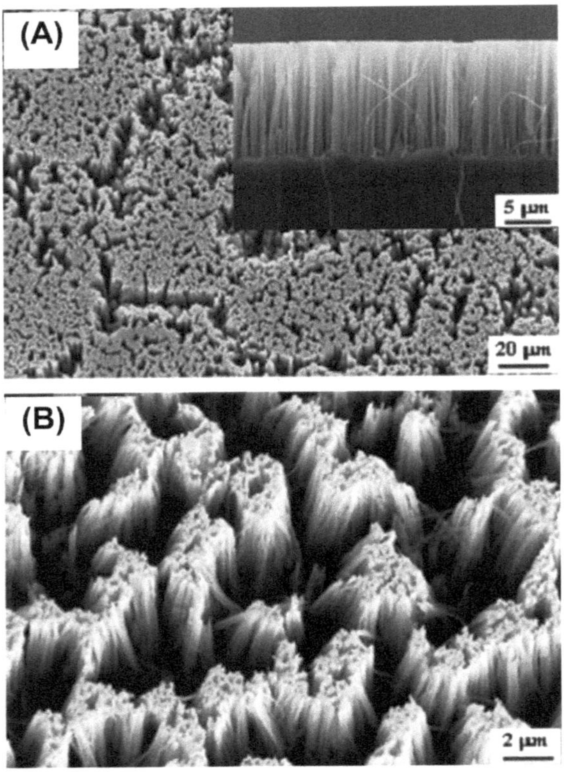

Figure 15 SEM images (Srivastava et al., 2010) of VA-SiNWA fabricated for 45 min in HF solution showing (A) top view (inset shows the cross-sectional view of Si wires of length ~12 μm) and (B) magnified tilted view of the sample surface.

goal of the reported work was the production of an anti-reflective coating for solar cells. A reflectivity of less than 2% was achieved for 12 micron nanowires in the spectral range of 300 to 600 nm.

Black silicon possessing different topography was obtained by Wang et al. (2013) using the same fabrication method of Ag-assisted chemical etching described by Srivastava et al. (2010). The authors deposited different mass thicknesses (3, 5, and 10 nm) of Ag to form NPs on a 30 nm SiN_x film that had been deposited on the black silicon surface as a passivation layer via PE chemical vapor deposition. Ag NPs created the surface plasmon resonance (SPR) effect and were shown to add to the decreased reflectivity in the near-IR wavelength range (Figure 16).

To demonstrate the existence of the SPR effect, which was shown to redshift with an increase in NP size, surface-enhanced Raman scattering (SERS) measurements using rhodamine-6G were performed. The study emphasized the efficiency of the resulting antireflection coating, which was fabricated for solar cell applications and was made of black silicon, a silicon nitride passivated layer together with a coating of Ag NPs.

A different wet etching approach was used by Toor et al. (2011) to roughen the surface of a silicon solar cell to provide it with better antireflective

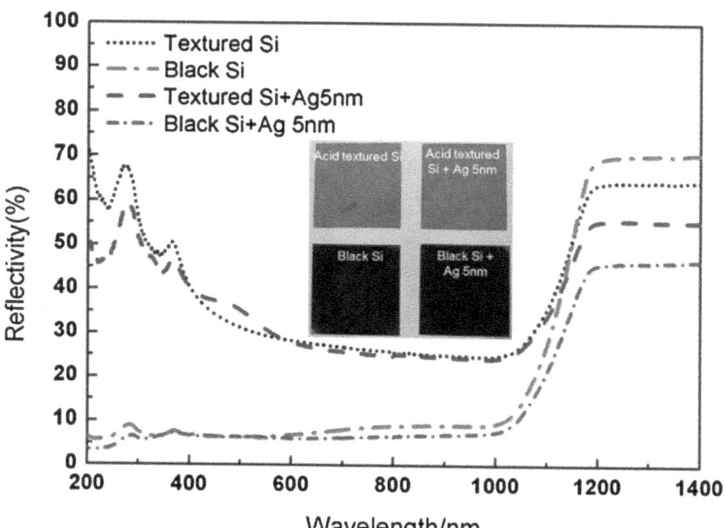

Figure 16 Graphs (Wang et al., 2013) demonstrating total hemispherical reflectance of acid textured Si, black Si, acid textured Si with 5 nm mass thickness of Ag, and black Si with 5 nm mass thickness of Ag. Inset is a photograph of these four samples.

properties. $HAuCl_4$ solution and an equal volume of 1:5:2 $HF:H_2O_2:H_2O$ mixture was used to form the Au NPs that catalyzed the metal-assisted silicon etch in order to form the nanoroughened surface features. The most noteworthy result in this work was the internal quantum efficiency response at 450 nm that was improved from 57% to 71%. The authors presented experimental results that suggested that most high-efficiency nanostructured solar cells would be limited in efficiency by a recombination of photoexcitation on their large area surface and would, therefore, benefit from the incorporation of light management strategies that minimized the nanostructured surface area without eliminating the benefits.

A femtosecond laser irradiation and rapid thermal annealing treatment was used to fabricate a black silicon sample (Figure 17) that showed strong IR PL as reported by Lu et al. (2013). The annealing temperature was varied between 500° C and 1200° C, and the observed PL peak position changed from 1.59 to 1.48 μm, respectively (no PL signal was observed for the unannealed sample). The reported x-ray photoelectron spectroscopy results from the black silicon before and after treatment in KOH solution suggested that the PL could be assigned to dislocation-related (D1) luminescence rather than sulfur-related impurity center–produced luminescence. The progress of research on black silicon in the area of Si-based optoelectronics is promising.

SERS Sensing

The versatility of Raman spectroscopy in detecting and analyzing chemical compounds by their unique spectra (or "fingerprints") has been advancing,

Figure 17 SEM images (Lu et al., 2013) of femtosecond laser fabrication performed in SF_6 gas atmosphere to create black silicon structures.

with its increased use in various areas such as biomedicine, material analysis, and environmental monitoring. The downside of this method is a low signal intensity when trace amounts of material are being investigated. The solution to this problem was found by Fleishman and McQuillan (1974) when a group of scientists were experimenting with a roughened silver electrode containing a pyridine monolayer on its surface. A huge increase in signal response was observed and this was attributed to the surface features on the silver sample that managed to enhance the signal. SERS is enabled by localized surface plasmon resonance that is created by an interaction between nanometer-sized metallic structures and a probing laser light. A range of promising applications was reviewed by Anker et al. (2008), who discussed various techniques for fabricating sensors based on a localized surface plasmon resonance effect.

A substrate for SERS was fabricated using a nano-roughened silicon surface containing a top coating layer of gold (Talian et al., 2009). The needle-like structures of black silicon or silicon grass were fabricated using a sulfur hexafluoride and oxygen plasma (Jansen et al., 2009). E-beam sputtering was used to deposit gold layers of up to 400 nm thickness. The increase in Au thickness was demonstrated to be beneficial for increases in the SERS signal response. Use of Ti or Ti together with Pt as an adhesion layer was shown to improve the extent of wetting, which assisted with adhering the gold to the silicon surface, although it produced a weaker SERS response. After the sample fabrication and analyte measurement conditions were fully optimized, the authors reported that an enhancement factor of 7.6×10^7 was achieved for rhodamine-6G using a 785 nm wavelength excitation laser.

Deng and Juang (2014) presented the results of their investigation into the SERS signal intensity dependence on the topography of black silicon and the thickness of the gold coating. The authors followed a procedure in order to fabricate nanometer-sized needle–structured samples using a cryogenic (−110 °C) etching process with a sulfur hexafluoride and oxygen mixed-gas plasma (Jansen et al., 2009). Three different morphologies with varying tip densities were obtained by changing the process gas ratios and total etching time. The authors suggested that a greater tip density and thicker gold layer produced a higher SERS signal. Although they tried to e-beam evaporate the gold layer to obtain thicknesses up to 600 nm as demonstrated in Figure 18, the lowest detection limit of 10 fmol/L for rhodamine-6G solution was demonstrated on 400 nm thick layer of gold deposited on the black silicon substrate that had a tip density of 30 tips/μm^2. It was found that the detection of the characteristic

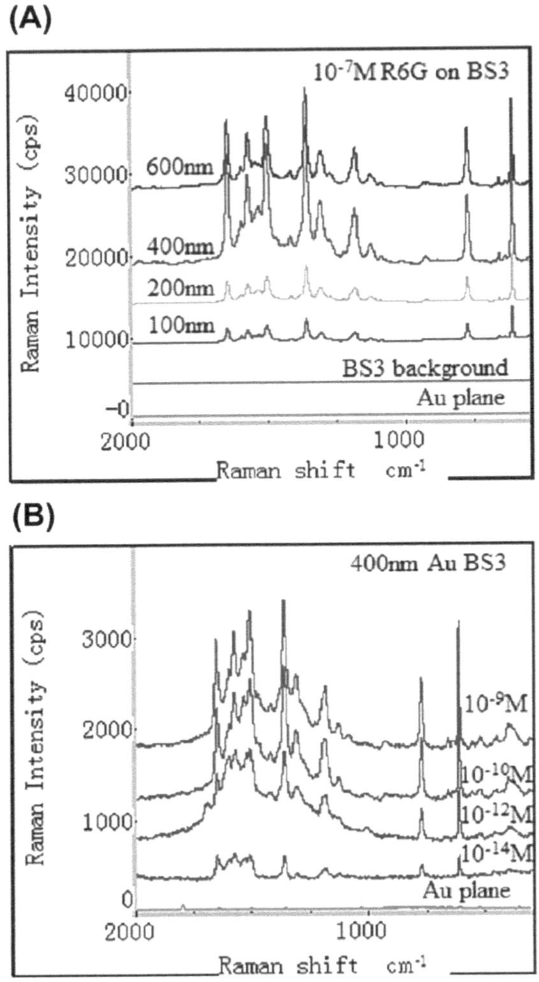

Figure 18 Graphs (Deng and Juang, 2014) demonstrating (A) SERS spectra of 10^{-7} mol/L concentration rhodamine-6G response intensity dependence on gold layer thickness that was deposited on the black silicon substrate. (B) The Raman spectra for 10^{-9} to 10^{-14} mol/L rhodamine-6G on the black silicon substrate deposited with 400-nm-thick gold layer.

lipid SERS spectrum from inside a single algal cell (*Chlorella vulgaris*) could be performed, with a response that was greatly enhanced in comparison to a comparable analysis using a planar gold substrate.

A straightforward method for structuring planar surfaces with simple structures was presented by Bowen et al. (2012). Gold-coated cone arrays (Figure 19) were fabricated by adhering PDMS decals to a hydroxylated substrate, warming the PDMS then peeling it off. If circular patterns on

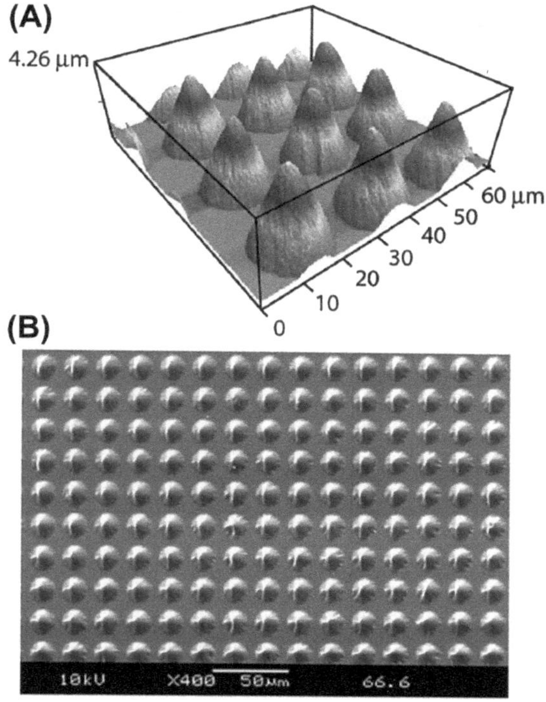

Figure 19 Images (Bowen et al., 2012) of (A) AFM height profile and (B) top-down SEM image of the SOG film imprinted with cones using a PDMS stamp containing conical reliefs (20-µm diameter, 4.26 µm height, 2 µm edge to edge spacing).

decals were defined, cone structures on the substrate were found to remain, and if line structures were used, triangular cross-sectional structures were formed on the sample surface. When the structures were coated with a 100 nm layer of gold, SERS benzenethiol detections could be successfully performed.

A different method for fabricating planar SERS substrates was suggested by Buividas et al. (2012). A femtosecond laser was used to create nanograting structures on the surface of sapphire wafers (Figure 20), which were then coated with gold or silver for subsequent analysis using SERS. This lithography-free fabrication technique is capable of fabricating surfaces with features of a few square millimeters in 1 hour or less when state-of-the-art scanning stages and a high repetition rate laser are used. Comparison with commercially available substrates, namely Klarite® and Solas, show that such substrates provide an order-of-magnitude higher SERS sensitivity. Integration of such a substrate into a handheld or lab-on-a-chip devices is promising as it requires much less power for the same SERS response.

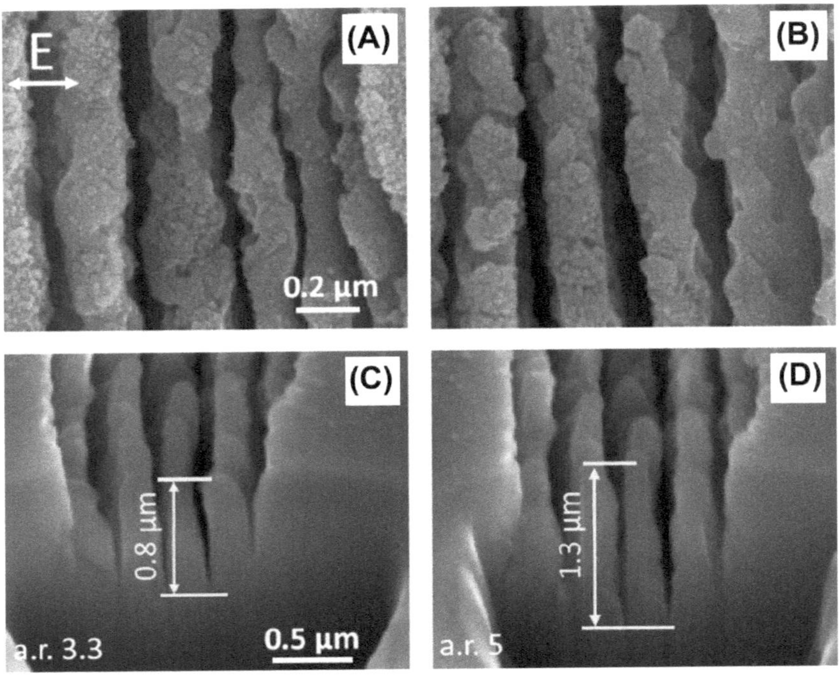

Figure 20 SEM images (Buividas et al., 2012) of ripple cross sections fabricated by different laser powers: (A,C) at 20 nJ and (B,D) at 25 nJ. The scale is the same for (A,B) and (C,D); cross-sectional views measured at an angle of 52° reveal the high aspect ratio (a. r.) of ripples.

Parisi et al. (2013) demonstrated a microfluidic platform for SERS measurements when the Raman response enhancing areas have vertical copper-core/carbon-sheath (Cu-core/C-sheath) nanowall structures synthesized using an electrodeposition technique. After structures were placed into a microfluidic channel, $AgNO_3$ solution was flowed through it to form Ag NP–decorated vertical nanowalls using the galvanic replacement of Cu with Ag+. The fluidic channel was formed on UV-lithography defined contacts out of PDMS. The influence of PDMS to a SERS response was investigated by removing the PDMS cover and measuring the signal directly from the synthesized areas, which proved to have no significant change in Raman signal. Crystal violet dye was finally used to evaluate the efficiency of this manufactured sensor, where a concentration as low as 50 pmol/L able to be reliably detected, demonstrating an enhancement factor of 10^9. The authors concluded that their concept (Figure 21) of combining the microfluidics with a high-sensitivity SERS platform has great potential in various chemical and biological sensing applications.

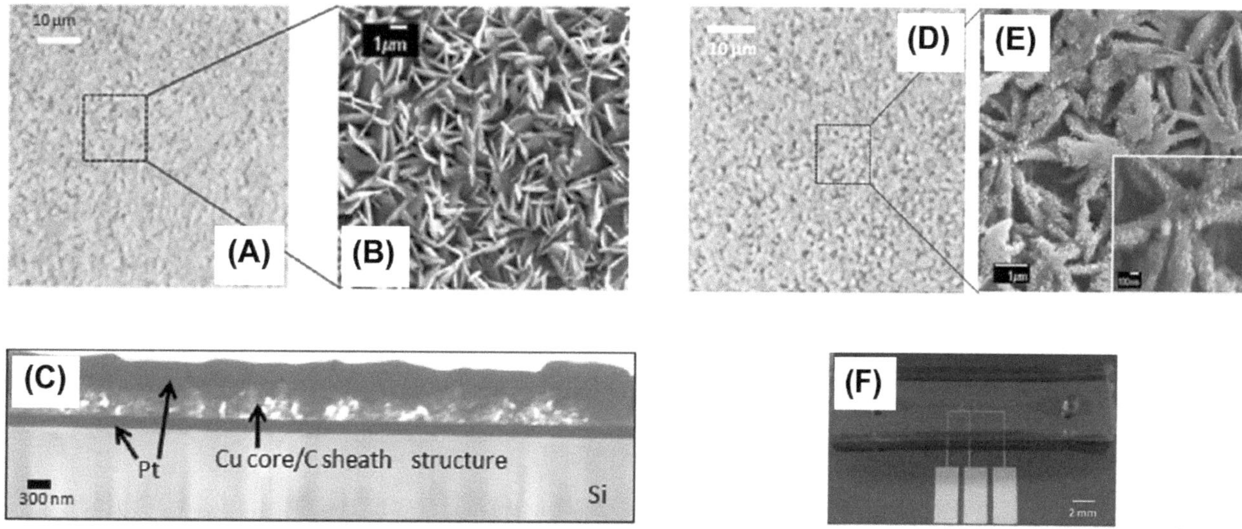

Figure 21 Assorted images (Parisi et al., 2013) showing (A) (optical) the Cu-core/C-sheath nanowalls grown on the working electrode, (B) (SEM) the vertical 3-D Cu-core/C-sheath nanowalls grown on the working electrode, (C) (STEM) the cross-section of the vertical 3-D Cu-core/C-sheath nanowalls grown on the working electrode (top Pt layer comes from Pt coating during the sample preparation), (D) (optical) the Cu-core/C-sheath nanowalls post Ag replacement, (E) (SEM) low and high magnification (inset) of Ag NP–decorated nanowalls, and (F) (optical) the as-prepared microfluidic device with three UV-patterned electrodes.

Another SERS sensor integration into a microfluidic chip was demonstrated by Tong et al. (2009). A focused near-IR laser tweezer beam was used to trap and optically aggregate Ag NPs in the flow at the bottom of the channel in order to form an efficient SERS substrate. An auxiliary separate green laser beam was used for the Raman excitation of sample molecules.

A self-aligning silver NP substrate was fabricated by Huang et al. (2011). A porous anodic aluminum oxide film was used for directing the silver NPs. Adenine was a test substance in SERS experiments and an enhancement of 10^6 was achieved.

A method for the fabrication of a low cost SERS substrate was proposed by Hoppmann et al. (2013). The printing ink was formed from concentrated gold colloid in water solution, glycerol, and ethanol. The ink was poured into a refillable cartridge, and a conventional ink-jet printer was used to print arbitrary shapes on chromatography paper. The dependence of the SERS response on the number of times a pattern was printed on the paper was investigated and showed the best response after 12 print cycles. Structured paper (Figure 22) was used as a dipstick and a swab. In the first case, it was dipped into malachite green (1 ppb in water) for 1 or 30 seconds and left to dry in ambient air; in the second case, the fungicide thiram was deposited in known quantities (10–300 ng) onto a glass surface and allowed to dry, after which the surface was swabbed with a paper stripe that had been pre-wet with acetone. Subsequent SERS measurements using these substrates highlighted the effectiveness with which they could be used for detecting analytes, which highlights their potential use in food safety, environmental pollution and security applications.

A disposable plastic SERS sensor capable of achieving a 10^7 Raman response enhancement was fabricated by Oo et al. (2013) using roll-to-roll and nanoimprint techniques. The geometry used was based on the industry standard Klarite® and was subsequently optimized for better performance. The original substrate consisted of an array of gold-coated inverted square pyramidal pits that were altered to form rectangular cones as well as varying their pitch; an 8-fold improvement of sensitivity was achieved.

An innovative idea for inexpensive large area SERS substrate fabrication was proposed by Akil-Jradi et al. (2012). A solution was prepared by dissolving $AgNO_3$ in ethanol and mixing it with PMMA dissolved in acetone. The manufacture of substrates was based on the vapor-induced phase separation and chemical reduction of a metal precursor-doped

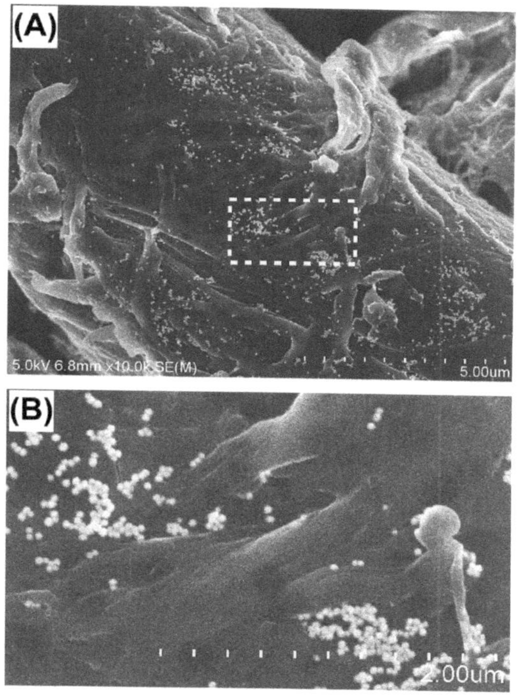

Figure 22 SEM images (Hoppmann et al., 2013) (A) of a printed gold nanoparticle region on cellulose paper and (B) of clustered gold nanoparticles on the cellulose fiber (magnified).

polymer dispersion. The silver NP–rich solution was spin-coated on different conductivity and wettability surfaces. It was left to dry, and the PMMA was then washed out in a subsequent step. Different substrates produced different results, varying from NP rings to flake-like patterns, which, when annealed at low temperatures, changed their morphology again. An enhancement factor exceeding 10^{13} was demonstrated for bipyridine ethylene (concentration 10^{-12} mol/L) deposited on Ag NP aggregations formed on indium tin oxide (ITO)-coated glass (Figure 23). The substrates could be used in biomedicine, safety, and environmental pollution applications for the detection of rare and dangerous chemicals.

Using nanoimprint lithography, Kostovski et al. (2009) replicated the nanostructures present on cicada wings onto the end faces of standard silica optical fibers. A PDMS mold was used to transfer the nanofeatures from a cicada wing segment onto the tip of an optical fiber. After the optical

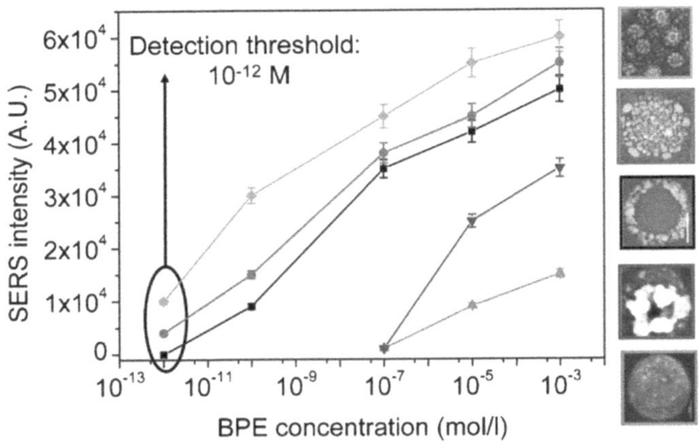

Figure 23 A graph (Akil-Jradi et al., 2012) demonstrating the evolution of the SERS intensity at 1605 cm^{-1} as a function of bipyridine ethylene (BPE) concentration for the different substrates.

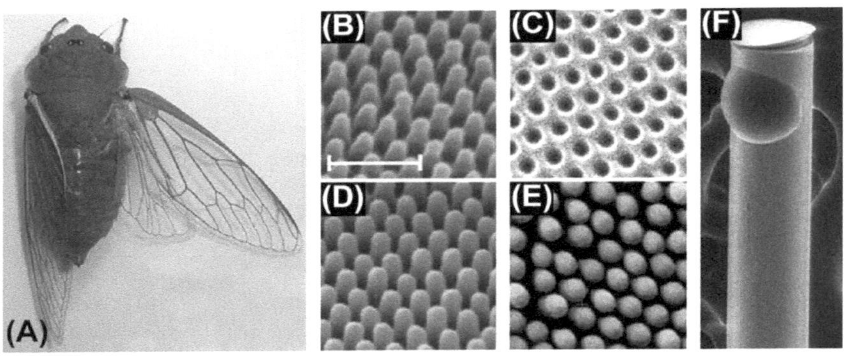

Figure 24 A photograph (Kostovski et al., 2009) of (A) the Australian greengrocer cicada is shown, together with SEM images of (B) a cicada wing nanotemplate, (C) the inverse h-PDMS mold, (D) the polymer replica on an optical fiber tip, (E) the silver-coated replica, and (F) a macroscopic view of the imprinted optical fiber. Scale bar is 500 nm.

fiber tip was patterned with nanopillars, a metallic silver coating was thermally deposited on top in order to enable a SERS response to be obtained (Figure 24). Successful measurements from the surface of the replica as well as through the fiber were compared with the measurements made directly on a cicada wing. By using a shorter fiber integrated into a filter subsystem, the authors were pursuing possible in vivo applications.

Xu et al. (2011) discuss the use of EBL-defined quasi-3D nanostructure arrays for the detection of molecules and large microorganisms. The structures used in SERS measurements consisted of an elevated nanohole-patterned gold film and gold nanodisks at the bottom of wells. After performing a series of finite difference time domain (FDTD) calculations, the authors suggested that the strongest electric field shifts from the nanodisk at the bottom of the well to the nanohole rim in the gold film as the diameter of wells was increased. Based on these results, they hypothesized that to obtain a high SERS signal response from large bacteria, the strongest field enhancement would need to be at the top of the well, as bacteria remain on the top surface of the substrate after being adsorbed. 4-Mercaptopyridine was chosen as a molecular analyte because it can form a self-assembled monolayer on a gold surface. Three Gram-negative strains, *Escherichia coli* K12, *Pseudomonas aeruginosa,* and *Acinetobacter calcoaceticus,* and one Gram-positive strain, *Staphylococcus epidermidis,* were selected for the experiments in order to investigate the possibility of distinguishing between different types of bacteria and a same type but different kind of microorganisms. Even though the spectra obtained were complicated, the authors managed to accomplish sufficient signal separation to distinguish between different bacterial samples. The authors are hopeful that their method will lead to a development of a new sensing platform with molecular specificity based on SERS with potential applications in areas such as environmental monitoring, health care, food safety, and security.

EMERGING 3D NANOSTRUCTURING TECHNOLOGIES

Methods for direct 3D structuring of materials are developing from several directions. Laser writing with tightly focused laser pulses/beams (Seet et al., 2006a) now reach 100 nm resolution in all three dimensions by exploiting the threshold of polymerization and precise control of thermal conditions at the small central part of the focal region and creation of new resist materials that are more suitable for polymerization reactions that are by avalanche and thermal processes (Juodkazis et al., 2005; Seet et al., 2005, 2006b; Sun et al., 2010). The 3D scaffolds can be infiltrated by different materials to expand this method of direct laser writing on surfaces (Juodkazis et al., 2009, 2008a, 2008b; Malinauskas et al., 2013; Seet et al., 2008).

FIB milling tools are other popular direct writing tools. Currently, Ga, Si, Au (Raith GmhH), and He ions are installed in commercial tools reaching sub-10 nm resolution. The highest resolution of milling with only 4 nm wide grooves and still at aspect ratios exceeding 100 are demonstrated with He ions. Together with a 3D stage with rotation functionality, FIB becomes a 3D structuring tool as demonstrated by the ability to mill slanted pore photonic crystal (Juodkazis et al., 2011) whose geometry and parameters are close to those required for photonic bandgap formation at visible wavelength (not yet demonstrated). Ion milling down to a resolution of a few tens of nanometers has been demonstrated (Gervinskas et al., 2013a; Rosa et al., 2010, 2011), and even higher resolution and faster writing speeds can be achieved by controlling the surface charging effects caused by impinging ions and secondary electrons. We demonstrated recently that deep-UV illumination by LED light at 250 to 270 nm wavelengths alleviates surface charging during ion milling (Gervinskas et al., 2013b). The photoelectrons generated by UV light are compensating strong charge gradients during ion milling and can help to solve substrate damage effects in the fabrication of EBL-defined NPs (Juodkazis and Rosa, 2010) and larger area ion milling applications for high throughput fabrication (Brasselet et al., 2013).

Recent technology for direct 3D writing was demonstrated using a heated AFM tip (Paul et al., 2011). An AFM tip heated to temperatures in excess of 700 °C evaporates monomers of molecular glass of unzip polymer resists with a resolution defined by the tip (~10 nm). This technology has already been demonstrated at writing speeds of 20 mm/s and pixel rates of 500 kHz. Since this is contact mode writing, it is not susceptible to vibrations and electromagnetic noises to the same degree as other electron and ion beam technologies. It also does not require vacuum enclosures and solves writing fields stitching errors by AFM image cross-correlation. Figure 25 shows Swinburne University's logo that has been patterned using a NanoFrazer from a SwissLitho setup, which commercialized this technology in 2012.

An interesting capability of writing 3D masks for plasma etching is shown in Figure 26. Ga^{+} ions were used to mill circular regions using ion beam lithography (Raith IonLine) forming nanometer-thick Ga-rich Si paddocks after removal of a Cr film coated onto a Si substrate. Ion beam lithography is a technique where ions impinge onto a sample surface at normal incidence. Since ion milling can be carried out at arbitrary angle (Juodkazis et al., 2011), it becomes possible to write complex masks that can withstand dry plasma etching (Figure 26). This opens new possibilities in 3D nanostructuring on the nanoscale.

Figure 25 Three-dimensional nanoengraving in polyphthalaldehyde (PPA) with a hot AFM tip with a 10 nm resolution defined by the tip's curvature (tool: NanoFrazor). *Courtesy of P. Paul and F. Holzner from SwissLitho Ltd.*

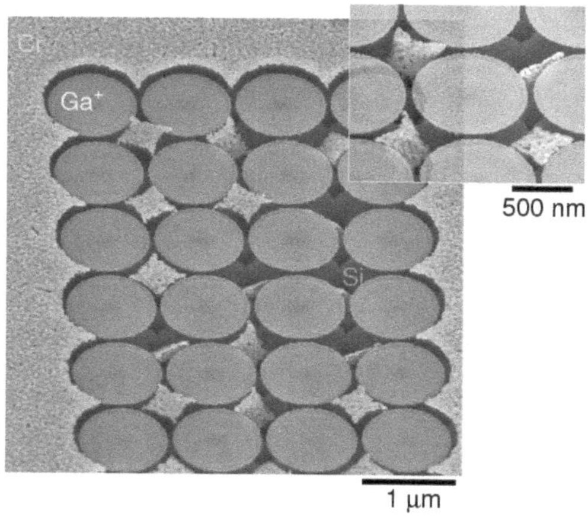

Figure 26 Ion beam lithography (IBL) direct write of a plasma etching mask by Ga^+ milling on Cr film–coated Si. Etching was carried out in SF_6 plasma at a 5 sccm flow, 2 Pa pressure, 10 W bias voltage for 20 min reaching a ~2 μm depth (tool: Samco RIE-101iPH). Region Ga^+ is Ga rich.

CONCLUSIONS AND OUTLOOK

In this chapter, several applications of nanotextured surfaces, many of which can exhibit superhydrophobic properties, were reviewed.

Fabrication of black silicon, using dry and wet etching techniques, was introduced and examples of possible applications of such surfaces were discussed. Black silicon surfaces can be used to control surface wettability. A wide and fast-growing field of applications of such unique surfaces is in the area of surface enhanced Raman spectroscopy (SERS), where black silicon has been shown to be a suitable platform for delivering large surface areas. A great variety of other fabrication techniques for the fabrication of low-cost substrates that achieve high SERS enhancement was also presented in detail.

ACKNOWLEDGMENTS

We are grateful for enlightening discussions on the nanofabrication of solar cells with Prof. Sajeev John, on Ga-ion mask writing with Dr. Warren McKenzie, and on AFM tip fabrication to Dr. Felix Holzner.

REFERENCES

Abate, A.R., Weitz, D.A., January 2011. Syringe-vacuum microfluidics: a portable technique to create monodisperse emulsions. Biomicrofluidics 5, 14107.

Abdulrahman, N.A., Fan, Z., Tonooka, T., Kelly, S.M., Gadegaard, N., Hendry, E., Govorov, A.O., Kadodwala, M., 2012. Induced chirality through electromagnetic coupling between chiral molecular layers and plasmonic nanostructures. Nano Lett. 12 (2), 977–983.

Anker, J.N., Hall, W.P., Lyandres, O., Shah, N.C., Zhao, J., Van Duyne, R.P., 2008. Biosensing with plasmonic nanosensors. Nat. Mater. 7 (6), 442–453.

Akil-Jradi, S., Jradi, S., Plain, J., Adam, P.-M., Bijeon, J.-L., Royer, P., Bachelot, R., 2012. Micro/nanoporous polymer chips as templates for highly sensitive SERS sensors. RSC Adv. 2 (20), 7837–7842.

Bowen, A.M., Motala, M.J., Lucas, J.M., Gupta, S., Baca, A.J., Mihi, A., Alivisatos, A.P., Braun, P.V., Nuzzo, R.G., 2012. Triangular elastomeric stamps for optical applications: near-field phase shift photolithography, 3d proximity field patterning, embossed antireflective coatings, and SERS sensing. Adv. Funct. Mater. 22 (14), 2927–2938.

Buividas, R., Stoddart, P.R., Juodkazis, S., 2012. Laser fabricated ripple substrates for surface-enhanced raman scattering. Ann. Phys. 524 (11), L5–L10.

Brasselet, E., Gervinskas, G., Seniutinas, G., Juodkazis, S., 2013. Topological shaping of light by closed-path nanoslits. Phys. Rev. Lett. 111 (19), 193901.

Cuennet, J.G., Vasdekis, A.E., Sio, L.D., Psaltis, D., De Sio, L., 2011. Optofluidic modulator based on peristaltic nematogen microflows. Nat. Photonics. 5 (4), 234–238.

Dietrich, K., Lehr, D., Helgert, C., Tünnermann, A., Kley, E.-B., 2012. Circular dichroism from chiral nanomaterial fabricated by on-edge lithography. Adv. Mater. 24 (44), OP321–OP325.

Deng, Y.-L., Juang, Y.-J., 2014. Black silicon SERS substrate: effect of surface morphology on SERS detection and application of single algal cell analysis. Biosens. Bioelectron. 53, 37–42.

Li, W.-D.D., Wu, W., Stanley Williams, R., 2012. Combined helium ion beam and nanoimprint lithography attains 4 nm half-pitch dense patterns. J. Vac. Sci. Technol., B: Microelectron. Nanometer Struct. 30 (6), 06F304.
Esfandiar, A., Savaloni, H., Placido, F., 2013. On the fabrication and characterization of graded slanted chiral nano sculptured silver thin films. Phys. E. 50, 88–96.
Lu, Q., Wang, J., Liang, C., Zhao, L., Jiang, Z., 2013. Strong infrared photoluminescence from black silicon made with femtosecond laser irradiation. Opt. Lett. 38 (8), 1274–1276.
Fleishman, H.P.M.J., McQuillan, A.J., 1974. Surface enhanced Raman scattering from silver particles on polymer-replica substrates. J. Chem Phys. Lett. 26, 163.
Garoli, D., Romanato, F., Zilio, P., Natali, M., Marinello, F., Ongarello, T., Sammito, D., De Salvador, D., August 2011. Fabrication of "nano-rocket-tips" for plasmonic nanofocusing. Microelectron. Eng. 88, 2530–2532.
Gervinskas, G., Seniutinas, G., Rosa, L., Juodkazis, S., 2013. Arrays of arbitrarily shaped nanoparticles: overlay errorless direct ion write. Adv. Opt. Mat. 1 (6), 456–459.
Gervinskas, G., Seniutinas, G., Juodkazis, S., 2013. Control of surface charge for high-fidelity nanostructuring of materials. Laser Phot. Rev. 7 (6), 1049–1053.
Han, C., Leung, H.M., Tam, W.Y., 2013. Chiral metamaterials by shadowing vapor deposition. J. Opt. 15 (7), 072101.
Huang, C.-H., Lin, H.-Y., Chen, S., Liu, C.-Y., Chui, H.-C., Tzeng, Y., 2011. Electrochemically fabricated self-aligned 2-d silver/alumina arrays as reliable sers sensors. Opt. Express 19 (12), 11441–11450.
Hoppmann, E.P., Yu, W.W., White, I.M., 2013. Highly sensitive and flexible inkjet printed SERS sensors on paper. Methods 63 (3), 219–224.
Jansen, H.V., de Boer, M.J., Unnikrishnan, S., Louwerse, M.C., Elwenspoek, M.C., March 2009. Black silicon method: X. A review on high speed and selective plasma etching of silicon with profile control: an in-depth comparison between Bosch and cryostat DRIE processes as a roadmap to next generation equipment. J. Micromech. Microeng. 19, 033001.
Juodkazis, S., Mizeikis, V., Seet, K.K., Miwa, M., Misawa, H., 2005. Two-photon lithography of nanorods in photoresist. Nanotechnology 16, 846–849.
Juodkazis, S., Mizeikis, V., Misawa, H., 2009. Three-dimensional microfabrication of materials by femtosecond lasers for photonics applications. J. Appl. Phys. 106 (5), 51101.
Juodkazis, S., Mizeikis, V., Misawa, H., 2008. Three-dimensional structuring of resists and resins by direct laser writing and holographic recording. Adv. Polym. Sci. 213, 157–206.
Juodkazis, S., Mizeikis, V., Matsuo, S., Ueno, K., Misawa, H., 2008. Three-dimensional micro- and nano structuring of materials by tightly focused laser radiation. Bull. Chem. Soc. Jpn. 81 (4), 411–448.
Juodkazis, S., Rosa, L., Bauerdick, S., Peto, L., ElGanainy, R., John, S., 2011. Sculpturing of photonic crystals by ion beam lithography: towards complete photonic bandgap at visible wavelength. Opt. Express 19 (7), 5802–5810.
Juodkazis, S., Rosa, L., 2010. Surface defect mediated electron hopping between nanoparticles separated by a nanogap. Phys. Status Solidi 4 (10), 244–246.
Kuzyk, A., Schreiber, R., Fan, Z., Pardatscher, G., Roller, E.M., Hogele, A., Simmel, F.C., Govorov, A.O., Liedl, T., 2012. DNA-based self-assembly of chiral plasmonic nanostructures with tailored optical response. Nature 483 (7389), 311–314.
Kostovski, G., White, D., Mitchell, A., Austin, M., Stoddart, P., 2009. Nanoimprinted optical fibres: biotemplated nanostructures for SERS sensing. Biosens. Bioelectron. 24 (5), 1531–1535.
Levy, U., Shamai, R., 2008. Tunable optofluidic devices. Microfluid. Nanofluid. 4 (1), 97–105.
Laser, D.J., Santiago, J.G., June 2004. A review of micropumps. J. Micromech. Microeng. 14, R35.
Lim, J.-M.M., Urbanski, J.P., Thorsen, T., Yang, S.-M.M., 2011. Pneumatic control of a liquid-core/liquid- cladding waveguide as the basis for an optofluidic switch. Appl. Phys. Lett. 98 (4), 44101.

Lilienthal, K., Fischer, M., Stubenrauch, M., Schober, a., May 2010. Self-organized nanostructures in silicon and glass for MEMS, MOEMS and BioMEMS. Mater. Sci. Eng., B. 169, 78–84.

Monat, C., Domachuk, P., Eggleton, B.J., 2007. Integrated optofluidics: a new river of light. Nat. Photonics 1 (2), 106–114.

Monat, C., Domachuk, P., Grillet, C., Collins, M., Eggleton, B.J., Cronin-Golomb, M., Mutzenich, S., Mahmud, T., Rosengarten, G., Mitchell, A., September 2008. Optofluidics: a novel generation of reconfigurable and adaptive compact architectures. Microfluid. Nanofluid. 4, 81–95.

Menard, L.D., Ramsey, J.M., 2010. Fabrication of sub-5 nm nanochannels in insulating substrates using focused ion beam milling. Nano Lett. 11 (2), 512–517.

Mehran, M., Sanaee, Z., Abdolahad, M., Mohajerzadeh, S., September 2011. Controllable silicon nano-grass formation using a hydrogenation assisted deep reactive ion etching. Mater. Sci. Semicond. Process 14, 199–206.

Murias, D., Reyes-Betanzo, C., Moreno, M., Torres, A., Itzmoyotl, A., Ambrosio, R., Soriano, M., Lucas, J., Cabarrocas, P.R.I., September 2012. Black silicon formation using dry etching for solar cells applications. Mater. Sci. Eng., B. 177, 1509–1513.

Malinauskas, M., Farsari, M., Piskarskas, A., Juodkazis, S., 2013. Ultrafast-laser micro/nano-structuring of photopolymers: a decade of advances. Phys. Rep. 533, 1–31.

Oo, S.Z., Chen, R., Siitonen, S., Kontturi, V., Eustace, D., Tuominen, J., Aikio, S., Charlton, M., 2013. Disposable plasmonic plastic SERS sensor. Opt. Express 21 (15), 18484–18491.

Psaltis, D., Quake, S.R., Yang, C., Others, July 2006. Developing optofluidic technology through the fusion of microfluidics and optics. Nature (London) 442, 381.

Pang, L., Chen, H.M., Ptasinski, J.N., Sun, P.-C.C., Fainman, Y., February 2011. Nanoplasmonic resonance integrated with optofluidics for biochemical sensing and identification. In: Kubby, J.A., Reed, G.T. (Eds.), Proc. SPIE, 7943. p. 794304.

Pezoldt, J., Kups, T., Stubenrauch, M., Fischer, M., 2011. Black luminescent silicon. Physica Status Solidi C. 8 (3), 1021–1026.

Parisi, J., Su, L., Lei, Y., 2013. In situ synthesis of silver nanoparticle decorated vertical nanowalls in a microfluidic device for ultrasensitive in-channel SERS sensing. Lab Chip 13 (8), 1501–1508.

Paul, P., Knoll, A., Holzner, F., Despont, M., Duerig, U., 2011. Rapid turnaround scanning probe nanolithography. Nanotechnology 23, 385307.

Ruffato, G., Garoli, D., Cattarin, S., Barison, S., Romanato, F., October 2012. FIB lithography of nanoporous gold slits for extraordinary transmission. Microelectron. Eng. 98, 419–423.

Rosa, L., Sun, K., Szymanska, J., Hudson, F.E., Dzurak, A., Linden, A., Bauerdick, S., Peto, L., Juodkazis, S., 2010. Tailoring spectral position and width of field enhancement by ion-beam trimming of plasmonic nanoparticles. Physica Status Solidi RRL. 4 (10), 262–264.

Rosa, L., Sun, K., Mizeikis, V., Bauerdick, S., Peto, L., Juodkazis, S., April 2011. 3D-tailored gold nanoparticles for light field enhancement and harvesting over visible-IR spectral range. J. Chem. Phys. C. 115, 5251–5256.

Song, W., Vasdekis, A.E., Li, Z., Psaltis, D., 2009. Optofluidic evanescent dye laser based on a distributed feedback circular grating. Appl. Phys. Lett. 94 (16), 161110.

Song, W., Psaltis, D., 2010. Pneumatically tunable optofluidic dye laser. Appl. Phys. Lett. 96 (8), 81101.

Song, W., Psaltis, D., 2011. Pneumatically tunable optofluidic 2 X 2 switch for reconfigurable optical circuit. Lab Chip 11, 2397–2402.

Staude, I., Decker, M., Renner, M., Waller, E., Neshev, D.N., von Freymann, G., Kivshar, Y.S., 2013. Hybrid fabrication of tapered gold double-helices for near infrared frequencies. In: CLEO: QELS-fundamental Science. Optical Society of America, p. QW3N-7.

Staude, I., Decker, M., Ventura, M.J., Jagadish, C., Neshev, D.N., Gu, M., Kivshar, Y.S., March 2013. Hybrid high resolution three-dimensional nanofabrication for meta materials and nanoplasmonics. Adv. Mater. (Deerfield Beach, Fla.) 25, 1260–1264.
Steglich, M., Lehr, D., Ratzsch, S., Kasebier, T., Schrempel, F., Kley, E.-B., Tunnermann, A., 2014. An ultra-black silicon absorber. Laser Photonics Rev. 82(2), L13–L17.
Srivastava, S.K., Kumar, D., Singh, P., Kar, M., Kumar, V., Husain, M., September 2010. Excellent antireflection properties of vertical silicon nanowire arrays. Sol. Energy Mater. Sol. Cells 94, 1506–1511.
Seet, K.K., Mizeikis, V., Juodkazis, S., Misawa, H., 2006. Three-dimensional horizontal circular spirals photonic crystals with stop gaps below 1 μm. Appl. Phys. Lett. 88 (22), 221101.
Seet, K.K., Juodkazis, S., Jarutis, V., Misawa, H., 2006. Feature-size reduction of photopolymerized structures by femtosecond optical curing of SU-8. Appl. Phys. Lett. 89, 24106.
Seet, K.K., Mizeikis, V., Matsuo, S., Juodkazis, S., Misawa, H., 2005. Three-dimensional spiral - architecture photonic crystals obtained by direct laser writing. Adv. Mat. 17 (5), 541–545.
Sun, Q., Juodkazis, S., Murazawa, N., Mizeikis, V., Misawa, H., 2010. Freestanding and movable photonic microstructures fabricated by photopolymerization with femtosecond laser pulses. J. Micromech. Microeng. 20, 035004/1–035004/5.
Seet, K.K., Mizeikis, V., Kannari, K., Juodkazis, S., Misawa, H., Tetreault, N., John, S., 2008. Templating and replication of spiral photonic crystals for silicon photonics. IEEE J. Sel. Top. Quantum Electron. 14 (4), 1064–1073.
Tripathi, S.K., Scanlan, D., O'Hara, N., Nadzeyka, A., Bauerdick, S., Peto, L., Cross, G.L.W., Hara, N.O., May 2012. Resolution, masking capability and throughput for direct-write, ion implant mask patterning of diamond surfaces using ion beam lithography. J. Micromech, Microeng. 22, 055005.
Toor, F., Branz, H.M., Page, M.R., Jones, K.M., Yuan, H.-C., 2011. Multi-scale surface texture to improve blue response of nanoporous black silicon solar cells. Appl. Phys. Lett. 99 (10), 103501.
Talian, I., Mogensen, K.B., Orinak, A., Kaniansky, D., Hubner, J., August 2009. Surface-enhanced Raman spectroscopy on novel black silicon-based nanostructured surfaces. J. Raman Spectrosc. 40, 982–986.
Tong, L., Righini, M., Gonzalez, M.U., Quidant, R., Kall, M., 2009. Optical aggregation of metal nanoparticles in a microfluidic channel for surface-enhanced raman scattering analysis. Lab Chip 9 (2), 193–195.
Vieweg, M., Gissibl, T., Pricking, S., Kuhlmey, B.T., Wu, D.C., Eggleton, B.J., Giessen, H., November 2010. Ultrafast nonlinear optofluidics in selectively liquid-filled photonic crystal fibers. Opt. Express 18, 25232–25240.
Whitesides, G.M., July 2006. The origins and the future of microfluidics. Nature 442, 368–373.
Winston, D., Manfrinato, V.R., Nicaise, S.M., Cheong, L.L., Duan, H., Ferranti, D., Marshman, J., McVey, S., Stern, L., Notte, J., Berggren, K.K., October 2011. Neon ion beam lithography (NIBL). Nano Lett. 11, 4343–4347.
Wang, Y., Liu, Y., Liang, H., Mei, Z., Du, X., 2013. Broadband antireflection on the silicon surface realized by ag nanoparticle-patterned black silicon. Phys. Chem. Chem. Phys. 15 (7), 2345–2350.
Xu, J., Zhang, L., Gong, H., Homola, J., Yu, Q., et al., 2011. Tailoring plasmonic nanostructures for optimal SERS sensing of small molecules and large microorganisms. Small 7 (3), 371–376.

CHAPTER EIGHT

Biological Interactions with Superhydrophobic Surfaces

GLOSSARY

LLSS Lotus leaf surface structure
Cassie-impregnating state Wetting state in which water droplets only partially penetrate between the surface features
Amphiphilic having the properties of hydrophilic (water-loving) and lipophilic (fat-loving)
Adipogenic Generation of adipose tissues

INTRODUCTION

The most promising of the potential applications of superhydrophobic surfaces revolve around interactions with biological systems (Bixler and Bhushan, 2013; Nishimoto and Bhushan, 2013; Zhang et al., 2013). From inhibiting the attachment of fouling bacteria to enhancing the integration of bioimplant materials by host tissues, the potential benefits of mediation of cellular adhesion by the substratum itself are substantial. Much research exists regarding the initial interactions between cells and various substrata; however, much of these data appear contradictory, and no universal rules have yet been identified (Luong-Van et al., 2013; Song and Mano, 2013). Most likely this is a result of the complexity of biological systems; even the simplest bacteria excrete numerous proteins and polysaccharides, which may form conditioning films, potentially altering the hydrophobicity and charge of the surface, and if these secretions are present in sufficient quantity, they can also change the surface topographical features (Costerton et al., 1995; O'Toole et al., 2000). In this chapter, a brief summary of what is known about biological interactions with superhydrophobic substrata will be presented, from the molecular level (i.e., protein adsorption) to the multicellular level of tissue adhesion. We will also attempt to reconcile the seemingly contradictory information available in the literature.

Superhydrophobic Surfaces
http://dx.doi.org/10.1016/B978-0-12-801109-6.00008-2

COMPLEXITY AND DYNAMICS

Many of the previous chapters in this book have addressed in great detail the thermodynamic theory behind the behavior of a majority of superhydrophobic surfaces. A majority of these surfaces are very rough, often containing hierarchical roughness, and hence they have the ability to trap large quantities of air (Truong et al., 2012; Papadopoulos et al., 2013). This entrapment of air makes the study of interfacial interactions of these materials with biological adsorbents difficult to study for several reasons. First, the thorough characterization of such substrate surfaces is no easy task, as it can be very difficult to map air bubbles at an interface (Bhushan and Jung, 2008; Bhushan et al., 2009). Second, the actual interface that is available to interact with the cells is then called into question, as in most cases cells require an aqueous environment and do not completely traverse the air–water interface to the substratum below (Habimana et al., 2014; Vasudevan et al., 2014). An additional complicating factor is the wetting phenomena that occur, such as the transition from the Cassie–Baxter wetting state to what is known as the Cassie-impregnating state (Feng et al., 2008; Su et al., 2010). These dynamic wetting factors, together with the natural dynamism of cells, form a system that is very difficult to characterise.

PROTEIN ADSORPTION ON SUPERHYDROPHOBIC SURFACES

As described, studying the adsorption behaviour of single molecules such as proteins on superhydrophobic surfaces is important due to their role in determining the conditions and subsequent success of attachment of prokaryotic or eukaryotic cells (Banerjee et al., 2011). The adsorption and adhesion of proteins are also important phenomena in their own right, as they can have a direct bearing on medical diagnostics and treatments if these proteins are present on biomedical devices (Hucknall et al., 2009). For example, the accumulation of proteins on in vitro diagnostic devices may lead to a decreased sensitivity of the device or, similarly, any build-up of a protein coating on medical implants may lead to the formation of thromboses (Amiji and Park, 1993).

The driving forces behind the adsorption of proteins on superhydrophobic surfaces are complex, dynamic, and diverse (Lee et al., 2010; Kolind et al., 2012). A wide range of noncovalent forces are at play, including hydrophobic interactions, electrostatic forces, hydrogen bonding, and van der Waals forces (Jeon et al., 1991; Tilton et al., 1991; Hlady and Buijs, 1996; Raffaini and Ganazzoli, 2010). As a consequence, every different protein will theoretically adsorb in a different way on any given substratum, as each protein has a unique sequence of amino acids. This means that the charge density and distribution, location of hydrophobic regions, and size and shape of each protein are different (Roach et al., 2005). In addition, the conformation and ionization of proteins are highly sensitive to the surrounding conditions, particularly the pH of the medium (Campbell and Reece, 2011).

At a superficial level, superhydrophobic surfaces can be considered to experience less protein adsorption than other, less-hydrophobic surfaces (Koc et al., 2008; Koh et al., 2010). The entrapped air pockets that serve to increase the observed water contact angles displayed by the surface effectively block portions of the surface from adsorption processes, decreasing their accessibility to incoming protein molecules (Lafuma and Quéré, 2003; Bhushan et al., 2009; Barthlott et al., 2010). This, however, is simply the first observable effect; a second complicating factor must also be considered. Due to their amphiphilic nature (the magnitude of which can vary greatly), proteins can act as surfactants and destabilize the solid–liquid–air interface (Koc et al., 2008). Essentially, this results in the release of some portion of the entrapped air, thereby increasing the protein accessibility to the surface. Provided that the air is not completely released, the net result is a decrease in protein adsorption relative to the amount that would take place on less-hydrophobic substrata; however, the effect is not as great as might be initially thought (Genzer and Efimenko, 2006; Wu et al., 2013).

Superhydrophobic surfaces invariably have complex or sophisticated surface topographies at the microscale and/or nanoscale. The influence with which the surface topography can affect protein adsorption is dependent on the presence or absence of shear forces within the system (Koc et al., 2008). In static systems, it has been shown that the initial protein adsorption on rougher substrata is higher than that found on smoother surfaces (Scopelliti et al., 2010). This is due to local accumulation of protein molecules within the pores of the surface, which increases the adsorption kinetics in these regions. Conversely, it has been shown that the introduction of a directional flow enabled the detachment of proteins to a greater degree from rough surfaces than from smoother ones, and in some cases almost complete protein removal was achieved (Luong-Van et al., 2013).

BACTERIAL INTERACTIONS WITH SUPERHYDROPHOBIC SURFACES

Historically, it has generally been thought that bacteria are typically slightly hydrophobic, and thus they are more likely to attach to substrates containing greater degrees of hydrophobicity (Donlan, 2002). This is a recurring theme throughout studies of biological interactions with surfaces, however, the actual case is substantially more complicated. First, the assumption that bacterial cells are hydrophobic is far from accurate; different bacteria can have varying degrees of hydrophobicity, depending on their external environment (e.g., availability of nutrients, external pH, salinity) (Truong et al., 2010). Two cells of the same bacterial species may even differ in the hydrophobicity of their outer layer, depending on the culturing conditions.

Determination of the role of surface topography in mediating bacterial cell attachment has been a popular objective among researchers for some time. There are several conflicting reports; however, a few collective observations can be made with some consistency. If a surface is sufficiently rough on the microscale (i.e., 1–100 μm average roughness), bacteria are able to accumulate in the "valley" regions between adjacent asperities and experience shelter from the turbulent flow of the surrounding medium (Bhushan, 2011; Bixler and Bhushan, 2013; Zhang et al., 2013). Roughness on the nanoscale tends to have the opposite effect, however, decreasing the number of cells that are able to effectively adhere to the surface (Truong et al., 2012; Webb et al., 2013) (Figure 1). It is generally accepted that the presence of nanoroughness limits the level of contact that occurs between the bacterium and

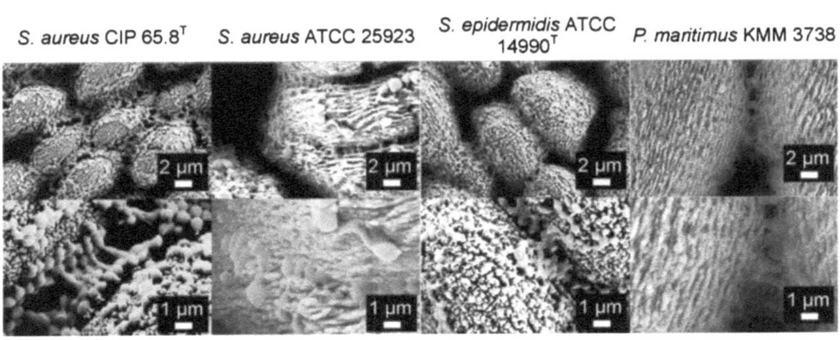

Figure 1 *Staphylococcus* spp. and *Planococcus maritimus* cells attached to the surface of titanium replicas of lotus leaf structures. The highly rough nanostructure on the substratum limited the contact that could be made by the bacterial cells, and thus they were only able to settle in the deeper, more sheltered regions within the surface structures. *Reproduced with permission from Truong et al. (2012), Taylor & Francis.*

the substratum; thus, there are fewer anchor points and, as a result, the aggregate force of adhesion is lower (Hsu et al., 2013; Habimana et al., 2014).

Superhydrophobic surfaces are thought to be strong candidates for the production of anti-biofouling materials that resist bacterial attachment. The rationale is that the self-cleaning effect (originally observed on lotus leaves) that repels any macroscale contaminating particles may extend to the nanoscale, helping colloidal particles such as bacteria to be repelled (Bhushan et al., 2010; Rothstein, 2010; Zhang et al., 2013). In some cases, this does appear to be true; air pockets trapped on superhydrophobic surfaces enable water to move very easily across the surface, and loosely attached cells are detached with minimal applied force. In some cases, superhydrophobic surfaces can have an opposite effect on bacterial cell adhesion. The wings of some insects (i.e., cicadae and dragonflies), while being superhydrophobic, do not repel bacterial cells. Instead, the cells experience adhesion forces that are of sufficient magnitude that they are ruptured by the action of the nanopillar asperities present on both types of wing (Hasan et al., 2012; Ivanova et al., 2012, 2013; Pogodin et al., 2013) (Figure 2).

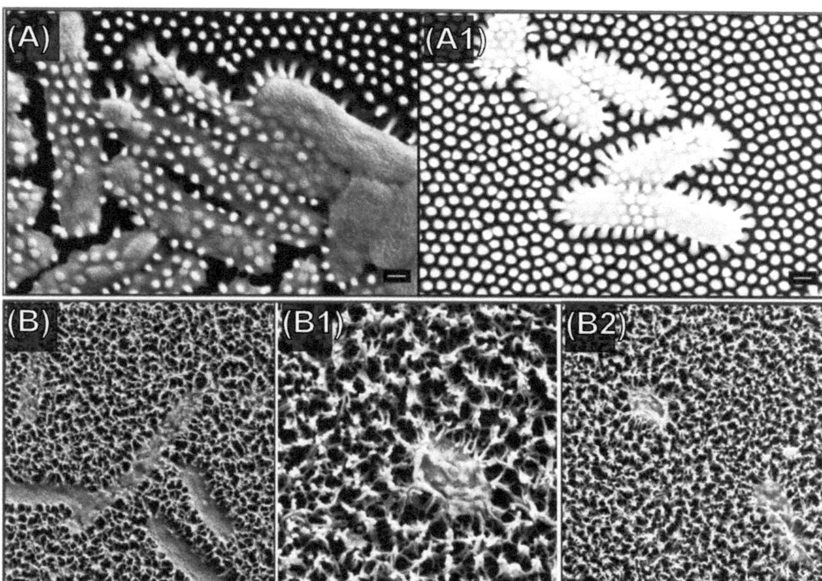

Figure 2 Bacterial cells adsorbed on the surface of cicada (A) and dragonfly (B) wings are subjected to the mechanobactericidal action of the surface. Cicada wings are highly effective against Gram-negative cell types such as *Pseudomonas aeruginosa* (A1) and retain this effect even when coated with a layer of gold (A2). Dragonfly wings, on the other hand have been shown to be effective against all cell types tested to date, including Gram-negative bacteria (B), Gram-positive bacteria (B1), and endospores (B2). *Adapted with permissions from Ivanova et al. (2012), Wiley, and Ivanova et al. (2013), Nature Publishing Group.*

Such surfaces are termed mechanobactericidal, due to their lethality to bacterial cells through a purely physical mechanism.

EUKARYOTIC CELL–TISSUE INTERACTIONS WITH SUPERHYDROPHOBIC SURFACES

The overarching trend throughout the literature on biological interactions with superhydrophobic surfaces is that as the biological system in question increases in complexity, so, too, do the interactions taking place. Accordingly, the diversity of data available on such systems is greater than even in the case of bacteria. Different cell types respond differently to the same surfaces, and the culture conditions play a role in determining the success of adhesion (Kikuchi et al., 2005; Luong-Van et al., 2013; Song and Mano, 2013).

Certainly, some similar effects have been observed for protein and bacterial interactions with superhydrophobic surfaces that have been reported for eukaryotic cells. One area of interest is in the prevention of platelet adhesion in blood, which is of interest as this process leads to the activation of the platelets, together with subsequent clotting of the blood (Sun et al., 2005; Milner et al., 2006; Zhou et al., 2008; Ye et al., 2009; Koh et al., 2010). Investigations of platelet adhesion to structured polydimethylsiloxane and microporous polypropylene have found that platelet adhesion and activation are decreased as the observed water contact angle on the substratum is increased (Hou et al., 2010). For cells such as these platelets, it is generally considered that nanoscopically rough substrata limit the extent of adhesion taking place, with relatively low shear forces required to actually prevent the adhesion. Much like bacterial cells, these eukaryotic cells cannot make extensive contact with the nanorough surface, which is exacerbated by the presence of trapped air pockets. An example is given in Figure 3, which depicts osteosarcoma cells (SaOs2) adsorbing on both smooth and nanorough superhydrophobic polystyrene.

In some cases, it is possible for superhydrophobic surfaces to stimulate the attachment of eukaryotic cells. The available literature on this topic tends to focus on the adhesion of stem cells derived from various tissues. Adipose-derived and mesenchymal stem cells have been show to attach in greater numbers to superhydrophobic polystyrene and modified TiO_2 nanotubes, respectively (Bauer et al., 2008; Cha et al., 2011). Cell attachment was also substantially enhanced; however, this attachment was lost after 3 days. When working with stem cells, it is also important to monitor the differentiation ability of the cells when the are attaching to the surface, rather than only their ability to adhere. Stem cells attached to superhydrophobic surfaces appear to vary in their ability to differentiate, depending on the type of differentiation. For example, Cha et al. (2011) demonstrated that adipogenic differentiation was higher when cells were attached to superhydrophobic surfaces, but chondrogenic and osteogenic differentiation was decreased when attaching onto similar surfaces (Figure 4).

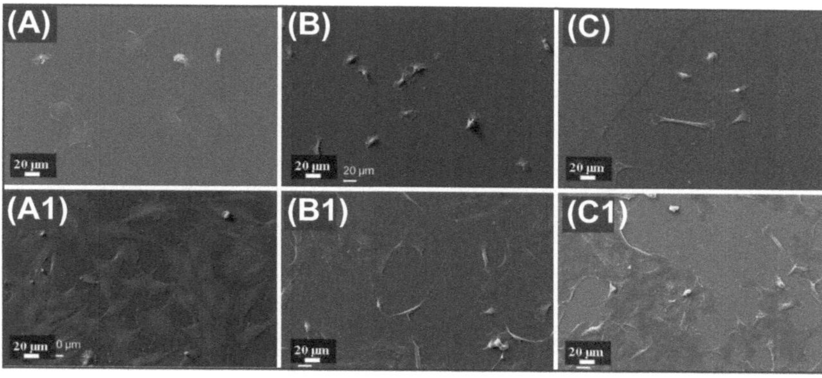

Figure 3 Osteosarcoma cells cultured on the surface of flat (A) and etched superhydrophobic polystyrene (B, C) for 24 h (top) and 8 days (bottom). The nanoroughness of the etched polystyrene caused the cells to attach in decreased number, which is particularly evident in the 8 day incubations. *Adapted with permission from Di Mundo et al. (2011), the American Chemical Society.*

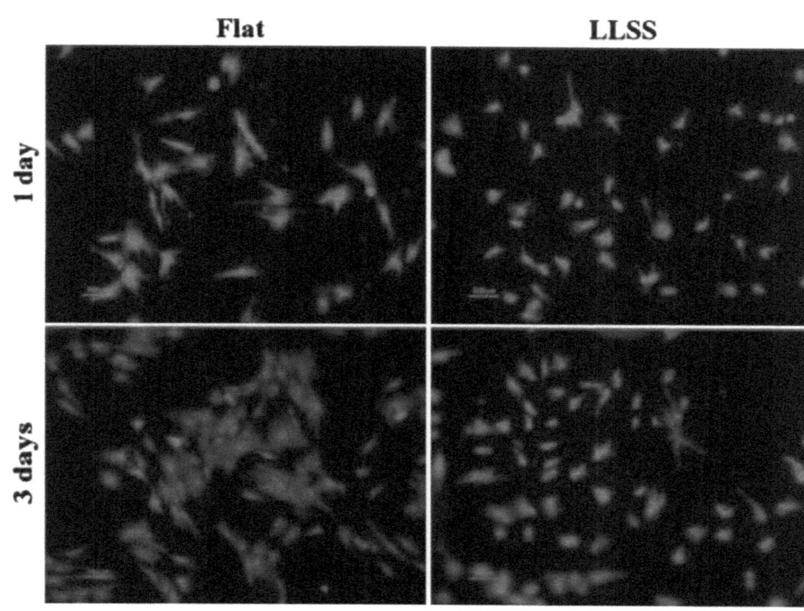

Figure 4 Adipose-derived stem cells adsorbed the surface of flat polystyrene and replicas of lotus leaf surface structure (LLSS) also fabricated from polystyrene, as visualized by fluorescence microscopy. The cell morphology after 3 days of incubation is clearly different, with cells being less capable of attaching onto the LLSS surfaces. *Reproduced with permission from Cha et al. (2011), Wiley.*

SUMMARY

Biological systems are complex by nature, and accordingly so, too, are their interactions with superhydrophobic materials. The dynamic nature of both the cells themselves and the wetting processes that occur on complex interfaces can lead to a range of observed effects that vary with time and conditions, and it cannot therefore be explicitly stated that superhydrophobic surfaces are beneficial or detrimental to cells in general. Rather, superhydrophobic surfaces are potentially useful in any application in which mediation of cell–substratum interactions is required; with proper and careful control of the conditions, superhydrophobic surfaces can both prevent and enhance cellular adhesion, and these surfaces may be able to exert an effect on the metabolism of the attaching organism.

REFERENCES

Amiji, M., Park, K., 1993. Surface modification of polymeric biomaterials with poly(ethylene oxide), albumin, and heparin for reduced thrombogenicity. J. Biomater. Sci. Polym. Ed. 4, 217.

Banerjee, I., Pangule, R.C., Kane, R.S., 2011. Antifouling coatings: recent developments in the design of surfaces that prevent fouling by proteins, bacteria, and marine organisms. Adv. Mater. 23, 690.

Barthlott, W., Schimmel, T., Wiersch, S., Koch, K., Brede, M., Barczewski, M., Walheim, S., Weis, A., Kaltenmaier, A., Leder, A., Bohn, H.F., 2010. The *salvinia* paradox: superhydrophobic surfaces with hydrophilic pins for air retention under water. Adv. Mater. 22, 2325.

Bauer, S., Park, J., von der Mark, K., Schmuki, P., 2008. Improved attachment of mesenchymal stem cells on super-hydrophobic TiO_2 nanotubes. Acta Biomater. 4, 1576.

Bhushan, B., 2011. Biomimetics inspired surfaces for drag reduction and oleophobicity/philicity. Beilstein J. Nanotechnol. 2, 66.

Bhushan, B., Jung, C.J., Nosonovsky, M., 2010. Lotus effect: surfaces with roughness-induced superhydrophobicity, self-cleaning, and low adhesion. In: Springer Handbook of Nanotechnology, third ed. Springer, New York, p. 1437.

Bhushan, B., Jung, Y.C., 2008. Wetting, adhesion and friction of superhydrophobic and hydrophilic leaves and fabricated micro/nanopatterned surfaces. J. Phys. Condens. Matter 20.

Bhushan, B., Jung, Y.C., Koch, K., 2009. Self-cleaning efficiency of artificial superhydrophobic surfaces. Langmuir 25, 3240.

Bixler, G.D., Bhushan, B., 2013. Fluid drag reduction and efficient self-cleaning with rice leaf and butterfly wing bioinspired surfaces. Nanoscale 5, 7685.

Campbell, N.A., Reece, J.B., 2011. Biology, ninth ed. Pearson, Boston.

Cha, K.J., Park, K.S., Kang, S.W., Cha, B.H., Lee, B.K., Han, I.B., Shin, D.A., Kim, D.S., Lee, S.H., 2011. Effect of replicated polymeric substrate with lotus surface structure on adipose-derived stem cell behaviors. Macromol. Biosci. 11, 1357.

Costerton, J.W., Lewandowski, Z., Caldwell, D.E., Korber, D.R., Lappin-Scott, H.M., 1995. Microbial biofilms. Annu. Rev. Microbiol. 49, 711.

Di Mundo, R., Nardulli, M., Milella, A., Favia, P., Dagostino, R., Gristina, R., 2011. Cell adhesion on nanotextured slippery superhydrophobic substrates. Langmuir 27, 4914.

Donlan, R.M., 2002. Biofilms: microbial life on surfaces. Emerg. Infect. Dis. 8, 881.

Feng, L., Zhang, Y., Xi, J., Zhu, Y., Wang, N., Xia, F., Jiang, L., 2008. Petal effect: a superhydrophobic state with high adhesive force. Langmuir 24, 4114.

Genzer, J., Efimenko, K., 2006. Recent developments in superhydrophobic surfaces and their relevance to marine fouling: a review. Biofouling 22, 339.

Habimana, O., Semião, A.J.C., Casey, E., 2014. The role of cell-surface interactions in bacterial initial adhesion and consequent biofilm formation on nanofiltration/reverse osmosis membranes. J. Membr. Sci. 454, 82.

Hasan, J., Webb, H.K., Truong, V.K., Pogodin, S., Baulin, V.A., Watson, G.S., Watson, J.A., Crawford, R.J., Ivanova, E.P., 2012. Selective bactericidal activity of nanopatterned superhydrophobic cicada *Psaltoda claripennis* wing surfaces. Appl. Microbiol. Biotechnol. 1.

Hlady, V., Buijs, J., 1996. Protein adsorption on solid surfaces. Curr. Opin. Biotechnol. 7, 72.

Hou, X., Wang, X., Zhu, Q., Bao, J., Mao, C., Jiang, L., Shen, J., 2010. Preparation of polypropylene superhydrophobic surface and its blood compatibility. Colloids Surf. B: Biointerfaces 80, 247.

Hsu, L.C., Fang, J., Borca-Tasciuc, D.A., Worobo, R.W., Moraru, C.I., 2013. Effect of micro- and nano-scale topography on the adhesion of bacterial cells to solid surfaces. Appl. Environ. Microbiol. 79, 2703.

Hucknall, A., Rangarajan, S., Chilkoti, A., 2009. In pursuit of zero: polymer brushes that resist the adsorption of proteins. Adv. Mater. 21, 2441.

Ivanova, E.P., Hasan, J., Webb, H.K., Truong, V.K., Watson, G.S., Watson, J.A., Baulin, V.A., Pogodin, S., Wang, J.Y., Tobin, M.J., Löbbe, C., Crawford, R.J., 2012. Natural bactericidal surfaces: mechanical rupture of *Pseudomonas aeruginosa* cells by cicada wings. Small 8, 2489.

Ivanova, E.P., Hasan, J., Webb, K.W., Gervinskas, G., Juodkazis, S., Truong, V.K., Wu, A.H.F., Lamb, R.N., Baulin, V.A., Watson, G.S., Watson, J.A., Mainwaring, D.E., Crawford, R.J., 2013. Bactericidal activity of black silicon. Nat. Commun. 2838.

Jeon, S.I., Lee, J.H., Andrade, J.D., De Gennes, P.G., 1991. Protein-surface interactions in the presence of polyethylene oxide. I. Simplified theory. J. Colloid Interface Sci. 142, 149.

Kikuchi, L., Park, J.Y., Victor, C., Davies, J.E., 2005. Platelet interactions with calcium-phosphate-coated surfaces. Biomaterials 26, 5285.

Koc, Y., De Mello, A.J., McHale, G., Newton, M.I., Roach, P., Shirtcliffe, N.J., 2008. Nanoscale superhydrophobicity: suppression of protein adsorption and promotion of flow-induced detachment. Lab Chip Miniaturisation Chem. Biol. 8, 582.

Koh, L.B., Rodriguez, I., Venkatraman, S.S., 2010. The effect of topography of polymer surfaces on platelet adhesion. Biomaterials 31, 1533.

Kolind, K., Leong, K.W., Besenbacher, F., Foss, M., 2012. Guidance of stem cell fate on 2D patterned surfaces. Biomaterials 33, 6626.

Lafuma, A., Quéré, D., 2003. Superhydrophobic states. Nat. Mater. 2, 457.

Lee, S.T., Yun, J.I., Jo, Y.S., Mochizuki, M., van der Vlies, A.J., Kontos, S., Ihm, J.E., Lim, J.M., Hubbell, J.A., 2010. Engineering integrin signaling for promoting embryonic stem cell self-renewal in a precisely defined niche. Biomaterials 31, 1219.

Luong-Van, E., Rodriguez, I., Low, H.Y., Elmouelhi, N., Lowenhaupt, B., Natarajan, S., Lim, C.T., Prajapati, R., Vyakarnam, M., Cooper, K., 2013. Review: micro-and nano-structured surface engineering for biomedical applications. J. Mater. Res. 28, 165.

Milner, K.R., Snyder, A.J., Siedlecki, C.A., 2006. Sub-micron texturing for reducing platelet adhesion to polyurethane biomaterials. J. Biomed. Mater. Res. Part A 76, 561.

Nishimoto, S., Bhushan, B., 2013. Bioinspired self-cleaning surfaces with superhydrophobicity, superoleophobicity, and superhydrophilicity. RSC Adv. 3, 671.

O'Toole, G., Kaplan, H.B., Kolter, R., 2000. Biofilm formation as microbial development. Annu. Rev. Microbiol. 54, 49.

Papadopoulos, P., Mammen, L., Deng, X., Vollmer, D., Butt, H.J., 2013. How superhydrophobicity breaks down. Proc. Natl. Acad. Sci. USA 110, 3254.
Pogodin, S., Hasan, J., Baulin, V.A., Webb, H.K., Truong, V.K., Nguyen, T.H.P., Boshkovikj, V., Fluke, C.J., Watson, G.S., Watson, J.A., Crawford, R.J., Ivanova, E.P., 2013. Biophysical model of bacterial cell interactions with nanopatterned cicada wing surfaces. Biophys. J. 104, 835.
Raffaini, G., Ganazzoli, F., 2010. Protein adsorption on a hydrophobic surface: a molecular dynamics study of lysozyme on graphite. Langmuir 26, 5679.
Roach, P., Farrar, D., Perry, C.C., 2005. Interpretation of protein adsorption: surface-induced conformational changes. J. Am. Chem. Soc. 127, 8168.
Rothstein, J.P., 2010. Slip on superhydrophobic surfaces. Annu. Rev. Fluid Mech. 42, 89.
Scopelliti, P.E., Borgonovo, A., Indrieri, M., Giorgetti, L., Bongiorno, G., Carbone, R., Podestà, A., Milani, P., 2010. The effect of surface nanometre-scale morphology on protein adsorption. PLoS One 3.
Song, W., Mano, J.F., 2013. Interactions between cells or proteins and surfaces exhibiting extreme wettabilities. Soft Matter 9, 2985.
Su, Y., Ji, B., Zhang, K., Gao, H., Huang, Y., Hwang, K., 2010. Nano to micro structural hierarchy is crucial for stable superhydrophobic and water-repellent surfaces. Langmuir 26, 4984.
Sun, T., Tan, H., Han, D., Fu, Q., Jiang, L., 2005. No platelet can adhere-largely improved blood compatibility on nanostructured superhydrophobic surfaces. Small 1, 959.
Tilton, R.D., Robertson, C.R., Gast, A.P., 1991. Manipulation of hydrophobic interactions in protein adsorption. Langmuir 7, 2710.
Truong, V.K., Lapovok, R., Estrin, Y.S., Rundell, S., Wang, J.Y., Fluke, C.J., Crawford, R.J., Ivanova, E.P., 2010. The influence of nano-scale surface roughness on bacterial adhesion to ultrafine-grained titanium. Biomaterials 31, 3674.
Truong, V.K., Webb, H.K., Fadeeva, E., Chichkov, B.N., Wu, A.H.F., Lamb, R., Wang, J.Y., Crawford, R.J., Ivanova, E.P., 2012. Air-directed attachment of coccoid bacteria to the surface of superhydrophobic lotus-like titanium. Biofouling 28, 539.
Vasudevan, R., Kennedy, A.J., Merritt, M., Crocker, F.H., Baney, R.H., 2014. Microscale patterned surfaces reduce bacterial fouling-microscopic and theoretical analysis. Colloids Surf. B: Biointerfaces 117, 225.
Webb, H.K., Boshkovikj, V., Fluke, C.J., Truong, V.K., Hasan, J., Baulin, V.A., Lapovok, R., Estrin, Y., Crawford, R.J., Ivanova, E.P., 2013. Bacterial attachment on sub-nanometrically smooth titanium substrata. Biofouling 29, 163.
Wu, A.H.F., Nakanishi, K., Cho, K.L., Lamb, R., 2013. Diatom attachment inhibition: limiting surface accessibility through air entrapment. Biointerphases 8, 1.
Ye, X., Shao, Y.l., Zhou, M., Li, J., Cai, L., 2009. Research on micro-structure and hemocompatibility of the artificial heart valve surface. Appl. Surf. Sci. 255, 6686.
Zhang, X., Wang, L., Levänen, E., 2013. Superhydrophobic surfaces for the reduction of bacterial adhesion. RSC Adv. 3, 12003.
Zhou, M., Yang, J., Ye, X., Zheng, A., Li, G., Yang, P., Zhu, Y., Cai, L., 2008. Blood platelet's behavior on nanostructured superhydrophobic surface. J. Nano Res. 129.

INDEX

Note: Page numbers followed by "f" and "t" indicate figures and tables respectively.

U

V

W

X

Z

Exact Methods for Nonlinear PDEs

The book describes effective analytical methods for finding exact solutions to nonlinear differential equations of mathematical physics and other partial differential equations and also demonstrates the practical applications of these methods. It covers the methods of generalized separation of variables, methods of functional separation of variables, the classical method of symmetry reductions, the direct method of symmetry reductions, the method of weak symmetry reductions, and the method of differential constraints.

The book presents several simple methods for finding exact solutions to nonlinear partial differential equations (PDEs). These methods do not require specialized knowledge and aim to minimize intermediate calculations. For the first time, it discusses the application of nonrigorous, intuitive reasoning in deriving exact solutions to nonlinear PDEs.

Each section provides numerous examples, problems, and exercises to help readers develop practical skills in applying the methods. The material is illustrated with equations of mass and heat transfer, hydrodynamics, wave theory, nonlinear optics, and other nonlinear equations of mathematical physics.

The key points that distinguish this book from others in the field include:

- it presents many methods in a simpler and more visual format;
- it describes a number of simple methods for constructing exact solutions to nonlinear PDEs and delay PDEs;
- it emphasizes and details the practical use of non-rigorous reasoning to derive exact solutions for nonlinear PDEs.

The book is intended for a diverse audience, including researchers, university professors, engineers, postgraduate students, and students specializing in applied mathematics, theoretical physics, and engineering sciences.

Andrei D. Polyanin, DSc, PhD, professor, is a well-known scientist of broad interests and is active in various areas of mathematics, mechanics, and chemical engineering sciences. Prof. Polyanin graduated with honors from the Department of Mechanics and Mathematics of Moscow State University in 1974. He earned his PhD in 1981 and DSc in 1986 at the Institute for Problems in Mechanics of the Russian (former USSR) Academy of Sciences. Since 1975, Professor Polyanin has been working at the Institute for Problems in Mechanics of the Russian Academy of Sciences. He is a member of the Russian National Committee on Theoretical and Applied Mechanics. Prof. Polyanin is an author of more than 30 books and over 300 articles and holds three patents.

Advances in Applied Mathematics
Series Editor: Daniel Zwillinger

Delay Ordinary and Partial Differential Equations
Andrei D. Polyanin
Vsevolod G. Sorkin
Alexi I. Zhurov

Clean Numerical Simulation
Shijun Liao

Multiplicative Partial Differential Equations
Svetlin Georgiev and Khaled Zennir

Engineering Statistics
A Matrix-Vector Approach with MATLAB®
Lester W. Schmerr Jr.

General Quantum Numerical Analysis
Svetlin Georgiev and Khaled Zennir

An Introduction to Partial Differential Equations with MATLAB®
Matthew P. Coleman and Vladislav Bukshtynov

Handbook of Exact Solutions to Mathematical Equations
Andrei D. Polyanin

Introducing Game Theory and its Applications, Second Edition
Elliott Mendelson and Dan Zwillinger

Modeling Operations Research and Business Analytics
William P. Fox and Robert E. Burks

Decision Analysis through Modeling and Game Theory
William P. Fox

Advances in High-Order Predictive Modelling
Dan Gabriel Cacuci

Introduction to Financial Mathematics, Second Edition
Kevin J. Hastings

General Quantum Variational Calculus
Svetlin G. Georgiev and Khaled Zennir

Advanced Engineering Mathematics with MATLAB®, Sixth Edition
Dean G. Duffy

Exact Methods for Nonlinear PDEs
Andrei D. Polyanin

https://www.routledge.com/Advances-in-Applied-Mathematics/book-series/CRCADVAPPMTH?pd=published,forthcoming&pg=1&pp=12&so=pub&view=list

Exact Methods for Nonlinear PDEs

Andrei D. Polyanin

CRC Press
Taylor & Francis Group
Boca Raton London New York

CRC Press is an imprint of the
Taylor & Francis Group, an **informa** business
A CHAPMAN & HALL BOOK

First edition published 2026
by CRC Press
2385 Executive Center Drive, Suite 320, Boca Raton, FL 33431, U.S.A.

and by CRC Press
4 Park Square, Milton Park, Abingdon, Oxon, OX14 4RN

CRC Press is an imprint of Taylor & Francis Group, LLC

Library of Congress Cataloging-in-Publication Dat
Names: Polyanin, A. D. (Andreĭ Dmitrievich) author.
Title: Exact methods for nonlinear PDEs / Andrei D. Polyanin.
Description: First edition. | Boca Raton : CRC Press/Taylor and Francis
Group, 2025. | Includes bibliographical references and index.
Identifiers: LCCN 2024062307 (print) | LCCN 2024062308 (ebook) | ISBN
9781032710518 (hbk) | ISBN 9781032716428 (pbk) | ISBN 9781032716367
(ebk)
Subjects: LCSH: Differential equations, Partial--Numerical solutions. |
Differential equations, Nonlinear--Numerical solutions.
Classification: LCC QA377 .P5677 2025 (print) | LCC QA377 (ebook) | DDC
518/.64--dc23/eng/20250318
LC record available at https://lccn.loc.gov/2024062307
LC ebook record available at https://lccn.loc.gov/2024062308

ISBN: 978-1-032-71051-8 (hbk)
ISBN: 978-1-032-71642-8 (pbk)
ISBN: 978-1-032-71636-7 (ebk)

DOI: 10.1201/9781032716367

Typeset in CMR10 font
by KnowledgeWorks Global Ltd.

Publisher's note: This book has been prepared from camera-ready copy provided by the authors.

Contents

Preface

The book describes effective analytical methods for finding exact solutions to nonlinear differential equations of mathematical physics and other partial differential equations and also demonstrates the practical applications of these methods.

Nonlinear partial differential equations (PDEs) of the second and higher orders, including nonlinear equations of mathematical physics, are commonly encountered in various fields of mathematics, physics, mechanics, chemistry, and biology and in numerous applications. Finding a general solution to these equations is only possible in exceptional cases. Therefore, one usually has to limit themselves to identifying and analyzing particular solutions, which are referred to as *exact solutions*.

A solution is exact if, when substituted into the mathematical equation, it turns the equation into an identity. No approximations or simplifications of the equation are allowed, and no a priori assumptions are used.

Exact solutions to equations of mathematical physics greatly enhance our understanding of the qualitative features of many phenomena and processes in various fields of natural science. Exact solutions of nonlinear equations elucidate the mechanisms of various complex nonlinear effects such as spatial localization of transport processes, the presence, multiplicity, or absence of stationary states under certain conditions, the existence of blow-up modes, the possible non-smoothness or discontinuity of the unknown quantities, and many more. Simple solutions of linear and nonlinear differential equations are widely used to illustrate theoretical material and various applications in university and college courses on applied and computational mathematics, asymptotic methods, theoretical physics, heat and mass transfer theory, fluid dynamics, wave theory, nonlinear optics, and other disciplines.

Notably, traveling-wave and self-similar exact solutions are often asymptotics of much broader classes of solutions that correspond to different initial and boundary conditions. This means that one can draw general conclusions and predict the dynamics of various phenomena and processes based on these properties of solutions.

Exact solutions of mathematical physics equations are crucial as 'mathematical standards' that are widely used to create test problems to develop various numerical, asymptotic, and approximate analytical methods and assess their accuracy. Exact solutions are also necessary to update and improve relevant sections of computer software for analytical calculations (such as computer algebra systems like Mathematica, Maple, MATLAB, and others).

This book understands exact solutions to nonlinear partial differential equations as follows:

(i) solutions expressed in terms of elementary functions;

(ii) solutions expressed through elementary functions, functions included in the equation (necessary when the equation depends on arbitrary or special functions), and indefinite integrals;

(iii) solutions expressed in terms of solutions to ordinary differential equations (ODEs) or systems of such equations.

Combinations of cases (i) and (iii), as well as (ii) and (iii) are also allowed.

The simplest case (i) is singled out from the more general case (ii) on purpose, since some authors limit themselves to searching only such exact solutions. In cases (i) and (ii), exact solutions can be represented explicitly, implicitly, or parametrically.

Exact methods for solving nonlinear partial differential equations are methods that allow obtaining exact solutions.

The book describes methods of generalized separation of variables (which include several methods varying in complexity and having different ranges of applicability), methods of functional separation of variables, the classical method of symmetry reductions (group analysis of differential equations), the direct method of symmetry reductions (Clarkson–Kruskal method), the method for finding weak symmetries, the method of differential constraints, and some other methods that allow one to find exact solutions to nonlinear equations of mathematical physics. The book outlines an elementary theory of invariants, which makes it possible, with minimal effort, to obtain the vast majority of exact solutions that one usually looks for using the much more complex and time-consuming classical method for seeking symmetries. To illustrate general ideas, enhance understanding of the following material, and simplify the presentation of certain sections, the book also includes auxiliary topics, which cover the symmetries of algebraic equations and systems, invariant solutions of ordinary differential equations, methods for solving quasilinear and nonlinear first-order partial differential equations, as well as solution methods for functional equations. Furthermore, the book provides numerous examples and exercises to help the reader better understand the material and develop practical skills in solving nonlinear differential equations.

The last two chapters focus on using non-rigorous (intuitive) reasoning based on the heuristic principles 'from simple to complex' and 'structural analogy of solutions' to find exact solutions to nonlinear PDEs. The book discusses several simple yet effective methods for constructing exact solutions of nonlinear mathematical physics equations and functional partial differential equations, including nonlinear PDEs with constant or variable delay. These methods minimize the need for intermediate calculations and are based on two main ideas: (i) exact solutions of given equations can be used to find more complex solutions of the same equations and (ii) exact solutions of some equations can be used as a basis for constructing solutions to more complex related equations with similar nonlinear terms. Through specific examples, the book shows how non-rigorous reasoning can help improve and develop new analytical methods. In some cases, the results obtained through non-rigorous reasoning can be generalized, and various statements and theorems can be rigorously proven. At present, non-rigorous reasoning is often the only feasible way to construct exact solutions of nonlinear functional partial differential equations.

When selecting the practical material (including examples and exercises), the author gave significant priority to two important types of equations:

- equations that are found in various applications, such as in heat and mass transfer theory, wave theory, fluid dynamics, nonlinear optics, biology, and other disciplines;
- equations of a general form, which depend on several free parameters or contain arbitrary functions; exact solutions of such equations are of particular interest for testing numerical and approximate analytical methods.

The main points that distinguish this book from the others in this area are:

- the book presents many methods in a simpler and more visual format;
- the book includes material that may benefit a wider audience of researchers than usual, and parts of it are accessible to readers with minimal training and no special mathematical background;
- the book describes a number of simple methods for constructing exact solutions to nonlinear partial differential equations (including delay PDEs), which require a small amount of intermediate calculations;
- the book details the practical use of non-rigorous reasoning to find exact solutions for nonlinear PDEs as well as functional PDEs.

The material is organized according to the 'from simple to complex' principle. The author aimed to avoid the use of special terms whenever possible to make the content accessible to readers with varying mathematical backgrounds. For that purpose, some results are described briefly and simply, without proof, but still adequately for most applications. Many sections can be read independently, making it easier to navigate through the material. Furthermore, a detailed table of contents is provided for quick reference.

The author thanks A. V. Aksenov and A. I. Zhurov for their valuable discussions and insightful comments.

The author hopes that the book will be helpful for a wide range of researchers, university teachers, and engineers, as well as graduate postgraduate and PhD students specializing in applied, computational, and engineering mathematics, theoretical physics, mechanics, control theory, biology, and chemical engineering. Individual sections of the book and exercises can be utilized for lecturing and conducting practical classes on applied mathematics and mathematical physics for students of physics and mathematics specialties at universities and colleges.

Andrei D. Polyanin
December 2024

Some Notations and Remarks

Latin Characters

$C_1, C_2, \ldots$	arbitrary constants;
t	time ($t \geq 0$), independent variable for PDEs;
u	unknown function (dependent variable) for PDEs;
x	space variable for PDEs, independent variable for ODEs and PDEs;
y	unknown function (dependent variable) for ODEs, space variable (independent variable) for PDEs;
$x_1, \ldots, x_n$	Cartesian coordinates in n-dimensional space (independent variables).

Short Notation for Derivatives and Differential Operators

Partial derivatives of a function $u = u(x, t)$:

$$u_x = \frac{\partial u}{\partial x}, \quad u_t = \frac{\partial u}{\partial t}, \quad u_{xx} = \frac{\partial^2 u}{\partial x^2}, \quad u_{xt} = \frac{\partial^2 u}{\partial x \partial t}, \quad u_{tt} = \frac{\partial^2 u}{\partial t^2}, \quad \ldots, \quad u_x^{(n)} = \frac{\partial^n u}{\partial x^n}.$$

Ordinary derivatives of a function $f = f(x)$:

$$f'_x = \frac{df}{dx}, \quad f''_{xx} = \frac{d^2 f}{dx^2}, \quad f'''_{xxx} = \frac{d^3 f}{dx^3}, \quad f''''_{xxxx} = \frac{d^4 f}{dx^4}, \quad f_x^{(n)} = \frac{d^n f}{dx^n} \quad \text{for} \quad n > 4.$$

Differential operators:

$$\partial_x = \frac{\partial}{\partial x}, \quad \partial_y = \frac{\partial}{\partial y}, \quad \partial_t = \frac{\partial}{\partial t}, \quad \partial_u = \frac{\partial}{\partial u}.$$

Remarks

1. The book often uses the abbreviations ODE and PDE, which stand for 'ordinary differential equation' and 'partial differential equation,' respectively.

2. If a formula or a solution involves a derivative of a function, the derivative is assumed to exist.

3. If a formula or a solution involves an indefinite or definite integral, the integral is assumed to exist.

4. Expressions of the form $\dfrac{f(x)}{a-2}$, appearing in formulas or solutions imply that $a \neq 2$; this is not usually stated explicitly.

5. The simple solutions that only involve a single independent variable appearing in the original equation are omitted throughout the book.

6. In general, the book adopts a simple and straightforward classification of the most common solutions by their appearance, which is unrelated to the type or appearance of the equations concerned (see the table below).

Table: Most common types of exact solutions for equations with two independent variables, x and t, and one unknown function, u.

No.	Type of solution	Solution structure (x and t can be swapped)
1	Additive separable solution	$u = \varphi(x) + \psi(t)$
2	Multiplicative separable solution	$u = \varphi(x)\psi(t)$
3	Traveling wave solution*	$u = U(z), \quad z = \alpha x + \beta t, \quad \alpha\beta \neq 0$
4	Self-similar solution	$u = t^{\alpha} F(z), \quad z = xt^{\beta}$
5	Generalized self-similar solution	$u = \varphi(t)F(z), \quad z = \psi(t)x$
6	Generalized traveling wave solution	$u = U(z), \quad z = \varphi(t)x + \psi(t)$
7	Generalized separable solution	$u = \varphi_1(x)\psi_1(t) + \cdots + \varphi_n(x)\psi_n(t)$
8	Functional separable solution (special case)	$u = U(z), \quad z = \varphi(x) + \psi(t)$
9	Functional separable solution	$u = U(z)$, $z = \varphi_1(x)\psi_1(t) + \cdots + \varphi_n(x)\psi_n(t)$

* Both independent variables can play the role of space coordinates.

Author

Andrei D. Polyanin, D.Sc., Ph.D., is a well-known scientist of broad interests and is active in various areas of mathematics, theory of heat and mass transfer, hydrodynamics, and chemical engineering sciences. He is one of the most prominent authors in the field of reference literature on mathematics. Professor Polyanin graduated with honors from the Department of Mechanics and Mathematics at the Lomonosov Moscow State University in 1974. Since 1975, Professor Polyanin has been working at the Ishlinsky Institute for Problems in Mechanics of the Russian (former USSR) Academy of Sciences, where he defended his Ph.D. in 1981 and D.Sc. degree in 1986.

Professor Polyanin has made important contributions to the theory of differential and integral equations, mathematical physics, applied and engineering mathematics, the theory of heat and mass transfer, and fluid dynamics. He is developing analytical methods for constructing solutions to mathematical equations of various types and has obtained a huge number of exact solutions of ordinary differential, partial differential, delay partial differential, integral, and functional equations.

Professor Polyanin is an author of more than 30 books and over 300 articles and holds three patents. His books include the following: V. F. Zaitsev and A. D. Polyanin, *Discrete-Group Methods for Integrating Equations of Nonlinear Mechanics*, CRC Press, 1994; A. D. Polyanin and V. V. Dilman, *Methods of Modeling Equations and Analogies in Chemical Engineering*, CRC Press/Begell House, Boca Raton, 1994; A. D. Polyanin and V. F. Zaitsev, *Handbook of Exact Solutions for Ordinary Differential Equations*, CRC Press, 1995 (2nd edition in 2003); A. D. Polyanin and A. V. Manzhirov, *Handbook of Integral Equations*, CRC Press, 1998 (2nd edition in 2008); A. D. Polyanin, *Handbook of Linear Partial Differential Equations for Engineers and Scientists*, Chapman & Hall/CRC Press, 2002 (2nd edition, co-authored with V. E. Nazaikinskii, in 2016); A. D. Polyanin, V. F. Zaitsev, and A. Moussiaux, *Handbook of First Order Partial Differential Equations*, Taylor & Francis, 2002; A. D. Polyanin, A. M. Kutepov, et al., *Hydrodynamics, Mass and Heat Transfer in Chemical Engineering*, Taylor & Francis, 2002; A. D. Polyanin and V. F. Zaitsev, *Handbook of Nonlinear Partial Differential Equations*, Chapman & Hall/CRC Press, 2004 (2nd edition in 2012); A. D. Polyanin and A. V. Manzhirov, *Handbook of Mathematics for Engineers and Scientists*, Chapman & Hall/CRC Press, 2007; A. D. Polyanin and V. F. Zaitsev, *Handbook of Ordinary Differential Equations: Exact Solutions, Methods, and Problems*, CRC Press, 2018; A. D. Polyanin and A. I. Zhurov, *Separation of Variables and Exact Solutions to Nonlinear PDEs*, CRC Press, 2022; A. D. Polyanin, V. G. Sorokin, and A. I. Zhurov, *Delay Ordinary and Partial Differential Equations*, CRC Press, 2023; and A. D. Polyanin, *Handbook of Exact Solutions to Mathematical Equations*, CRC Press, 2024.

Professor Polyanin is the editor-in-chief of the international scientific educational website 'EqWorld—The World of Mathematical Equations' and a member of the editorial boards for several journals.

1. Elementary Invariant Theory: Algebraic and Ordinary Differential Equations

► **Preliminary remarks.** This chapter describes a simple method based on the elementary theory of invariants. The method allows one to simplify algebraic equations (also known as polynomial equations) and systems of such equations, integrate first-order ordinary differential equations, reduce the order of second- and higher-order ordinary differential equations, as well as obtain exact solutions of nonlinear partial differential equations (this will be discussed further in Chapter 4 'Elementary Invariant Theory: Partial Differential Equations'). Invariants are constructed by searching for transformations that preserve the form of the equations under consideration. Numerous examples of solving specific algebraic and differential equations are given. It is important to note that using even simple linear translation or scaling transformations (or their combinations) allows one to describe, in a unified way, significantly more solvable (or order-reducing) ordinary differential equations than available in standard textbooks. When using this simple method, one must be able to solve only the simplest algebraic equations (or systems of such equations) and be able to differentiate.

1.1. Symmetries. General Scheme of Using Invariants to Solve Equations

1.1.1. Symmetries and Reductions. Transformations Preserving the Form of Mathematical Equations. Invariants

In nature and technology (and elementary geometry), one usually encounters either mirror symmetry, when one-half of an object is mirror-symmetrical to the other, or central symmetry, when an object transforms into itself when rotated about a center. However, the concept of symmetry is much broader than in the cases indicated. In general, symmetry is understood as the invariance of objects as well as physical phenomena, mathematical formulas, equations, and so on, under some procedure. Below, we will discuss how to use the property of symmetry to find solutions or simplify various classes of nonlinear algebraic and differential equations (ordinary or partial) and systems of such equations.

Symmetries of a mathematical equations will be understood as transformations that preserve the form of the equation. A reduction is a technique that simplifies a mathematical equation for easier analysis and solution or converts it into a more convenient form from some viewpoint.

Below are a few specific equations that preserve their form under simple transformations.

► Example 1.1. Let us consider the biquadratic equation

$$x^4 + ax^2 + 1 = 0. \tag{1.1.1.1}$$

The change of variable

$$x = -\bar{x}$$

leads to exactly the same equation

$$\bar{x}^4 + a\bar{x}^2 + 1 = 0.$$

This means that the biquadratic equation (1.1.1.1) preserves its form under the transformation $x = -\bar{x}$.

Two other transformations

$$x = \pm\frac{1}{\tilde{x}}$$

also preserve the form of equation (1.1.1.1) since multiplying by $\tilde{x}^4$ results in

$$\tilde{x}^4 + a\tilde{x}^2 + 1 = 0.$$

Suppose $x_1 = x_*$ is a solution of equation (1.1.1.1). Then the above transformations, which preserve the form of the equation, allow us to find three more solutions: $x_2 = -x_*$, $x_3 = 1/x_*$, and $x_4 = -1/x_*$. ◄

► Example 1.2. The form of the second-order linear ordinary differential equation

$$y''_{xx} - y'_x = 0 \tag{1.1.1.2}$$

will remain unchanged under any of the following three transformations:

$$\begin{aligned} &x = \bar{x} + a, && y = \bar{y} && (a \text{ is any number}); \\ &x = \bar{x}, && y = \bar{y} + b && (b \text{ is any number}); \\ &x = \bar{x}, && y = c\bar{y} && (c \text{ is any nonzero number}), \end{aligned} \tag{1.1.1.3}$$

since all of them lead to exactly the same equation

$$\bar{y}''_{\bar{x}\bar{x}} - \bar{y}'_{\bar{x}} = 0.$$

◄

► Example 1.3. The form of the first-order nonlinear ordinary differential equation

$$y'_x = e^y$$

will remain unchanged under either of the two transformations

$$\begin{aligned} &x = \bar{x} + a, && y = \bar{y} && (a \text{ is any number}); \\ &x = e^b\bar{x}, && y = \bar{y} - b && (b \text{ is any number}). \end{aligned}$$

◄

It will be shown below that transformations that preserve the form of equations allow us to 'multiply' solutions of equations.

A *invariant of a transformation* is a function (other than a constant) that preserves its form under the transformation. A transformation invariant can depend on the independent variable as well as the dependent variable and its derivatives (we deal with differential equations).

To clarify the concept of invariants that preserve their form under given transformations, let us consider a few simple examples.

► Example 1.4. The one-parameter translation in both axes

$$x = \bar{x} + a, \quad y = \bar{y} + a,$$

where a is any number, has the invariant

$$I = y - x = \bar{y} - \bar{x}.$$

If x is the independent variable and y is the dependent variable, then the derivatives

$$I_2 = y'_x = \bar{y}'_{\bar{x}}, \quad I_3 = y''_{xx} = \bar{y}''_{\bar{x}\bar{x}}, \quad \ldots$$

are also invariants of the translation transformation. ◄

► Example 1.5. The one-parameter uniform scaling in both axes

$$x = a\bar{x}, \quad y = a\bar{y},$$

where $a \neq 0$ is any number, has the invariant

$$I_1 = \frac{y}{x} = \frac{\bar{y}}{\bar{x}}.$$

If x is the independent variable and y is the dependent variable, then there are more complex invariants, dependent on derivatives, that remain unchanged under the above transformation. These are two examples

$$I_2 = y'_x = \bar{y}'_{\bar{x}}, \quad I_3 = xy''_{xx} = \bar{x}\bar{y}''_{\bar{x}\bar{x}}.$$

◄

► Example 1.6. The one-parameter transformation that represents scaling in one variable and translation in the other variable

$$x = a\bar{x}, \quad y = \bar{y} + k \ln a,$$

where $a > 0$ is any number and k is a given constant, has the invariant

$$I_1 = y - k \ln x = \bar{y} - k \ln \bar{x}.$$

There are two more invariants dependent on derivatives and remaining unchanged under the above transformation:

$$I_2 = xy'_x = \bar{x}\bar{y}'_{\bar{x}}, \quad I_3 = x^2 y''_{xx} = \bar{x}^2 \bar{y}''_{\bar{x}\bar{x}}.$$

◄

Remark 1.1. *The transformation invariants in examples 1.4–1.6 did not depend explicitly on the parameter a. Such invariants of one-parameter transformations or similar ones play a primary role in analyzing and solving ordinary differential equations (and partial differential equations).*

Remark 1.2. *An invariant of a transformation is not uniquely defined. If I is an invariant of some transformation, then $f(I)$, where f is an arbitrary function, is also an invariant of this transformation. In practice, it is preferable to use invariants of the simplest form, which lead to simple calculations.*

1.1.2. General Scheme for Using Invariants to Solve Mathematical Equations

Figure 1.1 displays a principal scheme for studying mathematical equations. The scheme focuses on identifying transformations that preserve the form of the equations, followed by a transition from the original variables to new variables, the invariants of the transformations.

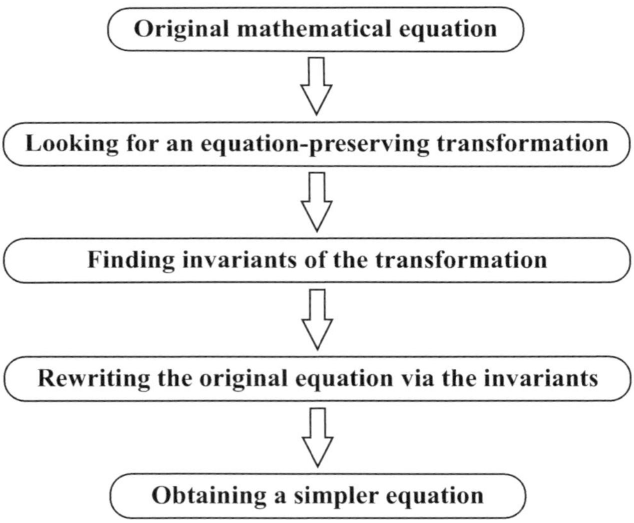

Figure 1.1. General scheme for using invariants to solve mathematical equations.

After following these steps, the equation is often simplified or transformed into a solvable form. It is important to note that this scheme can be effectively applied to find solutions for a diverse range of mathematical equations (e.g., see Kudryashov, 1998; Polyanin, 2008; and Polyanin & Zaitsev, 2012).

For better understanding and assimilating the concept of using invariants to solve mathematical equations, the author follows the principle of 'from simple to complex' in presenting the material. The presentation begins with relevant results for algebraic equations (see Section 1.2), followed by ordinary differential equations (see

Section 1.3), and concludes with nonlinear equations of mathematical physics (see Chapter 4).

Problems and Exercises for Section 1.1

1. Specify transformations that preserve the form of the following second-order linear ODEs:

a) $y''_{xx} = x$,
b) $y''_{xx} = xy$,
c) $xy''_{xx} = y'_x$,
d) $y''_{xx} = f(x)y'_x$.

2. Specify transformations that preserve the form of the following second-order nonlinear ODEs:

a) $y''_{xx} = e^y$,
b) $y''_{xx} = y^n$,
c) $y''_{xx} = (y'_x)^2$,
d) $yy''_{xx} = (y'_x)^2$,
e) $y''_{xx} = f(y)y'_x$,
f) $y''_{xx} = f(y'_x)$.

3. Prove that the rotation transformation on a plane

$$x = \bar{x}\cos\beta - \bar{y}\sin\beta, \qquad y = \bar{x}\sin\beta + \bar{y}\cos\beta \qquad (\beta = \text{const})$$

has the invariant $I = x^2 + y^2$.

4. Find a one-parameter linear transformation of the independent variables of the form

$$x = a_1\bar{x} + b_1\bar{y}, \qquad y = a_2\bar{x} + b_2$$

that would have the invariant $I = y^2 - x^2$.

5. For the nonuniform scaling transformation in two axes

$$x = a\bar{x}, \qquad y = a^k\bar{y},$$

where $a > 0$ is a free parameter and k is a given number, find the first three invariants of this transformation that involves the independent variable and the dependent variable with its first two derivatives.

1.2. Algebraic Equations and Systems of Algebraic Equations

1.2.1. Simple Symmetries and Transformations Reducing the Degree of Algebraic Equations

1°. Consider the algebraic (polynomial) equation

$$a_{2n}x^{2n} + a_{2n-2}x^{2n-2} + a_{2n-4}x^{2n-4} + \cdots + a_4x^4 + a_2x^2 + a_0 = 0, \quad (1.2.1.1)$$

that only involves even powers. A biquadratic equation is a special case of equations (1.2.1.1) with $n = 2$.

The change of variable

$$x = -\bar{x} \tag{1.2.1.2}$$

leads to exactly the same equation for $\bar{x}$. (Equation (1.2.1.1) is said to be invariant under transformation (1.2.1.2).) It follows that if equation (1.2.1.1) has a solution $x = x_1$, then $x = -x_1$ is also a solution to the equation.

Squaring both sides of (1.2.1.2) gives a simple algebraic function that remains unchanged under transformation (1.2.1.2):

$$x^2 = \bar{x}^2 \tag{1.2.1.3}$$

This function is an invariant of transformation (1.2.1.2). With the invariant (1.2.1.3) taken to be the new variable, $z = x^2$, equation (1.2.1.1) of degree $2n$ becomes an equation of degree n:

$$a_{2n}z^n + a_{2n-2}z^{n-1} + a_{2n-4}z^{n-2} + \cdots + a_4 z^2 + a_2 z + a_0 = 0.$$

Consequently, changing the variable x in the original equation to the invariant $z = x^2$ of transformation (1.2.1.3) simplifies the equation (reduces its degree by half).

2°. Now let us look at a more general algebraic equation of even degree $2n$:

$$P_{2n}(x) = 0, \tag{1.2.1.4}$$

which can include lower-order terms with odd degrees.

Proposition 1. *Suppose that for some λ and any x, the left-hand side of equation (1.2.1.4) satisfies the relation $P_{2n}(x) = P_{2n}(\lambda - x)$ identically. Then the original equation can be reduced with the substitution $z = (x - \frac{1}{2}\lambda)^2$ to a simpler algebraic equation of degree n.*

This proposition is easy to prove by using the relation

$$\begin{aligned} P_{2n}(x) &= \tfrac{1}{2}[P_{2n}(x) + P_{2n}(\lambda - x)] \\ &= \tfrac{1}{2}\big[P_{2n}\big(\tfrac{1}{2}\lambda + y\big) + P_{2n}\big(\tfrac{1}{2}\lambda - y\big)\big], \quad y = x - \tfrac{1}{2}\lambda. \end{aligned} \tag{1.2.1.5}$$

It follows from (1.2.1.5) that equation (1.2.1.4), rewritten in terms of the new variable y, remains unchanged when y is replaced with $-y$. Hence, its left-hand side can be represented as a linear combination of only even powers of y. Further, the change of variable $z = y^2$ (z is an invariant of the transformation $\bar{y} = -y$) reduces the degree of the equation by half.

► Example 1.7. Let us look at the equation

$$(x + a)^{2n} + (x + b)^{2n} - c = 0, \tag{1.2.1.6}$$

which is a special case of equation (1.2.1.4).

It is easy to verify that the left-hand side of (1.2.1.6) will remain unchanged if we replace x with $\lambda - x$, where $\lambda = -a - b$. Therefore, after expanding brackets and substituting $z = [x + \frac{1}{2}(a + b)]^2$, the original equation reduces to an equation of degree n. ◄

3°. Now consider an algebraic equation of odd degree $2n+1$:

$$P_{2n+1}(x) = 0. \tag{1.2.1.7}$$

Proposition 2. *Suppose that for some λ, the left-hand side of equation (1.2.1.7) satisfies the relation $P_{2n+1}(x) = -P_{2n+1}(\lambda - x)$ identically. The original equation has a root $x = \frac{1}{2}\lambda$, and hence, its left-hand side can be represented as*

$$P_{2n+1}(x) = (x - \tfrac{1}{2}\lambda)Q_{2n}(x). \tag{1.2.1.8}$$

The polynomial of odd degree $Q_{2n}(x)$ satisfies the relation $Q_{2n}(x) = Q_{2n}(\lambda - x)$ identically. By virtue of Proposition 1, the equation $Q_{2n}(x) = 0$ can be reduced with the substitution $z = (x - \frac{1}{2}\lambda)^2$ to a simpler algebraic equation of degree n.

► Example 1.8. Consider the equation

$$(x+a)^{2n+1} + (x+b)^{2n+1} + c(2x+a+b) = 0, \tag{1.2.1.9}$$

which is a special case of equation (1.2.1.7).

It is easy to verify that the sign of the left-hand side of (1.2.1.9) will change if we replace x with $\lambda - x$, where $\lambda = -a - b$. Therefore, equation (1.2.1.9) has a root $x = \frac{1}{2}\lambda = -\frac{1}{2}(a+b)$, and so its left-hand side can be represented as (1.2.1.8). ◄

1.2.2. Reciprocal Equations and Their Generalizations

1°. The algebraic equation of even degree of the form

$$a_0x^{2n} + a_1x^{2n-1} + a_2x^{2n-2} + \cdots + a_2x^2 + a_1x + a_0 = 0, \tag{1.2.2.1}$$

where $a_0 \neq 0$, is called a *reciprocal equation.* Its left-hand side is known as a *reciprocal polynomial.**

Let us change the variable in equation (1.2.2.1) as follows:

$$x = \frac{1}{\bar{x}}. \tag{1.2.2.2}$$

Multiplying by $\bar{x}^{2n}$ results in exactly the same equation. It follows that if equation (1.2.2.1) has a root $x = b$, then $x = 1/b$ is also a root of the equation.

The quadratic equation

$$a_0x^2 + a_1x + a_0 = 0$$

is the simplest reciprocal equation. Dividing it by x gives

$$a_0\left(x + \frac{1}{x}\right) + a_1 = 0.$$

This can be rewritten as the linear equation

$$a_0z + a_1 = 0,$$

where

$$z = x + \frac{1}{x} = \bar{x} + \frac{1}{\bar{x}} \tag{1.2.2.3}$$

is the simplest *invariant of transformation* (1.2.2.2).

* The coefficients of a reciprocal polynomial equidistant from its ends are equal.

Proposition 3. *In general, a reciprocal equation of degree $2n$ can be reduced to a simpler algebraic equation of degree n using the substitution (1.2.2.3); for example, see Leung et al. (1992) and Boltyanskii & Vilenkin (2002).*

► Example 1.9. Consider the fourth-degree reciprocal equation

$$ax^4 + bx^3 + cx^2 + bx + a = 0.$$

Dividing by x^2 gives

$$a\left(x^2 + \frac{1}{x^2}\right) + b\left(x + \frac{1}{x}\right) + c = 0. \tag{1.2.2.4}$$

Taking into account that

$$\left(x + \frac{1}{x}\right)^2 = x^2 + 2 + \frac{1}{x^2} \quad \Longrightarrow \quad x^2 + \frac{1}{x^2} = z^2 - 2,$$

we arrive at the second-degree equation

$$az^2 + bz + c - 2a = 0.$$

The substitution (1.2.2.3) used here is based on the invariant of transformation (1.2.2.2). ◄

Proposition 4. *In general, a reciprocal algebraic equation of an odd degree*

$$P_{2n+1}(x) = 0,$$

where

$$P_{2n+1}(x) \equiv a_0 x^{2n+1} + a_1 x^{2n} + a_2 x^{2n-1} + \cdots + a_2 x^2 + a_1 x + a_0,$$

has a root $x = -1$, and its left-hand side can be represented as

$$P_{2n+1}(x) = (x + 1)Q_{2n}(x),$$

where $Q_{2n}(x)$ is a reciprocal polynomial of degree $2n$.

► Example 1.10. The third-degree reciprocal equation

$$ax^3 + bx^2 + bx + a = 0$$

can be rewritten as

$$(x + 1)[ax^2 + (b - a)x + a] = 0.$$

◄

It follows from Proposition 4 that a reciprocal equation of degree $2n + 1$ can be reduced, by dividing by $(x + 1)$ and introducing the new variable (1.2.2.3), to an equation of degree n.

2°. Generalized reciprocal equations

$$a_0\beta^n x^{2n} + a_1\beta^{n-1}x^{2n-1} + \cdots + a_{n-1}\beta x^{n+1} + a_n x^n$$
$$+ \gamma a_{n-1}x^{n-1} + \gamma^2 a_{n-2}x^{n-2} + \cdots + \gamma^{n-1}a_1 x + \gamma^n a_0 = 0 \quad (1.2.2.5)$$

where $a_0, \beta, \gamma \neq 0$, remain unchanged under the transformation

$$x = \frac{\gamma}{\beta \bar{x}}.$$

The simplest invariant of this transformation is

$$z = \beta x + \frac{\gamma}{x} = \beta \bar{x} + \frac{\gamma}{\bar{x}}. \quad (1.2.2.6)$$

By changing from x in the original $2n$th-degree equation to the new variable z, we arrive at a simpler equation of degree n.

► Example 1.11. The equation

$$(x^2 + a)^n + bx^{n-m}(x^2 + a)^m + cx^n = 0 \qquad (n > m)$$

is a special case of the reciprocal equation (1.2.2.5). Using substitution (1.2.2.6) with $\beta = 1$ and $\gamma = a$, we can reduce it to a simpler equation: $z^n + bz^m + c = 0$. ◄

1.2.3. Symmetric Systems of Algebraic Equations

A bivariate polynomial $P_s(x, y)$ is called *symmetric* if it does not change when its arguments are interchanged: $P_s(x, y) = P_s(y, x)$.

Remark 1.3. *In the language of transformations, a symmetric polynomial is a polynomial that remains unchanged under the transformation* $x = \bar{y}$, $y = \bar{x}$.

The simplest symmetric polynomials

$$\sigma_1 = x + y, \quad \sigma_2 = xy \quad (1.2.3.1)$$

are called *elementary*. These are the simplest algebraic invariants that remain unchanged under the permutation of their arguments.

Let us use the notation

$$s_n = x^n + y^n, \quad (1.2.3.2)$$

where n is a positive integer.

The recurrence formula

$$s_k = \sigma_1 s_{k-1} - \sigma_2 s_{k-2}$$

holds true. It makes it possible to express sums of powers of the form (1.2.3.2) in terms of elementary symmetric polynomials sequentially. The *Waring formula* (Dickson, 1922):

$$\frac{1}{k}s_k = \frac{1}{k}\sigma_1^k - \frac{(k-2)!}{1!\,(k-2)!}\sigma_1^{k-2}\sigma_2 + \frac{(k-3)!}{2!\,(k-4)!}\sigma_1^{k-4}\sigma_2^2 - \frac{(k-4)!}{3!\,(k-6)!}\sigma_1^{k-6}\sigma_2^3 + \cdots$$

is more convenient to use. It allows one to express sums of powers immediately in an explicit form.

Proposition 5 *(fundamental theorem on symmetric polynomials).* Any symmetric bivariate polynomial can be uniquely expressed via elementary polynomials (1.2.3.1).

A proof of this proposition and historical background can be found in the works of Boltyanskii & Vilenkin (2002) and Blum-Smith & Coskey (2017).

To solve systems of algebraic equations

$$P_s(x,y)=0, \qquad Q_s(x,y)=0, \tag{1.2.3.3}$$

where P_s and Q_s are symmetric polynomials, it is convenient to use the elementary symmetric polynomials (1.2.3.1) as new variables. This approach simplifies the original system. It is noteworthy that systems of this kind possess the following property: if the system has a solution $x=x_0$, $y=y_0$, then $x=y_0$, $y=x_0$ is also a solution of this system.

If σ_1 and σ_2 solve the transformed system, then the unknowns x and y are determined from the simple system (1.2.3.1), whose solution is reduced to solving the quadratic equation

$$x^2-\sigma_1 x+\sigma_2=0. \tag{1.2.3.4}$$

Systems of equations of the form (1.2.3.3) and related algebraic systems with three or more unknowns—which remain unchanged under any permutation of their variables—will be referred to as *classical symmetric systems of algebraic equations* (or simply *symmetric systems of algebraic equations*).

For clarity, Figure 1.2 presents a principal scheme illustrating the solution of symmetric systems of algebraic equations by using invariants (1.2.3.1) as new variables.

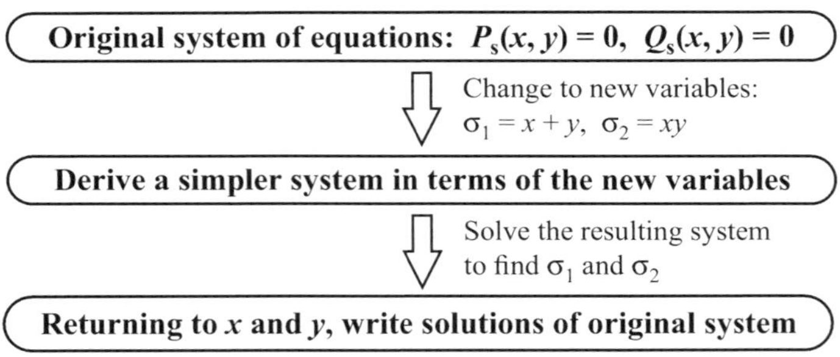

Figure 1.2. Solution scheme for classical symmetric systems of algebraic equations.

► Example 1.12. Let us look at the nonlinear system of algebraic equations

$$\begin{aligned} x^2+axy+y^2&=b,\\ x^4+cx^2y^2+y^4&=d. \end{aligned} \tag{1.2.3.5}$$

It remains unchanged under the permutation of its arguments.

By changing from x and y in system (1.2.3.5) to the new variables (1.2.3.1) and considering the formulas

$$x^2+y^2=(x+y)^2-2xy=\sigma_1^2-2\sigma_2,$$
$$x^4+y^4=(x^2+y^2)^2-2x^2y^2=(\sigma_1^2-2\sigma_2)^2-2\sigma_2^2=\sigma_1^4-4\sigma_1^2\sigma_2+2\sigma_2^2,$$

we obtain

$$\sigma_1^2+(a-2)\sigma_2=b,\qquad \sigma_1^4-4\sigma_1^2\sigma_2+(c+2)\sigma_2^2=d. \tag{1.2.3.6}$$

Eliminating σ_1 from this system yields a quadratic equation

$$(a^2+c-2)\sigma_2^2-2ab\sigma_2+b^2-d=0.$$

The subsequent solution is simple and omitted here. ◄

2°. Sometimes, algebraic equations can possess hidden symmetries that are not immediately apparent. By introducing an additional variable, these equations can be transformed into symmetric systems of equations. Let us illustrate this with a specific example.

► Example 1.13. Consider the sixth-degree algebraic equation

$$(a-x^2)^3=(b-x^3)^2, \tag{1.2.3.7}$$

which involves two free parameters, a and b. By expanding the brackets, we can rewrite this equation as

$$2x^6-3ax^4-2bx^3+3a^2x^2+b^2-a^3=0. \tag{1.2.3.8}$$

Equation (1.2.3.7) and its equivalent (1.2.3.8) do not have apparent symmetries. We will demonstrate that solutions to equation (1.2.3.7) can be expressed in terms of solutions of the symmetric algebraic system

$$x^2+y^2=a,\qquad x^3+y^3=b. \tag{1.2.3.9}$$

Indeed, the direct elimination of the additional variable y from system (1.2.3.9) leads to equation (1.2.3.7).

Changing to the variables (1.2.3.1) in (1.2.3.9) and considering that $x^3+y^3=(x+y)^3-3xy(x+y)$, we obtain a simpler system:

$$\sigma_1^2-2\sigma_2=a,\qquad \sigma_1^3-3\sigma_1\sigma_2=b.$$

Eliminating σ_2 yields the cubic equation

$$\sigma_1^3-3a\sigma_1+2b=0. \tag{1.2.3.10}$$

Since the roots of any cubic equation can be expressed in radicals (see, for example, Korn & Korn, 2000, Bronshtein & Semendyayev, 2015, and Polyanin, 2024), it is thus constructively proven that all roots of the two-parameter sixth-degree equation (1.2.3.7) can also be expressed in radicals. ◄

3°. In some cases, a bivariate system of two algebraic equations consisting of nonsymmetric equations can be reduced to a symmetric system by introducing new (auxiliary) unknowns using scaling transformation. Let us illustrate this with a specific example.

▶ Example 1.14. Let us look at the nonsymmetric system of algebraic equations

$$\begin{aligned} a_1x + b_1y &= c_1, \\ a_2x^n + b_2y^n &= c_2, \end{aligned} \tag{1.2.3.11}$$

where a_1, b_1, a_2, and b_2 are all nonzero. Using scaling transformation

$$x = \mu X, \qquad y = \lambda Y,$$

where μ and λ are parameters to be determined, we rewrite system (1.2.3.11) in the form

$$\begin{aligned} a_1\mu X + b_1\lambda Y &= c_1, \\ a_2\mu^n X^n + b_2\lambda^n Y^n &= c_2. \end{aligned} \tag{1.2.3.12}$$

For this system to be symmetric, we must set

$$a_1\mu = b_1\lambda, \qquad a_2\mu^n = b_2\lambda^n. \tag{1.2.3.13}$$

Eliminating μ and λ from equations (1.2.3.13) gives a single condition,

$$a_2b_1^n = b_2a_1^n, \tag{1.2.3.14}$$

which constrains four parameters of the original system. The transformed system (1.2.3.12) will be symmetric if condition (1.2.3.14) is satisfied. The resulting consistent solution of equations (1.2.3.13) is

$$\mu = kb_1, \qquad \lambda = ka_1, \tag{1.2.3.15}$$

where k is an arbitrary constant. Substituting (1.2.3.15) into (1.2.3.12) with $k = 1$ and taking into account (1.2.3.14), we arrive at the symmetric system

$$\begin{aligned} a_1b_1(X + Y) &= c_1, \\ a_2b_1^n(X^n + Y^n) &= c_2. \end{aligned}$$

◀

Remark 1.4. *The book by Boltyanskii & Vilenkin (2002) discusses systems of algebraic equations in three or more unknowns that preserve their form under permutations of the variables, along with methods for solving these equations. It also demonstrates how symmetric polynomials can be utilized to prove inequalities and solve certain irrational equations. For tables containing exact solutions to various algebraic equations, refer to the book by Polyanin (2024).*

1.2.4. Systems of Algebraic Equations Reducible to Symmetric Systems

1°. A bivariate polynomial $Q_{\mathrm{a}}(x, y)$ is called *antisymmetric* if it changes sign under permutation of the arguments, meaning that $Q_{\mathrm{a}}(x, y) = -Q_{\mathrm{a}}(y, x)$.

Here are two main properties of antisymmetric polynomials (Boltyanskii & Vilenkin, 2002):

$$\begin{aligned} &1. \quad Q_{\rm a}(x,x) = 0, \\ &2. \quad Q_{\rm a}(x,y) = (x-y)R_{\rm s}(x,y), \end{aligned} \qquad (1.2.4.1)$$

where $R_{\rm s}(x,y)$ is a symmetric polynomial.

2°. Consider the mixed algebraic system

$$\begin{aligned} P_{\rm s}(x,y) &= 0, \\ Q_{\rm a}(x,y) &= 0, \end{aligned} \qquad (1.2.4.2)$$

where the polynomial $P_{\rm s}(x,y)$ is symmetric, and the polynomial $Q_{\rm a}(x,y)$ is antisymmetric.

By virtue of the second property in (1.2.4.1), system (1.2.4.2) can be represented as

$$\begin{aligned} P_{\rm s}(x,y) &= 0, \\ (x-y)R_{\rm s}(x,y) &= 0. \end{aligned} \qquad (1.2.4.3)$$

The second equation in (1.2.4.3) can be satisfied by equating either factor on the left-hand side with zero. As a result, system (1.2.4.2) splits into two simpler subsystems

$$\begin{aligned} P_{\rm s}(x,x) &= 0, \quad y = x; \\ P_{\rm s}(x,y) &= 0, \quad R_{\rm s}(x,y) = 0. \end{aligned} \qquad (1.2.4.4)$$

The second subsystem in (1.2.4.4) is symmetric and so can be solved by the method described in Section 1.2.3.

► Example 1.15. Let us look at the system

$$\begin{aligned} x^2 + y^2 &= a, \\ bx^3 + cy &= by^3 + cx. \end{aligned} \qquad (1.2.4.5)$$

By moving all terms in (1.2.4.5) to the left-hand sides, we arrive at a special case of system (1.2.4.2) with

$$P_{\rm s}(x,y) = x^2 + y^2 - a, \qquad Q_{\rm a}(x,y) = b(x^3 - y^3) + c(y - x).$$

The second equation in (1.2.4.5) can be represented as

$$(x-y)(x^2 + xy + y^2 - c) = 0.$$

Hence, the original system (1.2.4.5) splits into two simple subsystems

$$\begin{aligned} 2x^2 - a &= 0, \quad y = x; \\ x^2 + y^2 - a &= 0, \quad x^2 + xy + y^2 - c = 0. \end{aligned}$$

Their solutions are easy to find. ◄

3°. Now let us look at a special system of algebraic equations:

$$P(x,y)+Q_{\mathrm{s}}(x,y)=0,\qquad P(y,x)+Q_{\mathrm{s}}(x,y)=0, \tag{1.2.4.6}$$

where $P(x,y)$ is a given bivariate polynomial ,and $Q_{\mathrm{s}}(x,y)$ is a symmetric polynomial. When the arguments are interchanged, the equations in (1.2.4.6) swap. We will refer to such systems as *nonclassical symmetric systems*.

Solving the system of equations (1.2.4.6) can be reduced to solving a simpler symmetric system and a single independent equation (Polyanin & Shingareva, 2024). To this end, we add the two equations together to obtain

$$P(x,y)+P(y,x)+2Q_{\mathrm{s}}(x,y)=0. \tag{1.2.4.7}$$

This equation remains the same when the arguments are interchanged. Then, by subtracting the second equation of (1.2.4.6) from the first one, we get

$$P(x,y)-P(y,x)=0. \tag{1.2.4.8}$$

It is easy to verify that the left-hand side of equation (1.2.4.8) is an antisymmetric polynomial. By virtue of the second property in (1.2.4.1), we can write

$$P(x,y)-P(y,x)=(x-y)R_{\mathrm{s}}(x,y), \tag{1.2.4.9}$$

where $R_{\mathrm{s}}(x,y)$ is a symmetric bivariate polynomial. Therefore, equation (1.2.4.8) can be rewritten as

$$(x-y)R_{\mathrm{s}}(x,y)=0. \tag{1.2.4.10}$$

It follows that the original system (1.2.4.6) is equivalent to the system of two equations (1.2.4.7) and (1.2.4.10). Equation (1.2.4.10) can be satisfied by equating either factor on the left-hand side with zero. As a result, the system of equations (1.2.4.7) and (1.2.4.10) splits into two simpler subsystems

$$\begin{aligned} &P(x,y)+P(y,x)+2Q_{\mathrm{s}}(x,y)=0, \quad y=x;\\ &P(x,y)+P(y,x)+2Q_{\mathrm{s}}(x,y)=0, \quad R_{\mathrm{s}}(x,y)=0. \end{aligned} \tag{1.2.4.11}$$

The first subsystem in (1.2.4.11) is simplified with the substitution $y=x$ and can be rewritten as

$$P(x,x)+Q_{\mathrm{s}}(x,x)=0, \qquad y=x.$$

The second subsystem in (1.2.4.11) is symmetric and can be solved using the method described in Section 1.2.3.

For clarity, Figure 1.3 presents a principal scheme for solving nonclassical symmetric systems of algebraic equations.

► Example 1.16. Consider the algebraic system

$$\begin{aligned} x&=ax^2+by^2+c,\\ y&=ay^2+bx^2+c, \end{aligned} \tag{1.2.4.12}$$

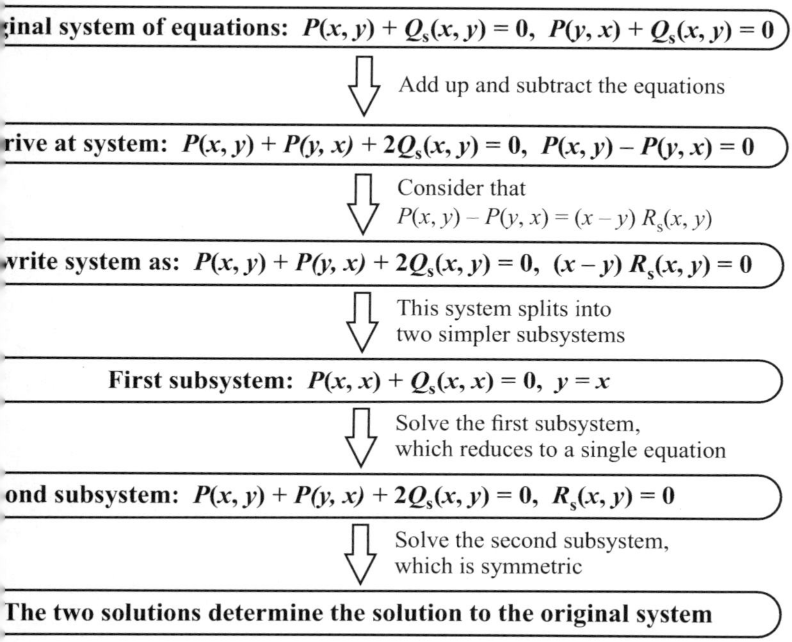

gure 1.3. A scheme for solving nonclassical symmetric algebraic equations.

$\neq \pm b$. By moving all terms to the right-hand side, we get a special case of 1.2.4.6) with

$$P(x,y) = ax^2 - x + by^2 + c, \qquad Q_s(x,y) \equiv 0.$$

dding and subtracting the equations of system (1.2.4.12) termwise, we arrive uivalent system of equations

$$\begin{aligned}(a+b)(x^2+y^2) - (x+y) + 2c &= 0,\\ (x-y)[(a-b)(x+y)-1] &= 0.\end{aligned}$$

nto two simple subsystems

$$\begin{aligned}(a+b)x^2 - x + c = 0, &\qquad y = x;\\ (a+b)(x^2+y^2) - (x+y) = 0, &\qquad (a-b)(x+y) - 1 = 0.\end{aligned}$$

ution is straightforward and can be omitted. ◂

Algebraic equations of the form

$$f(f(x)) = x, \tag{1.2.4.13}$$

eft-hand side representing the second iteration of the given polynomial $f(x)$, ible to systems (1.2.4.6) with $Q_s(x,y) \equiv 0$ (Polyanin & Shingareva, 2024).

The roots of the simpler equation $f(x) = x$ are also roots of the original equation (1.2.4.13). This property is also valid for more complex, transcendental equations, when $f(x)$ is any given continuous function.

Introducing the new variable $y = f(x)$ allows us to represent equation (1.2.4.13) as the equivalent algebraic system

$$y = f(x), \qquad x = f(y), \tag{1.2.4.14}$$

which is a special case of system (1.2.4.6). After moving all terms of the equations in (1.2.4.14) to the left-hand side, we can treat the system using the method outlined in Item 3°.

► Example 1.17. Let us explore the quartic equation

$$(x^2 + a)^2 + a = x, \tag{1.2.4.15}$$

which involves a free parameter a and can be represented as equation (1.2.4.13) with $f(x) = x^2 + a$.

With the introduction of the new variable $y = x^2 + a$, equation (1.2.4.15) can be reduced to a system of the form (1.2.4.6):

$$y = x^2 + a, \qquad x = y^2 + a. \tag{1.2.4.16}$$

By adding up and subtracting the equations in (1.2.4.16) as described in Item 3° followed by straightforward rearrangements and factorization, we obtain

$$x^2 + y^2 - x - y + 2a = 0, \qquad (y - x)(x + y + 1) = 0. \tag{1.2.4.17}$$

Equating the factors in the second equation with zero, we arrive at two independent quadratic equations:

$$\begin{aligned} x^2 - x + a = 0, &\qquad y = x; \\ x^2 + x + a + 1 = 0, &\qquad y = -x - 1. \end{aligned}$$

which determine four roots x of the original equation (1.2.4.15). ◄

► Example 1.18. Let us consider a more complicated ninth-degree algebraic equation of the form (1.2.4.13) with $f(x) = x^3 + a$:

$$(x^3 + a)^3 + a = x. \tag{1.2.4.18}$$

It contains a free parameter a and can be reduced to the nonclassical symmetric system

$$y = x^3 + a, \qquad x = y^3 + a. \tag{1.2.4.19}$$

By adding up and subtracting the equations in (1.2.4.16) as described in Item 3° followed by straightforward rearrangements and factorization, we obtain the equivalent system of equations

$$x^3 + y^3 - x - y + 2a = 0, \qquad (y - x)(x^2 + xy + y^2 + 1) = 0, \tag{1.2.4.20}$$

which splits into two simpler subsystems

$$\begin{aligned} x^3 - x + a = 0, \quad & y = x; \\ x^3 + y^3 - x - y + 2a = 0, \quad & x^2 + xy + y^2 + 1 = 0. \end{aligned} \tag{1.2.4.21}$$

The solutions to the first subsystem (1.2.4.21) can be found by determining the roots of a cubic equation, which leads us to identify three roots of the original equation (1.2.4.18). To solve the second subsystem (1.2.4.21), which is symmetric and defines the remaining six roots of the equation (1.2.4.18), we introduce new variables based on the formulas (1.2.3.1). This approach allows us to obtain a simpler system:

$$\sigma_1^3 - 3\sigma_1\sigma_2 - \sigma_1 + 2a = 0, \quad \sigma_1^2 - \sigma_2 + 1 = 0. \tag{1.2.4.22}$$

Eliminating σ_2 yields a cubic equation for σ_1:

$$\sigma_1^3 + 2\sigma_1 - a = 0. \tag{1.2.4.23}$$

The roots of this equation and the relationship $\sigma_2 = \sigma_1^2 + 1$, which derives from the second equation of (1.2.4.22), yield three pairs of real or complex numbers σ_{1k} and σ_{2k} ($k = 1, 2, 3$). By substituting these values into the quadratic equation (1.2.3.4), we can determine the six roots of the original equation (1.2.4.18).

Since the roots of any cubic equation can be expressed in radicals, this fact constructively proves that all roots of the one-parameter equation of the ninth degree (1.2.4.18) can be expressed in radicals.

It should be reminded that a general algebraic equation of degree greater than four, with arbitrary coefficients, cannot be solved in radicals (Ruffini–Abel theorem).

Remark 1.5. *If we are only interested in real roots of the algebraic equation (1.2.4.18), the second system in (1.2.4.21) should be discarded. This is because the inequality* $x^2 + xy + y^2 + 1 = (x + \frac{1}{2}y)^2 + \frac{3}{4}y^2 + 1 > 0$ *holds in this case.* ◄

5°. Let us demonstrate that algebraic equations of the form

$$f(af(x) + x + ab) + f(x) + 2b = 0, \tag{1.2.4.24}$$

where $f(x)$ is a given polynomial, and a and b are free parameters, possess a hidden symmetry and are reducible to systems (1.2.4.6) with $Q_s(x, y) \equiv 0$. It is noteworthy that the roots of the simpler equation $f(x) = -b$ are also roots of the original equation (1.2.4.24). This property also holds for more complicated, transcendental equations when $f(x)$ is any given continuous function.

We put

$$y = af(x) + x + ab \tag{1.2.4.25}$$

and rewrite equation (1.2.4.24) as

$$f(y) + f(x) + 2b = 0. \tag{1.2.4.26}$$

Eliminating $f(x)$ from (1.2.4.25) and (1.2.4.26) yields

$$x = af(y) + y + ab. \tag{1.2.4.27}$$

Equations (1.2.4.25) and (1.2.4.27) make up a system of the form (1.2.4.6):

$$y = af(x) + x + ab, \quad x = af(y) + y + ab. \tag{1.2.4.28}$$

► Example 1.19. Consider the one-parameter quartic equation

$$(x^2 + x + b)^2 + x^2 + 2b = 0, \qquad (1.2.4.29)$$

which is a special case of equation (1.2.4.24), with $f(x) = x^2$ and $a = 1$. It reduces to the nonclassical symmetric system

$$y = x^2 + x + b, \quad x = y^2 + y + b. \qquad (1.2.4.30)$$

Adding up and subtracting the equations (1.2.4.30), we obtain an equivalent system of equations, which can be represented as

$$x^2 + y^2 + 2b = 0, \quad (y - x)(x + y + 2) = 0. \qquad (1.2.4.31)$$

Equating the factors in the second equation with zero followed by simple rearrangements, we arrive at two independent quadratic equations for x:

$$x^2 + b = 0, \quad y = x;$$
$$x^2 + 2x + b + 2 = 0, \quad y = -x - 2.$$

The roots of these equations are

$$x_1 = -\sqrt{-b}, \quad x_2 = \sqrt{-b}, \quad x_3 = -1 - \sqrt{-b-1}, \quad x_4 = -1 + \sqrt{-b-1},$$

which determine the desired solutions to the original equation (1.2.4.31). ◄

► Example 1.20. It can be shown that the one-parameter ninth-degree algebraic equation

$$(x^3 + x + b)^3 + x^3 + 2b = 0,$$

which is a special case of equation (1.2.4.24) with $f(x) = x^3$ and $a = 1$, is solvable in radicals. ◄

Remark 1.6. *The more complicated algebraic system that derives from equation (1.2.4.6) through the replacement of the symmetric polynomial $Q_s(x, y)$ with an arbitrary polynomial $Q(x, y)$ can be simplified by subtracting one equation from the other. This process leads to an equation of the form (1.2.4.10), which should then be analyzed in conjunction with one of the equations from the original system.*

6°. An important property of system (1.2.4.6) is that its equations coincide when $y = x$. Let us explore a more general algebraic system comprising two equations

$$P(x, y) = 0, \quad Q(x, y) = 0, \qquad (1.2.4.32)$$

assuming that for some values of the constants λ and μ, the bivariate polynomials $P(x, y)$ and $Q(x, y)$ satisfy the condition

$$P(x, \lambda x) \equiv \mu Q(x, \lambda x). \qquad (1.2.4.33)$$

From (1.2.4.33) it follows that

$$P(x, y) - \mu Q(x, y) = (y - \lambda x)R(x, y), \qquad (1.2.4.34)$$

where $R(x, y)$ is some polynomial.

Instead of system (1.2.4.32), we consider an equivalent system by retaining the first equation and replacing the second equation with the linear combination $P(x,y) - \mu Q(x,y) = 0$. By taking into account the representation (1.2.4.34), we obtain the system

$$P(x,y) = 0, \quad (y - kx)R(x,y) = 0, \tag{1.2.4.35}$$

which splits into two simpler subsystems

$$\begin{aligned} P(x,y) &= 0, \quad y = kx; \\ P(x,y) &= 0, \quad R(x,y) = 0. \end{aligned} \tag{1.2.4.36}$$

► Example 1.21. Let us look at the algebraic system

$$ax^2 + by^2 + x + c = 0, \quad (a+b)x^2 - y + c = 0, \tag{1.2.4.37}$$

which is a special case of system (1.2.4.32). It satisfies condition (1.2.4.33) with $\lambda = -1$ and $\mu = 1$.

By retaining the first equation and replacing the second one with the equation difference, we obtain an equivalent system that can be represented as

$$ax^2 + by^2 + x + c = 0, \quad (x+y)[b(y-x)+1] = 0.$$

This system splits into two simple subsystems

$$\begin{aligned} ax^2 + by^2 + x + c &= 0, \quad y = -x; \\ ax^2 + by^2 + x + c &= 0, \quad b(y-x)+1 = 0, \end{aligned}$$

each reducible to a quadratic equation for x after eliminating y. ◄

7°. Irrational equations of the form

$$P_s\big(\sqrt[n]{a-x}, \sqrt[n]{b+x}\big) = 0, \tag{1.2.4.38}$$

where $P_s(y,z)$ is a symmetric polynomial, can be transformed by introducing the new variables

$$y = \sqrt[n]{a-x}, \quad z = \sqrt[n]{b+x} \tag{1.2.4.39}$$

into a symmetric system of algebraic equations:

$$\begin{aligned} P_s(y,z) &= 0, \\ y^n + z^n &= a+b. \end{aligned}$$

Remark 1.7. *If n is an even number, each radical in (1.2.4.38)–(1.2.4.39) can also have a minus sign in front of it. Let us consider a specific example to illustrate the aforesaid.*

► Example 1.22. The irrational equations

$$\sqrt[4]{a-x} \pm \sqrt[4]{b+x} = c$$

can be reduced through the introduction of the new variables

$$y = \sqrt[4]{a-x}, \quad z = \pm\sqrt[4]{b+x}$$

to a symmetric system of algebraic equations:

$$y + z = c,$$
$$y^4 + z^4 = a + b.$$

This system is solved using the standard method outlined in Section 1.2.3. ◀

Concluding remarks. The method outlined above, utilizing invariants, allows for the solution or simplification (lower the degree) of certain types of algebraic equations and symmetric systems of algebraic equations.

Problems and Exercises for Section 1.2

1. Find the general form of a fourth-degree algebraic equation that remains unchanged under the transformation $x = -1/\bar{x}$. Specify a change of variable that would reduce the order of the resulting equation.

2. Reduce the fifth-degree algebraic equation $x^5 + ax^4 + bx^3 + bx^2 + ax + x = 0$ to a quadratic equation. Find the roots of the original equation for $a = b = -1/2$.

3. Solve the system of equations

$$x + y = a, \quad x^3 + y^3 = b(x^2 + y^2).$$

4. Solve the system of equations

$$x^3 + y^3 = a, \quad x^2y + xy^2 = b.$$

5. Solve the system of equations

$$x + y = a, \quad x^4 + y^4 = b.$$

6. Solve the system of equations

$$x + y = a, \quad x^5 + y^5 = b.$$

7. Solve the system of equations

$$x + y = a, \quad x^7 + y^7 = a^7.$$

8. Solve the system of equations

$$x = ax^2 + by^2 + c(x+y)^3, \quad y = ay^2 + bx^2 + c(x+y)^3.$$

9. Solve the system of equations

$$x = ax^2 + by^3, \quad y = ay^2 + bx^3.$$

10. Solve the irrational equation

$$\sqrt[3]{a - x} + \sqrt[3]{b + x} = c.$$

11. For the system of equations

$$a_1x^3 + b_1y^3 = c_1, \quad a_2x^2y + b_2xy^2 = c_2,$$

use scaling transformation and find a condition for the coefficients that would allow for the reduction of the system to a symmetric form.

12. For the system of equations

$$a_1 x^m + b_1 y^m = c_1, \quad a_2 x^n + b_2 y^n = c_2,$$

use scaling transformation and find a condition for the coefficients that would allow for the reduction of the system to a symmetric form.

1.3. Ordinary Differential Equations

1.3.1. Transformations Preserving the Form of Equations. Transformation Invariants

An ordinary differential equation (ODE)

$$F(x, y, y'_x, \dots, y_x^{(n)}) = 0 \tag{1.3.1.1}$$

is said to be *invariant under an (invertible) transformation*

$$x = \varphi(\bar{x}, \bar{y}), \quad y = \psi(\bar{x}, \bar{y}) \tag{1.3.1.2}$$

if the equation acquires the same form

$$F(\bar{x}, \bar{y}, \bar{y}'_{\bar{x}}, \dots, \bar{y}_{\bar{x}}^{(n)}) = 0 \tag{1.3.1.3}$$

after substituting (1.3.1.2) into (1.3.1.1). The function F in equations (1.3.1.1) and (1.3.1.3) is the same.

Transformations that preserve the form of an equation allow one to 'multiply' its solutions. Indeed, suppose

$$y = g(x) \tag{1.3.1.4}$$

is a known solution of equation (1.3.1.1). Since the equation remains the same after changing to the new variables (1.3.1.2), the transformed equation (1.3.1.3) has a solution

$$\bar{y} = g(\bar{x}). \tag{1.3.1.5}$$

Using formulas (1.3.1.2) to return to the old variables (the formulas must first be solved for $\bar{x}$ and $\bar{y}$), we arrive at a solution of equation (1.3.1.1) that will, in general, be different to the original solution (1.3.1.4).

► Example 1.23. The third-order differential equation

$$y'''_{xxx} - y'_x = 0 \tag{1.3.1.6}$$

has a particular solution

$$y = e^x.$$

The transformation

$$x = \bar{x} + a, \quad y = \bar{y} + b \tag{1.3.1.7}$$

preserves the form of equation (1.3.1.6). Therefore, the transformed equation $\bar{y}'''_{\bar{x}\bar{x}\bar{x}} - \bar{y}'_{\bar{x}} = 0$ has a solution $\bar{y} = e^{\bar{x}}$. Inserting the old variables, obtained by inverting formulas (1.3.1.7), we arrive at a new solution of equation (1.3.1.6):

$$y = Ae^{x} + b, \qquad A = e^{-a}.$$

This solution involves two arbitrary constants, A and b. ◄

A function $I(x, y, y'_x)$, other than a constant, is called an *invariant of transformation* (1.3.1.2) if the function preserves its form under this transformation

$$I(x, y, y'_x) = I(\bar{x}, \bar{y}, \bar{y}'_{\bar{x}}).$$

Remark 1.8. *If $I = I(x, y, y'_x)$ is an invariant of transformation (1.3.1.2), then $\Psi(I)$, where Ψ is an arbitrary function, is also an invariant of this transformation.*

1.3.2. Order Reduction for ODEs with $n \geq 2$ (Reduction to a Solvable Form for $n = 1$)

Section 1.1.2 outlines a general approach to analyzing mathematical equations using invariants. Let us describe this approach as applied to ordinary differential equations.

First, we look for a *one-parameter transformation*

$$x = \varphi(\bar{x}, \bar{y}; c), \qquad y = \psi(\bar{x}, \bar{y}; c), \tag{1.3.2.1}$$

that would preserve the form of equation (1.3.1.1). Transformation (1.3.2.1) must depend on *one free parameter*, $c \in [c_1, c_2]$. The original equation (1.3.1.1) is independent of this parameter.

In the second step, for second- or higher-order equations ($n \geq 2$), we construct two functionally independent invariants of transformation (1.3.2.1):

$$I_1 = I_1(x, y), \qquad I_2 = I_2(x, y, y'_x). \tag{1.3.2.2}$$

These are explicitly independent of c.

In the third step, we choose the invariants (1.3.2.2) to be the new variables for equation (1.3.1.1). To this end, we use the transformation

$$u = I_2, \quad z = I_1, \quad u = u(z).$$

This procedure reduces the order of the equation by one.

For first-order equations ($n = 1$), the third step suggests making the change of variable

$$z = I_1, \quad z = z(x),$$

in (1.3.1.1). The resulting equation becomes solvable.

1.3.3. Transformations Employed. Determining Invariants

Here and henceforth, following Polyanin (2008) and Polyanin & Zaitsev (2012 and 2018), we will only use the simplest transformations of the form

$$x = \bar{x} + A, \qquad y = \bar{y} + B \qquad \textit{translation transformation};$$
$$x = A\bar{x}, \qquad y = B\bar{y} \qquad \textit{scaling transformation}$$

or their combinations

$$x = A_1\bar{x} + B_1, \quad y = A_2\bar{y} + B_2. \tag{1.3.3.1}$$

In the latter case, the derivatives are connected linearly to each other:

$$y'_x = \frac{A_2}{A_1}\bar{y}'_{\bar{x}}, \quad y''_{xx} = \frac{A_2}{A_1^2}\bar{y}''_{\bar{x}\bar{x}}, \quad y^{(n)}_x = \frac{A_2}{A_1^n}\bar{y}^{(n)}_{\bar{x}}. \tag{1.3.3.2}$$

The coefficients A_1, A_2, B_1, and B_2 of transformation (1.3.3.1) must depend on the free parameter c and are determined from the invariance condition of the equation.

The following proposition holds true. Suppose transformation (1.3.3.1) preserves the form of an equation that has a particular solution (1.3.1.4). Then

$$y = B_2 + A_2 g\left(\frac{x - B_1}{A_1}\right)$$

is also a solution of the equation.

The first invariant I_1 results from eliminating the parameter c from relations (1.3.3.1). The second invariant I_2 is obtained by eliminating the parameter c from one of the relations of (1.3.3.1) and the first relation in (1.3.3.2).

1.3.4. General Scheme of Applying Invariants. Analysis of Specific Ordinary Differential Equations

For clarity, Figure 1.4 displays a general scheme of applying invariants to simplify second-order ordinary differential equations.

Below are several examples of employing elementary invariant theory to simplify and integrate specific ordinary differential equations.

► Example 1.24. Let us look at a second-order ODE explicitly independent of y:

$$y''_{xx} = F(x, y'_x). \tag{1.3.4.1}$$

This equation remains unchanged under an arbitrary translation in the dependent variable, $y \Longrightarrow y + c$, where c is a free parameter. This fact is equivalent to the transformation

$$x = \bar{x}, \quad y = \bar{y} + c \quad (y'_x = \bar{y}'_{\bar{x}}). \tag{1.3.4.2}$$

Two out of three quantities x, y, and y'_x remain unchanged:

$$x = \bar{x}, \quad y'_x = \bar{y}'_{\bar{x}}.$$

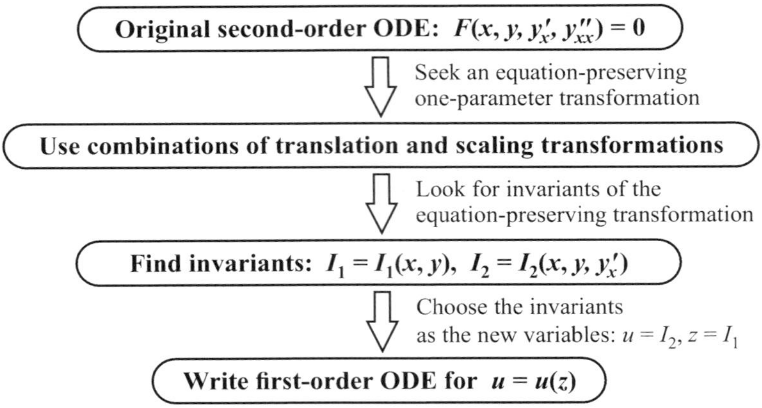

Figure 1.4. A scheme of using invariants to simplify second-order ordinary differential equations.

These are invariants of transformation (1.3.4.2): $I_1 = x$ and $I_2 = y'_x$. Let us choose them to be the new variables:

$$u = y'_x, \quad z = x, \quad u = u(z).$$

As a result, we arrive at a first-order ODE: $u'_x = F(x, u)$. ◄

► Example 1.25. Let us look at the following second-order autonomous ODE explicitly independent of x:

$$y''_{xx} = F(y, y'_x). \qquad (1.3.4.3)$$

This equation remains unchanged under an arbitrary translation in the independent variable, $x \Longrightarrow x + c$, where c is a free parameter. This fact is equivalent to the transformation

$$x = \bar{x} + c, \quad y = \bar{y} \quad (y'_x = \bar{y}'_{\bar{x}}). \qquad (1.3.4.4)$$

Two out of three quantities x, y, and y'_x remain unchanged:

$$y = \bar{y}, \quad y'_x = \bar{y}'_{\bar{x}}.$$

These are invariants of transformation (1.3.4.4): $I_1 = y$ and $I_2 = y'_x$. Let us choose them to be the new variables:

$$u = y'_x, \quad z = y, \quad u = u(z).$$

As a result, we arrive at a first-order ODE: $uu'_y = F(y, u)$. ◄

► Example 1.26. The second-order nonlinear equation

$$y''_{xx} = yF\Big(x, \frac{y'_x}{y}\Big) \qquad (1.3.4.5)$$

remains unchanged under scaling of the dependent variable, $y \Longrightarrow cy$, where $c \neq 0$ is a free parameter. This is equivalent to the transformation

$$x = \bar{x}, \quad y = c\bar{y} \quad (y'_x = c\bar{y}'_{\bar{x}}). \tag{1.3.4.6}$$

Two combinations of three quantities x, y, and y'_x remain unchanged:

$$x = \bar{x}, \quad \frac{y'_x}{y} = \frac{\bar{y}'_{\bar{x}}}{\bar{y}}.$$

These are invariants of transformation (1.3.4.6). We take them to be the new variables:

$$u = \frac{y'_x}{y}, \quad z = x, \quad u = u(z).$$

Differentiating u with respect to x yields

$$u'_x = \frac{y''_{xx}}{y} - \left(\frac{y'_x}{y}\right)^2 = \frac{y''_{xx}}{y} - u^2.$$

Using this relation to eliminate y''_{xx} from (1.3.4.5), we arrive at the first-order ODE:

$$u'_x = F(x, u) - u^2.$$

◄

Remark 1.9. *For $F(x, u) = g(x) + f(x)u$, equation (1.3.4.5) is a general second-order linear homogeneous ODE. The above transformation reduces it to a first-order ODE with a quadratic nonlinearity.*

► Example 1.27. Let us consider the second-order nonlinear ODE

$$yy''_{xx} - (y'_x)^2 = ky^3 e^{\lambda x}. \tag{1.3.4.7}$$

We look for an invariant transformation in the form

$$x = \bar{x} + b, \quad y = c\bar{y}. \tag{1.3.4.8}$$

Substituting (1.3.4.8) into (1.3.4.7) and cancelling by c, we obtain the equation

$$\bar{y}\bar{y}''_{\bar{x}\bar{x}} - (\bar{y}'_{\bar{x}})^2 = ce^{\lambda b} k\bar{y}^3 e^{\lambda \bar{x}}.$$

Requiring that it coincides with equation (1.3.4.7), we obtain an equation for the parameter b:

$$ce^{\lambda b} = 1 \quad \Longrightarrow \quad b = -\frac{1}{\lambda} \ln c. \tag{1.3.4.9}$$

The parameter c remains free ($c > 0$).

Substituting (1.3.4.9) into (1.3.4.8) and eliminating c from the second relation using the first relation, we obtain

$$y = \bar{y} e^{\lambda(\bar{x} - x)} \quad \Longrightarrow \quad ye^{\lambda x} = \bar{y} e^{\lambda \bar{x}}.$$

This gives us the first invariant of transformation (1.3.4.8)–(1.3.4.9):

$$I_1 = ye^{\lambda x}. \tag{1.3.4.10}$$

To find the second invariant, we calculate the derivative

$$y'_x = c\bar{y}'_{\bar{x}}.$$

Eliminating the parameter c with the help of the second relation of (1.3.4.8), we obtain the second invariant:

$$\frac{y'_x}{y} = \frac{\bar{y}'_{\bar{x}}}{\bar{y}} = I_2. \tag{1.3.4.11}$$

To reduce the order of the original equation, we choose the invariants (1.3.4.10) and (1.3.4.11) as the new variables:

$$z = e^{\lambda x}y, \quad u = \frac{y'_x}{y}, \quad u = u(z). \tag{1.3.4.12}$$

On the one hand

$$u'_x = \frac{y''_{xx}}{y} - \left(\frac{y'_x}{y}\right)^2 = \frac{y''_{xx}}{y} - u^2, \tag{1.3.4.13}$$

while on the other hand

$$u'_x = u_z z'_x = (\lambda e^{\lambda x}y + e^{\lambda x}y'_x)u'_z = \left(\lambda z + e^{\lambda x}y\frac{y'_x}{y}\right)u'_z = (\lambda z + zu)u'_z. \tag{1.3.4.14}$$

Equating the right-hand sides of relations (1.3.4.13) and (1.3.4.14), we get

$$\frac{y''_{xx}}{y} - u^2 = (\lambda z + zu)u'_z \quad \Longrightarrow \quad \frac{y''_{xx}}{y} = u^2 + (\lambda z + zu)u'_z. \tag{1.3.4.15}$$

Dividing equation (1.3.4.7) by y^2 and replacing the old variables with the new ones using (1.3.4.12) and (1.3.4.15), we obtain a first-order separable ODE

$$(\lambda + u)u'_z = k,$$

which is easy to integrate. ◄

Remark 1.10. *The more general second-order nonlinear ODE*

$$y''_{xx} = yF\Big(e^{\lambda x}y, \frac{y'_x}{y}\Big)$$

possesses similar properties. Transformation (1.3.4.12) reduces this equation to a first-order ODE:

$$u^2 + (\lambda z + zu)u'_z = F(z,u).$$

► Example 1.28. Now let us look at the first-order nonlinear ODE

$$y'_x = f\left(\frac{\alpha_1 x + \beta_1 y + \gamma_1}{\alpha_2 x + \beta_2 y + \gamma_2}\right). \tag{1.3.4.16}$$

Its left-hand side remains unchanged under transformations of the form

$$x = a\bar{x} + b, \quad y = a\bar{y} + c, \tag{1.3.4.17}$$

where a is an arbitrary constant, and the constants b and c will be defined below. Let us insert (1.3.4.17) into the argument of the right-hand side of equation (1.3.4.16):

$$\frac{\alpha_1 x + \beta_1 y + \gamma_1}{\alpha_2 x + \beta_2 y + \gamma_2} = \frac{a(\alpha_1 \bar{x} + \beta_1 \bar{y}) + \alpha_1 b + \beta_1 c + \gamma_1}{a(\alpha_2 \bar{x} + \beta_2 \bar{y}) + \alpha_2 b + \beta_2 c + \gamma_2}. \tag{1.3.4.18}$$

For equation (1.3.4.16) to be invariant under transformation (1.3.4.17), we must set

$$\begin{aligned} \alpha_1 b + \beta_1 c + \gamma_1 &= a\gamma_1, \\ \alpha_2 b + \beta_2 c + \gamma_2 &= a\gamma_2 \end{aligned} \tag{1.3.4.19}$$

in (1.3.4.18). Then, the numerator and denominator of the last fraction in (1.3.4.18) can be cancelled by a. Relations (1.3.4.19) can be treated as a system of two linear algebraic equations of first degree for the coefficients b and c. The coefficient a remains arbitrary.

To sum up, we have shown that equation (1.3.4.16) is invariant under transformation (1.3.4.17) subject to the constraints (1.3.4.19). Furthermore, the argument of the function f on right-hand side of equation (1.3.4.16) is the first invariant of this transformation. Therefore, we should make the change of variable

$$z = \frac{\alpha_1 x + \beta_1 y + \gamma_1}{\alpha_2 x + \beta_2 y + \gamma_2}, \quad \text{where} \quad z = z(x), \tag{1.3.4.20}$$

in equation (1.3.4.16). Let us solve (1.3.4.20) for y and then differentiate with respect to x. Now replacing y'_x with $f(z)$, which follows from (1.3.4.16) and (1.3.4.20), and performing simple rearrangements, we arrive at a first-order separable ODE:

$$\frac{(\alpha_2\beta_1 - \alpha_1\beta_2)x + \beta_1\gamma_2 - \beta_2\gamma_1}{(\beta_2 z - \beta_1)^2} z'_x = f(z) + \frac{\alpha_2 z - \alpha_1}{\beta_2 z - \beta_1}.$$

◀

Table 1.1 lists some second-order ODEs that admit order reduction through the use of the simplest invariant transformations. For first-order ODEs, when the function $F(u, v, w)$ does not include the third argument, the equations specified in Table 1.1 can be solved by changing from y to the new dependent variable $z = I_1(x, y)$, where I_1 is the first invariant.

The results presented in Table 1.1 are easy to generalize to nonlinear ODEs of any order.

► **Concluding remarks.** The described method of studying ODEs uses the ideas of group analysis of differential equations (e.g., Ovsiannikov, 1982; Ibragimov, 1991 and 1994; and Olver, 1993) but is much simpler than the latter. To apply this method, one must be able to solve only the simplest algebraic equations (and systems) and be able to differentiate. Whereas when using the group analysis method, one has to deal with partial differential equations at intermediate stages (i.e., one has to go beyond the standard course on ordinary differential equations). Another noteworthy advantage of this simple approach is that no new concepts have to be introduced here, which are abundant in the group analysis method. This method allows one to describe, in a unified way, significantly more solvable (or allowing order reduction) ordinary differential equations than presented in the overwhelming majority of existing textbooks.

Table 1.1. Some ordinary differential equations admitting order reduction with their invariant transformations and invariants.

Equation	Invariant transformation	First invariant	Second invariant
$F(x, y'_x, y''_{xx}) = 0$	$y = \bar{y} + c$	$I_1 = x$	$I_2 = y'_x$
$F(y, y'_x, y''_{xx}) = 0$	$x = \bar{x} + c$	$I_1 = y$	$I_2 = y'_x$
$F(\alpha x + \beta y + \gamma, y'_x, y''_{xx}) = 0$	$x = \bar{x} + c\beta,\ y = \bar{y} - c\alpha$	$I_1 = \alpha x + \beta y + \gamma$	$I_2 = y'_x$
$F\Big(x, \frac{y'_x}{y}, \frac{y''_{xx}}{y}\Big) = 0$	$y = c\bar{y}$	$I_1 = x$	$I_2 = \frac{y'_x}{y}$
$F(y, xy'_x, x^2 y''_{xx}) = 0$	$x = c\bar{x}$	$I_1 = y$	$I_2 = xy'_x$
$F\Big(e^{\lambda x} y, \frac{y'_x}{y}, \frac{y''_{xx}}{y}\Big) = 0$	$x = \bar{x} - \frac{1}{\lambda} \ln c,\ y = c\bar{y}$	$I_1 = e^{\lambda x} y$	$I_2 = \frac{y'_x}{y}$
$F(e^{\lambda x} y, e^{\lambda x} y'_x, e^{\lambda x} y''_{xx}) = 0$	$x = \bar{x} - \frac{1}{\lambda} \ln c,\ y = c\bar{y}$	$I_1 = e^{\lambda x} y$	$I_2 = e^{\lambda x} y'_x$
$F(xe^{\lambda y}, xy'_x, x^2 y''_{xx}) = 0$	$x = c\bar{x},\ y = \bar{y} - \frac{1}{\lambda} \ln c$	$I_1 = xe^{\lambda y}$	$I_2 = xy'_x$
$F(x^k y, x^{k+1} y'_x, x^{k+2} y''_{xx}) = 0$	$x = c\bar{x},\ y = c^{-k} \bar{y}$	$I_1 = x^k y$	$I_2 = x^{k+1} y'_x$
$F\Big(x^n y^m, \frac{xy'_x}{y}, \frac{x^2 y''_{xx}}{y}\Big) = 0$	$x = c^m \bar{x},\ y = c^{-n} \bar{y}$	$I_1 = x^n y^m$	$I_2 = \frac{xy'_x}{y}$

Problems and Exercises for Section 1.3

1. Reduce the order of equation (1.1.1.2) in three different ways by using transformation (1.1.1.3).

2. Find three transformations that preserve the form of the equation $xy''_{xx} - y'_x = 0$. Use these transformations to reduce the order of the equation in different ways.

3. Find three transformations that preserve the form of the equation $y''_{xx} - (y'_x)^2 = 0$. Use these transformations to reduce the order of the equation in different ways.

4. Solve the equation $y'_x = y(kx + \ln y)$, where $f(z)$ is an arbitrary function.

5. Solve the equation $xy'_x = f(ky + \ln x)$, where $f(z)$ is an arbitrary function.

6. Solve the generalized homogeneous equation $y'_x = ax^{k-1} \exp(yx^{-k})$.

7. Solve the generalized homogeneous equation $xy'_x = yf(x^n y^m)$.

8. Reduce the order of the Emden–Fowler equation $y''_{xx} = Ax^n y^m$.

9. Reduce the order of the equation $y''_{xx} = Ax^n e^{\lambda y}$.

10. Reduce the order of the equation $y''_{xx} = Ay^n e^{\lambda x}$.

11. Find a transformation that reduces the order of the generalized Emden–Fowler equation $y''_{xx} = Ax^n y^m (y'_x)^k$.

12. Find a transformation that reduces the order of the equation $y''_{xx} = Ay^n e^{\lambda x} (y'_x)^m$.

13. Find a transformation that reduces the order of the equation $y''_{xx} = Ax^n e^{\lambda y} (y'_x)^m$.

14. Write out generalizations of the equations presented in Table 1.1 for third-order ODEs.

15. Write out generalizations of the equations presented in Table 1.1 for arbitrary-order ODEs.

References for Chapter 1

Blum-Smith B., Coskey S. The fundamental theorem on symmetric polynomials: History first whiff of Galois theory. *College Math. Journal*, 2017, Vol. 48, No. 1, pp. 18–29.

Boltyanskii V.G., Vilenkin N.Ya. *Symmetry in Algebra, 2nd ed.* Nauka, Moscow, 2002 (in Russian).

Bronshtein I.N., Semendyayev K.A. *Handbook of Mathematics, 6th ed.* Springer, Berlin, 2015.

Dickson L.E. *First Course in the Theory of Equations.* John Wiley & Sons, New York, 1922.

Ibragimov N.H. *Experience in Group Analysis of Ordinary Differential Equations.* Znaniye, Moscow, No. 7, 1991 (in Russian).

Ibragimov N.H. (ed.). *CRC Handbook of Lie Group Analysis of Differential Equations, Vol. 1, Symmetries, Exact Solutions and Conservation Laws.* CRC Press, Boca Raton, 1994.

Korn G.A., Korn T.M. *Mathematical Handbook for Scientists and Engineers.* Dover Publ., New York, 2000.

Kudryashov N.A. Symmetry of algebraic and differential equations. *Soros Educational Journal*, 1998, No. 9, pp. 104–110 (in Russian).

Leung K.T., Mok I.A.C., Suen S.N. *Polynomials and Equations.* Hong Kong Univ. Press, Hong Kong, 1992.

Olver P.J. *Application of Lie Groups to Differential Equations.* Springer, New York, 1993.

Ovsiannikov L.V. *Group Analysis of Differential Equations.* Academic Press, New York, 1982.

Polyanin A.D. An elementary theory of using invariants to solve mathematical equations. *Vestnik Samarskogo Gosudarstvennogo Universiteta*, 2008, No. 6(65), pp. 152–176 (in Russian).

Polyanin A.D. *Handbook of Exact Solutions to Mathematical Equations.* CRC Press, Boca Raton, 2024.

Polyanin A.D., Shingareva I.K. Non-classical symmetries and reductions of algebraic equations and systems of equations. *Mathematical Education*, 2024, No. 2 (110), pp. 22–34 (in Russian).

Polyanin A.D., Zaitsev V.F. *Handbook of Nonlinear Partial Differential Equations, 2nd ed.* Chapman & Hall/CRC Press, Boca Raton, 2012.

Polyanin A.D., Zaitsev V.F. *Handbook of Ordinary Differential Equations: Exact Solutions, Methods, and Problems.* Boca Raton—London: CRC Press, 2018.

2. First-Order Partial Differential Equations

► **Preliminary remarks.** First-order partial differential equations (PDEs) are mathematical equations that contain two or more first-order partial derivatives of the unknown function.

This chapter provides a brief overview of the most common methods for integrating linear, quasilinear, and nonlinear first-order PDEs with two or more independent variables. The methods discussed rely on analyzing characteristic systems of ordinary differential equations (ODEs), using exact solutions of specific types, and applying separation of variables. The chapter also addresses the Cauchy problem for these equations and presents a procedure for solving it. Examples of solutions to specific first-order PDEs and initial value problems are also included.

Methods for solving first-order PDEs are further utilized to address certain classes of functional equations (see Section 3.2) and to construct exact solutions for second- and higher-order PDEs (see Sections 8.1.2 and 9.5.5).

2.1. Quasilinear First-Order Partial Differential Equations

2.1.1. Equations in Two Independent Variables. Characteristic ODE System. General Solution

Quasilinear first-order partial differential equations with two independent variables, x and y, are written as

$$f(x,y,u)u_x + g(x,y,u)u_y = h(x,y,u), \qquad (2.1.1.1)$$

where u is the unknown function. Such equations are common in various applications, including continuum mechanics, gas dynamics, hydrodynamics, wave theory, acoustics, filtration theory, mass and heat transfer theory, chemical engineering, and other fields.

In this chapter, we will consider only continuously differentiable solutions of equation (2.1.1.1), which in some domain $\Omega(x,y)$ turn this equation into an identity.

Equation (2.1.1.1) is *linear* if the functional coefficients f and g do not depend on u, and the right-hand side of the equation has the form $h(x,y,u) = h_1(x,y)u + h_0(x,y)$.

The general solution of equation (2.1.1.1) is constructed in several stages as described below.

1°. The quasilinear first-order partial differential equation (2.1.1.1) is associated to a *characteristic system of ordinary differential equations* (Kamke, 1965):

$$\frac{dx}{f(x,y,u)} = \frac{dy}{g(x,y,u)} = \frac{du}{h(x,y,u)}. \tag{2.1.1.2}$$

2°. Then, one finds two functionally independent integrals* of the characteristic ODE system (2.1.1.2):

$$Q_1(x,y,u) = C_1, \qquad Q_2(x,y,u) = C_2, \tag{2.1.1.3}$$

where C_1 and C_2 are arbitrary constants.

3°. One writes out the general solution of equation (2.1.1.1) in implicit form:

$$\Psi(Q_1, Q_2) = 0, \tag{2.1.1.4}$$

where Ψ is an arbitrary function of two arguments that has continuous first derivatives, and $Q_k = Q_k(x,y,u)$ $(k = 1,\ 2)$ are the functions appearing in the integrals (2.1.1.3).

Remark 2.1. *On solving (2.1.1.4) for Q_1 or Q_2, one rewrites the solution as*

$$Q_k = \Phi(Q_{3-k}), \qquad k = 1,\ 2, \tag{2.1.1.5}$$

where Φ is an arbitrary continuously differentiable function of a single argument.

Additional remarks. The characteristic ODE system is written in a *symmetric form* (2.1.1.2). It is equivalent to two ODEs

$$y'_x = \frac{g(x,y,u)}{f(x,y,u)}, \qquad u'_x = \frac{h(x,y,u)}{f(x,y,u)}. \tag{2.1.1.6}$$

Similarly, other equivalent systems consisting of two ODEs can be used: $x'_y = f/g$, $u'_y = h/g$ or $u'_x = h/f$, and $u'_y = h/g$. In practice, one chooses systems that are easier to integrate (for example, systems consisting of isolated equations).

Remark 2.2. *In the special case $h \equiv 0$, the second equation of the ODE system (2.1.1.6) has a trivial first integral*

$$u = C_2. \tag{2.1.1.7}$$

When integrating the first equation of (2.1.1.6), the function u is treated as a constant free parameter.

► Example 2.1. Consider the linear first-order partial differential equation with constant coefficients

$$u_x + a u_y = b, \tag{2.1.1.8}$$

which is a special case of equation (2.1.1.1) with $f(x,y,u) = 1$, $g(x,y,u) = a$, and $h(x,y,u) = b$.

*The Jacobian of the functions Q_1 and Q_2 is nonzero.

The characteristic ODE system is

$$\frac{dx}{1} = \frac{dy}{a} = \frac{du}{b}. \tag{2.1.1.9}$$

It is equivalent to two independent linear equations

$$y'_x = a, \qquad u'_x = b.$$

Integrating yields $y = ax + C_1$ and $u = bx + C_2$, where C_1 and C_2 are arbitrary constants. On solving these solutions for C_1 and C_2, we find the first integrals of the characteristic system (2.1.1.9):

$$y - ax = C_1, \qquad u - bx = C_2,$$

Using Remark 2.1 with $k = 2$ and moving the term bx to the right-hand side of (2.1.1.5), we obtain the general solution of equation (2.1.1.8) in the explicit form

$$u = bx + \Phi(y - ax), \tag{2.1.1.10}$$

where $\Phi(z)$ is an arbitrary function.

For $b = 0$, solution (2.1.1.10) is called a *traveling wave solution.* ◄

► Example 2.2. Consider the linear first-order partial differential equation with variable coefficients

$$u_x + axu_y + bu = 0, \tag{2.1.1.11}$$

which is a special case of equation (2.1.1.1) with $f(x, y, u) = 1$, $g(x, y, u) = ax$, and $h(x, y, u) = -bu$.

We write the characteristic ODE system:

$$\frac{dx}{1} = \frac{dy}{ax} = -\frac{du}{bu}.$$

It is equivalent to two independent linear equations

$$y'_x = ax, \qquad u'_x = -bu.$$

Integrating yields $y = \frac{1}{2}ax^2 + C_1$ and $u = C_2 e^{-bx}$. On solving these relations for C_1 and C_2, we find two first integrals of the characteristic system

$$y - \tfrac{1}{2}ax^2 = C_1, \qquad ue^{bx} = C_2.$$

Using Remark 2.1 with $k = 2$ and multiplying relation (2.1.1.5) by e^{-bx}, we obtain the general solution of equation (2.1.1.8) in the explicit form

$$u = e^{-bx}\Phi(y - \tfrac{1}{2}ax^2),$$

where $\Phi(z)$ is an arbitrary function. ◄

Remark 2.3. *The general solution of the linear first-order partial differential equation (2.1.1.1) can be represented in explicit form (solved for the unknown u).*

► Example 2.3. Now let us look at the nonlinear equation

$$u_x + auu_y = 1. \tag{2.1.1.12}$$

The associated characteristic ODE system

$$\frac{dx}{1} = \frac{dy}{au} = \frac{du}{1}$$

is equivalent to two independent equations (the second one is nonlinear):

$$u'_x = 1, \qquad u'_y = 1/(au).$$

Two independent integrals of this system are

$$u - x = C_1, \qquad au^2 - 2y = C_2.$$

The general solution of the original equation (2.1.1.12) can be written in implicit form as

$$\Psi(u - x,\, au^2 - 2y) = 0. \tag{2.1.1.13}$$

Remark 2.4. *Unlike the solutions obtained in Examples 2.1 and 2.2, solution (2.1.1.13) cannot be represented in a form solved for the unknown u. This property is typical of nonlinear first-order partial differential equations.*

◄

2.1.2. Equation in Any Number of Independent Variables. Characteristic ODE System. General Solution

A quasilinear first-order partial differential equation in n independent variables has the general form

$$f_1(x_1,\dots,x_n,u)\frac{\partial u}{\partial x_1}+\dots+f_n(x_1,\dots,x_n,u)\frac{\partial u}{\partial x_n}=g(x_1,\dots,x_n,u). \tag{2.1.2.1}$$

The general solution of equation (2.1.2.1) can be constructed as follows.

First, to equation (2.1.2.1) one associates a *characteristic system of ordinary differential equations*:

$$\frac{dx_1}{f_1(x_1,\dots,x_n,u)} = \dots = \frac{dx_n}{f_n(x_1,\dots,x_n,u)} = \frac{du}{g(x_1,\dots,x_n,u)}.$$

Then, one determines n independent integrals of this system:

$$Q_1(x_1,\dots,x_n,u) = C_1, \quad \dots, \quad Q_n(x_1,\dots,x_n,u) = C_n, \tag{2.1.2.2}$$

where C_1, ..., C_n are arbitrary constants. One writes out the general solution of equation (2.1.2.1) in implicit form:

$$\Psi(Q_1,\dots,Q_n) = 0, \tag{2.1.2.3}$$

where Ψ is an arbitrary function of n arguments, and $Q_m = Q_m(x_1,\dots,x_n,u)$ are the functions appearing in the integrals (2.1.2.2), $m = 1,\dots,n$.

► Example 2.4. Consider the first-order linear PDE with three independent variables

$$u_x + a u_y + b u_z = 0. \tag{2.1.2.4}$$

The associated characteristic ODE system is

$$\frac{dx}{1} = \frac{dy}{a} = \frac{dz}{b} = \frac{du}{0}.$$

It is equivalent to three independent linear ODEs

$$y'_x = a, \qquad z'_x = b, \qquad u'_x = 0.$$

Integrating these equations yields three first integrals of the characteristic system:

$$y - ax = C_1, \qquad z - bx = C_2, \qquad u = C_3.$$

Therefore, the general solution of the original PDE can be written in implicit form as

$$\Psi(y - ax, z - bx, u) = 0, \tag{2.1.2.5}$$

where $\Psi(\xi, \eta, u)$ is an arbitrary function of three arguments. On solving (2.1.2.5) for u, we can rewrite the general solution of equation (2.1.2.4) in explicit form as

$$u = \Phi(y - ax, z - bx),$$

where $\Phi(\xi, \eta)$ is an arbitrary function of two arguments. ◄

2.1.3. Cauchy Problem. Solution Construction. Existence and Uniqueness Theorem

► Two statements of the Cauchy problem for an equation with two independent variables.

1°. *Generalized Cauchy problem.* It is required to find a solution $u = u(x, y)$ of equation (2.1.1.1) that satisfies the initial conditions

$$x = h_1(\xi), \quad y = h_2(\xi), \quad u = h_3(\xi), \tag{2.1.3.1}$$

where ξ is a parameter ($\alpha \le \xi \le \beta$) and $h_k(\xi)$ are given functions ($k = 1, 2, 3$).

Geometric interpretation: it is required to find an integral surface of equation (2.1.1.1) that passes through the line (2.1.3.1) defined parametrically.

2°. *Classical Cauchy problem.* It is required to find a solution $u = u(x, y)$ of equation (2.1.1.1) that satisfies the initial condition

$$u = \varphi(y) \quad \text{at} \quad x = 0, \tag{2.1.3.2}$$

where $\varphi(y)$ is a given function.

Sometimes, it is helpful to represent the classical Cauchy problem as the generalized Cauchy problem by reformulating the initial condition (2.1.3.2) in parametric form:

$$x = 0, \quad y = \xi, \quad u = \varphi(\xi). \tag{2.1.3.3}$$

► Procedure for solving the Cauchy problem. Hopf equation problem.

The procedure for solving the Cauchy problem for equation (2.1.1.1) subject to the initial conditions (2.1.3.1) involves several stages. First, one identifies two independent integrals (2.1.1.3) from the characteristic system (2.1.1.2). Next, one substitutes the initial data (2.1.3.1) into the integrals (2.1.1.3) to determine the integration constants C_1 and C_2:

$$Q_1(h_1(\xi), h_2(\xi), h_3(\xi)) = C_1, \qquad Q_2(h_1(\xi), h_2(\xi), h_3(\xi)) = C_2. \tag{2.1.3.4}$$

Eliminating the constants C_1 and C_2 from (2.1.1.3) and (2.1.3.4) yields

$$\begin{aligned} Q_1(x, y, u) &= Q_1(h_1(\xi), h_2(\xi), h_3(\xi)), \\ Q_2(x, y, u) &= Q_2(h_1(\xi), h_2(\xi), h_3(\xi)). \end{aligned} \tag{2.1.3.5}$$

Formulas (2.1.3.5) represent a *parametric form of the solution* to the Cauchy problem (2.1.1.1), (2.1.3.1). In some cases, the solution can be represented in an explicit form by eliminating the parameter ξ from (2.1.3.5).

► Example 2.5. Consider the Cauchy problem for the *Hopf equation*

$$u_x + uu_y = 0 \tag{2.1.3.6}$$

subject to the initial condition

$$u = \varphi(y) \quad \text{at} \quad x = 0. \tag{2.1.3.7}$$

First, we rewrite the initial condition (2.1.3.7) in the parametric form (2.1.3.1):

$$x = 0, \quad y = \xi, \quad u = \varphi(\xi). \tag{2.1.3.8}$$

Solving the characteristic ODE system

$$\frac{dx}{1} = \frac{dy}{u} = \frac{du}{0}, \tag{2.1.3.9}$$

we identify two independent integrals

$$u = C_1, \qquad y - ux = C_2. \tag{2.1.3.10}$$

Employing the initial conditions (2.1.3.8), we determine the constant of integration: $C_1 = \varphi(\xi)$ and $C_2 = \xi$. Inserting these expressions into (2.1.3.10), we obtain a solution to the Cauchy problem (2.1.3.6), (2.1.3.7) in parametric form:

$$u = \varphi(\xi), \tag{2.1.3.11}$$
$$y = \xi + \varphi(\xi)x. \tag{2.1.3.12}$$

The characteristics (2.1.3.12) are straight lines in the plane (x, y) with slope $\varphi(\xi)$ and y-intercept ξ. The function u maintains the same value, $\varphi(\xi)$, at each characteristic; in general, the values of u are different of different characteristics.

For $\varphi'(\xi) > 0$, the distinct characteristics do not intersect, and the formulas (2.1.3.11) and (2.1.3.12) describe a single-valued solution. As an example, consider the initial profile

$$\varphi(\xi) = \begin{cases} u_1 & \text{for } \xi \le 0, \\ \dfrac{u_2\xi^2 + \varepsilon u_1}{\xi^2 + \varepsilon} & \text{for } \xi > 0, \end{cases} \tag{2.1.3.13}$$

where $u_1 < u_2$ and $\varepsilon > 0$. From formulas (2.1.3.11) and (2.1.3.12) we obtain a unique smooth solution in the entire half-plane $x > 0$. In the region filled by the characteristics $y = \xi + u_1 x$ (for $\xi \le 0$), the solution is constant:

$$u = u_1 \quad \text{for} \quad y/x \le u_1. \tag{2.1.3.14}$$

For $\xi > 0$, the solution is defined by formulas (2.1.3.11)–(2.1.3.13).

Let us examine the solution in the limit as $\varepsilon \to 0$, which corresponds to a piecewise continuous initial profile

$$\varphi(\xi) = \begin{cases} u_1 & \text{for } \xi \le 0, \\ u_2 & \text{for } \xi > 0, \end{cases} \quad \text{where} \quad u_1 < u_2. \tag{2.1.3.15}$$

In what follows, we assume that $\xi > 0$; for $\xi \le 0$, formula (2.1.3.14) holds. If $\xi = \text{const} \neq 0$ and $\varepsilon \to 0$, it follows from (2.1.3.13) that $\varphi(\xi) = u_2$. Hence, in the region filled by the characteristics $y = \xi + u_2 x$ (for $\xi > 0$), the solution is constant:

$$u = u_2 \quad \text{for} \quad y/x \ge u_2 \qquad (\text{for } \varepsilon \to 0). \tag{2.1.3.16}$$

For $\xi \to 0$, the function φ can assume any value between u_1 and u_2 depending on the relation between the two small parameters ε and ξ; the first term on the right-hand side of (2.1.3.12) can be neglected. As a result, formulas (2.1.3.11) and (2.1.3.12) give the desired asymptotic of the solution in explicit form:

$$u = y/x \quad \text{for} \quad u_1 \le y/x \le u_2 \qquad (\text{for } \varepsilon \to 0). \tag{2.1.3.17}$$

Combining formulas (2.1.3.14), (2.1.3.16), and (2.1.3.17), we obtain the following solution of the Cauchy problem for equation (2.1.3.6) subject to the initial condition (2.1.3.15):

$$u(x,y) = \begin{cases} u_1 & \text{for } y \le u_1 x, \\ y/x & \text{for } u_1 x \le y \le u_2 x, \\ u_2 & \text{for } y \ge u_2 x. \end{cases} \tag{2.1.3.18}$$

Figure 2.1 displays some characteristics of equation (2.1.3.6) under condition (2.1.3.15) as well as the u versus y for $u_1 = \frac{1}{2}$, $u_2 = 2$, and $x_0 = 1$. In applications, such a solution is known as a *centered rarefaction wave*.

Remark 2.5. *In the case where $\varphi'(\xi) < 0$, the characteristics will intersect in some region. At the intersection point of two characteristics defined by different values of the parameter, ξ_1 and ξ_2, the function u, according to the first formula (2.1.3.13), will yield two different values: $\varphi(\xi_1)$ and $\varphi(\xi_2)$. Consequently, the solution will be multivalued in the region where the characteristics intersect. This example illustrates the locality of the existence and uniqueness theorem, which will be discussed further below.*

◀

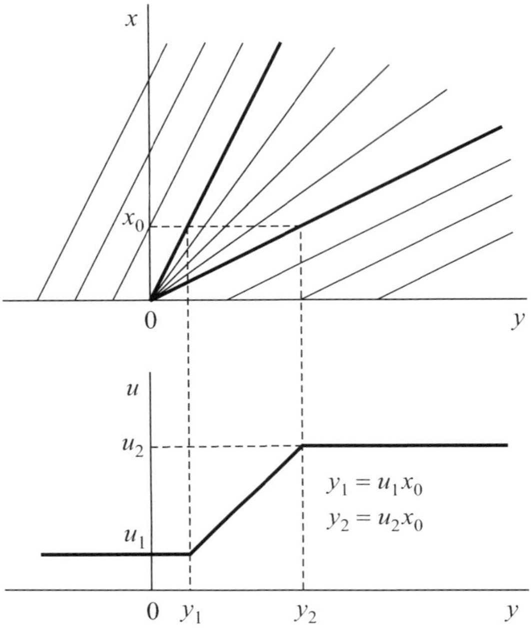

Figure 2.1. Characteristics of the Cauchy problem (2.1.3.6), (2.1.3.7) for the initial profile (2.1.3.15) and the graph of u as a function of the coordinate y, where $u_1 = \frac{1}{2}$, $u_2 = 2$, and $= 1$.

. The existence and uniqueness theorem for the Cauchy problem.

Suppose D_0 is a region of the plane (x, y) and D is a cylindrical region of the space $x, y, u)$ obtained from D_0 by adding the coordinate u, where $|u| < A_1$. Let the coefficients f, g, and h of equation (2.1.1.1) be continuously differentiable functions x, y, u in D and let $x = h_1(\xi)$, $y = h_2(\xi)$, and $u = h_3(\xi)$ be continuously differentiable functions of ξ for $|\xi| < A_2$ that define a curve C in D with a simple projection of C_0 on D_0 and satisfy the condition $(h_1')^2 + (h_2')^2 \neq 0$ (the prime stands for a derivative with respect to ξ). We assume that $fh_2' - gh_1' \neq 0$ on C. Then there exists a subdomain $\overline{D}_0$ of D_0 containing C_0 in which a continuously differentiable function $u = u(x, y)$ is defined satisfying the differential equation (2.1.1.1) in D_0 and the initial condition (2.1.3.1) on C_0. This function is defined uniquely.

It is important to note that this theorem is local in nature: the existence of a solution is only guaranteed in some sufficiently 'narrow' neighborhood, not fixed in advance, of the line C (see Remark 2.5 at the end of Example 2.5).

. Some remarks and comments.

For more information on the properties, geometric interpretation, and applications of quasilinear first-order PDEs, see Kamke (1965), Whitham (1974), Rhee et al. (1986),

Courant & Hilbert (1989), Bedrikovetsky (1993), Korn & Korn (2000), and Zauderer (2006). Extensive tables of exact solutions of these equations can be found in Kamke (1965), Polyanin et al. (2002), and Polyanin & Zaitsev (2012). Generalized solutions (nonsmooth shock waves type solutions) and jump and stability conditions, as well as viscosity solutions, are discussed in Oleinik (1957, 1959), Gelfand (1959), Rozhdestvenskii & Yanenko (1983), Rhee et al. (1986), Polyanin et al. (2002), and Polyanin & Zaitsev (2012).

Problems and Exercises for Section 2.1

1. Find the general solution of the linear equation

$$au_x + au_y = f(x).$$

2. Find the general solution of the linear equation

$$xu_x + ayu_y = b.$$

3. Find the general solution of the linear equation

$$axu_x + byu_y = cu.$$

4. Find the general solution of the linear equation

$$u_x + au_y = f(u).$$

5. Find the general solution to the model equation of gas dynamics

$$u_x + f(u)u_y = 0.$$

6. Find the general solution of the quasilinear equation

$$u_x + [f(u) + g(x)]u_y = 0.$$

7. Find the general solution of the quasilinear equation

$$[f_1(u) + g(y)]u_x + [f_2(u) + h(x)]u_y = 0.$$

8. Find the general solution of the quasilinear equation

$$u_x + f(u)u_y = g(u).$$

9. Find the general solution of the linear equation

$$axu_x + byu_y + czu_z = 0.$$

10. Find the general solution of the linear equation

$$u_x + au_y + bu_z = c.$$

2.2. First-Order Nonlinear Partial Differential Equations

2.2.1. Complete, General, and Singular Integrals

A *nonlinear first-order partial differential equation* with two independent variables has the general form

$$F(x, y, u, u_x, u_y) = 0. \tag{2.2.1.1}$$

Such equations are encountered in analytical mechanics, calculus of variations, optimal control, differential games, dynamic programming, geometric optics, differential geometry, and some other fields.

In this section, we consider only smooth solutions $u = u(x, y)$ of equation (2.2.1.1), which are continuously differentiable with respect to both arguments.

1°. Let an exact solution of equation (2.2.1.1),

$$u = \Theta(x, y, C_1, C_2), \tag{2.2.1.2}$$

depending on two free parameters C_1 and C_2, which are not included in the equation under consideration, be known. The two-parameter family of solutions (2.2.1.2) is called a *complete integral* of equation (2.2.1.1) if the rank of the matrix

$$\begin{pmatrix} \Theta_1 & \Theta_{x1} & \Theta_{y1} \\ \Theta_2 & \Theta_{x2} & \Theta_{y2} \end{pmatrix} \tag{2.2.1.3}$$

is equal to two in the domain being considered (for example, this is valid if $\Theta_{x1}\Theta_{y2} - \Theta_{x2}\Theta_{y1} \neq 0$). In matrix (2.2.1.3), Θ_n denotes the partial derivative of Θ with respect to C_n ($n = 1, 2$), Θ_{xn} is the second partial derivative with respect to x and C_n, and Θ_{yn} is the second partial derivative with respect to y and C_n.

In some cases, a complete integral can be found using the method of undetermined coefficients by presetting an appropriate structure of the particular solution sought. Note that the complete integral is determined by the differential equation nonuniquely.

► Example 2.6. Consider the equation

$$u_x = a u_y^n + b.$$

We seek a particular solution as the sum $u = C_1 y + C_2 + C_3 x$. Substituting this expression into the equation yields the relation $C_3 = aC_1^n + b$ for the coefficients C_1 and C_3. With this relation, we find a complete integral in the form $u = C_1 y + (aC_1^n + b)x + C_2$. ◄

A complete integral of equation (2.2.1.1) is often written in implicit form:*

$$\Theta(x, y, u, C_1, C_2) = 0. \tag{2.2.1.4}$$

*In equations (2.2.1.2) and (2.2.1.4), the symbol Θ denotes different functions.

2°. The *general integral* of equation (2.2.1.1) can be represented in parametric form by using the complete integral (2.2.1.2) [or (2.2.1.4)] and the two equations

$$\begin{aligned} &C_2 = f(C_1),\\ &\frac{\partial\Theta}{\partial C_1} + \frac{\partial\Theta}{\partial C_2} f'(C_1) = 0, \end{aligned} \tag{2.2.1.5}$$

where $f = f(z)$ is an arbitrary function and the prime stands for the derivative. In a sense, the general integral plays the role of the general solution depending on an arbitrary function (the question whether it describes all solutions calls for further analysis).

► Example 2.7. For the equation considered in Example 2.6, the general integral can be written in parametric form by using the relations

$$u = C_1 y + \big(aC_1^n + b\big)x + C_2, \quad C_2 = f(C_1), \quad y + anC_1^{n-1}x + f'(C_1) = 0.$$

Eliminating C_2 from these relations and renaming C_1 by C, one can represent the general integral in a more graphic manner in the form

$$\begin{aligned} u &= Cy + \big(aC^n + b\big)x + f(C),\\ y &= -anC^{n-1}x + f'(C). \end{aligned}$$

◄

3°. *Singular integrals* of equation (2.2.1.1) can be found without invoking a complete integral by eliminating $p = u_x$ and $q = u_y$ from the following system of three algebraic (or transcendental) equations:

$$F = 0, \quad F_p = 0, \quad F_q = 0,$$

where the first equation coincides with equation (2.2.1.1).

2.2.2. Method of Separation of Variables. Solution Construction Examples for Nonlinear First-Order PDEs

► Brief description of the method of separation of variables.

The method of separation of variables is one of the simplest yet effective techniques for constructing exact solutions and complete integrals of nonlinear partial differential equations of the first order. For equations involving two independent variables, x and y, where the desired function is $u = u(x, y)$, this method typically entails finding solutions in the form of a sum of functions, each depending on different arguments

$$u = \varphi(x) + \psi(y). \tag{2.2.2.1}$$

Additionally, exact solutions for certain classes of nonlinear equations that involve first-order partial derivatives are sometimes sought as products of functions of different arguments:

$$u = \varphi(x)\psi(y). \tag{2.2.2.2}$$

Solutions of the form (2.2.2.1) or (2.2.2.2) are referred to as *additive separable* or *multiplicative separable* solutions, respectively. For equations involving two independent variables, substituting (2.2.2.1) or (2.2.2.2) into the equation and performing simple rearrangements leads us to the equality of two expressions, each dependent on a different variable. The equality is possible only if both expressions are equal to the same constant value. As a result, we obtain two ordinary differential equations for the two desired functions, one for $\varphi = \varphi(x)$ and the other for $\psi = \psi(y)$.

► Example 2.8. Consider the first-order nonlinear PDE

$$u_x + a u_y^2 = by, \tag{2.2.2.3}$$

which governs the free vertical drop of a point body near the Earth's surface (y is the vertical coordinate measured downward, x time, $m = \frac{1}{2a}$ the mass of the body, and $g = 2ab$ the gravitational acceleration).

We look for a complete integral of this PDE as the sum of functions of different arguments. Substituting (2.2.2.1) into (2.2.2.3) and performing simple rearrangements, we get

$$\varphi'_x = by - a(\psi'_y)^2.$$

The left-hand side of this equality depends only on x, while the right-hand side depends on y alone. This is possible only if the conditions

$$\varphi'_x = C_1, \qquad by - a(\psi'_y)^2 = C_1 \tag{2.2.2.4}$$

are met, where C_1 is an arbitrary constant. Integrating ODEs (2.2.2.4) yields

$$\varphi = C_1 x + C_2, \qquad \psi = \pm\frac{2}{3\sqrt{a}\,b}(by - C_1)^{3/2} + C_3,$$

Substituting these expressions into (2.2.2.1), we obtain a complete integral in the form

$$u = -\tilde{C}_1 x \pm \frac{2}{3\sqrt{a}\,b}(by + \tilde{C}_1)^{3/2} + \tilde{C}_2,$$

where the arbitrary constants are renamed as $\tilde{C}_1 = -C_1$ and $\tilde{C}_2 = C_2 + C_3$. ◄

Remark 2.6. *Complicated nonlinear first-order PDEs may not have exact solutions of the form (2.2.2.1) or (2.2.2.2).*

► Complete integrals of some general first-order nonlinear PDEs.

Below are the structures of several complete integrals, which are solutions of the form (2.2.2.1) or (2.2.2.2), for various classes of first-order nonlinear PDEs that allow for separation of variables.

1°. If the equation does not depend explicitly on y and u, i.e.,

$$F(x, u_x, u_y) = 0,$$

then one can seek a complete integral in the form of the sum of two functions with different arguments

$$u = C_1 y + C_2 + \varphi(x).$$

The new unknown function φ is determined by solving the following ordinary differential equation:

$$F(x, \varphi'_x, C_1) = 0.$$

Expressing φ'_x from this equation in terms of x, one arrives at a separable differential equation for $\varphi = \varphi(x)$.

2°. Consider an equation with separated variables

$$F_1(x, u_x) = F_2(y, u_y).$$

Then one can seek a complete integral as the sum of two functions with different arguments,

$$u = \varphi(x) + \psi(y) + C_1,$$

which are determined by the following two ordinary differential equations:

$$F_1(x, \varphi'_x) = C_2,$$
$$F_2(y, \psi'_y) = C_2.$$

3°. Let the equation have the form (generalizes the equation of Item 2°):

$$F_1(x, u_x) + F_2(y, u_y) = au.$$

Then one can seek a complete integral as the sum of two functions with different arguments,

$$u = \varphi(x) + \psi(y) + C_1,$$

which are determined by the following two ordinary differential equations:

$$F_1(x, \varphi'_x) - a\varphi = aC_1 + C_2,$$
$$F_2(y, \psi'_y) - a\psi = -C_2,$$

where C_1 is an arbitrary constant.

4°. Suppose the equation can be rewritten in the form

$$F\big(G(x, u_x),\, y, u_y\big) = 0.$$

Then one can seek a complete integral as the sum of two functions with different arguments,

$$u(x, y) = \varphi(x) + \psi(y) + C_1,$$

which are determined by the following two ordinary differential equations:

$$G(x, \varphi'_x) = C_2,$$
$$F(C_2, y, \psi'_y) = 0,$$

where C_2 is an arbitrary constant.

5°. Let the equation have the form

$$F_1(x, u_x/u) = u^k F_2(y, u_y/u).$$

Then one can seek a complete integral in the form of the product of two functions with different arguments,

$$u = \varphi(x)\psi(y),$$

which are determined by the following two ordinary differential equations:

$$F_1(x, \varphi'_x/\varphi) = C_1\varphi^k,$$
$$F_2(y, \psi'_y/\psi) = C_1\psi^{-k},$$

where C_1 is an arbitrary constant.

6°. Table 2.1 lists complete integrals of the above and some other first-order nonlinear PDEs of general form involving arbitrary functions with several arguments.

Remark 2.7. *The handbooks by Kamke (1965), Polyanin et al. (2002), and Polyanin & Zaitsev (2012) provide complete integrals for many more nonlinear first-order partial differential equations with two independent variables than those listed in Table 2.1.*

2.2.3. Lagrange–Charpit Method Based on Using a Characteristic ODE System

► **Method description.**

One associates the nonlinear first-order partial differential equation (2.2.1.1) with the *characteristic system of ordinary differential equations* (Kamke, 1965):

$$\frac{dx}{F_p} = \frac{dy}{F_q} = \frac{du}{pF_p + qF_q} = -\frac{dp}{F_x + pF_u} = -\frac{dq}{F_y + qF_u}, \qquad (2.2.3.1)$$

where $p = u_x$ and $q = u_y$.

Suppose that a first integral,

$$\Phi(x, y, u, u_x, u_y) = C_1, \qquad (2.2.3.2)$$

of the characteristic ODE system (2.2.3.1) is known (here $\Phi \not\equiv F$).

We assume that the first integral (2.2.3.2) and the original PDE (2.2.1.1) can be solved for the derivatives u_x and u_y, i.e.,

$$u_x = \varphi_1(x, y, u, C_1), \quad u_y = \varphi_2(x, y, u, C_1). \qquad (2.2.3.3)$$

The first equation of this system can be treated as an ordinary differential equation with independent variable x and parameter y. On finding the general solution of this equation depending on an arbitrary function $\psi(y)$, one substitutes this solution into the second equation to arrive at an ordinary differential equation for ψ. On determining $\psi(y)$ and on substituting it into the solution of the first equation of (2.2.3.3), one finds a complete integral of equation (2.2.1.1). In a similar manner, one can start solving system (2.2.3.3) with the second equation, treating it as an ordinary differential equation with independent variable y and parameter x.

Table 2.1. Complete integrals for some special types of nonlinear first-order PDE; C_1 and C_2 are arbitrary constants.

No.	Equations and comments	Complete integrals	Auxiliary equations
1	$F(u_x, u_y)=0$, does not depend on x, y, and u implicitly	$u=C_1+C_2x+C_3y$	$F(C_2, C_3)=0$
2	$F(x, u_x, u_y)=0$, does not depend on y and u implicitly	$u=C_1y+C_2+\varphi(x)$	$F(x, \varphi'_x, C_1)=0$
3	$F(u, u_x, u_y)=0$, does not depend on x and y implicitly	$u=\varphi(z)$, $z=C_1x+C_2y$	$F(\varphi, C_1\varphi'_z, C_2\varphi'_z)=0$
4	$F_1(x, u_x)=F_2(y, u_y)$, separated equation	$u=\varphi(x)+\psi(y)+C_1$	$F_1(x, \varphi'_x)=C_2$, $F_2(y, \psi'_y)=C_2$
5	$F_1(x, u_x)+F_2(y, u_y)=au$, generalizes equation 4	$u=\varphi(x)+\psi(y)$	$F_1(x, \varphi'_x)-a\varphi=C_1$, $F_2(y, \psi'_y)-a\psi=-C_1$
6	$F_1(x, u_x)=e^{au}F_2(y, u_y)$, generalizes equation 4	$u=\varphi(x)+\psi(y)$	$F_1(x, \varphi'_x)=C_1e^{a\varphi}$, $F_2(y, \psi'_y)=C_1e^{-a\psi}$
7	$F_1(x, u_x/u)=u^kF_2(y, u_y/u)$, can be reduced to equation 6 by replacing $u=e^z$	$u=\varphi(x)\psi(y)$	$F_1(x, \varphi'_x/\varphi)=C_1\varphi^k$, $F_2(y, \psi'_y/\psi)=C_1\psi^{-k}$
8	$F_1(x, u_x/u)$ $+F_2(y, u_y/u)=a\ln u$	$u=\varphi(x)\psi(y)$	$F_1(x, \varphi'_x/\varphi)-a\ln\varphi=C_1$, $F_2(y, \psi'_y/\psi)-a\ln\psi=-C_1$
9	$u=xu_x+yu_y+F(u_x, u_y)$, Clairaut equation	$u=C_1x+C_2y$ $+F(C_1, C_2)$	NA
10	$F(x, u_x, u_y, u-yu_y)=0$, generalizes equation 2	$u=C_1y+\varphi(x)$	$F\big(x, \varphi'_x, C_1, \varphi\big)=0$
11	$F(u, u_x, u_y, xu_x+yu_y)=0$, generalizes equations 3 and 9	$u=\varphi(z)$, $z=C_1x+C_2y$	$F(\varphi, C_1\varphi'_z, C_2\varphi'_z, z\varphi'_z)=0$
12	$F_1\big(F_2(x, u_x), y, u_y\big)=0$, generalizes equation 4	$u=\varphi(x)+\psi(y)+C_1$	$F_2(x, \varphi'_x)=C_2$, $F_1(C_2, y, \psi'_y)=0$
13	$F_1\big(F_2(x, u_x/u), y, u_y/u\big)=0$	$u=C_1\varphi(x)\psi(y)$	$F_2(x, \varphi'_x/\varphi)=C_2$, $F_1(C_2, y, \psi'_y/\psi)=0$
14	$u_x=yF_1(x, u_y)+F_2(x, u_y)$	$u=\varphi(x)y$ $+\int F_2(x, \varphi(x))\,dx+C_1$	$\varphi'_x=F_1(x, \varphi)$
15	$u_x=uF(x, u_y)$ $+yG(x, u_y)+H(x, u_y)$	$u=\varphi(x)y+\psi(x)$	$\varphi'_x=F(x, \varphi)\varphi+G(x, \varphi)$, $\psi'_x=F(x, \varphi)\psi+H(x, \varphi)$

► Example 2.9. Consider the first-order nonlinear PDE

$$yuu_x^2 - u_y = 0. \tag{2.2.3.4}$$

In this case, the characteristic system (2.2.3.1) has the form

$$\frac{dx}{2yup} = -\frac{dy}{1} = \frac{du}{2yup^2 - q} = -\frac{dp}{yp^3} = -\frac{dq}{up^2 + yp^2 q}, \qquad p = u_x, \ q = u_y.$$

By making use of the original equation, we simplify the denominator of the third ratio to obtain an integrable combination: $du/(yup^2) = -dp/(yp^3)$. This yields the first integral $p = C_1/u$, that is $u_x = C_1/u$. Taking this relation into account and solving the original equation (2.2.3.4) for u_y, we obtain the system

$$u_x = \frac{C_1}{u}, \qquad u_y = \frac{C_1^2 y}{u}. \tag{2.2.3.5}$$

The general solution of the first equation has the form $u^2 = 2C_1 x + \psi(y)$, where $\psi(y)$ is an arbitrary function. With this solution, it follows from the second equation of the system (2.2.3.5) that $\psi'(y) = 2C_1^2 y$. Thus, $\psi(y) = C_1^2 y^2 + C_2$. Finally, we arrive at a complete integral of the form

$$u^2 = 2C_1 x + C_1^2 y^2 + C_2.$$

◄

Note that the general solution of the completely integrable *Pfaffian equation*

$$du = \varphi_1(x, y, u, C_1)\, dx + \varphi_2(x, y, u, C_1)\, dy \tag{2.2.3.6}$$

where the functions φ_1 and φ_2 are the same as in system (2.2.3.3), is a complete integral of PDE (2.2.1.1). For Pfaffian equations of type (2.2.3.6), see, for example, the books by Petrovskii (1966), Polyanin & Manzhirov (2007), and Polyanin & Zaitsev (2012).

Remark 2.8. *The relation $F(x, y, u, p, q) = C$ is an obvious first integral of the characteristic system (2.2.3.1). Hence, the function Φ determining the integral (2.2.3.2) must differ from F. However, the use of relation (2.2.1.1) makes it possible to reduce the order of system (2.2.3.1) by one.*

► **Case where the PDE does not depend on u explicitly.**

Suppose the original equation does not contain the unknown first explicitly, i.e., it has the form

$$F(x, y, u_x, u_y) = 0. \tag{2.2.3.7}$$

1°. Given a one-parameter family of exact solutions $u = \theta(x, y, C_1)$ such that $\partial\theta/\partial C_1 \not\equiv \text{const}$, a complete integral is given by $u = \theta(x, y, C_1) + C_2$.

2°. The first integral may be sought in the form

$$\Phi(x, y, u_x, u_y) = C_1$$

similar to that of equation (2.2.3.7). In this case, the characteristic system (2.2.3.1) is represented as

$$\frac{dx}{F_p} = \frac{dy}{F_q} = -\frac{dp}{F_x} = -\frac{dq}{F_y}, \qquad \text{where} \quad p = u_x, \quad q = u_y.$$

The corresponding Pfaffian equation (2.2.3.6) becomes

$$du = \varphi_1(x, y, C_1)\,dx + \varphi_2(x, y, C_1)\,dy.$$

One may integrate this equation in quadrature, thus arriving at the following expression for the complete integral:

$$u = \int_{x_0}^{x} \varphi_1(t, y, C_1)\,dt + \int_{y_0}^{y} \varphi_2(x_0, s, C_1)\,ds + C_2, \qquad (2.2.3.8)$$

where x_0 and y_0 are arbitrary numbers.

3°. Suppose that equation (2.2.3.7) can be solved for u_x or u_y; for example,

$$u_x = -\mathcal{H}(x, y, u_y).$$

Then, by differentiating this relation with respect to y, we obtain a quasilinear equation for the derivative $q = u_y$ in the form

$$q_x + \frac{\partial}{\partial y}\mathcal{H}(x, y, q) = 0.$$

This equation is simpler than the original one; qualitative features of it and solution methods can be found in Section 2.1.

▶ Hamilton–Jacobi equation.

Equation (2.2.1.1) solved for one of the derivatives, e.g.,

$$u_x + \mathcal{H}(x, y, u, u_y) = 0 \qquad (2.2.3.9)$$

is commonly referred to as the *Hamilton–Jacobi equation** and the function $\mathcal{H}$ as the *Hamiltonian*. Equations of the form (2.2.3.9) are frequently encountered in various fields of mechanics, control theory, and differential games, where the variable x usually plays the role of time and the variable y the role of the spatial coordinate. To the Hamilton–Jacobi equation (2.2.3.9) there corresponds the function $F(x, y, u, u_x, u_y) = u_x + \mathcal{H}(x, y, u, u_y)$ in equation (2.2.1.1).

*The Hamilton–Jacobi equation is often the one in which the function $\mathcal{H}$ is independent of u explicitly, i.e., the equation $u_x + \mathcal{H}(x, y, u_y) = 0$.

The characteristic system (2.2.3.1) for equation (2.2.3.9) can be reduced, by taking into account the relation $u_x = -\mathcal{H}$, to a simpler system consisting of three ODEs:

$$y'_x = \mathcal{H}_q, \quad u'_x = q\mathcal{H}_q - \mathcal{H}, \quad q'_x = -q\mathcal{H}_u - \mathcal{H}_y, \quad \text{where} \quad q = u_y, \tag{2.2.3.10}$$

which are independent of $p = u_x$; the left-hand sides of these equations are derivatives with respect to x.

2.2.4. Cauchy Problem. Solution Procedure. Some Examples

► **Statement of the problem. Solution procedure.**

Consider the Cauchy problem for equation (2.2.1.1) subject to the initial conditions

$$x = h_1(\xi), \quad y = h_2(\xi), \quad u = h_3(\xi), \tag{2.2.4.1}$$

where ξ is a parameter ($\alpha \le \xi \le \beta$) and the $h_k(\xi)$ are given functions.

The solution of this problem is carried out in several steps:

1°. First, one determines additional initial conditions for the derivatives,

$$u_x = p_0(\xi), \quad u_y = q_0(\xi). \tag{2.2.4.2}$$

To this end, one must solve the algebraic (or transcendental) system of equations

$$F\big(h_1(\xi), h_2(\xi), h_3(\xi), p_0, q_0\big) = 0, \tag{2.2.4.3}$$

$$p_0 h'_1(\xi) + q_0 h'_2(\xi) - h'_3(\xi) = 0 \tag{2.2.4.4}$$

for $p_0 = p_0(\xi)$ and $q_0 = q_0(\xi)$. Equation (2.2.4.3) results from substituting the initial data (2.2.4.1) into the original equation (2.2.1.1). Equation (2.2.4.4) is a consequence of the dependence of u on x and y and the relation $du = u_x\,dx + u_y\,dy$, where dx, dy, and du are calculated in accordance with the initial data (2.2.4.1), and conditions (2.2.4.2) are taken into account.

2°. One solves the autonomous system

$$\frac{dx}{F_p} = \frac{dy}{F_q} = \frac{du}{pF_p + qF_q} = -\frac{dp}{F_x + pF_u} = -\frac{dq}{F_y + qF_u} = d\tau, \tag{2.2.4.5}$$

where $p = u_x$ and $q = u_y$, which is obtained from (2.2.3.1) by introducing the additional variable τ (playing the role of time).

3°. Finally, one determines the constant of integration from the initial conditions

$$x = h_1(\xi), \quad y = h_2(\xi), \quad u = h_3(\xi), \quad u_x = p_0(\xi), \quad u_y = q_0(\xi) \quad \text{at} \quad \tau = 0, \tag{2.2.4.6}$$

obtained by combining conditions (2.2.4.1) and (2.2.4.2). This results in the three functions

$$x = x(\tau, \xi), \quad y = y(\tau, \xi), \quad u = u(\tau, \xi), \tag{2.2.4.7}$$

which give the solution of the stated Cauchy problem in parametric form (τ and ξ are parameters).

► Existence and uniqueness theorem.

Let the function $F = F(x, y, u, p, q)$, determining equation (2.2.1.1), be twice continuously differentiable with respect to all five arguments (in the domain being considered), with $F_p^2 + F_q^2 \neq 0$. Let the functions $h_1(\xi)$, $h_2(\xi)$, and $h_3(\xi)$, determining the initial conditions (2.2.4.1), be twice differentiable with respect to ξ, with $(h_1')^2 + (h_2')^2 \neq 0$. Assume that the functions $p_0(\xi)$ and $q_0(\xi)$, determining the additional initial condition (2.2.4.2), satisfy system (2.2.4.3), (2.2.4.4). Moreover, we adopt the *transversality condition*

$$\Delta \equiv F_p h_2' - F_q h_1' \neq 0,$$

where the functions h_1, h_2, p, and q are defined by equations (2.2.4.1) and (2.2.4.2), and the prime denotes the derivative with respect to ξ. Under the adopted assumptions, there exists a unique, twice continuously differentiable solution of equation (2.2.1.1) satisfying the initial conditions (2.2.4.1) and (2.2.4.2).

Remark 2.9. *This theorem has a local character, i.e., the existence of a unique solution of the Cauchy problem is merely guaranteed in some neighborhood of the line defined by the initial data (2.2.4.1) together with the additional conditions (2.2.4.2).*

Remark 2.10. *The algebraic (or transcendental) system (2.2.4.3), (2.2.4.4) can have several solutions (see Example 2.11 at the end of this section) that result in distinct additional conditions (2.2.4.2) for the derivatives p and q. Each pair of these additional conditions will generate a solution of the Cauchy problem (2.2.1.1), (2.2.4.1).*

Remark 2.11. *For nonlinear equations, the global solution of the Cauchy problem (2.2.1.1), (2.2.4.1) may be nonunique for another reason—because the characteristics in the xy-plane may intersect (see, for example, Melikyan, 1998 and Polyanin & Zaitsev, 2012).*

► Cauchy problem for the Hamilton–Jacobi equation.

The initial condition for the Hamilton–Jacobi equation (2.2.3.9) is usually stated in the form

$$u = \varphi(y) \quad \text{at} \quad x = L.$$

In this case, the solution of the Cauchy problem is reduced to the solution of the characteristic system (2.2.3.10) subject to the initial conditions

$$y = \xi, \quad u = \varphi(\xi), \quad q = \varphi'(\xi) \quad \text{at} \quad x = L,$$

where the prime stands for the derivative with respect to the parameter ξ.

► Examples of solving the Cauchy problem.

Consider a few specific examples.

► Example 2.10. Find a solution of the first-order nonlinear PDE

$$u_x u_y = au \tag{2.2.4.8}$$

satisfying the initial condition

$$u = f(y) \quad \text{at} \quad x = 0. \tag{2.2.4.9}$$

Rewrite the initial condition (2.2.4.9) in parametric form:

$$x=0,\quad y=\xi,\quad u=f(\xi). \tag{2.2.4.10}$$

The system of equations (2.2.4.3), (2.2.4.4) for $p_0=p_0(\xi)$ and $q_0=q_0(\xi)$ is written as

$$af(\xi)=p_0q_0,\qquad q_0-f'(\xi)=0.$$

It follows that

$$p_0=a\frac{f(\xi)}{f'(\xi)},\quad q_0=f'(\xi). \tag{2.2.4.11}$$

Since in this case $F=pq-au$, the characteristic system (2.2.4.5) takes the form

$$\frac{dx}{q}=\frac{dy}{p}=\frac{du}{2pq}=\frac{dp}{ap}=\frac{dq}{aq}=d\tau. \tag{2.2.4.12}$$

Its solution is given by (the last two equations are the easiest to integrate first):

$$\begin{aligned}&p=C_1e^{a\tau},\quad q=C_2e^{a\tau},\\&x=\frac{C_2}{a}e^{a\tau}+C_3,\quad y=\frac{C_1}{a}e^{a\tau}+C_4,\quad u=\frac{C_1C_2}{a}e^{2a\tau}+C_5.\end{aligned} \tag{2.2.4.13}$$

By using the initial conditions (2.2.4.10), (2.2.4.11), which must be satisfied at $\tau=0$, we find the constants of integration in the form

$$C_1=a\frac{f(\xi)}{f'(\xi)},\quad C_2=f'(\xi),\quad C_3=-\frac{f'(\xi)}{a},\quad C_4=\xi-\frac{f(\xi)}{f'(\xi)},\quad C_5=0.$$

Substituting theses values into (2.2.4.13) yields a solution of the Cauchy problem (2.2.4.8), (2.2.4.9) in the parametric form

$$x=\frac{1}{a}f'(\xi)(e^{a\tau}-1),\quad y=\frac{f(\xi)}{f'(\xi)}(e^{a\tau}-1)+\xi,\quad u=f(\xi)e^{2a\tau}.$$

◄

► **Example 2.11.** Find a solution of the equation

$$u_x^2+u_y^2=a^2, \tag{2.2.4.14}$$

passing through the circle

$$x^2+y^2=b^2,\quad u=0. \tag{2.2.4.15}$$

By introducing a parameter ξ, we rewrite the condition on the circle (2.2.4.15) in the form

$$x=b\sin\xi,\quad y=b\cos\xi,\quad u=0. \tag{2.2.4.16}$$

In this case, the equations (2.2.4.3), (2.2.4.4) for determining additional initial conditions are

$$p_0^2+q_0^2=a^2,\quad p_0\cos\xi-q_0\sin\xi=0.$$

Whence,

$$p_0 = \varepsilon a \sin\xi, \quad q_0 = \varepsilon a \cos\xi, \qquad \text{where} \quad \varepsilon = \pm 1. \tag{2.2.4.17}$$

Since in this case $F = p^2 + q^2 - a^2$, the characteristic system (2.2.4.5) is written as

$$\frac{dx}{2p} = \frac{dy}{2q} = \frac{du}{2(p^2+q^2)} = -\frac{dp}{0} = -\frac{dq}{0} = d\tau. \tag{2.2.4.18}$$

The general solution is given by (the last two equations are integrated first):

$$p = C_1, \quad q = C_2, \quad x = 2C_1\tau + C_3, \quad y = 2C_2\tau + C_4, \quad u = 2(C_1^2 + C_2^2)\tau + C_5. \tag{2.2.4.19}$$

Using the initial conditions (2.2.4.16), (2.2.4.17), which must be satisfied at $\tau = 0$, we find the constants of integration

$$C_1 = \varepsilon a \sin\xi, \quad C_2 = \varepsilon a \cos\xi, \quad C_3 = b\sin\xi, \quad C_4 = b\cos\xi, \quad C_5 = 0, \quad \varepsilon = \pm 1.$$

Substituting these values into (2.2.4.19) yields a solution of the Cauchy problem (2.2.4.14), (2.2.4.15) in the parametric form

$$x = (2\varepsilon a\tau + b)\sin\xi, \quad y = (2\varepsilon a\tau + b)\cos\xi, \quad u = 2a^2\tau.$$

On eliminating the parameters ξ and τ, we can rewrite this solution in a more graphic form,

$$a^2(x^2 + y^2) = (ab \pm u)^2. \tag{2.2.4.20}$$

Relation (2.2.4.20) describes two circular coaxial cones in the space (x, y, u). The circle (2.2.4.15) is a base of the cones. The common axis coincides with the u-axis. The vertices of the cones have the coordinates $u = \pm ab$.

It is significant that solution (2.2.4.20) is a many-valued function. ◄

► Additional remarks.

For more information on the properties, geometric interpretation, and applications of nonlinear first-order PDEs, see Bellman (1957), Pontryagin et al. (1962), Kamke (1965), and Courant & Hilbert (1989). Extensive tables of exact solutions of these equations can be found in Kamke (1965), Polyanin et al. (2002), and Polyanin & Zaitsev (2012). Generalized solutions (nonsmooth viscosity solutions) are discussed in Kruzhkov (1975), Lions (1982), Lions & Souganidis (1985), Subbotin (1995), Melikyan (1998), Polyanin et al. (2002), and Polyanin & Zaitsev (2012).

Problems and Exercises for Section 2.2

1. Find a complete integral of the nonlinear equation

$$u_x + a u_y^2 = f(x).$$

2. Find a complete integral of the nonlinear equation

$$u_x + a u_y^2 = bu.$$

3. Find a complete integral of the nonlinear equation

$$u_x^2 + a u_y^2 = f(x).$$

4. Prove that the tubular surface equation

$$u^2(u_x^2 + u_y^2 + 1) = 1$$

has a complete integral that can be written in implicit form as

$$(x - C_1)^2 + (y - C_2)^2 + u^2 = 1.$$

5. Find a complete integral of the nonlinear equation

$$u_x u_y = a u^2.$$

6. Find a complete integral of the nonlinear equation

$$u_x u_y^2 = f(x).$$

7. Find a complete integral of the nonlinear equation

$$u_x u_y = a e^u.$$

8. Find a complete integral of the nonlinear equation

$$u_x u_y = f(x) g(y) e^u.$$

9. Find a complete integral of the nonlinear equation

$$u_x = f(u_y).$$

10. Find a complete integral of the nonlinear equation

$$u_x = f(u_y) + g(x).$$

11. Find a complete integral of the nonlinear equation

$$u_x = f(x, u_y).$$

12. Find a complete integral of the nonlinear equation

$$u_y = f(x, u_x) + a u.$$

13. Find a solution to the Cauchy problem for the first-order nonlinear PDE

$$u_x u_y = a$$

with the initial condition

$$u = f(y) \quad \text{at} \quad x = 0.$$

14. Find a solution to the Cauchy problem for the first-order nonlinear PDE

$$u_x u_y = a$$

with the initial data

$$y = x \quad \text{at} \quad u = f(x).$$

References for Chapter 2

Bedrikovetsky P.G. *Mathematical Theory of Oil and Gas Recovery*. Kluwer, Dordrecht, 1993.

Bellman R. *Dynamic Programming*. Princeton Univ. Press, Princeton, 1957.

Courant R. and Hilbert D. *Methods of Mathematical Physics, Vol. 2*. Wiley–Interscience Publ., New York, 1989.

Gelfand I.M. Some problems of the theory of quasi-linear equations. *Uspekhi Matem. Nauk*, 1959, Vol. 14, No. 2, pp. 87–158 [*Amer. Math. Soc. Translation, Series 2*, 1963, pp. 295–381].

Kamke E. *Differentialgleichungen: Lösungsmethoden und Lösungen, II, Partielle Differentialgleichungen Erster Ordnung für eine gesuchte Funktion*. Akad. Verlagsgesellschaft Geest & Portig, Leipzig, 1965.

Korn G.A. and Korn T.M. *Mathematical Handbook for Scientists and Engineers*. Dover Publ., New York, 2000.

Kruzhkov S.N. Generalized solutions of the Hamilton–Jacobi equations of eikonal type. I. Formulation of the problems; existence, uniqueness and stability theorems; some properties of the solutions. *Mathematics of the USSR-Sbornik*, 1975, Vol. 27, No. 3, pp. 406–446.

Lions P.L. *Generalized Solutions of Hamilton–Jacobi Equations*, Pitman, Boston, 1982.

Lions P.L. and Souganidis P.E. Differential games, optimal control and directional derivatives of viscosity solutions of Bellman's and Isaacs' solutions. *SIAM J. Control and Optimization*, 1985, Vol. 23, No. 4, pp. 566–583.

Melikyan A.A. *Generalized Characteristics of First Order PDE's: Applications in Optimal Control and Differential Games*. Birkhäuser, Boston, 1998.

Oleinik O.A. Discontinuous solutions of nonlinear differential equations. *Uspekhi Matem. Nauk*, 1957, Vol. 12, No. 3, pp. 3–73 [*Amer. Math. Soc. Translation, Series 2*, 1963, Vol. 26, pp. 95–172].

Oleinik O.A. On uniqueness and stability of a general solution of Cauchy's problem of quasilinear equations (in Russian). *Uspekhi Matem. Nauk*, 1959, Vol. 14, No. 2, pp. 159–164.

Petrovskii I.G. *Ordinary Differential Equations*. Prentice-Hall, Englewood Cliffs, N.J., 1966.

Polyanin A.D. and Manzhirov A.V. *Handbook of Mathematics for Engineers and Scientists*. Chapman & Hall/CRC Press, Boca Raton–London, 2007.

Polyanin A.D. and Zaitsev V.F. *Handbook of Nonlinear Partial Differential Equations, 2nd ed.* Chapman & Hall/CRC Press, Boca Raton, 2012.

Polyanin A.D., Zaitsev V.F., and Moussiaux A. *Handbook of First Order Partial Differential Equations*. Taylor & Francis, London, 2002.

Pontryagin L.S., Boltyansky V.G., Gamkrelidze R.V., and Mishchenko E.F. *Mathematical Theory of Optimal Processes*. Wiley-Interscience, New York, 1962.

Rhee, H., Aris, R., and Amundson, N. R., *First Order Partial Differential Equations, Vol. I*. Prentice Hall, Englewood Cliffs, 1986.

Rozhdestvenskii B.L. and Yanenko N.N. *Systems of Quasilinear Equations and Their Applications to Gas Dynamics*. American Mathematical Society, Providence, 1983.

Subbotin A.I. *Generalized Solutions of First Order PDEs: the Dynamical Optimization Perspective*. Birkhäuser, Boston, 1995.

Whitham G.B. *Linear and Nonlinear Waves*. Wiley, New York, 1974.

Zauderer E. *Partial Differential Equations of Applied Mathematics, 3rd ed.* John Wiley & Sons, New York, 2006.

3. Solution Methods for Functional Equations

► **Preliminary remarks.** *Functional equations* are equations that involve unknown functions dependent on different arguments, while *functional differential equations* include both unknown functions of different arguments and their derivatives.

The most common and simplest functional equations are *difference equations*, which are used for numerically solving differential equations. The methods for solving these equations are well developed.

Functional equations other than difference equations are usually more complex than differential equations. There are no general methods for solving functional equations; specific techniques are applied to particular narrow classes of these equations. The general solutions of functional equations may involve arbitrary constants or depend on arbitrary functions. Furthermore, functional equations can have multiple solutions or, in some cases, no solutions at all.

This chapter discusses functional equations whose solution can be reduced to solving simpler, ODEs or linear equations with first-order partial derivatives. Examples of specific functional equations and their solutions are provided.

The method of differentiation with respect to independent variables, used to solve functional equations and described in Section 3.1, can be generalized. It is also employed to solve more complex functional differential equations that arise in the generalized or functional separation of variables in nonlinear PDEs; see Sections 5.4 and 6.4, respectively.

3.1. Method of Differentiation with Respect to Independent Variables

3.1.1. Brief Description of the Method

Sometimes, differentiating with respect to independent variables can help eliminate some arguments from the analyzed functional equation, thereby simplifying it to an ordinary differential equation. The solution thus obtained must then be substituted into the original equation to eliminate any unnecessary integration constants that may arise from the differentiation process used to derive the solution.

Remark 3.1. *In certain situations, the process of differentiating with respect to independent variables must be combined with multiplying (or dividing) the equation and its differential consequences by appropriate functions. Occasionally, it is beneficial to take the logarithm of the equation or its consequences beforehand.*

3.1.2. Solution of Specific Functional Equations by the Method of Differentiation with Respect to Independent Variables

▶ Example 3.1. Consider the *functional Cauchy equation*

$$f(x) + f(y) = f(x + y), \tag{3.1.2.1}$$

where x and y are independent variables, and $f(x)$ is the unknown function. It is noteworthy that equation (3.1.2.1) is linear.

Differentiating the functional equation (3.1.2.1) with respect to x and y, we arrive at the ordinary differential equation $f''_{zz}(z) = 0$, where $z = x + y$. Its solution is a linear function

$$f(z) = az + b, \tag{3.1.2.2}$$

where a and b are arbitrary constants.

On substituting the function (3.1.2.2) into (3.1.2.1), we find that $b = 0$. Hence, the Cauchy equation (3.1.2.1) has a one-parameter solution

$$f(x) = ax,$$

where a is an arbitrary constant. ◀

Remark 3.2. *In Example 3.1, it sufficed to differentiate the functional equation (3.1.2.1) once with respect to any independent variable. For example, after differentiating with respect to x, we obtain*

$$f'_x(x) = f'_z(z), \qquad z = x + y.$$

The left- and right-hand sides of this equation are functions of different arguments, which is possible only if $f'_x(x) = f'_z(z) = a$, where a is an arbitrary constant. This technique is known as separation of variables. Integrating $f'_z(z) = a$ yields formula (3.1.2.2). Next, we follow the same process as in Example 3.1.

▶ Example 3.2. Consider the nonlinear functional equation (known as the *logarithmic Cauchy equation*):

$$f(x)f(y) = f(x + y). \tag{3.1.2.3}$$

Assuming that $f > 0$, we take the logarithm of equation (3.1.2.3) and then differentiate the resulting expression with respect to x and y to obtain

$$[\ln f(x)f(y)]_{xy} = [\ln f(x) + \ln f(y)]_{xy} = 0 \qquad \text{(left-hand side of the equation)},$$
$$[\ln f(x + y)]_{xy} = [\ln f(z)]''_{zz}, \quad z = x + y \qquad \text{(right-hand side of the equation)}.$$

As a result, we arrive at an ordinary differential equation: $[\ln f(z)]''_{zz} = 0$, where $z = x + y$. Integrating twice yields $\ln f(z) = az + b$, or

$$f(z) = e^{az+b}, \tag{3.1.2.4}$$

where a and b are arbitrary constants.

Substituting the function (3.1.2.4) into (3.1.2.3) gives $b = 0$. Hence, the functional equation (3.1.2.1) has a one-parameter solution:

$$f(x) = e^{ax},$$

where a is an arbitrary constant.

It is noteworthy that equation (3.1.2.3) also admits a trivial solution $f = 0$ (we assumed that $f > 0$ when taking the logarithm). ◄

► Example 3.3. Consider the nonlinear functional equation

$$f(x+y) = f(x) + f(y) + af(x)f(y), \qquad a \neq 0, \tag{3.1.2.5}$$

which, for $a = -1$, occurs in probability theory.

Differentiating the equation with respect to x and y, we get

$$f''_{zz}(z) = af'_x(x)f'_y(y), \tag{3.1.2.6}$$

where $z = x + y$. Taking the logarithm of equation (3.1.2.6) and then differentiating with respect to x and y, we arrive at an ordinary differential equation

$$[\ln f''_{zz}(z)]''_{zz} = 0. \tag{3.1.2.7}$$

Integrating (3.1.2.7) twice with respect to z gives

$$f''_{zz}(z) = C_1 \exp(C_2 z), \tag{3.1.2.8}$$

where C_1 and C_2 are arbitrary constants. Substituting (3.1.2.8) into (3.1.2.6), we find that equation

$$C_1 \exp[C_2(x+y)] = af'_x(x)f'_y(y),$$

which allows for separation of variables. Integrating gives

$$f(x) = A\exp(C_2 x) + B, \qquad A = \pm\frac{1}{C_2}\sqrt{\frac{C_1}{a}}. \tag{3.1.2.9}$$

On substituting (3.1.2.9) into the original equation (3.1.2.5), we identify the constants: $A = -B = 1/a$ and $C_2 = \beta$ (arbitrary constant). As a result, we obtain the desired solution

$$f(x) = \frac{1}{a}\bigl(e^{\beta x} - 1\bigr).$$

◄

So far, we have considered functional equations with one unknown function. Functional equations involving multiple unknown functions are solved similarly.

► Example 3.4. Consider the *Pexider equation*:

$$f(x) + g(y) = h(x+y), \tag{3.1.2.10}$$

where x and y are independent variables, and $f(x)$, $g(y)$, and $h(z)$ are the unknown functions.

Differentiating the functional equation (3.1.2.10) with respect to x and y gives an ordinary differential equation $h''_{zz}(z) = 0$, where $z = x + y$. Its solution is a linear function:

$$h(z) = az + b. \tag{3.1.2.11}$$

Inserting this expression into (3.1.2.10), we get

$$f(x) + g(y) = ax + ay + b.$$

Separating the variables, we identify the functions f and g:

$$\begin{aligned} f(x) &= ax + b + c, \\ g(y) &= ay - c. \end{aligned} \tag{3.1.2.12}$$

Hence, the solution of the Pexider equation (3.1.2.10) is given by formulas (3.1.2.11) and (3.1.2.12), where a, b, and c are arbitrary constants. ◄

Remark 3.3. *The method of differentiation with respect to independent variables allows for various generalizations. This approach will be utilized further in Chapters 6 and 7 to construct exact solutions for nonlinear partial differential equations that can be expressed in terms of solutions to functional differential equations.*

Problems and Exercises for Section 3.1

1. Solve *Jensen's functional equation*

$$f\left(\frac{x+y}{2}\right) = \frac{f(x) + f(y)}{2}.$$

2. Solve the functional equation

$$f(xy) = f(x) + f(y).$$

3. Solve the functional equation

$$f(x) + f(y/x) = f(y).$$

4. Solve the functional equation

$$f\left(\frac{x+y}{2}\right) = \sqrt{f(x)f(y)}.$$

5. Solve the functional equation

$$f(xy) = af(x)f(y).$$

6. Solve *Lobachevsky's functional equation*

$$f(x+y)f(x-y) = f^2(x).$$

7. Solve *Gauss's functional equation*

$$f\left(\sqrt{x^2+y^2}\right) = f(x)f(y).$$

8. Solve the functional equation

$$f(x)g(y) = h(x+y).$$

9. Solve the functional equation

$$f(x)g(y) + h(y) = f(x+y).$$

10. Solve the functional equation

$$f(xy)g(y/x) = f^2(y).$$

11. Solve the functional equation

$$f(x+y) = a^{xy} f(x) f(y).$$

12. Solve the functional equation

$$f(xy) = [f(x)]^y.$$

13. Solve the functional equation

$$f(x+y) + f(x-y) = 2f(x) + 2f(y).$$

14. Solve the functional equation

$$f\big((x^n + y^n)^{1/n}\big) = af(x)f(y).$$

3.2. Method of Differentiation with Respect to a Parameter

3.2.1. Classes of Functional Equations. Description of the Method

► Preliminary remarks. Classes of equations under consideration.

In some cases, differentiating the considered equation with respect to a parameter can reduce it to a linear first-order partial differential equation. This linear equation can then be solved using the method outlined in Section 2.1. Once a solution is obtained, it must be substituted back into the original equation to eliminate any extraneous solutions that may have resulted from the differentiation process.

We will consider linear functional equations of the form

$$w(x,t) = \theta(x,t,a)\, w\big(\varphi(x,t,a), \psi(x,t,a)\big), \tag{3.2.1.1}$$

where x and t are independent variables, $w = w(x,t)$ is the unknown function, $\theta = \theta(x,t,a)$, $\varphi = \varphi(x,t,a)$, and $\psi = \psi(x,t,a)$ are given functions, and a is a free parameter, which can assume any values from a certain interval.

We assume that the relations

$$\theta(x,t,a_0) = 1, \quad \varphi(x,t,a_0) = x, \quad \psi(x,t,a_0) = t \tag{3.2.1.2}$$

hold at some $a = a_0$, suggesting that the functional equation (3.2.1.1) turns into an identity at $a = a_0$.

Remark 3.4. *Functional equations of the form (3.2.1.1) often arise when solving some inverse issues in mathematical physics. In these cases, given the properties of the solutions, it becomes necessary to determine the corresponding nonlinear equations. These types of problems will be discussed further in Section 4.3.*

► Description of the method.

Let us expand (3.2.1.1) in a power series in the parameter a at $a = a_0$ while taking into account expressions (3.2.1.2). Up to small terms of the order of $o(a-a_0)$, we get

$$\begin{aligned} \varphi(x,t,a) &\simeq x + \varphi^\circ_a(x,t)\Delta a, \\ \psi(x,t,a) &\simeq t + \psi^\circ_a(x,t)\Delta a, \\ \theta(t,x,a) &\simeq 1 + \theta^\circ_a(x,t)\Delta a, \\ w\big(\varphi(x,t,a),\psi(x,t,a)\big) &\simeq w\big(x+\varphi^\circ_a(x,t)\Delta a,\ t+\psi^\circ_a(x,t)\Delta a\big) \\ &\simeq w(x,t) + \varphi^\circ_a(x,t)\Delta a\, w_x + \psi^\circ_a(x,t)\Delta a\, w_t, \end{aligned} \tag{3.2.1.3}$$

where

$$\Delta a = a - a_0,\quad \varphi^\circ_a(x,t) = \frac{\partial\varphi}{\partial a}\bigg|_{a=a_0},\quad \psi^\circ_a(x,t) = \frac{\partial\psi}{\partial a}\bigg|_{a=a_0},\quad \theta^\circ_a(x,t) = \frac{\partial\theta}{\partial a}\bigg|_{a=a_0}.$$

Inserting the last two expressions of (3.2.1.3) into the right-hand side of equation (3.2.1.1), we obtain

$$\begin{aligned} &\theta(x,t,a)\, w\big(\varphi(x,t,a),\psi(x,t,a)\big) \\ &\qquad \simeq w(x,t) + [\varphi^\circ_a(x,t)w_x + \psi^\circ_a(x,t)w_t + \theta^\circ_a(x,t)w]\Delta a \end{aligned} \tag{3.2.1.4}$$

up to $o(\Delta a)$. Now we replace the right-hand side of equation (3.2.1.1) with the right-hand side of relation (3.2.1.4) and cancel out the function $w(x,t)$ in both sides. Then, we divide the resulting expression by Δa and pass to the limit as $\Delta a \to 0$. As a result, we arrive at a linear first-order partial differential equation for the function w (Polyanin & Manzhirov, 2007):

$$\varphi^\circ_a(x,t)w_x + \psi^\circ_a(x,t)w_t + \theta^\circ_a(x,t)w = 0. \tag{3.2.1.5}$$

To solve the PDE (3.2.1.5), we should consider the associated characteristic ODE system (see Section 2.1):

$$\frac{dx}{\varphi^\circ_a(x,t)} = \frac{dt}{\psi^\circ_a(x,t)} = -\frac{dw}{\theta^\circ_a(x,t)w}. \tag{3.2.1.6}$$

Suppose

$$Q_1(x,t) = C_1,\quad Q_2(x,t,w) = C_2 \tag{3.2.1.7}$$

are independent integrals of the characteristic system (3.2.1.6). Then the general solution of equation (3.2.1.5) can be expressed as

$$Q_2(x,t,w) = F\big(Q_1(x,t)\big), \tag{3.2.1.8}$$

where $F(z)$ is an arbitrary function.

We should express w from solution (3.2.1.8) and substitute into the original equation (3.2.1.1) for verification. This is necessary because extra solutions may have emerged or the solution of the partial differential equation (3.2.1.5) may happen not to satisfy the functional equation (3.2.1.1) at all; see Example 3.7 below.

Remark 3.5. *Equation (3.2.1.5) can be derived from (3.2.1.1) by differentiating with respect to a and setting $a = a_0$. In other words, equation (3.2.1.5) can be obtained using the formula*

$$\left[\frac{\partial}{\partial a}\theta(x,t,a)w\big(\varphi(x,t,a),\psi(x,t,a)\big)\right]_{a=a_0}=0,$$

which is easy to memorize.

Remark 3.6. *It is convenient to choose the second integral (3.2.1.7) to be linear with respect to w, or $Q_2(x,t,w)=\xi(x,t)w$, and to rewrite formula (3.2.1.8) in a form solved for w.*

Remark 3.7. *In the degenerate cases $\varphi_a^\circ(x,t)\equiv 0$ or $\psi_a^\circ(x,t)\equiv 0$, equation (3.2.1.5) is easy to integrate, since it is a linear ordinary differential equation, where one of the independent variables plays the role of a parameter.*

Remark 3.8. *The method of differentiation with respect to a parameter can also be used for solving more complex functional equations of the form*

$$w\big(\varphi_1(x,t,a),\psi_1(x,t,a)\big)=\theta(x,t,a)w\big(\varphi_2(x,t,a),\psi_2(x,t,a)\big)+\eta(x,t,a).$$

Such equations hold as identities for certain $a=a_0$. Several examples of this type are provided for practice at the end of Section 3.2.2.

3.2.2. Solution of Specific Functional Equations Using the Method of Differentiation with Respect to a Parameter

► Example 3.5. *Self-similar solutions*, which are often encountered in mathematical physics, can be defined as solutions that are invariant under a scaling transformation and, hence, satisfy the functional equation

$$w(x,t)=a^k w(a^m x, a^n t), \tag{3.2.2.1}$$

where k, m, and n are some constants, and a is an arbitrary positive constant.

Equation (3.2.2.1) becomes an identity at $a = 1$. Differentiating (3.2.2.1) with respect to a and then setting $a = 1$, we arrive at a first-order partial differential equation:

$$mxw_x+ntw_t+kw=0. \tag{3.2.2.2}$$

The first integrals of the associated characteristic system of ordinary differential equations

$$\frac{dx}{mx}=\frac{dt}{nt}=-\frac{dw}{kw}$$

are

$$xt^{-m/n}=C_1,\quad t^{k/n}w=C_2\qquad(n\neq 0).$$

Therefore, the general solution of the partial differential equation (3.2.2.2) is expressed as

$$w(x,t) = t^{-k/n} F(z), \qquad z = xt^{-m/n}, \tag{3.2.2.3}$$

where $F(z)$ is an arbitrary function.

A direct check can verify that expression (3.2.2.3) is a solution to the functional equation (3.2.2.1). ◄

► Example 3.6. Consider the functional equation

$$w(x,t) = a^k w(a^m x, t + \beta \ln a), \tag{3.2.2.4}$$

where k, m, and β are given constants, and a is an arbitrary positive constant.

Equation (3.2.2.4) becomes an identity when $a = 1$. By differentiating (3.2.2.4) with respect to a and then setting $a = 1$, we arrive at a first-order partial differential equation:

$$mxw_x + \beta w_t + kw = 0. \tag{3.2.2.5}$$

The associated characteristic system of ordinary differential equations

$$\frac{dx}{mx} = \frac{dt}{\beta} = -\frac{dw}{kw}$$

admits the first integrals

$$x \exp(-mt/\beta) = C_1, \quad w \exp(kt/\beta) = C_2.$$

Therefore, the general solution of the partial differential equation (3.2.2.5) is expressed as

$$w(x,t) = \exp(-kt/\beta) F(z), \quad z = x \exp(-mt/\beta), \tag{3.2.2.6}$$

where $F(z)$ is an arbitrary function.

A direct check can verify that expression (3.2.2.6) satisfies the functional equation (3.2.2.4). ◄

► Example 3.7. Now let us look at the functional equation

$$w(x,t) = a^k w\bigl(x + (1-a)t,\, a^n t\bigr), \tag{3.2.2.7}$$

where $a > 0$ is any and n is some constant.

Equation (3.2.2.4) becomes an identity when $a = 1$. By differentiating (3.2.2.7) with respect to a and then setting $a = 1$, we arrive at a first-order partial differential equation:

$$-tw_x + ntw_t + kw = 0. \tag{3.2.2.8}$$

The associated characteristic system

$$-\frac{dx}{t} = \frac{dt}{nt} = -\frac{dw}{kw}$$

has the first integrals

$$t + nx = C_1, \qquad wt^{k/n} = C_2.$$

Therefore, the general solution of the partial differential equation (3.2.2.8) is expressed as

$$w(x,t) = t^{-k/n} F(nx + t), \tag{3.2.2.9}$$

where $F(z)$ is an arbitrary function, and $n \neq 0$.

On substituting expression (3.2.2.9) into the original equation (3.2.2.7) and cancelling by $t^{-k/n}$, we obtain

$$F(nx + t) = F(nx + \sigma t), \qquad \sigma = (1 - a)n + a^n. \tag{3.2.2.10}$$

For $F(z) \neq \text{const}$, it follows that $\sigma = 1$, or

$$(1 - a)n + a^n = 1. \tag{3.2.2.11}$$

Since relation (3.2.2.10) must hold for any $a > 0$, relation (3.2.2.11) must also hold for any $a > 0$. This is possible only if

$$n = 1.$$

(The value $n = 1$ can be obtained by passing to the limit $a \to +0$ in (3.2.2.11).) In this case, the solution of equation (3.2.2.7) is defined by the formula (see (3.2.2.9) with $n = 1$):

$$w(x,t) = t^{-k} F(x + t), \tag{3.2.2.12}$$

where $F(z)$ is an arbitrary function.

If $n \neq 1$, equation (3.2.2.7) only has a degenerate solution, $w(x,t) = Ct^{-k/n}$, where C is an arbitrary constant (the degenerate solution corresponds to $F = \text{const}$ in (3.2.2.10)). ◀

•◆ *Problems and Exercises for Section 3.2*

1. Solve the functional equation

$$w(x,t) = w(x + ak_1, t + ak_2).$$

2. Solve the functional equation

$$w(x + ak_1, t) = w(x, t + ak_2).$$

3. Solve the functional equation

$$w(a^m x, t) = a^k w(x, a^n t).$$

4. Solve the functional equation

$$w(a^m x, t) = w(x, a^n t) + k \ln a.$$

5. Solve the functional equation

$$w(a^m x, t) = a^k w(x, t + \beta \ln a).$$

6. Solve the functional equation

$$w(x,t) = w(x + ak_1, t + ak_2) + ac.$$

7. Solve the functional equation

$$w(x,t) = e^{ac} w(x + ak_1, t + ak_2).$$

8. Solve the functional equation

$$w(x + ak_1, t) = e^{ac} w(x, t + ak_2).$$

Remark 3.9. *To solve equations 1–8, use the method of differentiation with respect to the parameter a (see Remark 3.8).*

3.3. Argument Elimination Method Using Test Functions

3.3.1. Classes of Equations. Description of the Method

We will investigate functional equations of the form (3.2.1.1) that satisfy conditions (3.2.1.2).

Instead of equation (3.2.1.1), we consider the more general auxiliary functional equation

$$w(x,t) = \theta(x,t,\xi)\, w\big(\varphi(x,t,\xi), \psi(x,t,\xi)\big), \tag{3.3.1.1}$$

where $\xi = \xi(x,t)$ is an arbitrary function.

The main idea of the method: if we obtain an exact solution to equation (3.3.1.1), this solution will also serve as a solution to the original functional equation (3.2.1.1). This is because equation (3.2.1.1) represents a special case of equation (3.3.1.1) with $\xi = a$.

Since the function $\xi = \xi(x,t)$ is arbitrary, we will now select a test (trial) function based on the following condition (Polyanin et al., 2005; Polyanin & Manzhirov, 2007):

$$\psi(x,t,\xi) = b, \tag{3.3.1.2}$$

where b is some constant (typically, setting $b = 1$ or $b = 0$ is convenient). By solving (3.3.1.2) for ξ and substituting the resulting test function $\xi = \xi(x,t)$ into (3.3.1.1), we obtain

$$w(x,t) = \theta(x,t,\xi(x,t))\, \Phi(\varphi(x,t,\xi(x,t))), \tag{3.3.1.3}$$

where $\Phi(\varphi) \equiv w(\varphi, b)$.

The expression in (3.3.1.3) serves as the foundation for constructing an exact solution to the original functional equation. One should substitute it into equation (3.2.1.1) and identify which functions $\Phi(\varphi)$ will satisfy this equation. In this process, some restrictions may arise concerning the forms of the determining functions θ, φ, and ψ.

Remark 3.10. *Condition (3.3.1.2) indicates the exclusion of the second argument, as it is replaced by a constant on the right-hand side of equation (3.3.1.1).*

Remark 3.11. *Rather than condition (3.3.1.2), the function $\xi = \xi(x,t)$ can be chosen based on a similar condition: $\varphi(x,t,\xi) = b$.*

3.3.2. Solution of Specific Functional Equations by Argument Elimination

► Example 3.8. Consider the functional equation (3.2.2.1), which is a special case of equation (3.2.1.1) where $\theta(x,t,a) = a^k$, $\varphi(x,t,a) = a^m x$, and $\psi(x,t,a) = a^n t$.

Following the scheme outlined in Section 3.3.1, we employ the auxiliary equation

$$w(x,t) = \xi^k w(\xi^m x, \xi^n t). \tag{3.3.2.1}$$

According to (3.3.1.2), we determine the test function ξ from the condition

$$\xi^n t = 1 \qquad (b = 1). \tag{3.3.2.2}$$

It follows that $\xi = t^{-1/n}$. Inserting this into (3.3.2.1), we get

$$w(x,t) = t^{-k/n}\Phi(t^{-m/n}x), \tag{3.3.2.3}$$

where $\Phi(\varphi) \equiv w(\varphi, 1)$.

It can be easily verified that expression (3.3.2.3) is a solution to the functional equation (3.2.2.1) for arbitrary Φ. Moreover, this solution coincides, up to renaming, with solution (3.2.2.3), which was obtained using the method of differentiation with respect to a parameter. ◄

Remark 3.12. *Instead of using 1 on the right-hand side of equation (3.3.2.2), one could use any non-zero constant b; the results would remain the same (up to a redefinition of the arbitrary function Φ).*

► Example 3.9. Let us look at the functional equation (3.2.2.4), which is a special case of equation (3.2.1.1) where $\theta(x,t,a) = a^k$, $\varphi(x,t,a) = a^m x$, and $\psi(x,t,a) = t + \beta \ln a$.

Following the above scheme, we use the more general auxiliary equation

$$w(x,t) = \xi^k w(\xi^m x, t + \beta \ln \xi). \tag{3.3.2.4}$$

We determined the test function ξ from the condition

$$t + \beta \ln \xi = 0 \qquad (b = 0).$$

It follows that $\xi = \exp(-t/\beta)$. Substituting this into (3.3.2.4), we get

$$w(x,t) = e^{-kt/\beta}\Phi(xe^{-mt/\beta}), \tag{3.3.2.5}$$

where $\Phi(\varphi) \equiv w(\varphi, 0)$. It is easy to verify that expression (3.3.2.5) is a solution to the functional equation (3.2.2.4) for arbitrary Φ. Moreover, this solution coincides with solution (3.2.2.6), which was obtained using the method of differentiation with respect to a parameter. ◄

► Example 3.10. Now consider the functional equation (3.2.2.7), which is a special case of equation (3.2.1.1) with $\theta(x,t,a)=a^k$, $\varphi(x,t,a)=x+(1-a)t$, and $\psi(x,t,a)=a^n t$.

Following the above scheme, we use the auxiliary equation

$$w(x,t)=\xi^k w\bigl(x+(1-\xi)t,\,\xi^n t\bigr). \tag{3.3.2.6}$$

According to (3.3.1.2), we determine the test function ξ from condition (3.3.2.2), which gives $\xi=t^{-1/n}$. Inserting this expression into (3.3.2.6) yields

$$w(x,t)=t^{-k/n}\Phi(z),\qquad z=x+t-t^{(n-1)/n}, \tag{3.3.2.7}$$

where $\Phi(\varphi)\equiv w(\varphi,1)$.

On substituting (3.3.2.7) into the original equation (3.2.2.7) and cancelling by $t^{-k/n}$, we obtain

$$\Phi\bigl(x+t-t^{(n-1)/n}\bigr)=\Phi\bigl(x+(1-a+a^n)t-a^{n-1}t^{(n-1)/n}\bigr). \tag{3.3.2.8}$$

Since this equality must hold for all $a>0$, we arrive at two possible outcomes:

$$\begin{aligned} &1)\quad n \text{ is any},\quad \Phi=C=\text{const};\\ &2)\quad n=1,\qquad \Phi \text{ is any}. \end{aligned} \tag{3.3.2.9}$$

In the second case, which corresponds to $n=1$ in the functional equation (3.2.2.7), the solution is written as

$$w(x,t)=t^{-k}F(x+t), \tag{3.3.2.10}$$

where $F(z)$ is an arbitrary function such that $F(z)=\Phi(z-1)$. It is clear that formula (3.3.2.10) matches solution (3.2.2.12) obtained through the method of differentiation with respect to a parameter. ◄

Remark 3.13. *The method of argument elimination is much simpler than the method of differentiation with respect to a parameter. This is because it only requires solving algebraic (or transcendental) equations of the form (3.3.1.2) for ξ, rather than deriving and solving first-order partial differential equations.*

Remark 3.14. *The argument elimination method using test functions is not mathematically rigorous. However, the solutions of specific functional equations obtained by the argument elimination method in Section 3.3.2 coincide with those obtained by the parameter differentiation method in Section 3.2.2. It should be noted that the intermediate results of solving equation (3.2.2.12) using these methods do not always match (compare formulas (3.2.2.9) and (3.3.2.7)).*

◆ *Problems and Exercises for Section 3.3*

See the problems and exercises provided at the end of Section 3.2.

► *For further information on various solution methods, properties, and applications of functional equations, the reader can refer to the following books: Kuczma (1968, 1985), Aczél & Dhombres (1989), Kuczma et al. (1990), Aczél (2002, 2006), Castillo et al. (2004), Polyanin & Manzhirov (2007), and Small (2007). Extensive tables of exact solutions for different functional equations can be found in the books by Polyanin & Manzhirov (2007) and Polyanin (2024). For functional equations that arise during the construction of exact solutions to nonlinear partial differential equations using the methods of generalized and functional separation of variables, see Sections 5.4 and 6.4. Additionally, refer to the books by Polyanin & Zaitsev (2012) and Polyanin & Zhurov (2022).*

References for Chapter 3

Aczél J. *Functional Equations: History, Applications and Theory*. Kluwer Academic, Dordrecht, 2002.

Aczél J. *Lectures on Functional Equations and Their Applications*, Dover Publications, New York, 2006.

Aczél J. and Dhombres J. *Functional Equations in Several Variables: With Applications to Mathematics, Information Theory and to the Natural and Social Sciences*. Cambridge University Press, Cambridge, 1989.

Castillo E., Iglesias A., and Ruiz-Cobo R. *Functional Equations in Applied Sciences*. Elsevier, New York, 2004.

Kuczma M. *An Introduction to the Theory of Functional Equations and Inequalities*, Polish Scientific Publishers, Warsaw, 1985.

Kuczma M. *Functional Equations in a Single Variable*, Polish Scientific Publishers, Warsaw, 1968.

Kuczma M., Choczewski B., and Ger R. *Iterative Functional Equations*, Cambridge University Press, Cambridge, 1990.

Polyanin A.D., Zaitsev V.F., and Zhurov A.I. *Solution Methods for Nonlinear Equations of Mathematical Physics and Mechanics*. Fizmatlit, Moscow, 2005.

Polyanin A.D. *Handbook of Exact Solutions to Mathematical Equations*. CRC Press, Boca Raton, 2024.

Polyanin A.D. and Manzhirov A.V. *Handbook of Mathematics for Engineers and Scientists (Chapters 17 and T12)*. Boca Raton–London: Chapman & Hall/CRC Press, 2007.

Polyanin A.D. and Zaitsev V.F. *Handbook of Nonlinear Partial Differential Equations, 2nd ed. (Sections 29.5 and 30.4)*. CRC Press, Boca Raton–London, 2012.

Polyanin A.D. and Zhurov A.I. *Separation of Variables and Exact Solutions to Nonlinear PDEs*. CRC Press, Boca Raton–London, 2022.

Small C.G. *Functional Equations and How to Solve Them*. Springer-Verlag, Berlin, 2007.

4. Elementary Invariant Theory: Partial Differential Equations

► **Preliminary remarks.** This chapter outlines a simple method for finding exact solutions to nonlinear equations of mathematical physics. The method utilizes linear translation and scaling transformations, as well as their combinations, and allows one to find the most common invariant solutions of nonlinear partial differential equations in a unified way.

The chapter provides examples of constructing self-similar and other exact solutions related to nonlinear equations in the fields of heat and mass transfer, wave theory, and hydrodynamics. In addition, it addresses some inverse problems, which require one to identify the form of equation that has solutions with prescribed properties.

It is important to highlight that this method does not require extensive training and involves only a minimal amount of intermediate calculations. Practically, one needs to be able to solve basic algebraic equations (or systems of such equations) and perform differentiation.

4.1. Description of the Method for Constructing Solutions Based on the Theory of Invariants

4.1.1. Transformations Preserving the Form of Equations. Transformation Invariants

We will consider reversible transformations of the form

$$x = X(\bar{x}, \bar{t}, \bar{u}), \quad t = T(\bar{x}, \bar{t}, \bar{u}), \quad u = U(\bar{x}, \bar{t}, \bar{u}). \tag{4.1.1.1}$$

A partial differential equation

$$F(x, t, u, u_x, u_t, u_{xx}, u_{xt}, u_{tt}, \dots) = 0 \tag{4.1.1.2}$$

is said to be *invariant under transformation* (4.1.1.1) if the substitution of (4.1.1.1) into (4.1.1.2) results in exactly the same equation

$$F(\bar{x}, \bar{t}, \bar{u}, \bar{u}_{\bar{x}}, \bar{u}_{\bar{t}}, \bar{u}_{\bar{x}\bar{x}}, \bar{u}_{\bar{x}\bar{t}}, \bar{u}_{\bar{t}\bar{t}}, \dots) = 0 \tag{4.1.1.3}$$

The function F in equations (4.1.1.1) and (4.1.1.2) is the same.

A function $I(x, t, u)$ (other than a constant) is called an *invariant of transformation* (4.1.1.1) if it remains unchanged under this transformation, or

$$I(x, t, u) = I(\bar{x}, \bar{t}, \bar{u}).$$

Remark 4.1. *If $I = I(x, t, u)$ is an invariant of transformation (4.1.1.1), then any function $\Psi(I)$ is also an invariant of the transformation.*

Note that the material presented later in this chapter is largely based on the works of Polyanin (2008) and Polyanin & Zaitsev (2012).

4.1.2. Procedure for Constructing Exact Solutions

Let us concretize the general scheme of using invariants for the analysis of mathematical equations described earlier in Chapter 1, 'Elementary Invariant Theory: Algebraic and Ordinary Differential Equations,' as applied to partial differential equations.

First, one seeks a *one-parameter transformation*

$$x = X(\bar{x}, \bar{t}, \bar{u}; c), \quad t = T(\bar{x}, \bar{t}, \bar{u}; c), \quad u = U(\bar{x}, \bar{t}, \bar{u}; c), \tag{4.1.2.1}$$

that preserves the form of equation (4.1.1.2). Transformation (4.1.2.1) includes a *free parameter* $c \in [c_1, c_2]$, whereas the original equation (4.1.1.2) does not feature this parameter.

In the second step, one identifies two functionally independent invariants of transformation (4.1.2.1):

$$I_1 = I_1(x, y, u), \qquad I_2 = I_2(x, y, u). \tag{4.1.2.2}$$

The invariants in equation (4.1.2.2) are independent of c and can be determined by eliminating c from the relations in (4.1.2.1).

In the third step, the invariants (4.1.2.2) are selected as the new variables, and one looks for an exact solution of equation (4.1.1.2) in the form

$$I_2 = \Phi(I_1). \tag{4.1.2.3}$$

The function $\Phi = \Phi(z)$ is defined by an ordinary differential equation, derived by substituting (4.1.2.3) into (4.1.1.2) (the corresponding solution is referred to as the *invariant solution*).

Remark 4.2. *Importantly, the invariant I_1 appearing in (4.1.2.2) and (4.1.2.3) is usually independent of u.*

As before (see Section 1.3), we will focus on the simplest one-parameter transformations (4.1.2.1),

$$x = \bar{x} + a_1, \quad t = \bar{t} + a_2, \quad u = \bar{u} + a_3 \quad (\textit{translation}); \tag{4.1.2.4}$$

$$x = b_1\bar{x}, \quad t = b_2\bar{t}, \quad u = b_3\bar{u} \quad (\textit{scaling}), \tag{4.1.2.5}$$

and their compositions. Here, a_n and b_n $(n = 1,\ 2,\ 3)$ are constant values dependent on the free parameter c.

For clarity, Figure 4.1 presents a general scheme for applying invariants to obtain invariant solutions of partial differential equations.

The example below illustrates how to determine invariants for the simplest transformations and the structure of the corresponding invariant solutions.

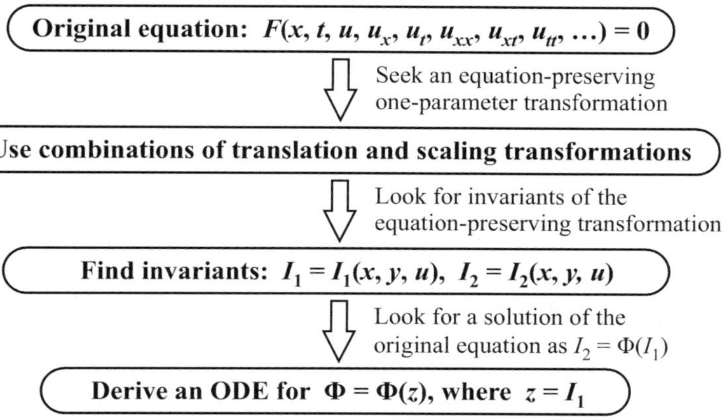

Figure 4.1. A scheme of using invariants to find invariant solutions for partial differential equations.

► Example 4.1. Let us consider a transformation that involves translation in the variable x and scaling in the variables t and u:

$$x = \bar{x} + k \ln c, \qquad t = c\bar{t}, \qquad u = c^m \bar{u}, \tag{4.1.2.6}$$

where k and m are some constants, and $c > 0$ is a free parameter.

We look for invariants of this transformation by eliminating the parameter c from relations (4.1.2.6). First, by expressing c from the second relation in (4.1.2.6), we obtain

$$c = t/\bar{t}. \tag{4.1.2.7}$$

By substituting (4.1.2.7) into the first relation in (4.1.2.6), we find that

$$x = \bar{x} + k \ln(t/\bar{t}) \implies x - k \ln t = \bar{x} - k \ln \bar{t}.$$

Therefore, the function

$$I_1 = x - k \ln t \tag{4.1.2.8}$$

is an invariant of transformation (4.1.2.6). Now by substituting (4.1.2.7) into the third relation of (4.1.2.6), we get

$$u = (t/\bar{t})^m \bar{u} \quad \implies \quad ut^{-m} = \bar{u}\bar{t}^{-m}.$$

Hence, the formula

$$I_2 = ut^{-m} \tag{4.1.2.9}$$

defines the second invariant of transformation (4.1.2.6). We seek an invariant solution in the form given by (4.1.2.3) or in the more convenient equivalent form solved for the unknown u:

$$u = t^m \Phi(z), \qquad z = x - k \ln t, \tag{4.1.2.10}$$

where the first invariant I_1 has been renamed to z for convenience. On substituting (4.1.2.10) into the partial differential equation in question, one can obtain an ordinary differential equation for $\Phi = \Phi(z)$. ◄

Remark 4.3. *In the given partial differential equation, if the order of the highest derivative with respect to x is greater than that with respect to t, then the first invariant $I_1 = z$ should be chosen to be linear in x. This choice makes it easier to calculate the partial derivatives with respect to x.*

Remark 4.4. *It is noteworthy that not every transformation has associated invariants (see Remark 4.10 following Example 4.9).*

4.2. Examples of Constructing Exact Solutions to Nonlinear Equations of Mathematical Physics

4.2.1. Traveling Wave Solutions (Obtained Through Translation)

Traveling wave solutions are exact solutions of the form

$$u = \Phi(z), \qquad z = kx - \lambda t,$$

where $\Phi = \Phi(z)$ is some function, and k and λ are some constants or free parameters. These are the simplest and very common solutions that remain unchanged under a one-parameter translation transformation of the independent variables described by the equations $t = \bar{t} + ck$ and $x = \bar{x} + c\lambda$, where c is an arbitrary constant.

► Example 4.2. Let us consider the nonlinear unsteady heat equation

$$u_t = [f(u)u_x]_x. \tag{4.2.1.1}$$

It remains unchanged under arbitrary translations in the independent variables:

$$t = \bar{t} + ck, \quad x = \bar{x} + c\lambda, \quad u = \bar{u}, \tag{4.2.1.2}$$

where c is a free parameter, and k and λ are given numbers (which can be selected arbitrarily). Eliminating c from the first two relations of (4.2.1.2), we find one invariant:

$$\frac{x - \bar{x}}{\lambda} = \frac{t - \bar{t}}{k} \quad \Longrightarrow \quad kx - \lambda t = k\bar{x} - \lambda\bar{t} = I_1.$$

The function $I_2 = u = \bar{u}$ is another invariant here. By virtue of (4.1.2.3), equation (4.2.1.1) admits solutions of the form

$$u = \Phi(z), \qquad z = kx - \lambda t. \tag{4.2.1.3}$$

On substituting (4.2.1.3) into (4.2.1.1), we arrive at an ordinary differential equation:

$$k^2[f(\Phi)\Phi'_z]'_z + \lambda\Phi'_z = 0.$$

Integrating twice gives an exact solution in implicit form

$$k^2 \int \frac{f(\Phi)\,d\Phi}{\lambda\Phi + C_1} = -z + C_2,$$

where C_1 and C_2 are arbitrary constants. ◄

Remark 4.5. *Traveling wave solutions are admitted by equations of a general form that do not explicitly depend on the independent variables:*

$$F(u, u_x, u_t, u_{xx}, u_{xt}, u_{tt}, \dots) = 0. \tag{4.2.1.4}$$

Substituting (4.2.1.3) into (4.2.1.4) yields an autonomous ordinary differential equation for $\Phi(z)$*:*

$$F(\Phi,\, k\Phi'_z,\, -\lambda\Phi'_z,\, k^2\Phi''_{zz},\, -k\lambda\Phi''_{zz},\, \lambda^2\Phi''_{zz},\, \dots) = 0. \tag{4.2.1.5}$$

► Example 4.3. Let us look at the linear wave equation

$$u_{tt} = a^2 u_{xx}. \tag{4.2.1.6}$$

We seek an exact solution in the traveling wave form (4.2.1.3). On substituting (4.2.1.3) into (4.2.1.6) and performing simple rearrangements, we obtain

$$(a^2k^2 - \lambda^2)\Phi''_{zz} = 0. \tag{4.2.1.7}$$

Equation (4.2.1.7) can be satisfied identically by setting the numerical factor in parentheses to zero. This yields two permissible values for the constants:

$$\lambda_1 = ak, \qquad \lambda_2 = -ak. \tag{4.2.1.8}$$

In both cases, the function $\Phi(z)$ in (4.2.1.3) remains arbitrary. Substituting the values (4.2.1.8) into (4.2.1.3), we get two solutions

$$u_1 = \Phi_1(kx - akt), \qquad u_2 = \Phi_2(kx + akt),$$

where $\Phi_1(z_1)$ and $\Phi_2(z_2)$ are arbitrary functions, and k is an arbitrary constant. Since the wave equation (4.2.1.6) is linear, the principle of linear superposition allows us to add up its solutions. Consequently, we derive a composite solution known as the *D'Alembert solution*:

$$u = \Phi_1(kx - akt) + \Phi_2(kx + akt). \tag{4.2.1.9}$$

It is the general solution of the equation in question. Without loss of generality, we can set $k = 1$ in (4.2.1.9), which results in a redefinition of the functions Φ_1 and Φ_2.

The linear solution $\Phi(z) = Az + B$, which corresponds to the vanishing of the second derivative Φ''_{zz} in (4.2.1.7), does not yield additional solutions. ◄

► Example 4.4. It is easy to verify that the homogeneous Monge–Ampère equation

$$u_{xx}u_{tt} - u_{xt}^2 = 0$$

has a traveling wave solution (4.2.1.3) where $\Phi(z)$ is an arbitrary function, and k and λ are arbitrary constants. ◄

4.2.2. Self-Similar Solutions (Obtained Through Scaling Transformations)

Self-similar solutions are exact solutions expressed in the form

$$u = t^k \Phi(z), \qquad z = t^m x,$$

where $\Phi = \Phi(z)$ is some function, and k and m are some functions. These solutions remain unchanged under one-parameter scaling transformations of the form $t = c\bar{t}$, $x = c^{-m}\bar{x}$, and $u = c^k \bar{u}$, where $c > 0$ is an arbitrary constant. Self-similar solutions, like traveling wave solutions, often arise in applications (e.g., see Sedov, 1993, and Barenblatt, 1996).

► Example 4.5. Let us consider the unsteady heat equation with a nonlinear power-law source

$$u_t = u_{xx} + \beta u^n. \tag{4.2.2.1}$$

We look for an invariant scaling transformation

$$x = a\bar{x}, \quad t = c\bar{t}, \quad u = b\bar{u}. \tag{4.2.2.2}$$

Substituting (4.2.2.2) into (4.2.2.1) and multiplying all terms by c/b, we get

$$\bar{u}_{\bar{t}} = \frac{c}{a^2}\bar{u}_{\bar{x}\bar{x}} + cb^{n-1}\beta\bar{u}^n.$$

Requiring that this equation matches (4.2.2.1), we arrive at two relationships

$$\frac{c}{a^2} = 1, \qquad cb^{n-1} = 1.$$

Expressing the parameters a and b via c gives

$$a = \sqrt{c}, \qquad b = c^{\frac{1}{1-n}}.$$

Substituting these expressions into (4.2.2.2) yields the invariant transformation

$$x = \sqrt{c}\,\bar{x}, \quad t = c\bar{t}, \quad u = c^{\frac{1}{1-n}}\,\bar{u}, \tag{4.2.2.3}$$

which involves a few instances of the parameters $c > 0$.

We will now determine the invariants. From the first two relations, we exclude c to obtain

$$x = \left(\frac{t}{\bar{t}}\right)^{1/2}\bar{x} \quad \Longrightarrow \quad xt^{-1/2} = \bar{x}\bar{t}^{-1/2} \quad \Longrightarrow \quad I_1 = xt^{-1/2}. \tag{4.2.2.4}$$

Eliminating c from the last two relationships of (4.2.2.3) yields

$$u = \left(\frac{t}{\bar{t}}\right)^{\frac{1}{1-n}}\bar{u} \quad \Longrightarrow \quad ut^{\frac{1}{n-1}} = \bar{u}\bar{t}^{\frac{1}{n-1}} \quad \Longrightarrow \quad I_2 = ut^{\frac{1}{n-1}}. \tag{4.2.2.5}$$

In view of (4.2.2.4) and (4.2.2.5), relation (4.1.2.3) defines the form of the self-similar solution:

$$u = t^{\frac{1}{1-n}}\Phi(z), \quad z = xt^{-1/2}. \tag{4.2.2.6}$$

Inserting (4.2.2.6) into (4.2.2.1) and performing simple rearrangements, we arrive at the ordinary differential equation

$$\frac{1}{1-n}\Phi - \frac{1}{2}z\Phi'_z = \Phi''_{zz} + \beta\Phi^n.$$

◀

Remark 4.6. *It is often more convenient to represent transformation (4.1.2.5) as*

$$t = c\bar{t}, \quad x = c^k\bar{x}, \quad u = c^m\bar{u} \qquad (c > 0) \tag{4.2.2.7}$$

and then determine the constants k and m at which the form of the equations remains unchanged. Here, c is a free parameter.

► Example 4.6. Let us illustrate how to apply scaling transformations (4.2.2.7) to the nonlinear heat equation

$$u_t = [f(u)u_x]_x, \tag{4.2.2.8}$$

where $f(u)$ is an arbitrary function.

Substituting (4.2.2.7) into (4.2.2.8) gives

$$\bar{u}_{\bar{t}} = c^{1-2k}[f(c^m\bar{u})\bar{u}_{\bar{x}}]_{\bar{x}}, \tag{4.2.2.9}$$

We require that the form of equations (4.2.2.8) and (4.2.2.9) coincide and considering the arbitrariness of the function $f(u)$, we determine the transformation parameters: $k = 1/2$ and $m = 0$. This implies that equation (4.2.2.8) has a self-similar solution (show this by reasoning similarly to Example 5.5):

$$u = \Phi(z), \quad z = xt^{-/2}. \tag{4.2.2.10}$$

Substituting (4.2.2.10) into (4.2.2.8) yields an ODE for $\Phi = \Phi(z)$:

$$[f(\Phi)\Phi'_z]'_z + \tfrac{1}{2}z\Phi'_z = 0. \tag{4.2.2.11}$$

The general solution of this nonlinear ODE can only be obtained for a few functions $f(\Phi)$, in particular, for $f(\Phi) = \text{const}$, which results in a linear ordinary differential equation. ◀

Remark 4.7. *The initial-boundary value problem for equations (4.2.2.8) for $x \geq 0$ and $t \geq 0$ subject to the initial and boundary conditions*

$$u(x, 0) = A, \quad u(t, 0) = B \qquad (A \text{ and } B \text{ are arbitrary constants})$$

has a self-similar solution (4.2.2.10) in which the function $\Phi = \Phi(z)$ is described by ODE (4.2.2.11) and the boundary conditions

$$\Phi(0) = B, \quad \Phi(\infty) = A.$$

Table 4.1. Some nonlinear equations of mathematical physics that allow for self-similar solutions.

Equation	Name	Form of solution	Final ODE
$u_t = au_{xx} + buu_x$	Burgers equation	$u = t^{-1/2}w(z)$, $z = xt^{-1/2}$	$aw'' + bww' + \frac{1}{2}zw' + \frac{1}{2}w = 0$
$u_t = au_{xx} + bu_x^2$	Potential Burgers equation	$u = u(z)$, $z = xt^{-1/2}$	$au'' + b(u')^2 + \frac{1}{2}zu' = 0$
$u_t = f(u_x)u_{xx}$	Equation of flow in porous medium	$u = t^{1/2}w(z)$, $z = xt^{-1/2}$	$2f(w')w'' + zw' - u = 0$
$u_{tt} = [f(u)u_x]_x$	Wave equation	$u = u(z)$, $z = x/t$	$(z^2u')' = [f(u)u']'$
$u_{xx} + u_{yy} = au^n$	Heat equations with source	$u = x^{\frac{2}{1-n}}w(z)$, $z = y/x$	$(1+z^2)w'' - \frac{2(1+n)}{1-n}zw$ $+ \frac{2(1+n)}{(1-n)^2}w - aw^n = 0$
$u_{xx} + au_yu_{yy} = 0$	Transonic gas flow equation	$u = x^{-3k-2}w(z)$, $z = x^ky$, k is any	$\frac{a}{k+1}w'w'' + \frac{k^2}{k+1}z^2w''$ $- 5kzw' + 3(3k+2)w = 0$
$u_t = au_{xxx} + buu_x$	Korteweg–de Vries equation	$u = t^{-2/3}w(z)$, $z = xt^{-1/3}$	$aw''' + bww' + \frac{1}{3}zw' + \frac{2}{3}w = 0$
$u_yu_{xy} - u_xu_{yy} = au_{yyy}$	Boundary layer equation	$u = x^{\lambda+1}w(z)$, $z = x^\lambda y$, λ is any	$(2\lambda+1)(w')^2 - (\lambda+1)ww''$ $= aw'''$

Some additional equations with self-similar solutions are listed in Table 4.1, where the primes denote derivatives with respect to z.

A similar approach can effectively be used to find self-similar solutions to systems of partial differential equations.

► Example 4.7. Let us consider the following system of equations of a steady laminar hydrodynamic boundary layer on a flat plate:

$$\begin{aligned} uu_x + vu_y &= au_{yy}, \\ u_x + v_y &= 0. \end{aligned} \tag{4.2.2.12}$$

Let us scale the independent and dependent variables in (4.2.2.12) as follows:

$$x = c\bar{x}, \qquad y = c^k\bar{y}, \qquad u = c^m\bar{u}, \qquad v = c^n\bar{v}, \tag{4.2.2.13}$$

where $c > 0$ is an arbitrary constant, and the constants k, m, and n are to be determined in the subsequent analysis. On substituting (4.2.2.13) into (4.2.2.12) and multiplying the resulting equations by suitable factors, we obtain

$$\begin{aligned} \bar{u}\bar{u}_{\bar{x}} + c^{n-m-k+1}\bar{v}\bar{u}_{\bar{y}} &= c^{-m-2k+1}a\bar{u}_{\bar{y}\bar{y}}, \\ \bar{u}_{\bar{x}} + c^{n-m-k+1}v_{\bar{y}} &= 0. \end{aligned} \tag{4.2.2.14}$$

We require that the transformed system (4.2.2.14) coincide with the original system (4.2.2.12). This condition results in two linear algebraic equations for the three constants k, m, and n:

$$\begin{aligned} n - m - k + 1 &= 0, \\ -2k - m + 1 &= 0. \end{aligned}$$

On solving these for m and n, we get

$$m = 1 - 2k, \qquad n = -k. \tag{4.2.2.15}$$

The exponent k can be chosen arbitrarily. Substituting relations (4.2.2.15) into transformation (4.2.2.13) yields

$$x = c\bar{x}, \quad y = c^k \bar{y}, \quad u = c^{1-2k}\bar{u}, \quad v = c^{-k}\bar{v}.$$

By eliminating c, we find the invariants

$$\begin{aligned} y &= \left(\frac{x}{\bar{x}}\right)^k \bar{y} &&\Longrightarrow && yx^{-k} = \bar{y}\bar{x}^{-k} = I_1 = \zeta; \\ u &= \left(\frac{x}{\bar{x}}\right)^{1-2k} \bar{u} &&\Longrightarrow && ux^{2k-1} = \bar{u}\bar{x}^{2k-1} = I_2; \\ v &= \left(\frac{x}{\bar{x}}\right)^k \bar{v} &&\Longrightarrow && vx^k = \bar{v}\bar{x}^k = I_3. \end{aligned}$$

We look for a solution in the form $I_2 = U(I_1)$, $I_3 = V(I_1)$, or

$$ux^{2k-1} = U(\zeta), \quad vx^k = V(\zeta).$$

On solving these relations for the unknown functions, we get

$$u(x,y) = x^{1-2k}U(\zeta), \quad v(x,y) = x^{-k}V(\zeta), \quad \zeta = yx^{-k}, \tag{4.2.2.16}$$

where k is an arbitrary constant. Substituting (4.2.2.16) into the original system (4.2.2.12), we arrive at a system of ordinary differential equations for $U = U(\zeta)$ and $V = V(\zeta)$:

$$\begin{aligned} U\big[(1-2k)U - k\zeta U'_\zeta\big] + VU'_\zeta &= aU''_{\zeta\zeta}, \\ (1-2k)U - k\zeta U'_\zeta + V'_\zeta &= 0. \end{aligned}$$

◀

Remark 4.8. *In various applications, self-similar solutions are often sought using similarity and dimensionality methods (e.g., see Birkhoff, 1960; Sedov, 1993; and Barenblatt, 1996).*

4.2.3. Other Invariant Solutions (Obtained Through Compositions of Translation and Scaling)

Specific examples are provided below to demonstrate how invariant solutions to nonlinear equations of mathematical physics can be found using compositions of translation and scaling.

► Example 4.8. Consider the Klein–Gordon equation with an exponential nonlinearity

$$u_{tt} = u_{xx} + \beta e^{\lambda u}. \tag{4.2.3.1}$$

We look for a invariant transformation as a composition of translation and scaling:

$$x = a\bar{x}, \quad t = c\bar{t}, \quad u = \bar{u} + b. \tag{4.2.3.2}$$

Inserting (4.2.3.2) into (4.2.3.1) and multiplying by c^2, we get

$$\bar{u}_{\bar{t}\bar{t}} = \frac{c^2}{a^2}\bar{u}_{\bar{x}\bar{x}} + c^2 e^{\lambda b}\beta e^{\lambda\bar{u}}.$$

For this equation to coincide with (4.2.3.1), we must set

$$\frac{c^2}{a^2} = 1, \quad c^2 e^{\lambda b} = 1.$$

It follows that

$$a = c, \quad b = -\frac{2}{\lambda}\ln c.$$

Then, transformation (4.2.3.2) becomes

$$x = c\bar{x}, \quad t = c\bar{t}, \quad u = \bar{u} - \frac{2}{\lambda}\ln c, \tag{4.2.3.3}$$

where $c > 0$ is an arbitrary constant.

On eliminating the parameter c from (4.2.3.2), we find the invariants

$$I_1 = \frac{x}{t} = \frac{\bar{x}}{\bar{t}}, \quad I_2 = u + \frac{2}{\lambda}\ln t = \bar{u} + \frac{2}{\lambda}\ln\bar{t}. \tag{4.2.3.4}$$

With allowance for (4.2.3.4), relation (4.1.2.3) gives an invariant solution in the form

$$u = \Phi(z) - \frac{2}{\lambda}\ln t, \quad z = \frac{x}{t}. \tag{4.2.3.5}$$

Substituting (4.2.3.5) into (4.2.3.1) yields an ordinary differential equation:

$$(z^2\Phi'_z)'_z + \frac{2}{\lambda} = \Phi''_{zz} + \beta e^{\lambda\Phi}.$$

◄

Remark 4.9. *It is often more convenient to represent transformation (4.2.3.2) in the form*

$$t = c\bar{t}, \quad x = c^k\bar{x}, \quad u = \bar{u} + m\ln c \qquad (c > 0) \tag{4.2.3.6}$$

and then determine the constants k and m that ensure that the form of the equation is preserved. Here, c is a free parameter.

► Example 4.9. Consider the nonlinear unsteady heat equation

$$u_t = (u^n u_x)_x. \tag{4.2.3.7}$$

We seek an invariant transformation as a composition of translation and scaling:

$$x = a\bar{x}, \quad t = \bar{t} + b, \quad u = c\bar{u}. \tag{4.2.3.8}$$

On substituting (4.2.3.8) into (4.2.3.7) and dividing by c, we get

$$\bar{u}_{\bar{t}} = \frac{c^n}{a^2}(\bar{u}^n \bar{u}_{\bar{x}})_{\bar{x}}. \tag{4.2.3.9}$$

The requirement that equations (4.2.3.9) and (4.2.3.7) coincide leads to one condition

$$\frac{c^n}{a^2} = 1 \quad \Longrightarrow \quad a = c^{n/2},$$

where $c > 0$ is an arbitrary constant. The other parameter, b, remains arbitrary. Hence the resulting invariant transformation

$$x = c^{n/2}\bar{x}, \quad t = \bar{t} + b, \quad u = c\bar{u} \tag{4.2.3.10}$$

is two-parameter.

Since the method we are using requires a one-parameter invariant transformation, we set

$$b = f(c). \tag{4.2.3.11}$$

The function f will be determined later. Eliminating the parameter c from the last two relations in (4.2.3.10) and considering (4.2.3.11), we obtain

$$t = \bar{t} + f\left(\frac{u}{\bar{u}}\right). \tag{4.2.3.12}$$

By adding a function $\varphi(u)$ to both sides of (4.2.3.12), we get

$$t + \varphi(u) = \bar{t} + \varphi(u) + f\left(\frac{u}{\bar{u}}\right).$$

This relation defines an invariant, $I(t,u) = I(\bar{t},\bar{u})$, if

$$\varphi(u) + f\left(\frac{u}{\bar{u}}\right) = \varphi(\bar{u}). \tag{4.2.3.13}$$

This equation can be treated as a functional equation. Assuming u and $\bar{u}$ to be independent variables, we can apply the differentiation method (see Section 3.1) to find a solution of equation (4.2.3.13):

$$f(c) = A \ln c, \tag{4.2.3.14}$$

where A is an arbitrary constant. On substituting (4.2.3.11) into (4.2.3.10) and taking into account (4.2.3.14), we obtain

$$x = c^{n/2}\bar{x}, \quad t = \bar{t} + A\ln c, \quad u = c\bar{u} \tag{4.2.3.15}$$

Assuming that $A \neq 0$, we express c via t from the second relation in (4.2.3.15) and substitute into the other relations to obtain the invariants

$$I_1 = xe^{knt} = \bar{x}e^{kn\bar{t}}, \quad I_2 = ue^{2kt} = \bar{u}e^{2k\bar{t}}, \quad \text{where} \quad k = -\frac{1}{2A}. \tag{4.2.3.16}$$

With the help of formula (4.1.2.3), we determine the form of exact solutions to equation (4.2.3.7):

$$u = e^{-2kt}\Phi(z), \qquad z = xe^{knt}. \tag{4.2.3.17}$$

On substituting (4.2.3.17) into (4.2.3.7), we arrive at an ordinary differential equation for $\Phi(z)$:

$$-2k\Phi + knz\Phi'_z = (\Phi^n\Phi'_z)'_z.$$

◀

Remark 4.10. *For $b = c$, transformation (4.2.3.10) preserves the form of equation (4.2.3.7) but does not have invariants. It follows that not every transformation that preserves the form of an equation has invariants.*

Remark 4.11. *Exact solutions of the form*

$$u = e^{kt}\Phi(z), \qquad z = xe^{mt},$$

where k and m are some constants, are called limiting self-similar solutions. Such solutions are invariant under compositions of translation in t and scaling in x and u.

Table 4.2 lists several examples of invariant solutions that can be obtained using compositions of translation and scaling that preserve the form of nonlinear equations of mathematical physics.

Table 4.2. Invariant solutions that can be obtained using compositions of translation and scaling that preserve the form of nonlinear equations of mathematical physics (c is an arbitrary constant, $c > 0$).

Invariant transformations	Form of invariant solutions	Examples of equations
$t = \bar{t} + ck,\ x = \bar{x} + c\lambda$	$u = u(z),\ z = kx - \lambda t$	$u_t = [f(u)u_x]_x$
$t = c\bar{t},\ x = c^k\bar{x},\ u = c^m\bar{u}$	$u = t^m w(z),\ z = xt^{-k}$	See Table 4.1 for equations
$t = \bar{t} + \ln c,\ x = c^k\bar{x},\ u = c^m\bar{u}$	$u = e^{mt}w(z),\ z = xe^{-kt}$	$u_t = a(u^n u_x)_x$; $k = \frac{1}{2}mn$, m is any
$t = c\bar{t},\ x = \bar{x} + k\ln c,\ u = c^m\bar{u}$	$u = t^m w(z),\ z = x - k\ln t$	$u_t = a(u^n u_x)_x$; $m = -1/n$, k is any
$t = c\bar{t},\ x = c^\beta\bar{x},\ u = \bar{u} + \alpha\ln c$	$u = w(z) + \alpha\ln t,\ z = xt^{-\beta}$	$u_t = (e^u u_x)_x$; $\alpha = 2\beta - 1$, β is any
$t = c\bar{t},\ x = \bar{x} + \beta\ln c,\ u = \bar{u} + \alpha\ln c$	$u = w(z) + \alpha\ln t,$ $z = x - \beta\ln t$	$u_{xt}^2 - u_{xx}u_{tt} = 0$; α and β are any
$t = \bar{t} + c,\ x = \bar{x} + c\lambda,\ u = \bar{u} + ck$	$u = w(z) + kt,\ z = x - \lambda t$	$u_t = f(u_x)u_{xx}$; k and λ are any
$t = \bar{t} + \ln c,\ x = \bar{x} + k\ln c,\ u = c^m\bar{u}$	$u = e^{mt}w(z),\ z = x - kt$	$u_{xt}^2 - u_{xx}u_{tt} = 0$; k and m are any

Using compositions of translation and scaling, one can also find invariant solutions for nonlinear systems of partial differential equations. We will illustrate this with a specific example.

► Example 4.10. Let us reconsider the system of equations of a steady-state laminar hydrodynamic boundary layer on a flat plate (4.2.2.12). We seek an invariant transformation as a composition of scaling and translation:

$$x = c\bar{x}, \quad y = \bar{y} + \beta \ln c, \quad u = c^k \bar{u}, \quad v = c^m \bar{v}, \tag{4.2.3.18}$$

where $c > 0$ and β are arbitrary constants, while the constants k and m are to be determined in the subsequent analysis. We substitute (4.2.3.18) into (4.2.2.12) and then multiply the resulting equations by suitable constant factors to obtain

$$\begin{aligned} \bar{u}\bar{u}_{\bar{x}} + c^{1-k+m}\bar{v}\bar{u}_{\bar{y}} &= c^{1-k}a\bar{u}_{\bar{y}\bar{y}}, \\ \bar{u}_{\bar{x}} + c^{1-k+m}v_{\bar{y}} &= 0. \end{aligned} \tag{4.2.3.19}$$

We require that the form of the transformed system (4.2.3.19) coincide with that of the original system (4.2.2.12). This condition leads to two simple linear algebraic equations, $1 - k + m = 0$ and $1 - k = 0$, whose solution is

$$k = 1, \quad m = 0. \tag{4.2.3.20}$$

On substituting the values of k and m from (4.2.3.20) into transformation (4.2.3.18), we get

$$x = c\bar{x}, \quad y = \bar{y} + \beta \ln c, \quad u = c\bar{u}, \quad v = \bar{v},$$

We eliminate the parameter c to obtain the invariants

$$\begin{aligned} y = \bar{y} + \beta \ln \frac{x}{\bar{x}} \quad &\Longrightarrow \quad y - \beta \ln x = \bar{y} - \beta \ln \bar{x} = I_1 = \zeta; \\ u = \frac{x}{\bar{x}}\bar{u} \quad &\Longrightarrow \quad \frac{u}{x} = \frac{u}{\bar{x}} = I_2; \\ v = \bar{v} \quad &\Longrightarrow \quad v = \bar{v} = I_3. \end{aligned}$$

We seek a solution in the form $I_2 = U(I_1)$, $I_3 = V(I_1)$, or

$$\frac{u}{x} = U(\zeta), \quad v = V(\zeta).$$

It follows that

$$u(x, y) = xU(\zeta), \quad v(x, y) = V(\zeta), \quad \zeta = y - \beta \ln x. \tag{4.2.3.21}$$

Substituting the expressions of (4.2.3.21) into the original system (4.2.2.12), we arrive at a system of ODEs for $U = U(\zeta)$ and $V = V(\zeta)$:

$$\begin{aligned} U\big(U - \beta U'_\zeta\big) + VU'_\zeta &= aU''_{\zeta\zeta}, \\ U - \beta U'_\zeta + V'_\zeta &= 0. \end{aligned}$$

◄

► **Concluding remarks.** The main limitations of constructing solutions based on invariant theory are as follows:

1) A transformation that preserves the form of the equation, other than the identity, does not always exist.

2) Not every transformation that preserves the form of the equation can be effectively utilized; it must depend on a free parameter.

3) An invariant for the given transformation does not always exist.

Problems and Exercises for Section 4.2

1. Find a self-similar solution, more general than (4.2.2.10), of the nonlinear heat equation (4.2.3.7).

2. Find a solution of the nonlinear heat equation (4.2.3.7) based on the invariant transformation

$$x = \bar{x} + a, \quad t = b\bar{t}, \quad u = c\bar{u}.$$

3. Find a self-similar solution of the nonlinear heat equation with a source

$$u_t = a(u^n u_x)_x + bu^k.$$

4. Find invariant solutions of the nonlinear heat equation

$$u_t = a(e^{\lambda u} u_x)_x.$$

5. Find invariant solutions of the steady-state combustion theory equation with exponential heat release

$$u_{xx} + u_{yy} = \beta e^{\lambda u}.$$

6. Find a self-similar solution and other invariant solutions of the nonlinear seepage equation

$$u_t = a u_x^k u_{xx}.$$

7. Find a self-similar solution and other invariant solutions of the nonlinear wave equation

$$u_{tt} = a(u^n u_x)_x.$$

8. Find invariant solutions of the nonlinear wave equation

$$u_{tt} = a(e^{\lambda u} u_x)_x.$$

9. Find invariant solutions of the transonic gas flow equation

$$u_{xx} + a u_y u_{yy} = 0$$

based on searching for invariant compositions of translation and scaling.

10. Find invariant solutions of the equation

$$u_{xx} + F(u_x, u_y) u_{yy} = 0.$$

11. Find invariant solutions of the equation

$$u_{xx} + F(u_x, u_y) u_{xy} = 0.$$

12. Find invariant solutions of the homogeneous Monge–Ampère equation

$$u_{xx} u_{yy} - u_{xy}^2 = 0.$$

13. Find invariant solutions of the Monge–Ampère type equation

$$u_{xx} u_{yy} - u_{xy}^2 = f(u).$$

14. Find invariant solutions of the Monge–Ampère type equation

$$u_{xx}u_{yy} - u_{xy}^2 = f(u_x).$$

15. Find invariant solutions of the Boussinesq equation

$$u_{tt} + a(uu_x)_x + bu_{xxxx} = 0.$$

16. Find invariant solutions of the equation

$$u_{tt} = au_{xxxx} + be^{\lambda u}.$$

17. Find a self-similar solution and other invariant solutions of the nth-order equation

$$u_t = a(u^k)_x^{(n)}.$$

18. Find a self-similar solution and other invariant solutions of the nth-order equation

$$u_{tt} = a(e^u)_x^{(n)}.$$

19. Find an invariant solution of the nonlinear system of reaction-diffusion type equations (Aksenov & Polyanin, 2021):

$$u_t = a(u^k u_x)_x + uf(u/v),$$
$$v_t = b(v^k v_x)_x + vg(u/v),$$

where a, b, and k are some constants ($k \neq 0$), and $f(z)$ and $g(z)$ are arbitrary functions.

20. Find an invariant solution of the nonlinear system of reaction-diffusion type equations (Aksenov & Polyanin, 2021):

$$u_t = a(u^k u_x)_x + u^{k+1} f(u/v),$$
$$v_t = b(v^k v_x)_x + v^{k+1} g(u/v),$$

where a, b, and k are some constants ($k \neq 0$), and $f(z)$ and $g(z)$ are arbitrary functions.

4.3. Inverse Problems (Determining Equation Structure from Its Properties)

4.3.1. Preliminary Remarks

Until now, we have focused on finding exact solutions to given nonlinear partial differential equations. However, there are times when it is necessary to identify the form of an equation that possesses certain properties or has specific solutions. These types of problems are referred to as *inverse problems*. In some cases, solving inverse problems can be reduced to solving functional equations; we discussed methods for solving functional equations earlier in Chapter 3. Below are several examples that illustrate how to formulate inverse problems as well as solution methods for these problems.

4.3.2. Examples of Inverse Problems and Their Solutions

► Example 4.11. Consider the nonlinear parabolic equation in a fairly general form

$$u_t = au_{xx} + F(u, u_x). \tag{4.3.2.1}$$

It is required to find the most general function $F = F(u, w)$ for which equation (4.3.2.1) admits self-similar solutions.

To have a self-similar solution, equation (4.3.2.1) must be invariant under the scaling transformation (4.2.2.7). Substituting (4.2.2.7) into (4.3.2.1) and performing simple rearrangements, we obtain

$$\bar{u}_{\bar{t}} = c^{1-2k} a\bar{u}_{\bar{x}\bar{x}} + c^{1-m} F(c^m \bar{u}, c^{m-k} \bar{u}_{\bar{x}}). \tag{4.3.2.2}$$

The requirement that equation (4.3.2.2) coincide with equation (4.3.2.1), up to renaming of variables, leads to $k = \frac{1}{2}$ and the following functional equation for $F = F(u, w)$:

$$F(u, w) = c^{1-m} F(c^m u, c^{m-1/2} w), \tag{4.3.2.3}$$

where $w = u_x$. Two solution methods for such functional equations were discussed in Sections 3.2 and 3.3, where u and w are treated as independent variables, and $c > 0$ is a free parameter. By applying any of these methods, we can derive the general solution of equation (4.3.2.3):

$$F(u, w) = u^{\frac{m-1}{m}} f\big(u^{\frac{1-2m}{2m}} w\big), \qquad m \neq 0, \tag{4.3.2.4}$$

where $f(z)$ is an arbitrary function. The reader is advised to obtain this solution by themselves.

Substituting (4.3.2.4) into (4.3.2.1) yields

$$u_t = au_{xx} + u^{\frac{m-1}{m}} f\big(u^{\frac{1-2m}{2m}} u_x\big). \tag{4.3.2.5}$$

This equation is invariant under the scaling transformation (4.2.2.7) at $k = \frac{1}{2}$ (m is a free parameter).

Using the scaling transformation (4.2.2.7) with the above parameters, one can find that equation (4.3.2.4) has a self-similar solution of the form (see Section 4.2.2):

$$w = t^m \varphi(z), \qquad z = xt^{-1/2},$$

where $\varphi = \varphi(z)$ is a function satisfying the ordinary differential equation

$$m\varphi - \tfrac{1}{2} z\varphi'_z = a\varphi''_{zz} + \varphi^{\frac{m-1}{m}} f\big(\varphi^{\frac{1-2m}{2m}} \varphi'_z\big).$$

Remark 4.12. *The case $m = 0$, in which equation (4.3.2.5) loses its meaning, should be treated separately. For $m = 0$, the functional equation (4.3.2.3) becomes*

$$F(u, w) = cF(u, c^{-1/2} w). \tag{4.3.2.6}$$

The general solution of this equation is given by

$$F(u, w) = f(u) w^2 \qquad (m = 0), \tag{4.3.2.7}$$

where $f(u)$ is an arbitrary function. (The reader is recommended to obtain this solution by themselves.) Substituting (4.3.2.7) into (4.3.2.1) yields the equation

$$u_t = au_{xx} + f(u)u_x^2, \tag{4.3.2.8}$$

which is invariant under the scaling transformation (4.2.2.7) at $k = \frac{1}{2}$ and $m = 0$.

Equation (4.3.2.8) has a self-similar solution of the form (see Section 4.2.2):

$$u = \varphi(z), \quad z = xt^{-1/2},$$

where $\varphi = \varphi(z)$ is a function satisfying the ordinary differential equation

$$-\tfrac{1}{2}z\varphi'_z = a\varphi''_{zz} + f(\varphi)(\varphi'_z)^2.$$

◀

► Example 4.12. Consider another nonlinear equation

$$u_{tt} = F(u, u_x)u_{xx}. \tag{4.3.2.9}$$

It is required to find the most general function $F = F(u, w)$ for which equation (4.3.2.9) admits limiting self-similar solutions (see Remark 4.11).

To have a limiting self-similar solution, equation (4.3.2.9) must be invariant under a composition translation and scaling of the form

$$t = \bar{t} + A\ln c, \quad x = c\bar{x}, \quad u = c^m\bar{u}, \tag{4.3.2.10}$$

where c is an arbitrary constant ($c > 0$), and A and m are some constants.

On substituting (4.3.2.10) into (4.3.2.9) and dividing by c^m, we obtain

$$\bar{u}_{\bar{t}\bar{t}} = c^{-2}F(c^m\bar{u}, c^{m-1}\bar{u}_{\bar{x}})\bar{u}_{\bar{x}\bar{x}}. \tag{4.3.2.11}$$

The requirement that equation (4.3.2.11) coincide with equation (4.3.2.9), up to renaming of variables, leads to the following functional equation for $F = F(u, w)$:

$$F(u, w) = c^{-2}F(c^m u, c^{m-1}w). \tag{4.3.2.12}$$

The general solution of the functional equation (4.3.2.12) is expressed as

$$F(u, w) = u^{\frac{2}{m}} f\big(u^{\frac{1-m}{m}} w\big), \quad m \neq 0, \tag{4.3.2.13}$$

where $f(z)$ is an arbitrary function. The reader is recommended to derive this solution by themselves (see Section 3.2 or 3.3).

Substituting (4.3.2.13) into (4.3.2.9) gives

$$u_{tt} = u^{\frac{2}{m}} f\big(u^{\frac{1-m}{m}} u_x\big)u_{xx}. \tag{4.3.2.14}$$

Using transformation (4.3.2.10), one can verify that equation (4.3.2.14) admits a limiting self-similar solution of the form

$$u = e^{kmt}\psi(y), \quad y = xe^{-kt},$$

where $k = 1/A$ and m are free parameters.

The reader is recommended to derive the equation for $\psi = \psi(y)$ by themselves.

Remark 4.13. *The case* $m = 0$ *should be treated separately. For* $m = 0$, *the functional equation (4.3.2.12) becomes*

$$F(u, w) = c^{-2} F(u, c^{-1} w). \tag{4.3.2.15}$$

The general solution of this equation (find it by yourself) is expressed as

$$F(u, w) = f(u) w^{-2} \qquad (m = 0), \tag{4.3.2.16}$$

where $f(u)$ *is an arbitrary function.*

Substituting (4.3.2.16) into (4.3.2.9) gives the equation

$$u_{tt} = f(u) u_x^{-2} u_{xx}, \tag{4.3.2.17}$$

which admits an exact solution of the form

$$u = \psi(y), \quad y = x e^{-kt}. \tag{4.3.2.18}$$

Here, the function $\psi = \psi(y)$ *is described by the ODE*

$$k^2 y (y \psi'_y)'_y = f(\psi)(\psi'_y)^{-2} \psi''_{yy}. \tag{4.3.2.19}$$

(The reader is recommended to demonstrate that equation (4.3.2.17) has a solution of the form (4.3.2.18) and to derive equation (4.3.2.19) for $\psi = \psi(y)$.*)* ◄

► Example 4.13. Now let us identify the general form of the function $F = F(u, w)$ for which equation (4.3.2.9) is invariant under the following composition scaling and translation:

$$t = c\bar{t}, \quad x = c^k \bar{x}, \quad u = \bar{u} + A \ln c, \tag{4.3.2.20}$$

where c is an arbitrary constant ($c > 0$), and A and k are some functions.

Substituting (4.3.2.20) into (4.3.2.9) gives

$$\bar{u}_{\bar{t}\bar{t}} = c^{2(1-k)} F(\bar{u} + A \ln c,\, c^{-k} \bar{u}_{\bar{x}}) \bar{u}_{\bar{x}\bar{x}}. \tag{4.3.2.21}$$

By comparing (4.3.2.21) and (4.3.2.9), we arrive at the functional equation

$$F(u, w) = c^{2(1-k)} F(u + A \ln c, c^{-k} w), \tag{4.3.2.22}$$

whose general solution is expressed as

$$F(u, w) = e^{2(k-1)\lambda u} f(e^{k\lambda u} w), \qquad \lambda = 1/A, \tag{4.3.2.23}$$

where $f(z)$ is an arbitrary function.

Equation (4.3.2.9) with function (4.3.2.23) admits invariant solutions of the form

$$u = \theta(r) + \frac{1}{\lambda} \ln t, \quad r = x t^{-k}.$$

The reader is recommended to derive the equation for $\theta(r)$ by themselves. ◄

Problems and Exercises for Section 4.3

1. For the equation

$$u_t = u_{xx} + F(x, u),$$

find the general form of the function $F(x, u)$ ensuring that the equation has a self-similar solution. Determine this solution.

2. For the equation

$$u_t = uu_{xx} + F(u, u_x),$$

find the general form of the function $F(u, v)$ ensuring that the equation has a self-similar solution. Determine this solution.

3. For the equation

$$u_t = u^2 F(u_x, u_{xx}),$$

find the general form of the function $F(v, w)$ ensuring that the equation has a self-similar solution. Determine this solution.

4. For the equation from Exercise 3, find the most general functions $F(v, w)$ for which the equation has other simple invariant solutions (obtainable using compositions of translation and scaling).

5. For the equation

$$u_{xt} = uF(u_x, u_{xx}),$$

find the general form of the function $F(v, w)$ ensuring that the equation has a self-similar solution. Determine this solution.

6. For the equation from Exercise 5, find the most general functions $F(v, w)$ for which the equation has other simple invariant solutions (obtainable using compositions of translation and scaling).

7. For the equation

$$u_t = [a(x)f(u)u_x]_x + b(x)g(u),$$

find the functions $a(x)$ and $b(x)$ ensuring that the equation has a self-similar solution; the functions $f(u)$ and $g(u)$ are considered arbitrary. Determine this solution.

8. Find the functions $a(x)$ and $b(x)$ for which the equation from Exercise 7 has other simple invariant solutions (obtainable using compositions of translation and scaling).

9. For the equation

$$u_t = [a(t)f(u)u_x]_x + b(x)g(u),$$

find the functions $a(t)$ and $b(x)$ ensuring that the equation has a self-similar solution; the functions $f(u)$ and $g(u)$ are considered arbitrary. Determine this solution.

10. For the equation

$$u_t = [a(x)f(u)u_x]_x + b(t)g(u),$$

find the functions $a(x)$ and $b(t)$ ensuring that the equation has a self-similar solution; the functions $f(u)$ and $g(u)$ are considered arbitrary. Determine this solution.

11. For the equation

$$u_t = [a(x)f(u)u_x]_x + b(x)g(u)u_x,$$

find the functions $a(x)$ and $b(x)$ ensuring that the equation has a self-similar solution; the functions $f(u)$ and $g(u)$ are considered arbitrary. Determine this solution.

12. For the equation

$$u_{xt} = [a(x)f(u)u_x]_x + b(x)g(u),$$

find the functions $a(x)$ and $b(x)$ ensuring that the equation has a self-similar solution; the functions $f(u)$ and $g(u)$ are considered arbitrary. Determine this solution.

4.4. Concluding Remarks

The method described in this chapter serves to find invariant solutions to nonlinear equations of mathematical physics. It employs concepts from the group analysis of differential equations (see Chapter 8 for more details and the books by Ovsiannikov, 1982; Ibragimov, 1994; and Olver, 1993). However, it is much simpler for practical application. The method relies on linear transformations of translation or scaling, or their compositions, and allows for the efficient identification of the most common invariant solutions to nonlinear PDEs using a unified approach. Notably, these solutions account for more than 80% of those obtainable through classical group analysis methods (see Section 8.6.2 for more details). It is important to emphasize that this method does not require extensive training and results in minimal intermediate calculations. Generally, users must be capable of solving basic algebraic equations (or systems of such equations) and performing differentiation.

It is also important to note that when using elementary invariant theory for higher-order PDEs, the technical difficulties remain almost the same as for second-order PDEs (for the vast majority of other methods, with an increase in the order of the PDE under consideration, the technical difficulties increase sharply).

The above circumstances suggest that it would be advisable to introduce elementary invariant theory into standard lecture courses on equations of mathematical physics and partial differential equations taught to students specializing in applied and engineering mathematics, theoretical physics and mechanics. Invariant theory can also be included in the relevant special courses taught at technical universities. Furthermore, introducing this simple method at an early stage would provide a helpful foundation before delving into more complex special topics related to group analysis methods for differential equations.

References for Chapter 4

Aksenov A.V. and Polyanin A.D. Methods for constructing complex solutions of nonlinear PDEs using simpler solutions. *Mathematics*, 2021, Vol. 9, No. 4, 345.

Barenblatt G.I. *Scaling, Self-Similarity, and Intermediate Asymptotics*. Cambridge University Press, Cambridge, 1996.

Birkhoff G. *Hydrodynamics, 2nd ed.* Princeton University Press, Princeton, NJ, 1960.

Ibragimov N.H. (ed.). *CRC Handbook of Lie Group Analysis of Differential Equations, Vol. 1, Symmetries, Exact Solutions and Conservation Laws*. CRC Press, Boca Raton, 1994.

Olver P.J. *Application of Lie Groups to Differential Equations*. Springer, New York, 1993.

Ovsiannikov L.V. *Group Analysis of Differential Equations*. Academic Press, New York, 1982.

Polyanin A.D. An elementary theory of using invariants for solving mathematical equations. *Vestnik Samarskogo Gosudarstvennogo Universiteta* [*Bulletin of Samara State University*], 2008, No. 6(65), pp. 152–176. (in Russian).

Polyanin A.D., Zaitsev V.F. *Handbook of Nonlinear Partial Differential Equations*. CRC Press, Boca Raton–London, 2012.

Sedov L.I. *Similarity and Dimensional Methods in Mechanics*. CRC Press, Boca Raton, 1993.

5. Methods of Generalized Separation of Variables

► **Preliminary remarks.** The methods of generalized separation of variables, which are a collection of several methods of different levels of complexity and different fields of application, are among the most constructive and effective methods for finding exact solutions to many nonlinear equations of mathematical physics and other PDEs. This chapter describes various techniques to obtaining generalized separable solutions following the simple to complex approach. These techniques include a simplified method based on presetting a system of coordinate functions, methods of reduction to a standard bilinear functional equation (followed by solving with the methods of differentiation and splitting), and the method of invariant subspaces. The focus is on nonlinear equations of heat and mass transfer, wave theory, and fluid dynamics, with particular attention to the most challenging equations for analysis that involve one or more arbitrary functions. The application of the methods is illustrated with numerous examples and tables. The chapter also discusses nonlinear PDEs with delay that admit generalized separable solutions. The solutions obtained can serve as test problems for numerical and approximate analytical methods.

5.1. Simple Separable Solutions

5.1.1. Multiplicative and Additive Separable Solutions

The method of separation of variables is the most commonly used approach to solving linear equations in mathematical physics. For equations involving two independent variables, x and t, and one unknown function, $u = u(x, t)$, this method relies on finding exact solutions in the form of the product of functions that depend on different arguments (see Tikhonov & Samarski, 1990, and Polyanin & Nazaikinskii, 2016):

$$u = \varphi(x)\psi(t). \tag{5.1.1.1}$$

The functions $\varphi = \varphi(x)$ and $\psi = \psi(t)$ are described by linear ordinary differential equations and determined in a subsequent analysis.

The integration of individual classes of nonlinear first-order partial differential equations involves seeking exact solutions as the sum of functions with different arguments (see Section 2.2.2):

$$u = \varphi(x) + \psi(t). \tag{5.1.1.2}$$

Nonlinear equations of mathematical physics of second or higher orders can also have exact solutions in the forms (5.1.1.1) or (5.1.1.2). These are referred to as *multiplicative separable solutions* and *additive separable solutions*, respectively.

5.1.2. Simple Separable Solutions for Nonlinear Equations of Mathematical Physics

In simple cases, the separation of variables in nonlinear partial differential equations with two independent variables is conducted similarly to how it is done in linear equations. An exact solution is sought as the product or sum of functions that depend on different variables. By substituting expression (5.1.1.1) or (5.1.1.2) into the equation and performing basic rearrangements, we obtain the equality of two expressions (for equations in two variables) that depend on different arguments. Such a situation can only occur if both of these expressions are equal to the same constant. Consequently, to determine the two unknown functions, $\varphi = \varphi(x)$ and $\psi = \psi(t)$, we derive two ordinary differential equations, one for φ and the other for ψ. We refer to these exact solutions as *simple separable solutions*.

Let us illustrate what has been said with a few specific examples.

► Example 5.1. The heat equation with a power-law nonlinearity

$$u_t = a(u^k u_x)_x \tag{5.1.2.1}$$

has an exact solution expressed as the product of functions with different arguments. On substituting (5.1.1.1) into equation (5.1.2.1) gives

$$\varphi\psi'_t = a\psi^{k+1}(\varphi^k \varphi'_x)'_x.$$

On dividing the result by $\varphi\psi^{k+1}$, we obtain

$$\frac{\psi'_t}{\psi^{k+1}} = \frac{a(\varphi^k \varphi'_x)'_x}{\varphi}.$$

The left-hand side of this equation only depends on t, while the right-hand side depends on x alone. This is possible only if

$$\frac{\psi'_t}{\psi^{k+1}} = C, \qquad \frac{a(\varphi^k \varphi'_x)'_x}{\varphi} = C, \tag{5.1.2.2}$$

where C is an arbitrary constant. The solution of the first ODE in (5.1.2.2) is expressed in terms of elementary functions, and that of the second ODE can be represented in an implicit form.

The process of constructing a separable solution of the form (5.1.1.1) for the nonlinear equation (5.1.2.1) when $k \neq 0$ is entirely analogous to the procedure for solving the linear heat equation, which corresponds to the case where $k = 0$ in (5.1.2.1). ◄

► Example 5.2. The wave equation with an exponential nonlinearity

$$u_{tt} = a(e^{\lambda u} u_x)_x \tag{5.1.2.3}$$

has an exact solution as the sum of functions with different arguments. Substituting expression (5.1.1.2) into equation (5.1.2.3) and dividing by $e^{\lambda\psi}$, we arrive at the equation

$$e^{-\lambda\psi}\psi''_{tt} = a(e^{\lambda\varphi}\varphi'_x)'_x, \tag{5.1.2.4}$$

whose left-hand side is dependent on only t, and the right-hand side on x alone. Equating the two sides of (5.1.2.4) with a constant, we get

$$e^{-\lambda\psi}\psi''_{tt} = C, \quad a(e^{\lambda\varphi}\varphi'_x)'_x = C. \tag{5.1.2.5}$$

Both ODEs in (5.1.2.5) are autonomous, meaning they do not explicitly depend on the independent variables. This characteristic allows us to reduce their order and transform them into simpler equations. Furthermore, after performing a single integration, the second equation in (5.1.2.5) simplifies to a first-order separable ODE. Consequently, we can derive a solution for equation (5.1.2.3) in the form (5.1.1.2); this solution is expressed in terms of elementary functions. ◄

► Example 5.3. The steady-state heat equation with a logarithmic source

$$u_{xx} + u_{yy} = au \ln u \tag{5.1.2.6}$$

has an exact solution expressed as the product of functions with different arguments:

$$u = \varphi(x)\psi(y). \tag{5.1.2.7}$$

Substituting expression (5.1.2.7) into equation (5.1.2.6), dividing by $\varphi\psi$, and moving some terms to the left- or right-hand side of the resulting equation, we obtain

$$\frac{\varphi''_{xx}}{\varphi} - a\ln\varphi = -\frac{\psi''_{yy}}{\psi} + a\ln\psi.$$

The left-hand side of this equation depends only on x, while the right-hand side depends solely on y. By equating both sides to a constant, we obtain two ordinary differential equations: one for $\varphi(x)$ and the other for $\psi(y)$. ◄

Table 5.1 illustrates several other simple separable solutions, whether additive or multiplicative, for various nonlinear equations in mathematical physics.

5.1.3. Examples of Nontrivial Separation of Variables in Nonlinear Equations

In many cases, the separation of variables in nonlinear partial differential equations occurs differently than in linear equations. To illustrate this point, we present several specific nonlinear equations that admit solutions of the form (5.1.1.1) or (5.1.1.2). However, it is essential to note that these solutions are not already simple separable solutions.

Table 5.1. Some nonlinear equations of mathematical physics that admit simple separable solutions.

Equation	Name	Form of solution	Determining equations
$u_t = au_{xx} + bu \ln u$	Heat equation with a source	$u = \varphi(x)\psi(t)$	$a\varphi''_{xx}/\varphi - b\ln\varphi = -\psi'_t/\psi + b\ln\psi = C$
$u_t = a(u^k u_x)_x + bu$	Heat equation with a source	$u = \varphi(x)\psi(t)$	$(\psi'_t - b\psi)/\psi^{k+1} = a(\varphi^k\varphi'_x)'_x/\varphi = C$
$u_t = a(u^k u_x)_x + bu^{k+1}$	Heat equation with a source	$u = \varphi(x)\psi(t)$	$a(\varphi^k\varphi'_x)'_x/\varphi + b\varphi^k = \psi'_t/\psi^{k+1} = C$
$u_t = a(e^{\lambda u}u_x)_x + b$	Heat equation with a source	$u = \varphi(x) + \psi(t)$	$e^{-\lambda\psi}(\psi'_t - b) = a(e^{\lambda\varphi}\varphi'_x)'_x = C$
$u_t = a(e^u u_x)_x + be^u$	Heat equation with a source	$u = \varphi(x) + \psi(t)$	$e^{-\psi}\psi'_t = a(e^\varphi\varphi'_x)'_x + be^\varphi = C$
$u_t = au_{xx} + bu_x^2$	Potential Burgers equation	$u = \varphi(x) + \psi(t)$	$\psi'_t = a\varphi''_{xx} + b(\varphi'_x)^2 = C$
$u_t = au_x^k u_{xx}$	Equation viscoplasticity	$u = \varphi(x) + \psi(t)$, $u = f(x)g(t)$	$\psi'_t = a(\varphi'_x)^k\varphi''_{xx} = C_1$, $g'_t/g^{k+1} = a(f'_x)^k f''_{xx}/f = C_2$
$u_{tt} = a(u^k u_x)_x$	Wave equation	$u = \varphi(x)\psi(t)$	$\psi''_{tt}/\psi^{k+1} = a(\varphi^k\varphi'_x)'_x/\varphi = C$
$u_{tt} = a(e^{\lambda u}u_x)_x$	Wave equation	$u = \varphi(x) + \psi(t)$	$e^{-\lambda\psi}\psi''_{tt} = a(e^{\lambda\varphi}\varphi'_x)'_x = C$
$u_{tt} = au_{xx} + bu \ln u$	Wave equation with a source	$u = \varphi(x)\psi(t)$	$\psi''_{tt}/\psi - b\ln\psi = a\varphi''_{xx}/\varphi + b\ln\varphi = C$
$u_{xx} + a(u^k u_y)_y = 0$	Steady-state anisotropic heat equation	$u = \varphi(x)\psi(y)$	$\varphi''_{xx}/\varphi^{k+1} = -a(\psi^k\psi'_y)'_y/\psi = C$
$u_{xx} + au_y u_{yy} = 0$	Steady-state transonic gas flow equation	$u = \varphi(x) + \psi(y)$, $u = f(x)g(y)$	$\varphi''_{xx} = -a\psi'_y\psi''_{yy} = C_1$, $f''_{xx}/f = -ag'_y g''_{yy}/g = C_2$

► Example 5.4. Let us consider the second-order equation with a cubic nonlinearity

$$u_t = u_{xx} + uu_x^2 - au^3, \tag{5.1.3.1}$$

where a is some constant.

We seek exact solutions in the product form with the multipliers dependent on different arguments. Substituting (5.1.1.1) into (5.1.3.1) and dividing by $\varphi(x)\psi(t)$, we get

$$\frac{\psi'_t}{\psi} = \frac{\varphi''_{xx}}{\varphi} + \psi^2[(\varphi'_x)^2 - a\varphi^2]. \tag{5.1.3.2}$$

The left-hand side of the functional differential equation (5.1.3.2) depends only on t, and the right-hand side is the sum of two terms, one dependent on only x and the other represented as the product of two functions with different arguments. Equation (5.1.3.2) cannot be represented as the equality of two functions with different arguments (unlike in the examples discussed above in Section 5.1.2). However, this does not mean that equation (5.1.3.1) has no solutions of the form (5.1.1.1).

1°. A direct verification shows that the functional differential equation (5.1.3.2) has two solutions when $a > 0$:

$$\varphi(x) = C\exp\bigl(\pm x\sqrt{a}\,\bigr), \quad \psi(t) = \exp(at), \tag{5.1.3.3}$$

where C is an arbitrary constant. With both solutions of the equation for φ in (5.1.3.3), the expression in square brackets in (5.1.3.2) becomes zero, allowing us to separate the variables.

2°. For $a > 0$, there is a more general solution to the functional differential equation (5.1.3.2):

$$\varphi(x) = C_1\exp\bigl(x\sqrt{a}\,\bigr) + C_2\exp\bigl(-x\sqrt{a}\,\bigr),$$
$$\psi(t) = \bigl[4C_1C_2 + C_3\exp(-2at)\bigr]^{-1/2},$$

where C_1, C_2, and C_3 are arbitrary constants. In this case, the function $\varphi = \varphi(x)$ is such that both x-dependent expressions in (5.1.3.2) will equal different constants:

$$\varphi''_{xx}/\varphi = a, \quad (\varphi'_x)^2 - a\varphi^2 = 4aC_1C_2.$$

This situation allows us to separate the variables. Notice that the function $\psi = \psi(t)$ satisfies the Bernoulli equation $\psi'_t = a\psi - 4aC_1C_2\psi^3$, which is also a separable ordinary differential equation.

3°. There is another solution to the functional differential equation (5.1.3.2) for $a < 0$:

$$\varphi(x) = C_1\sin\bigl(x\sqrt{-a}\,\bigr) + C_2\cos\bigl(x\sqrt{-a}\,\bigr),$$
$$\psi(t) = \bigl[C_1^2 + C_2^2 + C_3\exp(-2at)\bigr]^{-1/2}.$$

In this case, the function $\varphi = \varphi(x)$ is such that both x-dependent expressions in (5.1.3.2) will equal different constants. The function $\psi = \psi(t)$ is described by the Bernoulli equation $\psi'_t = a\psi - a(C_1^2 + C_2^2)\psi^3$, which is also a separable ODE. ◄

► Example 5.5. Consider the third-order partial differential equation with a quadratic nonlinearity

$$u_y u_{xx} + a u_x u_{yy} = b u_{xxx} + c u_{yyy}. \tag{5.1.3.4}$$

We will look for exact separable solutions to equation (5.1.3.4) represented as the sum of two functions with different arguments:

$$u = f(x) + g(y). \tag{5.1.3.5}$$

Substituting (5.1.3.5) into (5.1.3.4) gives a functional differential equation

$$g'_y f''_{xx} + a f'_x g''_{yy} = b f'''_{xxx} + c g'''_{yyy}, \tag{5.1.3.6}$$

which cannot be represented as an equality of two functions with different arguments.

In this case, it is easy to infer that the functional differential equation can hold

$$\begin{aligned} &\text{if } g'_y = C_1 \quad \Longrightarrow \quad g(y) = C_1 y + C_2, \quad f(x) = C_3 e^{C_1 x/b} + C_4 x \quad \text{(case 1)},\\ &\text{if } f'_x = C_1 \quad \Longrightarrow \quad f(x) = C_1 x + C_2, \quad g(y) = C_3 e^{aC_1 y/c} + C_4 y \quad \text{(case 2)}, \end{aligned}$$

where $C_1, \dots, C_4$ are arbitrary constants. In both cases, two of the four terms in (5.1.3.6) simultaneously vanish, allowing the variables to be separated.

Equation (5.1.3.4) also has a more complex solution of the form (5.1.3.5):

$$u = C_1 e^{-a\lambda x} + \frac{c\lambda}{a} x + C_2 e^{\lambda y} - ab\lambda y + C_3,$$

where λ is an arbitrary constant. The separation-of-variables mechanism is different here: both expressions on the left-hand side of (5.1.3.6) contain nonlinear terms that, although unrepresentable as the sum of functions with different arguments, are the same in magnitude but different in sign. When added together, the nonlinear terms cancel each other out, ultimately leading to the separation of variables:

$$\begin{array}{rl} g'_y f''_{xx} & = C_1 C_2 a^2 \lambda^3 e^{\lambda y - a\lambda x} - C_1 b(a\lambda)^3 e^{-a\lambda x} \\ + \quad a f'_x g''_{yy} & = -C_1 C_2 a^2 \lambda^3 e^{\lambda y - a\lambda x} + C_2 c \lambda^3 e^{\lambda y} \\ \hline g'_y f''_{xx} + a f'_x g''_{yy} & = -C_1 b(a\lambda)^3 e^{-a\lambda x} + C_2 c\lambda^3 e^{\lambda y} = b f'''_{xxx} + c g'''_{yyy} \end{array}$$ ◂

The examples above highlight key characteristics of multiplicative and additive separable solutions that are not found in simple separable solutions. The following sections outline a few fairly general methods for constructing similar and more complex exact solutions for nonlinear partial differential equations.

Problems and Exercises for Section 5.1

1. Find a simple separable solution to the nonlinear first-order PDE

$$u_t + a u_x^2 = f(x).$$

2. Find a simple separable solution to the nonlinear first-order PDE

$$u_t + a u_x^2 = f(t).$$

3. Find a simple separable solution to the nonlinear first-order PDE

$$u_x u_y = a u^2.$$

4. Find two simple separable solutions to the nonlinear first-order PDE

$$u_x u_y^2 = f(x) g(y).$$

5. Find a simple separable solution to the nonlinear first-order PDE

$$u_x^2 + a u_y^2 = f(x) + g(y).$$

6. Find a simple separable solution of the generalized potential Burgers equation

$$u_t = au_{xx} + f(u_x).$$

7. Find a simple separable solution for the anisotropic heat equation with a logarithmic source

$$[f(x)u_x]_x + [g(y)u_y]_y = au \ln u.$$

8. Find a simple separable solution for the nonlinear seepage equation

$$u_t = f(u_x)u_{xx}.$$

9. Find two simple separable solutions for the homogeneous Monge–Ampère equation

$$u_{xx}u_{yy} - u_{xy}^2 = 0.$$

10. Find a simple separable solution for the potential Korteweg–de Vries equation

$$u_t = au_{xxx} + bu_x^2.$$

11. Find two simple separable solutions for the boundary layer equation

$$u_y u_{xy} - u_x u_{yy} = au_{yyy}.$$

5.2. Structure of Generalized Separable Solutions

5.2.1. General Form of Solutions. Classes of Nonlinear Differential Equations

To simplify our presentation, we will focus on nonlinear equations of mathematical physics that involve two independent variables, x and y, and a single dependent variable, u. One of the independent variables may be treated as time.

Linear partial differential equations. Linear equations of mathematical physics with constant coefficients, as well as many linear equations with variable coefficients, have exact solutions that can be expressed as a sum of pairwise products of functions with different arguments:

$$u(x,y) = \varphi_1(x)\psi_1(y) + \varphi_2(x)\psi_2(y) + \cdots + \varphi_n(x)\psi_n(y), \qquad (5.2.1.1)$$

where $u_i = \varphi_i(x)\psi_i(y)$ are particular solutions. The functions $\varphi_i(x)$ for different i, as well as the functions $\psi_i(y)$, are not related to one another.

Furthermore, linear partial differential equations often admit exact solutions of the form (5.2.1.1), where pairwise products of functions with different arguments, $\varphi_i(x)\psi_i(y)$, are not particular solutions to these equations.

► Example 5.6. The linear heat equation

$$u_t = au_{xx}$$

admits the following solutions:

$$
\begin{aligned}
u &= x^2 + 2at,\\
u &= x^3 + 6atx,\\
u &= x^4 + 12atx^2 + 12a^2t^2,\\
u &= e^{-\mu x}\cos(\mu x)\cos(2a\mu^2 t) + e^{-\mu x}\sin(\mu x)\sin(2a\mu^2 t),\\
u &= e^{-\mu x}\sin(\mu x)\cos(2a\mu^2 t) - e^{-\mu x}\cos(\mu x)\sin(2a\mu^2 t),
\end{aligned}
$$

where μ is an arbitrary constant. The individual terms in these solutions are not solutions of the linear heat equation. ◀

Nonlinear partial differential equations. Nonlinear partial differential equations of mathematical physics with quadratic or power-law nonlinearities can be written as

$$f_1(x)g_1(y)\Pi_1[u] + f_2(x)g_2(y)\Pi_2[u] + \cdots + f_m(x)g_m(y)\Pi_m[u] = 0, \quad (5.2.1.2)$$

where $\Pi_i[u]$ are differential forms representing products of nonzero integer powers of u and its partial derivatives u_x, u_y, u_{xx}, u_{xy}, u_{yy}, u_{xxx}, ... Many of these equations also have exact solutions of the form (5.2.1.1). We will refer to such solutions as *generalized separable solutions*. Unlike linear equations, some functions $\varphi_i(x)$ (or functions $\psi_j(y)$) in solutions of nonlinear equations are usually connected with one another. In general, the functions $\varphi_i(x)$ and $\psi_j(y)$ are not known in advance and are to be determined in a subsequent analysis. Numerous examples of exact solutions to nonlinear equations of the form (5.2.1.1) for the simplest cases of $n = 1$ and $n = 2$ (with $\psi_1 = \varphi_2 = 1$) are provided in Sections 5.1.2 and 5.1.3.

The methods for finding exact solutions of the form (5.2.1.1) for nonlinear partial differential equations will be referred to as *methods of generalized separation of variables*.

In practice, when constructing generalized separable solutions to nonlinear equations of mathematical physics, the most frequently encountered solutions have the following special form that includes three undetermined functions:

$$u(x, y) = \varphi(x)\theta(y) + \psi(x). \quad (5.2.1.3)$$

(The independent variables on the right-hand side can be swapped.) In the special case $\psi(x) = 0$, this solution converts into a multiplicative separable solution. If $\varphi(x) = 1$, we get an additive separable solution.

Remark 5.1. *The function $\theta = \theta(y)$ in (5.2.1.3) is determined up to translation and scaling, because the linear transformation $\theta = C_1\bar{\theta} + C_2$, where C_1 and C_2 are arbitrary constants, leads to a similar relation: $u = \bar{\varphi}(x)\bar{\theta}(y) + \bar{\psi}(x)$, where $\bar{\varphi} = C_1\varphi$ and $\bar{\psi} = \psi + C_2\varphi$.*

Remark 5.2. *Expressions of the form (5.2.1.1) are frequently utilized in applied and computational mathematics to construct approximate analytical and numerical solutions for differential equations using Bubnov–Galerkin type projection methods.*

5.2.2. Functional Differential Equations Arising in Generalized Separation of Variables

In general, after substituting the expression (5.2.1.1) into the differential equation (5.2.1.2) for $\varphi_i(x)$ and $\psi_i(y)$, one obtains a functional differential equation

$$\Phi_1[X]\Psi_1[Y] + \Phi_2[X]\Psi_2[Y] + \cdots + \Phi_k[X]\Psi_k[Y] = 0. \tag{5.2.2.1}$$

The functionals $\Phi_j[X]$ and $\Psi_j[Y]$ are dependent on the variables x and y, respectively:

$$\begin{aligned} \Phi_j[X] &\equiv \Phi_j\big(x, \varphi_1, \varphi_1', \varphi_1'', \dots, \varphi_n, \varphi_n', \varphi_n''\big), \\ \Psi_j[Y] &\equiv \Psi_j\big(y, \psi_1, \psi_1', \psi_1'', \dots, \psi_n, \psi_n', \psi_n''\big). \end{aligned} \tag{5.2.2.2}$$

For clarity, the formulas are written for the case of a second-order equation (5.2.1.2). For higher-order equations, the right-hand sides of formulas (5.2.2.2) will include the respective higher-order derivatives of φ_i and ψ_i.

Below, we will describe several methods for solving functional differential equations of the form (5.2.2.1) with (5.2.2.2).

Remark 5.3. *Unlike ordinary differential equations, equations of the form (5.2.2.1) with (5.2.2.2) involve several functions (with their derivatives) that depend on different variables. This necessitates using specialized (non-standard) techniques to solve such equations.*

5.3. Simplified Method for Constructing Generalized Separable Solutions

5.3.1. Simplified Method Employing a Pre-defined System of Coordinate Functions. Description

To construct exact solutions for differential equations of the form (5.2.1.2) that involve quadratic or power-law nonlinearities and do not depend explicitly on y, implying that all g_i are constants, we can use a simplified approach. We will seek solutions represented as finite sums (5.2.1.1). Let us assume that the system of coordinate functions $\psi_i(y)$ is described by linear differential equations with constant coefficients. The most common solutions to such equations are as follows:

$$\psi_i(y) = y^i, \quad \psi_i(y) = e^{\lambda_i y}, \quad \psi_i(y) = \sin(\alpha_i y), \quad \psi_i(y) = \cos(\beta_i y). \tag{5.3.1.1}$$

Finite sets of these functions (in various combinations) can be used to find generalized separable solutions of the form (5.2.1.1), where λ_i, α_i, and β_i are treated as free parameters. The other system of functions, $\varphi_i(x)$, is determined by solving the nonlinear equations obtained by substituting expression (5.2.1.1) into the equation in question.

This approach lacks the generality of the methods discussed in Sections 5.4 and 5.5 below. However, the explicit specification of a system of coordinate functions, $\{\psi_i(y)\}$, significantly simplifies the process of constructing exact solutions.

It is important to note that, in doing so, individual solutions of the form (5.2.1.1) may be overlooked. Currently, the known generalized separable solutions to partial differential equations with quadratic nonlinearities are predominantly defined by coordinate functions of the form (5.3.1.1), most frequently for the case when $n = 2$.

Remark 5.4. *The simplified approach for finding generalized separable solutions is quite similar to the method of undetermined coefficients, commonly used to find particular solutions of linear nonhomogeneous ordinary differential equations with a special right-hand side, to calculate integrals, and in other situations.*

5.3.2. Examples of Constructing Exact Solutions for Nonlinear Equations in Two Independent Variables

Let us look at a few specific examples of using the simplified method to construct generalized separable solutions for nonlinear partial differential equations of the second or third order in two independent variables.

► Example 5.7. Consider the *Guderley equation*

$$u_{xx} = au_y u_{yy}, \tag{5.3.2.1}$$

which is used to describe transonic gas flows.

We immediately observe that equation (5.3.2.1) has the obvious degenerate generalized separable solution

$$u = (C_1 x + C_2)y + C_3 x + C_4, \tag{5.3.2.2}$$

where $C_1, \ldots, C_4$ are arbitrary constants. Its highest derivatives are zero: $u_{xx} = u_{yy} = 0$.

We will look for generalized separable solution in the form

$$u(x,y) = \varphi(x)y^k + \psi(x), \tag{5.3.2.3}$$

where the functions $\varphi(x)$ and $\psi(x)$ and constant $k \neq 0$ are to be determined. The degenerate solution (5.3.2.2) is a special case of solution (5.3.2.3). It is noteworthy that such two-term solutions are quite common in practice and are the simplest generalized separable solutions of PDEs, along with solutions in which the power-law function y^k is replaced by the exponential function $e^{\lambda y}$.

On substituting (5.3.2.3) into (5.3.2.1) and rearranging terms, we arrive at the relation

$$\varphi''_{xx} y^k - ak^2(k-1)\varphi^2 y^{2k-3} + \psi''_{xx} = 0, \tag{5.3.2.4}$$

which contains power-law function y^k and y^{2k-3} and must hold identically for any y.

Let us consider two cases: $\psi''_{xx} = 0$ and $\psi''_{xx} \neq 0$.

1°. *First case.* If $\psi''_{xx} = 0$, we get a two-term equation with separable variables, which can be satisfied by setting

$$k = 3, \quad \varphi''_{xx} - 18a\varphi^2 = 0. \tag{5.3.2.5}$$

The general solution of the autonomous equation for φ can be written in an implicit form as $x = \pm\int(12a\varphi^3 + C_1)^{-1/2}d\varphi + C_2$. In addition, ODE (5.3.2.5) admits a simple particular power-law solution $\varphi = \frac{1}{3a}(x + C_1)^{-2}$, which leads to a three-parameter exact solution of equation (5.3.2.1):

$$u = \frac{1}{3a}(x + C_1)^{-2}y^3 + C_2x + C_3, \qquad (5.3.2.6)$$

where C_1, C_2, and C_3 are arbitrary constants.

2°. *Second case.* To balance the function $\psi''_{xx} \neq 0$ with the second term in equation (5.3.2.4), we must set $k = 3/2$. As a result, we obtain a two-term equation, which can be satisfied if we set

$$\varphi''_{xx} = 0, \quad \psi''_{xx} = \tfrac{9}{8}a\varphi^2.$$

These equations are easy to integrate and lead to a four-parameter exact solution of equation (5.3.2.1):

$$u = (C_1x + C_2)y^{3/2} + \frac{3a}{32C_1^2}(C_1x + C_2)^4 + C_3x + C_4.$$

◄

► Example 5.8. A steady-state flow of a viscous incompressible fluid in a laminar boundary layer of a thin flat plate in a longitudinal flow is described by the following third-order nonlinear equation for the stream function:

$$u_yu_{xy} - u_xu_{yy} = \nu u_{yyy}, \qquad (5.3.2.7)$$

where x and y are a longitudinal and a transverse coordinate, and ν is the kinematic viscosity of the fluid.

We will seek a generalized separable solution for equation (5.3.2.7) in the form

$$u(x, y) = x^k\varphi(y) + \psi(y), \qquad (5.3.2.8)$$

where the functions $\varphi(y)$ and $\psi(y)$ and constants k are to be determined. On substituting (5.3.2.8) into (5.3.2.7) and rearranging the terms, we obtain

$$kx^{2k-1}[(\varphi'_y)^2 - \varphi\varphi''_{yy}] + kx^{k-1}(\varphi'_y\psi'_y - \varphi\psi''_{yy}) - \nu x^k\varphi'''_{yyy} - \nu\psi'''_{yyy} = 0. \quad (5.3.2.9)$$

This equation involved power-law functions x^{2k-1}, x^{k-1}, x^k and must hold identically for any x. To balance the last term in (5.3.2.9) if $\psi'''_{yyy} \not\equiv 0$, we must set one of the exponents of x to zero. Discarding the degenerate case $k = 0$, we find two admissible values: $k = 1$ and $k = 1/2$.

1°. *First case.* Substituting $k = 1$ into (5.3.2.9) and rearranging, we get

$$x[(\varphi'_y)^2 - \varphi\varphi''_{yy} - \nu\varphi'''_{yyy}] + [\varphi'_y\psi'_y - \varphi\psi''_{yy} - \nu\psi'''_{yyy}] = 0.$$

To make sure that this equation holds for any x, we must equate both expressions in square brackets with zero. This will result in a system of ordinary differential equations for $\varphi = \varphi(y)$ and $\psi = \psi(y)$:

$$(\varphi'_y)^2 - \varphi\varphi''_{yy} - \nu\varphi'''_{yyy} = 0,$$
$$\varphi'_y\psi'_y - \varphi\psi''_{yy} - \nu\psi'''_{yyy} = 0.$$

An exact solution of this system is given by

$$\varphi = \frac{6\nu}{y + C_1}, \quad \psi = \frac{C_2}{y + C_1} + \frac{C_3}{(y + C_1)^2} + C_4,$$

where $C_1, \ldots, C_4$ are arbitrary constants.

2°. *Second case.* With $k = 1/2$, we arrive at a degenerate solution, which holds no practical significance. ◀

► Example 5.9. Consider the nonlinear third-order PDE with a mixed derivative

$$u_{xt} + u_x^2 - uu_{xx} = \nu u_{xxx}, \qquad (5.3.2.10)$$

which arises in fluid dynamics.

We look for generalized separable solutions of the form

$$u = \varphi(t)e^{\lambda x} + \psi(t), \qquad \lambda \neq 0. \qquad (5.3.2.11)$$

Substituting (5.3.2.11) into (5.3.2.10) yields

$$\varphi_t' - \lambda\varphi\psi = \nu\lambda^2\varphi.$$

On solving this equation for ψ and inserting the resulting expression into (5.3.2.11), we obtain a solution of equation (5.3.2.10) in the form

$$u = \varphi(t)e^{\lambda x} + \frac{1}{\lambda}\frac{\varphi_t'(t)}{\varphi(t)} - \nu\lambda, \qquad (5.3.2.12)$$

where $\varphi(t)$ is an arbitrary function and λ is an arbitrary constant. ◀

► Example 5.10. Consider the Boussinesq equation with a quadratic nonlinearity

$$u_t = a(uu_x)_x. \qquad (5.3.2.13)$$

The equation describes unsteady groundwater flow through a porous medium with a free surface, where u represents the liquid pressure.

We seek an exact solution for equation (5.3.2.13) as the sum

$$u(x,t) = \varphi(t)x^2 + \psi(t)x^k, \qquad (5.3.2.14)$$

where the functions $\varphi(t)$ and $\psi(t)$ and the coefficient k are to be determined.

On substituting (5.3.2.14) into (5.3.2.13) and performing elementary rearrangements, we obtain

$$(\varphi_t' - 6a\varphi^2)x^2 + [\psi_t' - a(k+1)(k+2)\varphi\psi]x^k - ak(2k-1)\psi^2x^{2k-2} = 0. \quad (5.3.2.15)$$

Since this equation must hold identically for any x, the functional coefficients of the different powers of x in (5.3.2.15) must all be equal to zero. This results in two possible cases: $k = 0$ and $k = 1/2$, both of which make the coefficient of x^{2k-2} equal to zero. These cases will need to be addressed separately.

1°. *First case*. For $k=0$, the functions $\varphi=\varphi(t)$ and $\psi=\psi(t)$ satisfy the system of ODEs

$$\begin{aligned}\varphi_t' - 6a\varphi^2 &= 0,\\ \psi_t' - 2a\varphi\psi &= 0.\end{aligned}$$

Its general solution is defined by

$$\varphi(t) = -\frac{1}{6a(t+C_1)}, \quad \psi(t) = \frac{C_2}{|t+C_1|^{1/3}}, \tag{5.3.2.16}$$

where C_1 and C_2 are arbitrary constants.

2°. *Second case*. For $k=1/2$, the functions $\varphi=\varphi(t)$ and $\psi=\psi(t)$ are described by the system

$$\begin{aligned}\varphi_t' - 6a\varphi^2 &= 0,\\ \psi_t' - \tfrac{15}{4}a\varphi\psi &= 0.\end{aligned}$$

Its general solution is

$$\varphi(t) = -\frac{1}{6a(t+C_1)}, \quad \psi(t) = \frac{C_2}{|t+C_1|^{5/8}}. \tag{5.3.2.17}$$

Considering formulas (5.3.2.14), (5.3.2.16), and (5.3.2.17), we finally arrive at two generalized separable solutions to equation (5.3.2.13):

$$\begin{aligned}u &= -\frac{1}{6a(t+C_1)}(x+C_3)^2 + \frac{C_2}{|t+C_1|^{1/3}},\\ u &= -\frac{1}{6a(t+C_1)}(x+C_3)^2 + \frac{C_2}{|t+C_1|^{5/8}}(x+C_3)^{1/2}.\end{aligned}$$

For greater generality, an arbitrary translation in x has been added here; this can be done because the Boussinesq equation is explicitly independent of x. ◀

Remark 5.5. *The heat equation with thermal diffusivity linearly dependent on temperature,* $v_t = [(av+b)v_x]_x$*, can be reduced to equation (5.3.2.13) with the substitution* $u = v + (b/a)$*. This linear approximation of thermal diffusivity is applicable to many metals over a wide range of temperature variations.*

◆ *Problems and Exercises for Section 5.3*

1. Find a generalized separable solution of the first-order nonlinear PDE
$u_x = yF(x,u_y) + G(x,u_y)$.
Hint. Seek a solution in the form $u = \varphi(x)y + \psi(x)$.

2. Find a generalized separable solution of the first-order nonlinear equation $u_x = u_y^2 - au^2 + f(x)u$.
Hint. Seek a solution in the form $u = \varphi(x) + \psi(x)e^{\lambda y}$.

3. Find generalized separable solutions of the following nonlinear equations:

a) $u_t = a(uu_x)_x + b$,
b) $u_t = a(uu_x)_x + bu$,
c) $u_t = a(uu_x)_x + bu_x$,
d) $u_t = a(uu_x)_x + bu_x + cu$,
e) $u_t = au_{xx} + bu_x^2 + cu + s$.

Hint. Seek solutions in the forms $u = f(t)x + g(t)$ and $u = f(t)x^2 + g(t)x + h(t)$.

4. Find generalized separable solutions of the linear equation $u_t + f(t)x^{-1}u_x = au_{xx}$.

Hint. Look for solutions in the forms:

a) $u = x^2 + \varphi(t)$,
b) $u = x^4 + \varphi(t)x^2 + \psi(t)$.

5. For the nonlinear equation $u_t = uu_{xx} + au_x^2 + b$, find a value of the parameter a at which the equation has a generalized separable solution of the form $u = f(t)x^3 + g(t)x^2 + h(t)x + p(t)$.

6. Find generalized separable solutions of the nonlinear equation $u_t = u_{xx} + u_x^2 + aw^2$.

Hint. Look for solutions in the forms:

a) $u = \varphi(t) + \psi(t)e^{\lambda x}$,
b) $u = \varphi(t) + \psi(t)\sinh(\lambda x)$,
c) $u = \varphi(t) + \psi(t)\cosh(\lambda x)$,
d) $u = \varphi(t) + \psi(t)\sin(\lambda x + C)$.

7. Find generalized separable solutions of the following nonlinear wave equations:

a) $u_{tt} = a(uu_x)_x$,
b) $u_{tt} = a(uu_x)_x + b$,
c) $u_{tt} = a(uu_x)_x + bu$.

Hint. Look for solutions in the forms $u = f(t)x^2 + g(t)x + h(t)$ and $u = f(t)x^2 + g(t)x^k$.

8. Find a generalized separable solution of the form $u = \varphi(x) + \psi(x)e^{\lambda y}$ for the boundary layer equation (5.3.2.7).

9. Find generalized separable solutions of the nonhomogeneous Monge–Ampère equation $u_{xy}^2 = u_{xx}u_{yy} + f(x)$.

Hint. Look for equations in the form $u = \varphi(x)y^k + \psi(x)$.

5.4. Solution of Functional Differential Equation by the Method of Differentiation

5.4.1. Description of the Method

When searching for generalized separable solutions, it is common to encounter functional differential equations in the form (5.2.2.1)–(5.2.2.2). This section outlines a simple yet fairly general procedure for solving these equations, consisting of three consecutive steps as described below (see Polyanin et al., 2005; Polyanin & Zaitsev, 2012; and Polyanin & Zhurov, 2022).

1°. Assuming that $\Psi_k \not\equiv 0$, we divide equation (5.2.2.1) by Ψ_k and then differentiate with respect to y. As a result, we obtain a similar equation but with fewer terms:

$$\begin{gathered}\widetilde{\Phi}_1[X]\widetilde{\Psi}_1[Y] + \widetilde{\Phi}_2[X]\widetilde{\Psi}_2[Y] + \cdots + \widetilde{\Phi}_{k-1}[X]\widetilde{\Psi}_{k-1}[Y] = 0,\\ \widetilde{\Phi}_j[X] = \Phi_j[X], \quad \widetilde{\Psi}_j[Y] = (\Psi_j[Y]/\Psi_k[Y])'_y.\end{gathered} \tag{5.4.1.1}$$

By repeating the same procedure $(k-3)$ more times, we ultimately arrive at a two-term separable equation:

$$\widehat{\Phi}_1[X]\widehat{\Psi}_1[Y] + \widehat{\Phi}_2[X]\widehat{\Psi}_2[Y] = 0. \tag{5.4.1.2}$$

We need to consider two possible scenarios.

Nondegenerate case. Assuming that $\widehat{\Phi}_2[X] \not\equiv 0$ and $\widehat{\Psi}_1[Y] \not\equiv 0$, we move the second term of equation (5.4.1.2) to the right-hand side and the divide both sides by $\widehat{\Phi}_2[X]\widehat{\Psi}_1[Y]$. As a result, we arrive at an equality of two expressions dependent on different variables. As before, equating these expressions with a constant C, we obtain two ordinary differential equations

$$\widehat{\Phi}_1[X] + C\widehat{\Phi}_2[X] = 0, \quad C\widehat{\Psi}_1[Y] - \widehat{\Psi}_2[Y] = 0. \tag{5.4.1.3}$$

Degenerate cases. There are a few degenerate cases in which both terms in the sum (5.4.1.2) vanish simultaneously:

$$\begin{aligned}
&\widehat{\Phi}_1[X] \equiv 0, \quad \widehat{\Phi}_2[X] \equiv 0 &&\Longrightarrow \quad \widehat{\Psi}_1[Y] \text{ and } \widehat{\Psi}_2[Y] \text{ are any;}\\
&\widehat{\Psi}_1[Y] \equiv 0, \quad \widehat{\Psi}_2[Y] \equiv 0 &&\Longrightarrow \quad \widehat{\Phi}_1[X] \text{ and } \widehat{\Phi}_2[X] \text{ are any;}\\
&\widehat{\Phi}_1[X] \equiv 0, \quad \widehat{\Psi}_2[Y] \equiv 0 &&\Longrightarrow \quad \widehat{\Phi}_2[X] \text{ and } \widehat{\Psi}_1[Y] \text{ are any;}\\
&\widehat{\Phi}_2[X] \equiv 0, \quad \widehat{\Psi}_1[Y] \equiv 0 &&\Longrightarrow \quad \widehat{\Phi}_1[X] \text{ and } \widehat{\Psi}_2[Y] \text{ are any.}
\end{aligned} \tag{5.4.1.4}$$

Although the degenerate cases $\widehat{\Phi}_1 = 0$, $\widehat{\Psi}_2 = 0$ and $\widehat{\Phi}_2 = 0$, $\widehat{\Psi}_1 = 0$ follow directly from equations (5.4.1.3) in the limits as $C \to 0$ and $C \to \infty$, they can be treated separately because the respective limit solutions of equations (5.4.1.3) may not exist.

2°. Once a solution to the two-term equation (5.4.1.2) has been found, it should be substituted back into the original functional differential equation (5.2.2.1) with (5.2.2.2). This step is necessary to eliminate any extra constants of integration that may appear. These constants arise because equation (5.4.1.2) is obtained from (5.2.2.1) through differentiation.

3°. The case $\Psi_k \equiv 0$ should be considered separately, as the equation at the first stage was divided by Ψ_k. Similarly, we should investigate all other cases where the functionals used to divide the intermediate functional differential equations are identically zero.

Remark 5.6. *The functional differential equation (5.2.2.1) with the functionals (5.2.2.2) may have one or more solutions or no solutions at all.*

Remark 5.7. *At each stage, the number of terms in the functional differential equation being analyzed can be reduced by differentiating with respect to y or x. In the first step, for instance, we can assume that $\Phi_i \not\equiv 0$ for some i, where $1 \le i \le k$. By dividing equation (5.2.2.1) by Φ_i and then differentiating with respect to x, we arrive at a new equation of the same type but with fewer terms.*

In the practical application of the differentiation method, the terms containing the highest derivatives or the most complex nonlinearities should be removed first. This approach usually ultimately allows us to obtain simple two-term equations of the form (5.4.1.2).

5.4.2. Examples of Constructing Generalized Separable Solutions by the Differentiation Method

Below are some examples of using the differentiation method to construct generalized separable solutions for nonlinear partial differential equations.

► Example 5.11. Let us revisit the third-order nonlinear partial differential equation

$$u_y u_{xy} - u_x u_{yy} = \nu u_{yyy}. \qquad (5.4.2.1)$$

It describes a laminar boundary layer on a flat plate; see equation (5.3.2.7).

We look for generalized separable solutions of the form

$$u = \varphi(x)\theta(y) + \psi(x). \qquad (5.4.2.2)$$

Substituting (5.4.2.2) into (5.4.2.1) and dividing by φ, we arrive at the functional differential equation

$$\varphi'_x[(\theta'_y)^2 - \theta\theta''_{yy}] - \psi'_x\theta''_{yy} = \nu\theta'''_{yyy}. \qquad (5.4.2.3)$$

First, we eliminate the highest derivative. To this end, we differentiate equation (5.4.2.3) with respect to x to obtain

$$\varphi''_{xx}[(\theta'_y)^2 - \theta\theta''_{yy}] = \psi''_{xx}\theta''_{yy}. \qquad (5.4.2.4)$$

Nondegenerate case. Separating the variables in (5.4.2.4), we get

$$\psi''_{xx} = C_1\varphi''_{xx},$$
$$(\theta'_y)^2 - \theta\theta''_{yy} - C_1\theta''_{yy} = 0.$$

Integrating yields the undetermined functions

$$\varphi(x) \text{ is any}, \quad \psi(x) = C_1\varphi(x) + C_2x + C_3, \quad \theta(y) = C_4e^{\lambda y} - C_1, \qquad (5.4.2.5)$$

where C_1, ..., C_4, and λ are arbitrary constants. On substituting the functions (5.4.2.5) into (5.4.2.3), we establish a relationship between constants: $C_2 = -\nu\lambda$. Finally, considering the aforesaid and formulas (5.4.2.2) and (5.4.2.5), we obtain a solution to equation (5.4.2.1) of the form (5.4.2.2):

$$u = \varphi(x)e^{\lambda y} - \nu\lambda x + C,$$

where $\varphi(x)$ is an arbitrary function, while C and λ are arbitrary constants ($C = C_3$ and $C_4 = 1$).

Degenerate case. Equation (5.4.2.4) can be satisfied if we set

$$\varphi''_{xx} = 0, \quad \psi''_{xx} = 0, \quad \theta(y) \text{ is an arbitrary function.} \tag{5.4.2.6}$$

Integrating the first two equations in (5.4.2.6) twice yields

$$\varphi(x) = C_1 x + C_2, \quad \psi(x) = C_3 x + C_4. \tag{5.4.2.7}$$

Substituting (5.4.2.7) into (5.4.2.3), we arrive at an ODE for $\theta = \theta(y)$:

$$C_1(\theta'_y)^2 - (C_1\theta + C_3)\theta''_{yy} = \nu\theta'''_{yyy}. \tag{5.4.2.8}$$

Formulas (5.4.2.2) and (5.4.2.7) in conjunction with equation (5.4.2.8) define and exact solution of equation (5.4.2.1). ◀

► Example 5.12. The steady-state motion of a viscous incompressible fluid is described by the Navier–Stokes equations along with the continuity equation. In the two-dimensional case, these equations can be reduced to a single nonlinear stationary fourth-order equation by introducing the stream function u:

$$u_y(\Delta u)_x - u_x(\Delta u)_y = \nu\Delta\Delta u, \qquad \Delta u = u_{xx} + u_{yy}. \tag{5.4.2.9}$$

We note one property of equation (5.4.2.9): if $u(x, y)$ is a solution of the equation, then $-u(y, x)$ is also a solution of this equation. In the current example, we will omit solutions that can be obtained from the solutions considered below by using this property.

We will seek exact solutions of equation (5.4.2.9) in the form

$$u = \varphi(x) + \psi(y). \tag{5.4.2.10}$$

Substituting (5.4.2.10) into (5.4.2.9) gives

$$\psi'_y\varphi'''_{xxx} - \varphi'_x\psi'''_{yyy} = \nu\varphi''''_{xxxx} + \nu\psi''''_{yyyy}. \tag{5.4.2.11}$$

On differentiating both sides of (5.4.2.11) with respect to x and y, we eliminate the terms containing the highest derivatives to obtain

$$\psi''_{yy}\varphi''''_{xxxx} - \varphi''_{xx}\psi''''_{yyyy} = 0. \tag{5.4.2.12}$$

Nondegenerate case. Assuming that $\varphi''_{xx} \not\equiv 0$ and $\psi''_{yy} \not\equiv 0$, we separate the variables in (5.4.2.12) and arrive at two linear ordinary differential equations with constant coefficients

$$\varphi''''_{xxxx} = C\varphi''_{xx}, \tag{5.4.2.13}$$

$$\psi''''_{yyyy} = C\psi''_{yy}. \tag{5.4.2.14}$$

These equations have different solutions depending on the value of the integration constant C.

1°. For $C = 0$, equations (5.4.2.13) and (5.4.2.14) have the following solutions:

$$\begin{aligned} \varphi(x) &= A_1 + A_2x + A_3x^2 + A_4x^3, \\ \psi(y) &= B_1 + B_2y + B_3y^2 + B_4y^3, \end{aligned} \tag{5.4.2.15}$$

where A_k and B_k are arbitrary constants ($k = 1, 2, 3, 4$). On substituting (5.4.2.15) into (5.4.2.11), we obtain the values of the constants:

$$\begin{array}{lll} A_4 = B_4 = 0, & A_n, B_n \text{ are any} & (n = 1, 2, 3); \\ A_k = 0, & B_k \text{ are any} & (k = 1, 2, 3, 4); \\ B_k = 0, & A_k \text{ are any} & (k = 1, 2, 3, 4). \end{array}$$

The first two sets of constants define two known polynomial solutions of the second and third degree for the equations (5.4.2.9):

$$\begin{aligned} u &= C_1x^2 + C_2x + C_3y^2 + C_4y + C_5, \\ u &= C_1y^3 + C_2y^2 + C_3y + C_4, \end{aligned} \tag{5.4.2.16}$$

where $C_1, \ldots, C_5$ are arbitrary constants.

2°. For $C = \lambda^2 > 0$, equations (5.4.2.13) and (5.4.2.14) have the following solutions:

$$\begin{aligned} \varphi(x) &= A_1 + A_2x + A_3e^{\lambda x} + A_4e^{-\lambda x}, \\ \psi(y) &= B_1 + B_2y + B_3e^{\lambda y} + B_4e^{-\lambda y}. \end{aligned} \tag{5.4.2.17}$$

Substituting (5.4.2.17) into (5.4.2.11), dividing by λ^3, and collecting like terms, we obtain

$$A_3(\nu\lambda - B_2)e^{\lambda x} + A_4(\nu\lambda + B_2)e^{-\lambda x} + B_3(\nu\lambda + A_2)e^{\lambda y} + B_4(\nu\lambda - A_2)e^{-\lambda y} = 0.$$

Equating the coefficients of the exponential with zero, we find the values of the constants:

$$\begin{array}{ll} A_3 = A_4 = B_3 = 0,\ A_2 = \nu\lambda & \text{(case 1)}, \\ A_3 = B_3 = 0,\ A_2 = \nu\lambda,\ B_2 = -\nu\lambda & \text{(case 2)}, \\ A_3 = B_4 = 0,\ A_2 = -\nu\lambda,\ B_2 = -\nu\lambda & \text{(case 3)}. \end{array}$$

The other constants can be arbitrary. These sets of constants define three different exact solutions of the fluid dynamic equation (5.4.2.9) of the form (5.4.2.10):

$$\begin{aligned} u &= C_1e^{-\lambda y} + C_2y + C_3 + \nu\lambda x, \\ u &= C_1e^{-\lambda x} + \nu\lambda x + C_2e^{-\lambda y} - \nu\lambda y + C_3, \\ u &= C_1e^{-\lambda x} - \nu\lambda x + C_2e^{\lambda y} - \nu\lambda y + C_3, \end{aligned}$$

where C_1, C_2, C_3, and λ are arbitrary constants.

3°. For $C = -\lambda^2 < 0$, equations (5.4.2.13) and (5.4.2.14) have the solutions

$$\begin{aligned} \varphi(x) &= A_1 + A_2x + A_3\cos(\lambda x) + A_4\sin(\lambda x), \\ \psi(y) &= B_1 + B_2y + B_3\cos(\lambda y) + B_4\sin(\lambda y). \end{aligned} \tag{5.4.2.18}$$

Substituting (5.4.2.18) into (5.4.2.11) does not reveal any further real solutions.

Degenerate cases. In the cases where $\varphi''_{xx} \equiv 0$ and $\psi''_{yy} \equiv 0$, equation (5.4.2.12) becomes an identity for any function $\psi = \psi(y)$ and any function $\varphi = \varphi(x)$, respectively. These cases need to be investigated separately. For instance, if $\varphi''_{xx} \equiv 0$, we get $\varphi(x) = Ax + B$, where A and B are any. On inserting this function into (5.4.2.11), we arrive at the equation $-A\psi'''_{yyy} = \nu\psi''''_{yyyy}$. Its general solution is given by $\psi(y) = C_1 \exp(-Ay/\nu) + C_2 y^2 + C_3 y + C_4$. As a result, we find one more solution to equation (5.4.2.9) of the form (5.4.2.10):

$$u = C_1 e^{-\lambda y} + C_2 y^2 + C_3 y + C_4 + \nu\lambda x \quad (A = \nu\lambda,\ B = 0). \quad ◀$$

▶ Example 5.13. Consider the parabolic equation with a quadratic nonlinearity

$$u_t = a u u_{xx} + b u_x^2 + c. \tag{5.4.2.19}$$

We will seek separable of the form

$$u = \varphi(t)\theta(x) + \psi(t). \tag{5.4.2.20}$$

Substituting solution (5.4.2.20) into equation (5.4.2.19) and collecting like terms, we obtain

$$\psi'_t - c + \varphi'_t \theta = a\varphi\psi\theta''_{xx} + \varphi^2\left[a\theta\theta''_{xx} + b(\theta'_x)^2\right]. \tag{5.4.2.21}$$

First, we eliminate the most complex nonlinearity, which is in square brackets. On dividing (5.4.2.21) by φ^2 and differentiating with respect to t and x, we arrive at a two-term functional differential equation:

$$(\varphi'_t/\varphi^2)'_t \theta'_x = a(\psi/\varphi)'_t \theta'''_{xxx}. \tag{5.4.2.22}$$

Separating the variables yields the ordinary differential equations

$$\theta'''_{xxx} = K\theta'_x, \tag{5.4.2.23}$$

$$(\varphi'_t/\varphi^2)'_t = aK(\psi/\varphi)'_t, \tag{5.4.2.24}$$

where K is an arbitrary constant. The general solution of equation (5.4.2.23) is given by

$$\theta = \begin{cases} A_1 x^2 + A_2 x + A_3 & \text{for } K = 0, \\ A_1 e^{\lambda x} + A_2 e^{-\lambda x} + A_3 & \text{for } K = \lambda^2 > 0, \\ A_1 \sin(\lambda x) + A_2 \cos(\lambda x) + A_3 & \text{for } K = -\lambda^2 < 0, \end{cases} \tag{5.4.2.25}$$

where A_1, A_2, and A_3 are arbitrary constants. Integrating equation (5.4.2.24) gives

$$\begin{aligned} &\varphi = \frac{B}{t + C_1}, \quad \psi(t) \text{ is any} && \text{if } K = 0, \\ &\psi = B\varphi + \frac{1}{aK}\frac{\varphi'_t}{\varphi}, \quad \varphi(t) \text{ is any} && \text{if } K \neq 0, \end{aligned} \tag{5.4.2.26}$$

where B is an arbitrary constant. On substituting solutions (5.4.2.25) and (5.4.2.26) into equation (5.4.2.21), we can eliminate the extra constants and determine the functions ψ and φ. The results are presented below.

1°. Solution for $b \neq -a$ and $b \neq -\frac{1}{2}a$ (corresponds to $K = 0$):

$$u = \frac{c(a+2b)}{2(a+b)}(t+C_1) + C_2(t+C_1)^{-\frac{a}{a+2b}} - \frac{(x+C_3)^2}{2(a+2b)(t+C_1)},$$

where C_1, C_2, and C_3 are arbitrary constants.

2°. Solution for $b = -a$:

$$u = \frac{1}{a\lambda^2}\frac{\varphi'_t}{\varphi} + \varphi(A_1 e^{\lambda x} + A_2 e^{-\lambda x}) \quad \text{(corresponds to } K = \lambda^2 > 0\text{)},$$

where $\varphi = \varphi(t)$ is a function determined by the autonomous ordinary differential equation

$$Z''_{tt} = ac\lambda^2 + 4a^2\lambda^4 A_1 A_2 e^{2Z}, \qquad \varphi = e^Z,$$

whose solution can be represented in an implicit form. In the special case $A_1 = 0$ or $A_2 = 0$, we easily find that $\varphi = C_1 \exp\bigl(\frac{1}{2}ac\lambda^2 t^2 + C_2 t\bigr)$.

3°. Solution for $b = -a$ (corresponds to $K = -\lambda^2 < 0$):

$$u = -\frac{1}{a\lambda^2}\frac{\varphi'_t}{\varphi} + \varphi[A_1 \sin(\lambda x) + A_2 \cos(\lambda x)],$$

where $\varphi = \varphi(t)$ is a function satisfying the autonomous ordinary differential equation

$$Z''_{tt} = -ac\lambda^2 + a^2\lambda^4(A_1^2 + A_2^2)e^{2Z}, \qquad \varphi = e^Z,$$

whose solution can be obtained in an implicit form.

4°. For $b = -\frac{1}{2}a$, there is a degenerate solution $u = ct + A_1$.

The degenerate solution $\varphi = \text{const}$, $\psi = \text{const}$ of the two-term functional differential equation (5.4.2.22) leads to a steady-state of the equation in question. We will not consider such solutions here and in what follows. ◀

◆ *Problems and Exercises for Section 5.4*

1. Find generalized separable solutions for the following nonlinear first-order PDEs:
a) $u_x = au_y^2 + f(x)u$,
b) $u_x = au_y^2 + buu_y + f(x)$.
Hint. Look for solutions in the form $u = \varphi(x) + \psi(x)\theta(y)$.

2. Find generalized separable solutions for the following nonlinear heat-type equations:
a) $u_t = a(uu_x)_x$,
b) $u_t = a(uu_x)_x + b$,
c) $u_t = a(uu_x)_x + bu_x$.
Hint. Look for solutions in the form $u = f(t)\theta(x) + g(t)$.

3. Find generalized separable solutions for the following nonlinear wave equations:
a) $u_{tt} = a(uu_x)_x$,
b) $u_{tt} = a(uu_x)_x + b$.
Hint. Look for solutions in the form $u = f(t)\theta(x) + g(t)$.

4. Find generalized separable solutions of the boundary layer equation with a pressure gradient

$$u_y u_{xy} - u_x u_{yy} = \nu u_{yyy} + f(x).$$

Hint. Look for solutions in the form $u = \varphi(x)\theta(y) + \psi(x)$. The function f also needs to be identified here.

5. Find generalized separable solutions of the following nonlinear third-order equations:

a) $u_{xt} + u_x^2 - uu_{xx} = au_{xxx}$,

b) $u_t + auu_x + bu_{xtt} = 0$.

Hint. Look for solutions in the form $u = \varphi(t)\theta(x) + \psi(t)$.

6. Find generalized separable solutions of the following nonlinear equations:

a) $u_t = a\exp(\lambda u_{xx})$,

b) $u_{tt} = a\exp(\lambda u_{xx})$.

Hint. Look for solutions in the form $u = f(x)\theta(t) + g(x)$. The analysis should begin by taking the logarithm of both sides of the equation.

5.5. Solution of Functional Differential Equation by the Method of Splitting

5.5.1. Preliminary Remarks. Description of the Method. Splitting Principle

The reduction of the number of terms in the functional differential equation (5.2.2.1)–(5.2.2.2) through differentiation may lead to the appearance of extra integration constants, which need to be eliminated in the final step (refer to Section 5.4). Moreover, the order of the resulting equation could be higher than that of the original equation. To address these issues, one may find it helpful to simplify the solution of the functional differential equation by breaking it down into two more manageable tasks: solving a bilinear functional equation of the standard form and solving a system of ordinary differential equations. This approach effectively divides the original problem into two simpler ones (e.g., see Polyanin et al., 2005; Polyanin & Zaitsev, 2012; and Polyanin & Zhurov, 2022). Below is a brief overview of the main stages of this method.

1°. In the first step, we consider equation (5.2.2.1) as a bilinear functional equation

$$\sum_{n=1}^{k} \Phi_n \Psi_n = 0, \tag{5.5.1.1}$$

where $\Phi_n = \Phi_n[X]$ and $\Psi_n = \Psi_n[Y]$ are unknown quantities ($n = 1, \ldots, k$), and X and Y are independent variables.

The splitting principle. All solutions of the bilinear functional equation (5.5.1.1) can be represented as a set of linear combinations involving $\Phi_1, \ldots, \Phi_k$ and linear combinations involving $\Psi_1, \ldots, \Psi_k$:

$$\sum_{n=1}^{k} \alpha_{ni}\Phi_n = 0, \quad i = 1, \ldots, l;$$
$$\sum_{n=1}^{k} \beta_{nj}\Psi_n = 0, \quad j = 1, \ldots, m, \tag{5.5.1.2}$$

where $1 \le l \le k-1$ and $1 \le m \le k-1$. The constants α_{ni} and β_{nj} in (5.5.1.2) are selected such that the bilinear equality (5.5.1.1) holds identically. As demonstrated in Section 5.5.2, this selection can always be achieved. It is important to emphasize that the relationships in (5.5.1.2) are purely algebraic and do not rely on the specific expressions of the differential forms (5.2.2.2).

2°. In the second step, we successively substitute the differential forms $\Phi_i[X]$ and $\Psi_j[Y]$ from (5.2.2.2) into solutions (5.5.1.2). This process results in several systems of ordinary differential equations, which are often overdetermined, for determining the unknown functions $\varphi_p(x)$ and $\psi_q(y)$. By solving these systems, we can find generalized separable solutions of the form (5.2.1.1).

3°. The degenerate cases, where one or more differential forms Φ_n and Ψ_n vanish, must be treated separately. In these cases, linear relations of the form (5.5.1.2) must be used for the remaining forms.

Remark 5.8. *The splitting principle can be proved by the method of mathematical induction (Polyanin & Zhurov, 2022).*

Remark 5.9. *It is important to emphasize that the bilinear functional equation (5.5.1.1) (or (5.2.2.1)) used in the splitting method and its solutions for a fixed k are the same for different classes of nonlinear equations of mathematical physics.*

For greater clarity, Figure 5.1 illustrates the key stages in constructing solutions using the generalized separation of variables with the splitting method.

Remark 5.10. *The bilinear functional equation (5.5.1.1) and its solutions (5.5.1.2) play the key role in the method of functional separation of variables (see further Section 6.5.2 and the book by Polyanin & Zhurov, 2022).*

5.5.2. Solutions of Bilinear Functional Equations

Here, we provide non-degenerate solutions for three simple functional equations of the form (5.5.1.1) or (5.2.2.1). These solutions will be necessary later to construct exact solutions of specific nonlinear partial differential equations.

1°. The three-term bilinear functional equation

$$\Phi_1\Psi_1 + \Phi_2\Psi_2 + \Phi_3\Psi_3 = 0, \tag{5.5.2.1}$$

where Φ_i are all functions of the same argument, and all Ψ_i are functions of a different argument, have two solutions:

$$\Phi_1 = A_1\Phi_3, \quad \Phi_2 = A_2\Phi_3, \quad \Psi_3 = -A_1\Psi_1 - A_2\Psi_2;$$
$$\Psi_1 = A_1\Psi_3, \quad \Psi_2 = A_2\Psi_3, \quad \Phi_3 = -A_1\Phi_1 - A_2\Phi_2, \tag{5.5.2.2}$$

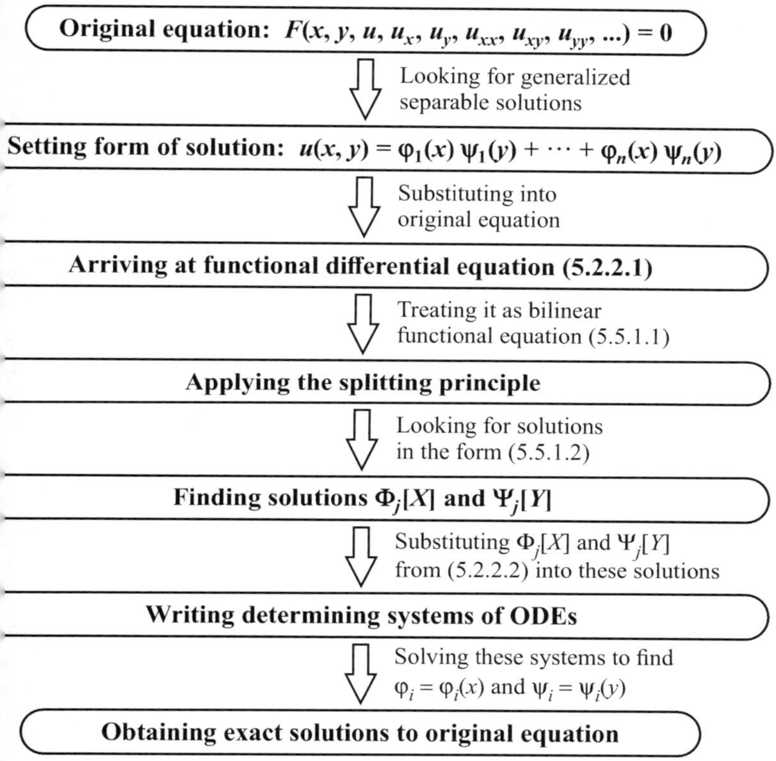

.1. General scheme for constructing generalized separable solutions using the nethod.

$_1$ and A_2 are arbitrary constants. The functions on the right-hand sides in are assumed arbitrary. The functions Φ_i in the first solution (5.5.2.2) or s Ψ_i in the second solution are all proportional to each other. It can be by direct substitution that solutions (5.5.2.2) satisfy the functional equation

The functional equation involving four terms

$$\Phi_1\Psi_1 + \Phi_2\Psi_2 + \Phi_3\Psi_3 + \Phi_4\Psi_4 = 0 \qquad (5.5.2.3)$$

solution

$$\begin{aligned} \Phi_1 &= A_1\Phi_3 + A_2\Phi_4, & \Phi_2 &= A_3\Phi_3 + A_4\Phi_4, \\ \Psi_3 &= -A_1\Psi_1 - A_3\Psi_2, & \Psi_4 &= -A_2\Psi_1 - A_4\Psi_2, \end{aligned} \qquad (5.5.2.4)$$

nt on four arbitrary constants A_m. The functions on the right-hand sides in are assumed arbitrary.

Equation (5.5.2.3) has two simpler solutions involving three arbitrary constants:

$$\begin{aligned}&\Phi_1=A_1\Phi_4,\quad \Phi_2=A_2\Phi_4,\quad \Phi_3=A_3\Phi_4,\quad \Psi_4=-A_1\Psi_1-A_2\Psi_2-A_3\Psi_3;\\&\Psi_1=A_1\Psi_4,\quad \Psi_2=A_2\Psi_4,\quad \Psi_3=A_3\Psi_4,\quad \Phi_4=-A_1\Phi_1-A_2\Phi_2-A_3\Phi_3.\end{aligned} \tag{5.5.2.5}$$

All functions Φ_i or all functions Ψ_i in solutions (5.5.2.5) are proportional to each other.

It can be verified by direct substitution that solutions (5.5.2.4) and (5.5.2.5) satisfy the functional equation (5.5.2.3).

Using solution (5.5.2.4) of the bilinear functional equation (5.5.2.3), one can more frequently find generalized separable solutions to nonlinear partial differential equations compared to using solutions (5.5.2.5).

Remark 5.11. *Due to the existence of discrete symmetries in bilinear functional equations with respect to various permutations of the quantities involved, the symbols Φ and Ψ in solutions (5.5.2.2), (5.5.2.4), and (5.5.2.5) can be interchanged: $\Phi_p \rightleftarrows \Psi_p$. Furthermore, simultaneous paired permutations, such as $\Phi_p \rightleftarrows \Phi_q$ and $\Psi_p \rightleftarrows \Psi_q$, are also possible.*

3°. In practice, one can follow these steps to solve the bilinear functional equation (5.5.1.1). First, select several first quantities, $\Phi_1, \ldots, \Phi_p$ ($p < k$) from the set $\Phi_1, \ldots, \Phi_k$; this can be done arbitrarily. Next, express these selected quantities as linear combinations of the remaining quantities in the set, $\Phi_{p+1}, \ldots, \Phi_k$. This procedure establishes the first group of relations (5.5.1.2). After substituting the quantities $\Phi_1, \ldots, \Phi_p$ with their linear combinations in equation (5.5.1.1) and collecting the terms proportional to Φ_q (for $q = p+1, \ldots, k$), one arrives at a relation of the form

$$\sum_{q=p+1}^{k} \Omega_q \Phi_q = 0, \qquad \Omega_q = \sum_{s=1}^{k} a_{qs} \Psi_s,$$

where a_{qs} are some constants. Equating the functional coefficients Ω_q ($q = p+1, \ldots, k$) with zero, one obtains the second group or relations (5.5.1.2).

Due to the symmetry of equation (5.5.1.1) with respect to the permutation $\Phi_n \rightleftarrows \Psi_n$, one can select elements from the set $\Psi_1, \ldots, \Psi_k$ in the first step.

► Example 5.14. Let us consider the five-term bilinear functional equation

$$\Phi_1\Psi_1 + \Phi_2\Psi_2 + \Phi_3\Psi_3 + \Phi_4\Psi_4 + \Phi_5\Psi_5 = 0. \tag{5.5.2.6}$$

We assume that the first three functional coefficients, Φ_1, Φ_2, and Φ_3, are linear combinations of the last two, Φ_4 and Φ_5:

$$\Phi_1 = A_1\Phi_4 + B_1\Phi_5, \quad \Phi_2 = A_2\Phi_4 + B_2\Phi_5, \quad \Phi_3 = A_3\Phi_4 + B_3\Phi_5, \tag{5.5.2.7}$$

where A_n and B_n are arbitrary constants. On substituting the expressions (5.5.2.7) into (5.5.2.6) and collecting the terms proportional to Φ_4 and Φ_5, we obtain

$$(A_1\Psi_1 + A_2\Psi_2 + A_3\Psi_3 + \Psi_4)\Phi_4 + (B_1\Psi_1 + B_2\Psi_2 + B_3\Psi_3 + \Psi_5)\Phi_5 = 0.$$

Equating the expressions in parentheses with zero, we get

$$\begin{aligned}\Psi_4 &= -A_1\Psi_1 - A_2\Psi_2 - A_3\Psi_3;\\ \Psi_5 &= -B_1\Psi_1 - B_2\Psi_2 - B_3\Psi_3.\end{aligned} \tag{5.5.2.8}$$

Formulas (5.5.2.7) and (5.5.2.8) provide one solution to equation (5.5.2.6). Other solutions can be found likewise. ◄

5.5.3. Examples of Constructing Generalized Separable Solutions by the Splitting Method

Below are a few examples of using the splitting method to construct generalized separable solutions for nonlinear PDEs.

► Example 5.15. Consider the nonlinear hyperbolic equation

$$u_{tt} = a(uu_x)_x + f(t)u + g(t), \tag{5.5.3.1}$$

where $f(t)$ and $g(t)$ are arbitrary functions.

We seek a generalized separable solution to this equation in the form

$$u = \varphi(t)\theta(x) + \psi(t). \tag{5.5.3.2}$$

Substituting (5.5.3.2) into (5.5.3.1) and rearranging, we obtain

$$a\varphi^2(\theta\theta'_x)'_x + a\varphi\psi\theta''_{xx} + (f\varphi - \varphi''_{tt})\theta + f\psi + g - \psi''_{tt} = 0.$$

This equation can be written as the bilinear functional equation (5.5.2.3) where

$$\begin{aligned}&\Phi_1 = (\theta\theta'_x)'_x, \quad \Phi_2 = \theta''_{xx}, \quad \Phi_3 = \theta; \qquad \Phi_4 = 1,\\ &\Psi_1 = a\varphi^2, \qquad \Psi_2 = a\varphi\psi, \quad \Psi_3 = f\varphi - \varphi''_{tt}, \quad \Psi_4 = f\psi + g - \psi''_{tt}.\end{aligned} \tag{5.5.3.3}$$

On inserting the expressions (5.5.3.3) into solution (5.5.2.4), we arrive at a system of ordinary differential equations for the undetermined functions $\theta = \theta(x)$, $\varphi = \varphi(t)$, and $\psi = \psi(t)$:

$$\begin{aligned}&(\theta\theta'_x)'_x = A_1\theta + A_2, \qquad\qquad \theta''_{xx} = A_3\theta + A_4;\\ &f\varphi - \varphi''_{tt} = -A_1 a\varphi^2 - A_3 a\varphi\psi, \quad f\psi + g - \psi''_{tt} = -A_2 a\varphi^2 - A_4 a\varphi\psi,\end{aligned} \tag{5.5.3.4}$$

where $A_1, \ldots, A_4$ are arbitrary constants. The first two equations in (5.5.3.4) form an overdetermined ODE system for a single function, θ. The second equation is linear and can be easily integrated. Depending on the value of A_3, its solution can be expressed in different forms: via trigonometric functions if $A_3 < 0$, via hyperbolic functions if $A_3 > 0$, or as a quadratic polynomial if $A_3 = 0$. By successively substituting the three solutions of the second equation into the first one, we conclude the quadratic polynomial is the only compatible solution for this system of ODEs:

$$\theta(x) = B_2x^2 + B_1x + B_0. \tag{5.5.3.5}$$

The constants B_0, B_1, and B_2 are expressed in terms of the constants A_n, appearing in system (5.5.3.4), as follows:

$$A_1 = 6B_2, \quad A_2 = B_1^2 - 4B_0B_2, \quad A_3 = 0, \quad A_4 = 2B_2, \tag{5.5.3.6}$$

Inserting the expressions (5.5.3.6) into the last two equations in (5.5.3.4), we obtain an ODE system for $\varphi(t)$ and $\psi(t)$:

$$\begin{aligned} \varphi''_{tt} &= 6aB_2\varphi^2 + f(t)\varphi, \\ \psi''_{tt} &= [2aB_2\varphi + f(t)]\psi + a(B_1^2 - 4B_0B_2)\varphi^2 + g(t). \end{aligned} \tag{5.5.3.7}$$

Formulas (5.5.3.2) and (5.5.3.5) and system (5.5.3.7) define a generalized separable solution of equation (5.5.3.1). The first equation in (5.5.3.7) can be solved independently; it is linear when $B_2 = 0$ and integrable by quadrature if $f(t) = \text{const}$. The second equation in (5.5.3.7) is linear in ψ (for known φ).

For $\theta \not\equiv 0$, $\varphi \not\equiv 0$, $\psi \not\equiv 0$, and arbitrary f and g, equation (5.5.3.1) does not have other solutions of the form (5.5.3.2). ◄

Remark 5.12. *Equation (5.5.3.1) admits a more general solution, quadratic in the space variable:* $u = x^2\psi_1(t) + x\psi_2(t) + \psi_3(t)$.

► Example 5.16. Consider the hydrodynamic nonlinear third-order equation

$$u_{xt} + u_x^2 - uu_{xx} = \nu u_{xxx}. \tag{5.5.3.8}$$

We look for exact solution to equation (5.5.3.8) in the form

$$u = \varphi(t)\theta(x) + \psi(t). \tag{5.5.3.9}$$

Substituting (5.5.3.9) into (5.5.3.8) yields

$$\varphi'_t\theta'_x - \varphi\psi\theta''_{xx} + \varphi^2\left[(\theta'_x)^2 - \theta\theta''_{xx}\right] - \nu\varphi\theta'''_{xxx} = 0. \tag{5.5.3.10}$$

This functional differential equation can be converted into the bilinear functional equation (5.5.2.3) if we set

$$\begin{aligned} &\Phi_1 = \varphi'_t, \quad \Phi_2 = \varphi\psi, \quad \Phi_3 = \varphi^2, \quad \Phi_4 = \nu\varphi; \\ &\Psi_1 = \theta'_x, \quad \Psi_2 = -\theta''_{xx}, \quad \Psi_3 = (\theta'_x)^2 - \theta\theta''_{xx}, \quad \Psi_4 = -\theta'''_{xxx}. \end{aligned} \tag{5.5.3.11}$$

First group of solutions. Inserting the expressions (5.5.3.11) into (5.5.2.4), we obtain the system of ordinary differential equations

$$\begin{aligned} &\varphi'_t = A_1\varphi^2 + A_2\nu\varphi, \qquad \varphi\psi = A_3\varphi^2 + A_4\nu\varphi; \\ &(\theta'_x)^2 - \theta\theta''_{xx} = -A_1\theta'_x + A_3\theta''_{xx}, \quad \theta'''_{xxx} = A_2\theta'_x - A_4\theta''_{xx}. \end{aligned} \tag{5.5.3.12}$$

The last two equations in (5.5.3.12) form an overdetermined ODE system for the function θ; the last equation is linear and has constant coefficients and, hence, is easy to integrate. It can be shown that these equations have a compatible solution only if the function θ is linearly related to its derivative:

$$\theta'_x = B_1\theta + B_2. \tag{5.5.3.13}$$

By eliminating the derivatives from the final equations in (5.5.3.12) using relation (5.5.3.13) and its consequences obtained through differentiation, we conclude that the six constants B_1, B_2, A_1, A_2, A_3, and A_4 must satisfy three conditions:

$$\begin{aligned} B_1(A_1+B_2-A_3B_1)&=0,\\ B_2(A_1+B_2-A_3B_1)&=0,\\ B_1^2+A_4B_1-A_2&=0. \end{aligned} \qquad (5.5.3.14)$$

Integrating equation (5.5.3.13) yields

$$\theta=\begin{cases} B_3\exp(B_1x)-\dfrac{B_2}{B_1} & \text{if } B_1\neq 0,\\ B_2x+B_3 & \text{if } B_1=0, \end{cases} \qquad (5.5.3.15)$$

where B_3 is an arbitrary constant.

The functions φ and ψ follow from the last two equations in (5.5.3.12):

$$\varphi=\begin{cases} \dfrac{A_2\nu}{C\exp(-A_2\nu t)-A_1} & \text{if } A_2\neq 0,\\ -\dfrac{1}{A_1t+C} & \text{if } A_2=0, \end{cases} \qquad \psi=A_3\varphi+A_4\nu. \qquad (5.5.3.16)$$

Formulas (5.5.3.15) and (5.5.3.16) and relations (5.5.3.14) allow us to find the following generalized separable solutions of equation (5.5.3.8):

$$\begin{aligned}
u&=\frac{C_1e^{-\lambda x}+1}{\lambda t+C_2}+\nu\lambda && \text{if}\quad A_2=0,\ B_1=-A_4,\ B_2=-A_1-A_3A_4;\\
u&=C_1e^{-\lambda(x+\beta\nu t)}+\nu(\lambda+\beta) && \text{if}\quad A_1=A_3=B_2=0,\ A_2=B_1^2+A_4B_1;\\
u&=\frac{\nu\beta+C_1e^{-\lambda x}}{1+C_2e^{-\nu\lambda\beta t}}+\nu(\lambda-\beta) && \text{if}\quad A_1=A_3B_1-B_2,\ A_2=B_1^2+A_4B_1;\\
u&=\frac{x+C_1}{t+C_2}+C_3 && \text{if}\quad A_2=B_1=0,\ B_2=-A_1,
\end{aligned}$$

where C_1, C_2, C_3, β, and λ are arbitrary constants; they can be expressed in terms of A_k and B_k.

Second group of solutions. Suppose that all x-dependent terms in equation (5.5.3.10) are proportional to θ'_x. Consequently, we get the following system of ordinary differential equations:

$$\begin{aligned} &\theta'''_{xxx}=A_1\theta'_x,\quad \theta''_{xx}=A_2\theta'_x,\quad (\theta'_x)^2-\theta\theta''_{xx}=A_3\theta'_x;\\ &\varphi'_t-A_2\varphi\psi+A_3\varphi^2-A_1\nu\varphi=0. \end{aligned} \qquad (5.5.3.17)$$

The overdetermined subsystem consisting of the first three equations of (5.5.3.17) admits two compatible solutions:

$$\begin{aligned} &\textit{Solution 1:}\quad \theta=e^{-\lambda x} && \text{if}\quad A_1=\lambda^2,\ A_2=-\lambda,\ A_3=0;\\ &\textit{Solution 2:}\quad \theta=x && \text{if}\quad A_1=A_2=0,\ A_3=1. \end{aligned} \qquad (5.5.3.18)$$

Solutions (5.5.3.18) in conjunction with solutions of the last equation in (5.5.3.17), which are easy to obtain, lead to two solutions of the original nonlinear differential equation (5.5.3.8):

$$u = \varphi(t)e^{-\lambda x} - \frac{\varphi_t'(t)}{\lambda\varphi(t)} + \nu\lambda,$$
$$u = \frac{x}{t+C} + \psi(t),$$

where $\varphi(t)$ and $\psi(t)$ are arbitrary functions, and λ is an arbitrary constant. ◀

► Example 5.17. Consider the following equation with an exponential nonlinearity in the highest derivative:

$$u_t = f(x)\exp(au_{xx}). \tag{5.5.3.19}$$

We look for exact solutions of the form

$$u = \varphi(x)\theta(t) + \psi(x). \tag{5.5.3.20}$$

Let us substitute (5.5.3.20) into (5.5.3.19), divide the resulting equation by $f(x)$, and take the logarithm of it. Assuming that $\varphi/f > 0$ and performing simple rearrangements, we obtain

$$a\psi_{xx}'' - \ln(\varphi/f) + a\theta\varphi_{xx}'' - \ln\theta_t' = 0.$$

This functional differential equation can be represented in the bilinear form (5.5.2.1) in which

$$\Phi_1 = a\psi_{xx}'' - \ln(\varphi/f),\quad \Phi_2 = \varphi_{xx}'',\quad \Phi_3 = 1;\quad \Psi_1 = 1,\quad \Psi_2 = a\theta,\quad \Psi_3 = -\ln\theta_t'.$$

On substituting these expressions into the first solution in (5.5.2.2), we arrive at the ordinary differential equations

$$a\psi_{xx}'' - \ln(\varphi/f) = A_1,\quad \varphi_{xx}'' = A_2,\quad \ln\theta_t' = A_1 + A_2 a\theta.$$

Integrating yields

$$\begin{aligned}
\varphi(x) &= \frac{1}{2}A_2x^2 + C_1x + C_2,\\
\psi(x) &= \frac{1}{2a}A_1x^2 + C_3x + C_4 + \frac{1}{a}\int_{x_0}^{x}(x-\xi)\ln\frac{\varphi(\xi)}{f(\xi)}\,d\xi,\\
\theta(t) &= -\frac{1}{A_2a}\ln\bigl(C_5 - A_2ae^{A_1}t\bigr).
\end{aligned} \tag{5.5.3.21}$$

Formulas (5.5.3.20) and (5.5.3.21) describe a generalized separable solution of equation (5.5.3.19). ◀

Problems and Exercises for Section 5.5

1. Find generalized separable solutions of the following nonlinear first-order PDEs:

a) $u_t = au_x^2 + f(t)u$,

b) $u_t = au_x^2 + buu_x + f(t)$.

Hint. Look for solutions in the form $u = \varphi(t)\theta(x) + \psi(t)$.

2. Find some functions $f(y)$ such that the nonlinear equations

a) $u_{xy} + ku_x u_y = f(y)$,

b) $u_{xy}^2 + ku_{xx}u_{yy} = f(y)$,

admit generalized separable solutions of the form $u = \varphi(x)\theta(y) + \psi(x)$.

3. Find generalized separable solutions of the nonlinear nth-order equation

$u_y u_{xy} - u_x u_{yy} = f(x)u_y^{(n)}$.

Hint. Look for solutions in the form $u = \varphi(x)\theta(y) + \psi(x)$.

4. Find generalized separable solutions of the form (5.4.2.10) for the nonlinear hydrodynamic equation (5.4.2.9) using the splitting method.

See also the problems and exercises for Section 5.4.

5.6. Invariant Subspace Method

5.6.1. Subspaces Invariant Under a Nonlinear Differential Operator. Method Description

This section outlines the *method of invariant subspaces* (Galaktionov, 1995; Galaktionov et al., 1995; Galaktionov & Svirshchevskii, 2007). It does not involve the analysis of functional differential equations and differs significantly from the methods discussed previously in Sections 5.4 and 5.5. Notably, the initial development of this method was significantly influenced by the works of Titov (1977, 1988) and Titov & Ustinov (1985).

Let us consider the partial differential equation

$$u_t = F[u], \tag{5.6.1.1}$$

where $F[u]$ is a nonlinear differential operator of the form

$$F[u] \equiv F(x, u, u_x, \dots, u_x^{(n)}). \tag{5.6.1.2}$$

Definition. A finite dimensional linear subspace

$$\mathscr{L}_k = \{\varphi_1(x), \dots, \varphi_k(x)\}, \tag{5.6.1.3}$$

whose elements are all possible linear combinations of the linearly independent functions $\varphi_1(x)$, …, $\varphi_k(x)$, is said to be *invariant under a differential operator* F if there exist functions f_1, …, f_k such that

$$F\left[\sum_{i=1}^{k} C_i\varphi_i(x)\right] = \sum_{i=1}^{k} f_i(C_1, \dots, C_k)\,\varphi_i(x), \tag{5.6.1.4}$$

where C_1, …, C_k are arbitrary constants.* The functions $\varphi_i(x)$ from (5.6.1.4) must be independent of C_1, …, C_k.

*Under the action of the operator F, the function $\sum_{i=1}^{k} C_i\varphi_i(x)$ converts into a similar function $\sum_{i=1}^{k} B_i\varphi_i(x)$ but with different coefficients, where $B_i = f_i(C_1, \dots, C_k)$.

Proposition 1. *Suppose the linear subspace* (5.6.1.3) *is invariant under a differential operator* F. *Then equation* (5.6.1.1) *has generalized separable solutions of the form*

$$u = \sum_{i=1}^{k} \psi_i(t)\varphi_i(x), \tag{5.6.1.5}$$

where $\psi_1(t)$, ..., $\psi_k(t)$ *are functions satisfying the autonomous system of ordinary differential equations (Galaktionov & Svirshchevskii, 2007):*

$$\psi_i' = f_i(\psi_1, \dots, \psi_k), \qquad i = 1, \dots, k. \tag{5.6.1.6}$$

The prime stands for a derivative with respect to t.

This proposition can be proved as follows. First, one substitutes the expression (5.6.1.5) into equation (5.6.1.1). Then, one employs relation (5.6.1.4) with the constants C_i replaced with the functions $\psi_i = \psi_i(t)$. After collecting the terms proportional to $\varphi_i = \varphi_i(x)$, one obtains

$$\sum_{i=1}^{k} [\psi_i' - f_i(\psi_1, \dots, \psi_k)]\varphi_i(x) = 0.$$

Since the functions φ_i are linearly independent, all the expressions in square brackets must be set equal to zero. This results in the ODE system (5.6.1.6).

The example below illustrates the application of the above method.

► Example 5.18. Consider the nonlinear heat equation with a linear source

$$u_t = (uu_x)_x + bu. \tag{5.6.1.7}$$

1°. Let us prove that the three-dimensional linear subspace of power-law functions $\mathscr{L}_3 = \{1, x, x^2\}$ is invariant under the nonlinear differential operator

$$F[u] = (uu_x)_x + bu, \tag{5.6.1.8}$$

which defines the right-hand side of equation (5.6.1.7).

Indeed, replacing u in (5.6.1.8) with $C_1 + C_2x + C_3x^2$, where C_1, C_2, and C_3 are arbitrary constants, we obtain (the intermediate calculations are simple, so the reader is recommended to carry them out by themselves):

$$F\big[C_1 + C_2x + C_3x^2\big] = 2C_1C_3 + C_2^2 + bC_1 + (6C_2C_3 + bC_2)x + (6C_3^2 + bC_3)x^2.$$

This relation demonstrates that any quadratic polynomial is transformed into another quadratic polynomial when acted by the operator (5.6.1.8). From Proposition 1 above, it follows that the nonlinear heat equation (5.6.1.7) possesses a generalized separable solution

$$u = \psi_1(t) + \psi_2(t)x + \psi_3(t)x^2, \tag{5.6.1.9}$$

where $\psi_i = \psi_i(t)$ $(i = 1, 2, 3)$ are functions satisfying the autonomous system of first-order ordinary differential equations

$$\begin{aligned}\psi_1' &= 2\psi_1\psi_3 + \psi_2^2 + b\psi_1,\\ \psi_2' &= 6\psi_2\psi_3 + b\psi_2,\\ \psi_3' &= 6\psi_3^2 + b\psi_3.\end{aligned}$$

This system can be successively integrated, starting from the last equation, which is a Bernoulli equation. As a result, one can obtain exact solutions, which are expressed in terms of elementary functions (not given here).

2°. Let us show that the two-dimensional linear subspace of power-law functions $\mathscr{L}_2 = \{\sqrt{x}, x^2\}$ is also invariant under the differential operator (5.6.1.8). Indeed, for arbitrary constants C_1 and C_2, the following relation holds (carry out the calculations yourself):

$$F\big[C_1\sqrt{x} + C_2x^2\big] = \big(\tfrac{15}{4}C_1C_2 + bC_1\big)\sqrt{x} + (6C_2^2 + bC_2)x^2.$$

Therefore, using Proposition 1, we conclude that the nonlinear equation (5.6.1.7) has a generalized separable solution different from (5.6.1.9):

$$u = \psi_1(t)\sqrt{x} + \psi_2(t)x^2,$$

where the functions $\psi_i = \psi_i(t)$ $(i = 1, 2)$ are described by the first-order ODEs

$$\begin{aligned}\psi_1' &= \tfrac{15}{4}\psi_1\psi_2 + b\psi_1,\\ \psi_2' &= 6\psi_2^2 + b\psi_2.\end{aligned}$$

This system can be successively integrated starting from the second equation, which is a Bernoulli equation. ◀

► Example 5.19. Now let us consider the nonlinear parabolic equation

$$u_t = au_{xx} + u_x^2 + ku^2 + bu + c. \tag{5.6.1.10}$$

We will show that for $k > 0$, the differential operator

$$F[u] = au_{xx} + u_x^2 + ku^2 + bu + c, \tag{5.6.1.11}$$

which defines the right-hand side of (5.6.1.10), has a two-dimensional invariant subspace $\mathscr{L}_2 = \{1, \cos(x\sqrt{k}\,)\}$. Indeed, for arbitrary C_1 and C_2, the relation

$$F\big[C_1 + C_2\cos(x\sqrt{k}\,)\big] = k(C_1^2 + C_2^2) + bC_1 + c + C_2(2kC_1 - ak + b)\cos(x\sqrt{k}\,)$$

holds. (Perform the calculations using the main trigonometric identity $\cos^2 z + \sin^2 z = 1$.) Therefore, equation (5.6.1.10) admits a generalized separable solution of the form

$$u = \psi_1(t) + \psi_2(t)\cos(x\sqrt{k}\,). \tag{5.6.1.12}$$

The functions $\psi_1(t)$ and $\psi_2(t)$ are described by the autonomous system of ordinary differential equations

$$\begin{aligned}\psi_1' &= k(\psi_1^2 + \psi_2^2) + b\psi_1 + c,\\ \psi_2' &= \psi_2(2k\psi_1 - ak + b).\end{aligned} \tag{5.6.1.13}$$

◀

Remark 5.13. *For $k > 0$, the nonlinear differential operator $F[u]$ defined in (5.6.1.11) admits a three-dimensional invariant subspace containing trigonometric functions* $\mathscr{L}_3 = \{1, \sin(x\sqrt{k}\,), \cos(x\sqrt{k}\,)\}$.

For $k < 0$, the nonlinear operator $F[u]$ defined in (5.6.1.11) admits a three-dimensional invariant subspace containing hyperbolic functions $\mathscr{L}_3 = \{1, \sinh(-x\sqrt{-k}\,), \cosh(-x\sqrt{-k}\,)\}$ *or an equivalent subspace with exponentials* $\mathscr{L}_3 = \{1, \exp(-x\sqrt{-k}\,), \exp(x\sqrt{-k}\,)\}$.

For $k = 0$, the nonlinear operator (5.6.1.11) admits a three-dimensional invariant subspace containing power-law functions $\mathscr{L}_3 = \{1, x, x^2\}$.

Table 5.2 lists some nonlinear differential operators and linear subspaces invariant under these operators; data from the books by Galaktionov & Svirshchevskii (2007) and Polyanin & Zhurov (2022) were used. Adding the linear operator $L[u] = \alpha u_{xx} + \beta u_x + \gamma u + \delta$ to the first seven nonlinear operators does not change the invariant subspaces (except for $\mathscr{L}_2$ for the third operator).

5.6.2. Some Generalizations

We will consider more general equations of the form

$$L_1[u] = L_2[w], \quad w = F[u], \tag{5.6.2.1}$$

where $L_1[u]$ and $L_2[w]$ are linear differential operators in the variable t:

$$L_1[u] \equiv \sum_{i=0}^{m_1} a_i(t)u_t^{(i)}, \quad L_2[w] \equiv \sum_{j=0}^{m_2} b_j(t)w_t^{(j)}, \tag{5.6.2.2}$$

and $F[u]$ is a nonlinear differential operator in the variable x:

$$F[u] \equiv F(t, x, u, u_x, \dots, u_x^{(n)}), \tag{5.6.2.3}$$

which can depend on t as a parameter.

Proposition 2. *Suppose a linear subspace* (5.6.1.3) *is invariant under nonlinear differential operator F in the sense that for arbitrary constants $C_1, \dots, C_k$, the equality*

$$F\left[\sum_{i=1}^{k} C_i\varphi_i(x)\right] = \sum_{i=1}^{k} f_i(t, C_1, \dots, C_k)\varphi_i(x) \tag{5.6.2.4}$$

holds. The functions $\varphi_i(x)$ are independent of t and the constants $C_1, \dots, C_k$. Then equation (5.6.2.1) *has a generalized separable solution of the form* (5.6.1.5), *where the functions $\psi_1(t), \dots, \psi_k(t)$ are described by the system of ordinary differential equations*

$$L_1\big[\psi_i(t)\big] = L_2\big[f_i(t, \psi_1, \dots, \psi_k)\big], \qquad i = 1, \dots, k. \tag{5.6.2.5}$$

Table 5.2. Some nonlinear differential operators and linear subspaces invariant under these operators (a, b, and c are constants).

No.	Nonlinear operator $F[u]$	Subspaces invariant under $F[u]$
1	$au_{xx}+bu_x^2$	$\mathscr{L}_3=\{1,x,x^2\}$
2	$au_{xx}+u_x^2+bu^2$	$\mathscr{L}_3=\{1,\sin(x\sqrt{b}),\cos(x\sqrt{b})\}$ if $b>0$, $\mathscr{L}_3=\{1,\sinh(x\sqrt{\lvert b\rvert}),\cosh(x\sqrt{\lvert b\rvert})\}$ if $b<0$
3	$auu_{xx}+bu_x^2+cu^2$	$\mathscr{L}_3=\{1,\sin(\lambda x),\cos(\lambda x)\}$ if $c/(a+b)=\lambda^2>0$, $\mathscr{L}_3=\{1,\sinh(\lambda x),\cosh(\lambda x)\}$ if $c/(a+b)=-\lambda^2<0$, $\mathscr{L}_3=\{1,x,x^2\}$ if $c=0$, $\mathscr{L}_2=\{x^2,x^\beta\}$, $\beta=a/(a+b)$ if $c=0,a\neq -b$
4	$uu_{xx}-u_x^2$ (special case of the third operator)	$\mathscr{L}_3=\{1,\sin(\lambda x),\cos(\lambda x)\}$, λ is an arbitrary constant, $\mathscr{L}_3=\{1,\sinh(\lambda x),\cosh(\lambda x)\}$, λ is an arbitrary constant, $\mathscr{L}_3=\{1,x,x^2\}$
5	$uu_{xx}-\frac{2}{3}u_x^2$ (special case of the third operator)	$\mathscr{L}_4=\{1,x,x^2,x^3\}$
6	$uu_{xx}-\frac{3}{4}u_x^2+au^2$ (special case of the third operator)	$\mathscr{L}_5=\{1,\cos(kx),\sin(kx),\cos(2kx),\sin(2kx)\}$ if $a=k^2>0$, $\mathscr{L}_5=\{1,\cosh(kx),\sinh(kx),\cosh(2kx),\sinh(2kx)\}$ if $a=-k^2<0$, $\mathscr{L}_5=\{1,x,x^2,x^3,x^4\}$ if $a=0$
7	$[(au^2+bu+c)u_x]_x$	$\mathscr{L}_2=\{1,x\}$
8	$u^2u_{xx}-\frac{1}{2}uu_x^2+au^3$	$\mathscr{L}_3=\{1,\cos(\sqrt{2a}\,x),\sin(\sqrt{2a}\,x)\}$ if $a>0$, $\mathscr{L}_3=\{1,\cosh(\sqrt{2\lvert a\rvert}\,x),\sinh(\sqrt{2\lvert a\rvert}\,x)\}$ if $a<0$, $\mathscr{L}_3=\{1,x,x^2\}$ if $a=0$
9	u_xu_{xx}	$\mathscr{L}_4=\{1,x,x^2,x^3\}$, $\mathscr{L}_3=\{1,x^{3/2},x^3\}$, $\mathscr{L}_2=\{1,\varphi(x)\}$, $\varphi'_x\varphi''_{xx}=p_1+p_2\varphi$, p_1 and p_2 are constants
10	$(u^2)_{xxxx}$	$\mathscr{L}_5=\{1,x,x^2,x^3,x^4\}$, $\mathscr{L}_3=\{x^{1/2},x^{3/2},x^4\}$
11	$(u^2)_x^{(n)}$	$\mathscr{L}_{n+1}=\{1,x,x^2,\dots,x^n\}$, $\mathscr{L}_3=\{x^{k/2},x^{m/2},x^n\}$, where $k<n$ and $m<n$; k, m, and $\frac{1}{2}(k+m)$ are nonnegative integers
12	$u_x^{(m)}u_x^{(n)}$	$\mathscr{L}_{m+n+1}=\{1,x,x^2,\dots,x^{m+n}\}$, $\mathscr{L}_3=\{1,x^{(m+n)/2},x^{n+m}\}$

► Example 5.20. Consider the generalized Guderley equation

$$u_{tt} + a_1(t)u_t = a_2(t)u_x u_{xx}, \tag{5.6.2.6}$$

which is used to describe transonic gas flows for $a_1(t) = 0$ and $a_2(t) = a$ and also for $a_1(t) = 1/t$ and $a_2(t) = a$. The t plays the role of a spatial variable, and $\gamma = a - 1$ is the adiabatic index.

Equation (5.6.2.6) is a special case of equation (5.6.2.1), where

$$L_1[u] = u_{tt} + a_1(t)u_t, \quad L_2[w] = a_2(t)w, \quad F[u] = u_x u_{xx}.$$

1°. It is straightforward to verify that the nonlinear differential operator $F[u] = u_x u_{xx}$ has a three-dimensional invariant subspace $\mathscr{L}_3 = \{1, x^{3/2}, x^3\}$. From Proposition 2 above, it follows that equation (5.6.2.6) has a generalized separable solution of the form

$$u = \psi_1(t) + \psi_2(t)x^{3/2} + \psi_3(t)x^3, \tag{5.6.2.7}$$

where the functions $\psi_i = \psi_i(t)$ ($i = 1,\ 2,\ 3$) are described by the system of ordinary differential equations

$$\begin{aligned}
\psi_1'' + a_1(t)\psi_1' &= \tfrac{9}{8}a_2(t)\psi_2^2,\\
\psi_2'' + a_1(t)\psi_2' &= \tfrac{45}{4}a_2(t)\psi_2\psi_3,\\
\psi_3'' + a_1(t)\psi_3' &= 18a_2(t)\psi_3^2.
\end{aligned}$$

2°. The operator $F[u] = u_x u_{xx}$ also admits a four-dimensional subspace, $\mathscr{L}_4 = \{1, x, x^2, x^3\}$. Hence, equation (5.6.2.6) also has a generalized separable solution of the form

$$u = \psi_1(t) + \psi_2(t)x + \psi_3(t)x^2 + \psi_4(t)x^3, \tag{5.6.2.8}$$

where the functions $\psi_i = \psi_i(t)$ ($i = 1,\ 2,\ 3,\ 4$) are described by the system of ordinary differential equations

$$\begin{aligned}
\psi_1'' + a_1(t)\psi_1' &= 2a_2(t)\psi_2\psi_3,\\
\psi_2'' + a_1(t)\psi_2' &= 2a_2(t)(3\psi_2\psi_4 + 2\psi_3^2),\\
\psi_3'' + a_1(t)\psi_3' &= 18a_2(t)\psi_3\psi_4,\\
\psi_4'' + a_1(t)\psi_4' &= 18a_2(t)\psi_4^2.
\end{aligned}$$

For arbitrary $a_1(t)$ and $a_2(t)$, this system admits exact solutions in explicit form when $\psi_3 = \text{const}$ and $\psi_4 = 0$, and reduces to one linear second-order ODE for $\psi_1 = \psi_1(t)$ when $\psi_3 \neq \text{const}$ and $\psi_4 = 0$.

3°. It is straightforward to demonstrate that the operator $F[u] = u_x u_{xx}$ has another two-dimensional invariant subspace, $\mathscr{L}_2 = \{1, \varphi(x)\}$, with the function $\varphi = \varphi(x)$ satisfying the autonomous ordinary differential equation

$$\varphi_x' \varphi_{xx}'' = p_1 + p_2\varphi, \tag{5.6.2.9}$$

where p_1 and p_2 are arbitrary constants. The method described below in Section 5.6.3 was used to derive this equation. The general solution of equation (5.6.2.9) can be written in implicit form:

$$x = \int \left(\tfrac{3}{2}p_2\varphi^2 + 3p_1\varphi + p_0\right)^{-1/3} d\varphi + p_3, \tag{5.6.2.10}$$

where p_0 and p_3 are arbitrary constants. Hence, equation (5.6.2.6) has a generalized separable solution of the form

$$u = \psi_1(t) + \psi_2(t)\varphi(x),$$

where the function $\varphi(x)$ is defined implicitly by relation (5.6.2.10), and the functions $\psi_1 = \psi_1(t)$ and $\psi_2 = \psi_2(t)$ are described by the system of ordinary differential equations

$$\psi_1'' + a_1(t)\psi_1' = p_1 a_2(t)\psi_2^2,$$
$$\psi_2'' + a_1(t)\psi_2' = p_2 a_2(t)\psi_2^2.$$

◄

5.6.3. Finding Linear Subspaces Invariant Under a Given Nonlinear Operator

Preliminary remarks. The primary challenge in using the method of invariant subspaces to construct exact solutions for specific equations lies in identifying linear subspaces that remain invariant under a given nonlinear operator. Until now, this problem has not been discussed in the book, and it has been implicitly assumed that the researcher knows in advance or has guessed from some considerations (for example, intuitive ones) what set of linearly independent functions should be used when searching for exact solutions of nonlinear partial differential equations. Therefore, Propositions 1 and 2 formulated in Sections 5.6.1 and 5.6.2 effectively reduce to finding exact solutions in the form of a bilinear sum (5.6.1.5), where the functions $\varphi_i(x)$ are predefined. This approach is in many ways similar to the simplified method for constructing generalized separable solutions discussed earlier in Section 5.3. The method described below theoretically enables identifying the functions $\varphi_i(x)$.

Description of the method. To determine the independent basis functions $\varphi_i = \varphi_i(x)$, we substitute the linear combination $\sum_{i=1}^k C_i\varphi_i(x)$ into the nonlinear differential operator (5.6.1.2). As a result, we obtain

$$F\Bigg[\sum_{i=1}^k C_i\varphi_i(x)\Bigg] = A_1(\mathbf{C})\Phi_1[X] + A_2(\mathbf{C})\Phi_2[X] + \cdots + A_m(\mathbf{C})\Phi_m[X]$$
$$+ B_1(\mathbf{C})\varphi_1(x) + B_2(\mathbf{C})\varphi_2(x) + \cdots + B_k(\mathbf{C})\varphi_k(x), \quad (5.6.3.1)$$

where $A_j(\mathbf{C})$ and $B_i(\mathbf{C})$ depend only on $C_1, \ldots, C_k$, while the functionals $\Phi_j[X]$ depend on x but are independent of $C_1, \ldots, C_k$:

$$\begin{aligned} A_j(\mathbf{C}) &\equiv A_j(C_1, \ldots, C_k), \quad j = 1, \ldots, m; \\ B_i(\mathbf{C}) &\equiv B_i(C_1, \ldots, C_k), \quad i = 1, \ldots, k; \\ \Phi_j[X] &\equiv \Phi_j\big(x, \varphi_1, \varphi_1', \varphi_1'', \ldots, \varphi_k, \varphi_k', \varphi_k''\big). \end{aligned} \quad (5.6.3.2)$$

For simplicity, the formulas are written for the case of a second-order differential operator. For a higher-order operator, the right-hand sides of the last relations in (5.6.3.2) will contain higher-order derivatives of φ_i. The functionals $\Phi_1[X]$,

..., $\Phi_m[X]$ are assumed to be linearly independent, and $A_j(\mathbf{C})$ are linearly independent functions of C_1, ..., C_k.

A finite-dimensional linear subspace (5.6.1.3) will be invariant under a nonlinear differential operator F of the form (5.6.1.2) if all functionals $\Phi_j[X]$ $(j=1,\dots,m)$ in (5.6.3.1) are linear combinations of the basis functions $\varphi_i(x)$ $(i=1,\dots,k)$. Thus, we arrive at the following system (usually overdetermined) of ordinary differential equations for the basis functions:

$$\begin{aligned}\Phi_j\big(x,\varphi_1,\varphi_1',\varphi_1'',\dots,\varphi_k,\varphi_k',\varphi_k''\big)\\ = p_{j,1}\varphi_1+p_{j,2}\varphi_2+\dots+p_{j,k}\varphi_k,\quad j=1,\dots,m,\end{aligned}\tag{5.6.3.3}$$

where $p_{j,i}$ are some constants independent of the parameters C_1, ..., C_k. If system (5.6.3.3) is solvable for some set of constants $p_{i,j}$ (in practice, it suffices to find a particular solution), then the functions $\varphi_i=\varphi_i(x)$ define a linear subspace invariant with respect to the nonlinear differential operator (5.6.1.2). In this case, the functions appearing on the right-hand side of equation (5.6.1.4) are expressed as

$$f_i(\mathbf{C})=p_{1,i}A_1(\mathbf{C})+p_{2,i}A_2(\mathbf{C})+\dots+p_{m,i}A_m(\mathbf{C})+B_i(\mathbf{C}),\quad i=1,\dots,k.\tag{5.6.3.4}$$

Remark 5.14. *It is worthwhile starting the analysis of nonlinear differential operators by looking for two-dimensional invariant subspaces of the form* $\mathscr{L}_2=\{1,\varphi(x)\}$.

Proposition 3. *Let a nonlinear differential operator* $F[u]$ *admit a two-dimensional invariant subspace of the form* $\mathscr{L}_2=\{1,\varphi(x)\}$, *where* $\varphi(x)=p\varphi_1(x)+q\varphi_2(x)$, *with* p *and* q *being arbitrary constants, and let the functions* 1, $\varphi_1(x)$, *and* $\varphi_2(x)$ *be linearly independent. Then the operator* $F[u]$ *also admits a three-dimensional invariant subspace* $\mathscr{L}_3=\{1,\varphi_1(x),\varphi_2(x)\}$.

Proposition 4. *Let two nonlinear differential operators* $F_1[u]$ *and* $F_2[u]$ *admit an invariant subspace* $\mathscr{L}_n=\{\varphi_1(x),\dots,\varphi_n(x)\}$. *Then the nonlinear operator* $pF_1[u]+qF_2[u]$, *where* p *and* q *are arbitrary constants, also admits the same invariant subspace.*

In particular, the operators $F[u]$ and $aF[u]+bu$ for any constant a and b have identical invariant subspaces ($a\neq 0$).

► Example 5.21. Let us consider the following differential operator (5.6.1.11):

$$F[u]=au_{xx}+u_x^2+ku^2+bu+c,$$

We seek its invariant subspaces in the form $\mathscr{L}_2=\{1,\varphi(x)\}$ to obtain

$$\begin{aligned}F[C_1+C_2\varphi(x)]&=C_2^2[(\varphi_x')^2+k\varphi^2]+C_2a\varphi_{xx}''\\&\quad+kC_1^2+bC_1+c+(bC_2+2kC_1C_2)\varphi.\end{aligned}$$

In this case,

$$\Phi_1[X]=(\varphi_x')^2+k\varphi^2,\qquad \Phi_2[X]=a\varphi_{xx}''.$$

Hence, the basis function $\varphi(x)$ must satisfy the overdetermined system of ordinary differential equations

$$(\varphi'_x)^2 + k\varphi^2 = p_1 + p_2\varphi, \qquad \varphi''_{xx} = p_3 + p_4\varphi, \tag{5.6.3.5}$$

where $p_1 = p_{1,1}$, $p_2 = p_{1,2}$, $p_3 = p_{2,1}/a$, and $p_4 = p_{2,2}/a$.

Let us investigate system (5.6.3.5) for consistency. To this end, we differentiate the first equation with respect to x and divide by φ'_x to obtain $\varphi''_{xx} = -k\varphi + p_2/2$. We utilize this relation to eliminate the second derivative from the second equation in (5.6.3.5). As a result, we get

$$(p_4 + k)\varphi + p_3 - \tfrac{1}{2}p_2 = 0.$$

For this equation to be satisfied identically, we must set

$$p_4 = -k, \quad p_3 = \tfrac{1}{2}p_2. \tag{5.6.3.6}$$

The solution of system (5.6.3.5) under condition (5.6.3.6) is expressed as

$$\begin{aligned}
&\varphi(x) = px^2 + qx && \text{if } k = 0 \quad (p_1 = q^2,\ p_2 = 4p),\\
&\varphi(x) = p\sin\bigl(x\sqrt{k}\,\bigr) + q\cos\bigl(x\sqrt{k}\,\bigr) && \text{if } k > 0 \quad (p_1 = kp^2 + kq^2,\ p_2 = 0),\\
&\varphi(x) = p\sinh\bigl(x\sqrt{-k}\,\bigr) + q\cosh\bigl(x\sqrt{-k}\,\bigr) && \text{if } k < 0 \quad (p_1 = -kp^2 + kq^2,\ p_2 = 0),
\end{aligned} \tag{5.6.3.7}$$

where p and q are arbitrary constants.

Since formulas (5.6.3.7) involve two arbitrary parameters, p and q, it follows from Proposition 3 that the nonlinear differential oprator (5.6.1.11) admits the following invariant subspaces:

$$\begin{aligned}
&\mathscr{L}_3 = \bigl\{1, x, x^2\bigr\} && \text{if } k = 0,\\
&\mathscr{L}_3 = \bigl\{1, \sin(x\sqrt{k}\,), \cos(x\sqrt{k}\,)\bigr\} && \text{if } k > 0,\\
&\mathscr{L}_3 = \bigl\{1, \sinh(x\sqrt{-k}\,), \cosh(x\sqrt{-k}\,)\bigr\} && \text{if } k < 0.
\end{aligned}$$

◄

► Example 5.22. Consider the heat equation with a quadratic nonlinearity

$$u_t = f(x)(uu_x)_x, \tag{5.6.3.8}$$

where the nonlinear differential operator is explicitly dependent on the space variable x. We have

$$F[u] = f(x)(uu_x)_x.$$

We look for invariant subspaces of the operator F in the form $\mathscr{L}_2 = \{1, \varphi(x)\}$ to obtain

$$F[C_1 + C_2\varphi(x)] = C_2^2 f(x)[\varphi\varphi''_{xx} + (\varphi'_x)^2] + C_1C_2 f(x)\varphi''_{xx}. \tag{5.6.3.9}$$

Hence,

$$\Phi_1[X] = f(x)[\varphi\varphi''_{xx} + (\varphi'_x)^2], \quad \Phi_2[X] = f(x)\varphi''_{xx},$$

and the basis function $\varphi(x)$ is described by the system of ordinary differential equations

$$\begin{aligned} f(x)[\varphi\varphi''_{xx} + (\varphi'_x)^2] &= p_1 + p_2\varphi, \\ f(x)\varphi''_{xx} &= p_3 + p_4\varphi. \end{aligned} \qquad (5.6.3.10)$$

Let us find some admissible functions $f(x)$ ensuring that the overdetermined system (5.6.3.10) is consistent.

In the nondegenerate case of $\varphi''_{xx} \neq 0$, eliminating $f(x)$ from (5.6.3.10) leads to an equation for φ:

$$(p_3 + p_4\varphi)[\varphi\varphi''_{xx} + (\varphi'_x)^2] = (p_1 + p_2\varphi)\varphi''_{xx}. \qquad (5.6.3.11)$$

Any solution of this equation generates a function $f(x)$ such that

$$f(x) = \frac{p_3 + p_4\varphi}{\varphi''_{xx}}. \qquad (5.6.3.12)$$

Substituting $(\varphi'_x)^2 = 2w(\varphi)$ converts (5.6.3.11) to the linear separable first-order equation

$$[p_4\varphi^2 + (p_3 - p_2)\varphi - p_1]w'_\varphi = -2(p_3 + p_4\varphi)w,$$

which is easy to integrate. Table 5.3 lists several cases where $\varphi(x)$ is expressed in terms of elementary functions.

Table 5.3. Determining functions $f(x)$ appearing in equation (5.6.3.8) and the respective basis functions $\varphi(x)$; a, b, n, and λ are arbitrary constants ($n \neq 2$ and $\lambda \neq 0$).

No.	Function $f(x)$	Function $\varphi(x)$	Parameters $p_1, \ldots, p_4$
1	a	$\frac{1}{2a}x^2 + bx$	$p_1 = ab^2$, $p_2 = 3$, $p_3 = 1$, $p_4 = 0$
2	ax^n	x^{2-n}	$p_1 = p_4 = 0$, $p_2 = a(2-n)(3-2n)$, $p_3 = a(1-n)(2-n)$
3	ax^2	$\ln x$	$p_1 = a$, $p_2 = p_3 = -a$, $p_4 = 0$
4	$ae^{\lambda x}$	$e^{-\lambda x}$	$p_1 = p_4 = 0$, $p_2 = 2a\lambda^2$, $p_3 = a\lambda^2$

Considering (5.6.3.10), we can represent relation (5.6.3.9) as

$$F[C_1 + C_2\varphi(x)] = C_2^2p_1 + C_1C_2p_3 + (C_2^2p_2 + C_1C_2p_4)\varphi(x).$$

The corresponding exact solutions of equation (5.6.3.8) are given by

$$u = \psi_1(t) + \psi_2(t)\varphi(x),$$

where the functions $\psi_m = \psi_m(t)$ are determined by solving the autonomous system of two ordinary differential equations

$$\begin{aligned} \psi_1' &= p_1\psi_2^2 + p_3\psi_1\psi_2, \\ \psi_2' &= p_2\psi_2^2 + p_4\psi_1\psi_2. \end{aligned}$$

The prime stands for a derivative with respect to t. By dividing the first equation by the second, we can transform this system to a single homogeneous first-order equation, which is easy to integrate. ◀

Problems and Exercises for Section 5.6

1. Show that the nonlinear differential operator

$F[u] = (a_1 u_x + a_2 u + a_3) u_{xx} + a_4 u_x^2 + a_5 u_x + a_6 u + a_7$

admits the three-dimensional linear invariant subspace $\mathscr{L}_3 = \{1, x, x^2\}$. Using this fact in view, find generalized separable solutions for the nonlinear equations:

a) $u_t = a u u_{xx} + bu$,

b) $u_t = a u u_{xx} + b u_x$,

c) $u_{tt} = a(u u_x)_x + bu + c$,

d) $u_{tt} = a(u u_x)_x + b u_x + cu$.

2. Find the conditions under which the nonlinear differential operator

$F[u] = (a_1 u_x + a_2 u + a_3) u_{xx} + a_4 u_x^2 + a_5 u_x + a_6 u^2 + a_7 u + a_8$

admits the following linear invariant subspaces:

a) $\mathscr{L}_2 = \{1, x\}$,

b) $\mathscr{L}_2 = \{x^2, x^\beta\}$,

c) $\mathscr{L}_2 = \{1, e^{\lambda x}\}$,

d) $\mathscr{L}_3 = \{1, \sinh(\lambda x), \cosh(\lambda x)\}$,

e) $\mathscr{L}_3 = \{1, \sin(\lambda x), \cos(\lambda x)\}$.

3. Find generalized separable solutions of the following equations:

a) $u_t = u_{xx} + a u_x^2 + b u^2$,

b) $u_t = u u_{xx} + b u^2 + c$,

c) $u_t = u u_{xx} + b u_x^2 + c$,

d) $u_t = a u u_{xx} + b u_x^2$,

e) $u_{tt} = (u u_x)_x + a u^2 + b$.

Hint. Take advantage of the results of solving the preceding problem.

4. Find the conditions under which the nonlinear differential operator

$F[u] = a_1 u_{xx}^2 + a_2 u_x u_{xx} + a_3 u u_{xx} + a_4 u_x^2 + a_5 u u_x + a_6 u^2 + a_7 u_x + a_8 u + a_9$

admits the following linear invariant subspaces:

a) $\mathscr{L}_3 = \{1, x, x^2\}$,

b) $\mathscr{L}_4 = \{1, x, x^2, x^3\}$,

c) $\mathscr{L}_5 = \{1, x, x^2, x^3, x^4\}$,

d) $\mathscr{L}_2 = \{1, e^{\lambda x}\}$,

e) $\mathscr{L}_3 = \{1, \sinh(\lambda x), \cosh(\lambda x)\}$,

f) $\mathscr{L}_3 = \{1, \sin(\lambda x), \cos(\lambda x)\}$.

5. Find generalized separable solutions of the following nonlinear equations:

a) $u_t = a u_{xx}^2$,

b) $u_{tt} = a u_{xx}^2 + b$,

c) $u_t = u_{xx}^2 + u^2$.

Hint. Take advantage of the results of solving the preceding problem.

6. Find generalized separable solutions of the following equations with a cubic nonlinearity:

a) $u_t = 2u^2 u_{xx} - u u_x^2$,

b) $u_{tt} = 2u^2 u_{xx} - u u_x^2 + a$.

Hint. Show that the nonlinear operator $F[u] = 2u^2 u_{xx} - u u_x^2$ admits the three-dimensional linear invariant subspace $\mathscr{L}_3 = \{1, x, x^2\}$.

7. Show that the nonlinear operator $F[u] = (u^2)_{xxxx}$ admits the following linear invariant subspaces:

a) $\mathscr{L}_5 = \{1, x, x^2, x^3, x^4\}$,

b) $\mathscr{L}_3 = \{x^{1/2}, x^{3/2}, x^4\}$.

8. Show that an equation of the form (5.6.1.10) with variable coefficients, where $a = a(t)$, $b = b(t)$, and $c = c(t)$ are arbitrary functions and $k = \text{const} > 0$, also has a generalized separable solution of the form (5.6.1.12), with the function $\psi_1(t)$ and $\psi_2(t)$ described by the system of ordinary differential equations (5.6.1.13) where $a = a(t)$, $b = b(t)$, and $c = c(t)$.

5.7. Nonlinear Partial Differential Equations with Delay

5.7.1. Preliminary Remarks. Reaction-Diffusion Equations with Delay

For the mathematical modeling of complex phenomena and processes, where the state depends not only on the present moment in time, but also on one or more moments in the past, differential equations with delay are utilized. These equations, which involve partial derivatives and a constant delay, include the unknown function $u = u(x,t)$ as well as the function $\bar{u} = u(x, t-\tau)$, where τ represents the delay time ($\tau > 0$).

Delay differential equations exhibit qualitative features that standard differential equations lack (see Wu, 1996; Polyanin et al., 2024). Among the nonlinear delay partial differential equations, the most frequently studied in the literature and applications are reaction-diffusion type equations (e.g., see Wu, 1996; Meleshko & Moyo, 2008; Sorokin & Vyazmin, 2022; and Polyanin et al., 2024):

$$u_t = [g(u)u_x]_x + f(u, \bar{u}), \qquad \bar{u} = u(x, t-\tau). \tag{5.7.1.1}$$

Here, the delay $\tau > 0$ only appears in the kinetic function $f(u, \bar{u})$.

The term *exact solution* in the context of nonlinear partial differential equations with delay refers to a solution that can be expressed in one of the following ways:

(a) in terms of elementary functions or in closed form through quadrature;

(b) in terms of solutions to one or more ordinary differential equations;

(c) in terms of solutions to one or more ordinary differential equations that include delay.

Combinations of cases (a), (b), and (c) are also permitted.

Remark 5.15. *This definition generalizes the concept of an exact solution for nonlinear partial differential equations without delay (see Preface).*

Remark 5.16. *If the equation involves special or arbitrary functions, those functions should be included with the elementary functions in Item (a).*

5.7.2. Simple Separable Solutions

Some nonlinear partial differential equations with delay allow for simple separable solutions. These solutions are typically expressed as products or sums of functions that depend on different variables. The method for finding these solutions is analogous to the technique used for identifying simple separable solutions in PDEs that do not involve delay, as explained in Section 5.1.2. To clarify this concept, we will provide a few specific examples.

► Example 5.23. Let us consider the nonlinear reaction-diffusion type PDE with delay

$$u_t = au_{xx} + uf(\bar{u}/u), \qquad \bar{u} = u(x, t-\tau), \tag{5.7.2.1}$$

where $f(z)$ is an arbitrary function.

Equation (5.7.2.1) admits a multiplicative separable solution. Substituting the expression (5.1.1.1), or

$$u = \varphi(x)\psi(t), \tag{5.7.2.2}$$

into equation (5.7.2.1), taking into account that $\bar{u}/u = \psi(t-\tau)/\psi(t)$, dividing all terms of the resulting equation by $\varphi\psi$, and moving the function f to the left-hand side, we obtain

$$\frac{\psi_t'}{\psi} - f(\bar{\psi}/\psi) = a\frac{\varphi_{xx}''}{\varphi}, \qquad \bar{\psi} = \psi(t-\tau). \tag{5.7.2.3}$$

The left-hand side of (5.7.2.3) depends only on t, while the right-hand side depends only on x. This is possible only if the conditions

$$\frac{\psi_t'}{\psi} - f(\bar{\psi}/\psi) = C, \quad a\frac{\varphi_{xx}''}{\varphi} = C \tag{5.7.2.4}$$

hold. These are equations for determining $\psi = \psi(t)$ and $\varphi = \varphi(x)$, where C is an arbitrary constant.

The general solution of the second equation in (5.7.2.4), which is a linear second-order ODE with constant coefficient, depends on the sign of the constant C and is given by

$$\varphi = \begin{cases} C_1\cos(\lambda x) + C_2\sin(\lambda x) & \text{if } C = -a\lambda^2 < 0, \\ C_1\exp(-\lambda x) + C_2\exp(\lambda x) & \text{if } C = a\lambda^2 > 0, \\ C_1 x + C_2 & \text{if } C = 0, \end{cases} \tag{5.7.2.5}$$

where C_1, C_2, and λ are arbitrary constants. Once multiplied by ψ, the first equation in (5.7.2.4), which is a nonlinear first-order ODEs with delay, becomes

$$\psi_t' = C\psi + \psi f(\bar{\psi}/\psi), \qquad \bar{\psi} = \psi(t-\tau). \tag{5.7.2.6}$$

Formulas (5.7.2.2) and (5.7.2.5), and the delay ODE (5.7.2.6) describe a multiplicative separable solution of the original delay PDE (5.7.2.1).

It is noteworthy that the delay ODE (5.7.2.6) has an exponential particular solution:

$$\psi(t) = C_3 \exp(\beta t),$$

where C_3 is an arbitrary constant, and β is determined by the algebraic (transcendental) equation

$$\beta = C + f(e^{-\beta\tau}). \qquad ◀$$

► Example 5.24. Consider the nonlinear PDE with delay

$$u_t = a u_{xx} + f(u - \bar{u}), \qquad \bar{u} = u(x, t - \tau), \tag{5.7.2.7}$$

where $f(z)$ is an arbitrary function.

This equation admits an additive separable solution (5.1.1.2). Substituting $u = \varphi(x) + \psi(t)$ into (5.7.2.7), taking into account that $u - \bar{u} = \psi(t) - \psi(t - \tau)$, and moving the function f to the other side of the equation, we obtain

$$\psi'_t - f(\psi - \bar{\psi}) = a\varphi''_{xx}, \qquad \bar{\psi} = \psi(t - \tau). \tag{5.7.2.8}$$

The left-hand side of relation (5.7.2.8) depends only on t, while the right-hand side depends on x alone. This is possible only if the conditions

$$\psi'_t - f(\psi - \bar{\psi}) = C, \quad a\varphi''_{xx} = C \tag{5.7.2.9}$$

hold. Here, C is an arbitrary constant. Integrating the second equation in (5.7.2.9), setting $C = 2aC_1$, and taking into account (5.1.1.2), we finally obtain

$$u = C_1 x^2 + C_2 x + \psi(t), \tag{5.7.2.10}$$

where C_1 and C_2 are arbitrary constants, and $\psi(t)$ is a function satisfying the delay ODE

$$\psi'_t = 2aC_1 + f(\psi - \bar{\psi}). \tag{5.7.2.11}$$

It is noteworthy that the delay ODE (5.7.2.11) has a particular solution linear in t:

$$\psi(t) = kt + C_3,$$

where C_3 is an arbitrary constant, and the constant k is determined from the algebraic (or transcendental) equation $k = 2aC_1 + f(k\tau)$. ◀

Table 5.4 gathers the above and a few other examples of simple (additive or multiplicative) separable solutions for several nonlinear PDEs with delay.

For all nonlinear delay PDEs specified in Table 5.4, the functions $\varphi(x)$ appearing in the solutions are either elementary or satisfy autonomous second-order ordinary differential equations. The functions $\psi(t)$ are defined by first-order ordinary differential equations that include delay.

Remark 5.17. *The procedure for constructing simple separable solutions to nonlinear Klein–Gordon type wave equations with delay,*

$$u_{tt} = [g(u)u_x]_x + f(u, \bar{u}), \qquad \bar{u} = u(x, t - \tau),$$

is entirely analogous to the method used for constructing simple separable solutions to nonlinear reaction-diffusion type delay PDEs (5.7.2.1). Specific examples of such equations that allow for multiplicative or additive separable solutions can be found in the exercises in Section 5.7. Additional references include Polyanin & Zhurov (2014b) and Polyanin et al. (2024).

Table 5.4. Reaction-diffusion equations with constant delay admitting additive or multiplicative separable solutions. Notations: $\bar{u} = u(x, t-\tau)$, $\bar{\psi} = \psi(t-\tau)$; C_1, C_2, and C_3 are arbitrary constants.

Original equation	Forms of solution	Determining equations
$u_t = au_{xx} + uf(\bar{u}/u)$	$u = [C_1\cos(\beta x) + C_2\sin(\beta x)]\psi(t)$; $u = [C_1\exp(-\beta x) + C_2\exp(\beta x)]\psi(t)$; $u = (C_1 x + C_2)\psi(t)$	$\psi'_t = -a\beta^2\psi + \psi f(\bar{\psi}/\psi)$; $\psi'_t = a\beta^2\psi + \psi f(\bar{\psi}/\psi)$; $\psi'_t = \psi f(\bar{\psi}/\psi)$
$u_t = au_{xx} + bu\ln u$ $+ uf(\bar{u}/u)$	$u = \varphi(x)\psi(t)$	$a\varphi''_{xx} = C_1\varphi - b\varphi\ln\varphi$, $\psi'_t = C_1\psi + b\psi\ln\psi$ $+ \psi f(\bar{\psi}/\psi)$
$u_t = au_{xx} + f(u - \bar{u})$	$u = C_2 x^2 + C_1 x + \psi(t)$	$\psi'_t = 2C_2 a + f(\psi - \bar{\psi})$
$u_t = au_{xx} + bu + f(u - \bar{u})$	$u = C_1\cos(\lambda x) + C_2\sin(\lambda x) + \psi(t)$, where $\lambda = \sqrt{b/a}$ (for $ab > 0$); $u = C_1\exp(-\lambda x) + C_2\exp(\lambda x) + \psi(t)$, where $\lambda = \sqrt{-b/a}$ (for $ab < 0$)	$\psi'_t = b\psi + f(\psi - \bar{\psi})$; $\psi'_t = b\psi + f(\psi - \bar{\psi})$
$u_t =$ $a(u^k u_x)_x + uf(\bar{u}/u)$	$u = \varphi(x)\psi(t)$	$a(\varphi^k\varphi'_x)'_x = C_1\varphi$, $\psi'_t = C_1\psi^{k+1} + \psi f(\bar{\psi}/\psi)$
$u_t = a(u^k u_x)_x + bu^{k+1}$ $+ uf(\bar{u}/u)$	$u = [C_1\cos(\beta x) + C_2\sin(\beta x)]^{\frac{1}{k+1}}\psi(t)$, where $\beta = \sqrt{b(k+1)/a}$, $b(k+1) > 0$; $u = (C_1 e^{-\beta x} + C_2 e^{\beta x})^{\frac{1}{k+1}}\psi(t)$, where $\beta = \sqrt{-b(k+1)/a}$, $b(k+1) < 0$; $u = C_1\exp\left(-\frac{b}{2a}x^2 + C_2 x\right)\psi(t)$ at $k = -1$; $u = \varphi(x)\psi(t)$ (generalizes preceding solutions)	$\psi'_t = \psi f(\bar{\psi}/\psi)$; $\psi'_t = \psi f(\bar{\psi}/\psi)$; $\psi'_t = \psi f(\bar{\psi}/\psi)$; $a(\varphi^k\varphi'_x)'_x + b\varphi^{k+1} = C_1\varphi$, $\psi'_t = C_1\psi^{k+1} + \psi f(\bar{\psi}/\psi)$
$u_t = a(e^{\lambda u}u_x)_x$ $+ f(u - \bar{u})$	$u = \dfrac{1}{\lambda}\ln(C_1\lambda x^2 + C_2 x + C_3) + \psi(t)$	$\psi'_t = 2aC_1 e^{\lambda\psi} + f(\psi - \bar{\psi})$
$u_t = a(e^{\lambda u}u_x)_x + be^{\lambda u}$ $+ f(u - \bar{u})$	$u =$ $\frac{1}{\lambda}\ln[C_1\cos(\beta x) + C_2\sin(\beta x)] + \psi(t)$, where $\beta = \sqrt{b\lambda/a}$, $b\lambda > 0$; $u = \frac{1}{\lambda}\ln(C_1 e^{-\beta x} + C_2 e^{\beta x}) + \psi(t)$, where $\beta = \sqrt{-b\lambda/a}$, $b\lambda < 0$; $u = \varphi(x) + \psi(t)$ (generalizes preceding solutions)	$\psi'_t = f(\psi - \bar{\psi})$; $\psi'_t = f(\psi - \bar{\psi})$; $a(e^{\lambda\varphi}\varphi'_x)'_x + be^{\lambda\varphi} = C_1$, $\psi'_t = C_1 e^{\lambda\psi} + f(\psi - \bar{\psi})$

5.7.3. Generalized Separable Solutions

To construct generalized separable solutions for nonlinear partial differential equations with delay, one can utilize the methods outlined in Sections 5.4 to 5.6.

Let us consider the nonlinear PDE involving a linear term with delay

$$u_t = F[u] + s\bar{u}, \qquad \bar{u} = u(x, t-\tau), \tag{5.7.3.1}$$

where s is some constant, and $F[u]$ is an nth-order nonlinear differential operator in x of the form (5.6.1.2).

Proposition 5. *Suppose the linear subspace* (5.6.1.3) *is invariant under an operator* F*, suggesting that relation* (5.6.1.4) *holds. Then equation* (5.7.3.1) *has a generalized separable solution of the form*

$$u(x,t) = \sum_{i=1}^{k} \psi_i(t)\varphi_i(x), \tag{5.7.3.2}$$

where $\psi_1(t)$, ..., $\psi_k(t)$ *are functions satisfying the system of ordinary differential equations with delay*

$$\psi_i' = f_i(\psi_1, \dots, \psi_n) + s\bar{\psi}_i, \qquad i = 1, \dots, k. \tag{5.7.3.3}$$

The prime stands for a derivative with respect to t*, and* $\bar{\psi}_i = \psi_i(t-\tau)$.

We will examine several applications of Proposition 5 to construct exact solutions for nonlinear reaction-diffusion equations of the form (5.7.1.1).

► Example 5.25. Consider the following delay reaction-diffusion equation of the form (5.7.3.1) with a quadratic nonlinearity:

$$u_t = [(a_1u + a_0)u_x]_x + b_1u + b_2\bar{u} + c. \tag{5.7.3.4}$$

For $b_2 = 0$, the differential operator on the right-hand side admits the linear invariant subspace $\mathscr{L}_3 = \{1, x, x^2\}$, since when acted by this operator, the polynomial $C_1 + C_2x + C_3x^2$ converts into a similar polynomial with other coefficients. It follows from Proposition 5 that the original equation (5.7.3.4) has a generalized separable solution of the form

$$u = \psi_1(t) + \psi_2(t)x + \psi_3(t)x^2. \tag{5.7.3.5}$$

The functions $\psi_i = \psi_i(t)$ ($i = 1, 2, 3$) are described by the ODE system with delay

$$\begin{aligned} \psi_1' &= 2a_1\psi_1\psi_3 + a_1\psi_2^2 + 2a_0\psi_3 + b_1\psi_1 + b_2\bar{\psi}_1 + c, \\ \psi_2' &= 6a_1\psi_2\psi_3 + b_1\psi + b_2\bar{\psi}_2, \\ \psi_3' &= 6a_1\psi_3^2 + b_1\psi_3 + b_2\bar{\psi}_3, \end{aligned} \tag{5.7.3.6}$$

where $\bar{\psi}_i = \psi_i(t-\tau)$. ◄

► Example 5.26. Consider a more complex delay reaction-diffusion equation of the form (5.7.3.1) with a quadratic nonlinearity:

$$u_t = [(a_1u + a_0)u_x]_x + ku^2 + b_1u + b_2\bar{u} + c, \quad k \neq 0. \tag{5.7.3.7}$$

1°. For $b_2 = 0$ and $a_1k < 0$, the differential operator on the right-hand side admits the linear invariant subspace $\mathscr{L}_3 = \{1, e^{-\lambda x}, e^{\lambda x}\}$, where $\lambda = \sqrt{-k/(2a_1)}$. In this case, it follows from Proposition 5 that equation (5.7.3.7) has a generalized separable solution of the form

$$u = \psi_1(t) + \psi_2(t)\exp(-\lambda x) + \psi_3(t)\exp(\lambda x), \quad \lambda = \sqrt{-\frac{k}{2a_1}}. \tag{5.7.3.8}$$

The functions $\psi = \psi_n(t)$ are described by the system of ODEs with delay

$$\psi_1' = k\psi_1^2 + 2k\psi_2\psi_3 + b_1\psi_1 + b_2\bar{\psi}_1 + c,$$
$$\psi_2' = (\tfrac{3}{2}k\psi_1 + a_0\lambda^2 + b_1)\psi_2 + b_2\bar{\psi}_2,$$
$$\psi_3' = (\tfrac{3}{2}k\psi_1 + a_0\lambda^2 + b_1)\psi_3 + b_2\bar{\psi}_3,$$

where $\bar{\psi}_i = \psi_i(t-\tau)$ $(i = 1,\ 2,\ 3)$.

2°. For $a_1k > 0$, it can be shown likewise that equation (5.7.3.7) admits the generalized separable solution

$$u = \psi_1(t) + \psi_2(t)\cos(\lambda x) + \psi_3(t)\sin(\lambda x), \quad \lambda = \sqrt{\frac{k}{2a_1}}, \tag{5.7.3.9}$$

where the functions $\psi = \psi_n(t)$ are described by the system of ODEs with delay

$$\psi_1' = k\psi_1^2 + \tfrac{1}{2}k(\psi_2^2 + \psi_3^2) + b_1\psi_1 + b_2\bar{\psi}_1 + c,$$
$$\psi_2' = (\tfrac{3}{2}k\psi_1 + b_1 - a_0\lambda^2)\psi_2 + b_2\bar{\psi}_2,$$
$$\psi_3' = (\tfrac{3}{2}k\psi_1 + b_1 - a_0\lambda^2)\psi_3 + b_2\bar{\psi}_3.$$

◀

◆ *Problems and Exercises for Section 5.7*

In the following text, the notation $\bar{u} = u(x, t-\tau)$ is consistently used throughout.

1. Find simple separable solutions for the nonlinear first-order PDEs with delay:
a) $u_t = au_x + uf(\bar{u}/u)$,
b) $u_t = au_x + f(u - \bar{u})$.

2. Find simple separable solutions for the nonlinear wave-type PDEs with delay:
a) $u_{tt} = au_{xx} + uf(\bar{u}/u)$,
b) $u_{tt} = au_{xx} + f(u - \bar{u})$,
c) $u_{tt} = au_{xx} + bu + f(u - \bar{u})$.

3. Find multiplicative separable solutions of the form $u = e^{\lambda t}\varphi(x)$ for the nonlinear PDE with delay

$$u_t = a(u^k u_x)_x + uf(\bar{u}/u) + u^{k+1}g(\bar{u}/u),$$

where $f(z)$ and $g(z)$ are arbitrary functions, and k is an arbitrary constant.

4. Find additive separable solutions of the form $u = \beta t + \varphi(x)$ for the nonlinear PDE with delay

$$u_t = a(e^{\lambda u}u_x)_x + f(u - \bar{u}) + e^{\lambda u}g(u - \bar{u}),$$

where $f(z)$ and $g(z)$ are arbitrary functions, and $\lambda \neq 0$ is an arbitrary constant.

5. Find multiplicative separable solutions of the form $u = \exp(\pm\lambda x)\psi(t)$ for the nonlinear PDE with delay

$$u_t = [(a\ln u + b)u_x]_x - cu\ln u + uf(\bar{u}/u),$$

where $f(z)$ is an arbitrary function.

6. Find multiplicative separable solutions of the form $u = e^{\lambda t}\varphi(x)$ for the nonlinear PDE with delay

$$u_t = au_{xx} + uf(u - k\bar{u}, \bar{u}/u), \qquad \bar{u} = u(x, t-\tau),$$

where $f(z_1, z_2)$ is an arbitrary function of two variable, and $k > 0$ is an arbitrary constant.

7. Prove that the nonlinear delay PDE (5.7.2.1) admits a mixed multiplicative separable solution:

$$u = \exp(\alpha x + \beta t)\theta(z), \qquad z = \gamma x + \delta t,$$

where α, β, γ, and δ are arbitrary constants.

8. Prove that the nonlinear delay PDE

$$u_t = au_{xx} + bu + f(u - \bar{u})$$

admits a mixed multiplicative separable solution:

$$u = \varphi(x) + \psi(z), \qquad z = \beta x + \gamma t,$$

where β and γ are arbitrary constants.

► **Concluding remarks.**

The methods for finding generalized separable solutions to nonlinear PDEs, as described in Sections 5.3 to 5.5, are explained in more detail in the books by Polyanin & Zhurov (2022) and Polyanin & Zaitsev (2012). Further information on various equations that allow for such solutions can also be found in these books. For extensive details on the method of invariant subspaces, discussed in Section 5.6, refer to the book by Galaktionov & Svirshchevskii (2007).

In addition to the methods described in Section 5.7, various modifications of the functional constraints method, along with other methods outlined in Polyanin & Zhurov (2014a, 2014c), Polyanin & Sorokin (2021), and Polyanin et al. (2024), can be employed to construct generalized separable solutions for delay reaction-diffusion equations (5.7.1.1), as well as for other nonlinear delay PDEs. A comprehensive collection of exact solutions for various delay PDEs can be found in the books by Polyanin et al. (2024) and Polyanin (2025).

References for Chapter 5

Galaktionov V.A. Invariant subspaces and new explicit solutions to evolution equations with quadratic nonlinearities. *Proc. Roy. Soc. Edinburgh, Sect. A*, 1995, Vol. 125, No. 2, pp. 225–246.

Galaktionov V.A., Posashkov S.A., and Svirshchevskii S.R. On invariant sets and explicit solutions of nonlinear evolution equations with quadratic nonlinearities. *Dif. & Integral Equations*, 1995, Vol. 8, No. 8, pp. 1997–2024.

Galaktionov V.A. and Svirshchevskii S.R. *Exact Solutions and Invariant Subspaces of Nonlinear Partial Differential Equations in Mechanics and Physics*. Chapman & Hall/CRC Press, Boca Raton, 2007.

Meleshko S.V. and Moyo S. On the complete group classification of the reaction–diffusion equation with a delay. *J. Math. Anal. Appl.*, 2008, Vol. 338, pp. 448–466.

Polyanin A.D. *Handbook of Exact Solutions to Mathematical Equations*. CRC Press, Boca Raton, 2025.

Polyanin A.D. and Nazaikinskii V.E. *Handbook of Linear Partial Differential Equations for Engineers and Scientists, 2nd ed.* CRC Press, Boca Raton, 2016.

Polyanin A.D. and Sorokin V.G. Construction of exact solutions to nonlinear PDEs with delay using solutions of simpler PDEs without delay. *Commun. Nonlinear Sci. Numer. Simul.*, 2021, Vol. 95, 105634.

Polyanin A.D., Sorokin V.G., and Zhurov A.I. *Delay Ordinary and Partial Differential Equations*. CRC Press, Boca Raton–London, 2024.

Polyanin A.D. and Zaitsev V.F. *Handbook of Nonlinear Partial Differential Equations, 2nd ed.* CRC Press, Boca Raton–London, 2012.

Polyanin A.D., Zaitsev V.F., and Zhurov A.I. *Solution Methods for Nonlinear Equations of Mathematical Physics and Mechanics* (in Russian). Fizmatlit, Moscow, 2005.

Polyanin A.D. and Zhurov A.I. Functional constraints method for constructing exact solutions to delay reaction-diffusion equations and more complex nonlinear equations. *Commun. Nonlinear Sci. Numer. Simul.*, 2014a, Vol. 19, No. 3, pp. 417–430.

Polyanin A.D. and Zhurov A.I. Generalized and functional separable solutions to nonlinear delay Klein–Gordon equations. *Commun. Nonlinear Sci. Numer. Simul.*, 2014b, Vol. 19, No. 8, pp. 2676–2689.

Polyanin A.D. and Zhurov A.I. The functional constraints method: Application to non-linear delay reaction-diffusion equations with varying transfer coefficients. *Int. J. Non-Linear Mechanics*, 2014c, Vol. 67, pp. 267–277.

Polyanin A.D. and Zhurov A.I. *Separation of Variables and Exact Solutions to Nonlinear PDEs*. CRC Press, Boca Raton–London, 2022.

Sorokin V.G. and Vyazmin A.V. Nonlinear reaction–diffusion equations with delay: Partial survey, exact solutions, test problems, and numerical integration. *Mathematics*, 2022, Vol. 10, No. 11, 1886.

Tikhonov A.N. and Samarskii A.A. *Equations of Mathematical Physics*. Dover Publ., New York, 1990.

Titov S.S. On solutions of nonlinear partial differential equations of the form of a simple variables (in Russian). *Chisl. Met. Mech. Sploshnoi Sredy*, 1977, Vol. 8, pp. 586–599.

Titov S.S. A method of finite-dimensional rings for solving nonlinear equations of mathematical physics (in Russian). In: *Aerodynamics* (ed. T.P. Ivanova), Saratov Univ., Saratov, 1988, pp. 104–110.

Titov S.S. and Ustinov V.A. Investigation of polynomial solutions to the equations of motion of gases through porous media with integer isentropic exponent (in Russian). In: *Approximate Methods of Solution of Boundary Value Problems in Solid Mechanics*, AN USSR, Ural. Otd-nie Inst. Matematiki i Mekhaniki, 1985, pp. 64–70.

Wu J. *Theory and Applications of Partial Functional Differential Equations*. Springer, New York, 1996.

6. Methods of Functional Separation of Variables

► **Preliminary remarks.** This chapter describes the methods of functional separation of variables that allow one to construct exact solutions to many nonlinear PDEs. Following the principle from simple to complex, several methods for finding solutions with a functional separation of variables are successively described. A simplified method for constructing functionally separable solutions of a special type is discussed. It is shown that to search for the generalized traveling wave solutions and some other exact solutions of a special type, methods of generalized separation of variables can be used. The differentiation method used to obtain a standard bilinear functional equation, which is then solved with the splitting method, is considered. Direct methods for constructing functional separable solutions in explicit or implicit form are described. The focus is on nonlinear equations of heat and mass transfer and wave theory, with particular attention to the most challenging equations for analysis that involve one or more arbitrary functions. The application of the methods is illustrated with numerous examples and tables. The chapter deals with various nonlinear PDEs, including those with variable coefficients and of arbitrary order.

6.1. Preliminary Remarks

6.1.1. Structure of Functional Separable Solutions

1°. Many linear equations of mathematical physics admit exact solutions that can be expressed as sums of products of functions with different arguments. When nonlinear partial differential equations are derived from these linear PDEs with the change of variable $u = U(z)$, they will have exact solutions in the form

$$u(x,t) = U(z), \quad \text{where} \quad z = \sum_{m=1}^{n} \varphi_m(x)\psi_m(t). \tag{6.1.1.1}$$

The function z is a solution of the original linear equation.

Many nonlinear partial differential equations that cannot be reduced to linear equations can also have exact solutions of the form (6.1.1.1). Such solutions will be called *functional separable solutions*. In the general case, the functions $\varphi_m(x)$, $\psi_m(t)$, and $U(z)$ in (6.1.1.1) are not known in advance and are to be determined in the course of a further analysis. The function U will be called the *external function*, and the functions φ_m and ψ_m will be called *internal functions*.

Remark 6.1. *A generalized separable solution (see Chapter 5) is a specific type of functional separable solution where $U(z) = z$. The presence of an external function U in (6.1.1.1) that must be determined significantly complicates the construction of functional separable solutions.*

The fundamental concept behind the method of functional separation of variables is as follows: when we substitute the expression (6.1.1.1) into the partial differential equation being analyzed, the resulting functional differential equation should be, if possible, transformed into the standard bilinear functional equation (5.5.1.1) from Section 5.5.1, or into the functional differential equation (5.2.2.1)–(5.2.2.2) from Section 5.2.2.

2°. Often, in a narrow sense, the term *functional separable solution* refers to simpler exact solutions of the form

$$u = U(z), \qquad z = \varphi(x) + \psi(t), \tag{6.1.1.2}$$

where all three functions $U(z)$, $\varphi(x)$, and $\psi(t)$ are to be determined. When looking for solutions (6.1.1.2), we assume that $\varphi \neq \text{const}$ and $\psi \neq \text{const}$.

Remark 6.2. *With the functional separation of variables, seeking solutions in either simplest form, $u = U(\varphi(x) + \psi(t))$ or $u = U(\varphi(x)\psi(t))$, leads to the same results. This is because the representation $U(\varphi(x)\psi(t)) = U_1(\varphi_1(x) + \psi_1(t))$ holds true, where $U_1(z) = U(e^z)$, $\varphi_1(x) = \ln \varphi(x)$, and $\psi_1(t) = \ln \psi(t)$.*

In what follows, we will not consider traveling wave solutions. These solutions are defined by substituting the linear functions $\varphi(x) = kx$ and $\psi(t) = \lambda t$, with constant k and λ, into (6.1.1.2). For these simple solutions, see Section 4.2.1.

6.1.2. Direct and Indirect Functional Separation of Variables

It is essential to distinguish between *direct* and *indirect functional separation of variables*, both of which utilize the same representation of solutions, (6.1.1.1) or (6.1.1.2). An implicit representation of functionally separable solutions can also be employed (refer to Section 6.5 for more details).

In the first stage of direct functional separation of variables, the chosen solution representation is substituted into the partial differential equation under consideration. The resulting functional differential equation is then analyzed.

Conversely, in the first stage of indirect functional separation of variables, the solution representation is replaced with an equivalent differential constraint (for further details, see Section 9.5.3). The overdetermined PDE system, which includes both the original equation and the differential constraint, is then examined for compatibility.

This chapter will employ direct methods to construct exact solutions of nonlinear partial differential equations. These methods utilize functional separation of variables, resulting in fewer intermediate calculations and a more straightforward approach than indirect methods relying on equivalent differential relations.

6.2. Simplified Method for Constructing Functional Separable Solutions

6.2.1. Description of the Simplified Method Based on Transformations of the Unknown

In certain instances, finding a solution in the form (6.1.1.1) can be approached in two stages. First, a transformation is applied to convert the original equation into one with a quadratic (or sometimes power-law) nonlinearity. Next, the solution to this resulting equation can be obtained using any of the methods outlined in Sections 5.3 through 5.6.

Unfortunately, there are no standard methods for reducing nonlinear partial differential equations of a specific type to PDEs with quadratic nonlinearity, and this conversion is not always feasible. However, equations with quadratic nonlinearity can sometimes be derived through transformations of the unknown function in the form $u = U(z)$. These are the most common transformations of this type:

$$u = z^\lambda \qquad \text{(for equations with power-law nonlinearity)},$$
$$u = \lambda \ln z \qquad \text{(for equations with exponential nonlinearity)},$$
$$u = e^{\lambda z} \qquad \text{(for equations with logarithmic nonlinearity)},$$

where λ is a constant that has to be determined. The proposed approach is similar to assigning the structure of the external function $U(z)$ a priori in expression (6.1.1.1); the success of its application largely relies on the researcher's experience and intuition.

6.2.2. Examples of Constructing Exact Solutions for Nonlinear Equations

Below, we give a few examples of utilizing the simplified method to construct functional separable solutions for nonlinear second-order partial differential equations.

► Example 6.1. Let us consider a three-parameter family of nonlinear diffusion equations with a power-law source:

$$u_t = a(u^k u_x)_x + bu^{k+1}, \qquad (6.2.2.1)$$

where a, b, and k are free parameters. The substitution $z = u^k$ converts (6.2.2.1) to an equation with quadratic nonlinearity:

$$z_t = azz_{xx} + \frac{a}{k} z_x^2 + bkz^2. \qquad (6.2.2.2)$$

This equation allows for various generalized separable solutions, with the specific form dependent on the coefficients of the nonlinear terms on the right-hand side of equation (6.2.2.2). Solutions to equation (6.2.2.2) can be easily found using Table 5.2 (refer to line 3). It is evident that when the inequality $ab(k+1) > 0$ holds, there will be solutions involving trigonometric functions. However, the solutions will involve exponential functions when $ab(k+1) < 0$.

As a result, we obtain the following solutions for the nonlinear PDE (6.2.2.1):

$$
\begin{aligned}
u &= \bigl\{\varphi(t)\bigl[C_1\cos(\beta x) + C_2\sin(\beta x)\bigr] + \psi(t)\bigr\}^{1/k} \quad &&\text{if } ab(k+1) > 0,\\
u &= \bigl\{\varphi(t)\bigl[C_1\cosh(\beta x) + C_2\sinh(\beta x)\bigr] + \psi(t)\bigr\}^{1/k} \quad &&\text{if } ab(k+1) < 0.
\end{aligned}
\tag{6.2.2.3}
$$

Here, C_1 and C_2 are arbitrary constants,

$$\beta = \sqrt{\frac{|b|k^2}{|a(k+1)|}},$$

and the functions $\varphi = \varphi(t)$ and $\psi = \psi(t)$ are described by the system of ordinary differential equations

$$
\begin{aligned}
\varphi'_t &= \frac{bk(k+2)}{k+1}\varphi\psi,\\
\psi'_t &= bk\psi^2 + \frac{bk}{k+1}(C_1^2 \pm C_2^2)\varphi^2.
\end{aligned}
\tag{6.2.2.4}
$$

The upper sign in the second equation corresponds to the first solution in (6.2.2.3), while the lower sign corresponds to the second solution in (6.2.2.3).

For $C_1 = C_2$, the general solution of equations (6.2.2.4), with the lower sign in the second ODE, can be expressed as

$$\varphi = C_3(t + C_4)^{-\frac{k+2}{k+1}}, \qquad \psi = -\frac{1}{bk(t+C_4)},$$

where C_3 and C_4 are arbitrary constants. ◄

► Example 6.2. Let us now examine the four-parameter family of nonlinear diffusion equations with an exponential source

$$u_t = a(e^{\lambda u}u_x)_x + be^{\lambda u} + c. \tag{6.2.2.5}$$

The substitution $z = e^{\lambda u}$ converts (6.2.2.5) to an equation with quadratic nonlinearity

$$z_t = azz_{xx} + b\lambda z^2 + c\lambda z. \tag{6.2.2.6}$$

Solutions to equation (6.2.2.6) can be found using Table 5.2 (see line 3). It is noteworthy that the last term on the right-hand side of (6.2.2.6) does not affect the structure of the solution. When the inequality $ab\lambda > 0$ is satisfied, equation (6.2.2.6) has solutions involving trigonometric functions. However, if $ab\lambda < 0$, the solutions are in terms of exponential functions. Furthermore, when $b = 0$, the solution is a quadratic polynomial in the variable x.

With the above change of variable, one can obtain, in particular, the following functional separable solutions for equation (6.2.2.5):

$$
\begin{aligned}
u &= \frac{1}{\lambda}\ln\bigl\{e^{c\lambda t}\bigl[C_1\cos\bigl(x\sqrt{b\lambda/a}\,\bigr) + C_2\sin\bigl(x\sqrt{b\lambda/a}\,\bigr)\bigr]\bigr\} \quad &&\text{if } ab\lambda > 0,\\
u &= \frac{1}{\lambda}\ln\bigl\{e^{c\lambda t}\bigl[C_1\cosh\bigl(x\sqrt{-b\lambda/a}\,\bigr) + C_2\sinh\bigl(x\sqrt{-b\lambda/a}\,\bigr)\bigr]\bigr\} \quad &&\text{if } ab\lambda < 0,
\end{aligned}
$$

where C_1 and C_2 are arbitrary constants. ◄

► **Example 6.3.** The nonlinear two-parameter diffusion equation with a logarithmic source

$$u_t = au_{xx} + bu \ln u \tag{6.2.2.7}$$

can be reduced with the substitution $u = e^z$ to an equation with quadratic nonlinearity:

$$z_t = az_{xx} + az_x^2 + bz. \tag{6.2.2.8}$$

Equation (6.2.2.8) admits a generalized separable solution in the form of a quadratic polynomial in the spatial variable (see row 1 in Table 5.2). The corresponding exact solution of the original equation (6.2.2.7) has the form

$$u = \exp[\psi_1(t)x^2 + \psi_2(t)x + \psi_3(t)],$$

where the functions $\psi_k = \psi_k(t)$ are described by the system of ordinary differential equations

$$\begin{aligned}\psi_1' &= 4a\psi_1^2 + b\psi_1,\\ \psi_2' &= 4a\psi_1\psi_2 + b\psi_2,\\ \psi_3' &= b\psi_3 + 2a\psi_1 + a\psi_2^2.\end{aligned}$$

The equations can be integrated sequentially with relative ease. The first equation is integrable because it is classified as a Bernoulli equation. By comparing the first and second equations, we find that $\psi_2 = C_1\psi_1$, where C_1 is an arbitrary constant. The final equation is a linear equation for ψ_3, which has a known nonhomogeneous term derived from the solutions of the first two equations. ◄

Table 6.1 presents examples of more complex diffusion or wave equations, specifically those involving power-law, exponential, or logarithmic nonlinearity. By applying a simple substitution of the form $u = U(z)$, these equations can be simplified to equations with quadratic or cubic nonlinearity. The last PDE in the table is the nonlinear telegraph equation, sometimes referred to in applications as a hyperbolic diffusion or heat equation.

Problems and Exercises for Section 6.2

1. Find functional separable solutions for the following heat equations with power-law nonlinearity:

a) $u_t = a(u^k u_x)_x + bu^{1-k}$,
b) $u_t = a(u^k u_x)_x + bu + cu^{1-k}$,
c) $u_t = a(u^k u_x)_x + bu^{1+k} + c + du^{1-k}$.

2. Find functional separable solutions for the following heat equations with exponential nonlinearity:

a) $u_t = a(e^{\lambda u} u_x)_x + b$,
b) $u_t = a(e^{\lambda u} u_x)_x + b + ce^{-\lambda u}$,
c) $u_t = a(e^{\lambda u} u_x)_x + be^{\lambda u} + c + de^{-\lambda u}$.

Table 6.1. Some nonlinear PDEs reducible to PDEs with quadratic or cubic nonlinearity using transformations of the form $u = U(z)$.

No.	Original equation	Transformation	Transformed equation	Solution structure for function z
1	$u_t = a(u^{2k}u_x)_x + bu^{1-k}$	$u = z^{1/k}$	$z_t = az^2 z_{xx} + a(1+k^{-1})zz_x^2 + bk$	$z = \varphi(t)x + \psi(t)$
2	$u_t = a(u^k u_x)_x + (bu^k + c)u_x$	$u = z^{1/k}$	$z_t = azz_{xx} + ak^{-1}z_x^2 + (bz+c)z_x$	$z = \varphi(t)x + \psi(t)$
3	$u_t = a(u^{2k}u_x)_x + bu^k u_x$	$u = z^{1/k}$	$z_t = az^2 z_{xx} + a(1+k^{-1})zz_x^2 + bzz_x$	$z = \varphi(t)x + \psi(t)$
4	$u_t = a(e^{2\lambda u}u_x)_x + b + ce^{-\lambda u}$	$u = \lambda^{-1}\ln z$	$z_t = az^2 z_{xx} + azz_x^2 + b\lambda z + c\lambda$	$z = \varphi(t)x + \psi(t)$
5	$u_t = au_{xx} + b(\ln u)u_x + cu$	$u = e^z$	$z_t = az_{xx} + az_x^2 + bzz_x + c$	$z = \varphi(t)x + \psi(t)$
6	$u_t = [(a\ln u + b)u_x]_x + cu$	$u = e^z$	$z_t = (az+b)z_{xx} + (az+a+b)z_x^2 + c$	$z = \varphi(t)x + \psi(t)$
7	$u_{tt} = a(u^{-1/2}u_x)_x + bu^{1/2}$	$u = z^2$	$zz_{tt} + z_t^2 = az_{xx} + \frac{1}{2}bz$	$z = \varphi(x)t^2 + \psi(x)t + \chi(x)$
8	$u_{tt} = a(u^{-2/3}u_x)_x + bu^{1/3}$	$u = z^3$	$z^2 z_{tt} + 2zz_t^2 = az_{xx} + \frac{1}{3}bz$	$z = \varphi(x)t + \psi(x)$
9	$u_{tt} + cu_t = au_{xx} + bu\ln u$	$u = e^z$	$z_{tt} + z_t^2 + cz_t = az_{xx} + az_x^2 + bz$	$z = \varphi(x) + \psi(t)$

3. Find functional separable solutions for the following heat equations with logarithmic nonlinearity:

a) $u_t = au_{xx} + bu\ln u + cu$,
b) $u_t = au_{xx} + bu\ln^2 u$,
c) $u_t = au_{xx} + bu\ln^2 u + cu\ln u + du$.

6.3. Generalized Traveling Wave Solutions

6.3.1. Generalized Traveling Wave Solutions and Other Special Solutions. Solution Construction Algorithm

To clarify the analysis, some internal functions in (6.1.1.1) can be defined a priori, while others, including the external function U, can be determined during the solution process. Such solutions will be referred to as *special functional separable solutions*.

eneralized traveling wave solution is the simplest but sufficiently common 'unctional separable solution. It can be represented as

$$= U(z), \quad z = \varphi(t)x + \psi(t) \quad \text{(argument } z \text{ is linear in } x\text{)}, \qquad (6.3.1.1)$$

and t can be swapped. For $\varphi(t) = k_1$ and $\psi(t) = k_2 t$, where k_1 and k_2 are ' constants, this solution becomes a regular *traveling wave solution.*

r substituting the expression (6.3.1.1) into the given PDE, we need to elim- : variable x using the expression for z; this involves making the substitution $- \psi)/\varphi$. As a result, we obtain a functional differential equation with two ıts: t and z. Its solution can be sought using the methods described in 5.

:larity, Figure 6.1 illustrates the general scheme for constructing generalized ; wave solutions for evolutionary equations.

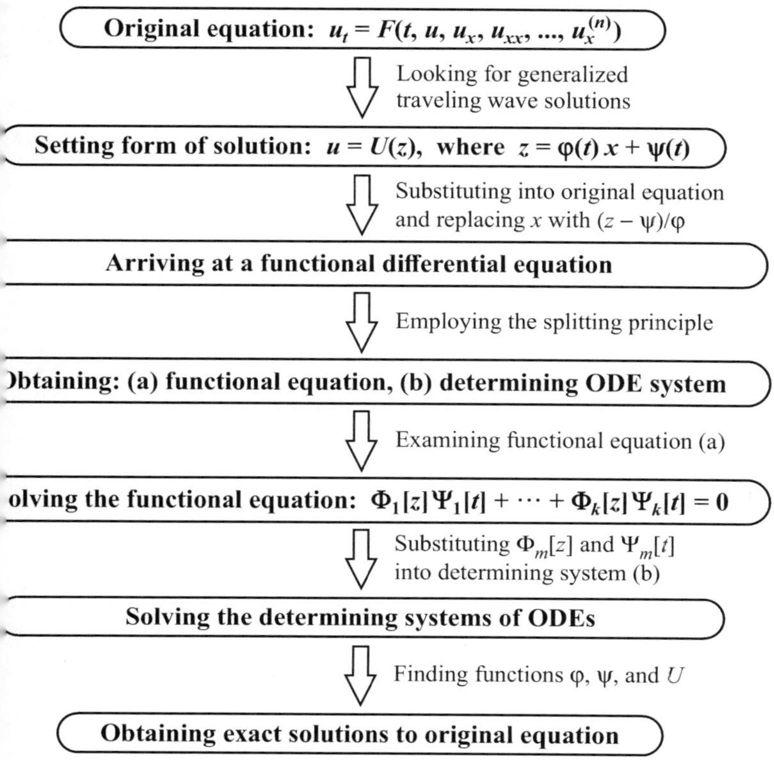

1. An algorithm for constructing generalized traveling wave solutions for evolution- ons.

Below, we specify two more complex special functional separable solutions:

$$u = U(z), \quad z = \varphi(t)x^2 + \psi(t) \quad \text{(argument } z \text{ is quadratic in } x\text{);} \tag{6.3.1.2}$$

$$u = U(z), \quad z = \varphi(t)e^{\lambda x} + \psi(t) \quad \text{(argument } z \text{ is exponential in } x\text{).} \tag{6.3.1.3}$$

After substituting the expression from (6.3.1.2) or (6.3.1.3) into the partial differential equation in question, we need to eliminate the variable x using the expression for z. This process results in a functional differential equation with two arguments: t and z. Solutions to this equation are sought using the methods described in Chapter 5.

It is noteworthy that $e^{\lambda x}$ in formulas (6.3.1.3) can be replaced with the simplest hyperbolic or trigonometric functions.

Remark 6.3. *The algorithm shown in Figure 6.1 can also be utilized to construct exact solutions of the more general form** $u = \sigma(t)F(z) + \varphi_1(t)x + \psi_2(t)$, *where* $z = \varphi_1(t)x + \psi_2(t)$. *Section 7.1.1 below discusses an example of such a solution (see Example 7.3).*

6.3.2. Examples of Constructing Generalized Traveling Wave Solutions

Some nonlinear PDEs that have generalized traveling wave solutions are given in Table 6.1 (see rows 1–6 and 8).

Let us examine several more complex examples of nonlinear equations of mathematical physics that allow for generalized traveling wave solutions of the form (6.3.1.1).

► Example 6.4. Consider the nonlinear third-order equation

$$u_y u_{xy} - u_x u_{yy} = a(u_{yy})^{n-1} u_{yyy}, \tag{6.3.2.1}$$

which describes the flow of a power-law fluid in a boundary layer on a flat plane. Here, u is the stream function, x and y are Cartesian coordinates (x is counted off along the plate), and n is a rheological parameter ($n = 1$ corresponds to a Newtonian fluid).

We will look for exact solutions to this equation in the form of a generalized traveling wave:

$$u = U(z), \qquad z = \varphi(x)y + \psi(x). \tag{6.3.2.2}$$

On substituting (6.3.2.2) into (6.3.2.1) and eliminating y using the change of variable $y = (z - \psi)/\varphi$, we arrive at the functional differential equation

$$\varphi'_x (U'_z)^2 = a\varphi^{2n} (U''_{zz})^{n-1} U'''_{zzz}, \tag{6.3.2.3}$$

*The specified class of solutions encompasses, as special cases, many of the most commonly encountered solutions, including traveling wave solutions, self-similar solutions, generalized self-similar solutions, additive separable solutions, and multiplicative separable solutions, as well as numerous invariant solutions.

which does not include the function ψ. On separating the variables in (6.3.2.3) and integrating, we get

$$\varphi(x) = \begin{cases} (ax+C)^{1/(1-2n)} & \text{for } n \neq 1/2, \\ C\exp(ax/\lambda) & \text{for } n = 1/2, \end{cases} \qquad \psi(x) \text{ is an arbitrary function,}$$

where C and λ are arbitrary constants. The function $U(z)$ satisfies the ordinary differential equations

$$\begin{aligned} &(U_z')^2 = (1-2n)(U_{zz}'')^{n-1}U_{zzz}''' && \text{for } n \neq 1/2, \\ &(U_z')^2 = \lambda(U_{zz}'')^{-1/2}U_{zzz}''' && \text{for } n = 1/2. \end{aligned}$$

The first equation admits, for $\frac{1}{2} < n < 2$, a power-law solution:

$$U(z) = Az^{\frac{2n-1}{n-2}}, \quad A = \left[\frac{3(2-n)}{2n-1}\right]^{\frac{1}{2-n}} \left[\frac{(2n-1)(n+1)}{(2-n)^2}\right]^{\frac{n}{2-n}}.$$

◀

▶ **Example 6.5.** Consider the unsteady heat equation with a nonlinear source

$$u_t = u_{xx} + f(u). \tag{6.3.2.4}$$

We look for exact solutions to equation (6.3.2.4) in the form of a generalized traveling wave*

$$u = u(z), \quad z = \varphi(t)x + \psi(t). \tag{6.3.2.5}$$

It is required to determine the functions $u(z)$, $\varphi(t)$, and $\psi(t)$ in (6.3.2.5) as well as the term $f(u)$ on the right-hand side of (6.3.2.4).

On substituting the expression (6.3.2.5) into (6.3.2.4) and dividing by u_z', we get

$$\varphi_t' x + \psi_t' = \varphi^2 \frac{u_{zz}''}{u_z'} + \frac{f(u)}{u_z'}. \tag{6.3.2.6}$$

On expressing x via z from (6.3.2.5), we obtain $x = (z-\psi)/\varphi$. Using this expression, we eliminate x from (6.3.2.6) to arrive at a functional differential equation in two variables, t and z:

$$-\psi_t' + \frac{\psi}{\varphi}\varphi_t' - \frac{\varphi_t'}{\varphi}z + \varphi^2 \frac{u_{zz}''}{u_z'} + \frac{f(u)}{u_z'} = 0.$$

It is convenient to represent this equation in the bilinear form

$$\Phi_1\Psi_1 + \Phi_2\Psi_2 + \Phi_3\Psi_3 + \Phi_4\Psi_4 = 0, \tag{6.3.2.7}$$

*In the following discussion, when we solve an equation using the method of functional separations of variables, and the equation includes a function $f(u)$ that also needs to be determined, we will use the notation $u = u(z)$ rather than $u = U(z)$. The argument $z = z(x,t)$ can be defined in various ways; refer to Section 6.1.1 for details.

where

$$\begin{aligned}
&\Phi_1 = -\psi'_t + \frac{\psi}{\varphi}\varphi'_t, \quad \Phi_2 = -\frac{\varphi'_t}{\varphi}, \quad \Phi_3 = \varphi^2, \quad \Phi_4 = 1,\\
&\Psi_1 = 1, \quad \Psi_2 = z, \quad \Psi_3 = \frac{u''_{zz}}{u'_z}, \quad \Psi_4 = \frac{f(u)}{u'_z}.
\end{aligned} \tag{6.3.2.8}$$

We substitute the expressions (6.3.2.8) into the formulas

$$\begin{aligned}
\Phi_1 &= A_1\Phi_3 + A_2\Phi_4, \quad &\Phi_2 &= A_3\Phi_3 + A_4\Phi_4,\\
\Psi_3 &= -A_1\Psi_1 - A_3\Psi_2, \quad &\Psi_4 &= -A_2\Psi_1 - A_4\Psi_2,
\end{aligned} \tag{6.3.2.9}$$

which satisfy equation (6.3.2.7) identically. (For these formulas, refer to solutions (5.5.2.4).) As a result, we obtain the following system of ordinary differential equations:

$$\begin{aligned}
-\psi'_t + \frac{\psi}{\varphi}\varphi'_t &= A_1\varphi^2 + A_2, \quad &-\frac{\varphi'_t}{\varphi} &= A_3\varphi^2 + A_4,\\
\frac{u''_{zz}}{u'_z} &= -A_1 - A_3 z, \quad &\frac{f(u)}{u'_z} &= -A_2 - A_4 z,
\end{aligned} \tag{6.3.2.10}$$

where $A_1, \dots, A_4$ are arbitrary constants.

Case 1. For $A_4 \neq 0$, the general solution of system (6.3.2.10) is

$$\begin{aligned}
\varphi(t) &= \pm\left(C_1 e^{2A_4 t} - \frac{A_3}{A_4}\right)^{-1/2},\\
\psi(t) &= -\varphi(t)\left[A_1\int\varphi(t)\,dt + A_2\int\frac{dt}{\varphi(t)} + C_2\right],\\
u(z) &= C_3\int\exp\bigl(-\tfrac{1}{2}A_3 z^2 - A_1 z\bigr)\,dz + C_4,\\
f(u) &= -C_3(A_4 z + A_2)\exp\bigl(-\tfrac{1}{2}A_3 z^2 - A_1 z\bigr),
\end{aligned} \tag{6.3.2.11}$$

where $C_1, \dots, C_4$ are arbitrary constants. The function $f = f(u)$ is defined by the last two expressions in parametric form (z serves as a parameters). For $A_3 \neq 0$, the source function $f(u)$ in (6.3.2.11) can be expressed in terms of elementary functions and the inverse of the error function.

In the special case of $A_3 = C_4 = 0$, $A_1 = -1$, and $C_3 = 1$, the source function can be represented in explicit form as

$$f(u) = -u(A_4 \ln u + A_2). \tag{6.3.2.12}$$

Case 2. For $A_4 = 0$ and $A_3 \neq 0$, the first two equations in (6.3.2.10) have the following solutions:

$$\varphi(t) = \pm\frac{1}{\sqrt{2A_3 t + C_1}}, \quad \psi(t) = \frac{C_2}{\sqrt{2A_3 t + C_1}} - \frac{A_1}{A_3} - \frac{A_2}{3A_3}(2A_3 t + C_1). \tag{6.3.2.13}$$

The other equations have solutions that are defined in the last two formulas of (6.3.2.11) with $A_4 = 0$.

For $A_4 = A_3 = 0$, the first two equations (6.3.2.10) have the following solutions:

$$\varphi = C_1, \quad \psi(t) = -(A_1 C_1^2 + A_2)t + C_2, \tag{6.3.2.14}$$

where C_1 and C_2 are arbitrary constants. ◀

► Example 6.6. Let us look at the following nonlinear convective heat equation with variable coefficients and a source:

$$u_t = a(t)u_{xx} + b(t)u_x + c(t)f(u), \tag{6.3.2.15}$$

where $a = a(t)$, $b = b(t)$, and $c = c(t)$ are arbitrary functions. This equation turns into equation (6.3.2.4) when $a = c = 1$ and $b = 0$.

We seek solutions of the form (6.3.2.5). Using reasoning similar to that in Example 6.5, we derive the system of ordinary differential equations

$$\begin{gathered} -\psi'_t + \frac{\psi}{\varphi}\varphi'_t + b\varphi = A_1 a\varphi^2 + A_2 c, \quad -\frac{\varphi'_t}{\varphi} = A_3 a\varphi^2 + A_4 c, \\ \frac{u''_{zz}}{u'_z} = -A_1 - A_3 z, \quad \frac{f(u)}{u'_z} = -A_2 - A_4 z. \end{gathered} \tag{6.3.2.16}$$

It is evident that, compared to system (6.3.2.10), only the first two equations for $\varphi = \varphi(t)$ and $\psi = \psi(t)$ have changed in system (6.3.2.16). Despite some added complexity, these equations can be fully integrated. This is because the second ODE, once multiplied by φ, simplifies to a Bernoulli equation, while the first ODE is linear in the function ψ. As a result, we obtain

$$\begin{gathered} \varphi(t) = \pm\left[C_1 E(t) + 2A_3 E(t)\int \frac{a(t)}{E(t)}\,dt\right]^{-1/2}, \quad E(t) = \exp\left(2A_4\int c(t)\,dt\right), \\ \psi(t) = -\varphi(t)\left[A_1\int a(t)\varphi(t)\,dt + A_2\int \frac{c(t)}{\varphi(t)}\,dt - \int b(t)\,dt + C_2\right]. \end{gathered}$$

As before, the functions $u(z)$ and $f(u)$ will be described by the last two formulas in (6.3.2.11). ◀

► Example 6.7. The nonlinear heat equation with a variable transfer coefficient and a source

$$u_t = [g(u)u_x]_x + f(u) \tag{6.3.2.17}$$

also has solutions of the form (6.3.2.5). In this case, the desired quantities are described by the ODE system

$$\begin{gathered} -\psi'_t + \frac{\psi}{\varphi}\varphi'_t = A_1\varphi^2 + A_2, \quad -\frac{\varphi'_t}{\varphi} = A_3\varphi^2 + A_4, \\ \frac{[g(u)u'_z]'_z}{u'_z} = -A_1 - A_3 z, \quad \frac{f(u)}{u'_z} = -A_2 - A_4 z. \end{gathered} \tag{6.3.2.18}$$

It is evident that as compared to system (6.3.2.10), only the third equation has changed in system (6.3.2.18). Consequently, the functions $\varphi(t)$ and $\psi(t)$ are determined by the first two formulas in (6.3.2.11). In equation (6.3.2.17), one of the two functions, $g(u)$ or $f(u)$, can be specified arbitrarily, while the other can be derived from the last two ordinary differential equations in (6.3.2.18).

The last two equations (6.3.2.18) involve three unknown functions: $f = f(u)$, $g = g(u)$, and $u = u(z)$. Therefore, we can specify one of these functions arbitrarily. In what follows, we will assume that the relationship $u = u(z)$ is given in implicit form as $z = Z(u)$, where $Z(u)$ is any function that is twice continuously differentiable. Utilizing the relation $u'_z = 1/Z'_u$, we can derive the following formulas from the last two equations in (6.3.2.18):

$$\begin{aligned} f(u) &= -[A_2 + A_4 Z(u)]\frac{1}{Z'_u(u)}, \\ g(u) &= -\left[A_1 u + A_3 \int Z(u)\,du + C\right] Z'_u(u). \end{aligned} \tag{6.3.2.19}$$

These formulas define the admissible structure of the functions $f = f(u)$ and $g = g(u)$ in equation (6.3.2.17). This equation has a solution of the form (6.3.2.5), where $u = u(z)$ is defined implicitly as $z = Z(u)$, while the functions $\varphi(t)$ and $\psi(t)$ are defined by the first two formulas in (6.3.2.11) if $A_4 \neq 0$, formulas (6.3.2.13) if $A_4 = 0$ and $A_3 \neq 0$, or formulas (6.3.2.14) if $A_4 = A_3 = 0$.

Let us explore some examples of how to use formulas (6.3.2.19).

1°. We set

$$Z(u) = u^k, \quad A_1 = -\frac{a_1}{k}, \quad A_2 = -b_2 k, \quad A_3 = -\frac{a_2(k+1)}{k}, \quad A_4 = -b_1 k, \quad C = 0$$

in (6.3.2.19) to obtain an equation of the form (6.3.2.17) with power-law nonlinearities:

$$u_t = [(a_1 u^k + a_2 u^{2k}) u_x]_x + b_1 u + b_2 u^{1-k}.$$

It admits the generalized traveling wave solution (6.3.2.5) for $u(z) = z^{1/k}$, where the functions $\varphi(t)$ and $\psi(t)$ are defined by the first two formulas in (6.3.2.11) if $b_1 \neq 0$, formulas (6.3.2.13) if $b_1 = 0$ and $a_2 \neq 0$, or formulas (6.3.2.14) if $b_1 = a_2 = 0$.

2°. Setting

$$Z(u) = e^{\lambda u}, \quad A_1 = -\frac{a_1}{\lambda}, \quad A_2 = -b_2 \lambda, \quad A_3 = -a_2, \quad A_4 = -b_1 \lambda, \quad C = 0$$

in (6.3.2.19), we arrive at an equation of the form (6.3.2.17) with exponential and exponential/power-law nonlinearities:

$$u_t = [(a_1 u e^{\lambda u} + a_2 e^{2\lambda u}) u_x]_x + b_1 + b_2 e^{-\lambda u}.$$

This equation admits the generalized traveling wave solution (6.3.2.5) for $u(z) = (\ln u)/\lambda$, where the functions $\varphi(t)$ and $\psi(t)$ are defined by the first two formulas in (6.3.2.11) is $b_1 \neq 0$, formulas (6.3.2.13) if $b_1 = 0$ and $a_2 \neq 0$, and formulas (6.3.2.14) if $b_1 = a_2 = 0$. ◀

► Example 6.8. The nonlinear nth-order PDE

$$u_t = u_x^{(n)} + f(u) \tag{6.3.2.20}$$

is treated likewise.

As before, we look for solutions in the form of a generalized traveling wave (6.3.2.5). By reasoning similarly to Example 6.5, we arrive at the system of ordinary differential equations

$$\begin{aligned} -\psi_t' + \frac{\psi}{\varphi}\varphi_t' &= A_1\varphi^n + A_2, & -\frac{\varphi_t'}{\varphi} &= A_3\varphi^n + A_4, \\ \frac{u_z^{(n)}}{u_z'} &= -A_1 - A_3 z, & \frac{f(u)}{u_z'} &= -A_2 - A_4 z, \end{aligned} \tag{6.3.2.21}$$

For $A_4 \neq 0$, the general solution of the first two equations in system (6.3.2.21) is expressed as

$$\begin{aligned} \varphi(t) &= \left(C_1 e^{A_4 nt} - \frac{A_3}{A_4}\right)^{-1/n}, \\ \psi(t) &= -\varphi(t)\left[A_1 \int \varphi^{n-1}(t)\,dt + A_2 \int \frac{dt}{\varphi(t)} + C_2\right]. \end{aligned} \tag{6.3.2.22}$$

For $A_4 = 0$, the solution of the second equation in (6.3.2.21) is

$$\varphi(t) = (A_3 nt + C_1)^{-1/n}.$$

Substituting this expression into the second formula in (6.3.2.22), we can obtain the function $\psi(t)$.

Let us explore some equations of the form (6.3.2.20) and find their solutions.

1°. Setting $A_1 = -1$, $A_3 = 0$, and $u(z) = e^z$ in the last two ODEs in (6.3.2.21), we arrive at an equation with a logarithmic nonlinearity:

$$u_t = u_x^{(n)} - A_4 u \ln u - A_2 u.$$

It has a solution of the form (6.3.2.5) with $u(z) = e^z$.

2°. For odd $n = 2m+1$ ($m = 1, 2, \ldots$), we set $A_1 = (-1)^{m+1}$, $A_3 = 0$, and $u(z) = \sin z$, to obtain

$$u_t = u_x^{(2m+1)} - (A_2 + A_4 \arcsin u)\sqrt{1-u^2}.$$

This equation has a solution of the form (6.3.2.5) with $u(z) = \sin z$. ◄

► Example 6.9. Consider the generalized nth-order Korteweg–de Vries equation

$$u_t = u_x^{(n)} + f(u)u_x. \tag{6.3.2.23}$$

Seeking its exact solutions in the form (6.3.2.5) leads to the following system of equations for the functions $\varphi(t)$, $\psi(t)$, $u(z)$, and $f(u)$:

$$\begin{aligned} -\psi_t' + \frac{\psi}{\varphi}\varphi_t' &= A_1\varphi^n + A_2\varphi, & -\frac{\varphi_t'}{\varphi} &= A_3\varphi^n + A_4\varphi, \\ \frac{u_z^{(n)}}{u_z'} &= -A_1 - A_3 z, & f(u) &= -A_2 - A_4 z. \end{aligned} \tag{6.3.2.24}$$

System (6.3.2.24) differs from system (6.3.2.21) in the first two equations and the last equation. The first two equations can be fully integrated because the second is separable, and the first is linear in ψ.

Next, we consider the case $A_3 = 0$, when the function $f(u)$ can be represented in explicit form. For $A_3 = 0$, the general solutions of the first two equations in (6.3.2.24) are expressed as follows:

$$\varphi(t) = \frac{1}{A_4 t + C_1}, \qquad \psi(t) = \frac{A_1}{A_4(n-2)}(A_4 t + C_1)^{1-n} - \frac{A_2 t + C_2}{A_4 t + C_1}.$$

Below, we present some equations of the form (6.3.2.23) along with their solutions.

$1°$. By setting $A_1 = -1$, $A_3 = 0$, and $u(z) = e^z$ in the last ODEs in (6.3.2.24), we arrive at an equation with a logarithmic nonlinearity:

$$u_t = u_x^{(n)} - (A_4 \ln u + A_2) u_x.$$

It has a solution of the form (6.3.2.5) with $u(z) = e^z$.

$2°$. For odd $n = 2m + 1$ ($m = 1, 2, \ldots$), we set $A_1 = (-1)^{m+1}$, $A_3 = 0$, and $u(z) = \sin z$ to obtain

$$u_t = u_x^{(2m+1)} - (A_2 + A_4 \arcsin u) u_x.$$

This equation admits a generalized traveling wave solution (6.3.2.5) with $u(z) = \sin z$. ◄

► Example 6.10. Consider the nonlinear PDE with an $(n+1)$st-order mixed derivative

$$\frac{\partial^{n+1} u}{\partial x^n \partial y} = f(u). \tag{6.3.2.25}$$

We look for a generalized traveling wave solution

$$u = u(z), \qquad z = \varphi(y)x + \psi(y). \tag{6.3.2.26}$$

By substituting (6.3.2.26) into (6.3.2.25) and eliminating x using the expression of z followed by dividing by $u_z^{(n+1)}$ and rearranging, we obtain a functional differential equation with two arguments:

$$\varphi^n \psi'_y - \varphi^{n-1} \psi \varphi'_y + \varphi^{n-1} \varphi'_y \left(z + n \frac{u_z^{(n)}}{u_z^{(n+1)}} \right) - \frac{f(u)}{u_z^{(n+1)}} = 0. \tag{6.3.2.27}$$

It reduces to a three-term bilinear functional equation (5.5.2.1), which has two solutions (5.5.2.2). Accordingly, we will consider two cases.

$1°$. In the first case, we equate the expression in parentheses and the last fraction in (6.3.2.27) with arbitrary constants, C_1 and C_2. As a result, after rearrangement, we get

$$\begin{aligned} (z - C_1) u_z^{(n+1)} + n u_z^{(n)} &= 0, \\ C_2 u_z^{(n+1)} - f(u) &= 0, \\ \varphi^n \psi'_y - \varphi^{n-1} \psi \varphi'_y + C_1 \varphi^{n-1} \varphi'_y - C_2 &= 0, \end{aligned}$$

Assuming that $C_1 = 0$, which corresponds to translation in z and renaming the function ψ, and integrating, we obtain

$$\begin{aligned} u &= A\ln|z| + B_{n-1}z^{n-1} + \cdots + B_1 z + B_0, \\ f(u) &= AC_2 n!\,(-1)^n z^{-n-1}, \\ \psi(y) &= C_2\varphi(y)\int \frac{dy}{[\varphi(y)]^{n+1}} + C_3\varphi(y), \end{aligned} \tag{6.3.2.28}$$

where A, B_m, and C_3 are arbitrary constants, and $\varphi(y)$ is an arbitrary function.

The first two formulas in (6.3.2.28) provide a parametric representation of the function $f(u)$. In the special case where $B_{n-1} = \cdots = B_0 = 0$, eliminating z leads us to the exponential dependence

$$f(u) = \alpha e^{\beta u}, \qquad \alpha = AC_2 n!\,(-1)^n, \quad \beta = -(n+1)/A.$$

By virtue of (6.3.2.28), the exact solution of equation (6.3.2.25) with an exponential right-hand side will exhibit functional arbitrariness.

2°. In the second case, in (6.3.2.27) we equate the difference of the first two terms and the functional factor in front of the expression in parentheses to constants. As a result, we arrive at three ordinary differential equations:

$$\begin{aligned} \varphi^{n-1}\varphi'_y &= C_1, \\ \varphi^n\psi'_y - \varphi^{n-1}\psi\varphi'_y &= C_2, \\ (C_1 z + C_2)u_z^{(n+1)} + C_1 n u_z^{(n)} - f(u) &= 0, \end{aligned} \tag{6.3.2.29}$$

where C_1 and C_2 are arbitrary constants. The solutions of the first two equations are

$$\varphi = (C_1 nt + C_3)^{1/n}, \qquad \psi = C_4(C_1 nt + C_3)^{1/n} - \frac{C_2}{C_1}.$$

These formulas, along with the last ODE in (6.3.2.29), define a generalized traveling wave solution of the form (6.3.2.26) for the nonlinear equation (6.3.2.25) with an arbitrary function $f(u)$. ◀

◆ *Problems and Exercises for Section 6.3*

The functions f, g, and h below are subject to determination.

1. Find special functional separable solutions of the nonlinear first-order PDE:
$u_t = f(u)u_x + g(u)$.
Hint. Look for solutions in the forms:
a) $u = u(z)$, $z = \varphi(t)x + \psi(t)$,
b) $u = u(z)$, $z = \varphi(x)t + \psi(x)$.

2. Find special functional separable solutions of the nonlinear first-order PDE:
$u_t = f(u)u_x^2 + g(u)$.
Hint. Look for solutions in the forms:
a) $u = u(z)$, $z = \varphi(t)x + \psi(t)$,
b) $u = u(z)$, $z = \varphi(t)x^2 + \psi(t)$.

3. Examine in detail and carry out all necessary calculations for Examples 6.8–6.10.

4. Find generalized traveling wave solutions of the nonlinear equations:
a) $u_t = [f(u)u_x]_x + g(u)u_x$,
b) $u_t = [f(u)u_x]_x + g(u)u_x + h(u)$.
Hint. Look for solutions in the form $u = u(z)$, where $z = \varphi(t)x + \psi(t)$.

5. Find a generalized traveling wave solution of the nonlinear heat equation:
$u_t = [f(u)u_x^n]_x + g(u)$.
Hint. Look for a solution in the form $u = u(z)$, where $z = \varphi(t)x + \psi(t)$.

6. Find generalized traveling wave solutions of the nonlinear third-order equations:
a) $u_t = f(u)u_{xxx} + g(u)$,
b) $u_t = f(u)u_{xxx} + g(u)u_x$,
c) $u_t = [f(u)u_{xx}]_x + g(u)$,
d) $u_t = [f(u)u_{xx}]_x + g(u)u_x$.

7. Find special functional separable solutions of the nonlinear heat equations:
a) $u_t = [f(u)u_x]_x + g(u)$,
b) $u_t = x^{-n}[x^n f(u)u_x]_x + g(u)$,
c) $u_t = au_{xx} + bxu_x + f(u)$.
Hint. Look for solutions in the form $u = u(z)$, where $z = \varphi(t)x^2 + \psi(t)$.

8. Find a generalized traveling wave solution of the nonlinear reaction-diffusion equation

$$u_t = u_{xx} + x^2 f(u),$$

where $f(u)$ is an arbitrary function (see Polyanin, 2019d).
Hint. Look for solutions in the form $u = u(z)$, where $z = t + \int \theta(x)\,dx$.

9. Find a generalized traveling wave solution of the nonlinear reaction-diffusion equation

$$u_t = [a(x)u_x]_x + \frac{x^2}{a(x)} f(u),$$

where $a(x)$ and $f(u)$ are arbitrary functions (see Polyanin, 2019d).
Hint. Look for solutions in the form $u = u(z)$, where $z = t + \int \theta(x)\,dx$.

10. Find a generalized traveling wave solution of the nonlinear reaction-diffusion equation

$$u_t = u_{xx} + \tanh^2 x\, f(u),$$

where $f(u)$ is an arbitrary function (see Polyanin, 2019d).
Hint. Look for solutions in the form $u = u(z)$, where $z = t + \int \theta(x)\,dx$.

11. Find a generalized traveling wave solution of the nonlinear convective diffusion equation

$$u_t = u_{xx} + x f(u)u_x,$$

where $f(u)$ is an arbitrary function (see Polyanin, 2019c).
Hint. Look for solutions in the form $u = u(z)$, where $z = t + \int \theta(x)\,dx$.

12. Find generalized traveling wave solution of the nonlinear convective diffusion equation

$$u_t = u_{xx} + \tanh x\, f(u)u_x,$$

where $f(u)$ is an arbitrary function (see Polyanin, 2019c).
Hint. Look for solutions in the form $u = u(z)$, where $z = t + \int \theta(x)\,dx$.

13. Find generalized traveling wave solution of the nonlinear convective diffusion equation

$$u_t = [a(x)u_x]_x + xf(u)u_x,$$

where $a(x)$ and $f(u)$ are arbitrary functions (see Polyanin, 2019c).

14. Find generalized traveling wave solution of the nonlinear Klein–Gordon equation

$$u_{tt} = [a(x)u_x]_x + \left[\frac{k}{a(x)} - 1\right]f(u),$$

where $a(x)$ and $f(u)$ are arbitrary functions (see Zhurov & Polyanin, 2020).

Hint. Look for solutions in the form $u = u(z)$, where $z = t + \int \theta(x)\,dx$.

6.4. Method of Differentiation

6.4.1. Brief Description of the Method. Reduction to a Standard Functional Differential Equation

The procedure for constructing functional separable solutions using the differentiation method typically involves several stages, which are briefly described below.

1°. Substitute the expression (6.1.1.2) into the nonlinear partial differential equation being studied. The result is a functional differential equation with three arguments: the first two arguments, x and t, are regular, and the third, z, is composite.

2°. Multiply the functional differential equation with three arguments by suitable functions and differentiate with respect to x or t (or both) to reduce it to a standard functional differential equation with two arguments, effectively eliminating the variable x or t. Two-argument functional differential equations are discussed in Section 5.2.

3°. Utilize the splitting method to construct a solution for the two-argument functional differential equation from Item 2°; use the appropriate formulas from Section 5.5.

4°. Insert the solution from Item 3° into the functional differential equation obtained in step 1°. This will allow us to establish relationships between the constants of integrations, remove extra constants (which may result from differentiation in Item 2°), and find all undetermined quantities.

5°. Examine all potential degenerate cases that may arise from violating the assumptions made during the solution.

Remark 6.4. *The second stage is the most difficult; its implementation may not always be feasible.*

6.4.2. Examples of Constructing Functional Separable Solutions by the Differentiation Method

The following examples illustrate how the differentiation method can be applied to construct functional separable solutions for nonlinear diffusion and wave equations. The discussion primarily focuses on the methodological aspects of deriving and utilizing the standard bilinear functional equation. As a result, some solutions are either not explored in detail or omitted altogether.

► Example 6.11. Consider the nonlinear heat equation

$$u_t = [f(u)u_x]_x. \tag{6.4.2.1}$$

In the general case, equation (6.4.2.1) admits a traveling wave solution $u = U_1(kx - \lambda t)$ and also a self-similar solution $u = U_2(xt^{-1/2})$.

We look for functional separable solutions to equation (6.4.2.1) in the form

$$u = u(z), \quad z = \varphi(x) + \psi(t). \tag{6.4.2.2}$$

Substituting (6.4.2.2) into (6.4.2.1) and dividing by u'_z, we obtain a functional differential equation in three variables:

$$\psi'_t = \varphi''_{xx} f(u) + (\varphi'_x)^2 H(z), \tag{6.4.2.3}$$

where

$$H(z) = f(u)\frac{u''_{zz}}{u'_z} + f'_z(u), \qquad u = u(z). \tag{6.4.2.4}$$

Differentiating (6.4.2.3) with respect to x yields

$$\varphi'''_{xxx} f(u) + \varphi'_x \varphi''_{xx}[f'_z(u) + 2H(z)] + (\varphi'_x)^3 H'_z = 0. \tag{6.4.2.5}$$

This functional differential equation in two variables can be interpreted as a three-term bilinear functional equation (5.5.2.1) from Section 5.5.2, which has two different solutions. Therefore, we must explore two cases.

Case 1. Solutions of the functional differential equation (6.4.2.5) can be determined from the system of ordinary differential equations

$$\begin{gathered} f'_z + 2H = 2A_1 f, \quad H'_z = A_2 f, \\ \varphi'''_{xxx} + 2A_1\varphi'_x\varphi''_{xx} + A_2(\varphi'_x)^3 = 0, \end{gathered} \tag{6.4.2.6}$$

where A_1 and A_2 are arbitrary constants.

The first two equations in (6.4.2.6) are linear and independent of the third one. Their general solution is given by

$$\begin{aligned} f &= \begin{cases} e^{A_1 z}(B_1 e^{kz} + B_2 e^{-kz}) & \text{for } A_1^2 > 2A_2, \\ e^{A_1 z}(B_1 + B_2 z) & \text{for } A_1^2 = 2A_2, \\ e^{A_1 z}[B_1 \sin(kz) + B_2\cos(kz)] & \text{for } A_1^2 < 2A_2, \end{cases} \quad k = |A_1^2 - 2A_2|^{1/2}, \\ H &= A_1 f - \tfrac{1}{2} f'_z, \end{aligned} \tag{6.4.2.7}$$

where B_1 and B_2 are constants of integration.

Substituting the expression of H from (6.4.2.7) into (6.4.2.4), we obtain a differential equation for $u = u(z)$. After integrating, we get

$$u = C_1 \int e^{A_1 z} |f(z)|^{-3/2} dz + C_2, \tag{6.4.2.8}$$

where C_1 and C_2 are arbitrary constants. Formula (6.4.2.7) for f together with expression (6.4.2.8) define the relationship $f = f(u)$ in parametric form.

Let us examine the case where $A_2 = 0$ and $A_1 \neq 0$ ($k = |A_1|$) in greater detail. From formulas (6.4.2.7) and (6.4.2.8), we obtain

$$\begin{aligned} &f(z) = B_1 e^{2A_1 z} + B_2, \quad H = A_1 B_2, \\ &u(z) = C_3 (B_1 + B_2 e^{-2A_1 z})^{-1/2} + C_2 \quad (C_1 = A_1 B_2 C_3). \end{aligned} \tag{6.4.2.9}$$

Eliminating z yields

$$f(u) = \frac{B_2 C_3^2}{C_3^2 - B_1 u^2}. \tag{6.4.2.10}$$

A first integral of the last equation in (6.4.2.6) with $A_2 = 0$ is expressed as

$$\varphi''_{xx} + A_1 (\varphi'_x)^2 = \text{const}.$$

Its general solution is given by

$$\begin{aligned} \varphi(x) &= -\frac{1}{2A_1} \ln\left[\frac{D_2}{D_1} \frac{1}{\sinh^2(A_1\sqrt{D_2}\,x + D_3)}\right] && \text{if} \quad D_1 > 0, \; D_2 > 0; \\ \varphi(x) &= -\frac{1}{2A_1} \ln\left[-\frac{D_2}{D_1} \frac{1}{\cos^2(A_1\sqrt{-D_2}\,x + D_3)}\right] && \text{if} \quad D_1 > 0, \; D_2 < 0; \\ \varphi(x) &= -\frac{1}{2A_1} \ln\left[-\frac{D_2}{D_1} \frac{1}{\cosh^2(A_1\sqrt{D_2}\,x + D_3)}\right] && \text{if} \quad D_1 < 0, \; D_2 > 0, \end{aligned} \tag{6.4.2.11}$$

where D_1, D_2, and D_3 are constants of integration. The relations

$$(\varphi'_x)^2 = D_1 e^{-2A_1\varphi} + D_2, \quad \varphi''_{xx} = -A_1 D_1 e^{-2A_1\varphi} \tag{6.4.2.12}$$

hold in all three cases.

Substituting the expressions (6.4.2.9) and (6.4.2.12) into the original functional differential equation (6.4.2.3) and considering the expression of z in (6.4.2.2), we obtain the following equation for $\psi = \psi(t)$:

$$\psi'_t = -A_1 B_1 D_1 e^{2A_1\psi} + A_1 B_2 D_2.$$

Integrating yields the solution

$$\psi(t) = \frac{1}{2A_1} \ln \frac{B_2 D_2}{D_4 \exp(-2A_1^2 B_2 D_2 t) + B_1 D_1}, \tag{6.4.2.13}$$

where D_4 is an arbitrary constant.

Formulas (6.4.2.2), (6.4.2.9) (for u), (6.4.2.11), and (6.4.2.13) define three solutions of the nonlinear equation (6.4.2.1) with $f(u)$ defined by (6.4.2.10). It should be reminded that these solutions apply to the special case $A_2 = 0$ in (6.4.2.7) and (6.4.2.8).

Case 2. Solutions of the functional differential equation (6.4.2.5) are determined from the system of ordinary differential equations

$$\begin{aligned} &\varphi'''_{xxx} = A_1(\varphi'_x)^3, \quad \varphi'_x\varphi''_{xx} = A_2(\varphi'_x)^3, \\ &A_1 f + A_2(f'_z + 2H) + H'_z = 0. \end{aligned} \tag{6.4.2.14}$$

The first two equations are compatible in two cases:

$$\begin{aligned} A_1 = A_2 = 0 \quad &\Longrightarrow \quad \varphi(x) = B_1 x + B_2, \\ A_1 = 2A_2^2 \quad &\Longrightarrow \quad \varphi(x) = -\frac{1}{A_2}\ln|B_1 x + B_2|. \end{aligned} \tag{6.4.2.15}$$

The first solution in (6.4.2.15) ultimately leads to a traveling wave solution of equation (6.4.2.1): $u = u(B_1 x + B_2 t)$. The second solution results in a self-similar solution $u = \widetilde{u}(x^2/t)$. In both cases, the function $f(u)$ in equation (6.4.2.1) is arbitrary. ◄

Remark 6.5. *The functional separable solutions (6.4.2.2) of equation (6.4.2.1) were obtained in a different way by Doyle & Vassiliou (1998).*

► Example 6.12. The more general nonlinear heat equation with a source

$$u_t = [f(u)u_x]_x + g(u) \tag{6.4.2.16}$$

also has a solution of the form (6.4.2.2). For the undetermined functions $\varphi(x)$ and $\psi(t)$, we obtain the following functional differential equation in three independent variables:

$$\psi'_t = \varphi''_{xx} f(u) + (\varphi'_x)^2 H(z) + g(u)/u'_z,$$

in which the function $H(z)$ is defined by (6.4.2.4). Differentiating this equation with respect to x, we get

$$\varphi'''_{xxx} f(u) + \varphi'_x\varphi''_{xx}[f'_z(u) + 2H(z)] + (\varphi'_x)^3 H'_z + \varphi'_x[g(u)/u'_z]'_z = 0.$$

This functional differential equation in two variables can be treated as a four-term bilinear functional equation (5.5.2.3) from Section 5.5.2, where $\Phi_1 = \varphi'''_{xxx}$, $\Phi_2 = \varphi'_x\varphi''_{xx}$, $\Phi_3 = (\varphi'_x)^3$, and $\Phi_4 = \varphi'_x$. ◄

Remark 6.6. *The study by Estévez et al. (2002) examined functional separable solutions for the equation $u_t = [f(u)u_x^n]_x + g(u)$, which is more general than equation (6.4.2.16).*

► Example 6.13. Consider the nonlinear Klein–Gordon equation

$$u_{tt} - u_{xx} = f(u). \tag{6.4.2.17}$$

We seek its exact solutions in the form

$$u = u(z), \qquad z = \varphi(x) + \psi(t). \tag{6.4.2.18}$$

Substituting the expression (6.4.2.18) into (6.4.2.17) gives

$$\psi''_{tt} - \varphi''_{xx} + \left[(\psi'_t)^2 - (\varphi'_x)^2\right] g(z) = h(z), \tag{6.4.2.19}$$

where

$$g(z) = u''_{zz}/u'_z, \quad h(z) = f\big(u(z)\big)/u'_z. \tag{6.4.2.20}$$

On differentiating equation (6.4.2.19) with respect to t and then x followed by dividing by $\psi'_t\varphi'_x$, we obtain

$$2(\psi''_{tt} - \varphi''_{xx})\, g'_z + \big[(\psi'_t)^2 - (\varphi'_x)^2\big]\, g''_{zz} = h''_{zz}.$$

Eliminating the difference $\psi''_{tt} - \varphi''_{xx}$ from this equation with the help of (6.4.2.19), we get

$$\big[(\psi'_t)^2 - (\varphi'_x)^2\big](g''_{zz} - 2gg'_z) = h''_{zz} - 2g'_z h. \tag{6.4.2.21}$$

The equality can hold in two cases:

$$\begin{aligned} &1) \quad g''_{zz} - 2gg'_z = 0, \qquad h''_{zz} - 2g'_z h = 0;\\ &2) \quad (\psi'_t)^2 = A\psi + B, \quad (\varphi'_x)^2 = -A\varphi + B - C,\\ &\quad\ \ h''_{zz} - 2g'_z h = (Az + C)(g''_{zz} - 2gg'_z), \end{aligned} \tag{6.4.2.22}$$

where A, B, and C are arbitrary constants.* Let us examine these cases in order.

Case 1. The first two equations in (6.4.2.22) allow us to find $g(z)$ and $h(z)$. Integrating the former one gives $g'_z = g^2 + \text{const}$. Integrating further, we obtain

$$g = k, \tag{6.4.2.23}$$

$$g = -1/(z + C_1), \tag{6.4.2.24}$$

$$g = -k\tanh(kz + C_1), \tag{6.4.2.25}$$

$$g = -k\coth(kz + C_1), \tag{6.4.2.26}$$

$$g = k\tan(kz + C_1), \tag{6.4.2.27}$$

where C_1 and k are arbitrary constants.

The second equation in (6.4.2.22) is a linear ODE for h and has a particular solution $h = g(z)$ because it coincides with the first equation when $h = g$. Its general solution is expressed as (prove by yourself):

$$h = C_2 g(z) + C_3 g(z)\int \frac{dz}{g^2(z)}, \tag{6.4.2.28}$$

where C_2 and C_3 are arbitrary constants.

It follows from relations (6.4.2.20) that the functions $u(z)$ and $f(u)$ are expressed as

$$u(z) = B_1\int G(z)\,dz + B_2, \quad f(u) = B_1 h(z)G(z), \quad G(z) = \exp\bigg[\int g(z)\,dz\bigg], \tag{6.4.2.29}$$

where B_1 and B_2 are arbitrary constants; the function f is defined parametrically.

*The first two ODEs in case 2 arise from differentiating the functional differential equation $(\psi'_t)^2 - (\varphi'_x)^2 = R(z)$ with respect to x and t, which is equation (6.4.2.21) divided by $(g''_{zz} - 2gg'_z)$. As a result, we get $R''_{zz} = 0$, which implies that $R(z) = Az + C$. Considering that $z = \varphi(x) + \psi(t)$ and $(\psi'_t)^2 - (\varphi'_x)^2 = R(z)$ leads us to the specified ODEs.

Let us examine the case (6.4.2.24) in more detail. Formula (6.4.2.28) yields

$$h = A_1(z + C_1)^2 + \frac{A_2}{z + C_1}, \tag{6.4.2.30}$$

where $A_1 = -C_3/3$ and $A_2 = -C_2$ are any. Substituting (6.4.2.24) and (6.4.2.30) into (6.4.2.29), we find that

$$u = B_1 \ln|z + C_1| + B_2, \quad f = A_1 B_1(z + C_1) + \frac{A_2 B_1}{(z + C_1)^2}.$$

Eliminating z from these relations, we derive the right-hand side of equation (6.4.2.17):

$$f(u) = A_1 B_1 e^w + A_2 B_1 e^{-2w}, \quad \text{where} \quad w = \frac{u - B_2}{B_1}. \tag{6.4.2.31}$$

For clarity, we set $C_1 = 0$, $B_1 = 1$, and $B_2 = 0$ and introduce the notation $A_1 = a$ and $A_2 = b$ to obtain

$$u(z) = \ln|z|, \quad f(u) = ae^u + be^{-2u}, \quad g(z) = -1/z, \quad h(z) = az^2 + b/z. \tag{6.4.2.32}$$

It remains to determine the functions $\psi(t)$ and $\varphi(x)$. Let us substitute the expressions (6.4.2.32) into the functional differential equation (6.4.2.19). Considering (6.4.2.18) and rearranging, we obtain

$$\begin{aligned}&[\psi\psi''_{tt} - (\psi'_t)^2 - a\psi^3 - b] - [\varphi\varphi''_{xx} - (\varphi'_x)^2 + a\varphi^3]\\ &\qquad + (\psi''_{tt} - 3a\psi^2)\varphi - \psi(\varphi''_{xx} + 3a\varphi^2) = 0.\end{aligned} \tag{6.4.2.33}$$

By differentiating (6.4.2.33) with respect to t and x, we remove the terms in square brackets. As a result, we arrive at a separable equation*

$$(\psi'''_{ttt} - 6a\psi\psi'_t)\varphi'_x - (\varphi'''_{xxx} + 6a\varphi\varphi'_x)\psi'_t = 0,$$

whose solution is described by the autonomous ordinary differential equations of the third order

$$\begin{aligned}\psi'''_{ttt} - 6a\psi\psi'_t &= A\psi'_t,\\ \varphi'''_{xxx} + 6a\varphi\varphi'_x &= A\varphi'_x,\end{aligned}$$

where A is the separation constant. Either equation can be integrated twice to obtain

$$\begin{aligned}(\psi'_t)^2 &= 2a\psi^3 + A\psi^2 + C_1\psi + C_2,\\ (\varphi'_x)^2 &= -2a\varphi^3 + A\varphi^2 + C_3\varphi + C_4,\end{aligned} \tag{6.4.2.34}$$

where $C_1, \ldots, C_4$ are arbitrary constants. On eliminating the derivatives from equation (6.4.2.33) using (6.4.2.34), we arrive at the following relationships between

*The easiest way to solve equation (6.4.2.33) is to take advantage of the solution results for equation (5.5.2.3) from Section 5.5; see formulas (5.5.2.4) and (5.5.2.5).

Table 6.2. Nonlinear wave-type equations $u_{tt} - u_{xx} = f(u)$ admitting functional separable solutions of the form $u = u(z)$, where $z = \varphi(x) + \psi(t)$.

No.	Right-hand side of equation $f(u)$	Solution $u(z)$	Equations for $\psi(t)$ and $\varphi(x)$
1	$au\ln u + bu$	e^z	$(\psi'_t)^2 = C_1e^{-2\psi} + a\psi - \frac{1}{2}a + b + A$, $(\varphi'_x)^2 = C_2e^{-2\varphi} - a\varphi + \frac{1}{2}a + A$
2	$ae^u + be^{-2u}$	$\ln\lvert z\rvert$	$(\psi'_t)^2 = 2a\psi^3 + A\psi^2 + C_1\psi + C_2$, $(\varphi'_x)^2 = -2a\varphi^3 + A\varphi^2 - C_1\varphi + C_2 + b$
3	$a\sin u + b(\sin u\ln\tan\frac{u}{4} + 2\sin\frac{u}{4})$	$4\arctan e^z$	$(\psi'_t)^2 = C_1e^{2\psi} + C_2e^{-2\psi} + b\psi + a + A$, $(\varphi'_x)^2 = -C_2e^{2\varphi} - C_1e^{-2\varphi} - b\varphi + A$
4	$a\sinh u + b(\sinh u\ln\tanh\frac{u}{4} + 2\sinh\frac{u}{2})$	$2\ln\left\lvert\coth\frac{z}{2}\right\rvert$	$(\psi'_t)^2 = C_1e^{2\psi} + C_2e^{-2\psi} - \sigma b\psi + a + A$, $(\varphi'_x)^2 = C_2e^{2\varphi} + C_1e^{-2\varphi} + \sigma b\varphi + A$
5	$a\sinh u + 2b(\sinh u\arctan e^{u/2} + \cosh\frac{u}{2})$	$2\ln\left\lvert\tan\frac{z}{2}\right\rvert$	$(\psi'_t)^2 = C_1\sin 2\psi + C_2\cos 2\psi + \sigma b\psi + a + A$, $(\varphi'_x)^2 = -C_1\sin 2\varphi + C_2\cos 2\varphi - \sigma b\varphi + A$
Notation: A, C_1, and C_2 are arbitrary constants; $\sigma = 1$ for $z > 0$, $\sigma = -1$ for $z < 0$.			

the constants: $C_3 = -C_1$ and $C_4 = C_2 + b$. Hence, the functions $\psi(t)$ and $\varphi(x)$ are described by the autonomous first-order ODEs with cubic nonlinearities

$$(\psi'_t)^2 = 2a\psi^3 + A\psi^2 + C_1\psi + C_2,$$
$$(\varphi'_x)^2 = -2a\varphi^3 + A\varphi^2 - C_1\varphi + C_2 + b.$$

The solutions of these equations are expressed in terms of elliptic functions.

The other cases specified in (6.4.2.23)–(6.4.2.27) are investigated similarly. The results are summarized in Table 6.2. (These solutions were initially obtained using a different method by Grundland & Infeld (1992); see also Zhdanov (1994) and Andreev et al. (1998).)

Case 2. Integrating the first two equations in (6.4.2.22) for the second case, we get two solutions

$$\psi = \pm\sqrt{B}\,t + D_1, \qquad \varphi = \pm\sqrt{B-C}\,t + D_2 \qquad \text{if} \quad A = 0;$$
$$\psi = \frac{1}{4A}(At + D_1)^2 - \frac{B}{A}, \quad \varphi = -\frac{1}{4A}(Ax + D_2)^2 + \frac{B-C}{A} \quad \text{if} \quad A \neq 0; \tag{6.4.2.35}$$

where D_1 and D_2 are arbitrary constants. In both cases, the function $f(u)$ in equation (6.4.2.17) is arbitrary. The first solution in (6.4.2.35) corresponds to a traveling wave, $u = u(kx + \lambda t)$, while the second solution leads to $u = u(x^2 - t^2)$. ◄

Table 6.3. Nonlinear heat-type equations $u_{xx} + u_{yy} = f(u)$ admitting functional separable solutions of the form $u = u(z)$, where $z = \varphi(x) + \psi(y)$.

No.	Right-hand side of equation $f(u)$	Solution $u(z)$	Equations for $\varphi(x)$ and $\psi(y)$
1	$au\ln u+bu$	e^z	$(\varphi'_x)^2=C_1e^{-2\varphi}+a\varphi-\frac{1}{2}a+b+A$, $(\psi'_y)^2=C_2e^{-2\psi}+a\psi-\frac{1}{2}a-A$
2	ae^u+be^{-2u}	$\ln\lvert z\rvert$	$(\varphi'_x)^2=2a\varphi^3+A\varphi^2+C_1\varphi+C_2$, $(\psi'_y)^2=2a\psi^3-A\psi^2+C_1\psi-C_2-b$
3	$a\sin u+b(\sin u\ln\tan\frac{u}{4}+2\sin\frac{u}{4})$	$4\arctan e^z$	$(\varphi'_x)^2=C_1e^{2\varphi}+C_2e^{-2\varphi}+b\varphi+a+A$, $(\psi'_y)^2=C_2e^{2\psi}+C_1e^{-2\psi}+b\psi-A$
4	$a\sinh u+b(\sinh u\ln\tanh\frac{u}{4}+2\sinh\frac{u}{2})$	$2\ln\left\lvert\coth\frac{z}{2}\right\rvert$	$(\varphi'_x)^2=C_1e^{2\varphi}+C_2e^{-2\varphi}-\sigma b\varphi+a+A$, $(\psi'_y)^2=-C_2e^{2\psi}-C_1e^{-2\psi}-\sigma b\psi-A$
5	$a\sinh u+2b(\sinh u\arctan e^{u/2}+\cosh\frac{u}{2})$	$2\ln\left\lvert\tan\frac{z}{2}\right\rvert$	$(\varphi'_x)^2=C_1\sin 2\varphi+C_2\cos 2\varphi+\sigma b\varphi+a+A$, $(\psi'_y)^2=C_1\sin 2\psi-C_2\cos 2\psi+\sigma b\psi-A$

Notation: A, C_1, and C_2 are arbitrary constants; $\sigma=1$ for $z>0$, $\sigma=-1$ for $z<0$.

► Example 6.14. The nonlinear steady-state heat (diffusion) equation with a source

$$u_{xx} + u_{yy} = f(u)$$

can be investigated in the same way as the nonlinear Klein–Gordon equation (see Example 6.13). The main results are summarized in Table 6.3. These solutions were initially obtained using a different method by Miller & Rubel, 1993. For any arbitrary function $f(u)$, there exists a traveling wave solution, $u = u(k_1x + k_2y)$, as well as a solution possessing radial symmetry, $u = u(x^2 + y^2)$. ◄

► Example 6.15. Consider a steady-state heat equation in an inhomogeneous anisotropic medium with a nonlinear source:

$$[a(x)u_x]_x + [b(y)u_y]_y = f(u). \qquad (6.4.2.36)$$

We will look for functional separable solutions

$$u = u(z), \qquad z = \varphi(x) + \psi(y). \tag{6.4.2.37}$$

Substituting (6.4.2.37) into (6.4.2.36) yields the functional differential equation

$$\big[a(\varphi'_x)^2 + b(\psi'_y)^2\big]u''_{zz} + \big[(a\varphi'_x)'_x + (b\psi'_y)'_y\big]u'_z = f(u), \tag{6.4.2.38}$$

which can be rewritten as

$$p_1(x) + p_2(y) + [q_1(x) + q_2(y)]R(z) + S(z) = 0, \tag{6.4.2.39}$$

where

$$p_1(x) = [a(x)\varphi'_x]'_x, \quad p_2(y) = [b(y)\psi'_y]'_y, \quad q_1(x) = a(x)(\varphi'_x)^2,$$
$$q_2(y) = b(y)(\psi'_y)^2, \quad R(z) = u''_{zz}/u'_z, \quad S(z) = -f(u)/u'_z, \quad u = u(z).$$

Let us differentiate equation (6.4.2.39) with respect to y. Next, we divide the result by $\psi'_y R'_z$ and then differentiate with respect to y. As a result, we arrive at a functional differential equation dependent on two variables, y and z. This equation can be reduced to a standard bilinear functional equation, whose solutions can be found using the method outlined in Section 5.5.2.

Without performing a complete analysis of equation (6.4.2.36), we will focus on examining generalized separable solutions that arise for an arbitrary kinetic function $f(u)$. In this context, the functions $a(x)$ and $b(y)$, which define the anisotropic properties of the medium, as well as the solution (6.4.2.37), are to be determined.

The functional differential equation (6.4.2.38) reduces to an ODE if we require that the expressions in square brackets be functions of z:

$$(a\varphi'_x)'_x + (b\psi'_y)'_y = M(z), \quad a(\varphi'_x)^2 + b(\psi'_y)^2 = N(z). \tag{6.4.2.40}$$

On differentiating either equation with respect to x and y, we arrive at the equations $M''_{zz} = 0$ and $N''_{zz} = 0$, whose general solutions are

$$M(z) = C_1 z + C_2, \qquad N(z) = C_3 z + C_4, \tag{6.4.2.41}$$

where $C_1, \ldots, C_4$ are arbitrary constants.

In view of (6.4.2.40) and (6.4.2.41), the functional differential equation (6.4.2.38) reduces to the ODE

$$(C_3 z + C_4)u''_{zz} + (C_1 z + C_2)u'_z = f(u). \tag{6.4.2.42}$$

Substituting (6.4.2.41) into (6.4.2.40) yields

$$(a\varphi'_x)'_x + (b\psi'_y)'_y = C_1(\varphi + \psi) + C_2, \quad a(\varphi'_x)^2 + b(\psi'_y)^2 = C_3(\varphi + \psi) + C_4.$$

By separating the variables, we obtain a system of ordinary differential equations for $\varphi(x)$, $a(x)$, $\psi(y)$, and $b(y)$:

$$(a\varphi'_x)'_x - C_1\varphi - C_2 = k_1, \qquad (b\psi'_y)'_y - C_1\psi = -k_1,$$
$$a(\varphi'_x)^2 - C_3\varphi - C_4 = k_2, \qquad b(\psi'_y)^2 - C_3\psi = -k_2.$$

This system can always be integrated by quadratures and can be converted to the form

$$\begin{aligned}
&(C_3\varphi + C_4 + k_2)\varphi''_{xx} + (C_1\varphi + C_2 + k_1 - C_3)(\varphi'_x)^2 = 0,\\
&(C_3\psi - k_2)\psi''_{yy} + (C_1\psi - k_1 - C_3)(\psi'_y)^2 = 0;\\
&a = (C_3\varphi + C_4 + k_2)(\varphi'_x)^{-2},\\
&b = (C_3\psi - k_2)(\psi'_y)^{-2}.
\end{aligned} \tag{6.4.2.43}$$

Here, the equations for φ and ψ are independent of a and b and so can be solved separately. Without performing a complete analysis of system (6.4.2.43), we note a simple particular case when the system is integrable in explicit form.

For $C_1 = C_2 = C_4 = k_1 = k_2 = 0$ and $C_3 = C \neq 0$, we have

$$a(x) = \alpha e^{\mu x}, \quad b(y) = \beta e^{\nu y}, \quad \varphi(x) = \frac{Ce^{-\mu x}}{\alpha\mu^2}, \quad \psi(y) = \frac{Ce^{-\nu y}}{\beta\nu^2},$$

where α, β, μ, and ν are arbitrary constants. Then, equation (6.4.2.42) for $u(z)$ becomes

$$Czu''_{zz} = f(u).$$

System (6.4.2.43) also has alternative solutions that yield different expressions for the functions $a(x)$ and $b(y)$. Table 6.4 presents the cases where these functions can be expressed explicitly (Polyanin & Zaitsev, 2012). Note that the traveling wave solution, which corresponds to $a = \text{const}$ and $b = \text{const}$ has been omitted from this table. For clarity, the equations are expressed using the auxiliary variable $\zeta = \sqrt{z} = \sqrt{\varphi(x) + \psi(y)}$; this variable is introduced to ensure that the functional coefficient of the highest derivative is equal to unity. In the general case, the solution of system (6.4.2.38) results in functions $a(x)$ and $b(y)$ that can be represented in parametric form (Polyanin & Zhurov, 1998). ◄

Remark 6.7. *The results detailed in Example 6.15 also apply to nonlinear hyperbolic equations, where the variable $y = t$ represents time, and the functional coefficients $a(x)$ and $b(y)$ have opposite signs, such that $a(x)b(y) < 0$. In this case, the functions $\varphi(x)$ and $\psi(y)$ provided for the first three equations in Table 6.4 also have opposite signs, meaning that $\varphi(x)\psi(y) < 0$. The sign of the parameter C is selected to ensure that the condition $\varphi(x) + \psi(y) \geq 0$ is satisfied in the region of interest.*

► Example 6.16. Let us consider the nonlinear Monge–Ampère type equation

$$u_{xx}u_{yy} - u_{xy}^2 = f(u). \tag{6.4.2.44}$$

Assuming that the function $f(u)$ is arbitrary, we look for solutions of equation (6.4.2.44) in the form

$$u = u(z), \qquad z = \varphi(x) + \psi(y). \tag{6.4.2.45}$$

Substituting (6.4.2.45) into (6.4.2.44) yields the functional differential equation

$$\big[(\varphi'_x)^2\psi''_{yy} + (\psi'_y)^2\varphi''_{xx}\big]u'_z u''_{zz} + \varphi''_{xx}\psi''_{yy}(u'_z)^2 = f(u). \tag{6.4.2.46}$$

Table 6.4. Functional separable solutions of the form $u = u(\zeta)$ with $\zeta = \sqrt{\varphi(x) + \psi(y)}$ for the heat equation for an inhomogeneous anisotropic medium and with an arbitrary nonlinear source.

Heat equation	Functions $\varphi(x)$ and $\psi(y)$	Equation for $u=u(\zeta)$
$(\alpha x^n u_x)_x + (\beta y^k u_y)_y = f(u)$	$\varphi(x)=\dfrac{Cx^{2-n}}{\alpha(2-n)^2}$, $\psi(y)=\dfrac{Cy^{2-k}}{\beta(2-k)^2}$	$u''_{\zeta\zeta} + \dfrac{4-nk}{(2-n)(2-k)}\dfrac{1}{\zeta}u'_\zeta = \dfrac{4}{C}f(u)$
$(\alpha e^{\mu x} u_x)_x + (\beta e^{\nu y} u_y)_y = f(u)$	$\varphi(x)=\dfrac{C}{\alpha\mu^2}e^{-\mu x}$, $\psi(y)=\dfrac{C}{\beta\nu^2}e^{-\nu y}$	$u''_{\zeta\zeta} - \dfrac{1}{\zeta}u'_\zeta = \dfrac{4}{C}f(u)$
$(\alpha e^{\mu x} u_x)_x + (\beta y^k u_y)_y = f(u)$	$\varphi(x)=\dfrac{C}{\alpha\mu^2}e^{-\mu x}$, $\psi(y)=\dfrac{Cy^{2-k}}{\beta(2-k)^2}$	$u''_{\zeta\zeta} + \dfrac{k}{2-k}\dfrac{1}{\zeta}u'_\zeta = \dfrac{4}{C}f(u)$
$(\alpha x^2 u_x)_x + (\beta y^2 u_y)_y = f(u)$	$\varphi(x)=\mu\ln\lvert x\rvert$, $\psi(y)=\nu\ln\lvert y\rvert$	Equation (6.4.2.38), both expressions in square brackets are constant
$\alpha u_{xx} + (\beta y^2 u_y)_y = f(u)$	$\varphi(x)=\mu x$, $\psi(y)=\nu\ln\lvert y\rvert$	Equation (6.4.2.38), both expressions in square brackets are constant
Notation: C, α, β, μ, ν, n, and k are free parameters ($C\neq0$, $\mu\neq0$, $\nu\neq0$, $n\neq2$, and $k\neq2$)		

If we set

$$\begin{aligned}(\varphi'_x)^2\psi''_{yy} + (\psi'_y)^2\varphi''_{xx} &= a(z),\\ \varphi''_{xx}\psi''_{yy} &= b(z),\end{aligned} \tag{6.4.2.47}$$

then equation (6.4.2.46) reduces to the ODE:

$$a(z)u'_z u''_{zz} + b(z)(u'_z)^2 = f(u). \tag{6.4.2.48}$$

Relations (6.4.2.47) represent a system of functional differential equations for $a = a(z)$, $b = b(z)$, $\varphi = \varphi(x)$, and $\psi = \psi(y)$, where $z = \varphi(x) + \psi(y)$.

Let us first examine the second equation of the system (6.4.2.47), which is easier to analyze. We take the logarithm of both sides and then differentiate once with respect to either independent variable. As a result, we obtain the following two relations:

$$\begin{aligned}\frac{\varphi'''_{xxx}}{\varphi''_{xx}} &= \varphi'_x\frac{b'_z}{b} \quad \text{(after differentiation in } x\text{)},\\ \frac{\psi'''_{yyy}}{\psi''_{yy}} &= \psi'_y\frac{b'_z}{b} \quad \text{(after differentiation in } y\text{)}.\end{aligned} \tag{6.4.2.49}$$

Separating the variables results in three ODEs:

$$\frac{\varphi'''_{xxx}}{\varphi'_x\varphi''_{xx}} = C, \qquad \frac{\psi'''_{yyy}}{\psi'_y\psi''_{yy}} = C, \qquad \frac{b'_z}{b} = C, \tag{6.4.2.50}$$

where C is an arbitrary constant. Integrating these equations yields

$$\varphi''_{xx} = A_1 e^{C\varphi}, \qquad \psi''_{yy} = A_2 e^{C\psi}, \qquad b = A_3 e^{Cz}, \tag{6.4.2.51}$$

where A_1, A_2, and A_3 are arbitrary constants.

We will examine the cases $C = 0$ and $C \neq 0$ separately.

1°. For $C = 0$, integrating the first two equations of (6.4.2.51) yields the functions appearing in (6.4.2.45):

$$\varphi = \tfrac{1}{2}A_1x^2 + B_1x, \qquad \psi = \tfrac{1}{2}A_2y^2 + B_2y, \tag{6.4.2.52}$$

where B_1 and B_2 are arbitrary constants; the inessential additive constants are omitted. Substituting (6.4.2.52) into (6.4.2.47) and rearranging taking into account the second relation in (6.4.2.45), we find the functional coefficients:

$$a(z) = 2A_1A_2z + A_1B_2^2 + A_2B_1^2, \qquad b(z) = A_1A_2 = \text{const}, \tag{6.4.2.53}$$

Inserting (6.4.2.53) into (6.4.2.48) yields an ODE for $u = u(z)$.

2°. For $C \neq 0$, the first two equations of (6.4.2.51) have the first integrals

$$(\varphi'_x)^2 = \frac{2A_1}{C}e^{C\varphi} + B_1, \qquad (\psi'_y)^2 = \frac{2A_2}{C}e^{C\psi} + B_2, \tag{6.4.2.54}$$

where B_1 and B_2 are arbitrary constants. Setting $B_1 = B_2 = 0$, substituting (6.4.2.51) and (6.4.2.54) into (6.4.2.47), and rearranging with the help of the second relation in (6.4.2.45), we obtain the functional coefficients

$$a(z) = \frac{4A_1A_2}{C}e^{Cz}, \qquad b(z) = A_1A_2e^{Cz}. \tag{6.4.2.55}$$

Inserting (6.4.2.55) into (6.4.2.48) yields an ODE for $u = u(z)$. Integrating (6.4.2.54) with $B_1 = B_2 = 0$, we determine the functions appearing in (6.4.2.45):

$$\varphi(x) = -\frac{2}{C}\ln\bigl(K_1 \pm \tfrac{1}{2}\sqrt{2A_1C}\,x\bigr), \qquad \psi(y) = -\frac{2}{C}\ln\bigl(K_2 \pm \tfrac{1}{2}\sqrt{2A_2C}\,y\bigr),$$

where K_1 and K_2 are arbitrary constants. ◀

❖ *Problems and Exercises for Section 6.4*

In all the problems presented below, exact solutions are sought in the form $u = u(z)$, where $z = \varphi(x) + \psi(t)$. The functional coefficients appearing in the equation are also subject to determination.

1. Find functional separable solutions of the following nonlinear first-order PDEs:
a) $u_t = f(u)u_x + g(u)$,
b) $u_t = f(u)u_x^2 + g(u)$,
c) $u_t = f(u)u_x^2 + g(x)$.

2. Find functional separable solutions of the following nonlinear heat equations:

a) $u_t = u_{xx} + f(u)$,
b) $u_t = a(x)u_{xx} + b(x)u_x + f(u)$,
c) $u_t = [f(u)u_x]_x + g(u)$,
d) $u_t = [f(u)u_x]_x + g(u)u_x$.

3. Find functional separable solutions of two nonlinear heat equations:

a) $u_t = x^{-n}[x^n f(u)u_x]_x$,
b) $u_t = x^{-n}[x^n f(u)u_x]_x + g(u)$.

4. Find functional separable solutions of the following nonlinear parabolic equations:

a) $u_t = [f(u)u_x^n]_x$,
b) $u_t = [f(u)u_x^n]_x + g(u)$.

5. Find functional separable solutions of the following nonlinear hyperbolic equations:

a) $u_{xt} = f(u)$,
b) $u_{xt} = u_x u_t f(u)$,
c) $u_{xt} = u_x^n f(u)$.

6. Prove by direct verification that the nonlinear Monge–Ampère type equation (6.4.2.44) admits a functional separable solution of the following form (Polyanin & Zaitsev, 2012):

$$u = u(z), \qquad z = \alpha_1 x^2 + \alpha_2 xy + \alpha_3 y^2 + \beta_1 x + \beta_2 y,$$

where α_1, α_2, α_3, β_1, and β_2 are arbitrary constants.

6.5. Construction of Functional Separable Solutions in Implicit Form

6.5.1. Preliminary Remarks. Traveling Wave Solutions in Implicit Form

The method for constructing functional separable solutions in implicit form is based on generalizing traveling wave solutions for various nonlinear partial differential equations. Before explaining this method in detail, we will present two simple examples that illustrate the existence of solutions in implicit form for nonlinear diffusion and wave equations.

► Example 6.17. Consider the nonlinear heat equation

$$u_t = [f(u)u_x]_x, \tag{6.5.1.1}$$

which involves an arbitrary function $f(u)$. The equation is explicitly independent of x and t and has a traveling wave solution

$$u = u(z), \qquad z = \lambda t + \kappa x, \tag{6.5.1.2}$$

where κ and λ are arbitrary constants. Substituting (6.5.1.2) into (6.5.1.1) yields the ODE $\lambda u'_z = \kappa^2[f(u)u'_z]'_z$. Integrating it, we obtain a solution in implicit form:

$$\kappa^2 \int \frac{f(u)\,du}{\lambda u + C_1} = \lambda t + \kappa x + C_2, \tag{6.5.1.3}$$

where C_1 and C_2 are arbitrary constants. The variable z on the right-hand side of solution (6.5.1.3) has been replaced with the initial variables using (6.5.1.2).

It is evident that even for simple functions such as $f(u) = u$, $f(u) = e^u$, $f(u) = \ln u$, $f(u) = \sin u$, and $f(u) = \cos u$, the solution (6.5.1.3) of equation (6.5.1.1) cannot be expressed explicitly in terms of elementary functions. Therefore, seeking exact solutions for more complex diffusion-type equations in an explicit form appears to be ineffective. ◄

► Example 6.18. The nonlinear Klein–Gordon equation

$$u_{tt} = u_{xx} + f(u), \tag{6.5.1.4}$$

where $f(u)$ is an arbitrary function, also admits traveling wave solutions $u = u(z)$ with $z = \lambda t + \kappa x$ (such that $\lambda \neq \pm\kappa$) that can be represented in implicit form:

$$\int \left[C_1 + \frac{2}{\lambda^2 - \kappa^2} F(u)\right]^{-1/2} du = C_2 \pm (\lambda t + \kappa x), \tag{6.5.1.5}$$
$$F(u) = \int f(u)\, du.$$

◄

Examples 6.17 and 6.18 demonstrate that the nonlinear equations (6.5.1.1) and (6.5.1.4), which are commonly encountered in various applications, possess traveling wave solutions. These solutions can be expressed in implicit form. It is important to emphasize that these solutions cannot be represented in explicit form in the general case with an arbitrary function $f(u)$.

Remark 6.8. *In practice, a number of methods are commonly employed to construct exact solutions of nonlinear partial differential equations by specifying the explicit form of the solutions a priori. These methods include the tanh-function method, the exp-function method, and the sin- and cos-function methods, among others. Free parameters are incorporated into these solutions, and their values are typically determined using the method of undetermined coefficients, often with the aid of computer algebra systems.*

However, these highly specialized direct methods have a very limited range of applicability because they prescribe the form of the solution 'blindly,' without considering the specific properties of the nonlinear equations being considered. This limitation is clearly illustrated by Examples 6.17 and 6.18, where, in general, exact solutions cannot be represented in explicit form at all.

Section 6.5.2 will describe a method for constructing exact solutions to nonlinear equations of mathematical physics. This method is based on a significant generalization of the traveling wave solutions considered in Examples 6.17–6.18.

6.5.2. Direct Method for Constructing Functional Separable Solutions in Implicit Form. Description

We will consider nonlinear partial differential equations of the general form

$$G(x, u_x, u_t, u_{xx}, u_{xt}, u_{tt}, \dots) = 0. \tag{6.5.2.1}$$

We will look for exact solutions in the implicit form

$$\int h(u)\,du = \xi(x)\omega(t) + \eta(x), \tag{6.5.2.2}$$

where $h = h(u)$, $\xi = \xi(x)$, $\eta = \eta(x)$, and $\omega = \omega(t)$ some functions that need to be determined in the subsequent analysis.

Remark 6.9. *The representation of solutions in the implicit form (6.5.2.2) is a significant generalization of the traveling wave solutions (6.5.1.3) and (6.5.1.5) of the nonlinear PDEs (6.5.1.1) and (6.5.1.4) that were discussed earlier. For instance, the representation of solutions in the form (6.5.2.2) derives from generalizing the solution in (6.5.1.3) of equation (6.5.1.1). This is implemented as follows:*

$$\frac{\kappa^2 f(u)}{\lambda u + C_1} \Longrightarrow h(u), \quad \lambda \Longrightarrow \xi(x), \quad t \Longrightarrow \omega(t), \quad \kappa x + C_2 \Longrightarrow \eta(x).$$

Let us outline the procedure for constructing exact solutions in the implicit form (6.5.2.2). First, we use (6.5.2.2) to calculate the partial derivatives u_x, u_t, u_{xx}, and so on, which are expressed in terms of the functions h, ξ, η, and ω, and their derivatives. Next, we substitute these partial derivatives into equation (6.5.2.1) and eliminate the variable t using the expression from (6.5.2.2). By making an appropriate choice of ω, we ultimately derive a bilinear functional differential equation of the form

$$\begin{gathered}\sum_{j=1}^{N} \Phi_j[x]\Psi_j[u] = 0,\\ \Phi_j[x] \equiv \Phi_j(x, \xi, \eta, \xi'_x, \eta'_x, \xi''_{xx}, \eta''_{xx} \dots),\\ \Psi_j[u] \equiv \Psi_j(u, h, h'_u, h''_{uu}, \dots).\end{gathered} \tag{6.5.2.3}$$

Here, $\Phi_j[x]$ and $\Psi_j[u]$ differential forms (or, sometimes, functional coefficients) that depend on only x and u, respectively. The following proposition holds true.

Splitting principle. Functional differential equations of the form (6.5.2.3) can have solutions only if the forms $\Psi_j[u]$ $(j = 1, \dots, N)$ are connected by some linear relations:

$$\sum_{j=1}^{m_i} k_{ij}\Psi_j[u] = 0, \qquad i = 1, \dots, n, \tag{6.5.2.4}$$

where k_{ij} are some constants, $1 \le m_i \le N - 1$, and $1 \le n \le N - 1$. Additionally, it is necessary to consider the degenerate cases where, besides linear relations, the individual differential forms $\Psi_j[u]$ may vanish.

The splitting principle is also valid for the forms $\Phi_j[x]$.

For a more detailed explanation of the splitting principle, refer to Sections 5.5.1 and 5.5.2.

The splitting principle will be employed further to construct solutions of certain functional differential equations of the form (6.5.2.3), which arise when seeking exact solutions to nonlinear diffusion equations.

Remark 6.10. *Constructing a solution in implicit form, with an integral term on the left-hand side of the equality (6.5.2.2), often results in lower-order differential equations for the function h compared to when seeking exact solutions in explicit form. Additionally, the implicit representation of solutions typically leads to simpler explicit expressions for the functions f and g in terms of h. In contrast, when searching for exact solutions in explicit form, the functions f and g are often expressed parametrically in terms of u. It is also important to note that, in general, different linear relations of the form (6.5.2.4) correspond to different solutions of the partial differential equation under consideration.*

6.5.3. Nonlinear Reaction-Diffusion Equations with Variable Coefficients

The class of nonlinear reaction-diffusion PDEs in question. We will consider one-dimensional nonlinear reaction-diffusion equations with variable coefficients of the form

$$c(x)u_t = [a(x)f(u)u_x]_x + b(x)g(u). \qquad (6.5.3.1)$$

It is noteworthy that equation (6.5.3.1), with $a(x) = b(x) = c(x) = x^n$, describes reaction-diffusion processes exhibiting radial symmetry in the two-dimensional ($n = 1$) and three-dimensional ($n = 2$) cases.

The arguments of the functions $a = a(x)$, $b = b(x)$, $c = c(x)$, $f = f(u)$, $g = g(u)$, $h = h(u)$, $\xi = \xi(x)$, $\eta = \eta(x)$, and $\omega = \omega(t)$ appearing in equation (6.5.3.1) and solution (6.5.2.2) will often be omitted below.

In what follows, the main focus will be on constructing exact solutions to nonlinear reaction-diffusion equations of the fairly general form (6.5.3.1) that depend on one or two arbitrary functions. Notably, exact solutions to equations of mathematical physics that involve arbitrary functions and have significant generality, are of considerable practical interest. They enable the evaluation of the accuracy of various numerical and approximate analytical methods used to solve the relevant initial-boundary value problems.

Derivation of the functional differential equation. We seek exact solutions for the class of reaction-diffusion equations (6.5.3.1) in the implicit form (6.5.2.2). Differentiating (6.5.2.2) with respect to t and x, we obtain

$$\begin{aligned}
hu_t &= \xi\omega'_t \quad \Longrightarrow \quad u_t = \frac{\xi\omega'_t}{h};\\
hu_x &= \xi'_x\omega + \eta'_x \quad \Longrightarrow \quad u_x = \frac{\xi'_x\omega + \eta'_x}{h};\\
(afu_x)_x &= \Big[(a\xi'_x\omega + a\eta'_x)\frac{f}{h}\Big]_x =\\
&= [(a\xi'_x)'_x\omega + (a\eta'_x)'_x]\frac{f}{h} + a(\xi'_x\omega + \eta'_x)^2\frac{1}{h}\Big(\frac{f}{h}\Big)'_u.
\end{aligned} \qquad (6.5.3.2)$$

Inserting these expressions into (6.5.3.1), we arrive at the functional differential equation

$$\omega'_t = Q_1(x,u)\omega^2 + Q_2(x,u)\omega + Q_3(x,u). \qquad (6.5.3.3)$$

The functions Q_n are explicitly independent of t and are defined by the formulas

$$\begin{aligned} Q_1(x,u) &= \frac{a(\xi'_x)^2}{c\xi}\Big(\frac{f}{h}\Big)'_u, \\ Q_2(x,u) &= \frac{1}{c\xi}\Big[(a\xi'_x)'_x f + 2a\xi'_x\eta'_x\Big(\frac{f}{h}\Big)'_u\Big], \\ Q_3(x,u) &= \frac{1}{c\xi}\Big[(a\eta'_x)'_x f + a(\eta'_x)^2\Big(\frac{f}{h}\Big)'_u + bgh\Big]. \end{aligned} \tag{6.5.3.4}$$

Equation (6.5.3.3) with the functions (6.5.3.4) depends on three variables: t, x, and u. It includes undetermined functions and their derivatives with different arguments, all connected by the additional relation (6.5.2.2). This equation is more complex than equations of the form (6.5.2.3).

The functional differential equation (6.5.3.3) with the functions (6.5.3.4) significantly simplifies in the following two cases:

$$\text{(i) } \xi'_x = 0, \qquad \text{(ii) } (f/h)'_u = 0. \tag{6.5.3.5}$$

Let us examine the first case in detail.

Generalized traveling wave solutions for $\omega(t) = kt$. When $\xi'_x = 0$, we can set $\xi = 1$ without loss of generality. Substituting $\xi = 1$ into formulas (6.5.3.4), we get $Q_1(x,u) = Q_2(x,u) = 0$. As a result, equation (6.5.3.3) becomes

$$\omega'_t = \frac{1}{c}\Big[(a\eta'_x)'_x f + a(\eta'_x)^2\Big(\frac{f}{h}\Big)'_u + bgh\Big]. \tag{6.5.3.6}$$

The functional differential equation (6.5.3.6) is separated; its left-hand side depends only on t, while the right-hand side depends on x and u. Therefore, we can set $\omega'_t = k = \text{const}$, which gives $\omega(t) = kt$. Substituting this function and $\xi(x) = 1$ into (6.5.2.2), we obtain a generalized traveling wave solution defined in parametric form:

$$\int h(u)\,du = kt + \int \theta(x)\,dx. \tag{6.5.3.7}$$

The functions $h(u)$ and $\theta(x) = \eta'_x(x)$ will be determined in the subsequent analysis from the functional differential equation

$$(a\theta)'_x f + a\theta^2\Big(\frac{f}{h}\Big)'_u + bgh - kc = 0, \tag{6.5.3.8}$$

which results from substituting $\omega(t) = kt$ and $\theta(x) = \eta'_x(x)$ into relation (6.5.3.6). Equation (6.5.3.8) represents a bilinear functional differential equation (6.5.2.3) with $N = 4$, which reduces to the functional equation (5.5.2.3); its solutions are defined by formulas (5.5.2.4) and (5.5.2.5).

Solution 1. First, consider the degenerate case in which the differential form $(f/h)'_u$ in equation (6.5.3.8) is zero. In this situation, according to the splitting principle (which will not be mentioned further), equation (6.5.3.8) has solutions under the following conditions:

$$h = f, \quad g = A + \frac{B}{f}, \quad (a\theta)'_x + Ab = 0, \quad Bb - kc = 0, \tag{6.5.3.9}$$

where A and B are arbitrary constants. It follows from relations (6.5.3.9) with $b(x) = c(x) = 1$ that the equation

$$u_t = [a(x)f(u)u_x]_x + A + \frac{k}{f(u)}, \tag{6.5.3.10}$$

which involves two arbitrary functions, $a(x)$ and $f(u)$, has a generalized traveling wave solution in implicit form

$$\int f(u)\,du = kt - A\int \frac{x\,dx}{a(x)} + C_1 \int \frac{dx}{a(x)} + C_2, \tag{6.5.3.11}$$

where C_1 and C_2 are arbitrary constants.

Solution 2. Equation (6.5.3.8) holds identically if we set

$$f = A, \quad g = \frac{1}{h}\left(\frac{f}{h}\right)'_u, \quad A(a\theta)'_x - kc = 0, \quad b = -a\theta^2, \tag{6.5.3.12}$$

where A is an arbitrary constant.

By utilizing relation (6.5.3.12) with $c(x) = 1$ and $A = k = 1$, one can obtain the nonlinear reaction-diffusion equation

$$u_t = [a(x)u_x]_x - \frac{x^2}{a(x)}g(u), \tag{6.5.3.13}$$

where $a(x)$ is an arbitrary function. The function $g(u)$ is expressed in terms of the arbitrary function $h = h(u)$ as

$$g(u) = -h^{-3}h'_u. \tag{6.5.3.14}$$

Equation (6.5.3.13) subject to condition (6.5.3.14) admits the exact solution

$$\int h(u)\,du = t + \int \frac{x\,dx}{a(x)} + C_1. \tag{6.5.3.15}$$

On solving relation (6.5.3.14) for h, we find two functions

$$h(u) = \pm\left(2\int g(u)\,du + C_2\right)^{-1/2}.$$

Using this relation to eliminate h from (6.5.3.15), we write generalized traveling solutions to equation (6.5.3.13) in implicit form:

$$\pm\int \left(2\int g(u)\,du + C_2\right)^{-1/2} du = t + \int \frac{x\,dx}{a(x)} + C_1. \tag{6.5.3.16}$$

Here, $a(x)$ and $g(u)$ are arbitrary functions, and C_1 and C_2 are arbitrary constants.

Solution 3. Equation (6.5.3.8) holds identically if we set

$$\left(\frac{f}{h}\right)'_u = A, \quad g = -\frac{f}{h}, \quad Aa\theta^2 = kc, \quad (a\theta)'_x = b, \tag{6.5.3.17}$$

where A is an arbitrary constant. Putting $c(x) = 1$, $A = 4k$, and $h(u) = f(u)/(4ku)$, we obtain two nonlinear reaction-diffusion equations

$$u_t = [a(x)f(u)u_x]_x \mp k\frac{a'_x(x)}{\sqrt{a(x)}}u, \tag{6.5.3.18}$$

which involve two arbitrary functions, $a(x)$ and $f(u)$, and an arbitrary constant k. Considering that $h(u) = f(u)/(4ku)$, we can represent exact solutions to these equations in implicit form:

$$\int \frac{f(u)}{u}\,du = 4k^2t \pm 2k\int \frac{dx}{\sqrt{a(x)}} + C_1, \tag{6.5.3.19}$$

where $C_1 = 4Ck$ is an arbitrary constant.

► Example 6.19. Substituting $a(x) = x^{2\beta}$ and $k = \mp\alpha/(2\beta)$ into (6.5.3.18) and (6.5.3.19) yields the equation

$$u_t = [x^{2\beta}f(u)u_x]_x + \alpha x^{\beta-1}u, \qquad \beta \neq 0,$$

which depends on an arbitrary function $f(u)$ and admits exact solutions in implicit form:

$$\int \frac{f(u)}{u}\,du = \begin{cases} \dfrac{\alpha^2}{\beta^2}t - \dfrac{\alpha}{\beta(1-\beta)}x^{1-\beta} + C_1 & \text{for } \beta \neq 1, \\ \alpha^2 t - \alpha\ln|x| + C_1 & \text{for } \beta = 1. \end{cases}$$ ◄

► Example 6.20. Substituting $a(x) = e^{2\beta x}$ and $k = \mp\alpha/(2\beta)$ into (6.5.3.18) and (6.5.3.19) yields the equation

$$u_t = [e^{2\beta x}f(u)u_x]_x + \alpha e^{\beta x}u, \qquad \beta \neq 0,$$

which depends on an arbitrary function $f(u)$ and admits exact solutions in implicit form:

$$\int \frac{f(u)}{u}\,du = \frac{\alpha^2}{\beta^2}t + \frac{\alpha}{\beta^2}e^{-\beta x} + C_1.$$ ◄

Other solutions of the form (6.5.3.7) for the diffusion equation (6.5.3.1). The nonlinear reaction-diffusion equation (6.5.3.1) admits some other generalized traveling wave solutions of the form (6.5.3.7), which corresponds to case (i) in (6.5.3.5). These solutions can be found in the studies by Polyanin (2019a) and Polyanin & Zhurov (2022), which also provide more complex solutions of the form (6.5.2.2) with $\xi(x) = 1$ and the functions $\omega(t) = ke^{\lambda t}$ and $\omega(t) = k\ln t$.

The reaction-diffusion equation (6.5.3.1) also admits exact solutions of the form (6.5.2.2) for $h(u) = f(u)$, which corresponds to case (ii) in (6.5.3.5). In particular, the studies cited above provide solutions (6.5.2.2) with $h(u) = f(u)$ and the functions $\omega(t) = t$, $\omega(t) = ke^{\lambda t}$, and $\omega(t) = kt^{\beta}$.

Functional separation of variables in general form. Implicit representation of solutions. To construct exact solutions of equation (6.5.2.1), one can first introduce a new variable ϑ, which relates to the original unknown function u through the nonlinear transformation

$$\vartheta = \int \zeta(u)\,du, \qquad (6.5.3.20)$$

The functions $\vartheta = \vartheta(x,t)$ and $\zeta = \zeta(u)$ are sought in the subsequent analysis using the generalized splitting method; these are described by overdetermined PDE and ODE systems, respectively. Once these functions are identified, the integral relation (6.5.3.20) will provide an implicit exact solution to the equation in question.

Notably, using the nonlinear transformation (6.5.3.20) in the first step further generalizes the approach based on the representation of solutions in the implicit form (6.5.2.2). Indeed, by setting

$$\vartheta = \xi(x)\omega(t) + \eta(x), \qquad \zeta(u) = h(u)$$

in (6.5.3.20), we arrive at (6.5.2.2).

The effectiveness of the nonlinear transformation defined by equation (6.5.3.20) was demonstrated by Polyanin (2020) and Polyanin & Zhurov (2020, 2022). The studies presented exact solutions for a more general reaction-diffusion equation that included an additional convective term; this equation is more complex than equation (6.5.3.1).

Problems and Exercises for Section 6.5

1. Derive an ODE that describes a traveling wave solution for the generalized Burgers equation

$$u_t + f(u)u_x = u_{xx},$$

where $f(u)$ is an arbitrary function. Demonstrate that the general solution of the resulting ODE can be expressed in implicit form. Find the simplest elementary functions $f(u)$ for which this solution can be represented explicitly.

2. Derive an ODE that describes a traveling wave solution for the nonlinear wave equation

$$u_{tt} = [f(u)u_x]_x,$$

where $f(u)$ is an arbitrary function. Demonstrate that the general solution of the resulting ODE can be expressed in implicit form. Find the simplest elementary functions $f(u)$ for which this solution can be represented explicitly.

3. Please perform detailed calculations for solutions 1–3 of equation (6.5.3.1) on your own.

4. Consider the nonlinear convective-diffusion equation

$$c(x)u_t = [a(x)f(u)u_x]_x + b(x)g(u)u_x.$$

Derive a functional differential equation that arises from searching for its generalized traveling wave solutions in the implicit form (6.5.3.7).

5. Find solutions to the functional differential equation from the previous problem and construct related exact solutions to the convective-diffusion equation discussed above.

6. Consider the nonlinear wave-type equation

$$c(x)u_{tt} = [a(x)f(u)u_x]_x + b(x)g(u)u_x.$$

Derive a functional differential equation that arises from searching for its generalized traveling wave solutions in the implicit form (6.5.3.7).

7. Find solutions to the functional differential equation from the previous problem with $h(u) = f(u)$ and construct related exact solutions to the wave-type equation discussed above.

► Concluding remarks.

The methods for finding functional separable solutions to nonlinear partial differential equations are discussed in the book by Polyanin & Zhurov (2022) much more comprehensively than in this chapter. Additional information can also be found in Polyanin & Zaitsev (2012). Various equations and their exact solutions employing functional separation of variables are presented, for example, in the studies by Grundland & Infeld (1992), Miller & Rubel (1993), Zhdanov (1994), Doyle & Vassiliou (1998), Pucci & Saccomandi (2000), Estévez et al. (2002), Galaktionov & Svirshchevskii (2007), Hu & Qu (2007), Polyanin & Zaitsev (2012), Polyanin (2019a, 2019b, 2019c, 2019d, 2020, 2025), Zhurov & Polyanin (2020), and Polyanin & Zhurov (2022).

Functional separable solutions for various nonlinear PDEs with delay are available in Polyanin & Zhurov (2014), Polyanin & Sorokin (2021), Polyanin et al. (2024), and Polyanin (2025).

References for Chapter 6

Andreev V.K., Kaptsov O.V., Pukhnachov V.V., and Rodionov A.A. *Applications of Group-Theoretical Methods in Hydrodynamics*. Kluwer, Dordrecht, 1998.

Doyle P.W. and Vassiliou P.J. Separation of variables for the 1-dimensional non-linear diffusion equation. *Int. J. Non-Linear Mech.*, 1998, Vol. 33, No. 2, pp. 315–326.

Estévez P.G., Qu C.Z., and Zhang S.L. Separation of variables of a generalized porous medium equation with nonlinear source. *J. Math. Anal. Appl.*, 2002, Vol. 275, pp. 44–59.

Galaktionov V.A. and Svirshchevskii S.R. *Exact Solutions and Invariant Subspaces of Nonlinear Partial Differential Equations in Mechanics and Physics*. Chapman & Hall/CRC Press, Boca Raton, 2007.

Grundland A.M. and Infeld E. A family of non-linear Klein–Gordon equations and their solutions. *J. Math. Phys.*, 1992, Vol. 33, pp. 2498–2503.

Hu J. and Qu C. Functionally separable solutions to nonlinear wave equations by group foliation method. *J. Math. Anal. Appl.*, 2007, Vol. 330, pp. 298–311.

Miller W. (Jr.) and Rubel L.A. Functional separation of variables for Laplace equations in two dimensions. *J. Phys. A.*, 1993, Vol. 26, pp. 1901–1913.

Polyanin A.D. Construction of exact solutions in implicit form for PDEs: New functional separable solutions of non-linear reaction-diffusion equations with variable coefficients. *Int. J. Non-Linear Mech.*, 2019a, Vol. 111, pp. 95–105.

Polyanin A.D. Construction of functional separable solutions in implicit form for non-linear Klein–Gordon type equations with variable coefficients. *Int. J. Non-Linear Mech.*, 2019b, Vol. 114, pp. 29–40.

Polyanin A.D. Functional separable solutions of nonlinear convection-diffusion equations with variable coefficients. *Commun. Nonlinear Sci. Numer. Simulat.*, 2019c, Vol. 73, pp. 379–390.

Polyanin A.D. Functional separable solutions of nonlinear reaction-diffusion equations with variable coefficients. *Applied Math. & Comput.*, 2019d, Vol. 347, pp. 282–292.

Polyanin A.D. Functional separation of variables in nonlinear PDEs: General approach, new solutions of diffusion-type equations. *Mathematics*, 2020, Vol. 8, No. 1, 90.

Polyanin A.D. *Handbook of Exact Solutions to Mathematical Equations*. CRC Press, Boca Raton, 2025.

Polyanin A.D. and Sorokin V.G. Construction of exact solutions to nonlinear PDEs with delay using solutions of simpler PDEs without delay. *Commun. Nonlinear Sci. Numer. Simul.*, 2021, Vol. 95, 105634.

Polyanin A.D., Sorokin V.G., and Zhurov A.I. *Delay Ordinary and Partial Differential Equations*. CRC Press, Boca Raton–London, 2024.

Polyanin A.D. and Zaitsev V.F. *Handbook of Nonlinear Partial Differential Equations, 2nd ed.,* Chapman & Hall/CRC Press, Boca Raton–London, 2012.

Polyanin A.D., Zaitsev V.F., and Zhurov A.I. *Solution Methods for Nonlinear Equations of Mathematical Physics and Mechanics* (in Russian). Fizmatlit, Moscow, 2005.

Polyanin A.D. and Zhurov A.I. The functional constraints method: Application to non-linear delay reaction-diffusion equations with varying transfer coefficients. *Int. J. Non-Linear Mechanics*, 2014, Vol. 67, pp. 267–277.

Polyanin A.D. and Zhurov A.I. Separation of variables in PDEs using nonlinear transformations: Applications to reaction-diffusion type equations. *Appl. Math. Lett.*, 2020, Vol. 100, 106055.

Polyanin A.D. and Zhurov A.I. Exact solutions to nonlinear equations of mechanics and mathematical physics. *Doklady Physics*, 1998, Vol. 43, No. 6, pp. 381–385.

Polyanin A.D. and Zhurov A.I. *Separation of Variables and Exact Solutions to Nonlinear PDEs*. CRC Press, Boca Raton–London, 2022.

Pucci E. and Saccomandi G. Evolution equations, invariant surface conditions and functional separation of variables. *Physica D*, 2000, Vol. 139, pp. 28–47.

Zhdanov R.Z. Separation of variables in the non-linear wave equation. *J. Phys. A*, 1994, Vol. 27, pp. L291–L297.

Zhurov A.I. and Polyanin A.D. Symmetry reductions and new functional separable solutions of nonlinear Klein–Gordon and telegraph type equations. *J. Nonlinear Math. Phys.*, 2020, Vol. 27, pp. 1–16.

7. Direct Method of Symmetry Reductions. Weak Symmetries

► **Preliminary remarks.** This chapter is devoted to the direct method of symmetry reductions and the direct method of weak symmetry reductions. As in other chapters, the presentation of the material is carried out in accordance with the principle 'from simple to complex'. To facilitate understanding of the method, the chapter begins with a description of a simplified scheme of the first method. In this case, the general structure of the solution is specified, including several desired functions, and it is required that one of them satisfy a single ordinary differential equation. The general form of reductions is further considered. Then the direct method of weak symmetry reductions is described, which combines the first method with the method of functional separation of variables. Unlike the direct method of symmetry reductions, using weak symmetry reductions assumes that the outer function can satisfy an overdetermined system of several ordinary differential equations rather than a single equation. This approach often allows one to obtain significantly more solutions, many of which can be represented in a closed form. The application of the methods is illustrated with numerous examples and tables, and the focus is, as before, on nonlinear equations of hydrodynamics, heat and mass transfer, and wave theory.

7.1. Direct Method of Symmetry Reductions

7.1.1. Simplified Scheme. Korteweg–de Vries and Boussinesq Equations

Before we proceed to describe the *direct method of symmetry reductions*[*] (also known as *Clarkson–Kruskal direct method*), let us first examine a simplified scheme.

The simplified scheme's basic idea is that exact solutions of partial differential equations in two independent variables, x and t, are sought in the form

$$u = f(t)w(z) + g(x,t), \qquad z = \varphi(t)x + \psi(t). \tag{7.1.1.1}$$

The functions $f(t)$, $g(x,t)$, $\varphi(t)$, and $\psi(t)$ are determined during the solution process. They are selected in such a way that the unknown function $w(z)$ ultimately satisfies *a single ordinary differential equation*.

Below, we examine specific examples of constructing exact solutions of the form (7.1.1.1) for nonlinear equations of mathematical physics.

[*] In this book, the short name *direct method of reductions* or *direct reduction method* will also be used sometimes.

► Example 7.1. Consider the unnormalized Korteweg–de Vries equation

$$u_t = au_{xxx} + buu_x. \tag{7.1.1.2}$$

We will look for its exact solution in the form (7.1.1.1). Substituting (7.1.1.1) into equation (7.1.1.2) yields

$$af\varphi^3 w'''_{zzz} + bf^2\varphi ww'_z + f(bg\varphi - \varphi'_t x - \psi'_t)w'_z + (bfg_x - f'_t)w + ag_{xxx} + bgg_x - g_t = 0. \tag{7.1.1.3}$$

Equating the functional coefficients of w_{zzz} and ww'_z in (7.1.1.3), we get

$$f = \varphi^2. \tag{7.1.1.4}$$

Next, equating the coefficient of w'_z with zero yields

$$g = \frac{1}{b\varphi}(\varphi'_t x + \psi'_t). \tag{7.1.1.5}$$

On substituting (7.1.1.4) and (7.1.1.5) into (7.1.1.3), we arrive at the relation

$$\varphi^5(aw'''_{zzz} + bww'_z) - \varphi\varphi'_t w + \frac{1}{b\varphi^2}\big\{\big[2(\varphi'_t)^2 - \varphi\varphi''_{tt}\big]x + 2\varphi'_t\psi'_t - \varphi\psi''_{tt}\big\} = 0.$$

Dividing by φ^5 and eliminating x using the formula $x = (z - \psi)/\varphi$, which follows from the second relation in (7.1.1.1), we obtain

$$aw'''_{zzz} + bww'_z - \varphi^{-4}\varphi'_t w + \frac{1}{b}\varphi^{-8}\big[2(\varphi'_t)^2 - \varphi\varphi''_{tt}\big]z + \frac{1}{b}\varphi^{-8}\big[\varphi\psi\varphi''_{tt} - \varphi^2\psi''_{tt} + 2\varphi\varphi'_t\psi'_t - 2\psi(\varphi'_t)^2\big] = 0. \tag{7.1.1.6}$$

We require that the functional coefficient of w and the last term be constant:

$$\varphi^{-4}\varphi'_t = -A, \qquad \varphi^{-8}\big[\varphi\psi\varphi''_{tt} - \varphi^2\psi''_{tt} + 2\varphi\varphi'_t\psi'_t - 2\psi(\varphi'_t)^2\big] = B,$$

where A and B are arbitrary constants. As a result, we arrive at a system of ordinary differential equations for φ and ψ:

$$\begin{aligned} \varphi'_t &= -A\varphi^4, \\ \psi''_{tt} + 2A\varphi^3\psi'_t - 2A^2\varphi^6\psi &= -B\varphi^6. \end{aligned} \tag{7.1.1.7}$$

Utilizing (7.1.1.6) and (7.1.1.7), we obtain an equation for $w(z)$:

$$aw_{zzz} + bww'_z - Aw - \frac{2A^2}{b}z + \frac{B}{b} = 0. \tag{7.1.1.8}$$

For $A \neq 0$, the genial solution of equations (7.1.1.7) is

$$\begin{aligned} \varphi(t) &= (3At + C_1)^{-1/3}, \\ \psi(t) &= C_2(3At + C_1)^{2/3} + C_3(3At + C_1)^{-1/3} + \frac{B}{2A^2}, \end{aligned} \tag{7.1.1.9}$$

where C_1, C_2, and C_3 arbitrary constants.

Formulas (7.1.1.1), (7.1.1.4), (7.1.1.5), and (7.1.1.9) along with equation (7.1.1.8) describe an exact solution to Korteweg–de Vries equation (7.1.1.2) (Clarkson & Kruskal, 1989). ◄

► Example 7.2. Now we consider the Boussinesq equation

$$u_{tt} + (uu_x)_x + au_{xxxx} = 0. \tag{7.1.1.10}$$

As in Example 7.1, we look for an exact solution in the form (7.1.1.1). The function $f(t)$, $g(x,t)$, $\varphi(t)$, and $\psi(t)$ will be determined in the solution process. On substituting (7.1.1.1) into (7.1.1.10), we get

$$\begin{aligned} af\varphi^4 w''''_{zzzz} &+ f^2\varphi^2 ww''_{zz} + f(z_t^2 + g\varphi^2)w''_{zz} \\ &+ f^2\varphi^2(w'_z)^2 + (fz_{tt} + 2fg_x\varphi + 2f'_t z_t)w'_z \\ &+ (fg_{xx} + f''_{tt})w + g_{tt} + gg_{xx} + g_x^2 + ag_{xxxx} = 0. \end{aligned} \tag{7.1.1.11}$$

Equating the functional coefficients of w''''_{zzzz} and ww''_{zz}, we find that

$$f = \varphi^2. \tag{7.1.1.12}$$

Equating the functional coefficient of w''_{zz} with zero and taking into account the second relation (7.1.1.1) and formula (7.1.1.12), we obtain

$$g = -\frac{1}{\varphi^2}(\varphi'_t x + \psi'_t)^2. \tag{7.1.1.13}$$

Substituting the expressions (7.1.1.12) and (7.1.1.13) into (7.1.1.11) yields

$$\begin{aligned} \varphi^6\big[aw''''_{zzzz} &+ ww''_{zz} + (w'_z)^2\big] + \varphi^2(x\varphi''_{tt} + \psi''_{tt})w'_z + 2\varphi\varphi''_{tt}w \\ &- \big[\varphi^{-2}(\varphi'_t x + \psi'_t)^2\big]_{tt} + 6\varphi^{-4}(\varphi'_t)^2(\varphi'_t x + \psi'_t)^2 = 0. \end{aligned}$$

We perform a double differentiation of the expression in square brackets in the second line and then divide all terms by φ^6. After eliminating x using the formula $x = (z - \psi)/\varphi$, we obtain

$$aw''''_{zzzz} + ww''_{zz} + (w'_z)^2 + \varphi^{-5}(\varphi''_{tt}z + \varphi\psi''_{tt} - \psi\varphi''_{tt})w'_z + 2\varphi^{-5}\varphi''_{tt}w + \cdots = 0. \tag{7.1.1.14}$$

We require that the functional coefficient of w'_z be a function of the single variable z:

$$\varphi^{-5}(\varphi''_{tt}z + \varphi\psi''_{tt} - \psi\varphi''_{tt}) = \varphi^{-5}\varphi''_{tt}z + \varphi^{-5}(\varphi\psi''_{tt} - \psi\varphi''_{tt}) \equiv Az + B,$$

where A and B are arbitrary constants. Consequently, we arrive at a system of ordinary differential equations for φ and ψ:

$$\begin{aligned} \varphi''_{tt} &= A\varphi^5, \\ \psi''_{tt} &= (A\psi + B)\varphi^4. \end{aligned} \tag{7.1.1.15}$$

With the help of (7.1.1.15), we eliminate the second and third derivatives of φ and ψ from (7.1.1.14) to arrive at an ordinary differential equation for $w = w(z)$:

$$aw''''_{zzzz} + ww''_{zz} + (w'_z)^2 + (Az + B)w'_z + 2Aw - 2(Az + B)^2 = 0. \tag{7.1.1.16}$$

Formulas (7.1.1.1), (7.1.1.12), and (7.1.1.13) together with equations (7.1.1.15) and (7.1.1.16) describe an exact solution to the Boussinesq equation (7.1.1.10).

The works by Clarkson & Kruskal (1989), Clarkson et al. (1997), and Polyanin & Zaitsev (2012) provide exact solutions for ODEs (7.1.1.15) and (7.1.1.16) as well as related solutions of the form (7.1.1.1) for the Boussinesq equation (7.1.1.10) expressible in terms of elementary or special functions. ◄

► Example 7.3. Let us consider the nonlinear third-order equation with a mixed derivative

$$u_{xt} + uu_{xx} - u_x^2 = \nu u_{xxx} + q(t)u_x + p(t). \tag{7.1.1.17}$$

It describes a wide class of exact solutions for the three-dimensional Navier–Stokes equations (Aristov & Polyanin, 2009); the constant ν is kinematic viscosity of the fluid. The functions $p = p(t)$ and $q = q(t)$, which appear in equation (7.1.1.17), can be chosen arbitrarily.

We look for exact solutions in the form

$$u = f(t)w(z) + g(t)x + h(t), \quad z = \varphi(t)x + \psi(t), \tag{7.1.1.18}$$

where the functions $f = f(t)$, $g = g(t)$, $h = h(t)$, $\varphi = \varphi(t)$, and $\psi = \psi(t)$ are to be determined into the subsequent analysis. On substituting (7.1.1.18) into (7.1.1.17), we get

$$[a(t)w + b(t)z + c(t)]w''_{zz} - a(t)(w'_z)^2 = \nu w'''_{zzz} + \widetilde{q}(t)w'_z + \widetilde{p}(t), \tag{7.1.1.19}$$

where

$$\begin{gathered} a = \frac{f}{\varphi}, \quad b = \frac{1}{\varphi^3}(g\varphi + \varphi'_t), \quad c = \frac{1}{\varphi^3}(h\varphi^2 - g\varphi\psi + \varphi\psi'_t - \psi\varphi'_t), \\ \widetilde{q} = \frac{1}{f\varphi^3}[fq\varphi + 2fg\varphi - (f\varphi)'_t], \quad \widetilde{p} = \frac{1}{f\varphi^3}(p + gq + g^2 - g'_t). \end{gathered} \tag{7.1.1.20}$$

By setting

$$a = \nu C_1, \quad b = \nu C_2, \quad c = \nu C_3, \quad \widetilde{q} = \nu C_4, \quad \widetilde{p} = \nu C_5, \tag{7.1.1.21}$$

where $C_1, \ldots, C_5$ are arbitrary constants, in (7.1.1.19), we obtain an ordinary differential equations for $w = w(z)$:

$$(\widetilde{C}_1 w + C_2 z + C_3)w''_{zz} - C_1(w'_z)^2 = w'''_{zzz} + C_4 w'_z + C_5.$$

In this case, the relations (7.1.1.20) subject to conditions (7.1.1.21) make up a mixed system of algebraic and ordinary differential equations for the functional parameters of solution (7.1.1.18) and functional coefficients $p = p(t)$ and $q = q(t)$ of equation (7.1.1.17). The functions $f = f(t)$ and $\psi = \psi(t)$ appearing in this system can be treated as arbitrary, while the other quantities, $\varphi = \varphi(t)$, $g = g(t)$, $h = h(t)$, $p = p(t)$, and $q = q(t)$ are expressed in terms of $f(t)$ and $\psi(t)$ without quadratures. ◄

Remark 7.1. *Numerous exact solutions to equation (7.1.1.17), along with related equations and hydrodynamic-type PDEs systems, can be found in Aristov et al. (2009), Polyanin & Zaitsev (2012), and Polyanin & Zhurov (2022).*

7.1.2. Special Reductions. Boussinesq and Wave Equations with Space Anisotropy

The process of finding exact solutions for nonlinear PDEs using the direct method of special reductions involves several successive steps, as described by Clarkson & Kruskal (1989), which are outlined below.

1°. Exact solutions of partial differential equations with two independent variables, x and t, are sought in the form

$$u(x,t) = f(x,t)w(z) + g(x,t), \qquad z = z(x,t). \tag{7.1.2.1}$$

Here the functions $f(x,t)$, $g(x,t)$, and $z(x,t)$ are to be determined in the solution process, which ensures that the function $w(z)$ is finally described by a *single ordinary differential equation.*

It is noteworthy that the functions u and w in formulas (7.1.1.1) and (7.1.2.1) are linearly related.

2°. On substituting the expression (7.1.2.1) into the nonlinear partial differential equation with a quadratic or power-law nonlinearity in question, we obtain

$$\Phi_1[x,t]\Psi_1[w] + \Phi_2[x,t]\Psi_2[w] + \cdots + \Phi_m[x,t]\Psi_m[w] = 0. \tag{7.1.2.2}$$

Here $\Psi_k[w]$ are differential forms consisting of products of nonnegative integer powers the function w and its derivatives, w'_z, w''_{zz}, and so on. The forms $\Phi_k[x,t]$ depend on the functions $f(x,t)$, $g(x,t)$, and $z(x,t)$ and their partial derivatives with respect to x and t. Suppose the differential form $\Psi_1[w]$ contains the highest derivative with respect to z. Then the form $\Phi_1[x,t]$ is utilized as the normalizing factor. This means that the relations

$$\Phi_k[x,t] = \Gamma_k(z)\,\Phi_1[x,t], \qquad k = 1,\ldots,m, \tag{7.1.2.3}$$

must hold. The functions $\Gamma_k(z)$ are to be determined; $\Gamma_1(z) \equiv 1$.

3°. In practice, to simplify the calculations when determining the functions f, g, z, u, and Γ_k, one can utilize the following properties:

a) if $f = f(x,t)$ can be represented as $f = f_0(x,t)\Omega(z)$, we can put $\Omega \equiv 1$, which is equivalent to the change of variable $w(z) \Rightarrow w(z)/\Omega(z)$;

b) if $g = g(x,t)$ is representable as $g = g_0(x,t) + f(x,t)\Omega(z)$, we can put $\Omega \equiv 0$, which is equivalent to the change of variable $w(z) \Rightarrow w(z) - \Omega(z)$;

c) if $z = z(x,t)$ is defined implicitly by an algebraic equation of the form $\Omega(z) = h(x,y)$, where $\Omega(z)$ is any invertible function, we can set $\Omega(z) = z$, which is equivalent to the change of variable $z \Rightarrow \Omega^{-1}(z)$.

4°. Once the functions $\Gamma_k(z)$ are determined, we insert (7.1.2.3) into (7.1.2.2) to obtain an ordinary differential equation for $w = w(z)$:

$$\Psi_1[w] + \Gamma_2(z)\Psi_2[w] + \cdots + \Gamma_m(z)\Psi_m[w] = 0. \tag{7.1.2.4}$$

Let us illustrate the main features of applying the direct reduction method with a specific example.

▶ Example 7.4. We will seek a solution to the Boussinesq equation (7.1.1.10) in the form (7.1.2.1). We have

$$afz_x^4 w''''_{zzzz} + a(6fz_x^2 z_{xx} + 4f_x z_x^3)w'''_{zzz} + f^2 z_x^2 ww''_{zz} + \cdots = 0. \qquad (7.1.2.5)$$

This equation displays only the first three terms and omits the arguments of the functions f and z. The functional coefficients of w''''_{zzzz} and ww''_{zz} must satisfy the following condition [see (7.1.2.3)]:

$$f^2 z_x^2 = afz_x^4 \Gamma_3(z),$$

where $\Gamma_3(z)$ is an undetermined function, which needs to be identified in the analysis. Then, taking advantage of property (a) from Item 3°, we choose

$$f = z_x^2, \qquad \Gamma_3(z) = 1/a. \qquad (7.1.2.6)$$

Similarly, the functional coefficients of w''''_{zzzz} and w'''_{zz} must satisfy the condition

$$6fz_x^2 z_{xx} + 4f_x z_x^3 = fz_x^4 \Gamma_2(z), \qquad (7.1.2.7)$$

where $\Gamma_2(z)$ is a new function that needs to be determined. Then, considering (7.1.2.6), we get

$$14\, z_{xx}/z_x = \Gamma_2(z) z_x.$$

Integrating with respect to x yields

$$\ln z_x = I(z) + \ln \widetilde{\varphi}(t), \qquad I(z) = \frac{1}{14}\int \Gamma_2(z)\, dz,$$

where $\widetilde{\varphi}(t)$ is an arbitrary function. Integrating once more, we obtain

$$\int e^{-I(z)}\, dz = \widetilde{\varphi}(t)x + \widetilde{\psi}(t),$$

where $\widetilde{\psi}(t)$ is an arbitrary function. The left-hand side is a function of z, and hence, taking advantage of property (c) from Item 3°, we get

$$z = x\varphi(t) + \psi(t). \qquad (7.1.2.8)$$

The functions $\varphi(t)$ and $\psi(t)$ are to be determined.

It follows from formulas (7.1.2.6)–(7.1.2.8) that

$$f = \varphi^2(t), \qquad \Gamma_2(z) = 0. \qquad (7.1.2.9)$$

On substituting the expressions (7.1.2.8) and (7.1.2.9) into (7.1.2.1), we obtain a solution in the form (7.1.1.1), where the function f is defined by (7.1.1.12). It follows that utilizing the general approach, based on representing a solution in the form (7.1.2.1), ultimately yields the same results as when using the simpler formula (7.1.1.1). ◀

Remark 7.2. *Similarly, constructing an exact solution of the Korteweg–de Vries equation (7.1.1.2) using formulas (7.1.1.1) and (7.1.2.1) can be shown to yield the same results.*

► Example 7.5. Let us consider the following nonlinear wave equation with two space variables and anisotropic in one of them:

$$u_{tt} = au_{xx} + [(bu + c)u_y]_y. \tag{7.1.2.10}$$

It is noteworthy that in the special case where $a = 1$, $b < 0$, and $c > 0$, equation (7.1.2.10) describes transonic spacial flows of an ideal polytropic gas (Pokhozhaev, 1989).

We look for an exact solution of equation (7.1.2.10) in the form (Polyanin & Zaitsev, 2012):

$$u = w(z) + g(x,t), \qquad z = y + h(x,t). \tag{7.1.2.11}$$

On substituting (7.1.2.11) into equation (7.1.2.10), we get

$$[(bw + ah_x^2 - h_t^2 + bg + c)w'_z]'_z + (ah_{xx} - h_{tt})w'_z + ag_{xx} - g_{tt} = 0.$$

Let the functions g and h satisfy the overdetermined system of equations

$$ag_{xx} - g_{tt} = C_1, \tag{7.1.2.12}$$
$$ah_{xx} - h_{tt} = C_2, \tag{7.1.2.13}$$
$$ah_x^2 - h_t^2 + bg = C_3, \tag{7.1.2.14}$$

where C_1, C_2, and C_3 are arbitrary constants. Then the function $w = w(z)$ is described by the autonomous ordinary differential equation

$$[(bw + c + C_3)w'_z]'_z + C_2 w'_z + C_1 = 0. \tag{7.1.2.15}$$

The general solutions of PDEs (7.1.2.12) and (7.1.2.13) are

$$\begin{aligned} g &= \varphi_1(\xi) + \psi_1(\eta) - \tfrac{1}{2}C_1 t^2,\\ h &= \varphi_2(\xi) + \psi_2(\eta) - \tfrac{1}{2}C_2 t^2,\\ \xi &= x + t\sqrt{a}, \quad \eta = x - t\sqrt{a}. \end{aligned}$$

Substituting these expressions into equation (7.1.2.14), eliminating t using the formula $t = \dfrac{\xi - \eta}{2\sqrt{a}}$, and performing simple rearrangements, we arrive at the functional differential equation with two arguments

$$\begin{aligned} &b\varphi_1(\xi) + C_2\xi\varphi_2'(\xi) - k\xi^2 - C_3 + b\psi_1(\eta) + C_2\eta\psi_2'(\eta) - k\eta^2\\ &\quad + \psi_2'(\eta)[4a\varphi_2'(\xi) - C_2\xi] + \eta[2k\xi - C_2\varphi_2'(\xi)] = 0, \end{aligned} \tag{7.1.2.16}$$

where

$$k = \frac{1}{8a}(bC_1 + 2C_2^2).$$

Equation (7.1.2.16) can be solved using the splitting method (see Section 5.5) if we set

$$\begin{aligned} b\varphi_1(\xi) + C_2\xi\varphi_2'(\xi) - k\xi^2 - C_3 &= A_1,\\ 4a\varphi_2'(\xi) - C_2\xi &= A_2,\\ 2k\xi - C_2\varphi_2'(\xi) &= A_3, \end{aligned} \tag{7.1.2.17}$$

where A_1, A_2, and A_3 are some constants. The compatible solution of the overdetermined system (7.1.2.17) is

$$\begin{aligned}\varphi_1(\xi) &= -\frac{C_2^2}{8ab}\xi^2 - \frac{BC_2}{b}\xi + \frac{A_1 + C_3}{b},\\ \varphi_2(\xi) &= \frac{C_2}{8a}\xi^2 + B\xi.\end{aligned} \tag{7.1.2.18}$$

It corresponds to the following values of the constants:

$$\begin{aligned}&A_1 \text{ is any}, \quad A_2 = 4aB, \quad A_3 = -BC_2, \quad B \text{ is any},\\ &C_1 = -\frac{C_2^2}{b}, \quad C_2 \text{ and } C_3 \text{ are any}, \quad k = \frac{C_2^2}{8a}.\end{aligned} \tag{7.1.2.19}$$

From relations (7.1.2.16) and (7.1.2.17), we obtain an equation that relates the functions ψ_1 and ψ_2:

$$A_1 + b\psi_1(\eta) + C_2\eta\psi_2'(\eta) - k\eta^2 + A_2\psi_2'(\eta) + A_3\eta = 0.$$

In view of (7.1.2.19), we find that

$$\psi_1(\eta) = -\frac{1}{b}(C_2\eta + 4aB)\psi_2'(\eta) + \frac{1}{b}\left(\frac{C_2^2}{8a}\eta^2 + BC_2\eta - A_1\right),$$

$\psi_2(\eta)$ is an arbitrary function.

Finally, we identify the functions determining solution (7.1.2.11):

$$\begin{aligned}g(x,t) &= -\frac{C_2^2}{2\sqrt{a}\,b}xt + \frac{C_2^2}{2b}t^2 - \frac{2\sqrt{a}\,BC_2}{b}t + \frac{C_3}{b} - \frac{1}{b}(C_2\eta + 4aB)\psi_2'(\eta),\\ h(x,t) &= \frac{C_2}{8a}(x^2 + 2\sqrt{a}\,xt - 3at^2) + B(x + \sqrt{a}\,t) + \psi_2(\eta),\end{aligned}$$

where $\eta = x - t\sqrt{a}$ and $\psi_2(\eta)$ is an arbitrary function. ◄

7.1.3. Special Form of Reductions. Nonlinear Reaction-Diffusion Equations

Class of nonlinear PDEs. We will consider reaction-diffusion equations with a nonlinear source and variable coefficients

$$c(x)u_t = [a(x)u_x]_x + p(x)f(u), \tag{7.1.3.1}$$

where $f(u)$ is an arbitrary function. Two of the three functional coefficients $a = a(x) > 0$, $c = c(x) > 0$, and $p = p(x)$ can be treated as free, and the remaining coefficient can be expressed in terms of them (this can be done in different ways as shown below). Without loss of generality, we assume that $p > 0$ (for $p < 0$, the functions p and f should be redefined by replacing them with $-p$ and $-f$).

Reductions of the reaction-diffusion equation to an ODE. Taking into account that $f(u)$ is arbitrary, we will follow Polyanin (2019b) to look for exact solutions of equation (7.1.3.1) as a superposition of functions

$$u = U(z), \qquad z = \varphi(x,t). \tag{7.1.3.2}$$

Formulas (7.1.3.2) follow from (7.1.2.1) if we set $f(x,t) = 1$ and $g(x,t) = 0$ and rename $w = U$ and $z(x,t) = \varphi(x,t)$. On substituting (7.1.3.2) into (7.1.3.1), we obtain

$$a(x)\varphi_x^2 U''_{zz} + \big\{[a(x)\varphi_x]_x - c(x)\varphi_t\big\}U'_z + p(x)f(U) = 0. \tag{7.1.3.3}$$

Suppose that the coefficients in equation (7.1.3.1) satisfy the relations

$$p(x) = a(x)s(\varphi)\varphi_x^2, \tag{7.1.3.4}$$

$$c(x)\varphi_t = [a(x)\varphi_x]_x + a(x)k(\varphi)\varphi_x^2, \tag{7.1.3.5}$$

where $s(\varphi)$ and $k(\varphi)$ are some functions ($s \not\equiv 0$). Then, equation (7.1.3.3) reduces to the ordinary differential equation

$$U''_{zz} - k(z)U'_z + s(z)f(U) = 0. \tag{7.1.3.6}$$

In the special case $k(z) \equiv 0$, when equation (7.1.3.5) is linear, the general solution of equation (7.1.3.6) with $s(z) = 1$ and arbitrary $f(U)$ can be represented in implicit form as

$$\int \left[C_1 - 2\int f(U)\,dU\right]^{-1/2} dU = C_2 \pm z, \tag{7.1.3.7}$$

where C_1 and C_2 are arbitrary constants.

Equations (7.1.3.4)–(7.1.3.6) allow one to construct exact solutions for a broad class of nonlinear reaction-diffusion equations of the form (7.1.3.1).

Analysis and solutions of the determining system of equations. A direct procedure for constructing exact solutions to nonlinear equations of the form (7.1.3.1) assumes that the functions $a(x)$, $c(x)$, and $f(u)$ are given, while the functions $u = u(x,t)$ and $p = p(x)$ are to be determined. In this context, if the functions $k(\varphi)$ and $s(\varphi)$ are predefined or known from certain considerations, it is required first to find particular solutions $p(x)$ and $\varphi = \varphi(x,t)$ for equations (7.1.3.4) and (7.1.3.5) (note that these equations can be linearized, as discussed below). Once these solutions are obtained, the solution to equation (7.1.3.1) can be determined using formula (7.1.3.2), in which the function $U(z)$ solves the ordinary differential equation (7.1.3.6).

In general, equations (7.1.3.4) and (7.1.3.5) for given functions $a = a(x)$, $c = c(x)$, $p = p(x)$, $k(\varphi)$, and $s(\varphi)$ form an overdetermined nonlinear system of equations for one function φ. This system will be referred to as the *determining system of equations*. We will examine the properties of equations (7.1.3.4) and (7.1.3.5) sequentially.

The nonlinear equation (7.1.3.4) reduces to two simple first-order separable equations: $\sqrt{s(\varphi)}\,\varphi_x = \pm\sqrt{p(x)/a(x)}$. Their general solutions are written as

$$\int \sqrt{s(\varphi)}\,d\varphi = \pm\int \sqrt{p(x)/a(x)}\,dx + \xi(t), \tag{7.1.3.8}$$

where $\xi(t)$ is an arbitrary function. Therefore, the function φ must have the general form

$$\varphi = G(y), \qquad y = \xi(t) + \theta(x). \tag{7.1.3.9}$$

It is noteworthy that solution (7.1.3.9) allows for another, equivalent representation: $\varphi = \bar{G}(\bar{y})$, $\bar{y} = \bar{\xi}(t)\bar{\theta}(x)$, where $\bar{y} = e^y$, $\bar{\xi} = e^\xi$, and $\bar{\theta} = e^\theta$.

Nonlinear transformations

$$\varphi = F(\psi) \tag{7.1.3.10}$$

preserve the form of equations (7.1.3.4) and (7.1.3.5), and the functional coefficients $k(\varphi)$ and $s(\varphi)$ are transformed as follows:

$$k(\varphi) \Longrightarrow k(F(\psi))F'_\psi(\psi) + \frac{F''_{\psi\psi}(\psi)}{F'_\psi(\psi)}, \qquad s(\varphi) \Longrightarrow s(F(\psi))[F'_\psi(\psi)]^2. \tag{7.1.3.11}$$

The degenerate case $k(\varphi) \equiv 0$ corresponds to a linear PDE with variable coefficients (7.1.3.5). If $k(\varphi) \not\equiv 0$, the nonlinear equation (7.1.3.5) is reduced by the substitution

$$\psi = C_1 \int K(\varphi)\,d\varphi + C_2, \qquad K(\varphi) = \exp\biggl[\int k(\varphi)\,d\varphi\biggr], \tag{7.1.3.12}$$

where C_1 and C_2 are arbitrary constants, to a linear PDE:

$$c(x)\psi_t = [a(x)\psi_x]_x. \tag{7.1.3.13}$$

In the special case $k(\varphi) = k = \text{const}$, we can use the substitution

$$\varphi = k^{-1}\ln|\psi|, \tag{7.1.3.14}$$

which follows from (7.1.3.12).

Transformations of the form (7.1.3.10) only affect the functional coefficients $k(\varphi)$ and $s(\varphi)$ in equations (7.1.3.4) and (7.1.3.5). Therefore, we can select the function F in such a way that it simplifies one of the equations. Below, we outline three approaches for simplifying these equations.

1°. For $s(\varphi) = 1$ and $k = k(\varphi)$, we find from formula (7.1.3.8) that

$$\varphi = \xi(t) + \theta(x), \tag{7.1.3.15}$$

which corresponds to $G(y) = y$ in (7.1.3.9). In this case, $p(x) = a(x)(\theta'_x)^2$.

2°. For $s(\varphi) = 1/\varphi$ and $k = k(\varphi)$, formula (7.1.3.8) gives

$$\varphi = \bar{\xi}(t)\bar{\theta}(x). \tag{7.1.3.16}$$

3°. For $s = s(\varphi)$ and $k(\varphi) = 0$, equation (7.1.3.5) is a linear autonomous PDE, whose solutions can be found by the method of separation of variables.

In what follows, we will use the simplest representation of the solution from Item 1°. On substituting (7.1.3.15) into equation (7.1.3.5), we arrive at the functional differential equation

$$c(x)\xi'_t = [a(x)\theta'_x]'_x + a(x)(\theta'_x)^2 k(\varphi), \qquad \varphi = \xi(t) + \theta(x). \tag{7.1.3.17}$$

We apply the method of differentiation (see Section 6.4) to identify admissible functions $k(\varphi)$ ensuring that this equation has solutions. To begin with, we divide equation (7.1.3.17) by c and rewrite as

$$\xi'_t = Q(x) + R(x)k(\varphi), \qquad \varphi = \xi(t) + \theta(x), \tag{7.1.3.18}$$

where $Q(x) = [(a\theta'_x)'_x]/c$ and $R(x) = a(\theta'_x)^2/c$. Then, by differentiating (7.1.3.18) with respect to t, we convert the resulting equation to $\xi''_{tt}/\xi'_t = R(x)k'_\varphi(\varphi)$. Taking the logarithm of both sides of this equation, differentiating with respect to t, and dividing by ξ'_t, we obtain $[\ln(\xi''_{tt}/\xi'_t)]'_t/\xi'_t = [\ln k'_\varphi(\varphi)]'_\varphi$. Further, differentiating with respect to x yields

$$[\ln k'_\varphi(\varphi)]''_{\varphi\varphi} = 0. \tag{7.1.3.19}$$

This ODE has the following solutions:

$$k(\varphi) = k_1\varphi + k_2 \qquad \text{(degenerate solution)}, \tag{7.1.3.20}$$

$$k(\varphi) = k_1 e^{-k_2\varphi} + k_3 \qquad \text{(nondegenerate solution)}, \tag{7.1.3.21}$$

where k_1, k_2, and k_3 are arbitrary constants. Formulas (7.1.3.20) and (7.1.3.21) determine all admissible functions $k(\varphi)$ that enable the functional differential equation (7.1.3.17) to have solutions.

Construction of exact solutions when $k(\varphi) = k$ and $s(\varphi) = 1$.

Direct method for constructing exact solutions. In the simplest case where $k(\varphi) = k = \text{const}$, which corresponds to $k_1 = 0$ and $k_2 = k$ in (7.1.3.20), substituting the expression (7.1.3.15) into equation (7.1.3.17) gives $\xi(t) = t$ (for convenience, the constant factor is set equal to 1). Therefore, in this case, the class of equations (7.1.3.1) admits functional separable solutions of the form (7.1.3.2), where

$$\varphi(x,t) = t + \int g(x)\,dx. \tag{7.1.3.22}$$

The function $g(x) = \theta'_x(x)$ can either be prescribed or determined in the process of the subsequent analysis, depending on the aim (see below). On substituting (7.1.3.22) into equation (7.1.3.4) with $s(\varphi) = 1$ and equation (7.1.3.5) with $k(\varphi) = k$, we obtain

$$p(x) = a(x)g^2(x), \tag{7.1.3.23}$$

$$c(x) = [a(x)g(x)]'_x + ka(x)g^2(x). \tag{7.1.3.24}$$

Relation (7.1.3.24) connects the first two functional coefficients of equation (7.1.3.1) and function $g = g(x)$ in (7.1.3.22). It is differential with respect to a and g and algebraic for c. Relation (7.1.3.23) is algebraic and serves to determine the functional coefficient $p(x)$.

If the functions $a(x)$ and $c(x)$ are assumed prescribed, relation (7.1.3.24) with $k \neq 0$ is a Riccati equation for $g = g(x)$. Let us rewrite it in the standard form:

$$a(x)g'_x + ka(x)g^2 + a'_x(x)g - c(x) = 0. \tag{7.1.3.25}$$

We examine two cases: degenerate and nondegenerate.

Degenerate case. If $k = 0$, the Riccati equation (7.1.3.25) degenerates into a linear equation, whose solution is expressed as

$$g(x) = \frac{1}{a(x)}\left[\int c(x)\,dx + C_1\right], \tag{7.1.3.26}$$

where C_1 is an arbitrary constant.

► Example 7.6. If the equation coefficients a and c are constant, $a = c = 1$, we can use formula (7.1.3.26) with $C_1 = 0$ to get $g(x) = x$. On substituting this function into (7.1.3.22) and (7.1.3.23), we find that $\varphi(x,t) = t + \frac{1}{2}x^2$ and $p(x) = x^2$. Hence, the nonlinear reaction-diffusion equation

$$u_t = u_{xx} + x^2 f(u) \tag{7.1.3.27}$$

admits, for arbitrary $f(u)$, a functional separable solution

$$u = U(z), \qquad z = t + \tfrac{1}{2}x^2. \tag{7.1.3.28}$$

The function $U(z)$ satisfies the autonomous ordinary differential equation

$$U''_{zz} + f(U) = 0, \tag{7.1.3.29}$$

obtained by substituting $k = 0$ and $s = 1$ into (7.1.3.6). Its general solution can be represented in the implicit form (7.1.3.7). ◄

► Example 7.7. Let us examine the more complicated scenario when one of the equation coefficients, $a = a(x)$, is an arbitrary function of the space variable, and the other coefficient is constant, $c(x) = 1$. Using formula (7.1.3.26) with $C_1 = 0$, we get $g(x) = x/a(x)$. On substituting this function into (7.1.3.22) and (7.1.3.23), we find that $\varphi(x,t) = t + \int \frac{x}{a(x)}\,dx$ and $p(x) = x^2/a(x)$. Therefore, the nonlinear reaction-diffusion equation

$$u_t = [a(x)u_x]_x + \frac{x^2}{a(x)} f(u), \tag{7.1.3.30}$$

dependent on two arbitrary functions, $a(x)$ and $f(u)$, admits a functional separable solution

$$u = U(z), \qquad z = t + \int \frac{x}{a(x)}\,dx. \tag{7.1.3.31}$$

The function $U(z)$ is described by the solvable ordinary differential equation (7.1.3.29). ◄

Nondegenerate case. If $k = \text{const}$ ($k \neq 0$), the substitution

$$g = \frac{1}{k}\frac{y'_x}{y} \tag{7.1.3.32}$$

converts the Riccati equation (7.1.3.25) into a linear second-order ODE:

$$a(x)y''_{xx} + a'_x(x)y'_x - kc(x)y = 0. \tag{7.1.3.33}$$

► Example 7.8. If the equation coefficients are constant, $a = c = 1$, the general solution of equation (7.1.3.33) is given by

$$y = \begin{cases} C_1 \cosh(mx) + C_2 \sinh(mx) & \text{for } k = m^2 > 0, \\ C_1 \cos(mx) + C_2 \sin(mx) & \text{for } k = -m^2 < 0, \end{cases} \tag{7.1.3.34}$$

where C_1 and C_2 are arbitrary constants. By setting $C_1 = 1$, $C_2 = 0$, and $k = 1$ in (7.1.3.34) and using formula (7.1.3.32), we find that

$$g(x) = \tanh x.$$

On substituting this function into (7.1.3.22) and (7.1.3.23), we obtain

$$\varphi(x,t) = t + \ln\cosh x, \qquad p(x) = \tanh^2 x.$$

It follows that the nonlinear reaction-diffusion equation

$$u_t = u_{xx} + \tanh^2 x f(u) \tag{7.1.3.35}$$

admits, for arbitrary $f(u)$, a functional separable solution

$$u = U(z), \qquad z = t + \ln\cosh x, \tag{7.1.3.36}$$

where the function $U(z)$ is described by the autonomous ordinary differential equation

$$U''_{zz} - U'_z + f(U) = 0. \tag{7.1.3.37}$$

The order of equation (7.1.3.37) can be reduced by one with the substitution $U'_z = \Phi(U)$, which leads to the canonical form of the second-kind Abel equation. ◄

Table 7.1 presents nonlinear reaction-diffusion type equations

$$u_t = u_{xx} + p(x)f(u),$$

where $f(u)$ is an arbitrary function, that admit functional separable solutions $u = U(z)$ with $z = \varphi(x,t)$; the function φ is defined up to an additive constant. In equations 1, 2, and 4–7, the function $\varphi(x,t)$ is the sum of functions of different arguments (7.1.3.22). The traveling wave solution (see equation 1) corresponds to the degenerate solution of equation (7.1.3.25) with $g = \alpha = \text{const}$. Some equations of this type with more complex functions $p(x)$ can have solutions derived using the

Table 7.1. Nonlinear equations $u_t = u_{xx} + p(x)f(u)$ admitting exact solutions of the form $u = U(z)$ with $z = \varphi(x,t)$.

No.	Function $p(x)$	Function $\varphi(x,t)$	Equation for $U = U(z)$
1	1	$t + \alpha x$	$\alpha^2 U''_{zz} - U'_z + f(U) = 0$
2	x^2	$t + \frac{1}{2}x^2$	$U''_{zz} + f(U) = 0$
3	x^{-2}	$xt^{-1/2}$	$U''_{zz} + \frac{1}{2}zU'_z + z^{-2}f(U) = 0$
4	$\tanh^2(\alpha x)$	$t + \alpha^{-2}\ln\cosh(\alpha x)$	$U''_{zz} - \alpha^2 U'_z + \alpha^2 f(U) = 0$
5	$\coth^2(\alpha x)$	$t + \alpha^{-2}\ln\lvert\sinh(\alpha x)\rvert$	$U''_{zz} - \alpha^2 U'_z + \alpha^2 f(U) = 0$
6	$\tan^2(\alpha x)$	$t - \alpha^{-2}\ln\lvert\cos(\alpha x)\rvert$	$U''_{zz} + \alpha^2 U'_z + \alpha^2 f(U) = 0$
7	$\cot^2(\alpha x)$	$t - \alpha^{-2}\ln\lvert\sin(\alpha x)\rvert$	$U''_{zz} + \alpha^2 U'_z + \alpha^2 f(U) = 0$

Here, $f(u)$ is an arbitrary function and α is an arbitrary constant ($\alpha \neq 0$).

formula (7.1.3.34) from Example 7.8. The solution of equation No. 3 is self-similar (refer to Example 7.10).

Other ways to construct exact solutions. Let us now explore alternative methods for constructing exact solutions of equations of the form (7.1.3.1) for $k(\varphi) = k$ and $s(\varphi) = 1$, without involving the integration of the Riccati equation (7.1.3.25). To this end, we will assume the function $g(x)$ and one of the two functions $a(x)$ or $c(x)$ to be given, and we will derive the other function from equation (7.1.3.25). Table 7.2 presents possible scenarios along with formulas for the desired functions. The final form of the nonlinear reaction-diffusion equation is determined by substituting $p(x) = a(x)g^2(x)$ into equation (7.1.3.1).

Table 7.2. Two different ways of defining the functional coefficients of equation (7.1.3.1) with $p(x) = a(x)g^2(x)$.

No.	Given functions	Desired functions
1	$a = a(x)$, $g = g(x)$	$c(x) = ag'_x + kag^2 + a'_x g$
2	$c = c(x)$, $g = g(x)$	$a(x) = g^{-1}E\bigl(\int cE^{-1}dx + C_1\bigr)$

Notation: k and C_1 are arbitrary constants, $g^{-1} = 1/g$, and $E = \exp\bigl(-k\int g\,dx\bigr)$.

► Example 7.9. We will employ the second scenario from Table 7.2 with $c = 1$ for an alternative representation of equations and their exact solutions. There are two possible cases to consider.

1°. *Degenerate case with* $k = 0$. Using row 2 in Table 7.2, we find that $a(x) = x/g(x)$ and $p(x) = x^2/a(x)$, which leads to PDE (7.1.3.30).

2°. *Nondegenerate case with* $k \neq 0$. From row 2 in Table 7.2 with $k \neq 0$ and $C_1 = 0$, we find that $a(x) = g^{-1}E\int E^{-1}dx$. We introduce a new function $h = h(x)$

defined as $h = \int E^{-1}dx$. Differentiating this expression and considering that $E = \exp\bigl(-k\int g\,dx\bigr)$, we express the function g via h. After simple rearrangements, we obtain $g = k^{-1}h''_{xx}/h'_x$, $a = kh/h''_{xx}$, and $p = k^{-1}hh''_{xx}/(h'_x)^2$. It follows that the nonlinear equation

$$u_t = [a(x)u_x]_x + p(x)f(u), \quad a(x) = k\frac{h}{h''_{xx}}, \quad p(x) = \frac{1}{k}\frac{hh''_{xx}}{(h'_x)^2}, \tag{7.1.3.38}$$

where $f(u)$ and $h = h(x)$ are arbitrary functions and $k \neq 0$ is an arbitrary constant, admits a generalized separable solution

$$u = U(z), \quad z = t + \frac{1}{k}\ln|h'_x|.$$

The function $U(z)$ is determined from the ordinary differential equation $U''_{zz} - kU'_z + f(U) = 0$.

By setting, for example, $h = \sinh(\alpha x)$ and $k = \alpha^2$ in (7.1.3.38), we obtain equation No. 4 from Table 7.1.

On substituting $h = \ln(\alpha x)$ and $k = -1$ into (7.1.3.38), we arrive at the equation

$$u_t = [x^2\ln(\alpha x)u_x]_x + \ln(\alpha x)f(u), \tag{7.1.3.39}$$

which admits an exact solution of the form $u = U(z)$, where $z = t + \ln x$. ◀

Direct construction of exact solutions for $k(\varphi) \neq$ const.

1°. *Case of $k(\varphi) = k_1\varphi$ and $s(\varphi) = 1$.* If $k(\varphi) = k_1\varphi$, which corresponds to $k_2 = 0$ in (7.1.3.20), substituting (7.1.3.15) into equation (7.1.3.17) yields $\xi(t) = e^{\lambda t}$. Then, the class of equations (7.1.3.1) admits functional separable solutions of the form (7.1.3.2), where

$$\varphi(x,t) = e^{\lambda t} + \theta(x). \tag{7.1.3.40}$$

On substituting the expression (7.1.3.40) into relation (7.1.3.4) with $s(\varphi) = 1$ and equation (7.1.3.17) with $k(\varphi) = k_1\varphi$, we obtain

$$c(x) = \frac{k_1}{\lambda}a(x)(\theta'_x)^2, \quad p(x) = a(x)(\theta'_x)^2. \tag{7.1.3.41}$$

In this case, the function $a(x)$ remains arbitrary, while the function $\theta = \theta(x)$ satisfies the following ODE with a cubic nonlinearity:

$$[a(x)\theta'_x]'_x + k_1a(x)\theta(\theta'_x)^2 = 0. \tag{7.1.3.42}$$

The substitution $\eta = \int\exp\bigl(\frac{1}{2}k_1\theta^2\bigr)\,d\theta$ converts equation (7.1.3.42) into the linear equation $[a(x)\eta'_x]'_x = 0$, whose general solution is expressed as $\eta = C_1\int\frac{1}{a}\,dx + C_2$.

2°. *Case of $k(\varphi) = k_1e^{-k_2\varphi} + k_3$ and $s(\varphi) = 1$.* For $k(\varphi) = k_1e^{-k_2\varphi} + k_3$, which corresponds to employing the relationship (7.1.3.21), we substitute (7.1.3.15) into equation (7.1.3.17) to obtain $\xi(t) = k_2^{-1}\ln t$. In this case, the class of equations (7.1.3.1) admits functional separable solutions of the form (7.1.3.2) where

$$\varphi(x,t) = \frac{1}{k_2}\ln t + \theta(x). \tag{7.1.3.43}$$

On substituting the expression (7.1.3.43) into relation (7.1.3.4) with $s(\varphi) = 1$ and equation (7.1.3.17) with $k(\varphi) = k_1 e^{-k_2\varphi} + k_3$, we find that

$$p(x) = a(x)(\theta'_x)^2, \quad c(x) = k_1 k_2 a(x) e^{-k_2\theta}(\theta'_x)^2. \tag{7.1.3.44}$$

The function $a(x)$ remains arbitrary, while the function $\theta = \theta(x)$ satisfies the following ODE with a quadratic nonlinearity:

$$[a(x)\theta'_x]'_x + k_3 a(x)(\theta'_x)^2 = 0. \tag{7.1.3.45}$$

This equation is easy to integrate because the substitution $\zeta(x) = \theta'_x$ converts it into a Bernoulli equation. In particular, for $k_3 = 0$, the general solution of equation (7.1.3.45) is expressed as

$$\theta(x) = C_1 \int \frac{1}{a}\,dx + C_2.$$

▶ Example 7.10. Let

$$a(x) = 1, \quad k_1 = -\tfrac{1}{2}, \quad k_2 = -2, \quad k_3 = 1. \tag{7.1.3.46}$$

Then, equation (7.1.3.45) has a solution $\theta = \ln x$. On substituting this function into formulas (7.1.3.43) and (7.1.3.44) and considering (7.1.3.46), we obtain

$$\varphi(x,t) = -\tfrac{1}{2}\ln t + \ln x, \quad p(x) = x^{-2}, \quad c(x) = 1. \tag{7.1.3.47}$$

Hence, equation

$$u_t = u_{xx} + x^{-2} f(u) \tag{7.1.3.48}$$

admits a self-similar solution

$$u = U(z), \quad z = -\tfrac{1}{2}\ln t + \ln x \equiv \ln(xt^{-1/2}), \tag{7.1.3.49}$$

where the function $U(z)$ satisfies the ordinary differential equation

$$U''_{zz} + (\tfrac{1}{2}e^{2z} - 1)U'_z + f(U) = 0. \tag{7.1.3.50}$$

It is important to note that a different representation of such solutions is often utilized in practical applications. This alternative approach involves defining a self-similar variable, $\bar{z} = e^z = xt^{-1/2}$, and transforming equation (7.1.3.48) into the form $U''_{\bar{z}\bar{z}} + \frac{1}{2}\bar{z}U'_{\bar{z}} + \bar{z}^{-2}U = 0$ (see equation No. 3 in Table 7.1). ◀

▶ Example 7.11. The relations (7.1.3.44) and equation (7.1.3.45) hold if we set

$$a(x) = 1, \quad c(x) = e^{-x}, \quad p(x) = 1, \quad \theta(x) = x, \quad k_1 = k_2 = 1, \quad k_3 = 0.$$

Therefore, the equation $e^{-x}u_t = u_{xx} + f(u)$ admits an solution $u = U(z)$ with $z = x + \ln t$. ◀

7.1.4. General Form of Reductions. Harry Dym Equation

The basic idea of the general scheme of using the direct reduction method is as follows: according to Clarkson & Kruskal (1989), exact solutions of partial differential equations with two independent variables x and t, are sought in the form

$$u(x,t) = F\big(x,t,w(z)\big), \qquad z = z(x,t). \tag{7.1.4.1}$$

The functions $F(x,t,w)$ and $z(x,t)$ should be selected in such a way that the function $w(z)$ satisfies a single ordinary differential equation. Unlike the representations of solutions in the form (7.1.1.1) or (7.1.2.1), the relationship between the functions w and u in equation (7.1.4.1) will generally be nonlinear.

Below we illustrate the main features of applying the direct reduction method to find exact solutions in the form (7.1.4.1).

► Example 7.12. Let us revisit the Boussinesq equation (7.1.1.10). Substituting (7.1.4.1) into (7.1.1.10) yields

$$aF_w z_x^4 w''''_{zzzz} + 4aF_{ww} z_x^4 w'_z w'''_{zzz} + a(4F_{xw} z_x^3 + 6F_w z_x^2 z_{xx}) w'''_{zzz} + \cdots = 0. \tag{7.1.4.2}$$

Only the three leading terms are written out here, and the arguments of the functions F and z have been omitted. For equation (7.1.4.2) to be reduced to an ordinary differential equation for $w = w(z)$, the ratios of the functional coefficients of $w'_z w'''_{zzz}$, w'''_{zzz}, and so forth, to the functional coefficient of the leading derivative w''''_{zzzz} must be functions of z and w, that is,

$$\frac{4aF_{ww} z_x^4}{aF_w z_x^4} = \Gamma_2(z,w), \qquad \frac{a(4F_{xw} z_x^3 + 6F_w z_x^2 z_{xx})}{aF_w z_x^4} = \Gamma_3(z,w), \qquad \ldots$$

Integrating the first relation,

$$4F_{ww}/F_w = \Gamma_2(z,w),$$

twice with respect to w yields

$$F(x,t,w) = f(x,t)\Theta(z,w) + g(x,t), \tag{7.1.4.3}$$

where $f(x,t)$ and $g(x,t)$ are arbitrary functions of two arguments, and

$$\Theta = \int \exp\biggl(\frac{1}{4}\int \Gamma_2\, dw\biggr)\, dw.$$

By setting $\Theta(z, w(z)) = U(z)$ in (7.1.4.3) and using the representation (7.1.4.1), we arrive at a solution that coincides, up to notation, with (7.1.2.1). Consequently, seeking exact solutions to the Boussinesq equation (7.1.1.10) using the general representation (7.1.4.1) results in a simpler special form of solution (7.1.2.1). ◄

► Example 7.13. Consider the Harry Dym equation

$$u_t + 2(u^{-1/2})_{xxx} = 0. \tag{7.1.4.4}$$

We will look for exact solutions in the form (7.1.4.1). On substituting this expression into equation (7.1.4.4), we obtain

$$-F^{-3/2}F_w z_x^3 w'''_{zzz} + \left(-3F^{-3/2}F_{ww} + \tfrac{9}{2}F^{-5/2}F_w^2\right) z_x^3 w'_z w''_{zz} + \cdots = 0.$$

The ratio of the functional coefficients of $w'_z w''_{zz}$ and w'''_{zzz} must be a function of z and w, that is,

$$3\frac{F_{ww}}{F_w} - \frac{9}{2}\frac{F_w}{F} = \Gamma(z, w).$$

Integrating twice yields

$$F^{-1/2}(x, t, w) = f(x, t)\Theta(z, w) + g(x, t), \qquad (7.1.4.5)$$

where $f(x, t)$ and $g(x, t)$ are arbitrary functions of two arguments, and

$$\Theta = -\int \exp\left(\frac{1}{3}\int \Gamma\, dw\right) dw.$$

It follows from formulas (7.1.4.1) and (7.1.4.5) that solutions to the Harry Dym equation (7.1.4.4) can be sought in the form

$$u^{-1/2}(x, t) = f(x, t)U(z) + g(x, t), \qquad z = z(x, t).$$

◀

◆ *Problems and Exercises for Section 7.1*

1. Find an exact solution for the unnormalized Burgers equation

$$u_t = au_{xx} + buu_x$$

of the form (7.1.1.1) by employing the direct reduction method.

2. Find an exact solution for the generalized nth-order Burgers–Korteweg–de Vries equation

$$u_t = au_x^{(n)} + buu_x$$

of the form (7.1.1.1) by employing the direct reduction method (Polyanin & Zaitsev, 2012).

3. Find an exact solution for the nonlinear convective-diffusion equation

$$u_t = u_{xx} + xf(u)u_x,$$

where $f(u)$ is an arbitrary function, by employing the direct reduction method.

Hint. Look for solutions in the form of a composition of functions (7.1.3.2) by using the expression (7.1.3.22) (Polyanin, 2019a).

4. Find an exact solution of the nonlinear convective-diffusion equation

$$u_t = u_{xx} + \tanh x f(u)u_x,$$

where $f(u)$ is an arbitrary function, by employing the direct reduction method.

Hint. Look for solutions in the form of a composition of functions (7.1.3.2) by using the expression (7.1.3.22) (Polyanin, 2019a).

5. Find an exact solution of the nonlinear convective-diffusion equation

$$u_t = [a(x)u_x]_x + xf(u)u_x,$$

where $a(x)$ and $f(u)$ are arbitrary functions, by employing the direct reduction method.

Hint. Look for solutions in the form of a composition of functions (7.1.3.2) by using the expression (7.1.3.22) (Polyanin, 2019a).

6. Find an exact solution of the form (7.1.3.2) for the nonlinear Klein–Gordon type equation

$$u_{tt} = [a(x)u_x]_x + \left[\frac{k}{a(x)} - 1\right] f(u),$$

where $a(x)$ and $f(u)$ are arbitrary functions, by employing the direct reduction method.

Hint. Look for solutions in the form of a composition of functions (7.1.3.2) by using the expression (7.1.3.22) (Zhurov & Polyanin, 2020).

7.2. Direct Method of Weak Symmetry Reductions

7.2.1. General Description of the Method. Steady-State Boundary Layer Equation

Preliminary remarks. In various modifications of the direct reduction method based on formulas (7.1.1.1), (7.1.2.1), and (7.1.4.1), the function $w = w(z)$ holds a privileged position. The other functions must be selected to ensure that it satisfies a *single ordinary differential equation*. This requirement imposes significant restrictions on the method's capabilities, which limits its effectiveness in finding many exact solutions that could be identified through other methods. In particular, the vast majority of generalized or functional separable solutions discussed in Chapters 6 and 7 cannot be obtained using this approach.

The efficiency of the direct reduction method can be significantly improved when combined with the ideas of the methods of generalized or functional separation of variables, where all determining functions are treated as equal, and the function $w(z)$ can be described by an *overdetermined system of several ODEs*.

Splitting principle. Direct method of weak symmetry reductions. We utilize the expression (7.1.2.1) as before to construct exact solutions of nonlinear partial differential equations. By substituting this expression into the nonlinear partial differential equation under investigation, we arrive at relation (7.1.2.2). In addition to the algorithm for constructing exact solutions described in Section 7.1.2, we will also assume that the function $w(z)$ can satisfy an overdetermined system of several ordinary differential equations. Moreover, the functions $f = f(x,t)$ and $g = g(x,t)$ can be described by overdetermined systems of PDEs. To obtain exact solutions of the resulting bilinear equation of the form (7.1.2.2), where $\Phi_n = \Phi_n[x,t]$ and $\Psi_n = \Psi_n[w]$, we will apply the generalized splitting principle, which is outlined below.

Generalized splitting principle. We seek solutions to the equation being studied as various combinations of elements from two sets, $\{\Phi_j\}$ and $\{\Psi_j\}$, which appear in equation (7.1.2.2):

$$\sum_{n=1}^{m} \alpha_{ni}\Phi_n = 0, \qquad i = 1, \dots, l;$$
$$\sum_{n=1}^{m} \beta_{nj}\Psi_n = 0, \qquad j = 1, \dots, k, \tag{7.2.1.1}$$

where $1 \le l \le m-1$ and $1 \le k \le m-1$. The constants α_{ni} and β_{nj} in (7.2.1.1) are selected to ensure that the bilinear equation (7.1.2.2) holds identically, which can always be done. It is important to emphasize that relations (7.2.1.1) are algebraic and independent of the specific structure of the differential forms Φ_n and Ψ_n.

Once relations (7.2.1.1) are obtained, we can insert them into the differential forms Φ_n and Ψ_n. As a result, we will arrive at systems of differential equations, often overdetermined, for the functions $f = f(x,t)$, $g = g(x,t)$, and $w = w(z)$ that appear in the expression (7.1.2.1).

Remark 7.3. *It is also necessary to investigate degenerate cases when, in addition to the linear relations of the form (7.2.1.1), one or more of the differential forms Ψ_n or Φ_n may vanish.*

Remark 7.4. *In the formulation of the generalized splitting principle, both sets Φ_n and Ψ_n depend on x and t, since $\Psi_n = \Psi_n[w]$, where $w = w(z)$ and $z = z(x,t)$.*

Following Polyanin & Zhurov (2022), we refer to the outlined combination of the direct reduction method and the generalized splitting principle as the *direct method of weak symmetry reductions*.

It is important to highlight a key qualitative difference between the final presentation of results obtained from the direct reduction method and those from the direct method of weak symmetry reductions. Solutions derived from the direct reduction method are typically expressed in terms of solutions to nonlinear ordinary differential equations. In contrast, solutions generated through the direct method of weak symmetry reductions often allow for a closed-form representation, typically expressed in quadratures.

► Example 7.14. Consider the equation of an extended axisymmetric steady laminar hydrodynamic boundary layer

$$u_y u_{xy} - u_x u_{yy} = a(y u_{yy})_y + \mathcal{F}(x), \tag{7.2.1.2}$$

where u is the stream function, x is the downstream coordinate measured along the symmetry axis, $y = \frac{1}{4}r^2$, r is the radial coordinate, and $\mathcal{F}(x)$ is a pressure function. The longitudinal and transverse components of the fluid velocity, v_1 and v_2, are expressed via the stream function as $v_1 = 2r^{-1}u_r$ and $v_2 = -2r^{-1}u_x$.

Following Burde (1994), we look for a solution to equation (7.2.1.2) in the form

$$u(x,y) = af(x)w(z) + ag(x), \qquad z = \varphi(x)y + \psi(x), \tag{7.2.1.3}$$

where a is a constant factor introduced for convenience. We substitute expression (7.2.1.3) into the original equation (7.2.1.2), eliminate y using the relationship

$\varphi(x)y = z - \psi(x)$, and divide by $a^2\varphi^2 f$ to arrive at a two-argument functional differential equation of the form (5.2.2.1)–(5.2.2.2) from Section 5.2.2 with $k = 6$:

$$(zw''_{zz})'_z - \psi w'''_{zzz} + f'_x ww''_{zz} + g'_x w''_{zz} - \frac{(f\varphi)'_x}{\varphi}(w'_z)^2 + \frac{\mathcal{F}}{a^2 f\varphi^2} = 0. \quad (7.2.1.4)$$

We will use a simplified approach to construct exact solutions. Assuming that the functional coefficients of ww''_{zz}, w''_{zz}, $(w'_z)^2$, and 1 are linear combinations of the coefficients 1 and ψ of the leading terms $(zw''_{zz})'_z$ and w'''_{zzz}, we find that

$$\begin{aligned} f'_x &= A_1 + B_1\psi, \\ g'_x &= A_2 + B_2\psi, \\ -(f\varphi)'_x/\varphi &= A_3 + B_3\psi, \\ \mathcal{F}/(a^2 f\varphi^2) &= A_4 + B_4\psi, \end{aligned} \quad (7.2.1.5)$$

where A_k and B_k are arbitrary constants. We substitute the expressions (7.2.1.5) into equation (7.2.1.4) and collect the terms proportional to ψ (we assume that $\psi \neq \text{const}$). Equating the functional coefficient of ψ with zero, we obtain an overdetermined system consisting of two ordinary differential equations for $w = w(z)$:

$$(zw''_{zz})'_z + A_1 ww''_{zz} + A_2 w''_{zz} + A_3(w'_z)^2 + A_4 = 0, \quad (7.2.1.6)$$

$$-w'''_{zzz} + B_1 ww''_{zz} + B_2 w''_{zz} + B_3(w'_z)^2 + B_4 = 0. \quad (7.2.1.7)$$

We will explore three cases.

Case 1. If we set

$$A_1 = A_3 = A_4 = 0, \quad A_2 = -n, \quad (7.2.1.8)$$

equation (7.2.1.6) will have the following solution:

$$w(z) = \frac{C_1}{n(n+1)} z^{n+1} + C_2 z + C_3, \quad (7.2.1.9)$$

where C_1, C_2, and C_3 constants of integration. The solution (7.2.1.9) of equation (7.2.1.6) is simultaneously a solution to equation (7.2.1.7) only if the conditions

$$n = -2, \quad B_1 = B_3, \quad C_1 = -\frac{4}{B_1}, \quad C_2^2 = -\frac{B_4}{B_1}, \quad C_3 = -\frac{B_2}{B_1} \quad (7.2.1.10)$$

hold. Inserting the coefficients (7.2.1.8) and (7.2.1.10) into system (7.2.1.5) and integrating, we obtain

$$g(x) = 2x - C_3 f, \quad \varphi = \frac{C_4}{f^2}, \quad \psi = -\frac{C_1}{4} f'_x, \quad \mathcal{F} = -(aC_2C_4)^2 \frac{f'_x}{f^3}, \quad (7.2.1.11)$$

where $f = f(x)$ is an arbitrary function.

Formulas (7.2.1.3), (7.2.1.9), and (7.2.1.11) provide an exact solution for the axisymmetric boundary-layer equation (7.2.1.2).

Case 2. If

$$A_2 = B_1 = B_3 = B_4 = 0, \quad B_2 = -\lambda, \quad A_3 = -A_1, \quad A_4 = \lambda^2/A_1 \quad (7.2.1.12)$$

a compatible solution to the system of equations (7.2.1.6) and (7.2.1.7) is given by

$$w(z) = \frac{1}{A_1}(C_1 e^{-\lambda z} + \lambda z - 3). \quad (7.2.1.13)$$

System (7.2.1.5) with coefficients (7.2.1.12) has the following solution:

$$f = A_1 x + C_2, \quad \varphi = C_3, \quad \psi = -\frac{1}{\lambda} g'_x, \quad \mathcal{F} = \frac{(aC_3\lambda)^2}{A_1}(A_1 x + C_2), \quad (7.2.1.14)$$

where C_1, C_2, and C_3 are arbitrary constants, and $g = g(x)$ is an arbitrary function.

Formulas (7.2.1.3), (7.2.1.13), and (7.2.1.14) provide an exact solution for the boundary layer equation (7.2.1.2).

Case 3. System (7.2.1.6)–(7.2.1.7) also admits solutions of the form

$$w(z) = C_1 z^2 + C_2 z + C_3,$$

where the constants C_1, C_2, and C_3 are related to A_n and B_n. The solution can be obtained more easily by working directly with the original equation (7.2.1.2), specifically by inserting $u = \varphi_2(x)y^2 + \varphi_1(x)y + \varphi_0(x)$, which corresponds to the method of generalized separation of variables. As a result, we find that

$$u(x, y) = C_1 y^2 + \varphi(x)y + \frac{1}{4C_1}\varphi^2(x) - \frac{1}{2C_1}\int \mathcal{F}(x)\,dx - x + C_3,$$

where $\mathcal{F}(x)$ and $\varphi(x)$ are arbitrary functions, while C_1 and C_3 are arbitrary constants. ◀

7.2.2. Burgers–Huxley Equation (Diffusion-Type Equation with Cubic Nonlinearity)

Let us consider the equation with a cubic nonlinearity

$$u_t + \sigma u u_x = a u_{xx} + b_3 u^3 + b_2 u^2 + b_1 u + b_0. \quad (7.2.2.1)$$

For $\sigma = 1$ and $b_0 = b_1 = b_2 = b_3 = 0$, it becomes a Burgers equation, which describes the propagation of waves in nonlinear dissipative systems. For $\sigma = b_0 = 0$, it coincides with the Huxley equation, which models nerve pulse propagation in nerve fibers. For $b_0 = 0$, equation (7.2.2.1) is an unnormalized Burgers–Huxley equation, which describes near-wall liquid flows in liquid crystals as well as population dynamics taking into account reproduction, mortality, nutrition, and diffusion.

We will look for solution of equation (7.2.2.1) in the form

$$u(x, t) = f(x, t)w(z) + \lambda, \quad z = z(x, t), \quad (7.2.2.2)$$

where the functions $f = f(x,t)$, $z = z(x,t)$, and $w = w(z)$ as well as the constant λ are to be determined. Substituting (7.2.2.2) into (7.2.2.1) yields an equation that can be conveniently represented in the bilinear form

$$\sum_{n=1}^{7} \Phi_n[x,t]\Psi_n[z] = 0. \tag{7.2.2.3}$$

The forms $\Phi_n = \Phi_n[x,t]$ depend on the coefficients of equation (7.2.2.1) as well as the functions (and their derivatives) appearing in solution (7.2.2.2). These are expressed as

$$\begin{aligned}
&\Phi_1 = b_3\lambda^3 + b_2\lambda^2 + b_1\lambda + b_0,\\
&\Phi_2 = 3b_3\lambda^2 f + 2b_2\lambda f + b_1 f + af_{xx} - \sigma\lambda f_x - f_t,\\
&\Phi_3 = 3b_3\lambda f^2 + b_2 f^2 - \sigma f f_x, \quad \Phi_4 = b_3 f^3,\\
&\Phi_5 = afz_{xx} + 2af_x z_x - \sigma\lambda f z_x - f z_t,\\
&\Phi_6 = -\sigma f^2 z_x, \quad \Phi_7 = afz_x^2.
\end{aligned} \tag{7.2.2.4}$$

The forms $\Psi_n = \Psi_n[z]$ are defined by

$$\begin{aligned}
&\Psi_1 = 1, \quad \Psi_2 = w, \quad \Psi_3 = w^2, \quad \Psi_4 = w^3,\\
&\Psi_5 = w'_z, \quad \Psi_6 = ww'_z, \quad \Psi_7 = w''_{zz}.
\end{aligned} \tag{7.2.2.5}$$

Further, we will utilize the direct method of weak symmetry reductions. It is easy to verify that if

$$w(z) = 1/z, \tag{7.2.2.6}$$

the following three linear relations between the functions (7.2.2.5) hold:

$$\Psi_7 = 2\Psi_4, \quad \Psi_6 = -\Psi_4, \quad \Psi_5 = -\Psi_3. \tag{7.2.2.7}$$

Substituting (7.2.2.7) into (7.2.2.3) yields

$$\Phi_1\Psi_1 + \Phi_2\Psi_2 + (\Phi_3 - \Phi_5)\Psi_3 + (\Phi_4 - \Phi_6 + 2\Phi_7)\Psi_4 = 0. \tag{7.2.2.8}$$

Equating the functional coefficient of Ψ_4 in (7.2.2.8) with zero and dividing by f, we arrive at the equation $b_3 f^2 + \sigma f z_x + 2az_x^2 = 0$, whose solution is given by

$$f = \beta z_x, \tag{7.2.2.9}$$

where β is a root of the quadratic equation

$$b_3\beta^2 + \sigma\beta + 2a = 0. \tag{7.2.2.10}$$

Equating the functional coefficients of Ψ_1, Ψ_2, and Ψ_3 in (7.2.2.8) with zero, considering (7.2.2.4) and (7.2.2.9), and performing simple rearrangements, we obtain

$$\begin{aligned}
z_t - (3a + \beta\sigma)z_{xx} + (\sigma\lambda + b_2\beta + 3b_3\beta\lambda)z_x &= 0 \quad \text{(coefficient of } \Psi_3\text{)},\\
z_{xt} - az_{xxx} + \sigma\lambda z_{xx} - (b_1 + 2\lambda b_2 + 3b_3\lambda^2)z_x &= 0 \quad \text{(coefficient of } \Psi_2\text{)},\\
b_3\lambda^3 + b_2\lambda^2 + b_1\lambda + b_0 &= 0 \quad \text{(coefficient of } \Psi_1\text{)}.
\end{aligned} \tag{7.2.2.11}$$

Here, the first two linear partial differential equations form an overdetermined system for $z = z(x,t)$. The last (cubic) equation serves to determine the constant λ.

Using (7.2.2.2), (7.2.2.6), and (7.2.2.9), we can represent a solution of equation (7.2.2.1) in the form

$$u(x,t) = \frac{\beta}{z}\frac{\partial z}{\partial x} + \lambda. \tag{7.2.2.12}$$

Let β be a square root of the equation (7.2.2.10), and let λ be a root of the last cubic equation in (7.2.2.11). Depending on the value of the constant b_3, we need to consider two cases.

1°. *Case* $b_3 \neq 0$. It follows from the first two equations in (7.2.2.11) that

$$\begin{aligned} z_t + p_1 z_{xx} + p_2 z_x &= 0, \\ z_{xxx} + q_1 z_{xx} + q_2 z_x &= 0, \end{aligned} \tag{7.2.2.13}$$

where

$$\begin{aligned} p_1 &= -\beta\sigma - 3a, \quad p_2 = \lambda\sigma + \beta b_2 + 3\beta\lambda b_3, \\ q_1 &= -\frac{\beta b_2 + 3\beta\lambda b_3}{\beta\sigma + 2a}, \quad q_2 = -\frac{3b_3\lambda^2 + 2b_2\lambda + b_1}{\beta\sigma + 2a}. \end{aligned} \tag{7.2.2.14}$$

Below are a few solutions to the overdetermined system of linear PDEs (7.2.2.13) with (7.2.2.14). Together with formula (7.2.2.12) and quadratic equation (7.2.2.10), these solutions allow us to find exact solutions of the original PDE (7.2.2.1).

Four main situations are possible here:

1.1. For $q_2 \neq 0$ and $q_1^2 \neq 4q_2$ ($q_1^2 > 4q_2$), we get

$$\begin{aligned} z(x,t) &= C_1 \exp(k_1 x + s_1 t) + C_2 \exp(k_2 x + s_2 t) + C_3, \\ k_n &= -\tfrac{1}{2}q_1 \pm \tfrac{1}{2}\sqrt{q_1^2 - 4q_2}, \quad s_n = -k_n^2 p_1 - k_n p_2, \end{aligned}$$

where C_1, C_2, and C_3 are arbitrary constants; $n = 1, 2$. For $q_1^2 < 4q_2$, we should separate the real and imaginary parts. This results in a solution that involves both exponential and trigonometric functions.

1.2. For $q_2 \neq 0$ and $q_1^2 = 4q_2$, we have

$$\begin{aligned} z(x,t) &= C_1 \exp(kx + s_1 t) + C_2(kx + s_2 t)\exp(kx + s_1 t) + C_3, \\ k &= -\tfrac{1}{2}q_1, \quad s_1 = -\tfrac{1}{4}p_1 q_1^2 + \tfrac{1}{2}p_2 q_1, \quad s_2 = -\tfrac{1}{2}p_1 q_1^2 + \tfrac{1}{2}p_2 q_1. \end{aligned}$$

1.3. For $q_2 = 0$ and $q_1 \neq 0$, we have

$$z(x,t) = C_1(x - p_2 t) + C_2 \exp[-q_1 x + q_1(p_2 - p_1 q_1)t] + C_3.$$

1.4. For $q_2 = q_1 = 0$, we have

$$z(x,t) = C_1(x - p_2 t)^2 + C_2(x - p_2 t) - 2C_1 p_1 t + C_3.$$

2°. *Case of* $b_3=0$ *and* $b_2\neq 0$. Solutions are defined by formula (7.2.2.12) with

$$z(x,t)=C_1+C_2\exp\left[Ax+A\left(\frac{b_1\sigma}{2b_2}+\frac{2ab_2}{\sigma}\right)t\right],$$
$$\beta=-\frac{2a}{\sigma},\qquad A=\frac{\sigma(b_1+2b_2\lambda)}{2ab_2},$$

where $\lambda=\lambda_{1,2}$ are roots of the quadratic equation $b_2\lambda^2+b_1\lambda+b_0=0$.

Remark 7.5. *Kawahara & Tanaka (1983), F. Cariello & Tabor (1989), Danilov & Subochev (1991), Nucci & Clarkson (1992), Kudryashov (1993), Clarkson & Mansfield (1994), and Estévez (1994) obtained various exact solutions to the nonlinear equation (7.2.2.1) and its special cases using different approaches.*

7.2.3. Equations of Unsteady Plane and Axisymmetric Boundary Layers

Unsteady plane boundary layer equation. Hydrodynamic boundary layer equations describe the flow of viscous fluids within thin layers near solid, liquid, and gaseous surfaces of various shapes at high Reynolds numbers.

The system of unsteady plane laminar boundary layer equations for a Newtonian incompressible fluid, which is the classical model of a viscous fluid, is written as follows (Schlichting, 1981 and Loitsyanskiy, 1996):

$$U_t+UU_x+VU_y=\nu U_{yy}+F(t,x),\tag{7.2.3.1}$$
$$U_x+V_y=0,\tag{7.2.3.2}$$

where t is time, x and y are the longitudinal and transverse coordinates ($y=0$ corresponds to the body surface), U and V are the longitudinal and transverse fluid velocity components, $F(t,x)=-p_x/\rho$ is a given function (proportional to the longitudinal pressure gradient), p is pressure, ρ is mass density, and ν is kinematic viscosity.

With the stream function W defined by

$$U=W_y,\qquad V=-W_x,\tag{7.2.3.3}$$

system (7.2.3.1)–(7.2.3.2) reduces to a single third-order PDE:

$$W_{ty}+W_yW_{xy}-W_xW_{yy}=\nu W_{yyy}+F(t,x).\tag{7.2.3.4}$$

Axisymmetric boundary layer equation. The system of unsteady axisymmetric laminar boundary layer equations is expressed as (Schlichting, 1981):

$$U_t+UU_x+VU_y=\nu U_{yy}+F(t,x),\tag{7.2.3.5}$$
$$(rU)_x+(rV)_y=0,\qquad r=r(x),\tag{7.2.3.6}$$

where x and y are the longitudinal and transverse coordinates (the respective unit vectors $\mathbf{e}_x$ and $\mathbf{e}_y$ point in the tangential and normal direction to the body of revolution, respectively), U and V are the longitudinal and transverse fluid velocity components, $F(t, x) = -p_x/\rho$ is a given function, $r = r(x)$ is the dimensionless cross-sectional radius perpendicular to the axis of revolution and defining the body shape. The other notations are the same as in equation (7.2.3.4). The choice of the unit length in the definition of $r = r(x)$ does not affect the form of the equation.

Introducing the stream function W defined by

$$U = W_y, \quad V = -W_x - \frac{r'_x}{r}W, \qquad r = r(x), \tag{7.2.3.7}$$

reduces system (7.2.3.5)–(7.2.3.6) to a single third-order PDE:

$$W_{ty} + W_y W_{xy} - W_x W_{yy} - \frac{r'_x}{r} W W_{yy} = \nu W_{yyy} + F(t, x). \tag{7.2.3.8}$$

If $r(x) = \text{const}$, equation (7.2.3.8) coincides with (7.2.3.4). If $r(x) \neq \text{const}$, equation (7.2.3.8) becomes much more complicated to analyze. For some specific $r(x)$, it was shown by Aksenov & Kozyrev (2022) that equation (7.2.3.8) with three independent variables can be reduced to an ordinary differential equation or a partial differential equation with two independent variables; the analysis was based on a modification of the Clarkson–Kruskal direct method.

Next, we will consider the general case of equation (7.2.3.8) for an arbitrary dependence $r = r(x)$. As usual, the admissible forms of the pressure function $F(t, x)$ will be determined during the analysis process.

Reduction to a plane boundary layer equation with varying viscosity. By introducing the new variables

$$z = r(x)y, \quad w = r(x)W, \tag{7.2.3.9}$$

we rewrite equation (7.2.3.8) in a form more convenient for analysis:

$$w_{tz} + w_z w_{xz} - w_x w_{zz} = \nu r^2(x) w_{zzz} + F(t, x). \tag{7.2.3.10}$$

In the special case $r(x) \equiv 1$, equation (7.2.3.10) coincides with the plane boundary layer equation (7.2.3.4).

General form of solutions. Determining equations. Reduction to an ODE. We look for solutions of equation (7.2.3.10) in the following form (Polyanin & Zhurov, 2015):

$$w = fu(\xi) + gz + h, \qquad \xi = \varphi z + \psi, \tag{7.2.3.11}$$

where the six functions $f = f(t, x)$, $g = g(t, x)$, $h = h(t, x)$, $\varphi = \varphi(t, x)$, $\psi = \psi(t, x)$, and $u = u(\xi)$ are to be determined in the solution process.

On substituting (7.2.3.11) into (7.2.3.10), we obtain an equation that is convenient to represent in a bilinear form:

$$\sum_{n=1}^{7} \Phi_n[x, t] \Psi_n[\xi] = 0. \tag{7.2.3.12}$$

The functions $\Phi_n = \Phi_n[x, t]$ depend on the functional coefficients (and their derivatives) that appear in equation (7.2.3.10) and solution (7.2.3.11) and are expressed as

$$\begin{gathered}\Phi_1 = g_t + gg_x - F, \quad \Phi_2 = (f\varphi)_t + (fg\varphi)_x, \quad \Phi_3 = f\varphi(f\varphi)_x,\\ \Phi_4 = f(\varphi\psi_t + g\varphi\psi_x + g_x\varphi\psi - \psi\varphi_t - g\varphi_x\psi - h_x\varphi^2),\\ \Phi_5 = f(\varphi_t + g\varphi_x - g_x\varphi), \quad \Phi_6 = -ff_x\varphi^2, \quad \Phi_7 = -\nu r^2 f\varphi^3.\end{gathered} \tag{7.2.3.13}$$

The functions $\Psi_n = \Psi_n[\xi]$ are given by

$$\begin{gathered}\Psi_1 = 1, \quad \Psi_2 = u'_\xi, \quad \Psi_3 = (u'_\xi)^2, \quad \Psi_4 = u''_{\xi\xi},\\ \Psi_5 = \xi u''_{\xi\xi}, \quad \Psi_6 = uu''_{\xi\xi}, \quad \Psi_7 = u'''_{\xi\xi\xi}.\end{gathered} \tag{7.2.3.14}$$

Equation (7.2.3.12) with the functionals (7.2.3.13) and (7.2.3.14) reduces to a single ODE for $u = u(\xi)$ if all Φ_n ($n \le 6$) are set proportional to Φ_7, that is,

$$\Phi_n = -a_n\Phi_7 \quad (n = 1, \ldots, 6), \tag{7.2.3.15}$$

where a_1, …, a_6 are free parameters. Substituting (7.2.3.13) into (7.2.3.15), we arrive at a nonlinear system of PDEs for the functions $f = f(t, x)$, $g = g(t, x)$, $h = h(t, x)$, $\varphi = \varphi(t, x)$, and $\psi = \psi(t, x)$:

$$g_t + gg_x - F = a_1\nu r^2 f\varphi^3, \tag{7.2.3.16}$$

$$(f\varphi)_t + (fg\varphi)_x = a_2\nu r^2 f\varphi^3, \tag{7.2.3.17}$$

$$(f\varphi)_x = a_3\nu r^2\varphi^2, \tag{7.2.3.18}$$

$$\varphi\psi_t + g\varphi\psi_x + g_x\varphi\psi - \varphi_t\psi - g\varphi_x\psi - h_x\varphi^2 = a_4\nu r^2\varphi^3, \tag{7.2.3.19}$$

$$\varphi_t + g\varphi_x - g_x\varphi = a_5\nu r^2\varphi^3, \tag{7.2.3.20}$$

$$f_x = -a_6\nu r^2\varphi. \tag{7.2.3.21}$$

As a result, equation (7.2.3.12) reduces to the following nonlinear ODE for $u = u(\xi)$:

$$a_1 + a_2u'_\xi + a_3(u'_\xi)^2 + a_4u''_{\xi\xi} + a_5\xi u''_{\xi\xi} + a_6uu''_{\xi\xi} = u'''_{\xi\xi\xi}. \tag{7.2.3.22}$$

Any consistent solution to the determining system of first-order PDEs (7.2.3.16)–(7.2.3.21), together with the corresponding solution to the third-order ODE (7.2.3.22), generates an exact solution of the form (7.2.3.11) for equation (7.2.3.10).

Analysis and solutions of the determining system (7.2.3.16)–(7.2.3.21) for an axisymmetric boundary layer with arbitrary $r(x)$. The system of equations (7.2.3.16)–(7.2.3.21) can be split into several simpler subsystems that can be analyzed independently.

The subsystem of equations (7.2.3.18) and (7.2.3.21) allows us to express f and φ in terms of $r = r(x)$. For given f and φ, the subsystem of equations (7.2.3.17) and (7.2.3.20) determines g and r. This indicates that, in general, the function $r = r(x)$ cannot be considered arbitrary. Once g and r are found, the pressure function can be calculated using the formula

$$F = g_t + gg_x - a_1\nu r^2 f\varphi^3, \tag{7.2.3.23}$$

which is a corollary of equation (7.2.3.16). The function ψ in equation (7.2.3.19) can be treated as an arbitrary given function. Integrating with respect to x yields

$$h = \int \frac{1}{\varphi^2}\left(\varphi\psi_t + g\varphi\psi_x + g_x\varphi\psi - \varphi_t\psi - g\varphi_x\psi - a_4\nu r^2\varphi^3\right) dx + p(t), \quad (7.2.3.24)$$

where $p(t)$ is an arbitrary function.

In what follows, the function $r = r(x)$ will be considered arbitrary.

We will begin the analysis with the subsystem consisting of equations (7.2.3.17) and (7.2.3.20). We isolate the derivative g_x and rewrite these equations in the form

$$g_x + \frac{(f\varphi)_x}{f\varphi} g = a_2\nu r^2\varphi^2 - \frac{(f\varphi)_t}{f\varphi}, \quad (7.2.3.25)$$

$$g_x - \frac{\varphi_x}{\varphi} g = -a_5\nu r^2\varphi^2 + \frac{\varphi_t}{\varphi}. \quad (7.2.3.26)$$

We will seek conditions under which equations (7.2.3.25) and (7.2.3.26) coincide. In this situation, the function $r = r(x)$ can be treated as arbitrary. Then, g is determined from (7.2.3.26) and, hence, is expressed in terms of φ and r.

Equating the functional coefficients of g and the right-hand sides of equations (7.2.3.25) and (7.2.3.26), we obtain

$$\frac{\varphi_x}{\varphi} + \frac{(f\varphi)_x}{f\varphi} = 0, \quad \frac{\varphi_t}{\varphi} + \frac{(f\varphi)_t}{f\varphi} = (a_2 + a_5)\nu r^2\varphi^2. \quad (7.2.3.27)$$

Integrating the first equation in (7.2.3.27) yields

$$f = \lambda(t)\varphi^{-2}, \quad (7.2.3.28)$$

where $\lambda(t)$ is an arbitrary function. Substituting the expression (7.2.3.28) into the second equation of (7.2.3.27), we find that

$$[\ln\lambda(t)]'_t = (a_2 + a_5)\nu r^2(x)\varphi^2. \quad (7.2.3.29)$$

Equation (7.2.3.29) holds identically in the following two cases:

$$(a)\quad a_5 = -a_2, \quad \lambda(t) = \lambda_0, \quad \varphi = \varphi(t,x) \text{ is any function}; \quad (7.2.3.30)$$

$$(b)\quad \varphi = \frac{\sigma(t)}{r(x)}, \quad \lambda(t) = \lambda_0 \exp\left[(a_2 + a_5)\,\nu \int \sigma^2(t)\,dt\right], \quad (7.2.3.31)$$

where λ_0 is an arbitrary constant, and $\sigma(t)$ is an arbitrary function.

Case (a). We can set $\lambda_0 = 1$ without loss of generality. In view of (7.2.3.28), we get $f = \varphi^{-2}$. Equations (7.2.3.18) and (7.2.3.21) become

$$\varphi'_x = -a_3\nu r^2\varphi^4, \quad \varphi'_x = \tfrac{1}{2}a_6\nu r^2\varphi^4. \quad (7.2.3.32)$$

These equations are compatible if $a_6 = -2a_3$. Integrating the first equation in (7.2.3.32) and then (7.2.3.26), we find an exact solution of equation (7.2.3.10) of the form (7.2.3.11), in which we should set

$$f = \left[3a_3\nu \int r^2(x)\,dx + b(t)\right]^{2/3},$$
$$\varphi = \left[3a_3\nu \int r^2(x)\,dx + b(t)\right]^{-1/3}, \qquad (7.2.3.33)$$
$$g = c(t)\varphi + \varphi \int \left(\frac{\varphi_t}{\varphi^2} + a_2\nu r^2\varphi\right) dx,$$

where $b = b(t)$ and $c = c(t)$ are arbitrary functions, h is determined by formula (7.2.3.24), $\psi = \psi(t, x)$ is an arbitrary function, and $u = u(\xi)$ is a function satisfying ODE (7.2.3.22) with $a_5 = -a_2$ and $a_6 = -2a_3$. The pressure function is defined by formula (7.2.3.23). It is noteworthy that if $a_3 \neq 0$, the function g can be represented as

$$g = c(t)\varphi - \frac{1}{3}b'_t(t)\varphi \int \varphi^2 dx + \frac{a_2}{2a_3}\varphi^{-1}.$$

Case (b). It follows from formulas (7.2.3.28) and (7.2.3.31) as well as equations (7.2.3.18) and (7.2.3.21) that the conditions $a_6 = -2a_3$ and $r(x) = \alpha x + \beta$ must hold, meaning that the function $r(x)$ is not arbitrary. This situation is not analyzed here.

Solutions of the determining system (7.2.3.16)–(7.2.3.21) for a plane boundary layer with $r(x) \equiv 1$. Now let us explore the case of the plane boundary layer equation (7.2.3.10) with $r(x) = 1$ and describe some of its exact solutions of the form (7.2.3.11). In all the cases discussed below, we only provide expressions of the functions $\varphi = \varphi(t, x)$, $f = f(t, x)$, and $g = g(t, x)$. The function $\psi = \psi(t, x)$ is arbitrary, the functions $h = h(t, x)$ and $F = F(t, x)$ are defined by formulas (7.2.3.24) and (7.2.3.23) with $r(x) = 1$, and the function $u = u(\xi)$ satisfies ODE (7.2.3.22).

$1°$. *Case of* $a_5 = -a_2$, $a_6 = -2a_3$, *and* $a_3 \neq 0$. We will obtain an exact solution of the form (7.2.3.11) by setting $r(x) = 1$ in (7.2.3.33), (7.2.3.23), and (7.2.3.24). As a result, we get

$$\varphi = \left[3a_3\nu x + b(t)\right]^{-1/3},$$
$$f = \left[3a_3\nu x + b(t)\right]^{2/3}, \qquad (7.2.3.34)$$
$$g = c(t)\varphi - \frac{1}{3a_3\nu}b'_t(t) + \frac{a_2}{2a_3}\varphi^{-1},$$

where $b = b(t)$ and $c = c(t)$ are arbitrary functions.

$2°$. *Case of* $a_5 = -a_2$ *and* $a_3 = a_6 = 0$. Formulas (7.2.3.33) with $a_3 = 0$ lead to an exact solution (7.2.3.11), in which we should set

$$\varphi = \varphi(t) \text{ is an arbitrary function,}$$
$$f = \varphi^{-2}, \qquad (7.2.3.35)$$
$$g = \left(\frac{\varphi_t}{\varphi} + a_2\nu\varphi^2\right)x + c(t)\varphi,$$

where $c = c(t)$ is an arbitrary function, and the other functions are described above.

3°. *Case of* $a_3 = a_6 = 0$. An exact solution of the form (7.2.3.11) is given by

$$\begin{aligned} &\varphi = \varphi(t) \text{ is an arbitrary function,} \\ &f = \frac{k}{\varphi^2} \exp\left[(a_2 + a_5)\,\nu \int \varphi^2\,dt\right], \\ &g = \left(\frac{\varphi_t}{\varphi} - a_5\nu\varphi^2\right)x + c(t)\varphi, \end{aligned} \qquad (7.2.3.36)$$

where $c(t)$ is an arbitrary function, and k is an arbitrary constant. Three functions (7.2.3.36) satisfy four equations (7.2.3.17), (7.2.3.18), (7.2.3.20), and (7.2.3.21) with $r = 1$ and $a_3 = a_6 = 0$. They acquire the form (7.2.3.35) when $a_5 = -a_2$ and $k = 1$.

Applying the method of weak symmetry reductions to construct exact solutions. The boundary layer equation (7.2.3.10) with arbitrary $r(x)$ has many more exact solutions of the form (7.2.3.11) than obtained previously by a direct reduction to the single ODE (7.2.3.22) with $a_5 = -a_2$ and $a_6 = -2a_3$. We outline below a procedure for constructing more exact solutions of the form (7.2.3.11).

The determining system (7.2.3.16)–(7.2.3.21) was derived under the assumption that the first six functions Ψ_i in (7.2.3.14) were linearly independent.

If some of the functions Ψ_i are linearly dependent, one should choose a linearly independent subsystem $\{\Psi_j\}$ and express the remaining functions in terms of the elements of this subsystem. In this situation, relation (7.2.3.12) must be rewritten as a linear combination of the functions Ψ_j. Next, the functional coefficients of Ψ_j should all be equated with zero. The resulting modified system of equations will differ from the original system (7.2.3.16)–(7.2.3.21) and will contain fewer equations. The smaller the number of linearly independent functions present in (7.2.3.14), the fewer equations will be included in the modified system. As a result, more arbitrary functions can be incorporated into its solutions.

Solutions of the form (7.2.3.11) other than those described by equations (7.2.3.16)–(7.2.3.22) can occur if there are two or more linearly independent subsystems among the functions Ψ_i one of which includes Ψ_7. For the functions $u = u(\xi)$, which define Ψ_i in (7.2.3.14), it is advisable to use particular solutions ODE (7.2.3.22) with appropriate values for the parameters a_i.

Table 7.3 presents twelve functions $u = u(\xi)$, each producing two or three linear relationships between the functions (7.2.3.14). The function in the first row results in $\Psi_7 = 0$, indicating a degenerate solution. The functions listed in the remaining rows generate nondegenerate solutions for equation (7.2.3.10).

Examples of constructing exact solutions. Here are a few examples of constructing exact solutions for the nonlinear third-order PDE (7.2.3.10) using some of the generating functions listed in Table 7.3.

► Example 7.15. Let us pick the function $u = \xi^3$ from row 2 in Table 7.3 and construct an exact solution. Considering formulas (7.2.3.14) and the linear relationships between Ψ_i (see Table 7.3), we rewrite equation (7.2.3.12), with functionals

Table 7.3. Generating functions u and respective linear relationships for Ψ_n. Notations: $\Psi_1 = 1$, $\Psi_2 = u'_\xi$, $\Psi_3 = (u'_\xi)^2$, $\Psi_4 = u''_{\xi\xi}$, $\Psi_5 = \xi u''_{\xi\xi}$, $\Psi_6 = uu''_{\xi\xi}$, and $\Psi_7 = u'''_{\xi\xi\xi}$.

No.	Generating functions u	Linear relationships for Ψ_n
1	$u=\xi^2$	$\Psi_4=2\Psi_1,\ \Psi_5=\frac{2}{3}\Psi_2,\ \Psi_6=\frac{1}{2}\Psi_3$
2	$u=\xi^3$	$\Psi_5=2\Psi_2,\ \Psi_6=\frac{2}{3}\Psi_3,\ \Psi_7=6\Psi_1$
3	$u=\xi^{-1}$	$\Psi_5=-2\Psi_2,\ \Psi_6=2\Psi_3,\ \Psi_7=-6\Psi_3$
4	$u=\exp\xi$	$\Psi_2=\Psi_4=\Psi_7,\ \Psi_6=\Psi_3$
5	$u=\cosh\xi$	$\Psi_6=\Psi_1+\Psi_3,\ \Psi_7=\Psi_2$
6	$u=\sinh\xi$	$\Psi_6=\Psi_3-\Psi_1,\ \Psi_7=\Psi_2$
7	$u=\cos\xi$	$\Psi_6=\Psi_3-\Psi_1,\ \Psi_7=-\Psi_2$
8	$u=\sin\xi$	$\Psi_6=\Psi_3-\Psi_1,\ \Psi_7=-\Psi_2$
9	$u=\tanh\xi$	$\Psi_6=-2\Psi_2+2\Psi_3,\ \Psi_7=-2\Psi_2-3\Psi_6$
10	$u=\coth\xi$	$\Psi_6=-2\Psi_2+2\Psi_3,\ \Psi_7=-2\Psi_2-3\Psi_6$
11	$u=\tan\xi$	$\Psi_6=-2\Psi_2+2\Psi_3,\ \Psi_7=2\Psi_2+3\Psi_6$
12	$u=\cot\xi$	$\Psi_6=2\Psi_2+2\Psi_3,\ \Psi_7=2\Psi_2-3\Psi_6$

defined by (7.2.3.13) and (7.2.3.14), in the form

$$\begin{aligned}&\big[g_t + gg_x - F - 6\nu r^2 f\varphi^3\big]\Psi_1\\ &\quad + \big[(f\varphi)_t + (fg\varphi)_x + 2f(\varphi_t + g\varphi_x - g_x\varphi)\big]\Psi_2\\ &\quad + \big[f\varphi(f\varphi)_x - \tfrac{2}{3}ff_x\varphi^2\big]\Psi_3\\ &\quad + f(\varphi\psi_t + g\varphi\psi_x + g_x\varphi\psi - \psi\varphi_t - g\varphi_x\psi - h_x\varphi^2)\Psi_4 = 0.\end{aligned}$$

Equating the functional coefficients of all Ψ_i ($i = 1, 2, 3, 4$) with zero, we obtain the following PDE system:

$$g_t + gg_x - F - 6\nu r^2 f\varphi^3 = 0, \tag{7.2.3.37}$$

$$(f\varphi)_t + (fg\varphi)_x + 2f(\varphi_t + g\varphi_x - g_x\varphi) = 0, \tag{7.2.3.38}$$

$$f\varphi_x + \tfrac{1}{3}f_x\varphi = 0, \tag{7.2.3.39}$$

$$\varphi\psi_t + g\varphi\psi_x + g_x\varphi\psi - \psi\varphi_t - g\varphi_x\psi - h_x\varphi^2 = 0. \tag{7.2.3.40}$$

Equation (7.2.3.39) has been reformulated for convenience, and the last two equations have been divided by the nonzero factors $f\varphi$ and f, respectively.

System (7.2.3.37)–(7.2.3.40) includes two fewer equations than system (7.2.3.16)–(7.2.3.21). In addition, the resulting equations do not require a consistency analysis. Equations (7.2.3.37)–(7.2.3.39) allow us to find f, g, and F for arbitrary $r(x)$. The functions $\varphi = \varphi(t, x)$ and $\psi = \psi(t, x)$ remain arbitrary, while the function h, as follows from (7.2.3.40), is defined by formula (7.2.3.24) with $a_4 = 0$.

The general solution of equation (7.2.3.39) is expressed as

$$f = a(t)\varphi^{-3}, \tag{7.2.3.41}$$

where $a(t)$ is an arbitrary function. After substituting this expression into (7.2.3.38) and integrating, we obtain

$$g = \frac{1}{a(t)}\left[a'_t(t)x + b(t)\right], \tag{7.2.3.42}$$

where $b(t)$ is an arbitrary function. Substituting formulas (7.2.3.41) and (7.2.3.42) into equation (7.2.3.37) yields the pressure function:

$$F(t,x) = -6\nu a(t)r^2(x) + \frac{1}{a(t)}\left[a''_{tt}(t)x + b'_t(t)\right].$$

For $\psi = h = 0$, by inserting (7.2.3.41) and (7.2.3.42) into (7.2.3.11), we find an exact solution:

$$u = a(t)z^3 + \frac{1}{a(t)}\left[a'_t(t)x + b(t)\right]z,$$

where $a(t)$ and $b(t)$ are arbitrary functions. Interestingly, the arbitrary function φ included in (7.2.3.41) does not appear in the final result, as it is cancelled out after substituting into (7.2.3.11). ◄

► Example 7.16. Now let us pick $u = \exp\xi$ from row 4 in Table 7.3. Using formulas (7.2.3.14) and the linear relationships between Ψ_i (see Table 7.3), we rewrite equation (7.2.3.12), with functionals defined by (7.2.3.13) and (7.2.3.14), in the form

$$\begin{aligned}(g_t + gg_x - F)\Psi_1 &+ \left[(f\varphi)_t + (fg\varphi)_x - f\varphi^2 h_x - \nu r^2 f\varphi^3\right]\Psi_2 \\ &+ f\varphi\left[(f\varphi)_x - f_x\varphi\right]\Psi_3 + f(\varphi_t + g\varphi_x - g_x\varphi)\Psi_5 = 0,\end{aligned}$$

where we have set $\psi = 0$ for simplicity. Equating the functional coefficients of Ψ_i $(i = 1,\ 2,\ 3,\ 5)$ with zero, we obtain

$$\begin{gathered} g_t + gg_x - F = 0, \\ (f\varphi)_t + (fg\varphi)_x - f\varphi^2 h_x - \nu r^2 f\varphi^3 = 0, \\ \varphi_x = 0, \quad \varphi_t + g\varphi_x - g_x\varphi = 0. \end{gathered} \tag{7.2.3.43}$$

It follows from the first, third, and fourth equations that

$$\varphi = \varphi(t), \quad g = \frac{1}{\varphi}\left[\varphi'_t x + b(t)\right], \quad F = \frac{1}{\varphi}\left[\varphi''_{tt}x + b'_t(t)\right], \tag{7.2.3.44}$$

where $\varphi = \varphi(t)$ and $b = b(t)$ are arbitrary functions. The second equation in (7.2.3.43) gives

$$h = \frac{2\varphi'_t}{\varphi^2}x + \frac{1}{\varphi}\int\left(\frac{f_t}{f} + g\frac{f_x}{f} - \nu r^2\varphi^2\right)dx + c(t), \tag{7.2.3.45}$$

where $c(t)$ is an arbitrary function.

Formulas (7.2.3.11), (7.2.3.44), and (7.2.3.45) for $\psi = 0$ and arbitrary $f = f(t,x)$ determine an exact solution of equation (7.2.3.10) for arbitrary function $r(x)$, meaning any shape of the body in the flow. ◄

Remark 7.6. *It is crucial to highlight the fact that the direct method of weak symmetry reductions, as outlined in Section 7.2.1, allows for determining significantly more exact solutions of equation (7.2.3.10) compared to the direct method of symmetry reductions, discussed in Section 7.1.*

Remark 7.7. *Polyanin and Zhurov (2016a, 2022) used the method of weak symmetry reductions to identify numerous exact solutions to the unsteady axisymmetric boundary layer equations for a Newtonian fluid on an extended body of revolution. Furthermore, in their studies (2016b, 2022), they successfully applied this method to construct solutions to both plane and unsteady axisymmetric boundary layer equations for power-law and other non-Newtonian fluids.*

Problems and Exercises for Section 7.2

1. Unsteady axisymmetric boundary-layer equations on an extensive body of revolution can be reduced to a single third-order PDE for the modified stream function:

$$w_{tz} + w_z w_{xz} - w_x w_{zz} = \nu z w_{zzz} + F(t,x). \tag{7.2.3.46}$$

Polyanin and Zhurov (2016a) showed that this equation admits exact solutions of the form

$$w = fu(\xi) + gz + h, \qquad \xi = \varphi z + \psi, \tag{7.2.3.47}$$

where the six functions $f = f(t,x)$, $g = g(t,x)$, $h = h(t,x)$, $\varphi = \varphi(t,x)$, $\psi = \psi(t,x)$, and $u = u(\xi)$ must be determined during the analysis process.

Using the direct method of symmetry reduction, it is required to:

a) derive a single third-order ODE for the function $u = u(\xi)$,

b) derive a determining PDE system for the remaining functions,

c) find simple particular solutions of the ODE from Item (a) (for suitable free coefficients of the ODE),

d) analyze and find solutions to the defining system of partial differential equations from Item (b).

2. Using the direct method of week symmetry reduction, it is required to:

a) find exact solutions of the form (7.2.3.47) with $u(\xi) = \xi^3$ of the nonlinear PDE (7.2.3.46),

b) find exact solutions of the form (7.2.3.47) with $u(\xi) = \xi^k$ of the nonlinear PDE (7.2.3.46),

c) find exact solutions of the form (7.2.3.47) with $u(\xi) = e^{\xi}$ of the nonlinear PDE (7.2.3.46).

3. Unsteady axisymmetric boundary-layer equations of power-law non-Newtonian fluids on a body of revolution can be reduced to a single third-order PDE for the stream function:

$$w_{tz} + w_z w_{xz} - w_x w_{zz} = \kappa r^{n+1}(x) w_{zz}^{n-1} w_{zzz} + F(t,x), \tag{7.2.3.48}$$

where n is a rheological parameter of the fluid, the function $r(x)$, determining the shape of the body, is assumed arbitrary. For $n = 1$, equation (7.2.3.48) coincides with equation (7.2.3.10) with $\kappa = \nu$.

Polyanin and Zhurov (2016b) showed that equation (7.2.3.48) admits exact solutions of the form (7.2.3.47).

Using the direct method of symmetry reduction, it is required to:

a) derive a single third-order ODE for the function $u = u(\xi)$,

b) derive a determining PDE system for the remaining functions,

c) find simple particular solutions of the ODE from Item (a) (for suitable free coefficients of the ODE),

d) analyze and find solutions to the defining system of partial differential equations from Item (b).

4. Using the direct method of week symmetry reduction, it is required to:

a) find exact solutions of the form (7.2.3.47) with $u(\xi) = c\xi^{\frac{2n+1}{n}}$ of the nonlinear PDE (7.2.3.48),

b) find exact solutions of the form (7.2.3.47) with $u(\xi) = e^{\xi}$ of the nonlinear PDE (7.2.3.48) with $n = 2$.

5. Show that the equation of the unsteady boundary layer of the general non-Newtonian fluid model

$$w_{tz} + w_z w_{xz} - w_x w_{zz} = [G(w_{zz})]_z + F(t, x),$$

where $G = G(u)$ is an arbitrary function, admits exact solutions of the form (7.2.3.47) with $f = \varphi^{-2}$ (Polyanin and Zhurov, 2016b).

References for Chapter 7

Aksenov A.V. and Kozyrev A.A. New reductions of the unsteady axisymmetric boundary layer equation to ODEs and simpler PDEs. *Mathematics*, 2022, Vol. 10, No. 10, 1673.

Aristov S.N. and Polyanin A.D. Exact solutions of unsteady three-dimensional Navier–Stokes equations. *Doklady Physics*, 2009, Vol. 54, No. 7, pp. 316–321.

Aristov S.N., Knyazev D.V., and Polyanin A.D. Exact solutions of the Navier–Stokes equations with the linear dependence of velocity components of two space variables. *Theor. Foundations of Chemical Engineering*, 2009, Vol. 43, No. 5, pp. 642–662.

Burde G.I. The construction of special explicit solutions of the boundary-layer equations. Steady flows. *Quart. J. Mech. Appl. Math.*, 1994, Vol. 47, No. 2, pp. 247–260.

Danilov V.G. and Subochev P.Yu. Wave solutions of semilinear parabolic equations. *Theor. Math. Phys.*, 1991, Vol. 89, No. 1, pp. 1029–1045.

Cariello F. and Tabor M. Painlevé expansions for nonintegrable evolution equations, *Physica D*, 1989, Vol. 39, No. 1, pp. 77–94.

Clarkson P.A. and Kruskal M.D. New similarity reductions of the Boussinesq equation. *J. Math. Phys.*, 1989, Vol. 30, No. 10, pp. 2201–2213.

Clarkson P.A., Ludlow D.K., and Priestley T.J. The classical, direct and nonclassical methods for symmetry reductions of nonlinear partial differential equations. *Methods Appl. Anal.*, 1997, Vol. 4, No. 2, pp. 173–195.

Clarkson P.A. and Mansfield E.L. Symmetry reductions and exact solutions of a class of nonlinear heat equations. *Physica D*, 1994, Vol. 70, No. 3, pp. 250–288.

Estévez P.G. Non-classical symmetry and the singular manifold: the Burgers and the Burgers–Huxley equation. *J. Phys. A: Math. Gen.*, 1994, Vol. 27, pp. 2113–2127.

Kawahara T. and Tanaka M. Interactions of traveling fronts: an exact solution of a nonlinear diffusion equations. *Phys. Lett. A*, 1983, Vol. 97, No. 8, pp. 311–314.

Kudryashov N.A. On exact solutions of families of Fisher equations. *Theor. Math. Phys.*, 1993, Vol. 94, No. 2, pp. 211–218.

Loitsyanskiy L.G. *Mechanics of Liquids and Gases*. Begell House, New York, 1996.

Nucci M.C. and Clarkson P.A. The nonclassical method is more general than the direct method for symmetry reductions. An example of the Fitzhugh–Nagumo equation. *Phys. Lett. A*, 1992 Vol. 164, pp. 49–56.

Pokhozhaev S.I. On a problem of L.V. Ovsyannikov. *J. Appl. Mech. Tech. Phys.*, 1989, Vol. 39, No. 2, pp. 169–174.

Polyanin A.D. Functional separable solutions of nonlinear convection-diffusion equations with variable coefficients. *Commun. Nonlinear Sci. Numer. Simulat.*, 2019a, Vol. 73, pp. 379–390.

Polyanin A.D. Functional separable solutions of nonlinear reaction-diffusion equations with variable coefficients. *Applied Math. & Comput.*, 2019b, Vol. 347, pp. 282–292.

Polyanin A.D. and Zaitsev V.F. *Handbook of Nonlinear Partial Differential Equations, 2nd ed.* Chapman & Hall/CRC Press, Boca Raton–London, 2012.

Polyanin A.D. and Zhurov A.I. Unsteady axisymmetric boundary-layer equations: Transformations, properties, exact solutions, order reduction and solution method. *Int. J. Non-Linear Mechanics*, 2015, Vol. 74, pp. 40–50.

Polyanin A.D. and Zhurov A.I. Direct functional separation of variables and new exact solutions to axisymmetric unsteady boundary-layer equations. *Commun. Nonlinear Science & Numer. Simulation*, 2016a, Vol. 31, No. 1-3, pp. 11–20.

Polyanin A.D. and Zhurov A.I. One-dimensional reductions and functional separable solutions to unsteady plane and axisymmetric boundary-layer equations for non-Newtonian fluids. *Int. J. Non-Linear Mechanics*, 2016b, Vol. 85, pp. 70–80.

Polyanin A.D. and Zhurov A.I. *Separation of Variables and Exact Solutions to Nonlinear PDEs*. CRC Press, Boca Raton–London, 2022.

Schlichting H. *Boundary Layer Theory*. McGraw-Hill, New York, 1981.

Zhurov A.I. and Polyanin A.D. Symmetry reductions and new functional separable solutions of nonlinear Klein–Gordon and telegraph type equations. *J. Nonlinear Math. Phys.*, 2020, Vol. 27, pp. 1–16.

8. Classical Method of Symmetry Reductions

► **Preliminary remarks.** The classical method of symmetry reductions of partial differential equations* presents a regular procedure that allows obtaining the following:

(i) transformations under which differential equations are invariant (such transformations bring the given equation to itself);

(ii) new variables (dependent and independent) in which differential equations become considerably simpler.

The transformations from Item (i) convert a solution of an equation to the same or another solution of this equation. In the former case, we obtain an *invariant solution*, which can be found passing in the equation under consideration to new variables that are fewer in number than the original variables (in this case, invariant solutions of PDEs with two independent variables are expressed through solutions of ODEs). In the latter case, we have noninvariant solutions, which may be 'multiplied' to obtain a family of solutions.

The application of the classical method of symmetry reductions is illustrated by examples of obtaining invariant solutions of nonlinear equations of heat and mass transfer, wave theory and hydrodynamics.

Remark 8.1. *The classical method of symmetry reductions for differential equations is a significant extension of the much simpler method based on elementary invariant theory and described earlier in Chapter 4.*

Remark 8.2. *The classical method of symmetry reductions is one of the most popular approaches for finding exact solutions to nonlinear equations. A comprehensive description of this method, along with numerous concrete examples of its application, can be found, for example, in the books by Bluman & Cole (1974), Ovsiannikov (1982), Bluman & Kumei (1989), Fushchich et al. (1993), Olver (1993), Ibragimov (1994), Hydon (2000), and Cantwell (2002).*

Remark 8.3. *Sections 8.1 to 8.3 of this book present a description of the classical method of symmetry reductions with minimal use of specialized (group) terminology, which makes the material easier to understand. In Section 8.4 an explanation of the origin of the frequently encountered term 'Lie group analysis' will be given.*

---

*It is also known as the *Lie group analysis of differential equations* or briefly *group analysis of differential equations*.

8.1. One-Parameter Transformations and Their Local Properties

8.1.1. One-Parameter Transformations. Infinitesimal Operator

We will consider invertible transformations of the form

$$\begin{aligned} \bar{x} &= \varphi_1(x, y, u, \varepsilon), & \bar{x}|_{\varepsilon=0} &= x, \\ \bar{y} &= \varphi_2(x, y, u, \varepsilon), & \bar{y}|_{\varepsilon=0} &= y, \\ \bar{u} &= \psi(x, y, u, \varepsilon), & \bar{u}|_{\varepsilon=0} &= u, \end{aligned} \qquad (8.1.1.1)$$

where φ_1, φ_2, and ψ are sufficiently smooth functions of their arguments, and ε is a real parameter. It is assumed that the successive application (composition) of two transformations of the form (8.1.1.1) with parameters ε and $\bar{\varepsilon}$ is equivalent to a single transformation of the form with parameter $\varepsilon+\bar{\varepsilon}$ (this means that such transformations have the group property).

Remark 8.4. *In the special case of transformations in the plane, the functions φ_1 and φ_2 in (8.1.1.1) are independent of u, and $\psi = u$ (i.e., $\bar{u} = u$).*

The expansion of (8.1.1.1) into truncated Taylor series in ε about $\varepsilon = 0$ to linear terms gives

$$\bar{x} \simeq x + \varepsilon\xi(x, y, u), \quad \bar{y} \simeq y + \varepsilon\eta(x, y, u), \quad \bar{u} \simeq u + \varepsilon\zeta(x, y, u), \qquad (8.1.1.2)$$

where

$$\xi(x, y, u) = \frac{\partial\varphi_1}{\partial\varepsilon}\bigg|_{\varepsilon=0}, \quad \eta(x, y, u) = \frac{\partial\varphi_2}{\partial\varepsilon}\bigg|_{\varepsilon=0}, \quad \zeta(x, y, u) = \frac{\partial\psi}{\partial\varepsilon}\bigg|_{\varepsilon=0}.$$

The linear first-order differential operator

$$\mathrm{X} = \xi(x, y, u)\frac{\partial}{\partial x} + \eta(x, y, u)\frac{\partial}{\partial y} + \zeta(x, y, u)\frac{\partial}{\partial u}, \qquad (8.1.1.3)$$

which corresponds to the infinitesimal transformation (8.1.1.2), is called an *infinitesimal operator.**

LIE'S THEOREM. Suppose the coordinates $\xi(x, y, u)$, $\eta(x, y, u)$, and $\zeta(x, y, z)$ of the infinitesimal operator (8.1.1.3) are known. Then the transformation (8.1.1.1), possessing the group property, can be completely recovered by solving the Lie equations

$$\frac{d\varphi_1}{d\varepsilon} = \xi(\varphi_1, \varphi_2, \psi), \quad \frac{d\varphi_2}{d\varepsilon} = \eta(\varphi_1, \varphi_2, \psi), \quad \frac{d\psi}{d\varepsilon} = \zeta(\varphi_1, \varphi_2, \psi)$$

subject to the initial conditions

$$\varphi_1|_{\varepsilon=0} = x, \quad \varphi_2|_{\varepsilon=0} = y, \quad \psi|_{\varepsilon=0} = u.$$

*In the literature, it is also known as an *infinitesimal generator* or a *group generator*. In this book, we call an operator (8.1.1.3) an infinitesimal operator, a group generator, or, for short, an operator.

8.1.2. Invariant of an Infinitesimal Operator. Transformations in the Plane

An *invariant of the operator* (8.1.1.3) is a function $I(x, y, u)$ that satisfies the condition

$$I(\bar{x}, \bar{y}, \bar{u}) = I(x, y, u).$$

Let us expand this relation in a Taylor series in the small parameter ε, divide the resulting relation by ε, and then proceed to the limit as $\varepsilon \to 0$ to obtain a linear partial differential equation for I:

$$\mathrm{X}I = \xi(x, y, u)I_x + \eta(x, y, u)I_y + \zeta(x, y, u)I_u = 0. \tag{8.1.2.1}$$

Let the associated characteristic system of ordinary differential equations (see Section 2.1.2)

$$\frac{dx}{\xi(x, y, u)} = \frac{dy}{\eta(x, y, u)} = \frac{du}{\zeta(x, y, u)} \tag{8.1.2.2}$$

have functionally independent integrals

$$I_1(x, y, u) = C_1, \qquad I_2(x, y, u) = C_2, \tag{8.1.2.3}$$

where C_1 and C_2 are arbitrary constants. Then the general solution of equation (8.1.2.1) has the form

$$I = \Psi(I_1, I_2), \tag{8.1.2.4}$$

where $\Psi(I_1, I_2)$ is an arbitrary function of two arguments, $I_1 = I_1(x, y, u)$ and $I_2 = I_2(x, y, u)$.

This means that the operator (8.1.1.3) has two functionally independent invariants, I_1 and I_2, and any function $\Phi(x, y, u)$ invariant under the operator (8.1.1.3) can be represented as a function of the two invariants.

Table 8.1 lists the most common transformations in the plane and the corresponding operator (8.1.1.3) and invariants; only one invariant is specified, with the other being the same: $I_2 = u$.

8.1.3. Formulas for Derivatives. Coordinates of the First and Second Prolongations

In terms of the new variables (8.1.1.1), the first derivatives become

$$\bar{u}_{\bar{x}} \simeq u_x + \varepsilon\zeta_1, \qquad \bar{u}_{\bar{y}} \simeq u_y + \varepsilon\zeta_2. \tag{8.1.3.1}$$

Here, ζ_1 and ζ_2 are the coordinates of the first prolongation, which are found as

$$\begin{aligned}
\zeta_1 &= D_x(\zeta) - u_x D_x(\xi) - u_y D_x(\eta) \\
&= \zeta_x + (\zeta_u - \xi_x)u_x - \eta_x u_y - \xi_u u_x^2 - \eta_u u_x u_y, \\
\zeta_2 &= D_y(\zeta) - u_x D_y(\xi) - u_y D_y(\eta) \\
&= \zeta_y - \xi_y u_x + (\zeta_u - \eta_y)u_y - \xi_u u_x u_y - \eta_u u_y^2,
\end{aligned} \tag{8.1.3.2}$$

Table 8.1. One-parameter transformations in the plane.

Name	Transformation	Operator	Invariant
Translation in the x-axis	$\bar{x} = x + \varepsilon,\ \bar{y} = y$	$\mathrm{X} = \frac{\partial}{\partial x}$	$I_1 = y$
Translation along the straight line $ax + by = 0$	$\bar{x} = x + b\varepsilon,\ \bar{y} = y - a\varepsilon$	$\mathrm{X} = b\frac{\partial}{\partial x} - a\frac{\partial}{\partial y}$	$I_1 = ax + by$
Rotation	$\bar{x} = x\cos\varepsilon + y\sin\varepsilon$, $\bar{y} = y\cos\varepsilon - x\sin\varepsilon$	$\mathrm{X} = y\frac{\partial}{\partial x} - x\frac{\partial}{\partial y}$	$I_1 = x^2 + y^2$
Galileo transformation	$\bar{x} = x + \varepsilon y,\ \bar{y} = y$	$\mathrm{X} = y\frac{\partial}{\partial x}$	$I_1 = y$
Lorentz transformation	$\bar{x} = x\cosh\varepsilon + y\sinh\varepsilon$, $\bar{y} = y\cosh\varepsilon + x\sinh\varepsilon$	$\mathrm{X} = y\frac{\partial}{\partial x} + x\frac{\partial}{\partial y}$	$I_1 = y^2 - x^2$
Uniform extension	$\bar{x} = xe^{\varepsilon},\ \bar{y} = ye^{\varepsilon}$	$\mathrm{X} = x\frac{\partial}{\partial x} + y\frac{\partial}{\partial y}$	$I_1 = y/x$
Nonuniform extension	$\bar{x} = xe^{a\varepsilon},\ \bar{y} = ye^{b\varepsilon}$	$\mathrm{X} = ax\frac{\partial}{\partial x} + by\frac{\partial}{\partial y}$	$I_1 = \lvert y\rvert^a \lvert x\rvert^{-b}$

where D_x and D_y are the total differential operators with respect to x and y:

$$D_x = \frac{\partial}{\partial x} + u_x\frac{\partial}{\partial u} + u_{xx}\frac{\partial}{\partial u_x} + u_{xy}\frac{\partial}{\partial u_y} + \cdots,$$
$$D_y = \frac{\partial}{\partial y} + u_y\frac{\partial}{\partial u} + u_{xy}\frac{\partial}{\partial u_x} + u_{yy}\frac{\partial}{\partial u_y} + \cdots. \tag{8.1.3.3}$$

Let us verify that the first formula in (8.1.3.1) holds. Obviously,

$$\bar{u}_x = \bar{u}_{\bar{x}}\bar{x}_x + \bar{u}_{\bar{y}}\bar{y}_x, \quad \bar{u}_y = \bar{u}_{\bar{x}}\bar{x}_y + \bar{u}_{\bar{y}}\bar{y}_y. \tag{8.1.3.4}$$

Differentiating (8.1.1.2) with respect to x and y and retaining all terms up to the first degree of ε, we have

$$\begin{aligned}\bar{x}_x &= 1 + \varepsilon D_x\xi, & \bar{x}_y &= \varepsilon D_y\xi,\\ \bar{y}_x &= \varepsilon D_x\eta, & \bar{y}_y &= 1 + \varepsilon D_y\eta,\\ \bar{u}_x &= u_x + \varepsilon D_x\zeta, & \bar{u}_y &= u_y + \varepsilon D_y\zeta.\end{aligned} \tag{8.1.3.5}$$

In order to calculate $\bar{u}_{\bar{x}}$, let us eliminate $\bar{u}_{\bar{y}}$ from (8.1.3.4) and then substitute the derivatives $\bar{x}_x$, $\bar{x}_y$, $\bar{y}_x$, $\bar{y}_y$, $\bar{u}_x$, $\bar{u}_y$ for their respective expression from (8.1.3.5) to obtain

$$\bar{u}_{\bar{x}} = \frac{u_x + \varepsilon(D_x\zeta + u_x D_y\eta - u_y D_x\eta) + \varepsilon^2(D_x\zeta D_y\eta - D_x\eta D_y\zeta)}{1 + \varepsilon(D_x\xi + D_y\eta) + \varepsilon^2(D_x\xi D_y\eta - D_x\eta D_y\xi)}.$$

Expanding into a series in ε, we have

$$\bar{u}_{\bar{x}} \simeq u_x + \varepsilon\zeta_1, \quad \zeta_1 = D_x\zeta - u_x D_x\xi - u_y D_x\eta,$$

which was to be proved. The coordinate ζ_2 is calculated likewise.

The second derivatives in terms of the new variables (8.1.1.1) are calculated as

$$\bar{u}_{\bar{x}\bar{x}} \simeq u_{xx} + \varepsilon\zeta_{11}, \quad \bar{u}_{\bar{x}\bar{y}} \simeq u_{xy} + \varepsilon\zeta_{12}, \quad \bar{u}_{\bar{y}\bar{y}} \simeq u_{yy} + \varepsilon\zeta_{22}. \qquad (8.1.3.6)$$

Here, the ζ_{ij} are the coordinates of the second prolongation and are found as (see, for example, Ovsiannikov, 1982 and Ibragimov, 1994):

$$\begin{aligned}
\zeta_{11} &= D_x(\zeta_1) - u_{xx}D_x(\xi) - u_{xy}D_x(\eta),\\
\zeta_{12} &= D_y(\zeta_1) - u_{xx}D_y(\xi) - u_{xy}D_y(\eta),\\
\zeta_{22} &= D_y(\zeta_2) - u_{xy}D_y(\xi) - u_{yy}D_y(\eta),
\end{aligned}$$

or, in a detailed form,

$$\begin{aligned}
\zeta_{11} &= \zeta_{xx} + (2\zeta_{xu} - \xi_{xx})u_x - \eta_{xx}u_y + (\zeta_{uu} - 2\xi_{xu})u_x^2 - 2\eta_{xu}u_xu_y\\
&\quad - \xi_{uu}u_x^3 - \eta_{uu}u_x^2u_y + (\zeta_u - 2\xi_x - 3\xi_uu_x - \eta_uu_y)u_{xx} - 2(\eta_x + \eta_uu_x)u_{xy},\\
\zeta_{12} &= \zeta_{xy} + (\zeta_{yu} - \xi_{xy})u_x + (\zeta_{xu} - \eta_{xy})u_y - \xi_{yu}u_x^2\\
&\quad - (\zeta_{uu} - \xi_{xu} - \eta_{yu})u_xu_y - \eta_{xu}u_y^2 - \xi_{uu}u_x^2u_y - \eta_{uu}u_xu_y^2\\
&\quad - (\xi_y + \xi_uu_y)u_{xx} + (\zeta_u - \xi_x - \eta_y - 2\xi_uu_x - 2\eta_uu_y)u_{xy} - (\eta_x + \eta_uu_x)u_{yy},\\
\zeta_{22} &= \zeta_{yy} - \xi_{yy}u_x + (2\zeta_{yu} - \eta_{yy})u_y - 2\xi_{yu}u_xu_y + (\zeta_{uu} - 2\eta_{yu})u_y^2\\
&\quad - \xi_{uu}u_xu_y^2 - \eta_{uu}u_y^3 - 2(\xi_y + \xi_uu_y)u_{xy} + (\zeta_u - 2\eta_y - \xi_uu_x - 3\eta_uu_y)u_{yy}.
\end{aligned} \qquad (8.1.3.7)$$

The above formulas for the first and second prolongation coordinates, (8.1.3.2) and (8.1.3.7), will be required later for the analysis of differential equations.

8.2. Symmetries of Nonlinear Second-Order Equations. Invariance Condition

8.2.1. Invariance Condition. Splitting in Derivatives

We will consider second-order partial differential equations in two independent variables

$$F\big(x, y, u, u_x, u_y, u_{xx}, u_{xy}, u_{yy}\big) = 0. \qquad (8.2.1.1)$$

The procedure for finding symmetries* of equation (8.2.1.1) consists of several stages. In the first stage, one requires that equation (8.2.1.1) be invariant (preserve its form) under transformations (8.1.1.1), so that

$$F\big(\bar{x}, \bar{y}, \bar{u}, \bar{u}_{\bar{x}}, \bar{u}_{\bar{y}}, \bar{u}_{\bar{x}\bar{x}}, \bar{u}_{\bar{x}\bar{y}}, \bar{u}_{\bar{y}\bar{y}}\big) = 0. \qquad (8.2.1.2)$$

Let us expand this equation into a series in ε about $\varepsilon = 0$, taking into account that the leading term vanishes, according to (8.2.1.1). Using formulas (8.1.1.2), (8.1.3.1),

*A symmetry of an equation is a transformation that preserves its form.

(8.1.3.6) and retaining the terms up to the first degree of ε, we obtain (see, for example, Ovsiannikov, 1982):

$$\underset{2}{\mathrm{X}}F\big(x, y, u, u_x, u_y, u_{xx}, u_{xy}, u_{yy}\big)\big|_{F=0} = 0, \tag{8.2.1.3}$$

where

$$\underset{2}{\mathrm{X}}F = \xi\frac{\partial F}{\partial x} + \eta\frac{\partial F}{\partial y} + \zeta\frac{\partial F}{\partial u} + \zeta_1\frac{\partial F}{\partial u_x} + \zeta_2\frac{\partial F}{\partial u_y} + \zeta_{11}\frac{\partial F}{\partial u_{xx}} + \zeta_{12}\frac{\partial F}{\partial u_{xy}} + \zeta_{22}\frac{\partial F}{\partial u_{yy}}. \tag{8.2.1.4}$$

The coordinates of the first and second prolongations, ζ_i and ζ_{ij}, are defined by formulas (8.1.3.2) and (8.1.3.7). Relation (8.2.1.3) is called the *invariance condition*, and the operator $\underset{2}{\mathrm{X}}$ is called the *second prolongation of the operator* (8.1.1.3).

In the second stage, the derivative u_{yy} is eliminated from (8.2.1.3) using equation (8.2.1.1); instead of u_{yy}, one can eliminate either of the first derivatives or another second derivative. The resulting relation is then written as a polynomial in the 'independent variables,' the various combinations of the products of the partial derivatives (involving various powers of u_x, u_y, u_{xx}, and u_{xy}):

$$\sum A_{k_1k_2k_3k_4}(u_x)^{k_1}(u_y)^{k_2}(u_{xx})^{k_3}(u_{xy})^{k_4} = 0, \tag{8.2.1.5}$$

where the functional coefficients $A_{k_1k_2k_3k_4}$ are dependent on x, y, u, ξ, η, and ζ and the derivatives of ξ, η, and ζ only and are independent of the derivatives of u. Equation (8.2.1.5) is satisfied if all $A_{k_1k_2k_3k_4}$ are zero. Thus, the invariance condition is split into an overdetermined determining system, resulting from equating all functional coefficients of the various products of the remaining derivatives with zero (recall that the unknowns ξ, η, and ζ are independent of u_x, u_y, u_{xx}, and u_{xy}).

In the third stage, one solves the determining system and finds admissible coordinates ξ, η, and ζ of the infinitesimal operator (8.1.1.3).

Remark 8.5. *It should be noted that the functional coefficients $A_{k_1k_2k_3k_4}$ and the determining system are linear in the unknowns ξ, η, and ζ.*

Remark 8.6. *An invariant I that is a solution of equation (8.1.2.1) is also a solution of the equation $\underset{2}{X}I = 0$.*

The procedure for finding symmetries of differential equations is illustrated below by specific examples.

8.2.2. Examples of Finding Symmetries of Nonlinear Equations

► Example 8.1. Consider the two-dimensional stationary heat equation with a nonlinear source

$$u_{xx} + u_{yy} = f(u). \tag{8.2.2.1}$$

The corresponding left-hand side of equation (8.2.1.1) is $F = u_{xx} + u_{yy} - f(u)$.

1°. Infinitesimal operators X admitted by the equations are sought in the form (8.1.2.1), where the coordinates $\xi = \xi(x, y, u)$, $\eta = \eta(x, y, u)$, $\zeta = \zeta(x, y, u)$ are yet unknown and are to be determined in the subsequent analysis. In view of identity $F = u_{xx} + u_{yy} - f(u)$, the invariance condition (8.2.1.3)–(8.2.1.4) is written as

$$\left[\zeta_{22} + \zeta_{11} - \zeta f'(u)\right]_{F=0} = 0.$$

Substituting here the coordinates of the second prolongation (8.1.3.7) and then replacing u_{yy} by $f(u) - u_{xx}$, which follows from equation (8.2.2.1), we obtain

$$\begin{aligned}
&-2\xi_u u_x u_{xx} + 2\eta_u u_y u_{xx} - 2\eta_u u_x u_{xy} - 2\xi_u u_y u_{xy} - 2(\xi_x - \eta_y) u_{xx} \\
&\quad - 2(\xi_y + \eta_x) u_{xy} - \xi_{uu} u_x^3 - \eta_{uu} u_x^2 u_y - \xi_{uu} u_x u_y^2 - \eta_{uu} u_y^3 + (\zeta_{uu} - 2\xi_{xu}) u_x^2 \\
&\quad - 2(\xi_{yu} + \eta_{xu}) u_x u_y + (\zeta_{uu} - 2\eta_{yu}) u_y^2 + (2\zeta_{xu} - \xi_{xx} - \xi_{yy} - f\xi_u) u_x \\
&\quad + (2\zeta_{yu} - \eta_{xx} - \eta_{yy} - 3f\eta_u) u_y + \zeta_{xx} + \zeta_{yy} + f(\zeta_u - 2\eta_y) - \zeta f' = 0,
\end{aligned}$$

where $f = f(u)$ and $f' = df/du$. Equating the coefficients of all combinations of the partial derivatives to zero, we get the overdetermined determining system

$$\begin{aligned}
u_x u_{xx}&: \quad \xi_u = 0, \\
u_y u_{xx}&: \quad \eta_u = 0, \\
u_{xx}&: \quad \xi_x - \eta_y = 0, \\
u_{xy}&: \quad \xi_y + \eta_x = 0, \\
u_x^2&: \quad \zeta_{uu} - 2\xi_{xu} = 0, \\
u_x u_y&: \quad \eta_{xu} + \xi_{yu} = 0, \\
u_y^2&: \quad \zeta_{uu} - 2\eta_{yu} = 0, \\
u_x&: \quad 2\zeta_{xu} - \xi_{xx} - \xi_{yy} - \xi_u f(u) = 0, \\
u_y&: \quad 2\zeta_{yu} - \eta_{xx} - \eta_{yy} - 3\eta_u f(u) = 0, \\
1&: \quad \zeta_{xx} + \zeta_{yy} - f'(u)\zeta + f(u)(\zeta_u - 2\eta_y) = 0.
\end{aligned} \tag{8.2.2.2}$$

Here, the left column indicates a combination of the derivatives and the right column gives the associated coefficient. The coefficients of $u_y u_{xy}$, $u_x u_{xy}$, u_x^3, $u_x^2 u_y$, $u_x u_y^2$, u_y^3 either are among those already listed or are their differential consequences, and therefore they are omitted. It follows from the first, second, and fifth equations and their consequences that

$$\xi = \xi(x, y), \qquad \eta = \eta(x, y), \qquad \zeta = a(x, y)u + b(x, y). \tag{8.2.2.3}$$

The third and fourth equations of system (8.2.2.2) give

$$\xi_{xx} + \xi_{yy} = 0, \qquad \eta_{xx} + \eta_{yy} = 0. \tag{8.2.2.4}$$

Substituting (8.2.2.3) into the eighth and ninth equations of (8.2.2.2) and using (8.2.2.4), we find that $a_x = a_y = 0$, whence

$$a(x, y) = a = \text{const.} \tag{8.2.2.5}$$

In view of (8.2.2.3) and (8.2.2.5), system (8.2.2.2) becomes

$$\begin{aligned} &\xi_x - \eta_y = 0, \\ &\xi_y + \eta_x = 0, \\ &b_{xx} + b_{yy} - auf'(u) - bf'(u) + f(u)(a - 2\eta_y) = 0. \end{aligned} \tag{8.2.2.6}$$

For arbitrary $f = f(u)$, it follows that $a = b = \eta_y = 0$, and then $\xi = C_1 y + C_2$, $\eta = -C_1 x + C_3$, and $\zeta = 0$. By setting one of the constants to unity and the others to zero, we establish that the original equation admits three different operators:

$$\begin{aligned} &\mathrm{X}_1 = \partial_x && (C_2 = 1,\ C_1 = C_3 = 0); \\ &\mathrm{X}_2 = \partial_y && (C_3 = 1,\ C_1 = C_2 = 0); \\ &\mathrm{X}_3 = y\partial_x - x\partial_y && (C_1 = 1,\ C_2 = C_3 = 0). \end{aligned} \tag{8.2.2.7}$$

The first two operators define a translation along the x- and y-axis. The third operator represents a rotational symmetry.

2°. Let us dwell on the third equation of system (8.2.2.6). If the relation

$$(au + b)f'(u) - f(u)(a - 2\eta_y) = 0 \tag{8.2.2.8}$$

holds, there may exist other solutions of system (8.2.2.6) that will result in operators other than (8.2.2.7). Let us study two cases: $a \neq 0$ and $a = 0$.

Case 1. If $a \neq 0$, the solution of equation (8.2.2.8) gives

$$f(u) = C(au + b)^{1 - \frac{2\gamma}{a}},$$

where $\gamma = \eta_y = \text{const}$ and $b = \text{const}$. Therefore, for $f(u) = u^k$, it follows from formulas (8.2.2.3) and equations (8.2.2.6) that the nonlinear PDE (8.2.2.1) admits another operator,

$$\mathrm{X}_4 = x\partial_x + y\partial_y + \frac{2}{1-k}u\partial_u,$$

that defines a nonuniform extension.

Case 2. If $a = 0$, we have

$$f(u) = Ce^{\lambda u},$$

where $\lambda = \text{const}$. Then $b = -2\eta_y/\lambda$, and the functions ξ and η satisfy the first two equations (8.2.2.6), which coincide with the Cauchy–Riemann conditions for analytic functions. The real and the imaginary part of any analytic function, $\Phi(z) = \xi(x,y) + i\eta(x,y)$, of the complex variable $z = x + iy$ satisfies the Cauchy–Riemann conditions. In particular, if $b = \text{const}$ and $f(u) = e^u$, the nonlinear PDE (8.2.2.1) admits another operator,

$$\mathrm{X}_4 = x\partial_x + y\partial_y - 2\partial_u,$$

which corresponds to extension in x and y with simultaneous translation in u. ◄

► Example 8.2. Consider the nonlinear unsteady heat equation

$$u_t = [f(u)u_x]_x. \tag{8.2.2.9}$$

1°. In the invariance condition (8.2.1.3)–(8.2.1.4), one should set

$$y = t, \quad F = u_t - f(u)u_{xx} - f'(u)u_x^2 \qquad (\zeta_{12} = \zeta_{22} = 0).$$

The coordinates of the first and second prolongations, ζ_1, ζ_2, and ζ_{11}, are expressed by (8.1.3.2) and (8.1.3.7) with $y = t$. In the resulting equation, one should replace u_t with the right-hand side of equation (8.2.2.9) and equate the coefficients of the various combinations of the remaining derivatives to zero, thus arriving at the determining system of equations

$$\begin{aligned}
u_x u_{xx}\colon &\quad 2f(u)[\eta_{xu} f(u) + \xi_u] + f'(u)\eta_x = 0,\\
u_x u_{xt}\colon &\quad f(u)\eta_u = 0,\\
u_{xx}\colon &\quad \zeta f'(u) - f^2(u)\eta_{xx} - f(u)(2\xi_x - \eta_t) = 0,\\
u_{xt}\colon &\quad f(u)\eta_x = 0,\\
u_x^4\colon &\quad f'(u)\eta_u + f(u)\eta_{uu} = 0,\\
u_x^3\colon &\quad 2[f'(u)]^2\eta_x + f(u)\xi_{uu} + f'(u)\xi_u + 2f(u)f'(u)\eta_{xu} = 0,\\
u_x^2\colon &\quad f(u)\zeta_{uu} + f''(u)\zeta - 2f(u)\xi_{xu} - f'(u)(2\xi_x - \eta_t)\\
&\qquad + f'(u)\zeta_u - f(u)f'(u)\eta_{uu} = 0,\\
u_x\colon &\quad 2f(u)\zeta_{xu} + 2f'(u)\zeta_x - f(u)\xi_{xx} + \xi_t = 0,\\
1\colon &\quad \zeta_t - f(u)\zeta_{xx} = 0.
\end{aligned} \tag{8.2.2.10}$$

Here, the left column indicates a combination of the derivatives, and the right column gives the associated equation (to a constant factor); identities and differential consequences are omitted. Since $f(u) \not\equiv 0$, it follows from the second and fourth equations of system (8.2.2.10) that

$$\eta = \eta(t). \tag{8.2.2.11}$$

Then from the first and third equations, we have

$$\xi = \xi(x,t), \quad \zeta = \frac{f(u)(2\xi_x - \eta_t)}{f'(u)}. \tag{8.2.2.12}$$

Substituting relations (8.2.2.11) and (8.2.2.12) into the remaining equations of system (8.2.2.10), after elementary transformations and cancellation to nonzero functions, we obtain

$$\begin{aligned}
&(f/f')''(2\xi_x - \eta_t) = 0,\\
&f[4ff'' - 7(f')^2]\xi_{xx} - (f')^2\xi_t = 0,\\
&2f\xi_{xxx} - 2\xi_{xt} + \eta_{tt} = 0.
\end{aligned} \tag{8.2.2.13}$$

In the general case, for arbitrary function $f = f(u)$, from the first equation (8.2.2.13) it follows that $2\xi_x - \eta_t = 0$, and from the second it follows that $\xi_t = 0$. The third equation (8.2.2.13) gives $\xi = C_1 x + C_2$, and then $\eta = 2C_1 t + C_3$. Therefore, for arbitrary f, equation (8.2.2.9) admits three operators

$$\begin{aligned} &\mathrm{X}_1 = \partial_x && (C_2 = 1,\ C_1 = C_3 = 0);\\ &\mathrm{X}_2 = \partial_t && (C_3 = 1,\ C_1 = C_2 = 0);\\ &\mathrm{X}_3 = 2t\partial_t + x\partial_x && (C_1 = 1,\ C_2 = C_3 = 0). \end{aligned}$$

2°. Let us now show that for some special types of functions $f(u)$ additional operators may appear.

The first equation (8.2.2.13) can be satisfied by setting $(f/f')'' = 0$. From here, integrating, we obtain two solutions

$$f(u) = a\exp(bu), \tag{8.2.2.14}$$
$$f(u) = a(u+b)^k, \tag{8.2.2.15}$$

where a, b, and k are arbitrary constants (the first solution is degenerate).

For function (8.2.2.14) as well as function (8.2.2.15) with $k \neq 0, -4/3$, from the second and third equations (8.2.2.13) we get $\xi_{xx} = \xi_t = \eta_{tt} = 0$, which gives

$$\xi = C_1 x + C_2, \qquad \eta = C_3 t + C_4, \tag{8.2.2.16}$$

where C_1, C_2, C_3, and C_4 are arbitrary constants.

Substituting functions (8.2.2.14) and (8.2.2.15) and coordinates (8.2.2.16) with $C_2 = C_3 = C_4 = 0$ into (8.2.2.12), we get additional operators

1. $f(u) = e^u$: $\quad \mathrm{X}_4 = x\partial_x + 2\partial_u \quad (a = b = 1,\ C_1 = 1)$;
2. $f(u) = u^k$, $k \neq 0, -4/3$: $\quad \mathrm{X}_4 = kx\partial_x + 2u\partial_u \quad (a = 1,\ b = 0,\ C_1 = k)$.

Solution (8.2.2.15) with $k = -4/3$ vanishes the first equation (8.2.2.13), as well as the expression in square brackets in the second equation. In this case, from the second and third equations (8.2.2.13) we have $\xi_t = \xi_{xxx} = \eta_{tt} = 0$, which gives

$$\xi = C_1 x^2 + C_2 x + C_3, \qquad \eta = C_4 t + C_5, \tag{8.2.2.17}$$

where $C_1, \ldots, C_5$ are arbitrary constants.

Substituting function (8.2.2.15) with $a = 1$, $b = 0$, $k = -4/3$ and the coordinates (8.2.2.17) with $C_3 = C_4 = C_5 = 0$ into (8.2.2.12), we get additional operators

3. $f(u) = u^{-4/3}$: $\quad \begin{cases} \mathrm{X}_4 = 2x\partial_x - 3u\partial_u & (C_1 = 0,\ C_2 = 2);\\ \mathrm{X}_5 = x^2\partial_x - 3xu\partial_u & (C_1 = 1,\ C_2 = 0). \end{cases}$

◄

► Example 8.3. Consider now the nonlinear wave equation

$$u_{tt} = [f(u)u_x]_x. \tag{8.2.2.18}$$

1°. In the invariance condition (8.2.1.3)–(8.2.1.4), one should set

$$y = t, \quad F = u_{tt} - f(u)u_{xx} - f'(u)u_x^2 \qquad (\zeta_2 = \zeta_{12} = 0),$$

and use the coordinates of the first and second prolongations, ζ_1, ζ_{11}, and ζ_{22}, expressed by (8.1.3.2) and (8.1.3.7) with $y = t$. In the resulting equation, one should replace u_{tt} with the right-hand side of equation (8.2.2.18) and equate the coefficients of the various combinations of the remaining derivatives to zero, thus arriving at the determining system of equations

$$\begin{aligned}
u_x u_{xx}&:\quad f(u)\xi_u = 0,\\
u_t u_{xx}&:\quad f(u)\eta_u = 0,\\
u_{xx}&:\quad f'(u)\zeta + 2f(u)(\eta_t - \xi_x) = 0,\\
u_{xt}&:\quad f(u)\eta_x - \xi_t = 0,\\
u_x^3&:\quad f'(u)\xi_u + f(u)\xi_{uu} = 0,\\
u_x^2 u_t&:\quad f(u)\eta_{uu} - f'(u)\eta_u = 0,\\
u_x^2&:\quad f(u)\zeta_{uu} + f'(u)\zeta_u + f''(u)\zeta - 2f(u)\xi_{xu} - 2f'(u)(\xi_x - \eta_t) = 0,\\
u_x u_t&:\quad 2f'(u)\eta_x + 2f(u)\eta_{xu} - 2\xi_{tu} = 0,\\
u_t^2&:\quad \zeta_{uu} - 2\eta_{tu} = 0,\\
u_x&:\quad 2f'(u)\zeta_x - f(u)\xi_{xx} + 2f(u)\zeta_{xu} + \xi_{tt} = 0,\\
u_t&:\quad f(u)\eta_{xx} + 2\zeta_{tu} - \eta_{tt} = 0,\\
1&:\quad \zeta_{tt} - f(u)\zeta_{xx} = 0.
\end{aligned}$$

Identities and differential consequences have been omitted. Since $f(u) \neq \text{const}$, it follows from the first two equations that

$$\xi = \xi(x,t), \qquad \eta = \eta(x,t).$$

The ninth equation of the system becomes $\zeta_{uu} = 0$, thus giving

$$\zeta = a(x,t)u + b(x,t).$$

As a result, the determining system is reduced to the form

$$\begin{aligned}
&uf'(u)a(x,y) + f'(u)b(x,y) + 2f(u)(\eta_t - \xi_x) = 0,\\
&f'(u)a(x,y) + uf''(u)a(x,y) + f''(u)b(x,y) - 2f'(u)(\xi_x - \eta_t) = 0,\\
&2f'(u)(a_x u + b_x) - f(u)\xi_{xx} + 2f(u)a_x = 0,\\
&2a_t - \eta_{tt} = 0,\\
&a_{tt}u + b_{tt} - f(u)(a_{xx}u + b_{xx}) = 0.
\end{aligned}$$

For arbitrary function $f(u)$, we have $a = b = 0$, $\xi_{xx} = 0$, $\eta_{tt} = 0$, and $\xi_x - \eta_t = 0$. Integrating yields three operators:

$$\mathrm{X}_1 = \partial_x, \quad \mathrm{X}_2 = \partial_t, \quad \mathrm{X}_3 = x\partial_x + t\partial_t.$$

2°. Similarly, it can be established that the following special forms of f result in additional operators:

1. $f = e^u$: $\quad \mathrm{X}_4 = x\partial_x + 2\partial_u$;
2. $f = u^k$, $k \neq 0, -4/3, -4$: $\quad \mathrm{X}_4 = kx\partial_x + 2u\partial_u$;
3. $f = u^{-4/3}$: $\quad \mathrm{X}_4 = 2x\partial_x - 3u\partial_u, \quad \mathrm{X}_5 = x^2\partial_x - 3xu\partial_u$;
4. $f = u^{-4}$: $\quad \mathrm{X}_4 = 2x\partial_x - u\partial_u, \quad \mathrm{X}_5 = t^2\partial_t + tu\partial_u$. ◂

The symmetries obtained with the procedure presented can be used to find exact solutions of the differential equations considered (see below).

Problems and Exercises for Section 8.2

1. Find admissible infinitesimal operators and invariants of the Burgers equation and potential Burgers equation:

a) $u_t + uu_x = au_{xx}$,
b) $u_t + au_x^2 = bu_{xx}$.

2. Find admissible infinitesimal operators and invariants of the generalized Burgers equation:

$u_t + f(u)u_x = au_{xx}$.

Classify the symmetries of the equation for all $f(u)$.

3. Find admissible infinitesimal operators and invariants of the unsteady heat equations with nonlinear sources:

a) $u_t = a(uu_x)_x + bu$,
b) $u_t = a(uu_x)_x + bu^2$,
c) $u_t = au_{xx} + f(u)$.

Classify the symmetries of the last equation for all $f(u)$.

4. Find admissible infinitesimal operators and invariants of the nonlinear equations:

a) $u_t = u_{xx} + au_x^2$,
b) $u_t = u_{xx} + auu_x^2$.

5. Find admissible infinitesimal operators and invariants of the nonlinear wave equation:

$u_{xt} = f(u)$.

Classify the symmetries of the equation for all $f(u)$.

6. Show that the transonic gas flow equation

$$u_x u_{xx} + u_{yy} = 0$$

admits the operators

$$\mathrm{X}_1 = \partial_x, \quad \mathrm{X}_2 = \partial_y, \quad \mathrm{X}_3 = \partial_u, \quad \mathrm{X}_4 = y\partial_u, \quad \mathrm{X}_5 = x\partial_x + 3u\partial_u, \quad \mathrm{X}_6 = y\partial_y - 2u\partial_u$$

and find their respective invariants.

7. Find admissible infinitesimal operators and invariants for the equation of motion of a nonlinear viscoplastic medium:

$u_t = f(u_x)u_{xx}$.

Classify the symmetries of the equation for all $f(u)$.

8. Find admissible infinitesimal operators and invariants of the nonlinear seepage theory equation:

$u_t = [f(u_x)u_x]_x$.

Classify the symmetries of the equation for all $f(u)$.

9. Find admissible infinitesimal operators and invariants of the nonlinear telegraph equation:

$u_{tt} + f(u)u_t = au_{xx}$.

Classify the symmetries of the equation for all $f(u)$.

10. Find admissible infinitesimal operators and invariants of the nonlinear heat equation in anisotropic media:

$$[f(x)u_x]_x + [g(y)u_y]_y = h(u).$$

Classify the symmetries of the equation for all $f(x)$ and $g(y)$ and arbitrary $h(u)$.

Hint. Look for the coordinates of the admissible operator as $\xi = \xi(x,y)$, $\eta = \eta(x,y)$, $\zeta = 0$.

11. Find admissible infinitesimal operators and invariants of the Calogero equation:

$$u_{xt} = uu_{xx} + f(u_x).$$

Classify the symmetries of the equation for all $f(u)$.

12. Find admissible infinitesimal operators and invariants of the nonlinear equation:

$$u_{tt} = f(u_{xx}).$$

Classify the symmetries of the equation for all $f(w)$.

13. Find admissible infinitesimal operators and invariants of the minimal surface equation:

$$(1 + u_y^2)u_{xx} - 2u_x u_y u_{xy} + (1 + u_x^2)u_{yy} = 0.$$

14. Find admissible infinitesimal operators and invariants of the Born–Infeld equation:

$$(1 - u_t^2)u_{xx} + 2u_x u_t u_{xt} - (1 + u_x^2)u_{tt} = 0.$$

15. Find admissible infinitesimal operators and invariants of the nonhomogeneous Monge–Ampère equations:

a) $u_{xy}^2 - u_{xx}u_{yy} = f(x)$,

b) $u_{xy}^2 - u_{xx}u_{yy} = yf(x)$,

c) $u_{xy}^2 - u_{xx}u_{yy} = y^2 f(x)$,

d) $u_{xy}^2 - u_{xx}u_{yy} = e^y f(x)$.

Classify the symmetries of the equations for all $f(x)$.

8.3. Using Symmetries of Equations for Finding Exact Solutions. Invariant Solutions

8.3.1. Using Symmetries of Equations for Constructing One-Parameter Solutions

Suppose a particular solution,

$$u = g(x,y), \qquad (8.3.1.1)$$

of a given equation is known. Let us show that any symmetry of the equation defined by a transformation of the form (8.1.1.1) generates a one-parameter family of solutions (except for the cases where the solution is not mapped into itself by the transformations; see Section 8.3.2).

Indeed, since equation (8.2.1.1) converted to the new variables (8.1.1.1) acquires the same form (8.2.1.2), then the transformed equation (8.2.1.2) has a solution

$$\bar{u} = g(\bar{x}, \bar{y}). \qquad (8.3.1.2)$$

In (8.3.1.2), going back to the old variables by formulas (8.1.1.1), we obtain a one-parameter solution of the original equation (8.2.1.1).

► Example 8.4. The two-dimensional heat equation with an exponential source

$$u_{xx} + u_{yy} = e^u \tag{8.3.1.3}$$

has a one-dimensional solution

$$u = \ln \frac{2}{x^2}. \tag{8.3.1.4}$$

Equation (8.3.1.3) admits the operator $X_3 = y\partial_x - x\partial_y$ (see Example 8.1 in Section 8.2.2), which defines rotation in the plane. The corresponding transformation is given in Table 8.1. Replacing x in (8.3.1.4) by $\bar{x}$ (from Table 8.1), we obtain a one-parameter solution of equation (8.3.1.3):

$$u = \ln \frac{2}{(x\cos\varepsilon + y\sin\varepsilon)^2},$$

where ε is a free parameter. ◄

8.3.2. Procedure for Constructing Invariant Solutions

Solution (8.3.1.1) of equation (8.2.1.1) is called *invariant* under transformations (8.1.1.1) if it coincides with solution (8.3.1.2), which must be rewritten in terms of the old variables using formulas (8.1.1.1). This means that an invariant solution is converted to itself under the given transformation. The basic stages of constructing invariant solutions are outlined below.

First, one finds a one-parameter transformation with operator (8.1.1.3) whose coordinates are determined from the invariance condition (8.2.1.3) following the procedure described in Section 8.2.1. Then, one looks for two functionally independent integrals (8.1.2.3) of the characteristic system of ordinary differential equations (8.1.2.2). The invariant solution of the partial differential equation (8.1.2.1) is sought in the form

$$I_2 = \Phi(I_1), \tag{8.3.2.1}$$

where the functions $I_1 = I_1(x, y, u)$ and $I_2 = I_2(x, y, u)$ are known,* and the function Φ is to be determined. Relation (8.3.2.1) is the basis for the construction of invariant solutions: solving (8.3.2.1) for u and substituting the resulting expression into (8.2.1.1), we arrive at an ordinary differential equation for Φ.

► Example 8.5. A well-known and very important special case of invariant solutions is the self-similar solutions (see Section 4.2.2); they are based on invariants of scaling groups. The corresponding infinitesimal operator and its invariants are

$$X = ax\frac{\partial}{\partial x} + by\frac{\partial}{\partial y} + cu\frac{\partial}{\partial u}; \qquad I_1 = |y|^a|x|^{-b}, \quad I_2 = |u|^a|x|^{-c}.$$

Substituting the invariants into (8.3.2.1) yields $|u|^a|x|^{-c} = \Phi\big(|y|^a|x|^{-b}\big)$. On solving this equation for u, we obtain the form of the desired solution, $u = |x|^{c/a}\Psi\big(y|x|^{-b/a}\big)$, where $\Psi(z)$ is an unknown function. ◄

*Usually, the invariant that is independent of u is taken to be I_1.

To make it clearer, the general scheme for constructing invariant solutions for evolution second-order equations is depicted in Fig. 8.1. The first-order partial differential equation (8.1.2.1) for finding group invariants is omitted, since the corresponding characteristic system of ordinary differential equations (8.1.2.2) can be immediately used.

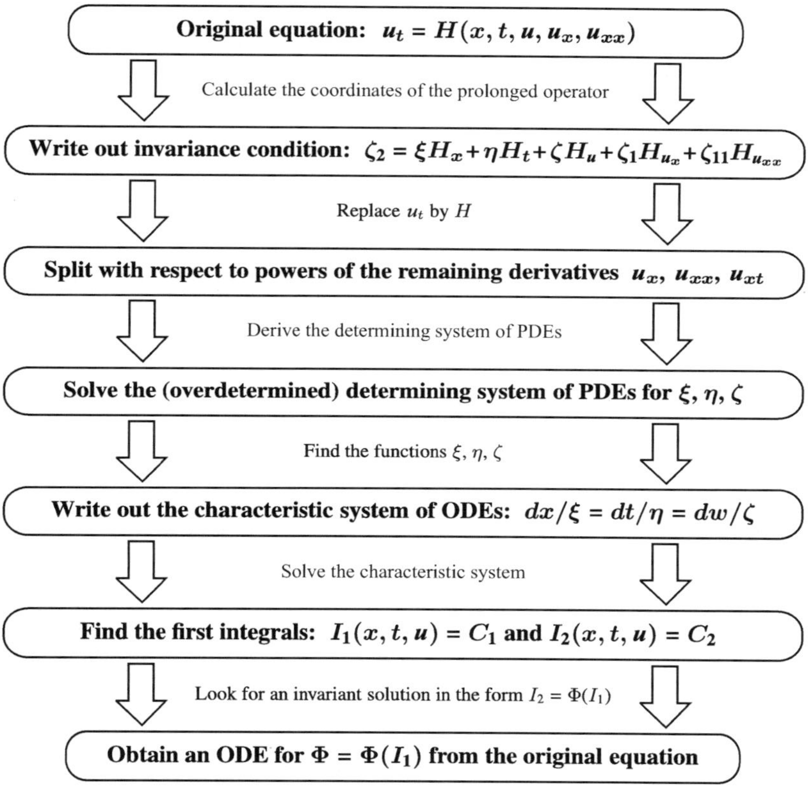

Figure 8.1. An algorithm for constructing invariant solutions for evolution second-order equations. Notation: ODE stands for ordinary differential equation, and PDE stands for partial differential equation; $\xi = \xi(x,t,u)$, $\eta = \eta(x,t,u)$, $\zeta = \zeta(x,t,u)$; ζ_1, ζ_2, and ζ_{11} are the coordinates of the prolonged operator, which are defined by formulas (8.1.3.2) and (8.1.3.7) with $y = t$.

8.3.3. Examples of Constructing Invariant Solutions to Nonlinear Equations

► Example 8.6. Consider once again the stationary heat equation with nonlinear source (8.2.2.1):

$$u_{xx} + u_{yy} = f(u).$$

1°. For arbitrary $f(u)$, this equation admits the operator (see Example 8.1 from Section 8.2.2):

$$\mathrm{X}_3 = y\partial_x - x\partial_y.$$

Invariants of X_3 are found for the linear first-order PDE $\mathrm{X}_3 I = 0$, or, in detailed form,

$$y\frac{\partial I}{\partial x} - x\frac{\partial I}{\partial y} + 0\cdot\frac{\partial I}{\partial u} = 0.$$

The associated characteristic system of ordinary differential equations,

$$\frac{dx}{y} = -\frac{dy}{x} = \frac{du}{0},$$

has the first integrals

$$x^2 + y^2 = C_1, \qquad u = C_2,$$

where C_1 and C_2 are arbitrary constants. Therefore, the functions $I_1 = x^2 + y^2$ and $I_2 = u$ are invariants of the operator X_3. Assuming that $I_2 = \Phi(I_1)$, we find the form of the invariant solution

$$u = \Phi(\xi), \qquad \xi = x^2 + y^2. \tag{8.3.3.1}$$

Substituting (8.3.3.1) into the original equation (8.2.2.1), we obtain the ODE

$$\xi\Phi''_{\xi\xi} + \Phi'_{\xi} = \tfrac{1}{4}f(\Phi),$$

that describes solutions invariant with respect to rotation.

Remark 8.7. *In applications, the polar radius* $r = \sqrt{x^2 + y^2}$ *is normally used as an invariant instead of* $\xi = x^2 + y^2$.

2°. Let us dwell on the special case $f(u) = u^k$, where the equation admits an additional operator (see Example 8.1 from Section 8.2.2):

$$\mathrm{X}_4 = x\partial_x + y\partial_y + \frac{2}{1-k}u\partial_u.$$

In order to find invariants of this operator, we have to consider the linear first-order partial differential equation $\mathrm{X}_4 I = 0$, or, in detailed form,

$$x\frac{\partial I}{\partial x} + y\frac{\partial I}{\partial y} + \frac{2}{1-k}u\frac{\partial I}{\partial u} = 0.$$

The corresponding characteristic system of ordinary differential equations,

$$\frac{dx}{x} = \frac{dy}{y} = \frac{1-k}{2}\frac{du}{u},$$

has the first integrals

$$y/x = C_1, \qquad x^{2/(k-1)}u = C_2,$$

where C_1, C_2 are arbitrary constants. Therefore, the functions $I_1 = y/x$ and $I_2 = x^{2/(k-1)}u$ are invariants of the operator X_4.

Assuming that $I_2 = \Psi(I_1)$ and expressing u, we find the form of the invariant (self-similar) solution:

$$u = x^{-2/(k-1)}\Psi(z), \qquad z = y/x. \tag{8.3.3.2}$$

Substituting (8.3.3.2) into the original PDE (8.2.2.1) yields a second-order ODE for $\Psi(z)$:

$$(k-1)^2(z^2+1)\Psi''_{zz} + 2(k^2-1)z\Psi'_z + 2(k+1)\Psi - (k-1)^2\Psi^k = 0.$$

3°. Let us now consider another special case $f(u) = e^u$, when the equation admits an additional operator (see Example 8.1 from Section 8.2.2):

$$X_4 = x\partial_x + y\partial_y - 2\partial_u,$$

To find the invariants of this operator, we need to consider the first-order linear PDE $X_4 I = 0$, which leads to the following characteristic ODE system:

$$\frac{dx}{x} = \frac{dy}{y} = -\frac{du}{2}.$$

The first integrals of this system have the form

$$y/x = C_1, \qquad u + 2\ln|x| = C_2.$$

Arguing further as in Item 2°, we find the form of the invariant solution

$$u = \theta(z) - 2\ln|x|, \qquad z = y/x. \tag{8.3.3.3}$$

Substituting (8.3.3.3) into the original PDE (8.2.2.1) yields a second-order ODE for $\theta(z)$:

$$(1+z^2)\theta''_{zz} + 2z\theta'_z + 2 = e^\theta.$$

4°. Table 8.2 summarizes the symmetries of the nonlinear stationary heat equation (8.2.2.1) (see Example 8.1 from Section 8.2.2 and Example 8.6 from Section 8.3.3).

Table 8.2. Operators, invariants, and solution structures admitted by the nonlinear stationary heat equation $u_{xx} + u_{yy} = f(u)$.

Function $f(u)$	Operators	Invariants	Solution structure
Arbitrary	$X_1 = \partial_x$, $X_2 = \partial_y$, $X_3 = y\partial_x - x\partial_y$	$I_1 = y$, $I_2 = u$, $I_1 = x$, $I_2 = u$, $I_1 = x^2 + y^2$, $I_2 = u$	$u = u(y)$, $u = u(x)$, $u = \Phi(\xi)$, $\xi = x^2 + y^2$
e^u	$X_4 = x\partial_x + y\partial_y - 2\partial_u$	$I_1 = y/x$, $I_2 = u + 2\ln\|x\|$	$u = \theta(y/x) - 2\ln\|x\|$
w^k $(k \neq 1)$	$X_4 = x\partial_x + y\partial_y + \frac{2}{1-k}u\partial_u$	$I_1 = y/x$, $I_2 = ux^{2/(k-1)}$	$u = x^{-2/(k-1)}\Psi(y/x)$

◀

► **Example 8.7.** Let us now consider the nonlinear unsteady heat equation (8.2.2.9),

$$u_t = [f(u)u_x]_x.$$

1°. For arbitrary $f(u)$, this equation admits the operator (see Example 8.2 from Section 8.2.2)

$$\mathrm{X}_3 = 2t\partial_t + x\partial_x.$$

Invariants of X_3 are found for the linear first-order PDE $\mathrm{X}_3 I = 0$, or

$$2t\frac{\partial I}{\partial t} + x\frac{\partial I}{\partial x} + 0\frac{\partial I}{\partial u} = 0.$$

The associated characteristic system of ordinary differential equations,

$$\frac{dx}{x} = \frac{dt}{2t} = \frac{du}{0},$$

has the first integrals

$$xt^{-1/2} = C_1, \qquad u = C_2,$$

where C_1 and C_2 are arbitrary constants. Therefore, the functions $I_1 = xt^{-1/2}$ and $I_2 = u$ are invariants of the operator X_3.

Assuming $I_2 = \Phi(I_1)$, we get

$$u = \Phi(z), \qquad z = xt^{-1/2}, \tag{8.3.3.4}$$

where $\Phi(z)$ is to be determined in the subsequent analysis. Substituting (8.3.3.4) in the original equation (8.2.2.9) yields the second-order ordinary differential equation

$$2[f(\Phi)\Phi'_z]'_z + z\Phi'_z = 0,$$

which describes an invariant (self-similar) solution.

2°. Let us dwell on the special case $f(u) = u^k$, where the equation admits the additional operator

$$\mathrm{X}_4 = kx\partial_x + 2u\partial_u.$$

The invariants are described by the first-order partial differential equation $\mathrm{X}_4 I = 0$, or

$$0\frac{\partial I}{\partial t} + kx\frac{\partial I}{\partial x} + 2u\frac{\partial I}{\partial u} = 0.$$

The associated characteristic system of ordinary differential equations,

$$\frac{dt}{0} = \frac{dx}{kx} = \frac{du}{2u},$$

has the first integrals

$$t = C_1, \qquad |x|^{-2/k}u = C_2,$$

where C_1, C_2 are arbitrary constants. Therefore, $I_1 = t$ and $I_2 = |x|^{-2/k}u$ are invariants of the operator X_4.

Assuming $I_2 = \theta(I_1)$ and expressing u, we get

$$u = |x|^{2/k}\theta(t), \qquad (8.3.3.5)$$

where $\theta(t)$ is to be determined in the process of the subsequent analysis. Substituting (8.3.3.5) in the original equation (8.2.2.9) with $f(u) = u^k$ gives the first-order ordinary differential equation

$$k^2\theta'_t = 2(k+2)\theta^{k+1}.$$

Integrating yields

$$\theta(t) = \left[A - \frac{2(k+2)}{k}t\right]^{-1/k},$$

where A is an arbitrary constant. Hence, the solution of equation (8.2.2.9) with $f(u) = u^k$, which is invariant under scaling, has the form

$$u = |x|^{2/k}\left[A - \frac{2(k+2)}{k}t\right]^{-1/k}.$$

3°. Table 8.3 summarizes the symmetries of the nonlinear unsteady heat equation (8.2.2.9) (see Example 8.2 from Section 8.2.2 and Example 8.7 from Section 8.3.3), which were described by Ovsiannikov (1959, 1982).

Table 8.3. Operators, invariants, and solution structures admitted by the nonlinear unsteady heat equation $u_t = [f(u)u_x]_x$.

Function $f(u)$	Operators	Invariants	Solution structure
Arbitrary	$X_1 = \partial_x$, $X_2 = \partial_t$, $X_3 = 2t\partial_t + x\partial_x$	$I_1 = t$, $I_2 = u$, $I_1 = x$, $I_2 = u$, $I_1 = x^2/t$, $I_2 = u$	$u = u(t) = \text{const}$, $u = u(x)$, $u = \Phi(z)$, $z = x^2/t$
e^u	$X_4 = x\partial_x + 2\partial_u$	$I_1 = t$, $I_2 = u - 2\ln\|x\|$	$u = 2\ln\|x\| + \theta(t)$
w^k $(k \neq 0, -\frac{4}{3})$	$X_4 = kx\partial_x + 2u\partial_u$	$I_1 = t$, $I_2 = u\|x\|^{-2/k}$	$u = \|x\|^{2/k}\theta(t)$
$w^{-4/3}$	$X_4 = 2x\partial_x - 3u\partial_u$, $X_5 = x^2\partial_x - 3xu\partial_u$	$I_1 = t$, $I_2 = u\|x\|^{3/2}$, $I_1 = t$, $I_2 = ux^3$	$u = \|x\|^{-3/2}\theta(t)$, $u = x^{-3}\theta(t)$, $\theta(t) = \text{const}$

◀

► Example 8.8. Consider the nonlinear wave equation (8.2.2.18),

$$u_{tt} = [f(u)u_x]_x.$$

1°. For arbitrary $f(u)$, this equation admits the following operator (see Example 8.3 from Section 8.2.2):

$$X_3 = t\partial_t + x\partial_x.$$

The invariants are found from the linear first-order PDE $X_3 I = 0$, or

$$t\frac{\partial I}{\partial t} + x\frac{\partial I}{\partial x} + 0\frac{\partial I}{\partial u} = 0.$$

The associated characteristic system of ordinary differential equations

$$\frac{dx}{x} = \frac{dt}{t} = \frac{du}{0}$$

admits the first integrals

$$xt^{-1} = C_1, \qquad u = C_2,$$

where C_1, C_2 are arbitrary constants. Therefore, $I_1 = xt^{-1}$ and $I_2 = u$ are invariants of the operator X_3.

Taking $I_2 = \Phi(I_1)$, we get

$$u = \Phi(z), \qquad z = xt^{-1}. \tag{8.3.3.6}$$

The function $\Phi(z)$ is found by substituting (8.3.3.6) in the original equation (8.2.2.18). This results in the ordinary differential equation

$$[f(\Phi)\Phi'_z]'_z = (z^2\Phi'_z)'_z,$$

which defines an invariant (self-similar) solution. This equation has the obvious first integral $f(\Phi)\Phi'_z = z^2\Phi'_z + C$.

2°. Table 8.4 summarizes the symmetries of the nonlinear wave equation (8.2.2.18) (see Example 8.3 from Section 8.2.2 and Example 8.8 from Section 8.3.3), which were described in the work of Ames et al. (1981).

Table 8.4. Operators, invariants, and solution structures admitted by the nonlinear wave equation $u_{tt} = [f(u)u_x]_x$.

Functions $f(u)$	Operators	Invariants	Solution structures
Arbitrary	$X_1 = \partial_x$, $X_2 = \partial_t$, $X_3 = t\partial_t + x\partial_x$	$I_1 = t$, $I_2 = u$, $I_1 = x$, $I_2 = u$, $I_1 = x/t$, $I_2 = u$	$u = u(t)$, $u = u(x)$, $u = \Phi(z)$, $z = x/t$
e^u	$X_4 = x\partial_x + 2\partial_u$	$I_1 = t$, $I_2 = u - 2\ln\lvert x\rvert$	$u = 2\ln\lvert x\rvert + \theta(t)$
w^k ($k \neq 0, -\frac{4}{3}, -4$)	$X_4 = kx\partial_x + 2u\partial_u$	$I_1 = t$, $I_2 = u\lvert x\rvert^{-2/k}$	$u = \lvert x\rvert^{2/k}\theta(t)$
$w^{-4/3}$	$X_4 = 2x\partial_x - 3u\partial_u$, $X_5 = x^2\partial_x - 3xu\partial_u$	$I_1 = t$, $I_2 = u\lvert x\rvert^{3/2}$, $I_1 = t$, $I_2 = ux^3$	$u = \lvert x\rvert^{-3/2}\theta(t)$, $u = x^{-3}\theta(t)$, $\theta''_{tt} = 0$
w^{-4}	$X_4 = 2x\partial_x - u\partial_u$, $X_5 = t^2\partial_t + tu\partial_u$	$I_1 = t$, $I_2 = u\lvert x\rvert^{1/2}$, $I_1 = x$, $I_2 = u/t$	$u = \lvert x\rvert^{-1/2}\theta(t)$, $u = t\theta(x)$, $\theta^{-4}\theta'_x = \text{const}$

◀

8.3.4. Solutions Induced by Linear Combinations of Admissible Operators

If a given equation admits N operators, then we have N associated different invariant solutions. However, when dealing with operators individually, one may overlook solutions that are invariant under a linear superposition of the operators; such solutions may have a significantly different form. In order to find all types of invariant solutions, one should study all possible linear combinations of the admissible operators.

► Example 8.9. Let us consider once again the nonlinear unsteady heat equation (8.2.2.9).

1°. For arbitrary $f(u)$, this equation admits three operators (refer to Table 8.3):

$$\mathrm{X}_1 = \partial_t, \quad \mathrm{X}_2 = \partial_x, \quad \mathrm{X}_3 = 2t\partial_t + x\partial_x.$$

The respective invariant solutions are

$$u = \Phi(x), \quad u = \Phi(t), \quad u = \Phi(x^2/t).$$

However, various linear combinations of the above operators give some other operators; for example,

$$\mathrm{X}_{1,2} = \mathrm{X}_1 + a\mathrm{X}_2 = \partial_t + a\partial_x, \tag{8.3.4.1}$$

where $a \neq 0$ is an arbitrary constant. The solution invariant under this operator is written as

$$u = \Phi(x - at). \tag{8.3.4.2}$$

It is apparent that solutions of this type (traveling waves) are not contained in the invariant solutions associated with the individual operators X_1, X_2, and X_3.

2°. If $f(u) = e^u$, apart from the above three operators, there is another one, $\mathrm{X}_4 = x\partial_x + 2\partial_u$ (see Table 8.3). In this case, the linear combination

$$\mathrm{X}_{3,4} = \mathrm{X}_3 + a\mathrm{X}_4 = 2t\partial_t + (a+1)x\partial_x + 2a\partial_u$$

gives another invariant solution,

$$u = \Phi(\xi) + a\ln t, \quad \xi = xt^{-\frac{a+1}{2}},$$

where the function $\Phi = \Phi(\xi)$ satisfies the ordinary differential equation

$$(e^\Phi \Phi'_\xi)'_\xi + \tfrac{1}{2}(a+1)\xi\Phi'_\xi = a.$$

3°. If $f(u) = u^k$ $(k \neq 0, -4/3)$, apart from the three operators from 1°, there is another one $\mathrm{X}_4 = kx\partial_x + 2u\partial_u$. The linear combination

$$\mathrm{X}_{3,4} = \mathrm{X}_3 + a\mathrm{X}_4 = 2t\partial_t + (ak+1)x\partial_x + 2au\partial_u$$

generates the invariant (self-similar) solution

$$u = t^a\Phi(\zeta), \quad \zeta = xt^{-\frac{ak+1}{2}},$$

where the function $\Phi = \Phi(\zeta)$ satisfies the ordinary differential equation

$$(\Phi^k \Phi'_\zeta)'_\zeta + \tfrac{1}{2}(ak+1)\zeta\Phi'_\zeta = a\Phi.$$

The invariant solutions presented in Items 1°–3° are not listed in Table 8.3. It is clearly important to consider solutions induced by linear combinations of admissible operators. ◄

► Example 8.10. Now consider the nonlinear wave equation (8.2.2.18).

1°. For arbitrary $f(u)$, this equation admits three operators (refer to Table 8.4):

$$X_1 = \partial_t, \quad X_2 = \partial_x, \quad X_3 = t\partial_t + x\partial_x.$$

The corresponding invariant solutions are

$$u = \Phi(x), \quad u = \Phi(t), \quad u = \Phi(x/t).$$

Going over the various linear combinations yields one more operator (8.3.4.1). There is a traveling-wave solution (8.3.4.2) which is invariant under this operator and not contained among the solutions corresponding to the 'pure' operators X_1, X_2, and X_3.

2°. If $f(u) = e^u$, a fourth operator, $X_4 = x\partial_x + 2\partial_u$, arises in addition to the above three (see Table 8.4). In this case, the linear combination

$$X_{3,4} = X_3 + aX_4 = t\partial_t + (a+1)x\partial_x + 2a\partial_u$$

($a \neq 0$ is an arbitrary constant) gives an invariant solution

$$u = \Phi(\xi) + 2a\ln t, \quad \xi = xt^{-a-1},$$

where the function $\Phi = \Phi(\xi)$ satisfies the ordinary differential equation

$$(e^\Phi \Phi'_\xi)'_\xi = (a+1)^2\xi^2\Phi''_{\xi\xi} + (a+1)(a+2)\xi\Phi'_\xi - 2a.$$

3°. If $f(u) = u^k$ ($k \neq 0, -4/3, -4$), there arises a fourth operator $X_4 = kx\partial_x + 2u\partial_u$. The linear combination

$$X_{3,4} = X_3 + aX_4 = t\partial_t + (ak+1)x\partial_x + 2au\partial_u$$

gives an invariant (self-similar) solution

$$u = t^{2a}\Phi(\zeta), \quad \zeta = xt^{-ak-1},$$

where the function $\Phi = \Phi(\zeta)$ satisfies the ordinary differential equation

$$(\Phi^k\Phi'_\zeta)'_\zeta = (ak+1)^2\zeta^2\Phi''_{\zeta\zeta} + (ak+1)(ak+2-4a)\zeta\Phi'_\zeta + 2a(2a-1)\Phi.$$ ◀

Problems and Exercises for Section 8.3

1. Find invariant solutions of the parabolic (evolutionary) equations:

a) $u_t + uu_x = au_{xx}$,
b) $u_t = u_{xx} + au_x^2$,
c) $u_t = u_{xx} + auu_x^2$,
d) $u_t = a(uu_x)_x + bu$,
e) $u_t = a(uu_x)_x + bu^2$,
f) $u_t + f(u)u_x = au_{xx}$,
g) $u_t = au_{xx} + f(u)$,
h) $u_t = f(u_x)u_{xx}$,
i) $u_t = [f(u_x)u_x]_x$.

Hint. Take advantage of the results of the previous symmetry analysis for these equations (see the problems and exercises for Section 8.2).

2. Find invariant solutions of the equations:

a) $xu_x + yu_y = au_{xx} + f(u)$,

b) $xu_x + yu_y = u_{xx} + u_{yy} + f(u)$.

3. Show that equation 2b has a noninvariant, traveling wave solution $u = u(z)$, where $z = k_1 x + k_2 y$, with k_1 and k_2 being arbitrary constants. This solution can be verified to exist by a direct substitution into the equation.

4. Find invariant solutions of the following equations:

a) $u_x u_{xx} + u_{yy} = 0$,

b) $u_{tt} + f(u)u_t = au_{xx}$,

c) $u_{xt} = f(u)$,

d) $u_{xt} = uu_{xx} + f(u_x)$,

e) $u_{tt} = f(u_{xx})$,

f) $(1 + u_y^2)u_{xx} - 2u_x u_y u_{xy} + (1 + u_x^2)u_{yy} = 0$,

g) $(1 - u_t^2)u_{xx} + 2u_x u_t u_{xt} - (1 + u_x^2)u_{tt} = 0$,

h) $u_{xy}^2 - u_{xx}u_{yy} = f(x)$.

Hint. Take advantage of the results of the previous symmetry analysis for these equations (see the problems and exercises for Section 8.2).

8.4. Some Generalizations. Higher-Order PDEs

8.4.1. One-Parameter Lie Groups of Point Transformations. Group Generator

This section will deal with functions dependent on $n+1$ variables, $x_1, \ldots, x_n$ and u. The brief notation $x = (x_1, \ldots, x_n)$ will be used.

The set of invertible transformations of the form

$$\mathrm{T}_\varepsilon = \begin{cases} \bar{x}_i = \varphi_i(x, u, \varepsilon), & \bar{x}_i|_{\varepsilon=0} = x_i, \\ \bar{u} = \psi(x, u, \varepsilon), & \bar{u}|_{\varepsilon=0} = u, \end{cases} \qquad (8.4.1.1)$$

where φ_i and ψ are sufficiently smooth functions of their arguments ($i = 1, \ldots, n$) and ε is a real parameter, is called a *one-parameter continuous point group* of transformations G if for any ε_1 and ε_2 the relation $\mathrm{T}_{\varepsilon_1} \circ \mathrm{T}_{\varepsilon_2} = \mathrm{T}_{\varepsilon_1+\varepsilon_2}$ holds, that is, the successive application (composition) of two transformations of the form (8.4.1.1) with parameters ε_1 and ε_2 is equivalent to a single transformation of the same form with parameter $\varepsilon_1 + \varepsilon_2$.

Further on, we consider local one-parameter continuous Lie groups of point transformations (or, for short, point groups), corresponding to the infinitesimal transformation (8.4.1.1) as $\varepsilon \to 0$. The expansion of (8.4.1.1) into Taylor series in the parameter ε about $\varepsilon = 0$ to the first order gives

$$\bar{x}_i \simeq x_i + \varepsilon\xi_i(x, u), \quad \bar{u} \simeq u + \varepsilon\zeta(x, u), \qquad (8.4.1.2)$$

where

$$\xi_i(x, u) = \frac{\partial\varphi_i(x, u, \varepsilon)}{\partial\varepsilon}\bigg|_{\varepsilon=0}, \quad \zeta(x, u) = \frac{\partial\psi(x, u, \varepsilon)}{\partial\varepsilon}\bigg|_{\varepsilon=0}.$$

The linear first-order differential operator

$$\mathrm{X} = \xi_i(x,u)\frac{\partial}{\partial x_i} + \zeta(x,u)\frac{\partial}{\partial u} \qquad (8.4.1.3)$$

corresponding to the infinitesimal transformation (8.4.1.2) is called a *group generator* (or an *infinitesimal operator*). In formula (8.4.1.3), summation is assumed over the repeated index i.

LIE'S THEOREM. Suppose the coordinates $\xi_i(x,u)$ and $\zeta(x,u)$ of the group generator (8.4.1.3) are known. Then the one-parameter group of transformations (8.4.1.1) can be completely recovered by solving the Lie equations

$$\frac{d\varphi_i}{d\varepsilon} = \xi_i(\varphi,\psi), \quad \frac{d\psi}{d\varepsilon} = \zeta(\varphi,\psi) \qquad (i=1,\dots,n)$$

with the initial conditions

$$\varphi_i|_{\varepsilon=0} = x_i, \quad \psi|_{\varepsilon=0} = u.$$

Here, the short notation $\varphi = (\varphi_1,\dots,\varphi_n)$ has been used.

Remark 8.8. *The widely known terms 'Lie group analysis of differential equations,' 'group-theoretic methods,' and others are due to the prevailing concept of a local one-parameter Lie group of point transformations. However, in this book, we prefer the widely used term 'classical method of symmetry reductions.'*

8.4.2. Group Invariants. Local Transformations of Derivatives

A *universal invariant* (or, for short, an *invariant*) of a group (8.4.1.1) and a operator (8.4.1.3) is a function $I(x,u)$ that satisfies the condition $I(\bar{x},\bar{u}) = I(x,u)$. The expansion in a series in powers of the small parameter ε gives rise to the linear partial differential equation for I:

$$\mathrm{X}I = \xi_i(x,u)\frac{\partial I}{\partial x_i} + \zeta(x,u)\frac{\partial I}{\partial u} = 0. \qquad (8.4.2.1)$$

Here summation is assumed over the repeated index i. From the theory of first-order partial differential equations it follows that the group (8.4.1.1) and the operator (8.4.1.3) have n functionally independent universal invariants. On the other side, this means that any function $F(x,u)$ that is invariant under the group (8.4.1.1) can be written as a function of n invariants.

In the new variable (8.4.1.1), the derivatives are transformed as follows:

$$\begin{gathered}\frac{\partial\bar{u}}{\partial\bar{x}_i} \simeq \frac{\partial u}{\partial x_i} + \varepsilon\zeta_i, \quad \bar{u}_{\bar{x}_i\bar{x}_j} \simeq u_{x_ix_j} + \varepsilon\zeta_{ij}, \\ \frac{\partial^3\bar{u}}{\partial\bar{x}_i\partial\bar{x}_j\partial\bar{x}_k} \simeq \frac{\partial^3 u}{\partial x_i\partial x_j\partial x_k} + \varepsilon\zeta_{ijk}, \quad \dots.\end{gathered} \qquad (8.4.2.2)$$

The coordinates ζ_i, ζ_{ij}, and ζ_{ijk} of the first three prolongations are expressed as

$$\begin{aligned}\zeta_i &= D_i(\zeta) - p_s D_i(\xi_s),\\ \zeta_{ij} &= D_j(\zeta_i) - q_{is} D_j(\xi_s),\\ \zeta_{ijk} &= D_k(\zeta_{ij}) - r_{ijs} D_k(\xi_s),\end{aligned} \tag{8.4.2.3}$$

where summation is assumed over the repeated index s and the following short notation partial derivatives are used:

$$p_i = \frac{\partial u}{\partial x_i}, \quad q_{ij} = \frac{\partial^2 u}{\partial x_i \partial x_j}, \quad r_{ijk} = \frac{\partial^3 u}{\partial x_i \partial x_j \partial x_k},$$
$$D_i = \frac{\partial}{\partial x_i} + p_i \frac{\partial}{\partial u} + q_{ij}\frac{\partial}{\partial p_j} + r_{ijk}\frac{\partial}{\partial q_{jk}} + \cdots,$$

with D_i being the total differential operator with respect to x_i.

8.4.3. Invariant Condition. Splitting Procedure. Invariant Solutions

We will consider partial differential equations of order m in n independent variables

$$F\left(x, u, \frac{\partial u}{\partial x_i}, \frac{\partial^2 u}{\partial x_i \partial x_j}, \frac{\partial^3 u}{\partial x_i \partial x_j \partial x_k}, \dots\right) = 0, \tag{8.4.3.1}$$

where $i, j, k = 1, \dots, n$.

The group analysis of equation (8.4.3.1) consists of several stages. At the first stage, let us require that equation (8.4.3.1) be invariant (must preserve its form) under transformations (8.4.1.1), so that

$$F\left(\bar{x}, \bar{u}, \frac{\partial \bar{u}}{\partial \bar{x}_i}, \frac{\partial^2 \bar{u}}{\partial x_i \partial \bar{x}_j}, \frac{\partial^3 \bar{u}}{\partial x_i \partial \bar{x}_j \partial \bar{x}_k}, \dots\right) = 0. \tag{8.4.3.2}$$

Let us expand this expression into a series in powers of ε about $\varepsilon = 0$ taking into account that the leading term (8.4.3.1) vanishes. Using formulas (8.4.1.1) and (8.4.2.2) and retaining terms to the first order of ε, we get

$$\underset{m}{\mathrm{X}} F\left(x, u, \frac{\partial u}{\partial x_i}, \frac{\partial^2 u}{\partial x_i \partial x_j}, \frac{\partial^3 u}{\partial x_i \partial x_j \partial x_k}, \dots\right)\bigg|_{F=0} = 0, \tag{8.4.3.3}$$

where

$$\underset{m}{\mathrm{X}} F = \xi_i \frac{\partial F}{\partial x_i} + \zeta \frac{\partial F}{\partial u} + \zeta_i \frac{\partial F}{\partial u_{x_i}} + \zeta_{ij}\frac{\partial F}{\partial u_{x_i x_j}} + \zeta_{ijk}\frac{\partial F}{\partial u_{x_i x_j x_k}} + \cdots. \tag{8.4.3.4}$$

The coordinates ζ_i, ζ_{ij}, and ζ_{ijk} of the first three prolongations are defined by formulas (8.4.2.2)–(8.4.2.3). Summation is assumed over repeated indices. Relation (8.4.3.3) is called the *invariance condition*, and the operator $\underset{m}{\mathrm{X}}$ is called the mth prolongation of the group generator; the partial derivatives of F with respect to all m derivatives of u appear last in (8.4.3.4).

At the second stage, one of the highest mth-order derivatives is eliminated from (8.4.3.3) using equation (8.4.3.1). The resulting relation is then represented as a polynomial in 'independent variables,' the various combinations of the remaining derivatives, which are the products of different powers of u_x, u_y, u_{xx}, u_{xy}, ... All the coefficients of this polynomial—they depend on x, u, ξ_i, and ζ only and are independent of the derivatives of u—are further equated to zero. As a result, the invariance condition is split into an overdetermined linear determining system.

The third stage involves solving the determining system and finding admissible coordinates ξ_i and ζ of the group generator (8.4.1.3).

For mth-order equations in two independent variables, the invariant solutions are defined in a similar way as for second-order equations. In this case, the procedure of constructing invariant solutions (for known coordinates of the group generator) is identical to that described in Section 8.3.2.

It is important to note that with an increase in the order of the PDEs under consideration, the technical difficulties in using the classical method of symmetry reductions increase sharply.

More complete information and examples on this topic can be found, for example, in the books of Ovsiannikov (1982), Bluman & Kumei (1989), Olver (1993), and Ibragimov (1994). In these books one can find solutions to many of the problems and exercises listed below.

Problems and Exercises for Section 8.4

1. Find admissible infinitesimal operators and invariant solutions of the Korteweg–de Vries equation:

$$u_t + u_{xxx} - 6uu_x = 0.$$

2. Find admissible infinitesimal operators and invariant solutions of the modified Korteweg–de Vries equation:

$$u_t + u_{xxx} - 6u^2u_x = 0.$$

3. Find admissible infinitesimal operators and invariant solutions of the Korteweg–de Vries type equations:

a) $u_t + u_{xxx} + auu_x + bt^{-1}u = 0,$

b) $u_t + u_{xxx} + au_x^2 = 0,$

c) $u_t + u_{xxx} + au_x^3 = 0.$

4. Find admissible infinitesimal operators and invariant solutions of the generalized Korteweg–de Vries equation:

$$u_t + u_{xxx} + f(u)u_x = 0.$$

Classify the symmetries of the equation for all $f(u)$.

5. Find admissible infinitesimal operators and invariant solutions of the following equation, which arises in viscous fluid dynamics:

$$u_{xt} + u_x^2 - uu_{xx} = \nu u_{xxx}.$$

6. Find admissible infinitesimal operators and invariant solutions of the following equation, which arises in viscous fluid dynamics:

$$u_{xt} + u_x^2 - uu_{xx} = \nu u_{xxx} + f(t).$$

Classify the symmetries of the equation for all $f(t)$.

7. Find admissible infinitesimal operators and invariant solutions of the steady-state gradientless hydrodynamic boundary layer equation:

$$u_y u_{xy} - u_x u_{yy} = u_{yyy}.$$

Remark 8.9. *The third-order nonlinear PDEs given above in Exercises 1–6 can be reduced to PDE systems by introducing a second unknown function $v = u_x$ (Exercise 7 uses the second unknown function $v = u_y$). Then the procedure described below in Section 8.5 can be employed.*

8.5. Symmetries of Systems of Equations of Mathematical Physics

8.5.1. Basic Relations Used in Symmetry Analysis of Systems of Equations

Symmetry analysis (group analysis) of systems of partial differential equations differs from the analysis of individual equations by only the fact that all dependent and independent variables must be taken into account in the operator (8.1.1.3) and its prolongations. Below are some formulas that are frequently used in practical symmetry analysis of simultaneous equations.

For system consisting of two equations with two independent variables, x and y, and two unknown functions, $u = u(x,y)$ and $v = v(x,y)$, admissible operators are sought in the form

$$\mathrm{X} = \xi(x,y,u,v)\frac{\partial}{\partial x} + \eta(x,y,u,v)\frac{\partial}{\partial y} + \zeta(x,y,u,v)\frac{\partial}{\partial u} + \chi(x,y,u,v)\frac{\partial}{\partial v}. \tag{8.5.1.1}$$

Generally, the second prolongation of (8.5.1.1) to the 'independent variables' u_x, u_y, v_x, v_y, u_{xx}, u_{xy}, u_{yy}, v_{xx}, v_{xy}, and v_{yy} is written as

$$\begin{aligned}\underset{2}{\mathrm{X}} = {}& \xi\frac{\partial}{\partial x} + \eta\frac{\partial}{\partial y} + \zeta\frac{\partial}{\partial u} + \chi\frac{\partial}{\partial v} + \zeta_1\frac{\partial}{\partial u_x} + \zeta_2\frac{\partial}{\partial u_y} + \chi_1\frac{\partial}{\partial v_x} + \chi_2\frac{\partial}{\partial v_y}\\ &+ \zeta_{11}\frac{\partial}{\partial u_{xx}} + \zeta_{12}\frac{\partial}{\partial u_{xy}} + \zeta_{22}\frac{\partial}{\partial u_{yy}} + \chi_{11}\frac{\partial}{\partial v_{xx}} + \chi_{12}\frac{\partial}{\partial v_{xy}} + \chi_{22}\frac{\partial}{\partial v_{yy}}.\end{aligned} \tag{8.5.1.2}$$

The coordinates of the first prolongation are calculated by formulas similar to (8.1.3.2) from Section 8.1.3,

$$\begin{aligned}\zeta_1 &= D_x(\zeta) - u_x D_x(\xi) - u_y D_x(\eta), & \chi_1 &= D_x(\chi) - v_x D_x(\xi) - v_y D_x(\eta),\\ \zeta_2 &= D_y(\zeta) - u_x D_y(\xi) - u_y D_y(\eta), & \chi_2 &= D_y(\chi) - v_x D_y(\xi) - v_y D_y(\eta);\end{aligned} \tag{8.5.1.3}$$

the coordinates of the second prolongation are calculated by

$$\begin{aligned}
\zeta_{11} &= D_x(\zeta_1) - u_{xx}D_x(\xi) - u_{xy}D_x(\eta), & \chi_{11} &= D_x(\chi_1) - v_{xx}D_x(\xi) - v_{xy}D_x(\eta),\\
\zeta_{12} &= D_y(\zeta_1) - u_{xx}D_y(\xi) - u_{xy}D_y(\eta), & \chi_{12} &= D_y(\chi_1) - v_{xx}D_y(\xi) - v_{xy}D_y(\eta),\\
\zeta_{22} &= D_y(\zeta_2) - u_{xy}D_y(\xi) - u_{yy}D_y(\eta), & \chi_{22} &= D_y(\chi_2) - v_{xy}D_y(\xi) - v_{yy}D_y(\eta),
\end{aligned} \tag{8.5.1.4}$$

where D_x and D_y are total differential operators in the independent variables:

$$\begin{aligned}
D_x &= \frac{\partial}{\partial x} + u_x\frac{\partial}{\partial u} + v_x\frac{\partial}{\partial v} + u_{xx}\frac{\partial}{\partial u_x} + u_{xy}\frac{\partial}{\partial u_y} + v_{xx}\frac{\partial}{\partial v_x} + v_{xy}\frac{\partial}{\partial v_y} + \cdots,\\
D_y &= \frac{\partial}{\partial y} + u_y\frac{\partial}{\partial u} + v_y\frac{\partial}{\partial v} + u_{xy}\frac{\partial}{\partial u_x} + u_{yy}\frac{\partial}{\partial u_y} + v_{xy}\frac{\partial}{\partial v_x} + v_{yy}\frac{\partial}{\partial v_y} + \cdots.
\end{aligned}$$

In the symmetry analysis of simultaneous partial differential equations, it is required that each equation must satisfy the invariance condition generated by the prolonged operator (8.5.1.2).

8.5.2. Symmetries of Equations of Steady Hydrodynamic Boundary Layer

The technique of symmetry analysis of nonlinear simultaneous partial differential equations is illustrated below by a specific example using formulas (8.5.1.1) to (8.5.1.4).

► Example 8.11. Consider the system of equations of a steady hydrodynamic boundary layer

$$\begin{aligned}
uu_x + vu_y + f(x) &= u_{yy},\\
u_x + v_y &= 0.
\end{aligned} \tag{8.5.2.1}$$

Determining relations for the coordinates of the admissible operator. We look for an operator admitted by system (8.5.2.1) in the form (8.5.1.1). The second prolongation is given by formula (8.5.1.2) in which

$$\chi_1 = 0, \quad \zeta_{11} = \zeta_{12} = 0, \quad \chi_{11} = \chi_{12} = \chi_{22} = 0. \tag{8.5.2.2}$$

(It has been taken into account that the only derivatives appearing in the system are u_x, u_y, v_y, and u_{yy}.) The nonzero coordinates of the prolonged operator (8.5.1.2) are calculated by formulas (8.5.1.3) and (8.5.1.4):

$$\begin{aligned}
\zeta_1 &= \zeta_x + \zeta_u u_x + \zeta_v v_x - u_x(\xi_x + \xi_u u_x + \xi_v v_x) - u_y(\eta_x + \eta_u u_x + \eta_v v_x),\\
\zeta_2 &= \zeta_y + \zeta_u u_y + \zeta_v v_y - u_x(\xi_y + \xi_u u_y + \xi_v v_y) - u_y(\eta_y + \eta_u u_y + \eta_v v_y),\\
\chi_2 &= \chi_y + \chi_u u_y + \chi_v v_y - v_x(\xi_y + \xi_u u_y + \xi_v v_y) - v_y(\eta_y + \eta_u u_y + \eta_v v_y),\\
\zeta_{22} &= \zeta_{yy} + 2\zeta_{vy}v_y + \zeta_{vv}(v_y)^2 + [2\zeta_{uy} - \eta_{yy} + 2(\zeta_{uv} - \eta_{vy})v_y - \eta_{vv}(v_y)^2]u_y\\
&\quad - [\xi_{yy} + 2(\xi_{uy} + \xi_{uv}v_y)u_y + 2\xi_{vy}v_y + \xi_{uu}(u_y)^2 + \xi_{vv}(v_y)^2]u_x +\\
&\quad - (\zeta_{uu} - 2\eta_{uy} - 2\eta_{uv}v_y)(u_y)^2 - \eta_{uu}(u_y)^3 - 2(\xi_y + \xi_u u_y + \xi_v v_y)u_{xy}\\
&\quad - [2\eta_y - \zeta_u - \zeta_v + (\xi_u + \xi_v)u_x + (3\eta_u + \eta_v)u_y + 2\eta_v v_y]u_{yy}.
\end{aligned} \tag{8.5.2.3}$$

Taking into account (8.5.2.2) and (8.5.2.3), we apply the prolonged operator (8.5.1.2) to system (8.5.2.1). Eliminating u_{yy} and v_y from the resulting relations using the equations of system (8.5.2.1) yields two invariance conditions. By collecting similar terms and rearranging, we arrive at the determining equations

$$\begin{aligned}
&[-3\xi_v u_x+(2\xi_u-\eta_v)u_y+2\xi_y+\zeta_v]u_{xy}+\xi_{vv}u_x^3-(2\xi_{uv}-\eta_{vv})u_x^2u_y\\
&\quad+(\xi_{uu}-2\eta_{uv})u_xu_y^2+\eta_{uu}u_y^3+(v\xi_v-2u\eta_v-\zeta_{vv}-2\xi_{yv})u_x^2\\
&\quad+(2\zeta_{uv}+2u\eta_u-2\eta_{yv}-v\eta_v+2\xi_{yu})u_xu_y+(2v\eta_u+2\eta_{yu}-\zeta_{uu})u_y^2\\
&\quad-u\xi_vu_xv_x-u\eta_vu_yv_x+(-v\xi_y-v\zeta_v-u\xi_x-2f\eta_v+2u\eta_y+f\xi_u\\
&\quad+\zeta+2\zeta_{yv}+\xi_{yy})u_x+(v\eta_y-u\eta_x+3f\eta_u-2\zeta_{yu}+\chi+\eta_{yy})u_y\\
&\quad+u\zeta_vv_x+f'_x\xi-\zeta_{yy}+u\zeta_x+v\zeta_y-f\zeta_u+2f\eta_y=0 \quad (\textit{first equation}),\\
&-(\xi_u+\eta_v)u_x^2+(\zeta_u-\chi_v+\eta_y-\xi_x)u_x-(\eta_v+\xi_u)u_yv_x+\zeta_vv_x\\
&\quad+(\chi_u-\eta_x-\xi_y)u_y+\zeta_x+\chi_y=0 \qquad (\textit{second equation}).
\end{aligned}$$

Splitting these equations into 'independent' variables results in two overdetermined systems, which must hold simultaneously:

First system:

$$\begin{aligned}
u_xu_{xy}&: \ 3\xi_v=0, && (1)\\
u_yu_{xy}&: \ 2\xi_u-\eta_v=0, && (2)\\
u_{xy}&: \ 2\xi_y+\zeta_v=0, && (3)\\
u_x^3&: \ \xi_{vv}=0, && (4)\\
u_x^2u_y&: \ 2\xi_{uv}-\eta_{vv}=0, && (5)\\
u_xu_y^2&: \ \xi_{uu}-2\eta_{uv}=0, && (6)\\
u_y^3&: \ \eta_{uu}=0, && (7)\\
u_x^2&: \ v\xi_v-2u\eta_v-\zeta_{vv}-2\xi_{yv}=0, && (8)\\
u_xu_y&: \ 2\zeta_{uv}+2u\eta_u-2\eta_{yv}-v\eta_v+2\xi_{yu}=0, && (9)\\
u_y^2&: \ 2v\eta_u+2\eta_{yu}-\zeta_{uu}=0, && (10)\\
u_xv_x&: \ u\xi_v=0, && (11)\\
u_yv_x&: \ u\eta_v=0, && (12)\\
u_x&: \ \xi_{yy}-v\xi_y-v\zeta_v-u\xi_x-2f\eta_v+2u\eta_y+f\xi_u+\zeta+2\zeta_{yv}=0, && (13)\\
u_y&: \ \eta_{yy}+v\eta_y-u\eta_x+3f\eta_u-2\zeta_{yu}+\chi=0, && (14)\\
v_x&: \ u\zeta_v=0, && (15)\\
1&: \ f'_x\xi-\zeta_{yy}+u\zeta_x+v\zeta_y-f\zeta_u+2f\eta_y=0. && (16)
\end{aligned}$$

Second system:

$$\begin{aligned}
u_x^2&: \ \xi_u+\eta_v=0, && (1)\\
u_yv_x&: \ \eta_v+\xi_u=0, && (2)\\
u_x&: \ \zeta_u-\chi_v+\eta_y-\xi_x=0, && (3)\\
u_y&: \ \chi_u-\eta_x-\xi_y=0, && (4)\\
v_x&: \ \zeta_v=0, && (5)\\
1&: \ \zeta_x+\chi_y=0. && (6)
\end{aligned}$$

It follows from the 11th and 12th equations of the first system that $\xi=\xi(x,y,u)$ and $\eta=\eta(x,y,u)$; using equation 2, it can be concluded that $\xi=\xi(x,y)$. From the 15th equation, it follows that $\zeta = \zeta(x, y, u)$, and equation 3 then gives $\xi = \xi(x)$. Equations 1 and 4–6 are satisfied identically. If follows from equation 9 that $\eta_u = 0$, and hence $\eta = \eta(x, y)$ (equation 7 is satisfied identically). So we have

$$\xi = \xi(x), \quad \eta = \eta(x, y), \quad \zeta = \zeta(x, y, u). \tag{8.5.2.4}$$

The remaining equations become

$$\begin{aligned} u_y^2\colon &\quad \zeta_{uu} = 0, \\ u_x\colon &\quad -u\xi'_x + 2u\eta_y + \zeta = 0, \\ u_y\colon &\quad \eta_{yy} + v\eta_y - u\eta_x - 2\zeta_{yu} + \chi = 0, \\ 1\colon &\quad f'_x\xi - \zeta_{yy} + u\zeta_x + v\zeta_y - f\zeta_u + 2f\eta_y = 0. \end{aligned} \tag{8.5.2.5}$$

By virtue of (8.5.2.4), equations 1, 2, and 5 in the second system are satisfied identically and the remaining equations become

$$\begin{aligned} u_x\colon &\quad \zeta_u - \chi_v + \eta_y - \xi_x = 0, \\ u_y\colon &\quad \chi_u - \eta_x = 0, \\ 1\colon &\quad \zeta_x + \chi_y = 0. \end{aligned} \tag{8.5.2.6}$$

The first equation in (8.5.2.5) implies that $\zeta = \zeta_1(x, y)u + \zeta_0(x, y)$. By substituting this expression into the second equation in (8.5.2.5) and splitting in u, one finds that $\zeta_0 \equiv 0$. Likewise, from the fourth equation in (8.5.2.5), one finds that $\zeta_1 = C_1 = \text{const}$, or

$$\zeta = C_1 u. \tag{8.5.2.7}$$

By virtue of this, the second equation in (8.5.2.5) and the last equation in (8.5.2.6) lead to

$$\eta = \tfrac{1}{2}(\xi_x - C_1)y + h(x), \qquad \chi = \chi(x, u, v), \tag{8.5.2.8}$$

where $h = h(x)$ is an arbitrary function. By substituting (8.5.2.7) and (8.5.2.8) into the third and fourth equations of system (8.5.2.5) and substituting the second equation of (8.5.2.6), one arrives at the following system of three equations:

$$\begin{aligned} &\chi = \tfrac{1}{2}\xi''_{xx}yu + h'_x u - \tfrac{1}{2}(\xi'_x - C_1)v, \\ &f'_x\xi + f(\xi'_x - 2C_1) = 0, \\ &\chi_u - \eta_x = 0. \end{aligned} \tag{8.5.2.9}$$

Note that the first equation of (8.5.2.6) coincides with the first equation of (8.5.2.9) differentiated with respect to v.

Since χ is independent of y, the first equation of (8.5.2.9) implies $\xi''_{xx} = 0$, or

$$\xi = C_2 x + C_3.$$

Substituting this expression into the first and last equations of system (8.5.2.9), taking into account (8.5.2.7) and (8.5.2.8), and renaming the constants of integration, one obtains the coordinates of the operator (8.5.1.1):

$$\xi = C_1 + (C_2 + 2C_3)x, \quad \eta = C_3 y + h(x), \quad \zeta = C_2 u, \quad \chi = -C_3 v + h'(x)u. \tag{8.5.2.10}$$

In addition, the condition

$$\big[C_1 + (C_2 + 2C_3)x\big] f'(x) = (C_2 - 2C_3) f(x) \tag{8.5.2.11}$$

must be satisfied, which follows from the second equation of (8.5.2.9) and contains the function $f(x)$ and the constants of integration C_1, C_2, and C_3.

Symmetries of the hydrodynamic boundary layer equations. Using condition (8.5.2.11), let us classify the symmetries of the boundary layer system (8.5.2.1) for all $f(x)$.

1°. In order that condition (8.5.2.11) is satisfied for arbitrary $f(x)$, one must set $C_1 = C_2 = C_3 = 0$. Then, from (8.5.2.10) it follows that $\xi = \zeta = 0$, $\eta = h(x)$, and $\chi = h'(x)u$, so the corresponding admissible operator is

$$\mathrm{X}_0 = h(x)\partial_y + h'(x)u\partial_v. \tag{8.5.2.12}$$

Recall that here $h(x)$ is an arbitrary function.

2°. If $f(x) = 0$, condition (8.5.2.11) is satisfied identically for any C_1, C_2, and C_3. Taking into account (8.5.2.10), we obtain the following three additional operators:

$$\begin{aligned} &\mathrm{X}_1 = \partial_x && (C_1 = 1,\ C_2 = C_3 = 0,\ h = 0);\\ &\mathrm{X}_2 = x\partial_x + u\partial_u && (C_2 = 1,\ C_1 = C_3 = 0,\ h = 0);\\ &\mathrm{X}_3 = 2x\partial_x + y\partial_y - v\partial_v && (C_3 = 1,\ C_1 = C_2 = 0,\ h = 0). \end{aligned} \tag{8.5.2.13}$$

The operator (8.5.2.12) exists in all cases.

3°. If $f(x) \equiv 1$, condition (8.5.2.11) implies $C_2 = 2C_3$. Then, along with X_0, there are two more admissible operators: X_1 and $2\mathrm{X}_2 + \mathrm{X}_3$, where the X_n are given by (8.5.2.13).

4°. If $f(x) = \pm e^x$, condition (8.5.2.11) implies $C_1 = -4C_3$ and $C_2 = -2C_3$, which corresponds to a single operator $4\mathrm{X}_1 + 2\mathrm{X}_2 - \mathrm{X}_3$.

5°. If $f(x) = \pm x^n$ $(n \neq 0)$, we have $C_1 = 0$ and $C_2 = -\frac{2(1+n)}{(1-n)}$, which results in a single operator $2(n+1)\mathrm{X}_2 - (n-1)\mathrm{X}_3$.

These results and the corresponding invariants of the system of boundary layer equations (8.5.2.1) are gathered in Table 8.5. The invariants I_n are determined by the first integrals of the characteristic system of ordinary differential equations

$$\frac{dx}{\xi} = \frac{dy}{\eta} = \frac{du}{\zeta} = \frac{dv}{\chi},$$

which corresponds to the linear first-order partial differential equation $XI = 0$.

Remark 8.10. *The first invariant was taken linear in y (wherever this was possible). This simplified the calculations, since the equations contain the highest-order derivative in y, while the derivatives with respect to x have a lower order. Recall that the invariants I_n can be replaced by $F_n(I_n)$, where the functions F_n are chosen for reasons of convenience.*

Table 8.5. Symmetries of the system of boundary layer equations (8.5.2.1); $a=$ const.

Function $f(x)$	Operator	Invariant 1	Invariant 2	Invariant 3
Arbitrary	$\mathrm{X}_0=h(x)\partial_y+h'(x)u\partial_v$	x	u	$v-(\ln h)'_x yu$
0	$\mathrm{X}_1=\partial_x$ $\mathrm{X}_2=x\partial_x+u\partial_u$ $\mathrm{X}_3=2x\partial_x+y\partial_y-v\partial_v$	y y $yx^{-1/2}$	u u/x u	v v $x^{1/2}v$
a	$\mathrm{X}_1=\partial_x$ $\mathrm{X}_2=4x\partial_x+y\partial_y+2u\partial_u-v\partial_v$	y $yx^{-1/4}$	u $ux^{-1/2}$	v $x^{1/4}v$
ae^x	$\mathrm{X}_1=2x\partial_x-y\partial_y+2u\partial_u-v\partial_v$	$yx^{1/2}$	u/x	$x^{1/2}v$
$ax^n,\ n\neq 0$	$\mathrm{X}_1=4x\partial_x-(n-1)y\partial_y$ $+2(n+1)u\partial_u+(n-1)v\partial_v$	$yx^{\frac{n-1}{4}}$	$x^{-\frac{n+1}{2}}u$	$x^{\frac{1-n}{4}}v$

Remark 8.11. *The symmetry* X_0 *indicates the invariance of system (8.5.2.1) under the transformation*

$$x=\bar{x},\quad y=\bar{y}+h(x),\quad u=\bar{u},\quad v=\bar{v}-h'_x u,$$

where $h=h(x)$ *is an arbitrary function.*

Exact solutions to the hydrodynamic boundary layer equations. We look for solutions in the form

$$I_2=\Phi(I_1),\quad I_3=\Psi(I_1),\tag{8.5.2.14}$$

where the I_n are invariants, and the functions Φ and Ψ are determined from the system of ordinary differential equations resulting from the substitution of (8.5.2.14) into the original system of equations (8.5.2.1).

1°. For arbitrary $f(x)$, exact solutions are sought in the form (see the first row in Table 8.5)

$$u=\Phi(x),\quad v=(\ln h)'_x yu+\Psi(x).\tag{8.5.2.15}$$

Substituting (8.5.2.15) into (8.5.2.1) yields a system of ordinary differential equations, which is easy to integrate. We have

$$\Phi(x)=\pm\Big[C_1-2\int f(x)\,dx\Big]^{1/2},\quad h(x)=\frac{C_1}{\Phi(x)},\quad \Psi(x)\text{ is arbitrary},\tag{8.5.2.16}$$

where C_1 and C_2 are arbitrary constants. Solution (8.5.2.15)–(8.5.2.16) corresponds to inviscid fluid flow, since $u_{yy}=0$.

2°. For $f(x)=0$, the invariants of the first operator in Table 8.5 give a one-dimensional solution independent of the coordinate x. Substituting $u=u(y)$ and $v=v(y)$ into system (8.5.2.1) and integrating, we obtain

$$\begin{aligned}&u=C_2\exp(C_1y)+C_3, && v=C_1 && \text{if } C_1\neq 0;\\ &u=C_2y+C_3, && v=0 && \text{if } C_1=0.\end{aligned}$$

The invariants of the second operator in Table 8.5 correspond to solutions of the form

$$u = x\varphi(y), \qquad v = \psi(y), \tag{8.5.2.17}$$

where the functions $\varphi(y)$ and $\psi(y)$ are described by the system of ordinary differential equations

$$\varphi^2 + \psi\varphi'_y = \varphi''_{yy},$$
$$\varphi + \psi'_y = 0,$$

which results from substituting (8.5.2.17) into the original system (8.5.2.1).

The invariants of the third operator (see Table 8.5) lead to the self-similar solution

$$u = U(\theta), \quad v = x^{-1/2}V(\theta), \quad \theta = yx^{-1/2},$$

where the functions $U(\theta)$ and $V(\theta)$ are determined by the system of ordinary differential equations

$$-\tfrac{1}{2}\theta UU'_\theta + VU'_\theta = U''_{\theta\theta},$$
$$-\tfrac{1}{2}\theta U'_\theta + V'_\theta = 0.$$

3°. Exact solutions for the other $f(x)$ presented in Table 8.5 are sought in a similar manner. ◀

Remark 8.12. *The above results of the group analysis of the system of boundary layer equations (8.5.2.1) were obtained in the work of Pavlovskii (1962) (see also Ovsiannikov, 1982).*

Problems and Exercises for Section 8.5

1. Find invariant solutions of the system of steady-state hydrodynamic boundary layer equations (8.5.2.1) in the following cases:

a) $f(x) = a$,
b) $f(x) = ae^x$,
c) $f(x) = ax^n$.

Hint. Take advantage of the results of the symmetry classification for this system, shown in Table 8.5.

2. Find admissible operators and invariant solutions for the system of steady-state transonic gas flow equations:

$u_y = v_x, \quad v_y = -uu_x.$

3. Find admissible operators and invariant solutions for the nonlinear system of equations describing one-dimensional long-wave vibrations of an elastic bar:

$u_t - v_x = 0, \quad v_t = -f(u)u_x.$

Classify the symmetries of the system of equations for all $f(u)$.

4. Find admissible operators and invariant solutions for the nonlinear system of equations describing one-dimensional isentropic motions of an ideal gas:

$\rho_t + (\rho u)_x = 0, \quad u_t + uu_x + f(\rho)\rho_x = 0.$

Classify the symmetries of the system of equations for all $f(\rho)$.

5. Find admissible operators and invariant solutions for the nonlinear system of equations describing two-dimensional steady-state flows of an ideal incompressible fluid:

$$uu_x + vu_y = -p_x, \quad uv_x + vv_y = -p_y, \quad u_x + v_y = 0.$$

Note that this system can be reduced to a system of two equations by eliminating pressure p. If we then introduce the stream function ψ defined by the formulas $u = \psi_y$ and $v = -\psi_x$, we can further reduce the resulting system to a single third-order equation.

6. Find admissible operators and invariant solutions for the following systems of equations:

a) $u_t = u_{xx} + u^k f(u/v), \quad v_t = u^m g(u/v);$

b) $u_t = u_{xx} + u^k f(u/v), \quad v_t + av_x = u^m g(u/v).$

Classify the symmetries of the systems for all $f(z)$ and $g(z)$.

7. Find admissible operators and invariant solutions for the system of equations of mass and heat transfer with volume chemical reactions:

$$u_t = au_{xx} + u^k f(u^n v^m), \quad v_t = bv_{xx} + u^p g(u^n v^m)$$

Classify the symmetries of the system of equations for all $f(z)$, $g(z)$.

8. Find invariant solutions of the Korteweg–de Vries equation

$$u_t + u_{xxx} - 6uu_x = 0$$

by reducing it to a PDE system, introducing a second unknown function $v = u_x$.

8.6. General Comments and Conclusions

8.6.1. General Comments

The classical method of symmetry reductions, based on the search for continuous groups of transformations that preserve the form of PDEs under consideration, allows one to find all invariant solutions of these equations. As a result of its application, nonlinear PDEs are reduced to significantly simpler ODEs (the solutions of which, unfortunately, usually cannot be expressed in terms of elementary functions or quadratures).

It is important to note that the use of the classical method is based on an algorithm that requires special training and leads to a fairly large amount of routine intermediate calculations. In addition, it is necessary to be able to analyze and solve overdetermined systems of differential equations, which is a complicating factor.

Note also that the classical method of symmetry reductions does not allow finding numerous non-invariant exact solutions of nonlinear PDEs, which can be obtained by the methods of generalized and functional separation of variables (see Chapters 5 and 6), the direct method of symmetry reductions (Chapter 7) and the method of differential constraints (Chapter 10).

8.6.2. Comparison of the Effectiveness of the Classical Method of Symmetry Reductions and the Elementary Theory of Invariants

Let us compare the effectiveness of the classical method of symmetry reductions and the elementary theory of invariants (see Chapter 4) using examples of constructing exact solutions of three different nonlinear PDEs,

$$\begin{aligned} u_{xx} + u_{yy} &= f(u), \\ u_t &= [f(u)u_x]_x, \\ u_{tt} &= [f(u)u_x]_x, \end{aligned} \tag{8.6.2.1}$$

which are discussed in detail in Sections 8.2.2, 8.3.3, and 8.3.4.

First, we recall that the elementary invariant theory relies on two fundamental transformations: translation and scaling, along with their combinations. This approach results in a relatively small number of intermediate calculations. It is much simpler than the classical symmetry reduction method because it only requires the ability to solve linear algebraic equations and perform differentiation.

Out of the five invariant solutions of the first equation in (8.6.2.1), listed in Table 8.2, only the third solution cannot be obtained using the elementary invariant theory.

Out of the seven invariant solutions of the second equation in (8.6.2.1), given in Table 8.3, only the last one cannot be obtained using invariant theory (note that the last solution is degenerate, for which $u_t = 0$). Three more complex solutions to this equation, induced by linear combinations of admissible operators, discussed in Example 8.9, are also easily constructed using the elementary invariant theory.

Out of the nine invariant solutions of the third equation in (8.6.2.1), given in Table 8.4, only the seventh and last solutions cannot be obtained using the elementary invariant theory. Note that the seventh and last solutions are degenerate, for which $u_{tt} = 0$ and $(u^{-4}u_x)_x = 0$ (that is, for these solutions, either the left- or right-hand side of the equation vanishes).

Out of the eight invariant solutions of the hydrodynamic system (8.5.2.1), which can be obtained using Table 8.5, only the first solution cannot be found using the elementary invariant theory.

The comparison of the effectiveness of the two methods, demonstrated using three nonlinear PDEs (8.6.2.1) and one PDE system (8.5.2.1), indicates that the elementary invariant theory allows one to find the vast majority of exact solutions obtainable using the much more complex classical method of symmetry reductions. This conclusion also holds true for most nonlinear PDEs encountered in mathematical physics.

Let us also recall that with an increase in the order of the PDEs under consideration, the technical difficulties in using the classical method of symmetry reduction increase sharply. On the contrary, when using the elementary theory of invariants for PDEs of the second and higher orders, the technical difficulties remain almost the same.

References for Chapter 8

Ames W.F., Lohner J.R., and Adams E. Group properties of $u_{tt} = [f(u)u_x]_x$. *Int. J. Non-Linear Mech.*, 1981, Vol. 16, No. 5–6, pp. 439–447.

Bluman G.W. and Cole J.D. *Similarity Methods for Differential Equations*. Springer-Verlag, New York, 1974.

Bluman G.W. and Kumei S. *Symmetries and Differential Equations*. Springer-Verlag, New York, 1989.

Cantwell B.J. *Introduction to Symmetry Analysis*. Cambridge Univ. Press, Cambridge, 2002.

Fushchich W.I., Shtelen W.M., and Serov N.I. *Symmetry Analysis and Exact Solutions of the Equations of Mathematical Physics*. Kluwer, Dordrecht, 1993.

Hydon P.E. *Symmetry Methods for Differential Equations: A Beginner's Guide*. Cambridge Univ. Press, Cambridge, 2000.

Ibragimov N.H. (ed.). *CRC Handbook of Lie Group Analysis of Differential Equations, Vol. 1, Symmetries, Exact Solutions and Conservation Laws*. CRC Press, Boca Raton, 1994.

Olver P.J. *Application of Lie Groups to Differential Equations*. Springer-Verlag, New York, 1993.

Ovsiannikov L.V. Group properties of nonlinear heat equations (in Russian). *Doklady Acad. Nauk U.S.S.R.*, 1959, Vol. 125, No. 3, pp. 492–495.

Ovsiannikov L.V. *Group Analysis of Differential Equations*. Academic Press, New York, 1982.

Pavlovskii Yu.N. Research into some invariant solutions of boundary layer equations. *U.S.S.R. Comput. Math. Math. Phys.*, 1962, Vol. 1, No. 2, pp. 321–339.

9. Differential Constraints Method

► **Preliminary remarks.** The method of differential constraints differs significantly from the methods covered in the previous chapters. It focuses on analyzing the compatibility of the given partial differential equation along with one or more additional, typically simpler, differential equations known as differential constraints.

The differential constraints method, on the one hand, is one of the most general methods (if not the most general method) for constructing exact solutions of nonlinear partial differential equations, and on the other hand, it is not sufficiently formalized since the previous experience and intuition of the researcher largely determines the choice of differential constraints. Most other methods can be treated as special cases of the differential constraints method.

The fundamental concepts of the method, along with a general overview, were initially introduced by Yanenko (1964). This chapter provides a concise description of the key characteristics of the method of differential constraints and includes specific examples of how it can be applied to derive exact solutions for nonlinear partial differential equations. The material is organized following the principle of moving 'from simple to complex.'

9.1. Differential Constraints Method for Ordinary Differential Equations

9.1.1. Description of the Method. First-Order Differential Constraints

Description of the method. Before discussing the differential constraints method for partial differential equations, let us first examine its characteristic features as applied to simpler ordinary differential equations.

The method's basic idea is to find exact solutions of a complex (non-integrable) equation by jointly analyzing this equation with a simpler auxiliary equation, usually integrable, known as a *differential constraint*.

The *order of a differential constraint* coincides with that of its highest derivative. Typically, the order of a differential constraint is lower than that of the equation being analyzed. The simplest and most commonly used differential constraints are first-order constraints. Both the equation and the differential constraint must include a set of free parameters (sometimes arbitrary functions) whose values are determined during the analysis to ensure compatibility between the equation and the constraint. After the compatibility analysis, all solutions derived from integrating the differential constraint will also serve as solutions to the original equation. This method allows one to find particular solutions of the original equation for certain values of the determining parameters.

For simplicity, we first consider autonomous ordinary differential equations of the form*

$$F(y, y'_x, \dots, y_x^{(n)}; \mathbf{a}) = 0, \tag{9.1.1.1}$$

which do not explicitly include the independent variable x and depend on the vector of free parameters $\mathbf{a} = \{a_1, \dots, a_k\}$. We will supplement equation (9.1.1.1) with an autonomous first-order differential constraint of the form

$$y'_x = \varphi(y; \mathbf{b}). \tag{9.1.1.2}$$

Its right-hand side depends on a vector of free parameters, $\mathbf{b} = \{b_1, \dots, b_s\}$.

On differentiating constraint (9.1.1.2) successively sufficiently many times, we express the leading derivatives through the function $\varphi = \varphi(y; \mathbf{b})$:

$$y''_{xx} = (y'_x)'_y y'_x = \varphi\varphi'_y, \quad y'''_{xxx} = (y''_{xx})'_y y'_x = \varphi(\varphi\varphi'_y)'_y, \quad \dots \tag{9.1.1.3}$$

On eliminating the derivatives from equation (9.1.1.1) using relations (9.1.1.2) and (9.1.1.3), we arrive at an algebraic (or transcendental) equation

$$P(y; \mathbf{a}, \mathbf{b}) = 0. \tag{9.1.1.4}$$

Further, we look for the values of the parameters **a** and **b** such that equation (9.1.1.4) holds identically for any y, which usually leads to certain restrictions on the components of the vector **a**. After that, we express the vector **b** in terms of **a**, leading to a dependence $\mathbf{b} = \mathbf{b}(\mathbf{a})$, and substitute back into the differential constraint (9.1.1.2). As a result, we obtain a first-order ordinary differential equation

$$y'_x = g(y; \mathbf{a}) \qquad (g = \varphi|_{\mathbf{b}=\mathbf{b}(\mathbf{a})}), \tag{9.1.1.5}$$

which is compatible with the original equation (9.1.1.1). By construction, any solution satisfying equation (9.1.1.5) is also a solution of the original equation (9.1.1.1). The general solution of equation (9.1.1.5) is also an exact solution of the original equation (9.1.1.1).

Remark 9.1. *Sometimes, one may use differential constraints that are unsolved for the derivative (see Example 9.3).*

Examples of nonlinear ODEs and their differential constraints. The form of nonlinearity of the differential constraint (9.1.1.2) can often be chosen to be similar to that of the original equation (9.1.1.1), albeit with different determining parameters. To clarify this point, we will illustrate it with examples of specific ordinary differential equations of the second, third, fourth, and higher orders.

*Such equations frequently arise in mathematical physics when exact solutions to nonlinear PDEs are sought as traveling waves.

► Example 9.1. Let us examine the second-order ordinary differential equation with a power-law nonlinearity

$$y''_{xx} - cy'_x = ay + by^n. \tag{9.1.1.6}$$

It arises in the theory of chemical reactors, combustion theory, and mathematical biology.* The simple cases where $n = 0$, $n = 1$, or $b = 0$, which lead to a linear constant-coefficient ODE, will not be considered.

We supplement equation (9.1.1.6) with the first-order differential constraint

$$y'_x = \alpha y + \beta y^m, \tag{9.1.1.7}$$

which is a first-order separable ODE and can be easily integrated. The right-hand side of equation (9.1.1.7) has been selected similar to that of the original equation (9.1.1.6), albeit with different parameters.

The equation and differential constraint include seven parameters: a, b, c, n, m, α, and β. The subsequent analysis aims to identify the parameters α, β, and m of the differential constraint, which must be expressed in terms of a, b, c, and n. In addition, we will determine restrictions on the equation parameters and construct an exact solution to the equation.

Differentiating (9.1.1.7) and replacing the first derivative with the right-hand side of (9.1.1.7), we find that

$$\begin{aligned} y''_{xx} &= (\alpha + m\beta y^{m-1})y'_x = (\alpha + m\beta y^{m-1})(\alpha y + \beta y^m) \\ &= \alpha^2 y + \alpha\beta(m+1)y^m + m\beta^2 y^{2m-1}. \end{aligned} \tag{9.1.1.8}$$

Eliminating the first and second derivatives from (9.1.1.6) with the help of (9.1.1.7) and (9.1.1.8) and performing straightforward rearrangements, we obtain

$$(\alpha^2 - \alpha c - a)y + \beta[\alpha(m+1) - c]y^m + m\beta^2 y^{2m-1} - by^n = 0. \tag{9.1.1.9}$$

For this relation to hold identically for any y, we must set

$$\begin{aligned} &\alpha^2 - \alpha c - a = 0, \\ &\alpha(m+1) - c = 0, \\ &2m - 1 = n, \\ &m\beta^2 - b = 0. \end{aligned} \tag{9.1.1.10}$$

The third equation helps balance the final two terms of relation (9.1.1.9). The other three algebraic equations of (9.1.1.10) result from setting the coefficients of the different powers of y in (9.1.1.9) to zero.

If conditions (9.1.1.10) are satisfied, any solution of equation (9.1.1.7) is also a solution of the more complex equation (9.1.1.6). The determining system comprised by the four equations (9.1.1.10) includes seven parameters: a, b, c, n, m, α, and β.

*Equation (9.1.1.6) describes traveling wave solutions of the Kolmogorov–Petrovskii–Piskunov equation $u_t = u_{zz} - f(u)$ for the special kinetic function $f(u) = au + bu^n$. In this case, we have $u = y(x)$, where $x = z + ct$.

The parameters b, c, and n of the original equation can be treated as arbitrary, while the others are expressed as

$$a = -\frac{2c^2(n+1)}{(n+3)^2}, \quad m = \frac{n+1}{2}, \quad \alpha = \frac{2c}{n+3}, \quad \beta = \pm\sqrt{\frac{2b}{n+1}}, \tag{9.1.1.11}$$

where $n \neq -1$, $n \neq -3$, and $b(n+1) > 0$. It is apparent that for equations (9.1.1.6) and (9.1.1.7) to be compatible, the parameter a of the original equation must be related to the other two parameters, c and n, in a certain way. In this case, there are two families of parameters (9.1.1.11) of equation (9.1.1.7), which bring about distinct one-parameter solutions of equations (9.1.1.6) and (9.1.1.7).

Integrating the differential constraint (9.1.1.7), which represents a separable first-order ODE, and considering formulas (9.1.1.11), we ultimately arrive at the following two exact solutions for equation (9.1.1.6) where $a = -2c^2(n+1)/(n+3)^2$:

$$y = \left[Ce^{\alpha(1-m)x} - (\beta/\alpha)\right]^{\frac{1}{1-m}},$$

where C is an arbitrary constant, and the constants m, α, and β are expressed in terms of the original equation parameters by formulas (9.1.1.11). ◄

► Example 9.2. Consider the nonlinear third-order ODE

$$y'''_{xxx} = ay^4 + by^2 + c \tag{9.1.1.12}$$

in conjunction with the first-order differential constraint

$$y'_x = \alpha y^2 + \beta. \tag{9.1.1.13}$$

Using (9.1.1.13), we find the derivatives

$$\begin{aligned} y''_{xx} &= 2\alpha y y'_x = 2\alpha y(\alpha y^2 + \beta) = 2\alpha^2 y^3 + 2\alpha\beta y, \\ y'''_{xxx} &= (6\alpha^2 y^2 + 2\alpha\beta)y'_x = (6\alpha^2 y^2 + 2\alpha\beta)(\alpha y^2 + \beta) \\ &= 6\alpha^3 y^4 + 8\alpha^2\beta y^2 + 2\alpha\beta^2. \end{aligned} \tag{9.1.1.14}$$

For the third derivative in (9.1.1.14) to coincide with the right-hand side of (9.1.1.12), the relations

$$a = 6\alpha^3, \quad b = 8\alpha^2\beta, \quad c = 2\alpha\beta^2 \tag{9.1.1.15}$$

must hold. On solving the first two equations for α and β and inserting the resulting expressions into the last equation, we obtain

$$\alpha = \left(\frac{a}{6}\right)^{1/3}, \quad \beta = \left(\frac{a}{6}\right)^{-2/3}\frac{b}{8}, \quad c = \frac{3b^2}{16a}. \tag{9.1.1.16}$$

It follows that the third-order equation (9.1.1.12) with $c = 3b^2/(16a)$ has a particular solution, which can be determined by integrating the separable first-order equation (9.1.1.13), whose parameters are related to those of the original equation by the first two relations in (9.1.1.16). ◄

► Example 9.3. Let us look at the fourth-order ODE with a cubic nonlinearity

$$y''''_{xxxx} = ay^3 + b \tag{9.1.1.17}$$

in conjunction with the first-order differential constraint

$$(y'_x)^2 = \alpha y^m + \beta. \tag{9.1.1.18}$$

Differentiating (9.1.1.18) successively, we express the derivatives as

$$\begin{aligned} y''_{xx} &= \tfrac{1}{2}\alpha m y^{m-1} \qquad \text{(after cancelling by } y'_x\text{)},\\ y'''_{xxx} &= \tfrac{1}{2}\alpha m(m-1)y^{m-2}y'_x,\\ y''''_{xxxx} &= \tfrac{1}{2}\alpha m(m-1)y^{m-2}y''_{xx} + \tfrac{1}{2}\alpha m(m-1)(m-2)y^{m-3}(y'_x)^2 =\\ &= \tfrac{1}{2}\alpha\beta m(m-1)(m-2)y^{m-3} + \tfrac{1}{4}\alpha^2 m(m-1)(3m-4)y^{2m-3}. \end{aligned} \tag{9.1.1.19}$$

Comparing the right-hand side equation (9.1.1.17) with that of the last relation in (9.1.1.19), we arrive at two pairs of determining parameters of the differential constraint (9.1.1.18):

$$m = 3, \qquad \alpha = \pm\sqrt{\frac{2a}{15}}, \qquad \beta = \pm b\sqrt{\frac{5}{6a}}. \tag{9.1.1.20}$$

One should choose either the upper or the lower signs here. Substituting (9.1.1.20) into (9.1.1.18) yields an equation reducible to two separable first-order ODEs. ◄

► Example 9.4. We will demonstrate that the nth-order ODE with an exponential nonlinearity

$$y_x^{(n)} = ae^{\lambda y} \tag{9.1.1.21}$$

admits a first-order differential constraint

$$y'_x = be^{\mu y}. \tag{9.1.1.22}$$

Indeed, differentiating relation (9.1.1.22) successively yields

$$y''_{xx} = b\mu e^{\mu y}y'_x = b^2\mu e^{2mu}, \qquad \ldots, \qquad y_x^{(m)} = b^m\mu^{m-1}(m-1)!\,e^{m\mu y}. \tag{9.1.1.23}$$

Substituting the final expression from (9.1.1.23) with $m = n$ into (9.1.1.21), we find the coefficients of the differential constraint:

$$\mu = \frac{\lambda}{n}, \qquad b = \left[\frac{an^{n-1}}{\lambda^{n-1}(n-1)!}\right]^{\frac{1}{n}}.$$

The general solution of the separable equation (9.1.1.22) and the corresponding exact solution of the original equation (9.1.1.21) are expressed as

$$y = -\frac{1}{\mu}\ln(-b\mu x + C).$$

◄

9.1.2. Differential Constraints of Arbitrary Order. General Method for Analyzing the Compatibility of Two Equations

In general, a differential constraint is an ordinary differential equation of arbitrary order. Therefore, it is necessary to be able to analyze overdetermined systems of two ordinary differential equations for compatibility. Below is a general algorithm that is used to analyze such systems.

1°. *ODEs of the same order.* Let us first consider two ordinary differential equations of the same order

$$F_1(x, y, y'_x, \dots, y_x^{(n)}) = 0, \qquad (9.1.2.1)$$

$$F_2(x, y, y'_x, \dots, y_x^{(n)}) = 0. \qquad (9.1.2.2)$$

Here and henceforth, we assume that the equations depend on free parameters, which are omitted for brevity. To eliminate the highest derivative, we solve either equation for $y_x^{(n)}$, which is assumed feasible. Then, we substitute the resulting expression into the other equation to obtain an $(n-1)$st-order equation:

$$G_1(x, y, y'_x, \dots, y_x^{(n-1)}) = 0. \qquad (9.1.2.3)$$

Differentiating (9.1.2.3) with respect to x and eliminating the derivative $y_x^{(n)}$ from the resulting equation using either equation (9.1.2.1) or (9.1.2.2), we arrive at another $(n-1)$st-order equation:

$$G_2(x, y, y'_x, \dots, y_x^{(n-1)}) = 0. \qquad (9.1.2.4)$$

Hence, the analysis of two nth-order equations, (9.1.2.1) and (9.1.2.2), reduces to the analysis of two $(n-1)$st-order equations, (9.1.2.3) and (9.1.2.4). By progressively reducing the order of the equations, we eventually arrive at a single algebraic (or transcendental) equation. This is because two first-order differential equations can be simplified to one algebraic equation. Analyzing the resulting algebraic equation is relatively straightforward and can be done similarly to the technique described earlier in Section 9.1.1 for the case of a first-order differential constraint.

► Example 9.5. Let us examine the overdetermined system of two ODEs obtained in Example 5.15 (refer to the first two equations in (5.5.3.4)):

$$\begin{aligned} (\theta\theta'_x)'_x &= A_1\theta + A_2, \\ \theta''_{xx} &= A_3\theta + A_4. \end{aligned} \qquad (9.1.2.5)$$

We will show how to find a consistent solution of this system through a compatibility analysis without using the solutions of the second linear equation in (9.1.2.5) for different values of the determining parameters A_n.

We expand the brackets in the first equation of (9.1.2.5) and eliminate the second derivative using the second equation to obtain a first-order ODE:

$$(\theta'_x)^2 + A_3\theta^2 = (A_1 - A_4)\theta + A_2.$$

Differentiating this equation and cancelling by θ'_x, we arrive at the second-order ODE:

$$2\theta''_{xx} + 2A_3\theta = A_1 - A_4.$$

Eliminating the second derivative with the help the second equation of (9.1.2.5), we obtain a linear algebraic equation for θ:

$$4A_3\theta + 3A_4 - A_1 = 0.$$

For this equation to hold identically for any θ, we must set

$$A_3 = 0, \qquad A_4 = \frac{1}{3}A_1. \tag{9.1.2.6}$$

With these values, the overdetermined system (9.1.2.5) has a compatible solution as a quadratic polynomial. First, one solves the second ODE of (9.1.2.5) with parameters (9.1.2.6) and then substitutes the solution into the first ODE to obtain

$$\theta = \frac{1}{6}A_1x^2 + Cx + \frac{3}{2A_1}(C^2 - A_2),$$

where C is an arbitrary constant. ◄

2°. *ODEs of different orders.* Suppose there are two ordinary differential equations of different orders:

$$F_1(x, y, y'_x, \dots, y_x^{(n)}) = 0, \tag{9.1.2.7}$$

$$F_2(x, y, y'_x, \dots, y_x^{(m)}) = 0, \tag{9.1.2.8}$$

where $m < n$. Differentiating ODE (9.1.2.8) $n - m$ times, we reduce system (9.1.2.7)–(9.1.2.8) to a system of the form (9.1.2.1)–(9.1.2.2) in which both equations are of the same order n.

► Example 9.6. Let us look at the fourth-order ODE with a quadratic nonlinearity

$$y''''_{xxxx} = a(y''_{xx})^2 - by^2 + c \tag{9.1.2.9}$$

in conjunction with the linear second-order differential constraint

$$y''_{xx} = \alpha y + \beta. \tag{9.1.2.10}$$

Differentiating (9.1.2.10) twice yields $y''''_{xxxx} = \alpha^2 y + \alpha\beta$. We use this expression and the differential constraint (9.1.2.10) to eliminate the derivatives from equation (9.1.2.9). As a result, arrive at a quadratic equation for y that is satisfied identically, provided that the conditions

$$a\alpha^2 - b = 0, \quad \alpha - 2a\beta = 0, \quad c = \alpha\beta - a\beta^2$$

hold (check that by yourself).

Two parameters of the original equation, a and b, can be considered arbitrary, while the others are expressed in terms of them as follows:

$$c = \frac{b}{4a^2}, \quad \alpha = \pm\sqrt{\frac{b}{a}}, \quad \beta = \pm\frac{1}{2a}\sqrt{\frac{b}{a}}.$$

◄

► Example 9.7. Let us explore another fourth-order ODE with a quadratic nonlinearity

$$y''''_{xxxx} = a(y''_{xx})^2 + b(y'_x)^2 + c. \tag{9.1.2.11}$$

We supplement (9.1.2.11) with the second-order differential constraint

$$y''_{xx} = \alpha(y'_x)^2 + \beta, \tag{9.1.2.12}$$

which represents an autonomous equation integrable by quadrature.

Differentiating (9.1.2.12) successively, we find the derivatives

$$\begin{aligned} y'''_{xxx} &= 2\alpha y'_x y''_{xx} = 2\alpha y'_x[\alpha(y'_x)^2 + \beta] = 2\alpha^2(y'_x)^3 + 2\alpha\beta y'_x, \\ y''''_{xxxx} &= [6\alpha^2(y'_x)^2 + 2\alpha\beta]y''_{xx} = [6\alpha^2(y'_x)^2 + 2\alpha\beta][\alpha(y'_x)^2 + \beta] \\ &= 6\alpha^3(y'_x)^4 + 8\alpha^2\beta(y'_x)^2 + 2\alpha\beta^2. \end{aligned} \tag{9.1.2.13}$$

Substituting (9.1.2.12) and (9.1.2.13) into (9.1.2.11) yields a biquadratic equation for the derivative:

$$(6\alpha^3 - a\alpha^2)(y'_x)^4 + (8\alpha^2\beta - 2a\alpha\beta - b)(y'_x)^2 + 2\alpha\beta^2 - c = 0.$$

It holds identically if we set

$$6\alpha^3 - a\alpha^2 = 0, \quad 8\alpha^2\beta - 2a\alpha\beta - b = 0, \quad 2\alpha\beta^2 - c = 0.$$

Two parameters of the original equation, a and b, can be considered arbitrary, while the others are expressed in terms of them as follows:

$$c = 27\,a^{-3}b^2, \quad \alpha = \tfrac{1}{6}a, \quad \beta = -9\,a^{-2}b.$$

◄

Problems and Exercises for Section 9.1

1. Find particular solutions of the second-order ODE

$y''_{xx} = ay^n + by^{2n+1}$

using the first-order differential constraint

$y'_x = \alpha + \beta y^{n+1}$.

2. Find particular solutions of the second-order ODE

$y''_{xx} - cy'_x = a + be^{\lambda y}$

using a first-order differential constraint.

Hint. The right-hand side of the differential constraint should be chosen in the same form as that of the original equation.

3. Find particular solutions of the fourth-order ODE

$y''''_{xxxx} = ay^n + by^{2n+3}$

using the first-order differential constraint

$(y'_x)^2 = \alpha y^m + \beta$.

Find the relationships between the parameters of the original equation and differential constraint as well as conditions when the equation and constraint are compatible.

4. Find particular solutions of the fourth-order ODE

$$y''''_{xxxx} = ae^{\lambda y} + be^{2\lambda y}$$

using the first-order differential constraint

$$(y'_x)^2 = \alpha + \beta e^{\lambda y}.$$

5. Find particular solutions of the nth-order ODE

$$y_x^{(n)} = ae^{\lambda y} y'_x$$

using the first-order differential constraint

$$y'_x = be^{\mu y}.$$

6. Show that the autonomous second-order differential constraint

$$y''_{xx} = f(y)(y'_x)^2 + g(y)y'_x$$

is equivalent to a suitable autonomous first-order differential constraint $y'_x = h(y)$.

9.2. Description of the Differential Constraints Method for Partial Differential Equations

9.2.1. Preliminary Remarks. Simple Example

The main idea of the method is to find exact solutions of a complex partial differential equation by analyzing it alongside a simpler auxiliary equation called a *differential constraint*.

Section 5.1 discussed several examples of nonlinear PDEs admitting additive separable solutions of the form

$$u(x, t) = \varphi(x) + \psi(t). \tag{9.2.1.1}$$

In the initial stage, the functions $\varphi(x)$ and $\psi(t)$ are assumed arbitrary and subject to determination in the process of the further analysis.

Differentiating the expression (9.2.1.1) with respect to t yields

$$u_t = f(t) \qquad (f = \psi'_t). \tag{9.2.1.2}$$

The converse is also true: relation (9.2.1.2) implies the representation of solutions in the form (9.2.1.1).

Further, differentiating (9.2.1.2) with respect to x, we find that

$$u_{xt} = 0. \tag{9.2.1.3}$$

Conversely, the representation of solutions in the form (9.2.1.1) can be derived from (9.2.1.3).

Thus, the problem of finding exact solutions of the form (9.2.1.1) for a specific partial differential equation can be replaced by an equivalent problem of finding exact solutions of this equation that satisfy an additional condition (9.2.1.2) or (9.2.1.3). Such additional conditions, written as one or more differential equations, are usually called *differential constraints*.

Before providing a general description of the differential constraints method, let us illustrate its key features with a simple example.

► Example 9.8. Let us explore the second-order PDE

$$u_y u_{xy} + a u_x u_{yy} = bu. \tag{9.2.1.4}$$

We will look for solutions of equation (9.2.1.4) that satisfy the linear first-order differential constraint

$$u_x = \varphi(y). \tag{9.2.1.5}$$

The function $\varphi(y)$ here cannot generally be arbitrary; it must satisfy a compatibility condition for equations (9.2.1.4) and (9.2.1.5). The compatibility condition represents a differential equation for $\varphi(y)$, which derives form equations (9.2.1.4) and (9.2.1.5) and their corollaries obtained by differentiation.

Differentiating constraint (9.2.1.5) successively with respect to x and y, we obtain

$$u_{xx} = 0, \quad u_{xy} = \varphi'_y, \quad u_{xxy} = 0, \quad u_{xyy} = \varphi''_{yy}. \tag{9.2.1.6}$$

Differentiating equation (9.2.1.4) with respect to x given

$$u_{xy}^2 + u_y u_{xxy} + a u_{xx} u_{yy} + a u_x u_{xyy} = b u_x. \tag{9.2.1.7}$$

On substituting the derivatives (9.2.1.5) and (9.2.1.6) into (9.2.1.7), we arrive at a second-order ODE for φ:

$$(\varphi'_y)^2 + a\varphi\varphi''_{yy} = b\varphi. \tag{9.2.1.8}$$

This equation represents a compatibility condition for equations (9.2.1.4) and (9.2.1.5).

We integrate equation (9.2.1.5) to obtain the exact solution

$$u = \varphi(y)x + \psi(y). \tag{9.2.1.9}$$

To find $\psi(y)$, we substitute (9.2.1.9) into (9.2.1.4) and take into account (9.2.1.8). As a result, we arrive at an ordinary differential equation for $\psi(y)$:

$$\varphi'_y \psi'_y + \varphi\psi''_{yy} = b\psi. \tag{9.2.1.10}$$

To sum up, we have derived an exact solution of the form (9.2.1.9), where the functions φ and ψ satisfy ODEs (9.2.1.8) and (9.2.1.10). These equations can be integrated by quadrature (the reader is encouraged to perform this themselves).

Remark 9.2. *The above solution is easier to obtain by direct substitution of (9.2.1.9) into the original equation (9.2.1.4).* ◄

9.2.2. General Description of the Differential Constraints Method

In general, the procedure for constructing exact solutions to nonlinear equations of mathematical physics using the differential constraints method is significantly more complex than that for ordinary differential equations. It involves several successive stages, which are described below.

1°. Seeking exact solutions to an equation

$$F(x, t, u, u_x, u_t, u_{xx}, u_{xt}, u_{tt}, \dots) = 0 \tag{9.2.2.1}$$

is carried out by supplementing it with a differential constraint (auxiliary differential equation)

$$G(x, t, u, u_x, u_t, u_{xx}, u_{xt}, u_{tt}, \dots) = 0. \tag{9.2.2.2}$$

The structure of the differential constraint (9.2.2.2) can be specified in two main ways:

- based on a priori considerations (for instance, one may require that the constraint be a solvable or simpler equation);
- based on properties of the given equation (these could include its symmetries or conservation laws).

2°. In general, the overdetermined system defined by equations (9.2.2.1) and (9.2.2.2) for a single unknown function u requires a compatibility analysis of its equations. If the differential constraint (9.2.2.2) is specified based on a priori considerations, it must incorporate sufficient functional arbitrariness, meaning it should include arbitrary determining functions. The compatibility analysis of system (9.2.2.1)–(9.2.2.2) must result in conditions that specify the form of these defining functions. These conditions, referred to as *compatibility conditions*, are typically expressed as a system of ordinary differential equations.

In simple cases,* the compatibility analysis is performed by (multiply) differentiating equations (9.2.2.1) and (9.2.2.2) with respect to x and t and eliminating the highest derivatives from the resulting differential relations as well as equations (9.2.2.1) and (9.2.2.2). This approach leads to an equation that only includes powers of the lower derivatives. Equating the coefficients of all powers of the derivatives with zero allows us to derive compatibility conditions that connect the functional coefficients of equations (9.2.2.1) and (9.2.2.2).

3°. We solve the system of differential equations for the determining functions obtained in stage 2°. Substituting the resulting functions into the differential constraint (9.2.2.2), we arrive at an equation

$$g(x, t, u, u_x, u_t, u_{xx}, u_{xt}, u_{tt}, \dots) = 0. \tag{9.2.2.3}$$

The differential constraint (9.2.2.3), which is compatible with equation (9.2.2.1) is referred to as a *invariant manifold* for equation (9.2.2.1).

*In more complex cases, it is advisable to use general methods for analyzing overdetermined systems of PDEs. These methods are based on (i) the Cartan algorithm or (ii) the Janet–Spenser–Kuranishi algorithm. Detailed descriptions of these algorithms, along with additional information on the theory of overdetermined systems of PDEs, can be found, for example, in the works of Kuranishi (1967), Pommaret (1978), and Sidorov et al. (1984). However, it is important to note that these algorithms are pretty intricate and labor-intensive; they are primarily of academic interest and have very limited practical application.

4°. We look for the general solution of (i) equation (9.2.2.3) or (ii) an equation that derives from (9.2.2.1) and (9.2.2.3). The resulting solution will include some arbitrary functions $\{\varphi_m\}$, which can be dependent on x and t as well as u. It is noteworthy that particular solutions of equation (9.2.2.3) or its corollaries can sometimes be utilized instead of the general solution.

5°. We substitute the solution found in stage 4° into the original PDE (9.2.2.1) to obtain an equation that allows us to identify the functions $\{\varphi_m\}$. Once $\{\varphi_m\}$ are determined, we substitute them into the solution from stage 4° to obtain an exact solution of the original equation (9.2.2.1).

Remark 9.3. *If the differential constraint is chosen poorly, equations (9.2.2.1) and (9.2.2.2) may turn out to be incompatible (not having common solutions).*

Remark 9.4. *The last three stages of the differential constraints method involve solving various equations (system of equations). If at least one of these stages fails to yield a solution, then no exact solution to the original equation can be constructed.*

Remark 9.5. *To find exact solutions of nonlinear partial differential equations or systems of partial differential equations, one can simultaneously use several differential constraints (see, for example, Polyanin & Zaitsev, 2012; Polyanin & Zhurov, 2022; and Section 9.4.3).*

For better clarity, Figure 9.1 shows a general scheme for using the differential constraints method with a single differential constraint.

9.3. First-Order Differential Constraints for Partial Differential Equations

9.3.1. Second-Order Evolution Equations

First-order differential constraint. Compatibility condition. We will consider second-order evolution equations of the form

$$u_t = \Phi(x, t, u, u_x, u_{xx}). \tag{9.3.1.1}$$

Before applying the differential constraints method, we will first solve equation (9.3.1.1) for the highest derivative:

$$u_{xx} = F(x, t, u, u_x, u_t). \tag{9.3.1.2}$$

Next, we supplement equation (9.3.1.1) with a general first-order differential constraint

$$u_t = G(x, t, u, u_x). \tag{9.3.1.3}$$

A compatibility condition for equations (9.3.1.2) and (9.3.1.3) can be derived by differentiating (9.3.1.2) once with respect to t and (9.3.1.3) twice with respect to x, followed by equating the resulting mixed third-order derivatives, u_{xxt} and u_{txx}:

$$\mathrm{D}_t F = \mathrm{D}_x^2 G. \tag{9.3.1.4}$$

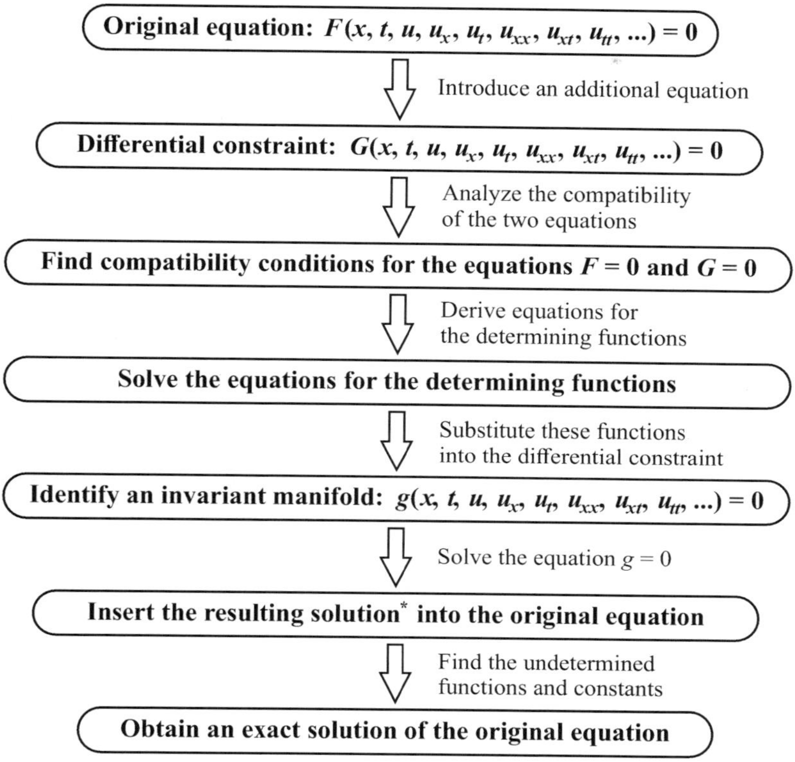

Figure 9.1. An algorithm for constructing exact solutions using the differential constraints method. (*Usually, the resulting solution includes some arbitrary functions and constants.)

Here D_t and D_x are total differentiation operators with respect to t and x:

$$\begin{aligned}\mathrm{D}_t &= \frac{\partial}{\partial t} + u_t\frac{\partial}{\partial u} + u_{xt}\frac{\partial}{\partial u_x} + u_{tt}\frac{\partial}{\partial u_t},\\ \mathrm{D}_x &= \frac{\partial}{\partial x} + u_x\frac{\partial}{\partial u} + u_{xx}\frac{\partial}{\partial u_x} + u_{xt}\frac{\partial}{\partial u_t}.\end{aligned} \tag{9.3.1.5}$$

The partial derivatives u_t, u_{xx}, u_{xt}, and u_{tt} in (9.3.1.5) must be expressed in terms of x, t, u, and u_x with the help of equations (9.3.1.2) and (9.3.1.3) and the relations resulting from differentiating these equations. As a result, the compatibility condition (9.3.1.4) becomes $R(x, t, u, u_x) = 0$. Often, the left-hand side of this relation can be represented as a polynomial in the derivative u_x:

$$\sum_{m=1}^{M} R_m(x, t, u)u_x^m = 0.$$

The splitting procedure with respect to the derivative results in a system of determining equations

$$R_m(x,t,u)=0, \qquad m=1,\dots,M.$$

An illustrative example. Below is an example illustrating the application of the differential constraints method to construct exact solutions for nonlinear PDEs. For simplicity and clarity, the formulas for the total differentiation operators (9.3.1.4) and (9.3.1.5) will not be used.

► Example 9.9. To seek exact solutions for the nonlinear heat equation with a source

$$u_t=[f(u)u_x]_x+g(u), \tag{9.3.1.6}$$

we use a simple general differential constraint

$$u_t=\varphi(u). \tag{9.3.1.7}$$

The functions $f(u)$, $g(u)$, and $\varphi(u)$ in equations (9.3.1.6) and (9.3.1.7) are not known in advance and will need to be identified during the subsequent analysis. The process for constructing a solution involves several stages, which are described in detail below.

1°. *Transformation of the original PDE into a more convenient form.* Solving equation (9.3.1.6) for the highest derivative yields

$$u_{xx}=\frac{u_t-f'_u(u)u_x^2-g(u)}{f(u)}=\frac{\varphi(u)-g(u)-f'_u(u)u_x^2}{f(u)}.$$

Here, in the second step, we eliminated the derivative u_t using differential constraint (9.3.1.7). This relation can be rewritten in the more convenient form

$$u_{xx}=h(u)-\frac{f'_u(u)}{f(u)}u_x^2, \quad h(u)=\frac{\varphi(u)-g(u)}{f(u)}. \tag{9.3.1.8}$$

In what follows, for brevity, we will omit the arguments of the functions $f=f(u)$, $g=g(u)$, and $\varphi=\varphi(u)$ in the intermediate results.

2°. *Derivation of the determining system of equations.* Differentiating the constraint (9.3.1.7) successively with respect to x, we find the third-order mixed derivative u_{txx}:

$$\begin{aligned} &u_t=\varphi, \quad u_{tx}=\varphi'_u u_x,\\ &u_{txx}=\varphi'_u u_{xx}+\varphi''_{uu}u_x^2=\varphi'_u\left(h-\frac{f'_u}{f}u_x^2\right)+\varphi''_{uu}u_x^2=\\ &\quad=h\varphi'_u+\left(\varphi''_{uu}-\varphi'_u\frac{f'_u}{f}\right)u_x^2. \end{aligned} \tag{9.3.1.9}$$

Here, the second derivative u_{xx} has been eliminated using relation (9.3.1.8).

Differentiating (9.3.1.8) with respect to t, we calculate the third-order mixed derivative u_{xxt}:

$$\begin{aligned} u_{xxt}&=h'_u u_t-\left(\frac{f'_u}{f}\right)'_u u_t u_x^2-2\frac{f'_u}{f}u_x u_{xt}=\\ &=h'_u\varphi-\left[\varphi\left(\frac{f'_u}{f}\right)'_u+2\varphi'_u\frac{f'_u}{f}\right]u_x^2. \end{aligned} \tag{9.3.1.10}$$

Here, the derivatives u_t and u_{xt} have been eliminated in the final step using the differential constraint (9.3.1.7) and the second relation in (9.3.1.9).

Equating the third-order derivatives u_{xxt} and u_{txx}, defined by formulas (9.3.1.9) and (9.3.1.10), we obtain

$$\left[\varphi''_{uu} + \varphi'_u \frac{f'_u}{f} + \varphi\left(\frac{f'_u}{f}\right)'_u\right] u_x^2 + h\varphi'_u - h'_u\varphi = 0. \tag{9.3.1.11}$$

Next, equating the functional coefficients of the different powers of u_x with zero (this implies utilizing the splitting procedure with respect to u_x), we arrive at the determining system of equations

$$\begin{aligned} \varphi''_{uu} + \varphi'_u \frac{f'_u}{f} + \varphi\left(\frac{f'_u}{f}\right)'_u &= 0, \\ h\varphi'_u - h'_u\varphi &= 0. \end{aligned} \tag{9.3.1.12}$$

3°. *Solving the determining system of equations.* System (9.3.1.12) can be integrated since the first equation can be represented as $[(f\varphi)'_u/f]'_u = 0$, and the second equation is separable. As a result, we find that

$$\begin{aligned} \varphi &= \frac{1}{f}\left(A\int f\,du + B\right), \\ g &= (1 - Cf)\varphi, \end{aligned} \tag{9.3.1.13}$$

where A, B, and C are arbitrary constants. In deriving the formula for g in (9.3.1.13), we replaced h with other functions by considering its definition in (9.3.1.8).

Further, we explore the nondegenerate case when $A \neq 0$ in (9.3.1.13). Assuming that $f = f(u)$ is an arbitrary given function, we use (9.3.1.13) to find the functions $g(u)$ and $\varphi(u)$:

$$g(u) = \frac{a + cf}{f}\left(\int f\,du + b\right), \quad \varphi(u) = \frac{a}{f}\left(\int f\,du + b\right), \tag{9.3.1.14}$$

where $a = A$, $b = B/A$, and $c = -AC$ are arbitrary constants.

4°. *Determining the general form of the solution using the differential constraint.* On substituting the function $\varphi(u)$ from (9.3.1.14) into the differential constraint (9.3.1.7), we arrive at the equation

$$u_t = \frac{a}{f}\left(\int f\,du + b\right). \tag{9.3.1.15}$$

Substituting $w = \int f\,du + b$ converts (9.3.1.15) into a simple linear ODE: $w_t = aw$. Integrating it yields the general solution of equation (9.3.1.15) in implicit form:

$$\int f(u)\,du = \theta(x)e^{at} - b. \tag{9.3.1.16}$$

The function $\theta = \theta(x)$ serves as a constant of integration dependent on x. This is because w is a function of both x and t, and the equation $w_t = aw$ does not explicitly involve x. At this stage, θ is treated as arbitrary.

5°. *Identifying solutions of the original PDE.* Differentiating (9.3.1.16) with respect to x and t, we get $u_t = ae^{at}\theta/f$ and $u_x = e^{at}\theta'_x/f$. Substituting these expressions into the original equation (9.3.1.6) and considering (9.3.1.14), we arrive at a linear second-order ODE:

$$\theta''_{xx} + c\theta = 0. \tag{9.3.1.17}$$

Its general solution is expressed as

$$\theta = \begin{cases} C_1 \sin\bigl(x\sqrt{c}\bigr) + C_2 \cos\bigl(x\sqrt{c}\bigr) & \text{for } c > 0, \\ C_1 \sinh\bigl(x\sqrt{-c}\bigr) + C_2 \cosh\bigl(x\sqrt{-c}\bigr) & \text{for } c < 0, \\ C_1 x + C_2 & \text{for } c = 0, \end{cases} \tag{9.3.1.18}$$

where C_1 and C_2 are arbitrary constants. Formulas (9.3.1.16) and (9.3.1.18) describe exact solutions of equation (9.3.1.6) in implicit form for arbitrary $f(u)$, and the function $g(u)$ given by formula (9.3.1.14). ◄

Remark 9.6. *In the degenerate case, substituting $A = 0$ into (9.3.1.13), we obtain the following expressions of the functions $g(u)$ and $\varphi(u)$ (as above, f is assumed to be an arbitrary given function):*

$$g(u) = \frac{b}{f} + c, \quad \varphi(u) = \frac{b}{f},$$

where $b = B$ and $c = -BC$ are arbitrary constants. This solution can be derived from (9.3.1.14) by renaming constants, $b \to b/a$ and $c \to ac/b$, followed by going to the limit as $a \to 0$. After simple calculations, we find the corresponding solution of equation (9.3.1.6) in implicit form:

$$\int f(u)\,du = bt - \frac{1}{2}cx^2 + C_1 x + C_2.$$

9.3.2. Hyperbolic Second-Order Equations

1°. Hyperbolic second-order equations of the form

$$u_{xt} = F(x, t, u, u_x, u_t) \tag{9.3.2.1}$$

supplemented with a first-order differential constraint (9.3.1.3) can be investigated in a similar fashion. We assume that $G_{u_x} \neq 0$.

A compatibility condition for these equations results from differentiating (9.3.2.1) once with respect to t and twice (9.3.1.3) with respect to t and x followed by equating the mixed third-order derivatives u_{xtt} and u_{ttx}:

$$\mathrm{D}_t F = \mathrm{D}_x[\mathrm{D}_t G]. \tag{9.3.2.2}$$

Here D_t and D_x are total differential operators (9.3.1.5) in which the partial derivatives u_t, u_{xx}, u_{xt}, and u_{tt} must be expressed in terms of x, t, u, and u_x with the help of (9.3.2.1) and (9.3.1.3) and their differential corollaries.

Let us illustrate the aforesaid with a specific example.

► Example 9.10. Consider the nonlinear equation

$$u_{xt} = f(u). \tag{9.3.2.3}$$

We supplement (9.3.2.3) with a first-order quasilinear differential constraint

$$u_x = \varphi(t)g(u). \tag{9.3.2.4}$$

Differentiating (9.3.2.3) with respect to x and replacing the first derivative with respect to x by the right-hand side of relation (9.3.2.4), we find that

$$u_{xxt} = \varphi g f'_u. \tag{9.3.2.5}$$

Differentiating (9.3.2.4) further with respect to x and t, we obtain two relations

$$u_{xx} = \varphi g'_u u_x = \varphi^2 g g'_u, \tag{9.3.2.6}$$

$$u_{xt} = \varphi'_t g + \varphi g'_u u_t. \tag{9.3.2.7}$$

After eliminating the mixed derivative from (9.3.2.7) using equation (9.3.2.3), we find the first derivative with respect to t as

$$u_t = \frac{f - \varphi'_t g}{\varphi g'_u}. \tag{9.3.2.8}$$

Differentiating (9.3.2.6) with respect to t and replacing u_t with the right-hand side or relation (9.3.2.8), we obtain

$$u_{xxt} = 2\varphi\varphi'_t g g'_u + \varphi^2 (g g'_u)'_u u_t = 2\varphi\varphi'_t g g'_u + \varphi (g g'_u)'_u \frac{f - \varphi'_t g}{g'_u}. \tag{9.3.2.9}$$

Now, equating the mixed third-order derivatives (9.3.2.5) and (9.3.2.9), cancelling by φ, and performing simple rearrangements, we arrive at the determining equation

$$\varphi'_t g[(g'_u)^2 - g g''_{uu}] = g g'_u f'_u - f(g g'_u)'_u, \tag{9.3.2.10}$$

which has two distinct solutions.

Solution 1. Equation (9.3.2.10) holds identically for any $\varphi = \varphi(t)$ if we set

$$(g'_u)^2 - g g''_{uu} = 0,$$

$$g g'_u f'_u - f(g g'_u)'_u = 0.$$

The general solution of this system is expressed as

$$f(u) = a e^{\lambda u}, \quad g(u) = b e^{\lambda u/2}, \tag{9.3.2.11}$$

where a, b, and λ are arbitrary constants. To simplify the calculations, we will further examine the case

$$a = b = 1, \quad \lambda = -2. \tag{9.3.2.12}$$

Substituting the function $g(u)$ from (9.3.2.11)–(9.3.2.12) into the differential constraint (9.3.2.4) and integrating the resulting equation, we get

$$u = \ln[\varphi(t)x + \psi(t)], \tag{9.3.2.13}$$

where $\psi(t)$ is an arbitrary function. Substituting (9.3.2.13) into the original equation (9.3.2.3) with the right-hand side (9.3.2.11)–(9.3.2.12), we arrive at the following linear ordinary differential equation for $\psi(t)$:

$$\psi\varphi'_t - \varphi\psi'_t = 1.$$

Its general solution is written as

$$\psi(t) = C\varphi(t) - \varphi(t)\int\frac{dt}{\varphi^2(t)}, \tag{9.3.2.14}$$

where C is an arbitrary constant.

To sum up, formulas (9.3.2.13) and (9.3.2.14), where $\varphi(t)$ is an arbitrary function, provide an exact solution of the nonlinear PDE with an exponential nonlinearity $u_{xt} = e^{-2u}$.

Solution 2. The second solution corresponds to the linear function

$$\varphi(t) = at + b, \tag{9.3.2.15}$$

where a and b are arbitrary constants. In this case, the functions $f(u)$ and $g(u)$ are related by equation (9.3.2.10), where $\varphi'_t = a$. Integrating (9.3.2.4) taking into account (9.3.2.15), we find the general form of solutions:

$$u = u(z), \quad z = (at + b)x + \psi(t), \tag{9.3.2.16}$$

where $\psi(t)$ is an arbitrary function. Inserting this relationship into the original equation (9.3.2.3) and replacing x with z with the help of (9.3.2.16), we obtain

$$[az + (at + b)\psi'_t - a\psi]u''_{zz} + au'_z = f(u). \tag{9.3.2.17}$$

For this relation to be an ordinary differential equation for $u = u(z)$, we must set

$$(at + b)\psi'_t - a\psi = \text{const}.$$

Integrating yields the function $\psi(t)$:

$$\psi(t) = ct + d, \tag{9.3.2.18}$$

where c and d are arbitrary constants.

Formulas (9.3.2.16) and (9.3.2.18) define a solution to equation (9.3.2.3) for arbitrary $f(u)$. The function $u(z)$ is described by the nonlinear ODE

$$(az + bc - ad)u''_{zz} + au'_z = f(u),$$

which follows from (9.3.2.17) and (9.3.2.18). The special case of $a = d = 0$ corresponds to a traveling wave solution, and the case $b = c = d = 0$ corresponds to a self-similar solution. ◄

2°. The process for constructing exact solutions to general hyperbolic equations, as well as other second-order partial differential equations, using first-order differential constraints is detailed in the books by Polyanin & Zaitsev (2012) and Polyanin & Zhurov (2022). To clarify this complex and laborious procedure, we will provide a simple illustrative example here, omitting the intricate details.

▶ Example 9.11. To find exact solutions for the nonlinear wave equation

$$u_{tt} = [f(u)u_x]_x \tag{9.3.2.19}$$

with arbitrary function $f(u)$, we utilize the first-order differential constraint

$$u_x = \varphi(u)u_t, \tag{9.3.2.20}$$

in which the function $\varphi(u)$ needs to be determined during the subsequent analysis.

Eliminating u_x from (9.3.2.19) with the help of (9.3.2.20), we transform the right-hand side of the equation as follows:

$$[f(u)u_x]_x = [f(u)\varphi(u)u_t]_x = [f(u)\varphi(u)u_x]_t = [f(u)\varphi^2(u)u_t]_t. \tag{9.3.2.21}$$

Using (9.3.2.21) and performing simple rearrangements, we can rewrite equation (9.3.2.19) as

$$\{[1 - f(u)\varphi^2(u)]u_t\}_t = 0. \tag{9.3.2.22}$$

This equation holds identically if we set the expression in square brackets to zero. As a result, we obtain two admissible determining functions:

$$\varphi(u) = \pm\frac{1}{\sqrt{f(u)}}. \tag{9.3.2.23}$$

Substituting (9.3.2.23) into the differential constraint (9.3.2.20) yields two quasilinear first-order PDEs, whose solutions, resulting from integrating the characteristic system of ODEs (see Section 2.1.1), can be represented in implicit form:

$$\begin{aligned} x + t\sqrt{f(u)} &= \Psi_1(u), \\ x - t\sqrt{f(u)} &= \Psi_2(u), \end{aligned} \tag{9.3.2.24}$$

where $\Psi_1(u)$ and $\Psi_2(u)$ are arbitrary functions.

It can be verified by a direct check that both implicit expressions (9.3.2.24) are exact solutions of the original equation (9.3.2.19). ◀

Problems and Exercises for Section 9.3

1. Find first-order and second-order differential constraints equivalent to additive or multiplicative separable solutions of the form

$$u(x,t) = \varphi(x)\psi(t),$$

where the functions $\varphi(x)$ and $\psi(t)$ are treated as arbitrary.

2. Find exact solutions of the normalized hydrodynamic boundary layer equation

$$u_y u_{xy} - u_x u_{yy} = u_{yyy}$$

by using the linear first-order differential constraint $u_x = \varphi(y)$.

3. Investigate the compatibility of the nonlinear reaction-diffusion equation

$$u_t = [f(u)u_x]_x + g(u) \tag{9.3.2.25}$$

and the first-order differential constraint

$$u_x = \varphi(u). \tag{9.3.2.26}$$

4. Taking advantage of the results from the analysis of the previous problem, find exact solutions to the nonlinear reaction-diffusion equation (9.3.2.25) using the first-order differential constraint (9.3.2.26).

5. Investigate the compatibility of the nonlinear diffusion equation

$$u_u = u_{xx} + f(u)u_x + g(u) \tag{9.3.2.27}$$

and the quasilinear first-order differential constraint

$$u_t = \varphi(u)u_x + \psi(u). \tag{9.3.2.28}$$

6. Taking advantage of the results from the analysis of the previous problem, find exact solutions to the nonlinear diffusion equation (9.3.2.27) using the quasilinear first-order differential constraint (9.3.2.26) (refer to Polyanin & Zhurov, 2022).

7. Investigate the compatibility of the nonlinear reaction-diffusion equation (9.3.2.25) and the nonlinear first-order differential constraint

$$u_t = \varphi(u)u_x^2 + \psi(u). \tag{9.3.2.29}$$

8. Taking advantage of the results from the analysis of the previous problem, find exact solutions to the nonlinear reaction-diffusion equation (9.3.2.25) using the first-order differential constraint (9.3.2.29) (refer to Polyanin & Zhurov, 2022).

9.4. Second-Order Differential Constraints for PDEs. Utilizing Several Constraints

9.4.1. Second-Order Differential Constraints and Equivalent Simpler Differential Constraints

Let us discuss the problem of compatibility for the evolution second-order equation

$$u_t = F_1(x, t, u, u_x, u_{xx}) \tag{9.4.1.1}$$

and a similar differential constraint

$$u_t = F_2(x, t, u, u_x, u_{xx}). \tag{9.4.1.2}$$

First of all, this problem can be reduced to a simpler problem with a first-order differential constraint, which was discussed in Section 9.3.1. To this end, we start with eliminating the second derivative u_{xx} from equations (9.4.1.1) and (9.4.1.2).

As a result, we obtain a first-order equation (9.3.1.3), which can be treated as a simpler differential constraint, and then we need to analyze it for compatibility with the original equation (9.4.1.1).

Secondly, on eliminating the derivative with respect to t from (9.4.1.1) and (9.4.1.2), we arrive at an equation

$$H(x,t,u,u_x,u_{xx}) = 0, \tag{9.4.1.3}$$

which can be treated as an ordinary differential equation in the variable x with a free parameter t. The constants of integration that arise in solving equation (9.4.1.3) will be arbitrary functions of t, that is, $C_1 = C_1(t)$ and $C_2 = C_2(t)$.

Thus, the general evolution second-order differential constraint (9.4.1.1) is equivalent to an appropriate first-order constraint of the form (9.3.1.3) and, on the other hand, to a second-order differential constraint in the form of the ordinary differential equation (9.4.1.3). Such equivalent constraints are technically easier to work with than the original differential constraint (9.4.1.2).

Remark 9.7. *Instead of* u_t, *the left-hand sides of equation (9.4.1.1) and differential constraint (9.4.1.2) can include the derivative* u_{tt} or u_{tx}. *Then, eliminating* u_{tt} *(or* u_{tx}*) also leads to a second-order ordinary differential equation of the form (9.4.1.3), which is more practical to use than the original differential constraint.*

9.4.2. Illustrative Examples of Using Second-Order Differential Constraints

► Example 9.12. Let us examine the nonlinear heat equation with a source

$$u_t = [f_1(u)u_x]_x + f_2(u). \tag{9.4.2.1}$$

To seek its exact solutions, we will use an autonomous second-order differential constraint quadratically nonlinear in the second derivative:

$$u_{xx} = g_1(u)u_x^2 + g_2(u). \tag{9.4.2.2}$$

This constraint can be treated as an ODE in the variable x. The functions $f_2(u)$, $f_1(u)$, $g_2(u)$, and $g_1(u)$ appearing in (9.4.2.1) and (9.4.2.2) are to be determined in the process of the subsequent analysis.

On eliminating the second derivative from (9.4.2.1) and (9.4.2.2), we arrive at the first-order PDE

$$u_t = \varphi(u)u_x^2 + \psi(u), \tag{9.4.2.3}$$

where

$$\varphi(u) = f_1(u)g_1(u) + f_1'(u), \quad \psi(u) = f_1(u)g_2(u) + f_2(u). \tag{9.4.2.4}$$

Here and henceforth, a prime stands for a derivative with respect to u.

Differentiating (9.4.2.2) with respect to t yields

$$u_{xxt} = 2g_1u_xu_{xt} + g_1'u_x^2u_t + g_2'u_t.$$

Eliminating the derivatives u_{xxt}, u_{xt}, u_t from this relation using equations (9.4.2.2) and (9.4.2.3) and their corollaries obtained through differentiation, we obtain

$$(2\varphi g_1^2+3\varphi' g_1+\varphi g_1'+\varphi'')u_x^4 \\ +(4\varphi g_1 g_2+5\varphi' g_2+\varphi g_2'-g_1\psi'-\psi g_1'+\psi'')u_x^2+2\varphi g_2^2+\psi' g_2-\psi g_2'=0.$$

Equating the functional coefficients of the different powers of u_x with zero, we get three equations, which can be written as

$$\begin{gathered}(\varphi'+\varphi g_1)'+2g_1(\varphi'+\varphi g_1)=0,\\ 4g_2(\varphi'+\varphi g_1)+(\varphi g_2-\psi g_1)'+\psi''=0,\\ \varphi=-\tfrac{1}{2}(\psi/g_2)'.\end{gathered} \qquad (9.4.2.5)$$

The first equation can be satisfied by setting $\varphi'+\varphi g_1=0$. The corresponding particular solution of system (9.4.2.5) is

$$\varphi=-\frac{1}{2}\mu', \quad \psi=\mu g_2, \quad g_1=-\frac{\mu''}{\mu'}, \quad g_2=\left(2C_1+\frac{C_2}{\sqrt{|\mu|}}\right)\frac{1}{\mu'}, \qquad (9.4.2.6)$$

where $\mu=\mu(u)$ is an arbitrary function.

Taking into account (9.4.2.4), we find the functional coefficients of the original equation (9.4.2.1) and differential constraint (9.4.2.2):

$$\begin{gathered}f_1=\left(C_3-\frac{1}{2}u\right)\mu', \quad f_2=(\mu-f_1)g_2,\\ g_1=-\frac{\mu''}{\mu'}, \quad g_2=\left(2C_1+\frac{C_2}{\sqrt{|\mu|}}\right)\frac{1}{\mu'}.\end{gathered} \qquad (9.4.2.7)$$

In view of (9.4.2.7), equation (9.4.2.2) admits the first integral

$$u_x^2=\left[4C_1\mu+4C_2\sqrt{|\mu|}+2\sigma_t'(t)\right]\frac{1}{(\mu')^2}, \qquad (9.4.2.8)$$

where $\sigma(t)$ is an arbitrary function. We eliminate u_x^2 from (9.4.2.3) with the help of (9.4.2.8) and insert the functions φ and ψ from (9.4.2.6) to obtain the equation

$$\mu' u_t=-C_2\sqrt{|\mu|}-\sigma_t'(t). \qquad (9.4.2.9)$$

Let us dwell in more detail on the special case $C_2=C_3=0$. Integrating equation (9.4.2.9) and considering that $\mu_t=\mu' u_t$, we find that

$$\mu=-\sigma(t)+\theta(x), \qquad (9.4.2.10)$$

where $\theta(x)$ is an arbitrary function. Substituting (9.4.2.10) into (9.4.2.8) and utilizing the relation $\mu_x=\mu' u_x$ yields $\theta_x^2-4C_1\theta=2\sigma_t-4C_1\sigma$. Equating both sides of this equation with zero and integrating the resulting ordinary differential equations, we find the functions on the right-hand side of (9.4.2.10):

$$\sigma(t)=A\exp(2C_1 t), \quad \theta(x)=C_1(x+B)^2, \qquad (9.4.2.11)$$

where A and B are arbitrary constants. It follows that an exact solution of equation (9.4.2.1) with the function f_1 and f_2 from (9.4.2.7) with $C_2 = C_3 = 0$ can be written in implicit form as

$$\mu(u) = -A\exp(2C_1t) + C_1(x+B)^2.$$

The function $\mu(u)$ in the solution and determining relations (9.4.2.7) can be defined arbitrarily. ◀

The differential constraints method can also be used to construct exact solutions for nonlinear systems of equations. Let us illustrate this with a specific example.

► Example 9.13. Let us look at the hydrodynamic-type system comprised by two nonlinear PDEs

$$\begin{aligned} F_{tx} + FF_{xx} - F_x^2 &= \nu F_{xxx} + q(t)F_x + p(t), \\ G_t + FG_x - GF_x &= \nu G_{xx}. \end{aligned} \qquad (9.4.2.12)$$

The first equation is independent of G, while the other is linear G. The functions $p = p(t)$ and $q = q(t)$ in the first equation can be defined arbitrarily. It is noteworthy that exact solutions of system (9.4.2.12) generate exact solutions of the unsteady three-dimensional Navier–Stokes equations (refer to Aristov & Polyanin, 2009 and Aristov et al., 2009).

Let us supplement system (9.4.2.12) with a second-order differential constraint

$$G = a(t) + b(t)F_x + c(t)F_{xx}. \qquad (9.4.2.13)$$

We eliminate G from the second equation of system (9.4.2.12) with the help of (9.4.2.13) and compare the resulting equation with the first equation in (9.4.2.12), as well as with the equations arising from differentiating the first equation in (9.4.2.12) with respect to x. As a result, we identify compatibility conditions for system (9.4.2.12) and differential constraint (9.4.2.13):

$$\begin{aligned} a'_t + bp &= 0, \\ b'_t + bq - a &= 0, \\ c'_t + qc &= 0. \end{aligned} \qquad (9.4.2.14)$$

Integrating the last equation in (9.4.2.14), we find that $c = C_1\exp(-\int q\,dt)$, where C_1 is an arbitrary constant. The first two equations in (9.4.2.14) lead to a linear second-order ODE for b:

$$b''_{tt} + qb'_t + (p + q'_t)b = 0. \qquad (9.4.2.15)$$

Once this equation is solved, the function $a = a(t)$ can be obtained from the second equation in (9.4.2.14) without integrating.

The result can be rephrased as follows: if an exact solution $F = F(t, x)$ to the first equation in (9.4.2.12) is known, then the corresponding exact solution to the second equation (9.4.2.12) can be obtained using formula (9.4.2.13); the functions $a = a(t)$, $b = b(t)$, and $c = c(t)$ are determined by solving the system of ordinary differential equations (9.4.2.14). ◀

9.4.3. Using Several Differential Constraints

Remark 9.5 from Section 9.2.2 noted that one can use several differential constraints with one equation. In general, all differential constraints must be analysed for compatibility with the original equation.

We will illustrate this with a few specific examples.

► Example 9.14. Consider the two-dimensional nonlinear anisotropic wave equation

$$u_{tt} = [f(u)u_x]_x + [g(u)u_x]_y + [h(u)u_y]_y, \tag{9.4.3.1}$$

where $f = f(u)$, $g = g(u)$, and $h = h(u)$ are arbitrary twice differentiable functions.

We look for exact solutions of equation (9.4.3.1) using the differential constraints method and assuming that

$$u_x = \varphi(u)u_t, \qquad u_y = \psi(u)u_t, \tag{9.4.3.2}$$

where the functions $\varphi(u)$ and $\psi(u)$ are subject to determination in the process of the further analysis. Using the constraints (9.4.3.2), we transform the terms on the right-hand side of equation (9.4.3.1):

$$\begin{aligned} {}_x &= [f(u)\varphi(u)u_t]_x = [f(u)\varphi(u)u_x]_t = [f(u)\varphi^2(u)u_t]_t,\\ [g(u)u_x]_y &= [g(u)\varphi(u)u_t]_y = [g(u)\varphi(u)u_y]_t = [g(u)\varphi(u)\psi(u)u_t]_t,\\ [h(u)u_y]_y &= [h(u)\psi(u)u_t]_y = [h(u)\psi(u)u_y]_t = [h(u)\psi^2(u)u_t]_t. \end{aligned} \tag{9.4.3.3}$$

In view of (9.4.3.3), we can represent equation (9.4.3.1) after simple rearrangements as

$$\{[1 - f(u)\varphi^2(u) - g(u)\varphi(u)\psi(u) - h(u)\psi^2(u)]u_t\}_t = 0. \tag{9.4.3.4}$$

This equation will hold identically if we set the expression in square brackets to zero. It is evident that one of the functions $\varphi(u)$ and $\psi(u)$ can be treated as arbitrary, while the other is found from the quadratic equation

$$f(u)\varphi^2(u) + g(u)\varphi(u)\psi(u) + h(u)\psi^2(u) = 1. \tag{9.4.3.5}$$

The function $u = u(x, y, t)$ can be found from the first-order equations (9.4.3.2). Integrating these equations separately by the method of characteristics, we obtain

$$t + \varphi(u)x = \Phi(y, u), \qquad t + \psi(u)y = \Psi(x, u), \tag{9.4.3.6}$$

where $\Phi(y, u)$ and $\Psi(x, u)$ are arbitrary functions of two variables. It follows that a compatible solution of equations (9.4.3.2) can be written in implicit form as

$$t + \varphi(u)x + \psi(u)y = \theta(u), \tag{9.4.3.7}$$

where $\theta(u)$ is an arbitrary function.

To sum up, the original nonlinear second-order PDE (9.4.3.1) has an exact solution that can be represented in implicit form as (9.4.3.7). It includes two arbitrary functions, $\psi(u)$ and $\theta(u)$, and the function $\varphi(u)$ is any root of the quadratic equation (9.4.3.5); it is assumed that the inequality $g^2\psi^2 - 4f(h\psi^2 - 1) > 0$ holds, suggesting that the roots of the quadratic equation are real. ◄

► Example 9.15. In three dimensions, the unsteady motion of a viscous incompressible fluid is described by the Navier–Stokes equations and the continuity equation, which can be briefly expressed as follows (see Schlichting, 1981, and Loitsyanskiy, 1996):

$$\begin{aligned} \mathbf{V}_t + (\mathbf{V}\cdot\nabla)\mathbf{V} &= -\nabla p + \nu\Delta\mathbf{V}, \\ \nabla\cdot\mathbf{V} &= 0, \end{aligned} \tag{9.4.3.8}$$

where $\mathbf{V} = (V_1, V_2, V_3)$ is the fluid velocity, t is time, p is pressure divided by the fluid density, ν is kinematic viscosity, and Δ and ∇ are the Laplace and gradient operators with respect to the space variable x, y, and z.

Equations (9.4.3.8) admits an exact solution of the form (Aristov & Polyanin, 2009 and Aristov et al., 2009):

$$\begin{aligned} &V_1 = x(-\tfrac{1}{2}F_z + w) + yv, \quad V_2 = xu - y(\tfrac{1}{2}F_z + w), \quad V_3 = F, \\ &p = p_0 - \tfrac{1}{2}\alpha x^2 - \tfrac{1}{2}\beta y^2 - \gamma xy - \tfrac{1}{2}F^2 + \nu F_z - \int F_t\,dz, \end{aligned} \tag{9.4.3.9}$$

where $p_0 = p_0(t)$, $\alpha = \alpha(t)$, $\beta = \beta(t)$, and $\gamma = \gamma(t)$ are arbitrary functions, and $F = F(t,x)$, $u = u(t,x)$, $v = v(t,x)$, and $w = w(t,x)$ are the unknown functions described by the nonlinear system of four PDEs:

$$F_{tx} + FF_{xx} - \tfrac{1}{2}F_x^2 = \nu F_{xxx} + 2(uv + w^2) - \alpha - \beta, \tag{9.4.3.10}$$

$$u_t + Fu_x - uF_x = \nu u_{xx} + \gamma, \tag{9.4.3.11}$$

$$v_t + Fv_x - vF_x = \nu v_{xx} + \gamma, \tag{9.4.3.12}$$

$$w_t + Fw_x - wF_x = \nu w_{xx} + \tfrac{1}{2}(\alpha - \beta). \tag{9.4.3.13}$$

Following Aristov & Polyanin (2009) and Aristov et al. (2009), we will analyze equations (9.4.3.10)–(9.4.3.13) in conjunction with the first-order differential constraints

$$u = mF_x + A, \quad v = nF_x + B, \quad w = kF_x + C, \tag{9.4.3.14}$$

where m, n, k, A, B, and C are undetermined functions dependent on t. We require that all four equations (9.4.3.10)–(9.4.3.13) coincide after inserting relations (9.4.3.14). As a result, we arrive at a nonlinear system for the undetermined functions, which consists of one algebraic equation and six ordinary differential equations:

$$mn + k^2 = \frac{1}{4}, \tag{9.4.3.15}$$

$$\frac{A - m'_t}{m} = \frac{B - n'_t}{n} = \frac{C - k'_t}{k} = 2(An + Bm + 2Ck), \tag{9.4.3.16}$$

$$\frac{\gamma - A'_t}{m} = \frac{\gamma - B'_t}{n} = \frac{\alpha - \beta - 2C'_t}{2k} = -\alpha - \beta + 2AB + 2C^2. \tag{9.4.3.17}$$

This system comprises seven equations for nine unknowns: six functions m, n, k, A, B, and C from (9.4.3.14) and three functions α, β, and γ from (9.4.3.10)–(9.4.3.13). In our case, the last three functions are also undetermined and need to be identified. It can be shown that the last equation in (9.4.3.16) follows from the other three equations of (9.4.3.15) and (9.4.3.16). Therefore, three undetermined functions in system (9.4.3.15)–(9.4.3.17) can generally be treated as arbitrary.

By virtue of (9.4.3.14)–(9.4.3.17), system (9.4.3.10)–(9.4.3.13) reduces to a single third-order partial differential equation:

$$F_{tx} + FF_{xx} - F_x^2 = \nu F_{xxx} + qF_x + p, \tag{9.4.3.18}$$

where the functions $p = p(t)$ and $q = q(t)$ are given by

$$p = \frac{\gamma - A_t'}{m}, \quad q = \frac{A - m_t'}{m}. \tag{9.4.3.19}$$

For $m = n$, the general solution of system (9.4.3.15)–(9.4.3.17) can be represented as

$$\begin{aligned}
&m = n = \frac{1}{2}\sin\varphi, \quad k = \frac{1}{2}\cos\varphi,\\
&A = B = \frac{1}{2}(q\sin\varphi + \varphi_t'\cos\varphi), \quad C = \frac{1}{2}(q\cos\varphi - \varphi_t'\sin\varphi),\\
&\alpha = \frac{1}{4}q^2 + \frac{1}{4}(\varphi_t')^2 - \frac{1}{2}p(1-\cos\varphi) + C_t',\\
&\beta = \frac{1}{4}q^2 + \frac{1}{4}(\varphi_t')^2 - \frac{1}{2}p(1+\cos\varphi) - C_t',\\
&\gamma = \frac{1}{2}p\sin\varphi + A_t',
\end{aligned} \tag{9.4.3.20}$$

where $p = p(t)$, $q = q(t)$, and $\varphi = \varphi(t)$ are arbitrary functions. For convenience, the free functions p and q in (9.4.3.20) are selected so as to ensure that system (9.4.3.10)–(9.4.3.13) together with the differential constraints (9.4.3.14) and formulas (9.4.3.20) reduce a single equation (9.4.3.18) with the same functions $p = p(t)$ and $q = q(t)$.

Hence, the following statement has been proven: any solution of equation (9.4.3.18) for any $p = p(t)$ and $q = q(t)$ generates a solution to the system of PDEs (9.4.3.10)–(9.4.3.13).

For the case $m \neq n$ and numerous exact solutions of equation (9.4.3.18), see Aristov & Polyanin (2009), Aristov et al. (2009), and Polyanin & Zaitsev (2012). ◄

Remark 9.8. *For utilizing differential constraints to construct exact solutions for gas dynamics equations, see Sidorov et al. (1984) and Meleshko (2005).*

Problems and Exercises for Section 9.4

1. Investigate the compatibility of the nonlinear second-order equation

$$u_t = u_{xx} + au_x^2 + bu^2$$

and the linear second-order differential constraint

$$u_{xx} = cu + \varphi(t).$$

2. Find exact solutions of the equation from the previous problem by using the specified linear differential constraint.

3. Investigate the compatibility of the nonlinear second-order equation

$$u_t = f(u)u_{xx} + g(u)u_x + h(u)$$

and the nonlinear second-order differential constraint (refer to Polyanin & Zaitsev, 2012):

$$u_{xx} = \varphi(u)u_x + \psi(u).$$

4. Find exact solutions of the three-dimensional anisotropic wave equation

$$u_{tt} = [f(u)u_x]_x + [g(u)u_y]_y + [h(u)u_z]_z,$$

using the first-order differential constraints

$$u_x = \varphi(u)u_t, \qquad u_y = \psi(u)u_t, \qquad u_z = \theta(u)u_t.$$

9.5. Relationship Between the Differential Constraints Method and Other Methods

9.5.1. Preliminary Remarks

The differential constraints method is grounded in the theory of compatibility between two or more partial differential equations. It is one of the most general approaches for deriving exact solutions to nonlinear partial differential equations. Many other methods can be viewed as specific cases of this approach.

The most challenging and informal issue that arises when applying this method is determining an appropriate differential constraint for a given nonlinear partial differential equation. For example, if the differential constraint lacks sufficient functional flexibility, it may not yield any solution. Conversely, if the constraint is too general, the analysis required to ensure compatibility among the nonlinear partial differential equations can become excessively complex, making it difficult to find exact solutions. Successfully addressing this problem extends beyond the formal description of the differential constraints method and is typically influenced by the experience and intuition of the researcher in each specific case.

Additionally, when applying the differential constraints method, one has to solve, at several stages, various differential equations or systems of equations that include arbitrary functions. This difficulty significantly complicates the analysis. If no solution can be found in at least one of these stages, it becomes impossible to construct an exact solution to the original equation. As a result, applying the differential constraints method is generally more challenging than using the other methods described in this book.

These circumstances explain why simpler, though less general, methods are often preferable for constructing exact solutions to nonlinear PDEs in practice.

Notably, less general methods often provide a constructive approach to deriving important relationships that can be treated as differential constraints. These relationships cannot be obtained directly through the differential constraints method. The concepts and algorithms that underlie these specialized methods often make them more effective* than the differential constraints method when it comes to constructing exact solutions for specific classes of nonlinear partial differential equations.

Thus, the methods of generalized and functional separation of variables, direct methods of symmetry reductions and weak symmetry reductions, and the differential constraints method are, in combination, complementary. Each of these methods has its own advantages and disadvantages, and their effectiveness can vary depending on the specific classes of nonlinear partial differential equations they are applied to.

We will now discuss the relationship between the differential constraints method and the other methods described in earlier chapters of this book.

9.5.2. Methods of Generalized Separation of Variables and the Differential Constraints Method

The methods of generalized separation of variables can be reformulated in terms of the differential constraints method, in the sense that any generalized separable solution can be assigned an equivalent differential constraint.

Generalized separable solutions of a given form can be reduced into differential constraints by eliminating the unknown functions through differentiation. Table 9.1 provides examples of some first- and second-order differential constraints that are equivalent to the most common separable solutions. The functions f and g can be expressed in terms of the original functions φ, ψ, and χ. It is evident that each simple or generalized separable solution can be associated with several equivalent differential constraints.

Table 9.1. First- and second-order differential constraints corresponding to some classes of separable solutions.

No.	Type of solution	Solution structure	Differential constraints
1	Additive separable solution	$u=\varphi(x)+\psi(t)$	$u_x=f(x)$ $u_t=g(t)$ $u_{xt}=0$
2	Multiplicative separable solution	$u=\varphi(x)\psi(t)$	$u_x=f(x)u$ $u_t=g(t)u$ $uu_{xt}-u_xu_t=0$
3	Generalized separable solution	$u=\varphi(t)x^2+\psi(t)x+\chi(t)$	$u_x=f(t)x+g(t)$ $xu_x-u=f(t)x^2+g(t)$ $xu_x-2u=f(t)x+g(t)$ $u_{xx}=f(t)$ $xu_{xx}-u_x=f(t)$
4	Generalized separable solution	$u=\varphi(t)\psi(x)+\chi(t)$	$u_x=f(t)g(x)$ $u_t=f(t)u+g(t)$ $u_x=f(x)[u+g(t)]$ $u_{xx}-f(x)u_x=0$ $u_{xt}-g(t)u_x=0$

*This suggests that these methods achieve the desired results more quickly, with fewer intermediate calculations, and they are much simpler.

The converse is also true: by integrating the differential constraints listed in the last column of Table 9.1, one can derive the general forms of both simple and generalized separable solutions, presented in the table's penultimate column.

For the solutions listed in Table 9.1, it is generally more efficient to utilize the methods of generalized separation of variables, as these methods involve fewer steps where intermediate differential equations need to be solved.

Remark 9.9. *The differential relations obtained with the splitting method (see Section 5.5) when studying bilinear functional differential equations, which arise after substituting generalized separable solutions into the PDEs in question, can be treated as differential constraints. Thus, the splitting method enables the construction of differential constraints during the solution process. It is important to emphasize that these constraints are not known beforehand and cannot be prescribed from a priori considerations in the initial stage of the analysis.*

9.5.3. Methods of Functional Separation of Variables and the Differential Constraints Method

The methods of functional separation of variables can also be reformulated in the language of the differential constraints method, in the sense that any functional separable solution can be associated with an equivalent differential constraint.

Table 9.2. First- and second-order differential constraints corresponding to two classes of functional separable solutions.

No.	Type of solution	Solution structure	Differential constraints
1	Generalized separable solution (generalized traveling-wave solution)	$u=U(z)$, $z=\varphi(t)x+\psi(t)$	$u_x=f(t)g(u)$ $u_t=[f(t)x+g(t)]u_x$ $u_{xx}-f(u)u_x^2=0$ $u_xu_{xt}-u_tu_{xx}=f(t)u_x^2$
2	Generalized separable solution	$u=U(z)$, $z=\varphi(x)+\psi(t)$	$u_x=f(x)g(u)$ $u_t=f(t)g(u)$ $u_t=f(x)g(t)u_x$ $uu_{xt}-f(u)u_xu_t=0$ $u_xu_{xt}-u_tu_{xx}=f(x)u_xu_t$ $u_xu_{tt}-u_tu_{xt}=g(t)u_xu_t$

Functional separable solutions of a specific form can be transformed into differential constraints by eliminating arbitrary functions through differentiation. Table 9.2 presents examples of first- and second-order differential constraints that correspond to the two most common types of functional separable solutions. The functions f and g can be expressed in terms of the original functions U, φ, and ψ. It is evident that each generalized separable solution can be associated with multiple equivalent differential constraints.

The converse is also true: by integrating the differential constraints listed in the last column of Table 9.2, one can derive the general forms of functional separable solutions, presented in the table's penultimate column.

The works of Polyanin (2019, 2020) provide specific examples demonstrating that using the method of functional separation of variables (in implicit form) is sometimes preferable to the differential constraints method.

9.5.4. Direct Method of Symmetry Reductions and Differential Constraints

The direct method of symmetry reductions, also called the direct reduction method, suggests looking for exact solutions in the form

$$u(x,t) = F\big(x,t,w(z)\big), \quad z = z(x,t), \tag{9.5.4.1}$$

where the functions $F(x,t,w)$ and $z(x,t)$ are selected to ensure that the function $w(z)$ satisfy a single ordinary differential equation; see Section 7.1.4.

Following the arguments of Olver (1994), we will demonstrate that looking for a solution in the form (9.5.4.1) is equivalent to seeking a solution with the differential constraints method using a quasilinear first-order differential constraint

$$\xi(x,t)u_t + \eta(x,t)u_x = \zeta(x,t,u). \tag{9.5.4.2}$$

Indeed, the characteristic system of ordinary differential equations

$$\frac{dt}{\xi(x,t)} = \frac{dx}{\eta(x,t)} = \frac{du}{\zeta(x,t,u)}$$

has the first integrals that can be written in the form

$$z(x,t) = C_1, \quad \varphi(x,t,u) = C_2, \tag{9.5.4.3}$$

where C_1 and C_2 are arbitrary constants. Therefore, the general solution of equation (9.5.4.2) can be represented as (see Section 2.1.1):

$$\varphi(x,t,u) = w\big(z(x,t)\big), \tag{9.5.4.4}$$

where $w(z)$ is an arbitrary function. On solving (9.5.4.4) for u, we obtain a solution in the form (9.5.4.1).

9.5.5. Nonclassical Method of Symmetry Reductions and Differential Constraints

The fundamental concepts of the nonclassical method for symmetry reductions were initially presented in the work of Bluman & Cole (1969).

For simplicity, we will further consider a second-order PDE in two independent variables of the general form

$$F(x,y,u,u_x,u_y,u_{xx},u_{xy},u_{yy}) = 0. \tag{9.5.5.1}$$

The results of the classical group analysis (see Chapter 8) can be substantially extended if, instead of finding invariants of an admissible infinitesimal operator X by means of solving the characteristic system of ODEs (8.1.2.2), which has the form

$$\frac{dx}{\xi(x,y,u)} = \frac{dy}{\eta(x,y,u)} = \frac{du}{\zeta(x,y,u)},$$

we supplement the original equation with a quasilinear first-order PDE (see Bluman & Cole, 1969):

$$\xi(x,y,u)u_x + \eta(x,y,u)u_y = \zeta(x,y,u), \qquad (9.5.5.2)$$

which is referred to as an *invariant surface condition*. The characteristic ODE system for (9.5.5.2) coincides with (8.1.2.2). Equation (9.5.5.1) and condition (9.5.5.2) are supplemented by the invariance condition

$$\underset{2}{\mathrm{X}} F\big(x,y,u,u_x,u_y,u_{xx},u_{xy},u_{yy}\big) = 0, \qquad (9.5.5.3)$$

which coincides with equation (8.2.1.3); the operator $\underset{2}{\mathrm{X}}$ is defined in (8.2.1.4).

All three equations (9.5.5.1)–(9.5.5.3) are used for the construction of exact solutions of the original equation (9.5.5.1), where the added auxiliary equations (9.5.5.2) and (9.5.5.3) play the role of differential constraints. It should be observed that in this case, the determining equations obtained for the unknown functions $\xi(x,y,u)$, $\eta(x,y,u)$, and $\zeta(x,y,u)$ by the splitting procedure are nonlinear. The symmetries determined by the invariant surface (9.5.5.2) are called *nonclassical*.

The nonclassical method of symmetry reductions is applied below to illustrate its practical use in constructing exact solutions of nonlinear PDEs through a specific example.

► Example 9.16. Consider the *Newell–Whitehead equation*

$$u_t = u_{xx} + u - u^3, \qquad (9.5.5.4)$$

which corresponds to the left-hand side $F = -u_t + u_{xx} + u - u^3$ of equation (9.5.5.1) with $y = t$.

Without loss of generality, we set $\eta = 1$ in the invariant surface condition (9.5.5.2) with $y = t$, thus assuming that $\eta \neq 0$. We have

$$u_t + \xi(x,t,u)u_x = \zeta(x,t,u). \qquad (9.5.5.5)$$

The invariance condition is obtained by a procedure similar to the classical algorithm (see Sections 8.1–8.2). Namely, we apply the operator

$$\underset{2}{\mathrm{X}} = \xi\partial_x + \eta\partial_t + \zeta\partial_u + \zeta_1\partial_{u_x} + \zeta_2\partial_{u_t} + \zeta_{11}\partial_{u_{xx}} + \zeta_{12}\partial_{u_{xt}} + \zeta_{22}\partial_{u_{tt}} \qquad (9.5.5.6)$$

to equation (9.5.5.4). Taking into account that $\partial_x = \partial_t = \partial_{u_x} = \partial_{u_{xt}} = \partial_{u_{tt}} = 0$, since the equation is explicitly independent of x, t, u_x, u_{xt}, u_{tt}, we get the invariance condition in the form

$$\zeta_2 = \zeta(3u^2 - 1) + \zeta_{11}.$$

Substituting here the expressions (8.1.3.2) and (8.1.3.7) for the coordinates ζ_2 and ζ_{11} of the first and second prolongations, with $y = t$ and $\eta = 1$, we obtain

$$\begin{aligned}\zeta_t - \xi_t u_x + \zeta_u u_t - \xi_u u_x u_t &= \zeta(-3u^2 + 1) + \zeta_{xx} + (2\zeta_{xu} - \xi_{xx})u_x \\ &+ (\zeta_{uu} - 2\xi_{xu})u_x^2 - \xi_{uu}u_x^3 + (\zeta_u - 2\xi_x - 3\xi_u u_x)u_{xx}.\end{aligned} \tag{9.5.5.7}$$

Let us express the derivatives u_t and u_{xx} with the help of (9.5.5.4) and (9.5.5.5) via the other quantities:

$$u_t = \zeta - \xi u_x, \qquad u_{xx} = \zeta - \xi u_x + u(u^2 - 1). \tag{9.5.5.8}$$

Inserting these expressions into the invariance condition (9.5.5.7), we arrive at a cubic polynomial in the remaining 'independent' derivative u_x. Equating all functional coefficients of the various powers of this polynomial to zero, we get the determining system

$$\begin{aligned} u_x^3:&\quad \xi_{uu} = 0,\\ u_x^2:&\quad \zeta_{uu} - 2(\xi_{xu} - \xi\xi_u) = 0,\\ u_x:&\quad 2\zeta_{xu} - 2\xi_u\zeta - 3u(u^2 - 1)\xi_u - \xi_{xx} + 2\xi\xi_x + \xi_t = 0,\\ 1:&\quad \zeta_t - \zeta_{xx} + 2\xi_x\zeta + (2\xi_x - \zeta_u)u(u^2 - 1) + (3u^2 - 1)\zeta = 0,\end{aligned} \tag{9.5.5.9}$$

which consists of only four equations.

The analysis of the system (9.5.5.9) allows us to conclude that

$$\xi = \xi(x,t), \qquad \zeta = -\xi_x u, \tag{9.5.5.10}$$

where the function $\xi(x,t)$ satisfies the nonlinear overdetermined PDE system

$$\begin{aligned}\xi_t - 3\xi_{xx} + 2\xi\xi_x &= 0,\\ \xi_{xt} - \xi_{xxx} + 2\xi_x^2 + 2\xi_x &= 0.\end{aligned} \tag{9.5.5.11}$$

After inserting the expressions (9.5.5.5), the associated invariant surface condition (9.5.5.5) becomes

$$u_t + \xi u_x = -\xi_x u. \tag{9.5.5.12}$$

We look for a stationary particular solution to equation (9.5.5.11) in the form $\xi = \xi(x)$. Let us differentiate the first equation of (9.5.5.11) with respect to x and then eliminate the third derivative, using the second equation of (9.5.5.11), from the resulting expression. This will give a second-order ODE. Eliminating from it the second derivative, using the first equation of (9.5.5.11) with $\xi_t \equiv 0$, after simple rearrangements we obtain

$$(6\xi_x' - 2\xi^2 + 9)\xi_x' = 0. \tag{9.5.5.13}$$

Equating the expression in parentheses in (9.5.5.13) to zero, we get a separable first-order ODE. Its general solution can be written as

$$\xi = -\frac{3}{\sqrt{2}}\,\frac{A\exp(\sqrt{2}\,x) + B}{A\exp(\sqrt{2}\,x) - B} = -\frac{3}{\sqrt{2}}\,\frac{A\exp(\frac{\sqrt{2}}{2}\,x) + B\exp(-\frac{\sqrt{2}}{2}\,x)}{A\exp(\frac{\sqrt{2}}{2}\,x) - B\exp(-\frac{\sqrt{2}}{2}\,x)}, \tag{9.5.5.14}$$

where A and B are arbitrary constants.

The characteristic system corresponding to PDE (9.5.5.12),

$$\frac{dt}{1} = \frac{dx}{\xi} = -\frac{du}{\xi'_x u}, \tag{9.5.5.15}$$

admits the first integrals

$$t + \tfrac{2}{3}\ln\bigl|A\exp\bigl(\tfrac{\sqrt{2}}{2}x\bigr) + B\exp\bigl(-\tfrac{\sqrt{2}}{2}x\bigr)\bigr| = C_1, \quad \xi u = C_2, \tag{9.5.5.16}$$

where C_1 and C_2 are arbitrary constants. For convenience, we introduce the new constant $\overline{C}_1 = \exp\bigl(\frac{3}{2}C_1\bigr)$ and look for a solution in the form $C_2 = -\frac{3}{\sqrt{2}}\overline{C}_1 F(\overline{C}_1)$. Inserting (9.5.5.16) here and taking into account (9.5.5.14), we obtain the solution structure

$$\begin{aligned} u(x,t) &= \bigl\{A\exp\bigl[\tfrac{1}{2}(\sqrt{2}x + 3t)\bigr] - B\exp\bigl[\tfrac{1}{2}(-\sqrt{2}x + 3t)\bigr]\bigr\}F(z),\\ z &= A\exp\bigl[\tfrac{1}{2}(\sqrt{2}x + 3t)\bigr] + B\exp\bigl[\tfrac{1}{2}(-\sqrt{2}x + 3t)\bigr]. \end{aligned} \tag{9.5.5.17}$$

Substituting (9.5.5.17) in the original equation (9.5.5.4), we find an equation for the function $F = F(z)$:

$$F''_{zz} = 2F^3. \tag{9.5.5.18}$$

The general solution of this ODE is expressed in terms of elliptic functions. Equation (9.5.5.18) admits the following particular solutions: $F = \pm\dfrac{1}{z + C}$, where C is an arbitrary constant. ◀

Remark 9.10. *The degenerate case* $\xi'_x = 0$ *in (9.5.5.13) is of little interest, since it corresponds to a traveling-wave solution.*

Remark 9.11. *Instead of solving the characteristic ODE system (9.5.5.15), after finding the coordinates* ξ *and* ζ*, it was possible to eliminate the derivative* u_t *from PDEs (9.5.5.4) and (9.5.5.5). Then the resulting equation is solved, which can be treated as an ordinary differential equation for* x *with parameter* t*.*

Remark 9.12. *The invariant surface condition (9.5.5.2) gives much more freedom in determining the coordinates* ξ, η, ζ *as compared with the classical scheme presented in Sections 8.1–8.3. This stems from the fact that in the classical scheme, the invariance condition is split with respect to two derivatives,* u_x *and* u_{xx}*, which are considered to be independent (see Examples 8.3 and 8.8 in Sections 8.2.2 and 8.3.3). In the nonclassical scheme, the derivatives* u_x *and* u_{xx} *are related by the second equation of (9.5.5.8) and the invariance conditions are split in only one derivative,* u_x*. This is why in the classical scheme the determining system consists of a larger number of equations, which impose additional constraints on the unknown quantities, as compared with the nonclassical scheme. In particular, the classical scheme fails to find exact solutions to equation Newell–Whitehead (9.5.5.4) in the form (9.5.5.17).*

► For additional information on the differential constraints method, its special cases (including the nonclassical method of symmetry reductions), and applications, see Meleshko (1983), Levi & Winternitz (1989), Nucci & Clarkson (1992), Arrigo et al. (1993), Galaktionov (1994), Clarkson et al. (1997), Kaptsov (1995, 1998), Ludlow et al. (1998), Kaptsov & Verevkin (2003), Martínez & Shabat (2003), Kruglikov (2008), Cherniha et al. (2018), and Polyanin & Zhurov (2022).

References for Chapter 9

Aristov S.N. and Polyanin A.D. Exact solutions of unsteady three-dimensional Navier–Stokes equations. *Doklady Physics*, 2009, Vol. 54, No. 7, pp. 316–321.

Aristov S.N., Knyazev D.V., and Polyanin A.D. Exact solutions of the Navier–Stokes equations with the linear dependence of velocity components of two space variables. *Theor. Foundations of Chemical Engineering*, 2009, Vol. 43, No. 5, pp. 642–662.

Arrigo D., Broadbridge P., and Hill J.M. Nonclassical symmetry solutions and the methods of Bluman–Cole and Clarkson–Kruskal. *J. Math. Phys.*, 1993, Vol. 34, pp. 4692–4703.

Bluman G.W. and Cole J.D. The general similarity solution of the heat equation. *J. Math. Mech.*, 1969, Vol. 18, pp. 1025–1042.

Cherniha R., Serov M., Pliukhin O. *Nonlinear Reaction-Diffusion-Convection Equations: Lie and Conditional Symmetry, Exact Solutions and Their Applications.* Chapman & Hall/CRC Press, Boca Raton, 2018.

Clarkson P.A., Ludlow D.K., and Priestley T.J. The classical, direct and nonclassical methods for symmetry reductions of nonlinear partial differential equations, *Methods and Applications of Analysis*, 1997, Vol. 4, No. 2, pp. 173–195.

Ludlow D.K., Clarkson P.A., and Bassom P. Nonclassical symmetry reductions of the three-dimensional incompressible Navier–Stokes equations. *J. Phys. A: Math. Gen.*, 1998, Vol. 31, No. 39, p. 7965.

Galaktionov V.A. Quasilinear heat equations with first-order sign-invariants and new explicit solutions. *Nonlinear Anal. Theor. Meth. Appl.*, 1994, Vol. 23, pp. 1595–621.

Kaptsov O.V. Determining equations and differential constraints. *Nonlinear Math. Phys.*, 1995, Vol. 2, No. 3-4, pp. 283–291.

Kaptsov O.V. Linear determining equations for differential constraints. *Sbornik: Mathematics*, 1998, Vol. 189, No. 12, pp. 1839–1854.

Kaptsov O.V. and Verevkin I.V. Differential constraints and exact solutions of nonlinear diffusion equations. *J. Phys. A: Math. Gen.*, 2003, Vol. 36, pp. 1401–1414.

Kruglikov B. Symmetry approaches for reductions of PDEs, differential constraints and Lagrange–Charpit method. *Acta Appl. Math.*, 2008, Vol. 101, pp. 145–161.

Kuranishi M. *Lectures on Involutive Systems on Partial Differential Equations*. Publ. Soc. Math., Sao Paulo, 1967.

Levi D. and Winternitz P. Non-classical symmetry reduction: example of the Boussinesq equation. *J. Phys. A*, 1989, Vol. 22, pp. 2915–2924.

Loitsyanskiy L.G. *Mechanics of Liquids and Gases*. Begell House, New York, 1996.

Martínez A.L. and Shabat A.B. Towards a theory of differential constraints of a hydrodynamic hierarchy. *J. Nonlin. Math. Phys.*, 2003, Vol. 10, pp. 229–242.

Meleshko S.V. Differential constraints and one-parameter Lie–Bäcklund groups. *Sov. Math. Dokl.*, 1983, Vol. 28, pp. 37–41.

Meleshko S.V. *Methods for Constructing Exact Solutions of Partial Differential Equations*. Springer-Verlag, New York, 2005.

Nucci M.C. and Clarkson P.A. The nonclassical method is more general than the direct method for symmetry reductions. An example of the Fitzhugh–Nagumo equation, *Phys. Lett. A*, 1992, Vol. 164, pp. 49–56.

Olver P.J. Direct reduction and differential constraints. *Proc. R. Soc. Lond. Ser. A*, 1994, Vol. 444, pp. 509–523.

Polyanin A.D. Comparison of the effectiveness of different methods for constructing exact solutions to nonlinear PDEs. Generalizations and new solutions. *Mathematics*, 2019, Vol. 7, No. 5, p. 386.

Polyanin A.D. Functional separation of variables in nonlinear PDEs: General approach, new solutions of diffusion-type equations. *Mathematics*, 2020, Vol. 8, No. 1, p. 90.

Polyanin A.D. and Zaitsev V.F. *Handbook of Nonlinear Partial Differential Equations, 2nd ed.* Chapman & Hall/CRC Press, Boca Raton–London, 2012.

Polyanin A.D. and Zhurov A.I. *Separation of Variables and Exact Solutions to Nonlinear PDEs*. CRC Press, Boca Raton–London, 2022.

Pommaret J.F. *Systems in Partial Differential Equations and Lie Pseudogroups*. Gordon & Breach Sci. Publ., New York, 1978.

Sidorov A.F., Shapeev V.P., and Yanenko N.N. *Method of Differential Constraints and its Applications in Gas Dynamics* (in Russian). Nauka, Novosibirsk, 1984.

Schlichting H. *Boundary Layer Theory*. McGraw-Hill, New York, 1981.

Yanenko N.N. Compatibility theory and methods of integration of systems of nonlinear partial differential equations, In: *Proceedings of the Fourth All-Union Mathematics Congress*, Vol. 2, pp. 613–621. Nauka, Leningrad, 1964.

10. Transformations of Equations of Mathematical Physics

► **Preliminary remarks.** This chapter discusses the transformations commonly used to simplify, linearize, reduce the order of, or construct exact solutions for nonlinear equations of mathematical physics and other partial differential equations. Consistent with previous chapters, the material is presented according to the principle of 'from simple to complex.'

The chapter begins with an overview of the simplest and most frequently used point transformations and their applications. Special attention is given to hodograph transformations, which are an important special case of point transformations. These transformations allow the linearization of wave-type PDEs and nonlinear systems of gas dynamics equations.

Next, the chapter explores more complex contact transformations, including Legendre and Euler transformations, which involve both the original variables and their first derivatives. It also describes Bäcklund transformations, which relate two different (or identical) PDEs through a suitable system of lower-order PDEs.

Furthermore, the chapter examines special hydrodynamic transformations of the von Mises and Crocco types, as well as transformations based on conservation laws. The application of these various transformations is illustrated with numerous examples, with a focus on nonlinear equations in hydrodynamics, heat and mass transfer, wave theory, and field theory.

10.1. Point Transformations

10.1.1. General Form of Point Transformations

Let x and y be independent variables and let $u = u(x, y)$ be a function of these variables. In general, a point transformation is defined as

$$x = X(\xi, \eta, w), \qquad y = Y(\xi, \eta, w), \qquad u = U(\xi, \eta, w), \tag{10.1.1.1}$$

where ξ and η are new independent variables, $w = w(\xi, \eta)$ is the new dependent variable, and X, Y, and U are some functions, prescribed or desired.

Point transformations preserve the original order of the equations to which they are applied and do not significantly alter the underlying structure of the equations. This is because the highest derivatives of the new variables are linearly dependent on the highest derivatives of the original variables.

Transformation (10.1.1.1) is invertible if

$$\begin{vmatrix} X_x & X_y & X_u \\ Y_x & Y_y & Y_u \\ U_x & U_y & U_u \end{vmatrix} \neq 0.$$

In general, a second-order equation in two independent variables

$$F(x, y, u, u_x, u_y, u_{xx}, u_{xy}, u_{yy}) = 0 \qquad (10.1.1.2)$$

can be reduced with an invertible point transformation (10.1.1.1) to the form

$$G(\xi, \eta, w, w_\xi, w_\eta, w_{\xi\xi}, w_{\xi\eta}, w_{\eta\eta}) = 0. \qquad (10.1.1.3)$$

If $w = w(\xi, \eta)$ is some solution of equation (10.1.1.3), then formulas (10.1.1.1) define an associated solution of equation (10.1.1.2) in a parametric form.

Point transformations are employed to simplify equations and convert them to known forms. They sometimes can reduce nonlinear equations to linear ones.

10.1.2. Linear Transformations and Classification of Second-Order Quasilinear PDEs in Two Independent Variables

Linear transformations of PDEs. *Linear point transformations* (or simply *linear transformations*),

$$x = X(\xi, \eta), \quad y = Y(\xi, \eta), \quad u = f_1(\xi, \eta)w + f_0(\xi, \eta), \qquad (10.1.2.1)$$

are most commonly used.

The simplest linear transformations of the independent variables aretranslation transformation

$$\begin{aligned} &x = \xi + x_0, && y = \eta + y_0 && (\textit{translation transformation}), \\ &x = k_1\xi, && y = k_2\eta && (\textit{scaling transformation}), \\ &x = \xi\cos\alpha - \eta\sin\alpha, && y = \xi\sin\alpha + \eta\cos\alpha && (\textit{rotation transformation}). \end{aligned}$$

These transformations correspond to the translation of the origin of coordinates to the point (x_0, y_0), scaling (extension or contraction) along the x- and y-axis, and the rotation of the coordinate system through the angle α, respectively. These transformations do not affect the dependent variable ($u = w$).

Linear transformations (10.1.2.1) are employed to simplify linear and nonlinear equations as well as to reduce second-order quasilinear PDEs to the canonical forms (see below in this section).

► Example 10.1. The nonlinear diffusion-type PDE

$$u_t = a(uu_x)_x + [xf(t) + g(t)]u_x,$$

where $f(t)$ and $g(t)$ are arbitrary functions, can be reduced with the linear point transformation

$$u(x,t) = w(z,\tau), \qquad z = xF(t) + \int g(t)F(t)\,dt,$$
$$\tau = \int F^2(t)\,dt, \qquad F(t) = \exp\left[\int f(t)\,dt\right].$$

to a simpler equation:

$$w_\tau = a(ww_z)_z.$$

This PDE admits, for example, a traveling wave solution as well as a self-similar solution. ◀

Types of quasilinear PDEs. Characteristic equation. Let us consider a second-order quasilinear partial differential equation in two independent variables of the special form

$$a(x,y)u_{xx} + 2b(x,y)u_{xy} + c(x,y)u_{yy} = F(x,y,u,u_x,u_y), \tag{10.1.2.2}$$

where $a = a(x,y)$, $b = b(x,y)$, and $c = c(x,y)$ are some functions of x and y that have continuous derivatives up to the second-order inclusive.

This equation is classified according to the sign of the discriminant

$$\delta = b^2 - ac, \tag{10.1.2.3}$$

where the arguments of the equation coefficients are omitted for brevity. Given a point (x,y), equation (10.1.2.2) is said to be

$$\begin{array}{ll} \textit{parabolic} & \text{if } b^2 - ac = 0,\\ \textit{hyperbolic} & \text{if } b^2 - ac > 0,\\ \textit{elliptic} & \text{if } b^2 - ac < 0 \end{array}$$

at this point (e.g., see Courant & Hilbert, 1989 and Tikhonov & Samarskii, 1990).

In order to reduce equation (10.1.2.2) to a canonical form, one should first write out the characteristic equation

$$a\,dy^2 - 2b\,dx\,dy + c\,dx^2 = 0,$$

which splits into two ODEs

$$a\,dy - \left(b + \sqrt{b^2 - ac}\right)dx = 0, \tag{10.1.2.4}$$

and

$$a\,dy - \left(b - \sqrt{b^2 - ac}\right)dx = 0, \tag{10.1.2.5}$$

and find their general integrals.

Remark 10.1. *The equations of characteristics (10.1.2.4)–(10.1.2.5) can be used if* $a \not\equiv 0$. *If* $a \equiv 0$, *one should use the simpler equations*

$$dx = 0,$$
$$2b\,dy - c\,dx = 0.$$

The former equation has the obvious general solution $x = C$.

Remark 10.2. *The classification of quasilinear PDEs of the form (10.1.2.2) does not depend on their solutions and is determined solely by the coefficients of the highest derivatives on the left-hand side.*

Canonical form of parabolic PDEs (case $b^2 - ac = 0$). In this case, equations (10.1.2.4) and (10.1.2.5) coincide and have a common general integral,

$$\varphi(x, y) = C.$$

By passing from x, y to new independent variables ξ, η in accordance with the relations

$$\xi = \varphi(x, y), \qquad \eta = \eta(x, y),$$

where $\eta = \eta(x, y)$ is any twice differentiable function that satisfies the condition of nondegeneracy of the Jacobian $\frac{D(\xi,\eta)}{D(x,y)}$ in a given domain, we reduce equation (10.1.2.2) to the canonical form

$$u_{\eta\eta} = F_1(\xi, \eta, u, u_\xi, u_\eta). \tag{10.1.2.6}$$

As η, one can take $\eta = x$ or $\eta = y$. Often $\eta = \eta(x, y)$ is selected so as to simplify the right-hand side of equation (10.1.2.6) as much as possible. In the special case $F_1 = u_\xi$, we have the classical linear heat equation.

Note that the transformed equation (10.1.2.6) has only one highest-derivative term.

Remark 10.3. *In the degenerate case where the function* F_1 *is independent of the derivative* u_ξ, *equation (10.1.2.6) is an ordinary differential equation for* η, *in which* ξ *serves as a parameter.*

► Example 10.2. Consider the nonlinear equation

$$u_{xx} - 2u_{xy} + u_{yy} = f(u).$$

It is a special case of equation (10.1.2.2) with $a = c = 1$, $b = -1$, and $F = f(u)$. In this case, characteristic equations (10.1.2.4) and (10.1.2.5) coincide and have a common integral,

$$x + y = C.$$

By changing to the new variables

$$\xi = x + y, \qquad \eta = x,$$

we arrive at the equation

$$u_{\eta\eta} = f(u).$$

It can be treated as an ordinary differential equation. Its general solution involves two constants of integration, which are arbitrary functions of the argument ξ, can be represented in the implicit form

$$\int\left[C_1+\int f(u)\,du\right]^{-1/2}du=C_2\pm\eta,$$

where $C_1=C_1(\xi)$ and $C_2=C_2(\xi)$ are arbitrary functions, with $\xi=x+y$, and $\eta=x$. ◄

Canonical form of hyperbolic PDEs (case $b^2-ac>0$). The general integrals

$$\varphi(x,y)=C_1,\qquad \psi(x,y)=C_2$$

of equations (10.1.2.4) and (10.1.2.5) are real and different. These integrals determine two different families of real characteristics.

By passing from x, y to new independent variables ξ, η in accordance with the relations

$$\xi=\varphi(x,y),\qquad \eta=\psi(x,y),$$

we reduce equation (10.1.2.2) to

$$u_{\xi\eta}=F_2(\xi,\eta,u,u_\xi,u_\eta). \tag{10.1.2.7}$$

This is the so-called *first canonical form of a hyperbolic equation.*

If the right-hand side of equation (10.1.2.7) is independent of u and u_ξ (or u and u_η), the equation admits an order reduction with the substitution $w=u_\eta$ (or $w=u_\xi$).

► Example 10.3. Consider the nonlinear equation

$$u_{xx}-u_{yy}=f(u)(u_x^2-u_y^2). \tag{10.1.2.8}$$

It is a special case of equation (10.1.2.2) with $a=1$, $b=0$, $c=-1$, and $F=f(u)(u_x^2-u_y^2)$. In this case, characteristic equations (10.1.2.4) and (10.1.2.5) have integrals $\varphi=y+x$ and $\psi=y-x$, so

$$\xi=y+x,\qquad \eta=y-x \tag{10.1.2.9}$$

can be taken as the new variables. In terms of the variables (10.1.2.9), equation (10.1.2.8) becomes

$$u_{\xi\eta}=f(u)u_\xi u_\eta. \tag{10.1.2.10}$$

This PDE coincides, up to obvious renaming, with the integrable equation (10.1.3.6). The general solution of the nonlinear PDE (10.1.2.10) with $f(u)=\alpha=\text{const}$ is given by

$$u=-\frac{1}{\alpha}\ln[\varphi(y+x)+\psi(y-x)],$$

where $\varphi(z_1)$ and $\psi(z_2)$ are arbitrary functions. ◄

The transformation

$$\xi = t + z, \qquad \eta = t - z$$

brings equation (10.1.2.7) to another canonical form,

$$u_{tt} - u_{zz} = F_3(t, z, u, u_t, u_z), \tag{10.1.2.11}$$

where $F_3 = 4F_2$. This is the so-called *second canonical form of a hyperbolic equation*.hyperbolic equation, second canonical form

In the special case $F_3 = 0$, equation (10.1.2.11) is the classical linear wave equation.

Canonical form of elliptic PDEs (case $b^2 - ac < 0$). In this case, the general integrals of equations (10.1.2.4) and (10.1.2.5) are complex conjugate; these determine two families of complex characteristics.

Let the general integral of equation (10.1.2.4) have the form

$$\varphi(x, y) + i\psi(x, y) = C, \qquad i^2 = -1,$$

where $\varphi(x, y)$ and $\psi(x, y)$ are real-valued functions.

By passing from x, y to new independent variables ξ, η in accordance with the relations

$$\xi = \varphi(x, y), \qquad \eta = \psi(x, y),$$

we reduce equation (10.1.2.2) to the canonical form

$$u_{\xi\xi} + u_{\eta\eta} = F_4(\xi, \eta, u, u_\xi, u_\eta).$$

In the special case $F_4 = 0$, we have the linear Laplace equation.

10.1.3. Simple Nonlinear Point Transformations

Point transformations can be used for the reduction of some nonlinear equations to linear ones. The simplest nonlinear transformations have the form

$$u = U(w) \tag{10.1.3.1}$$

and do not affect the independent variables ($x = \xi$ and $y = \eta$). Combinations of transformations (10.1.2.1) and (10.1.3.1) are also used quite often. Sometimes it is more convenient to write transformation (10.1.3.1) in an implicit form $w = W(u)$.

► Example 10.4. Consider the nonlinear equation

$$u_t = au_{xx} + bu_x^2. \tag{10.1.3.2}$$

With the substitution (10.1.3.1), this equation becomes

$$w_t = aw_{xx} + \frac{1}{U'_w}[aU''_{ww} + b(U'_w)^2]w_x^2. \tag{10.1.3.3}$$

The function $U(w)$ is chosen so that the expression in square brackets vanishes:

$$aU''_{ww} + b(U'_w)^2 = 0. \tag{10.1.3.4}$$

Integrating (10.1.3.4) yields the solution

$$U(w) = \frac{a}{b} \ln(C_1 w + C_2), \tag{10.1.3.5}$$

where C_1 and C_2 are arbitrary constants. Hence, the transformation determined by (10.1.3.1) and (10.1.3.5) with $C_1 \neq 0$ reduces the original equation (10.1.3.2) to the linear heat equation

$$w_t = aw_{xx}.$$

◀

► Example 10.5. Consider another nonlinear PDE

$$u_{xy} = f(u)u_x u_y, \tag{10.1.3.6}$$

where $f(u)$ is an arbitrary function.

Substitution (10.1.3.1) leads this equation to the form

$$U'_w w_{xy} + [U''_{ww} - f(U)(U'_w)^2] w_x w_y = 0. \tag{10.1.3.7}$$

Equating the expression in square brackets to zero, we get an ODE to define the function U:

$$U''_{ww} - f(U)(U'_w)^2 = 0. \tag{10.1.3.8}$$

Integrating, we have

$$\int \exp\Big[-\int f(U)\,dU\Big]\,dU = C_1 w + C_2, \tag{10.1.3.9}$$

where C_1 and C_2 are arbitrary constants. For $C_1 \neq 0$, formulae (10.1.3.9) defines the transformation (10.1.3.1), given in an implicit form. It follows from (10.1.3.7) and (10.1.3.8) that the original nonlinear PDE (10.1.3.6) is reduced to the simplest linear PDE, $w_{xy} = 0$, the general solution of which has the form $w = \varphi(x) + \psi(y)$, where $\varphi(x)$ and $\psi(y)$ are arbitrary functions. As a result, without limiting generality, assuming $C_1 = 1$ and $C_2 = 0$, the general solution of equation (10.1.3.6) can be written implicitly

$$\int \exp\Big[-\int f(u)\,du\Big]\,du = \varphi(x) + \psi(y).$$

◀

Problems and Exercises for Section 10.1

1. Find a transformation that reduces the unnormalized heat equation with a nonlinear source $u_t = au_{xx} + bu^n$ to the canonical form $w_\tau = w_{xx} + w^n$.

2. Find a transformation that reduces the unnormalized Burgers equation $u_t + auu_x = bu_{xx}$ to the canonical form $w_\tau + ww_z = w_{zz}$.

3. Find a transformation that reduces the unnormalized Boussinesq equation $u_{tt}+a(uu_x)_x+bu_{xxxx}=0$ to the canonical form $w_{\tau\tau}+(w_z)_z+w_{zzzz}=0$.

4. Find point transformations that reduce the nonlinear equations

a) $u_t=u_{xx}+auu_x^2$,

b) $u_t=u_{xx}+f(u)u_x^2$

to the linear heat equation $w_t=w_{xx}$.

Hint. Seek the transformations in the implicit form $w=F(u)$.

5. Find a point transformation reducing the equation $u_t=u_{xx}+au\ln u+f(t)u$ to the simpler equation $w_t=w_{xx}+aw\ln w$.

Hint. Seek the transformation in the form $u=\varphi(t)w$, where $w=w(x,t)$.

6. Find a point transformation reducing the equation $u_t=u_{xx}+au_x^2+f(t)$ to the simpler equation $w_t=w_{xx}+aw_x^2$.

Hint. Seek the transformation in the form $u=\varphi(t)+w$, where $w=w(x,t)$.

7. Show that the nonlinear equation $u_t=u_{xx}+au_x^2+f(x,t)$ can be reduced with the change of variable $w=\exp(au)$ to a linear equation for $w=w(x,t)$:

$$w_t=w_{xx}+af(x,t)w.$$

8. Find a point transformation reducing the equation $u_t=u_{xx}+au_x^k+bu+f(t)$ to the simpler equation $w_t=w_{xx}+aw_x^k+bw$.

Hint. Seek the transformations in the form $u=\varphi(t)+w$, where $w=w(x,t)$.

9. Show that the nonlinear equation

$$u_t=a(u^ku_x)_x+[xf(t)+g(t)]u_x+h(t)u$$

where $f(t)$, $g(t)$, and $h(t)$ are arbitrary functions, can be reduced, using the point transformation (Polyanin & Zaitsev, 2012)

$$u(x,t)=w(z,\tau)H(t),\quad z=xF(t)+\int g(t)F(t)\,dt,\quad \tau=\int F^2(t)H^k(t)\,dt,$$

where the functions F and H are defined by

$$F(t)=\exp\Bigl[\int f(t)\,dt\Bigr],\quad H(t)=\exp\Bigl[\int h(t)\,dt\Bigr],$$

to the simpler equation

$$w_\tau=a(w^kw_z)_z.$$

10. Show that the nonlinear equation

$$u_t=a(e^{\lambda u}u_x)_x+[xf(t)+g(t)]u_x+h(t)$$

where $f(t)$, $g(t)$, and $h(t)$ are arbitrary functions, can be reduced, using the point transformation

$$u(x,t)=w(z,\tau)+H(t),\quad z=xF(t)+\int g(t)F(t)\,dt,\quad \tau=\int F^2(t)e^{\lambda H(t)}dt,$$

where the functions F and H are defined by

$$F(t)=\exp\Bigl[\int f(t)\,dt\Bigr],\quad H(t)=\int h(t)\,dt,$$

to the simpler equation

$$w_\tau=a(e^{\lambda w}w_z)_z.$$

11. Show that the nonlinear equation

$$u_t = [f(u)u_x]_x + [xg(t) + h(t)]u_x$$

where $f(u)$, $g(t)$, and $h(t)$ are arbitrary functions, can be reduced, using a point transformation of the form

$$u(x,t) = U(z,\tau), \quad z = \varphi(t)x + \psi(t), \quad \tau = \theta(t),$$

to the simpler equation (Polyanin & Zaitsev, 2012):

$$U_\tau = [f(U)U_z]_z.$$

Identify the functions $\varphi(t)$, $\psi(t)$, and $\theta(t)$ determining the point transformation.

12. Show that the homogeneous Monge–Ampère equation $u_{xx}u_{yy} - u_{xy}^2 = 0$ is invariant under the linear transformation

$$\bar{x} = a_1x + b_1y + c_1, \quad \bar{y} = a_2x + b_2y + c_2, \quad \bar{u} = ku + a_3x + b_3y + c_3,$$

meaning that it reduces to an equation of exactly the same form. The parameter k needs to be determined.

13. Find a linear transformation of the form

$$\tilde{t} = f(t), \quad \tilde{x} = g(t)x, \quad \tilde{u} = h(t)u + p(t)x$$

connecting the cylindrical Korteweg–de Vries equation

$$\tilde{u}_{\tilde{t}} + \tilde{u}_{\tilde{x}\tilde{x}\tilde{x}} - 6\tilde{u}\tilde{u}_{\tilde{x}} + \frac{\tilde{u}}{2\tilde{t}} = 0$$

and the classical Korteweg–de Vries equation

$$u_t + u_{xxx} - 6uu_x = 0.$$

10.2. Hodograph Transformations (Special Point Transformations)

► **Preliminary remarks.** Hodograph transformations are significant special cases of point transformations. For ODEs of the form $F(x, u, u'_x, u''_{xx}, \dots) = 0$, the hodograph transformation suggests that a solution is sought in the implicit form $x = x(u)$. This means that u is considered the independent variable, while x is treated as the dependent variable. The hodograph transformation does not alter the order of the ODE and can also be expressed in the equivalent form: $x = \tilde{u}$, $u = \tilde{x}$. When employing this transformation, the derivatives are expressed as follows:

$$u'_x = \frac{1}{x'_u}, \quad u''_{xx} = -\frac{x''_{uu}}{x'^3_u}, \quad u'''_{xxx} = -\frac{1}{x'_u}\frac{d}{du}\left(\frac{x''_{uu}}{x'^3_u}\right), \quad \dots$$

The hodograph transformation is applied to linearize and simplify certain ordinary differential equations, and it also serves as an intermediate step in finding solvable nonlinear ODEs (Polyanin & Zaitsev, 2018).

In fluid mechanics, gas dynamics, solid mechanics, and other fields of natural and applied sciences, hodograph transformations are also used for linearization and simplification of nonlinear equations and systems of partial differential equations, when some dependent and independent variables are swapped (e.g., see Whitham, 1974; Rozhdestvenskii & Yanenko, 1983; Courant & Friedrichs, 1985; Siddiqui et al. 1985; and Chakrabarty, 1987); also refer to Kumei & Bluman (1982), Clarkson at al. (1989), and Polyanin & Zaitsev (2012).

10.2.1. One PDE: One of the Independent Variables Is Taken to Be the Unknown

For equations with two independent variables, x and t, and one unknown function $u = u(x,t)$, using a hodograph transformation suggests that a solution is sought in the implicit form

$$x = x(t,u), \qquad (10.2.1.1)$$

(The independent variables x and t can be swapped.) This means that t and u are taken as the independent variables and x as the dependent one. The hodograph transformation (10.2.1.1) does not alter the order of the equation and is a special case of a point transformation; it can be represented in the equivalent form: $x = \widetilde{u}$, $t = \widetilde{t}$, $u = \widetilde{x}$.

Differentiating relation (10.2.1.1) with respect to both variables as an implicit function and taking into account that $u = u(x,t)$, we get

$$\begin{aligned}
1 &= u_x x_u && \text{(differentiation in } x\text{)},\\
0 &= x_t + u_t x_u && \text{(differentiation in } t\text{)},\\
0 &= u_x^2 x_{uu} + u_{xx} x_u && \text{(double differentiation in } x\text{)},\\
0 &= u_t u_x x_{uu} + u_x x_{tu} + u_{tx} x_u && \text{(differentiation in } x \text{ and } t\text{)},\\
0 &= x_{tt} + 2u_t x_{tu} + u_t^2 x_{uu} + u_{tt} x_u && \text{(double differentiation in } t\text{)},
\end{aligned}$$

where the subscripts indicate the corresponding partial derivatives. We solve these relations to express the "old" derivatives through the "new" ones,

$$\begin{gathered}
u_x = \frac{1}{x_u}, \quad u_t = -\frac{x_t}{x_u}, \quad u_{xx} = -\frac{u_x^2 x_{uu}}{x_u} = -\frac{x_{uu}}{x_u^3},\\
u_{tx} = -\frac{x_u x_{tu} - x_t x_{uu}}{x_u^3}, \quad u_{tt} = \frac{-x_u^2 x_{tt} + 2x_t x_u x_{tu} - x_t^2 x_{uu}}{x_u^3}, \qquad (10.2.1.2)\\
\frac{\partial}{\partial x} = u_x \frac{\partial}{\partial u} = \frac{1}{x_u}\frac{\partial}{\partial u}, \quad \frac{\partial}{\partial t} = \frac{\partial}{\partial t} + u_t \frac{\partial}{\partial u} = \frac{\partial}{\partial t} - \frac{x_t}{x_u}\frac{\partial}{\partial u}.
\end{gathered}$$

Below are examples of using the hodograph transformation (10.2.1.1) and formulas (10.2.1.2) to simplify some nonlinear PDEs.

► Example 10.6. Consider the *generalized Hopf equation*

$$u_t = f(u) u_x, \qquad (10.2.1.3)$$

which is a *model equation of gas dynamics*. The hodograph transformation (10.2.1.1) leads PDE (10.2.1.3) to a simple first-order linear ODE

$$x_t = -f(u), \tag{10.2.1.4}$$

in which the variable u is included as a parameter. Integrating equation (10.2.1.4), we find the general solution of the original PDE in implicit form

$$x = -tf(u) + \varphi(u),$$

where $\varphi(u)$ is an arbitrary function. ◀

▶ Example 10.7. Consider the nonlinear second-order PDE

$$u_t = au_x^{-2} u_{xx}. \tag{10.2.1.5}$$

Let us seek its solution in implicit form (10.2.1.1). Substituting expressions (10.2.1.2) into (10.2.1.5), we obtain the linear heat equation

$$x_t = ax_{uu}.$$

◀

▶ Example 10.8. The equation of a steady-state boundary layer of a Newtonian fluid at a flat plate has the form

$$u_y u_{xy} - u_x u_{yy} = \nu u_{yyy}, \tag{10.2.1.6}$$

where u is the stream function, x and y are the coordinates along and across the flow, and ν is the kinematic viscosity of the fluid. In equation (10.2.1.6), let us perform the hodograph transformation; by doing so, we assume that $y = y(x, u)$. Using the formulas (10.2.1.2) with the renaming $x \Rightarrow y$ and $t \Rightarrow x$, we get

$$u_y u_{xy} - u_x u_{yy} = -\frac{y_{xu}}{y_u^3}, \qquad u_{yyy} = -\frac{1}{y_u}\frac{\partial}{\partial u}\left(\frac{y_{uu}}{y_u^3}\right).$$

As a result, equation (10.2.1.6) becomes

$$\frac{y_{xu}}{y_u^2} = \nu \frac{\partial}{\partial u}\left(\frac{y_{uu}}{y_u^3}\right). \tag{10.2.1.7}$$

This equation is simpler than the original one, since it admits an order reduction with the obvious substitution $y_u = z(x, u)$:

$$\frac{1}{z^2}\frac{\partial z}{\partial x} = \nu \frac{\partial}{\partial u}\left(\frac{1}{z^3}\frac{\partial z}{\partial u}\right).$$

The substitution $z = 1/U$ leads to the heat equation with quadratic nonlinearity

$$U_t = \nu(UU_u)_u.$$

◀

10.2.2. One PDE: The First Hodograph Transformation Based on Using an Equivalent PDE System

In order to investigate equations with one unknown function $u = u(x,t)$, it may be useful to convert the original PDE to an equivalent system of PDEs for two unknown functions

$$u = u(x,t), \qquad v = v(x,t) \tag{10.2.2.1}$$

(the elimination of v from the system results in the original equation).

The first hodograph transformation: in the equivalent system of PDEs, one of the independent variables is taken to be the dependent one. The hodograph transformation consists of representing the solution in implicit form

$$x = x(u,t), \qquad v = v(u,t). \tag{10.2.2.2}$$

Thus, u and t are treated as independent variables, while x and v are taken to be the dependent variables. The hodograph transformation (10.2.2.2) does not change the order of the equation and belongs to the class of point transformations (equivalently, it can be represented as $x = \widetilde{u}$, $t = \widetilde{t}$, $u = \widetilde{x}$, $v = \widetilde{v}$).

From (10.2.2.1) and (10.2.2.2) we get

$$du = u_x\,dx + u_t\,dt, \qquad dv = v_x\,dx + v_t\,dt; \tag{10.2.2.3}$$

$$dx = x_u\,du + x_t\,dt, \qquad dv = v_u\,du + v_t\,dt. \tag{10.2.2.4}$$

Solving (10.2.2.4) for du and dv, we have

$$du = \frac{1}{x_u}\,dx - \frac{x_t}{x_u}dt, \qquad dv = \frac{v_u}{x_u}\,dx + \left(v_t - \frac{x_t}{x_u}v_u\right)dt. \tag{10.2.2.5}$$

The comparison between (10.2.2.3) and (10.2.2.5) allows us to express the "old" derivatives in terms of the "new" ones

$$u_x = \frac{1}{x_u}, \quad u_t = -\frac{x_t}{x_u}, \quad v_x = \frac{v_u}{x_u}, \quad v_t = v_t - \frac{x_t}{x_u}v_u. \tag{10.2.2.6}$$

For the hodograph transformation (10.2.2.2), the highest derivatives of $u = u(x,t)$ are given in (10.2.1.2) and are independent of the auxiliary function v.

Remark 10.4. *In formulas (10.2.2.2) and (10.2.2.6), the variables x and t can be swapped.*

Remark 10.5. *The first transformation (10.2.2.2) will also be called a nonclassical hodograph transformation.*

Let us illustrate the hodograph transformation (10.2.2.2) by some examples.

► Example 10.9. Consider the second-order nonlinear equation

$$u_t + \frac{\partial}{\partial x}\left[\frac{f(u)}{u_x}\right] = 0. \tag{10.2.2.7}$$

Assuming that $v = f(u)/u_x$, let us represent equation (10.2.2.7) as an equivalent system of two equations

$$u_t + v_x = 0, \quad vu_x = f(u). \tag{10.2.2.8}$$

Applying to (10.2.2.8) the hodograph transformation (10.2.2.2) and taking into account (10.2.2.6), we get the linear system of equations

$$x_t - v_u = 0, \quad f(u)x_u = v.$$

Eliminating v yields the linear heat equation with the variable thermal conductivity coefficient

$$x_t = [f(u)x_u]_u. \tag{10.2.2.9}$$

Remark 10.6. *In equations (10.2.2.7)–(10.2.2.9), the function $f(u)$ can be replaced by an arbitrary function of two variables, $f(t,u)$.* ◄

► Example 10.10. Consider the *generalized Calogero equation*

$$v_{xt} = vf(v_x)v_{xx} + g(v_x). \tag{10.2.2.10}$$

By letting $u = v_x$, we rewrite equation (10.2.2.10) as the system of PDEs

$$u = v_x, \quad u_t = vf(u)u_x + g(u).$$

Using the hodograph transformation (10.2.2.2), we obtain the linear system of PDEs

$$ux_u = v_u, \quad x_t + g(u)x_u + f(u)v = 0.$$

Remark 10.7. *In equation (10.2.2.10), the functions $f(u)$ and $g(u)$ can be replaced by arbitrary functions of two variables, $f(t,u)$ and $g(t,u)$.* ◄

10.2.3. One PDE: The Second Hodograph Transformation Based on Using an Equivalent PDE System

First, as before, one rewrites the original PDE with one unknown function $u = u(x,t)$ as an equivalent system of PDEs for two unknown functions (10.2.2.1).

The second hodograph transformation: in the equivalent system of PDEs, both independent variables are taken to be the dependent ones. In this case, one seeks a solution in implicit form using the following hodograph transformation

$$x = x(u,v), \quad t = t(u,v), \tag{10.2.3.1}$$

where u, v are treated as the independent variables and x, t as the dependent variables. The hodograph transformation (10.2.3.1) does not change the order of the equation and belongs to the class of point transformations (equivalently, it can be represented as $x = \widetilde{u}$, $t = \widetilde{v}$, $u = \widetilde{x}$, $v = \widetilde{t}$).

From (10.2.2.1) we get the relations (10.2.2.3), and from (10.2.3.1) we have

$$dx = x_u\,du + x_v\,dv, \quad dt = t_u\,du + t_v\,dv. \tag{10.2.3.2}$$

Solving (10.2.3.2) for du and dv, we get

$$du = \frac{t_v}{J}\,dx - \frac{x_v}{J}\,dt, \qquad dv = -\frac{t_u}{J}\,dx + \frac{x_u}{J}\,dt, \qquad J = x_u t_v - x_v t_u. \quad (10.2.3.3)$$

The comparison between (10.2.2.3) and (10.2.3.3) allows us to express the "old" derivatives in terms of the "new" ones

$$u_x = \frac{t_v}{J}, \quad u_t = -\frac{x_v}{J}, \quad v_x = -\frac{t_u}{J}, \quad v_t = \frac{x_u}{J}, \qquad J = x_u t_v - x_v t_u = \frac{1}{u_x v_t - u_t v_x}. \quad (10.2.3.4)$$

Remark 10.8. *Transformation (10.2.3.1) will also be called the classical hodograph transformation (since it is used more frequently and was suggested long before the nonclassical hodograph transformation (10.2.2.2)).*

Let us illustrate the classical hodograph transformation (10.2.3.1) by examples of specific equations of mathematical physics.

► Example 10.11. Rewrite the *nonlinear wave equation*

$$u_{tt} = a(uu_x)_x \quad (10.2.3.5)$$

as the system of first-order PDEs

$$u_t = v_x, \qquad auu_x = v_t. \quad (10.2.3.6)$$

We now take advantage of the hodograph transformation (10.2.3.1), which amounts to taking u and v as the independent variables and x and t as the dependent variables. Using (10.2.3.4), we substitute expressions of the derivatives u_x, u_t, v_x, and v_t into (10.2.3.6) to arrive at the system

$$x_v = t_u, \qquad aut_v = x_u. \quad (10.2.3.7)$$

Let us differentiate the first equation in u and the second in v, and then eliminate the mixed derivative x_{uv}. As a result, we obtain the following linear equation for the function $t = t(u, v)$:

$$t_{uu} - aut_{vv} = 0. \quad (10.2.3.8)$$

Similarly, from system (10.2.3.7) one can obtain another second-order linear PDE for the function $x = x(u, v)$, which is not given here.

Given a particular solution $t = t(u, v)$ of equation (10.2.3.8), we substitute this solution into system (10.2.3.7) and find $x = x(u, v)$ by straightforward integration. Eliminating v from (10.2.3.1), we obtain an exact solution $u = u(x, t)$ of the nonlinear equation (10.2.3.5).

Remark 10.9. *Equation (10.2.3.8) with an arbitrary a admits a simple particular solution, namely,*

$$t = C_1 uv + C_2 u + C_3 v + C_4, \quad (10.2.3.9)$$

where $C_1, \dots, C_4$ are arbitrary constants. Substituting this solution into system (10.2.3.7), we obtain

$$x_v = C_1 v + C_2, \quad x_u = a(C_1 u + C_3)u. \tag{10.2.3.10}$$

Integrating the first equation in (10.2.3.10) yields $x = \frac{1}{2}C_1 v^2 + C_2 v + \varphi(u)$. Substituting this solution into the second equation in (10.2.3.10), we find the function $\varphi(u)$, and consequently

$$x = \tfrac{1}{2}C_1 v^2 + C_2 v + \tfrac{1}{3}aC_1 u^3 + \tfrac{1}{2}aC_3 u^2 + C_5. \tag{10.2.3.11}$$

Formulas (10.2.3.9) and (10.2.3.11) define an exact solution of equation (10.2.3.5) in parametric form (v is the parameter).

In a similar way, one can construct more complex solutions of equation (10.2.3.5) in parametric form.

◀

▶ Example 10.12. Consider the *Born–Infeld equation*

$$(1 - u_t^2)u_{xx} + 2u_x u_t u_{xt} - (1 + u_x^2)u_{tt} = 0, \tag{10.2.3.12}$$

which is used in nonlinear electrodynamics (field theory).

By introducing the new variables

$$\xi = x - t, \quad \eta = x + t, \quad v = u_\xi, \quad w = u_\eta,$$

equation (10.2.3.12) can be rewritten as the equivalent system (Whitham, 1974):

$$\begin{aligned} v_\eta - w_\xi &= 0, \\ w^2 v_\xi - (1 + 2vw)v_\eta + v^2 w_\eta &= 0. \end{aligned}$$

The hodograph transformation, where v, w are taken to be the independent variables and ξ, η the dependent ones, leads to the linear PDE system

$$\begin{aligned} \xi_w - \eta_v &= 0, \\ w^2 \eta_w + (1 + 2vw)\xi_w + v^2 \xi_v &= 0. \end{aligned} \tag{10.2.3.13}$$

Eliminating η yields the second-order linear PDE

$$v^2 \xi_{vv} + (1 + 2vw)\xi_{vw} + w^2 \xi_{ww} + 2v\xi_v + 2w\xi_w = 0.$$

Assuming that the solution of interest is in the domain of hyperbolicity, we write out the equation of characteristics (see Section 10.1.2):

$$v^2\, dw^2 - (1 + 2vw)\, dv\, dw + w^2\, dv^2 = 0.$$

This ODE has the integrals $r = C_1$ and $s = C_2$, where

$$r = \frac{\sqrt{1 + 4vw} - 1}{2w}, \qquad s = \frac{\sqrt{1 + 4vw} - 1}{2v}. \tag{10.2.3.14}$$

Passing in (10.2.3.13) to the new variables (10.2.3.14), we obtain

$$\begin{aligned} r^2 \xi_r + \eta_r &= 0, \\ \xi_s + s^2 \eta_s &= 0. \end{aligned} \tag{10.2.3.15}$$

Eliminating η yields the simple PDE (Whitham, 1974):

$$\xi_{rs} = 0,$$

whose general solution is the sum of two arbitrary functions with different arguments: $\xi = f(r) + g(s)$. The function η is determined from system (10.2.3.15) and can be represented as

$$\eta = -\int r^2 f_r'(r)\, dr - \int \frac{g_s'(s)}{s^2}\, ds + C,$$

where C is an arbitrary constant.

Remark 10.10. *The Cauchy problem for equation (10.2.3.12) was solved in the work of Barbashov & Chernikov (1966).* ◄

10.2.4. System of Two Coupled PDEs: One of the Independent Variables Is Taken to Be the Dependent One

Consider systems consisting of two PDEs with the unknown functions $u = u(x,t)$ and $v = v(x,t)$. We use the first hodograph transformation (10.2.2.2):

$$x = x(u,t), \qquad v = v(u,t)$$

where u, t are treated as the independent variables and x, v as the dependent variables. The expressions of the partial derivatives are given in (10.2.2.6).

► Example 10.13. The quasilinear PDE system

$$f_1(t,u)u_t + [f_2(t,u) + f_3(t,u)x + f_4(t,u)v]u_x + f_5(t,u)v_x = f_6(t,u),$$
$$g_1(t,u)u_t + [g_2(t,u) + g_3(t,u)x + g_4(t,u)v]u_x + g_5(t,u)v_x = g_6(t,u)$$

is reduced by the hodograph transformation (10.2.2.2) to the linear PDE system

$$f_1(t,u)x_t + f_6(t,u)x_u - f_5(t,u)v_u = f_2(t,u) + f_3(t,u)x + f_4(t,u)v,$$
$$g_1(t,u)x_t + g_6(t,u)x_u - g_5(t,u)v_u = g_2(t,u) + g_3(t,u)x + g_4(t,u)v.$$

◄

10.2.5. System of Two Coupled PDEs: Both of the Independent Variables Are Taken to Be the Dependent Ones

Consider systems consisting of two PDEs with the unknown functions $u = u(x,t)$ and $v = v(x,t)$. We use the second hodograph transformation (10.2.3.1):

$$x = x(u,v), \qquad t = t(u,v),$$

where u, v are treated as the independent variables and x, t as the dependent variables. The expressions of the partial derivatives are given in (10.2.3.4).

► Example 10.14. Consider the *system of gas dynamic type equations*

$$f_1(u,v)u_t + f_2(u,v)u_x + f_3(u,v)v_t + f_4(u,v)v_x = 0,$$
$$g_1(u,v)u_t + g_2(u,v)u_x + g_3(u,v)v_t + g_4(u,v)v_x = 0.$$

Treating u, v as the independent variables and x, t as the dependent ones, we arrive at the following system of linear PDEs (the calculations are similar to those performed in Example 10.11):

$$-f_1(u,v)x_v + f_2(u,v)t_v + f_3(u,v)x_u - f_4(u,v)t_u = 0,$$
$$-g_1(u,v)x_v + g_2(u,v)t_v + g_3(u,v)x_u - g_4(u,v)t_u = 0.$$

◄

Remark 10.11. *Hodograph transformations can be utilized to linearize certain PDE systems in three or more dependent and independent variables (see Lei & Chang, 1996).*

◆ *Problems and Exercises for Section 10.2*

1. Find the general solutions in implicit form for the following nonlinear first-order equations:
a) $u_t = f(t,u)u_x$,
b) $u_t = f(x,u)u_x$.
Hint. Use the hodograph transformation described in Example 10.6.

2. Transform the nonlinear first-order equation

$$u_t = f(u)u_x + g(u)$$

into a linear equation.

Hint. Use the hodograph transformation described in Example 10.6.

3. Transform the nonlinear second-order equation

$$u_t u_x^2 = f(t,u)u_{xx}$$

into a linear equation.

Hint. Use the hodograph transformation described in Example 10.7.

4. Show that the equation of motion of a nonlinear viscoplastic medium

$$u_t = f(u_x)u_{xx}$$

can be reduced, using the hodograph transformation (10.2.1.1), to a similar equation with a different function $\tilde{f}(z)$.

5. Transform the nonlinear wave equation

$$u_{tt} = [f(u)u_x]_x$$

into a linear equation.

Hint. Rewrite the equation as a system of equations and apply the hodograph transformation as in Example 10.11.

It is noteworthy that Zaitsev & Polyanin (2001) and Polyanin & Zaitsev (2012) present parametric exact solutions of the original equation, obtained using a hodograph transformation.

6. Transform the nonlinear steady-state heat equation in an anisotropic medium

$$[f(u)u_x]_x + [g(u)u_y]_y = 0$$

into a linear equation.

Hint. By renaming the variable $y \Rightarrow t$, rewrite the equation as a system of equations and apply the hodograph transformation as in Example 10.11.

7. Using the hodograph transformation (10.2.1.1), show that the general solution of nonlinear second-order equation

$$u_x u_{xt} - u_t u_{xx} = 0$$

can be represented in the implicit form $x = \varphi(t) + \psi(u)$, where $\varphi(t)$ and $\psi(u)$ are arbitrary functions.

It is noteworthy that on solving this solution for u, one can rewrite the general solution in explicit form as $u = F(x - \varphi(t))$, where $F(z)$ and $\varphi(t)$ are arbitrary functions (see also Zwillinger, 1997).

8. Using the hodograph transformation (10.2.1.1), find the general solution of nonlinear second-order equation

$$u_x u_{xt} - u_t u_{xx} = f(u).$$

9. Reduce the order of the boundary layer equation with a pressure gradient

$$u_y u_{xy} - u_x u_{yy} = \nu u_{yyy} + f(x), \qquad \nu = \text{const}.$$

Hint. Use the hodograph transformation described in Example 10.8.

10. Reduce the order of the boundary layer equation for a non-Newtonian fluid

$$u_y u_{xy} - u_x u_{yy} = [f(u_{yy})]_y,$$

where $f(z)$ is an arbitrary function.

Hint. Use the hodograph transformation described in Example 10.8 (see also Polyanin & Zaitsev, 2012).

11. Transform the nonlinear system of equations

$$\begin{aligned} \rho_t + (\rho v)_x &= 0, \\ (\rho v)_t + [\rho v^2 + p(\rho)]_x &= 0, \end{aligned}$$

which describes one-dimensional adiabatic flows of an ideal gas, into a linear system. Here, v and ρ are the unknowns and $p = p(\rho)$ is a given function (for example, $p = A\rho^\gamma$ for an ideal polytropic gas).

Hint. Take v and ρ to be the independent variables and x and t as the dependent ones. This system is a special case of the system considered in Example 10.14 with $u = \rho$.

10.3. Contact Transformations. Legendre and Euler Transformations

10.3.1. Preliminary Remarks. Contact Transformations for Ordinary Differential Equations

Prior to describing contact transformations for partial differential equations, it is helpful to first consider their application to ordinary differential equations, as they are simpler.

1°. Consider functions of a single variable, $u = u(x)$. In general, a contact transformation has the form

$$x = X(\xi, w, w'_\xi), \qquad u = U(\xi, w, w'_\xi), \tag{10.3.1.1}$$

where the functions $X(\xi, w, z)$ and $U(\xi, w, z)$ are chosen such that the derivative u'_x does not depend on $w''_{\xi\xi}$:

$$u'_x = \frac{u'_\xi}{x'_\xi} = \frac{U_\xi + U_w w'_\xi + U_z w''_{\xi\xi}}{X_\xi + X_w w'_\xi + X_z w''_{\xi\xi}} = H(\xi, w, w'_\xi). \tag{10.3.1.2}$$

The subscripts ξ, w, and z after X and U denote the respective partial derivatives (it is assumed that $X_z \not\equiv 0$ and $U_z \not\equiv 0$).

It follows from (10.3.1.2) that the relation

$$U_z(X_\xi + zX_w) - X_z(U_\xi + zU_w) = 0, \tag{10.3.1.3}$$

which connects the functions X and U, must hold; the derivative is calculated by

$$u'_x = \frac{U_z}{X_z}, \tag{10.3.1.4}$$

where $U_z/X_z \not\equiv \text{const}$.

2°. Suppose the function $X = X(\xi, w, z)$ in the contact transformation (10.3.1.1) is specified. Then relation (10.3.1.3) can be viewed as a linear partial differential equation for the second function U. The corresponding characteristic system of ordinary differential equations (see Section 2.1.1),

$$\frac{d\xi}{1} = \frac{dw}{z} = -\frac{X_z\,dz}{X_\xi + zX_w}, \tag{10.3.1.5}$$

admits the obvious first integral

$$X(\xi, w, z) = C_1, \tag{10.3.1.6}$$

which is a corollary of the second equation in (10.3.1.5); C_1 is an arbitrary constant. It follows that, to obtain the general representation of the function $U = U(\xi, w, z)$, one has to deal with the ODE

$$w'_\xi = z, \tag{10.3.1.7}$$

which follows from the first equation in (10.3.1.5) and whose right-hand side is defined implicitly by (10.3.1.6). Let the first integral of equation (10.3.1.7) have the form

$$\Phi(\xi, w, C_1) = C_2.$$

Then the general representation of $U = U(\xi, w, z)$ in transformation (10.3.1.1) is given by (Polyanin & Zaitsev, 2018):

$$U = \Psi(X, \Phi),$$

where $\Psi(X, \Phi)$ is an arbitrary function of two variables, $X = X(\xi, w, z)$, and $\Phi = \Phi(\xi, w, X)$.

The application of contact transformations preserves the order of differential equations. The inverse of a contact transformation can be obtained by solving system (10.3.1.1) and (10.3.1.4) for ξ, w, and w'_ξ.

3°. Below, we will present several examples of contact transformations and provide expressions for the derivatives of u'_x and the corresponding inverse transformations.

► Example 10.15. Legendre transformation:

$$\begin{aligned} &x = w'_\xi, \quad u = \xi w'_\xi - w, \quad u'_x = \xi \quad \text{(direct transformation)};\\ &\xi = u'_x, \quad w = x u'_x - u, \quad w'_\xi = x \quad \text{(inverse transformation)}. \end{aligned} \tag{10.3.1.8}$$

The second derivative is calculated as follows:

$$u''_{xx} = (u'_x)'_x = \frac{(u'_x)'_\xi}{x'_\xi} = \frac{1}{w''_{\xi\xi}}.$$

The Legendre transformation allows simplifying some ODEs. For example, the first-order nonlinear equation

$$x f(u'_x) + u g(u'_x) + h(u'_x) = 0$$

is reduced by the Legendre transformation (10.3.1.8) to the first-order linear ODE:

$$[f(\xi) + \xi g(\xi)] w'_\xi - g(\xi) w + h(\xi) = 0.$$

The Legendre transformation (10.3.1.8) allows us to rewrite a general second-order ODE:

$$X(x,\, u,\, u'_x,\, u''_{xx}) = 0 \tag{10.3.1.9}$$

in the form

$$X\Big(w'_\xi,\, \xi w'_\xi - w,\, \xi,\, \frac{1}{w''_{\xi\xi}}\Big) = 0. \tag{10.3.1.10}$$

Sometimes equation (10.3.1.10) may be simpler than the original ODE (10.3.1.9); see Exercise 2 at the end of Section 10.3.

It is important to note that the Legendre transformation, like any other contact transformations, does not depend on the type and form of ODEs, for the analysis and solution of which they can be used. ◄

► Example 10.16. Contact transformation ($a \neq 0$):

$$x = w'_\xi + aw, \quad u = be^{a\xi} w'_\xi, \quad u'_x = be^{a\xi} \quad \text{(direct transformation);}$$
$$\xi = \frac{1}{a} \ln \frac{u'_x}{b}, \quad w = \frac{1}{a}\Big(x - \frac{u}{u'_x}\Big), \quad w'_\xi = \frac{u}{u'_x} \quad \text{(inverse transformation).}$$

◄

► Example 10.17. Contact transformation:

$$x = w'_\xi + a\xi, \quad u = \tfrac{1}{2}(w'_\xi)^2 + aw, \quad u'_x = w'_\xi \quad \text{(direct transformation);}$$
$$\xi = \frac{1}{a}(x - u'_x), \quad w = \frac{1}{2a}\big[2u - (u'_x)^2\big], \quad w'_\xi = u'_x \quad \text{(inverse transformation).}$$

◄

Remark 10.12. *For examples of other contact transformations for ODEs, see Zwillinger (1997) and Polyanin & Zaitsev (2018).*

10.3.2. General Form of Contact Transformations for Partial Differential Equations

We consider functions of two variables: $u = u(x, y)$. A common property of contact transformations is the dependence of the original variables on the new variables and their first derivatives:

$$x = X(\xi, \eta, w, w_\xi, w_\eta), \quad y = Y(\xi, \eta, w, w_\xi, w_\eta), \quad u = U(\xi, \eta, w, w_\xi, w_\eta). \tag{10.3.2.1}$$

The functions X, Y, and U cannot be arbitrary and are selected so as to ensure that the first derivatives of the original variables depend only on the transformed variables and, possibly, their first derivatives:

$$u_x = V(\xi, \eta, w, w_\xi, w_\eta), \quad u_y = W(\xi, \eta, w, w_\xi, w_\eta). \tag{10.3.2.2}$$

The differential relations that connect the functions X, Y, and U in (10.3.2.1) and the corresponding functions V and W in (10.3.2.2) can be obtained in a similar way to that used for ODEs in Section 10.3.1 (for details, see Polyanin & Zaitsev, 2012).

Basic properties of contact transformations:

(i) Contact transformations do not increase the order of the equations to which they are applied.

(ii) Contact transformations do not depend on the type or form of PDEs for the analysis of which they can be used.

In general, a contact transformation (10.3.2.1)–(10.3.2.2) reduces a second-order PDE in two independent variables

$$F(x, y, u, u_x, u_y, u_{xx}, u_{xy}, u_{yy}) = 0 \tag{10.3.2.3}$$

to a second-order PDE of the form

$$G(\xi, \eta, w, w_\xi, w_\eta, w_{\xi\xi}, w_{\xi\eta}, w_{\eta\eta}) = 0. \tag{10.3.2.4}$$

In some cases, equation (10.3.2.4) turns out to be more simple than (10.3.2.3). If $w = w(\xi, \eta)$ is a solution of equation (10.3.2.4), then formulas (10.3.2.1) define the corresponding solution of equation (10.3.2.3) in parametric form.

The next two sections discuss the two most important contact transformations and provide examples of their application to specific nonlinear PDEs.

10.3.3. Euler Transformation

The *Euler transformation* belongs to the class of contact transformations and is defined by the relations (Kamke, 1965):

$$x = w_\xi, \quad y = \eta, \quad u = \xi w_\xi - w. \tag{10.3.3.1}$$

Differentiating the last relation in (10.3.3.1) with respect to x and y and taking into account the other two relations, we find that

$$u_x = \xi, \quad u_y = -w_\eta. \tag{10.3.3.2}$$

Differentiating these expressions in x and y, we find the second derivatives:

$$u_{xx} = \frac{1}{w_{\xi\xi}}, \quad u_{xy} = -\frac{w_{\xi\eta}}{w_{\xi\xi}}, \quad u_{yy} = \frac{w_{\xi\eta}^2 - w_{\xi\xi}w_{\eta\eta}}{w_{\xi\xi}}. \tag{10.3.3.3}$$

The subscripts indicate the corresponding partial derivatives.

The Euler transformation (10.3.3.1)–(10.3.3.3) is employed in finding solutions and linearization of certain nonlinear partial differential equations.

Remark 10.13. *When applying Euler transformations, certain integrals may vanish if, in some subdomain, the second derivative u_{xx} (or u_{yy}) is identically zero.*

Remark 10.14. *For functions of one variable, $u = u(x)$, the Euler transformation (10.3.3.2) coincides with the Legendre transformation (10.3.1.8), which is applied in the theory of ordinary differential equations.*

► Example 10.18. The first-order nonlinear PDE

$$xf(y, u_x) + u_y g(y, u_x) + h(y, u_x) = 0$$

is reduced by the Euler transformation (10.3.3.1) to the linear first-order PDE

$$f(\eta, \xi)w_\xi - g(\eta, \xi)w_\eta + h(\eta, \xi) = 0.$$ ◄

The Euler transformation (10.3.3.1) allows us to reduce a general second-order PDE with two independent variables

$$F(x, y, u, u_x, u_y, u_{xx}, u_{xy}, u_{yy}) = 0 \tag{10.3.3.4}$$

to the equation

$$F\left(w_\xi, \eta, \xi w_\xi - w, \xi, -w_\eta, \frac{1}{w_{\xi\xi}}, -\frac{w_{\xi\eta}}{w_{\xi\xi}}, \frac{w_{\xi\eta}^2 - w_{\xi\xi}w_{\eta\eta}}{w_{\xi\xi}}\right) = 0. \qquad (10.3.3.5)$$

In some cases, equation (10.3.3.5) may become simpler than (10.3.3.4).

Let $w = w(\xi, \eta)$ be a solution of PDE (10.3.3.5). Then formulas (10.3.3.1) define the corresponding solution of the original PDE (10.3.3.4) in parametric form.

► Example 10.19. The second-order PDE with a quadratic nonlinearity

$$u_y u_{xx} + a = 0$$

using the Euler transformation (10.3.3.1)–(10.3.3.3) is reduced to the linear heat equation

$$w_\eta = a w_{\xi\xi}.$$

◄

► Example 10.20. A more complex second-order PDE with quadratic nonlinearity

$$u_{xy} = a u_y u_{xx}, \qquad (10.3.3.6)$$

can be linearized by the Euler transformation (10.3.3.1)–(10.3.3.3) to obtain

$$w_{\xi\eta} = a w_\eta.$$

This PDE is reduced with the substitution $v = w_\eta$ to a first-order linear ODE. Considering the aforesaid and integrating, we find that

$$w = f(\xi) + g(\eta) e^{a\xi}, \qquad (10.3.3.7)$$

where $f(\xi)$ and $g(\eta)$ are arbitrary functions.

Using formulas (10.3.3.1) and (10.3.3.7), we obtain the general solution of the original equation (10.3.3.6) in parametric form:

$$u = x\xi - f(\xi) - g(y)e^{a\xi},$$
$$x = f'_\xi(\xi) + a g(y) e^{a\xi}.$$

◄

► Example 10.21. The potential filtration equation

$$u_t = f(u_{xx})$$

can be reduced with the Euler transformation

$$x = w_\xi, \quad t = \eta, \quad u = x\xi - w \qquad (u_x = \xi, \quad u_t = -w_\eta)$$

to a similar equation (Ibragimov, 1994):

$$w_\eta = -f(1/w_{\xi\xi}).$$

◄

10.3.4. Legendre Transformation

An important special case of contact transformations is the *Legendre transformation* defined by the relations (see Kamke, 1965 and Courant & Hilbert, 1989):

$$x = w_\xi, \quad y = w_\eta, \quad u = \xi w_\xi + \eta w_\eta - w, \tag{10.3.4.1}$$

where w is the new dependent variable and ξ, η are the new independent variables.

Differentiating the last relation in (10.3.4.1) with respect to x and y and taking into account the other two relations, we obtain the first derivatives:

$$u_x = \xi, \quad u_y = \eta. \tag{10.3.4.2}$$

With (10.3.4.1)–(10.3.4.2), we find the second derivatives

$$u_{xx} = Jw_{\eta\eta}, \quad u_{xy} = u_{yx} = -Jw_{\xi\eta}, \quad u_{yy} = Jw_{\xi\xi},$$

where

$$J = u_{xx}u_{yy} - u_{xy}^2, \quad \frac{1}{J} = w_{\xi\xi}w_{\eta\eta} - w_{\xi\eta}^2.$$

The Legendre transformation (10.3.4.1), with $J \neq 0$, allows us to reduce a general second-order PDE with two independent variables

$$F(x, y, u, u_x, u_y, u_{xx}, u_{xy}, u_{yy}) = 0 \tag{10.3.4.3}$$

to the following equation (Courant & Hilbert, 1989):

$$F(w_\xi, w_\eta, \xi w_\xi + \eta w_\eta - w, \xi, \eta, Jw_{\eta\eta}, -Jw_{\xi\eta}, Jw_{\xi\xi}) = 0. \tag{10.3.4.4}$$

Sometimes equation (10.3.4.4) may be simpler than (10.3.4.3).

Let $w = w(\xi, \eta)$ be a solution of equation (10.3.4.4). Then the formulas (10.3.4.1) define the corresponding solution of equation (10.3.4.3) in parametric form.

Remark 10.15. *When applying the Legendre transformation, individual solutions may be lost if, in some subdomain, the Jacobian* $J = u_{xx}u_{yy} - u_{xy}^2$ *is identically zero.*

► Example 10.22. The first-order nonlinear PDE

$$xf(u_x, y_y) + yg(u_x, y_y) + h(u_x, y_y) = 0$$

is reduced by the Legendre transformation (10.3.4.1) to the linear first-order PDE:

$$f(\xi, \eta)w_\xi + g(\xi, \eta)w_\eta + h(\xi, \eta) = 0.$$ ◄

► Example 10.23. The equation of steady-state transonic gas flow

$$au_x u_{xx} + u_{yy} = 0$$

is reduced by the Legendre transformation (10.3.4.1) to the second-order linear PDE with variable coefficients

$$a\xi w_{\eta\eta} + w_{\xi\xi} = 0.$$ ◄

10.3.5. Euler–Legendre Transformation with Many Variables

In general, for a function of many variables $u = u(x_1, \dots, x_n)$, the Euler–Legendre transformation and its inverse are defined by the following formulas (Kamke, 1965):

Euler–Legendre transformation

$$x_1 = X_1, \quad \dots, \quad x_{k-1} = X_{k-1},$$
$$x_k = \frac{\partial U}{\partial X_k}, \quad \dots, \quad x_n = \frac{\partial U}{\partial X_n},$$
$$u(\mathbf{x}) = \sum_{i=k}^{n} X_i \frac{\partial U}{\partial X_i} - U(\mathbf{X}),$$

Inverse Euler–Legendre transformation

$$X_1 = x_1, \quad \dots, \quad X_{k-1} = x_{k-1},$$
$$X_k = \frac{\partial u}{\partial x_k}, \quad \dots, \quad X_n = \frac{\partial u}{\partial x_n},$$
$$U(\mathbf{X}) = \sum_{i=k}^{n} x_i \frac{\partial u}{\partial x_i} - u(\mathbf{x}),$$

where $\mathbf{x} = \{x_1, \dots, x_n\}$, $\mathbf{X} = \{X_1, \dots, X_n\}$, and the derivatives are related by

$$\frac{\partial u}{\partial x_1} = -\frac{\partial U}{\partial X_1}, \quad \dots, \quad \frac{\partial u}{\partial x_{k-1}} = -\frac{\partial U}{\partial X_{k-1}}.$$

The transformation described above generalizes the Euler and Legendre transformations, which are discussed in Sections 10.3.3 and 10.3.4.

❖ *Problems and Exercises for Section 10.3*

1. Show that the first-order nonlinear ODE

$$xf(y'_x) + yg(y'_x) + (xy' - y)^k h(y'_x) = 0$$

can be reduced with the Legendre transformation (10.3.1.8) to the Bernoulli equation.

2. Show that the second-order nonlinear ODE

$$y''_{xx} = \big[xf(y'_x) + yg(y'_x) + h(y'_x)\big]^{-1}$$

can be reduced with the Legendre transformation (10.3.1.8) to the second-order linear ODE.

3. Reduce the nonlinear second-order PDE

$$u_y u_{xx} = f(y, u_x)$$

to a linear equation by using the Euler transformation.

4. Convert the nonlinear second-order PDE

$$u_{xy} = f(y, u_x) u_y u_{xx}$$

into a linear equation by using the Euler transformation. By reducing the order of the resulting equation, find its general solution.

5. Show that the nonhomogeneous Monge–Ampère equation

$$u_{xx} u_{yy} - u_{xy}^2 = -a^2$$

can be reduced to a linear wave equation using the Euler transformation. Find the general solution of the original equation in parametric form.

6. Show that the Monge–Ampère type equation

$$u_{xx}u_{yy} - u_{xy}^2 = au_{xx}$$

is reducible to the linear ordinary differential equation $w_{\eta\eta} = -a$ using the Euler transformation. Find the general solution of the original equation in parametric form.

7. Using formulas (10.3.3.1) and (10.3.3.2) find the inverse Euler transform.

8. Using formulas (10.3.4.1) and (10.3.4.2) find the inverse Legendre transform.

9. Reduce the nonlinear second-order PDE

$$u_{xx} + au_x u_y u_{yy} = 0.$$

to a linear equation using the Legendre transformation.

10. Reduce the nonlinear second-order PDE

$$u_{tt} = f(u_x)u_{xx}$$

to a linear equation using the Legendre transformation.

11. Reduce the minimal surface equation (describing, for example, the shape of a soap film bounded by a given contour)

$$(1 + u_y^2)u_{xx} - 2u_x u_y u_{xy} + (1 + u_x^2)u_{yy} = 0$$

to a linear equation using the Legendre transformation.

12. Reduce the Born–Infeld equation

$$(1 - u_t^2)u_{xx} + 2u_x u_t u_{xt} - (1 + u_x^2)u_{tt} = 0$$

to a linear equation using the Legendre transformation.

13. Show that the nonlinear potential hyperbolic heat equation

$$u_{tt} + u_t = f(u_x)u_{xx}, \qquad a \neq 0$$

can be reduced by the contact transformation

$$x = w + w_\eta, \quad t = \eta + \ln|w_\xi|, \quad u = \xi + w_\eta/w_\xi \quad (u_x = 1/w_\xi, \quad u_t = -w_\eta/w_\xi)$$

leads to an equation of a similar form (Svirshchevskii, 1986 and Ibragimov, 1994):

$$w_{\eta\eta} + w_\eta = F(w_\xi)w_{\xi\xi}, \qquad \text{where} \qquad F(w) = w^{-2}f(1/w).$$

14. Reduce the nonlinear second-order PDE

$$f(u_x, u_y)u_{xx} + g(u_x, u_y)u_{xy} + h(u_x, u_y)u_{yy} = 0$$

to a linear equation using the Legendre transformation.

15. Show that the Euler–Legendre contact transformation

$$t = T, \quad x = U_X, \quad y = U_Y, \quad u = XU_X + YU_Y - U$$

reduces the parabolic Monge–Ampère equation

$$u_t = u_{xx}u_{yy} - u_{xy}^2$$

to the equation

$$U_T(U_{XX}U_{YY} - U_{XY}^2) = -1.$$

Hint. Use the contact transformation with $n = 3$ and $k = 2$ from Section 10.3.5, setting $x_1 = t$, $x_2 = x$, $x_3 = y$, and $X_1 = T$ (refer to Aksenov & Polyanin, 2025).

10.4. Bäcklund Transformations. RF Pairs and Their Applications. Differential Substitutions

10.4.1. General Description and Examples of Bäcklund Transformations

For simplicity and clarity, we first describe the Bäcklund transformations as applied to second-order PDEs.

Let $u = u(x, y)$ be a solution of the PDE

$$F_1(x, y, u, u_x, u_y, u_{xx}, u_{xy}, u_{yy}) = 0, \qquad (10.4.1.1)$$

and let $w = w(x, y)$ be a solution of another PDE

$$F_2(x, y, w, w_x, w_y, w_{xx}, w_{xy}, w_{yy}) = 0. \qquad (10.4.1.2)$$

Equations (10.4.1.1) and (10.4.1.2) are said to be related by the *Bäcklund transformation*

$$\begin{aligned} &\Phi_1(x, y, u, u_x, u_y, w, w_x, w_y) = 0, \\ &\Phi_2(x, y, u, u_x, u_y, w, w_x, w_y) = 0 \end{aligned} \qquad (10.4.1.3)$$

if the compatibility of the pair (10.4.1.3) and PDE (10.4.1.1), implies equation (10.4.1.2), and the compatibility of the pair (10.4.1.3) and PDE (10.4.1.2) implies (10.4.1.1). If, for some specific solution $w = w(x, y)$ of equation (10.4.1.2), one succeeds in solving equations (10.4.1.3) for $u = u(x, y)$, then this function $u = u(x, y)$ will be a solution of equation (10.4.1.1).

Bäcklund transformations may preserve the form of equations* (such transformations are used for obtaining new solutions) or establish relations between solutions of different equations (such transformations are used for obtaining solutions of one equation from solutions of another equation).

If a solution w to equation (10.4.1.2) is known, the corresponding solution u to equation (10.4.1.1) can be obtained by solving the overdetermined system of first-order partial differential equations (10.4.1.3). This involves solving the simplest equation first and then eliminating any resulting excess functional arbitrariness by substituting the obtained solution into another equation. Similarly, a solution w of equation (10.4.1.2) can be constructed from a given solution u of equation (10.4.1.1).

► Example 10.24. The *Burgers equation*

$$u_t + uu_x = u_{xx} \qquad (10.4.1.4)$$

is related to the linear heat equation

$$w_t = w_{xx} \qquad (10.4.1.5)$$

* In such cases, these are referred to as *auto-Bäcklund transformations*.

by the Bäcklund transformation (e.g., see Ibragimov, 1994):

$$\begin{aligned} &w_x + \tfrac{1}{2}wu = 0, \\ &w_t + \tfrac{1}{2}(wu)_x = 0. \end{aligned} \tag{10.4.1.6}$$

Eliminating u from (10.4.1.6), we obtain equation (10.4.1.5).

Conversely, let $w(x,t)$ be a nonzero solution of the heat equation (10.4.1.5). Dividing (10.4.1.5) by w, differentiating the resulting equation with respect to x, and taking into account that $(w_t/w)_x = (w_x/w)_t$, we obtain

$$\left(\frac{w_x}{w}\right)_t = \left(\frac{w_{xx}}{w}\right)_x. \tag{10.4.1.7}$$

From the first equation in (10.4.1.6) we have

$$\frac{w_x}{w} = -\frac{u}{2} \quad\Longrightarrow\quad \frac{w_{xx}}{w} - \left(\frac{w_x}{w}\right)^2 = -\frac{u_x}{2} \quad\Longrightarrow\quad \frac{w_{xx}}{w} = -\frac{u_x}{2} + \frac{1}{4}u^2. \tag{10.4.1.8}$$

Replacing the expressions in parentheses in (10.4.1.7) with the right-hand sides of the first and the last relation (10.4.1.8), we obtain the Burgers equation (10.4.1.4).

The first relation in (10.4.1.6) is the *Hopf–Cole transformation* (see also Example 10.34). It can be rewritten as a differential substitution by solving (10.4.1.6) for u, which leads to formula 10.4.3.9. ◄

► Example 10.25. Let us demonstrate that *Liouville's equation*

$$u_{xy} = e^{\lambda u} \tag{10.4.1.9}$$

is connected to the linear wave equation

$$w_{xy} = 0 \tag{10.4.1.10}$$

by the Bäcklund transformation (e.g., see Ibragimov, 1994):

$$\begin{aligned} w_x - u_x &= \frac{2k}{\lambda}\exp\left[\frac{1}{2}\lambda(u+w)\right], \\ w_y + u_y &= -\frac{1}{k}\exp\left[\frac{1}{2}\lambda(u-w)\right], \end{aligned} \tag{10.4.1.11}$$

where $k \neq 0$ is an arbitrary constant.

Indeed, let us differentiate the first relation of (10.4.1.11) with respect to y and the second one with respect to x. Then, taking into account that $w_{yx} = w_{xy}$ and $u_{yx} = u_{xy}$ and eliminating the combinations of the first derivatives using (10.4.1.11), we obtain

$$\begin{aligned} w_{xy} - u_{xy} &= k\exp\left[\frac{1}{2}\lambda(u+w)\right](w_y + u_y) = -\exp(\lambda u), \\ w_{xy} + u_{xy} &= \frac{\lambda}{2k}\exp\left[\frac{1}{2}\lambda(u-w)\right](w_x - u_x) = \exp(\lambda u). \end{aligned} \tag{10.4.1.12}$$

Adding relations (10.4.1.12) together, we get the linear PDE (10.4.1.10). Subtracting the latter equation from the former gives the nonlinear equation (10.4.1.9). ◄

More complex Bäcklund transformations exist than (10.4.1.3), which include second-order derivatives and connect third-order PDEs. To illustrate this, let us consider a specific example.

► Example 10.26. The Korteweg–de Vries equation

$$u_t - 6uu_x + u_{xxx} = 0 \tag{10.4.1.13}$$

and the modified Korteweg–de Vries equation

$$w_t - 6w^2 w_x + w_{xxx} = 0 \tag{10.4.1.14}$$

are related by the Bäcklund transformation (Ibragimov, 1994):

$$\begin{aligned} w_x &= w^2 - u, \\ w_t &= u_{xx} + 2(uw)_x. \end{aligned} \tag{10.4.1.15}$$

Eliminating u from (10.4.1.15) yields equation (10.4.1.14).

Now, differentiating the equations (10.4.1.15) with respect to t and x, eliminating the mixed derivative $w_{xt} = w_{tx}$ from the resulting relations, and performing simple rearrangements, we obtain

$$u_t + u_{xxx} + 2uw_{xx} + 4u_x w_x - 4uww_x - 4w^2 u_x = 0. \tag{10.4.1.16}$$

Differentiating the first relation in (10.4.1.15) with respect to x, we find that

$$w_{xx} = 2ww_x - u_x = 2w(w^2 - u) - u_x = 2w^3 - 2uw - u_x. \tag{10.4.1.17}$$

On eliminating the derivatives w_x and w_{xx} from (10.4.1.16) with the help of the first relation in (10.4.1.15) and expression (10.4.1.17), we arrive at the Korteweg–de Vries equation (10.4.1.13).

The first relation in (10.4.1.15) is the *Miura transformation* (Miura, 1976a). It can be rewritten as a differential substitution by solving (10.4.1.15) for u, which leads to formula $u = w^2 - w_x$. ◄

Remark 10.16. *It is significant that unlike the contact transformations, the Bäcklund transformations are determined by the specific equations (a Bäcklund transformation that relates given equations does not always exist).*

Remark 10.17. *For additional information, examples, and applications of Bäcklund transformations, for instance, see Nimmo & Crighton (1982), Rogers & Shadwick (1982), Ibragimov (1994), Rogers & Schief (2002), and Polyanin & Zaitsev (2012); also refer to Section 10.4.2.*

10.4.2. RF Pairs and Their Use for Constructing Bäcklund Transformations

Outlined below is the method for constructing Bäcklund transformations for partial differential equations based on using RF pairs.

1°. Prior to describing RF pairs for partial differential equations, we describe them for simpler, ordinary differential equations. This order of presentation is chosen for better understanding of the subsequent material.

An RF pair is an operation of consecutive raising and lowering of the order of an equation (RF stands for "Rise" and "Fall"). The nth-order ordinary differential equation of general form

$$F(x,u,u'_x,u''_{xx},\dots,u_x^{(n)})=0 \tag{10.4.2.1}$$

admits two basic RF pairs, which are considered below.

The first RF pair. On solving equation (10.4.2.1) for the dependent variable (which is assumed to be feasible), we get

$$\Phi(x,u'_x,u''_{xx},\dots,u_x^{(n)})=u. \tag{10.4.2.2}$$

The differentiation with respect to x takes (10.4.2.2) to an $(n+1)$st-order equation, which does not depend on u explicitly. By lowering the order of the resulting equation using the standard substitution

$$\xi(x)=u'_x, \tag{10.4.2.3}$$

we arrive at the nth-order equation

$$\frac{d}{dx}\Phi(x,\xi,\xi'_x,\dots,\xi_x^{(n-1)})=\xi. \tag{10.4.2.4}$$

Equations (10.4.2.2) and (10.4.2.4) are connected by the Bäcklund transformation

$$\begin{aligned}\Phi(x,\xi,\xi'_x,\dots,\xi_x^{(n-1)})&=u,\\ u'_x&=\xi.\end{aligned} \tag{10.4.2.5}$$

If we know a solution $u=u(x)$ to equation (10.4.2.2) (or (10.4.2.1)), then we can find a solution $\xi=\xi(x)$ to equation (10.4.2.4) using the second relation in (10.4.2.5). If we know a solution $\xi=\xi(x)$ to equation (10.4.2.4), then we obtain a solution $u=u(x)$ to equation (10.4.2.2) using the first relation in (10.4.2.5).

The second RF pair. On solving equation (10.4.2.1) for the independent variable (which is assumed to be feasible), we get

$$\Psi(u,u'_x,u''_{xx},\dots,u_x^{(n)})=x. \tag{10.4.2.6}$$

Differentiating (10.4.2.6) with respect to x yields an $(n+1)$st-order autonomous equation, whose order is further reduced using the standard substitution

$$\eta(u)=u'_x.$$

As a result, we arrive at the nth-order equation

$$\eta\frac{d}{du}\Psi(u,\eta,\eta\eta'_u,\dots,\eta_x^{(n-1)})=1. \tag{10.4.2.7}$$

The derivatives with respect to x are found successively using the following rule: $\eta_x^{(k)} = \eta(\eta_x^{(k-1)})_u$. The derivative with respect to u on the left-hand side of (10.4.2.7) is calculated by the chain rule taking into account that F depends on u, η, η_u, ...

Equations (10.4.2.6) and (10.4.2.7) are connected by the Bäcklund transformation

$$\begin{aligned}\Psi(u, \eta, \eta'_x, \dots, \eta_x^{(n-1)}) &= x,\\ u'_x &= \eta.\end{aligned} \qquad (10.4.2.8)$$

A considerable number of new integrable ordinary differential equations have been found in a systematic manner using RF pair operations and associated Bäcklund transformations (Zaitsev & Polyanin, 1994 and Polyanin & Zaitsev, 2018).

The above ideas and results for ordinary differential equations turn out to be transferrable, when appropriately modified, to nonlinear partial differential equations.

2°. Consider the nth-order evolution equation

$$u_t = s(t)u + F\big(t, x, u_x, u_{xx}, \dots, u_x^{(n)}\big). \qquad (10.4.2.9)$$

PDE (10.4.2.9) has the following general property: if $\widetilde{u}(t, x)$ is a solution to this equation, then the function

$$u = \widetilde{u}(t, x) + C\exp\biggl(\int s(t)\,dt\biggr),$$

where C is an arbitrary constant, is also a solution to (10.4.2.9).

Equation (10.4.2.9) does not depend on u explicitly if $s(t) \equiv 0$. If the unknown function u as well as the functions s and F on the right-hand side are independent of t, then equation (10.4.2.9) becomes (10.4.2.2) (after F has been renamed $-s\Phi$).

By analogy with ordinary differential equations, let us apply the first RF pair to equation (10.4.2.9).

Differentiating equation (10.4.2.9) with respect to x yields the $(n + 1)$st-order equation

$$u_{xt} = s(t)u_x + \frac{\partial}{\partial x}F\big(t, x, u_x, u_{xx}, \dots, u_x^{(n)}\big), \qquad (10.4.2.10)$$

which does not depend on u explicitly. In equation (10.4.2.10), let us change from the old variables to the new ones (analogue of transformation (10.4.2.3)):

$$t,\ x,\ u = u(t, x) \quad\Longrightarrow\quad t,\ x,\ \xi = u_x, \quad \text{where} \quad \xi = \xi(t, x). \qquad (10.4.2.11)$$

The evolution nth-order equation for $\xi = \xi(x, t)$ has the form

$$\xi_t = s(t)\xi + \frac{\partial}{\partial x}F\big(t, x, \xi, \xi_x, \dots, \xi_x^{(n-1)}\big). \qquad (10.4.2.12)$$

The derivative with respect to x on the right-hand side of (10.4.2.12) is obtained by the chain rule taking into account that F depends on x, ξ, ξ_x, ...

Equations (10.4.2.9) and (10.4.2.12) are connected by the Bäcklund transformation

$$u_t = s(t)u + F(t, x, \xi, \xi_x, \dots, \xi_x^{(n-1)}),\\ u_x = \xi. \qquad (10.4.2.13)$$

If a solution to equation (10.4.2.9) is known, one can find a solution to equation (10.4.2.12) using the second relation in (10.4.2.13). Given a solution to equation (10.4.2.12), one obtains a solution to equation (10.4.2.9) by solving the linear first-order system (10.4.2.13) for u (by virtue of (10.4.2.12), the right-hand sides of the system satisfy the solvability condition).

Remark 10.18. *By specifying an integrable equation of the form (10.4.2.9) using transformation (10.4.2.13), one can obtain an integrable equation of the form (10.4.2.12).*

► Example 10.27. Consider the nonlinear equation

$$u_t = au_{xx} + bu_x^2 + f(x,t)u_x + g(x,t). \qquad (10.4.2.14)$$

With the transformation

$$Z = \exp\Bigl(\frac{b}{a}u\Bigr) \qquad (10.4.2.15)$$

it can be reduced to the linear equation

$$Z_t = aZ_{xx} + f(x,t)Z_x + \frac{b}{a}g(x,t)Z. \qquad (10.4.2.16)$$

Note that if $f(x,t) = f_1(t)x + f_0(t)$ and $g(x,t) = g_1(t)x + g_0(t)$, the linear PDE (10.4.2.16) can be reduced to the classical heat equation (see Polyanin & Nazaikinskii, 2016, p. 383).

The Bäcklund transformation (10.4.2.13) corresponding to equation (10.4.2.14) has the form

$$u_t = a\xi_x + b\xi^2 + f(x,t)\xi + g(x,t),\\ u_x = \xi. \qquad (10.4.2.17)$$

Eliminating u from (10.4.2.17), we arrive at the nonlinear equation

$$\xi_t = a\xi_{xx} + 2b\xi\xi_x + [f(x,t)\xi]_x + g_x(x,t), \qquad (10.4.2.18)$$

which generalizes the Burgers equation (corresponding to the special case where $f = g = 0$). Replacing u in (10.4.2.17) by Z using (10.4.2.15), we obtain the Bäcklund transformation

$$Z_t = \frac{b}{a}Z[a\xi_x + b\xi^2 + f(x,t)\xi + g(x,t)],\\ Z_x = \frac{b}{a}Z\xi$$

that connects the nonlinear PDE (10.4.2.18) to the linear PDE (10.4.2.16). ◄

► Example 10.28. The nonlinear third-order equation

$$u_t + u_{xxx} - 3u_x^2 + \frac{1}{2t}u = 0$$

can be reduced, with a Bäcklund transformation of the form (10.4.2.13),

$$u_t = -\xi_{xx} + 3\xi^2 - \frac{1}{2t}u,$$
$$u_x = \xi$$

to the cylindrical Korteweg–de Vries equation

$$\xi_t + \xi_{xxx} - 6\xi\xi_x + \frac{1}{2t}\xi = 0.$$

◀

► Example 10.29. Consider the nonlinear equation

$$u_t = f(u_x)u_{xx}.$$

It arises in the nonlinear theory of flows in porous media and also governs the motion of a nonlinear viscoplastic medium.

The Bäcklund transformation of the form (10.4.2.13),

$$u_t = f(\xi)\xi_x,$$
$$u_x = \xi,$$

reduces the original equation to the nonlinear heat equation

$$\xi_t = [f(\xi)\xi_x]_x. \tag{10.4.2.19}$$

For the special case $f(\xi) = a\xi^{-2}$, the original PDE can be reduced with the hodograph transformation to the linear heat equation (see Example 10.7). Therefore, the resulting PDE (10.4.2.19) also admits linearization. ◀

3°. Consider the following nth-order evolution equation of a quite general form:

$$u_t = s(t)xu_x + F\big(t, u, u_x, u_{xx}, \ldots, u_x^{(n)}\big). \tag{10.4.2.20}$$

Equation (10.4.2.20) includes as special cases the Burgers, Korteweg–de Vries, and Harry Dym equations as well as nonlinear heat equations and other nonlinear equations of mathematical physics. If the unknown u and the right-hand side functions s and F are independent of t, then equation (10.4.2.20) reduces to (10.4.2.6) (with F renamed $-su_x\Psi$).

Equation (10.4.2.20) has the following general property: if $\widetilde{u}(t,x)$ is a solution of the equation, then

$$u = \widetilde{u}(t, x + \psi(t)), \quad \psi(t) = C\exp\left(-\int s(t)\,dt\right),$$

where C is an arbitrary constant, is also a solution.

By analogy with ordinary differential equations, let us apply the second RF pair to equation (10.4.2.20). The formula

$$\frac{\partial}{\partial x}\left(\frac{u_t}{u_x}\right) \equiv \frac{u_x u_{tx} - u_t u_{xx}}{u_x^2} = \frac{\eta_t}{\eta} \tag{10.4.2.21}$$

will be required later; it is obtained by differentiating u_t/u_x with respect to x followed by changing from t, x, and u to the von Mises variables (for more information about this transformation, see in Section 10.5.1):

$$t,\ x,\ u=u(t,x) \quad\Longrightarrow\quad t,\ u,\ \eta=\eta(t,u), \quad \text{where} \quad \eta=u_x. \tag{10.4.2.22}$$

Dividing equation (10.4.2.20) by u_x and differentiating with respect to x, we obtain an $(n+1)$st-order equation whose left-hand side coincides with (10.4.2.21). Changing to the von Mises variables followed by elementary rearrangements leads to the nth-order equation

$$\eta_t = s(t)\eta + \eta^2 \frac{\partial}{\partial u}\left(\frac{F}{\eta}\right), \qquad F = F(t,u,\eta,\eta\eta_u,\dots,\eta_x^{(n-1)}), \tag{10.4.2.23}$$

where the partial derivatives with respect to x are calculated successively with the formula $\eta_x^{(k)} = \eta(\eta_x^{(k-1)})_u$. On the right-hand side of (10.4.2.23), the derivative with respect to u is found by the chain rule taking into account the fact that F depends on u, η, η_u, ...

It follows that equations (10.4.2.20) and (10.4.2.23) are connected by the Bäcklund transformation (Polyanin & Zhurov, 2012a):

$$\begin{aligned} u_t &= s(t)x\eta + F(t,u,\eta,\eta_x,\dots,\eta_x^{(n-1)}),\\ u_x &= \eta. \end{aligned} \tag{10.4.2.24}$$

Any solution $u=g(t,x)$ to equation (10.4.2.20) determines a solution to equation (10.4.2.23), which can be written in parametric form as $u=g(t,x)$, $\eta=g_x(t,x)$.

Suppose $\eta=\eta(t,u)$ is a solution to equation (10.4.2.23). Let us calculate the spatial derivatives $\eta_x=\eta\eta_u$, $\eta_{xx}=\eta(\eta\eta_u)_u$, ... and substitute them into the right-hand side of equation (10.4.2.24). As a result, the first equation in (10.4.2.24) becomes $u_t=G(t,x,u)$, where $G(t,x,u)$ is a known function of three variables. This means that system (10.4.2.24) has been transformed into a system of first-order equations for $u=u(t,x)$.

Remark 10.19. *Instead of solving the above system of first-order equations, one can proceed in a simpler way. Substituting the solution $\eta=\eta(t,u)$ of equation (10.4.2.23) into the second equation in (10.4.2.24) and integrating, one obtains*

$$\int \frac{du}{\eta(t,u)} = x + \varphi(t), \tag{10.4.2.25}$$

where $\varphi(t)$ is an arbitrary function. Further, on solving (10.4.2.25) for $u=u(t,x)$, one should substitute the resulting expression into the original equation (10.4.2.20) and find the function $\varphi(t)$.

New integrable equations can be obtained as follows. Starting from an integrable equation of the form (10.4.2.20), one can use the Bäcklund transformation (10.4.2.24) to arrive at another integrable equation (10.4.2.23). Conversely: if equation (10.4.2.23) is integrable for some function F, then the original equation (10.4.2.20) is also integrable.

Remark 10.20. *Sometimes, it may be helpful to use additional point transformations of equation (10.4.2.23). In particular, it may be convenient to represent equation (10.4.2.23) in the form*

$$\zeta_t = -s(t)\zeta - \frac{\partial}{\partial u}(\zeta F), \qquad \zeta = \frac{1}{\eta}.$$

► Example 10.30. Consider the generalized Burgers equation

$$u_t + f(u)u_x = au_{xx}. \tag{10.4.2.26}$$

The Bäcklund transformation (10.4.2.24) with $F = au_{xx} - f(u)u_x = a\eta\eta_u - f(u)\eta$ and $s(t) \equiv 0$ takes (10.4.2.26) to the equation $\eta_t = a\eta^2\eta_{uu} - f'_u(u)\eta^2$, which can be reduced, with the substitution $\eta = 1/\zeta$, to the nonlinear heat equation with a source (Polyanin & Zhurov, 2012a):

$$\zeta_t = a(\zeta^{-2}\zeta_u)_u + f'_u(u). \tag{10.4.2.27}$$

Assuming that $f(u) = bu$ in (10.4.2.27), we have an integrable equation (it is reduced to the classical Burgers equation (10.4.2.26)), which was derived by Dorodnitsyn & Svirshchevskii (1983) using a different method. ◄

► Example 10.31. Consider the nonlinear third-order equation

$$u_t + f(u)u_x = au_{xxx}. \tag{10.4.2.28}$$

The Bäcklund transformation (10.4.2.24) with $F = au_{xxx} - f(u)u_x = a\eta(\eta\eta_u)_u - f(u)\eta$ and $s(t) \equiv 0$ reduces (10.4.2.28) to the equation $\eta_t = a\eta^2(\eta\eta_u)_{uu} - f'_u(u)\eta^2$. With the substitution $\eta = 1/\zeta$, this equation can be rewritten as (Polyanin & Zhurov, 2012a):

$$\zeta_t = a(\zeta^{-3}\zeta_u)_{uu} + f'_u(u). \tag{10.4.2.29}$$

For $f(u) \equiv 0$, the original equation (10.4.2.28) is linear. For $f(u) = bu$ and $f(u) = bu^2$, we have the Korteweg–de Vries and modified Korteweg–de Vries equations, respectively. By substituting these functions $f(u)$ into (10.4.2.29), we obtain new integrable equations. ◄

Remark 10.21. *By applying again a Bäcklund transformation (10.4.2.24) based on the von Mises transformation to equations (10.4.2.27) with $f(u) = bu$ and (10.4.2.29) with $f(u) = bu$, one obtains other integrable equations.*

Table 10.1 lists integrable equations obtained in Polyanin & Zhurov (2012a) by using the Bäcklund transformation (10.4.2.24) (in some cases, followed by an additional point transformation, specified in the last column). In the fifth row, the original PDE is linearized using the Euler transformation (see formulas (10.3.3.1)–(10.3.3.3), in which y must be replaced by t); the function $g(\xi)$ is defined parametrically.

Table 10.2 lists nonlinear equations obtained in Polyanin & Zaitsev (2012) by using the Bäcklund transformation (10.4.2.24) from some common classes of equations of mathematical physics. Whenever solutions to the original equations are known, one can construct exact solutions to the transformed equations.

Table 10.1. Some integrable nonlinear PDEs obtained using the Bäcklund transformation (10.4.2.24).

No.	Original integrable PDE (10.4.2.20)	Resulting integrable PDE (10.4.2.23)	Additional notation and remarks
1	$u_t = au_{xx} + bu$	$\zeta_t + b(u\zeta)_u = a(\zeta^{-2}\zeta_u)_u$	$\zeta = 1/\eta$, original PDE is linear
2	$u_t + buu_x = au_{xx}$	$\zeta_t = a(\zeta^{-2}\zeta_u)_u + b$	$\zeta = 1/\eta$, original PDE is the Burgers PDE
3	$u_t = u_{xx} + f(u)u_x^2$	$\eta_t = \eta^2\eta_{uu} + \eta^2[f(u)\eta]_u$	original PDE reduces to the linear heat equation by the substitution $v = \int \exp\bigl[\int f(u)\,du\bigr]\,du$
4	$u_t = au_{xx} - au_x^2 + be^u$	$\eta_t = a\eta^2\eta_{uu} - (a\eta^2 + be^u)\eta_u + be^u\eta$	original PDE reduces to a linear one by the substitution $v = e^{-u}$
5	$u_t = au_{xxx} + bu$	$\zeta_t + b(u\zeta)_u = a(\zeta^{-3}\zeta_u)_{uu}$	$\zeta = 1/\eta$, original PDE is linear
6	$u_t = au_{xxx} + bu_{xx}$	$\zeta_t = a(\zeta^{-3}\zeta_u)_{uu} + b(\zeta^{-2}\zeta_u)_u$	$\zeta = 1/\eta$, original PDE is linear
7	$u_t + buu_x = au_{xxx}$	$\zeta_t = a(\zeta^{-3}\zeta_u)_{uu} + b$	$\zeta = 1/\eta$, original PDE is the KdV equation
8	$u_t + bu^2u_x = au_{xxx}$	$\zeta_t = a(\zeta^{-3}\zeta_u)_{uu} + 2bu$	$\zeta = 1/\eta$, original PDE is the modified KdV equation
9	$u_t = au_{xxx} + bu_x^2$	$\theta_t = a\theta^{3/2}\theta_{uuu} + b\theta\theta_u$	$\theta = \eta^2$, original PDE is the potential KdV equation
10	$u_t = au_{xxx} + bu_x^3$	$\theta_t = a\theta^{3/2}\theta_{uuu} + 2b\theta^{3/2}\theta_u$	$\theta = \eta^2$, original PDE is the potential modified KdV equation
11	$u_t = au^3u_{xxx}$	$\theta_t = a\theta^{3/2}(u^3\theta_{uu})_u$	$\theta = \eta^2$, original PDE is the Harry Dym PDE
12	$u_t = au_{xxxx}$	$\theta_t = a\theta^{3/2}(\theta^{1/2}\theta_{uu})_{uu}$	$\theta = \eta^2$, original PDE is linear

4°. Transformations of the forms (10.4.2.13) and (10.4.2.24) are called *canonic Bäcklund transformations of evolution equations*.

One of the methods for finding other Bäcklund transformations involves the following. First, one looks for a point transformation that reduces the equation in question to an equation of the form (10.4.2.9) or (10.4.2.20). Then a canonical

Table 10.2. Some nonlinear equations obtained using the Bäcklund transformation (10.4.2.24) and related equations. Notation for auxiliary variables: $\zeta = 1/\eta$ and $\theta = \eta^{1/2}$.

Name of original equation	Original equation of the form (10.4.2.20)	Resulting PDE (10.4.2.23) and some other equations
Model equation of gas dynamics (with $g \equiv 0$)	$u_t + f(u)u_x = g(u)$	$\zeta_t + [g(u)\zeta]_u = f'_u(u)$ (linear equation)
Nonlinear heat equation with source	$u_t = [f(u)u_x]_x + g(u)$	$\eta_t = \eta^2[f(u)\eta]_{uu} + \eta^2[g(u)/\eta]_u$; $\zeta_t = [f(u)/\zeta]_{uu} - [g(u)\zeta]_u$
Generalized Burgers equation (case 1)	$u_t + f(u)u_x = g(u)u_{xx}$	$\eta_t = \eta^2[g(u)\eta_u]_u - \eta^2 f'_u(u)$; $\zeta_t = [g(u)\zeta^{-2}\zeta_u]_u + f'_u(u)$
Generalized Burgers equation (case 2)	$u_t + f(u)u_x = [g(u)u_x]_x$	$\eta_t = \eta^2[g(u)\eta]_{uu} - \eta^2 f'_u(u)$; $\zeta_t = [g(u)/\zeta]_{uu} + f'_u(u)$
Generalized equation of nonlinear theory of fluid flows through porous media	$u_t = f(u, u_x)u_{xx}$	$\eta_t = \eta^2[f(u,\eta)\eta_u]_u$
Heat equation for rheologically complex media	$u_t = [f(u, u_x)]_x$	$\eta_t = \eta^2[f(u,\eta)]_{uu}$; $\zeta_t = -[f(u, 1/\zeta)]_{uu}$
Korteweg–de Vries–Harry Dym type equation	$u_t + f(u)u_x = g(u)u_{xxx}$	$\eta_t = \eta^2[g(u)(\eta\eta_u)_u]_u - \eta^2 f'_u(u)$; $\theta_t = \theta^{3/2}[g(u)\theta_{uu}]_u - 2\theta^{3/2} f'_u(u)$
Generalized modified Korteweg–de Vries equation	$u_t + f(u_x) = au_{xxx}$	$\zeta_t = a(\zeta^{-3}\zeta_u)_{uu} + [\zeta f(1/\zeta)]_u$
Generalized modified Harry Dym equation	$u_t = [f(u, u_x)]_{xx}$	$\eta_t = \eta^2\{\eta[f(u,\eta)]_u\}_{uu}$

Bäcklund transformation of the form (10.4.2.13) or (10.4.2.24) is rewritten in terms of the new variables. The composition of these transformations determines the Bäcklund transformation of the original equation.

► Example 10.32. Consider the following evolution equation, which is homogeneous in the unknown function u:

$$u_t = uF\left(t, x, \frac{u_x}{u}, \frac{u_{xx}}{u}, \ldots, \frac{u_x^{(n)}}{u}\right). \qquad (10.4.2.30)$$

This equation has the characteristic property of remaining unchanged under the scaling transformation of the dependent variable $u = C\bar{u}$, where $C \neq 0$ is an arbitrary constant.

It is not difficult to verify that equation (10.4.2.30) is reduced to an equation of the form (10.4.2.9) with the substitution $u = e^w$. The second equation in (10.4.2.13), rewritten in terms of w and ξ, has the form $w_x = \xi$.

Returning back to the original variable $w = \ln u$, we obtain $u_x = u\xi$. Therefore the desired Bäcklund transformation of equation (10.4.2.30) has the form

$$u_t = uF(t, x, \xi, \xi_x + \xi^2, \dots),\\ u_x = u\xi. \tag{10.4.2.31}$$

The arguments omitted in (10.4.2.31) are calculated using the recurrence relations

$$\frac{u_x}{u} = \xi, \quad \frac{u_x^{(k)}}{u} = \frac{\partial}{\partial x}\left[\frac{u_x^{(k-1)}}{u}\right] + \xi\frac{u_x^{(k-1)}}{u}, \qquad k = 2, \dots, n,$$

which follow from the second equation in (10.4.2.31). In particular, for $n = 3$ we have $u_{xxx}/u = \xi_{xx} + 3\xi\xi_x + \xi^3$.

Dividing (10.4.2.31) by u and eliminating u by cross-differentiation with the aid of the relation

$$\frac{\partial}{\partial t}\left(\frac{u_x}{u}\right) = \frac{\partial}{\partial x}\left(\frac{u_t}{u}\right),$$

we arrive at the nth-order equation

$$\xi_t = \frac{\partial}{\partial x}F(t, x, \xi, \xi_x + \xi^2, \dots).$$

◄

10.4.3. Differential Substitutions

1°. In mathematical physics, apart from point transformations, contact transformations, and the Bäcklund transformations, one sometimes resorts to the so-called *differential substitutions*. First-order differential substitutions have the form

$$u = \Psi(x, y, w, w_x, w_y). \tag{10.4.3.1}$$

A differential substitution increases the order of an equation (if it is inserted into an equation for u) and allows us to obtain solutions of one equation from those of another. The relationship between the solutions of the two equations is generally not invertible and is, in a sense, unilateral. A differential substitution may decrease the order of an equation (if it is inserted into an equation for w).

► Example 10.33. The nonlinear second-order PDE

$$w_{xy} + 2aww_x + bw_y + abw^2 = 0 \tag{10.4.3.2}$$

reduces with the help of the differential substitution

$$u = w_y + aw^2 \tag{10.4.3.3}$$

to a simple linear first-order PDE:

$$u_x + bu = 0. \tag{10.4.3.4}$$

Integrating yields its general solution

$$u = f(y)e^{-bx}, \tag{10.4.3.5}$$

where $f(y)$ is an arbitrary function. On substituting (10.4.3.5) into (10.4.3.3), we obtain a Riccati equation for $w = w(x, y)$: $w_y + aw^2 = f(y)e^{-bx}$; this ODE includes x as a free parameter. ◀

2°. Below is a common definition of the term 'differential substitution' in a narrower sense (see Kuznetsova, 2012, and Kuznetsova et al., 2012).

Definition. Relation (10.4.3.1) is called a differential substitution from the equation

$$F(x, y, w, w_x, w_y, w_{xx}, \dots) = 0 \tag{10.4.3.6}$$

to the equation

$$G(x, y, u, u_x, u_y, u_{xx}, \dots) = 0 \tag{10.4.3.7}$$

if the function (10.4.3.1) satisfied equation (10.4.3.7) for every solution $w = w(x, y)$ of equation (10.4.3.6).

► Example 10.34. Consider the Burgers equation

$$u_t + uu_x = u_{xx}. \tag{10.4.3.8}$$

Let us make the differential substitution (the *Hopf–Cole transformation*)

$$u = -\frac{2w_x}{w}. \tag{10.4.3.9}$$

Substituting (10.4.3.9) into (10.4.3.8), we obtain the third-order equation

$$-\frac{2w_{tx}}{w} + \frac{2w_t w_x}{w^2} = -\frac{2w_{xxx}}{w} + \frac{2w_x w_{xx}}{w^2},$$

which can be converted to

$$[(w_t - w_{xx})/w]_x = 0. \tag{10.4.3.10}$$

Thus, using formula (10.4.3.9), one can transform each solution of the linear heat equation

$$w_t = w_{xx} \tag{10.4.3.11}$$

into a solution of the Burgers equation (10.4.3.8). The converse is not generally true. Indeed, it follows from (10.4.3.10) that a solution of equation (10.4.3.8) generates a solution of the more general equation

$$w_t - w_{xx} = f(t)w,$$

where $f(t)$ is a function of t. ◀

► Example 10.35. Consider the equation with an exponential nonlinearity

$$u_{xy} = -e^u u_x. \tag{10.4.3.12}$$

Let us use the following differential substitution (Kuznetsova, 2012):

$$u = \ln(w_y/w). \tag{10.4.3.13}$$

Substituting (10.4.3.13) into (10.4.3.12) and cancelling identical terms, we obtain the third-order PDE

$$\frac{w_{xyy}}{w_y} - \frac{w_{xy}w_{yy}}{w_y^2} = 0,$$

which can be rewritten as

$$(w_{xy}/w_y)_y = 0. \tag{10.4.3.14}$$

It follows that any solution of the linear equation

$$w_{xy} = 0 \tag{10.4.3.15}$$

generates, using formula (10.4.3.13), an exact solution of the original equation (10.4.3.12).

Considering that the general solution of equation (10.4.3.15) has the form $w = f(x)+g(y)$, where $f(x)$ and $g(y)$ are arbitrary functions, we use formula (10.4.3.13) to find an exact solution of equation (10.4.3.12):

$$u = \ln \frac{g'_y(y)}{f(x) + g(y)}. \tag{10.4.3.16}$$

◄

Remark 10.22. *The general solution of equation (10.4.3.14) is expressed as* $w = f(x) + h(x)g(y)$, *where* $f(x)$, $h(x)$, *and* $g(y)$ *are arbitrary functions. On substituting this function into (10.4.3.13), setting* $f(x) = \bar{f}(x)h(x)$, *and cancelling by* $h(x)$, *we arrive at solution (10.4.3.16), in which* f *is renamed* $\bar{f}$. *Therefore, the function (10.4.3.16) is the general solution of the original equation (10.4.3.12).*

► Example 10.36. Consider the Korteweg–de Vries equation

$$u_t + 6uu_x + u_{xxx} = 0. \tag{10.4.3.17}$$

Let us make the differential substitution

$$u = w_x - w^2, \tag{10.4.3.18}$$

which is referred to as the *Miura transformation*. This results in the fourth-order equation (Miura, 1976b):

$$\begin{aligned} &\frac{\partial}{\partial x} M[w] - 2wM[w] = 0, \\ &M[w] \equiv w_t - 6w^2 w_x + w_{xxx}. \end{aligned} \tag{10.4.3.19}$$

It follows from (10.4.3.18)–(10.4.3.19) that any solution $w = w(x,t)$ to the modified Korteweg–de Vries equation

$$M[w] \equiv w_t - 6w^2 w_x + w_{xxx} = 0 \tag{10.4.3.20}$$

generates, with the use of (10.4.3.18), an exact solution to the Korteweg–de Vries equation (10.4.3.17). ◄

In the last three examples discussed above, the differential substitutions (10.4.3.9), (10.4.3.13), and (10.4.3.18) increased the order of equations (cf. the orders of (10.4.3.8) and (10.4.3.10), the orders of (10.4.3.12) and (10.4.3.14), and those of (10.4.3.17) and (10.4.3.19)) on the one hand and, on the other hand, allowed us to obtain exact solutions to some equations using solutions to other equations of the same order (cf. the orders of (10.4.3.8) and (10.4.3.11), the orders of (10.4.3.12) and (10.4.3.15), and those of (10.4.3.17) and (10.4.3.20)).

Remark 10.23. *A differential substitution may be obtained as a consequence of a Bäcklund transformation (although this is not always the case); see Examples 10.24 and 10.26 in Section 10.4.1.*

Problems and Exercises for Section 10.4

1. Show that two Laplace equations

$$u_{xx} + u_{yy} = 0 \quad \text{and} \quad w_{xx} + w_{yy} = 0$$

are connected by the auto-Bäcklund transformation

$$u_x = w_y, \qquad w_x = -u_y.$$

Note that in the theory of functions of a complex variable, the last relations are called the *Cauchy–Riemann conditions*.

2. Show that nonlinear heat equation with an exponential source

$$u_{xx} + u_{yy} = ae^{\beta u}$$

is connected to the Laplace equation

$$w_{xx} + w_{yy} = 0,$$

which is a linear equation, by the Bäcklund transformation

$$\begin{aligned} w_x + \tfrac{1}{2}\beta u_y &= \left(\tfrac{1}{2}a\beta\right)^{1/2} \exp\left(\tfrac{1}{2}\beta u\right) \sin w, \\ w_y - \tfrac{1}{2}\beta u_x &= \left(\tfrac{1}{2}a\beta\right)^{1/2} \exp\left(\tfrac{1}{2}\beta u\right) \cos w. \end{aligned}$$

Hint. Apply differentiation with respect to appropriate variables and use the conditions $w_{xy} = w_{yx}$ and $u_{xy} = u_{yx}$.

3. Show that two Bonnet equations

$$u_{xy} = \sin u \quad \text{and} \quad w_{xy} = \sin w$$

are connected by the auto-Bäcklund transformation (the Darboux–Bianchi transformation):

$$\begin{aligned} w_x &= u_x + 2k \sin\left[\tfrac{1}{2}(u + w)\right], \\ w_y &= -u_y - 2k^{-1} \sin\left[\tfrac{1}{2}(u - w)\right], \end{aligned}$$

where $k \neq 0$ is an arbitrary constant; see Ibragimov (1994).

Hint. Apply differentiation with respect to appropriate variables and use the conditions $w_{xy} = w_{yx}$ and $u_{xy} = u_{yx}$.

4. Show that the nonlinear equations

$$u_{xx} + u_{yy} = \sinh u \cosh u,$$
$$w_{xx} + w_{yy} = \sin w \cos w$$

are connected by the Bäcklund transformation

$$w_x + u_y = \sin w \cosh u,$$
$$w_y - u_x = \cos w \sinh u.$$

Hint. Apply differentiation with respect to appropriate variables and use the conditions $w_{xy} = w_{yx}$ and $u_{xy} = u_{yx}$.

5. Show that the nonlinear Schrödinger equation

$$iu_t + u_{xx} + |u|^2 u = 0,$$

where $i^2 = -1$ and u is a complex-valued function of real variables x and t, is invariant under the Bäcklund transformation

$$u_x - \widetilde{u}_x = iaf_1 - \tfrac{1}{2} if_2 g_1,$$
$$u_t - \widetilde{u}_t = \tfrac{1}{2} g_1 (u_x + \widetilde{u}_x) - ag_2 + \tfrac{1}{4} if_1 \big(|f_1|^2 + |f_2|^2\big).$$

Notation:

$$f_1 = u - \widetilde{u}, \quad f_2 = u + \widetilde{u}, \quad g_1 = i\varepsilon\big(b - 2|f_1|^2\big)^{1/2}, \quad g_2 = i\big(af_1 - \tfrac{1}{2} f_2 g_1\big),$$

where a and b are arbitrary real constants and $\varepsilon = \pm 1$.

6. Show that the linear third-order equation

$$u_t = au_{xxx} + bu_{xx} + sxu_x + cu,$$

where $a = a(t)$, $b = b(t)$, $c = c(t)$, and $s = s(t)$ are arbitrary functions, can be reduced with the Bäcklund transformation (10.4.2.24), in which $F = au_{xxx} + bu_{xx} + cu = a\eta(\eta\eta_u)_u + b\eta\eta_u + cu$, followed by the substitution $\eta = 1/\zeta$, to the nonlinear PDE

$$\zeta_t + c(u\zeta)_u + s\zeta = a(\zeta^{-3}\zeta_u)_{uu} + b(\zeta^{-2}\zeta_u)_u.$$

7. Show that the nonlinear second-order PDE

$$u_{xy} + ae^u u_x + bu_y + abe^u = 0$$

is reducible with the differential substitution

$$w = u_y + ae^u$$

to the simple first-order linear PDE

$$w_x + bw = 0,$$

which can be treated as an ODE implicitly dependent on the parameter y.

8. Find the general solution of the second-order nonlinear PDE from the preceding exercise.

9. Show that two second-order nonlinear PDEs

$$u_{xy} = \exp u, \tag{10.4.3.21}$$

$$w_{xy} = f(w) w_x w_y \tag{10.4.3.22}$$

are connected by the differential substitution (Kuznetsova, 2012):

$$u = \ln(w_x w_y) + g(w), \tag{10.4.3.23}$$

where the functions $f = f(w)$ and $g = g(w)$ satisfy the ODE:

$$2f'_w + 2f^2 + g''_{ww} + f g'_w - \exp g = 0. \tag{10.4.3.24}$$

10. By setting $f \equiv 0$ in (10.4.3.22) and using (10.4.3.23) and (10.4.3.24), find a solution of equation (10.4.3.21) in the form

$$u = \ln(w_x w_y) + g(w), \qquad w = \varphi(x) + \psi(y),$$

where $\varphi(x)$ and $\psi(y)$ are arbitrary functions, and the function $g = g(w)$ satisfies the autonomous ODE: $g''_{ww} - \exp g = 0$. Determine $g(w)$ in implicit form by integrating this ODE.

10.5. Hydrodynamic Type Transformations. Transformations Based on Conservation Laws

10.5.1. Von Mises Hydrodynamic Type Transformation

The *von Mises transformation* is a significant special case of transformations, particularly in fluid dynamics, where it is employed to reduce the order of boundary layer equations. In this transformation, the unknown function is treated as a new independent variable, while an appropriate partial derivative is considered the new dependent variable.

1°. Prior to describing the von Mises transformation for partial differential equations, let us first consider a simple analogue of this transformation that frequently arises in the theory of ordinary differential equations.

Let $u = u(x)$ be a function of a single variable. The nth-order autonomous ordinary differential equation

$$F(u, u'_x, u''_{xx}, \ldots, u^{(n)}_x) = 0, \tag{10.5.1.1}$$

which does not depend on x explicitly, can be reduced, with the substitution

$$\eta(u) = u'_x \qquad (\textit{analogue of the von Mises transformation}), \tag{10.5.1.2}$$

to an $(n-1)$st-order equation for $\eta = \eta(u)$ where the derivatives with respect to x must be substituted by derivatives with respect to u by successively applying the formula $\eta^{(k)}_x = \eta(\eta^{(k-1)}_x)_u$.

Remark 10.24. *Equation (10.5.1.1) has the characteristic property that it is invariant under translation in the independent variable $x = \bar{x} + C$, where C is an arbitrary constant.*

2°. Now consider a nonlinear nth-order partial differential equation of the special form (Polyanin & Zhurov, 2012a):

$$u_x u_{tx} - u_t u_{xx} = F(t, u, u_x, u_{xx}, \dots, u_x^{(n)}). \tag{10.5.1.3}$$

If u and F are independent of t, then equation (10.5.1.3) becomes (10.5.1.1).

In equation (10.5.1.3), let us change from the old variables to the von Mises variables (analogue of transformation (10.5.1.2)):

$$t,\ x,\ u = u(t,x) \quad\Longrightarrow\quad t,\ u,\ \eta = \eta(t,u), \quad \text{where} \quad \eta = u_x. \tag{10.5.1.4}$$

The derivatives are transformed as follows:

$$\begin{gathered} \frac{\partial}{\partial x} = u_x \frac{\partial}{\partial u} = \eta \frac{\partial}{\partial u}, \qquad \frac{\partial}{\partial t} = \frac{\partial}{\partial t} + u_t \frac{\partial}{\partial u}, \\ u_{xx} = \eta\eta_u, \quad u_{xxx} = \eta(\eta\eta_u)_u, \quad \dots, \quad u_{tx} = \eta_t + u_t\eta_u. \end{gathered} \tag{10.5.1.5}$$

Hence, the von Mises transformation (10.5.1.4) reduces the nth-order PDE (10.5.1.3) to the $(n-1)$st-order PDE

$$\eta\eta_t = F(t, u, \eta, \eta\eta_u, \dots, \eta_x^{(n-1)}), \tag{10.5.1.6}$$

where the third- and higher-order derivatives are calculated by the formulas

$$u_x^{(k)} = \eta_x^{(k-1)} = \eta[\eta_x^{(k-2)}]_u, \qquad \frac{\partial}{\partial x} = \eta \frac{\partial}{\partial u}, \qquad k = 3, \dots, n.$$

Remark 10.25. *Equation (10.5.1.3) has the following characteristic property. Suppose $\widetilde{u}(t,x)$ is a solution of the equation in question. Then the function*

$$u(t,x) = \widetilde{u}\bigl(t, x + \varphi(t)\bigr),$$

where $\varphi(t)$ is an arbitrary function, is also a solution of the equation.

► Example 10.37. The nonlinear second-order PDE

$$u_x u_{tx} - u_t u_{xx} = f(t,u)u_{xx} + g(t,u)u_x^2 + h(t,u)u_x,$$

which is a special case of equation (10.5.1.3) with $F = f(t,u)u_{xx} + g(t,u)u_x^2 + h(t,u)u_x$, can be reduced with the von Mises transformation (10.5.1.4) to a linear first-order PDE:

$$\eta_t - f(t,u)\eta_u = g(t,u)\eta + h(t,u).$$

◄

► Example 10.38. The equation of a steady laminar hydrodynamic boundary layer on a flat plate has the form (see Schlichting, 1981):

$$u_y u_{xy} - u_x u_{yy} = a u_{yyy}, \tag{10.5.1.7}$$

where u is the stream function, x and y are the coordinates along and across the flow, and a is the kinematic viscosity of the fluid. Performing the renaming $x \Rightarrow t$ and $y \Rightarrow x$ in (10.5.1.7), one obtains a special case of equation (10.5.1.3).

The von Mises transformation

$$x,\ y,\ u = u(x,y) \quad \Longrightarrow \quad x,\ u,\ \eta = \eta(x,u), \quad \text{where} \quad \eta = u_y,$$

decreases the order of equation (10.5.1.7) and brings it to the simpler nonlinear heat equation

$$\eta_x = a(\eta\eta_u)_u.$$

◀

Any solution $u = g(t,x)$ to equation (10.5.1.3) determines a solution to equation (10.5.1.6), which can be written in the parametric form $u = g(t,x)$, $\eta = g_x(t,x)$.

Suppose $\eta = \eta(t,u)$ is a solution to equation (10.5.1.6). Then the function $u = u(t,x)$ is determined from the simple first-order equation $u_x = \eta(t,u)$, the integration of which gives a solution in implicit form:

$$\int_{u_0}^{u} \frac{ds}{\eta(t,s)} = x + \varphi(t),$$

where $\varphi(t)$ is an arbitrary function, and u_0 is an arbitrary constant. Thus, the inverse von Mises transformation involves the nonlocal operation of integration of a function that depends on the solution. Such transformations are called *nonlocal*. In the general case, nonlocal transformations can be defined by differential equations.

3°. The von Mises transformation (10.5.1.4) can be generalized to the case of greater number of independent variables.

Consider the nonlinear nth-order equation

$$F(t,x,u,u_y,u_{yy},\dots,u_y^{(n)},u_yu_{ty}-u_tu_{yy},u_yu_{xy}-u_xu_{yy}) = 0, \qquad (10.5.1.8)$$

which involves three independent variables t, x, and y. The von Mises transformation

$$t,\ x,\ y,\ u = u(t,x,y) \quad \Longrightarrow \quad t,\ x,\ u,\ \eta = \eta(t,x,u), \quad \text{where} \quad \eta = u_y, \qquad (10.5.1.9)$$

reduces the order of equation (10.5.1.8) and takes it to the form

$$F(t,x,u,\eta,\eta\eta_u,\dots,\eta_y^{(n-1)},\eta\eta_t,\eta\eta_x) = 0, \qquad (10.5.1.10)$$

where the derivatives with respect to y are calculated successively by the formula $\eta_y^{(k)} = \eta\big(\eta_y^{(k-1)}\big)_u$.

► Example 10.39. Consider the Prandtl system

$$\begin{aligned} u_t + uu_x + vu_y &= \nu u_{yy} + f(t,x), \\ u_x + v_y &= 0, \end{aligned} \qquad (10.5.1.11)$$

which describes a flat unsteady boundary layer with pressure gradient, with u and v being the fluid velocity components. On dividing the first equation in (10.5.1.11) by u_y, we differentiate the resulting expression with respect to y and eliminate v_y using the second equation of the system. As a result, we arrive at a third-order PDE of the form (10.5.1.8):

$$u_y u_{ty} - u_t u_{yy} + u(u_y u_{xy} - u_x u_{yy}) = \nu(u_y u_{yyy} - u_{yy}^2) - f(t,x)u_{yy}. \quad (10.5.1.12)$$

Applying the von Mises transformation (10.5.1.9), we obtain the second-order PDE (Polyanin & Zhurov, 2012a):

$$\eta_t + u\eta_x + f(t,x)\eta_u = \nu\eta^2\eta_{uu}. \quad (10.5.1.13)$$

Once a solution of this equation has been found, the second velocity component v can be obtained directly from the first equation in (10.5.1.11) without integration.

Note that PDE (10.5.1.13) can be reduced, with the change of variable $\eta = 1/\zeta$, to the nonlinear heat type equation

$$\zeta_t + u\zeta_x + f(t,x)\zeta_u = \nu(\zeta^{-2}\zeta_u)_u. \quad (10.5.1.14)$$

Below are two special cases when equation (10.5.1.14) admits linearization.

1. Consider the special case $f(t,x) = f(t)$. We look for exact solutions to (10.5.1.14) in the form

$$\zeta = Z(\xi,\tau), \qquad \xi = x - ut + \int tf(t)\,dt, \qquad \tau = \frac{1}{3}t^3.$$

By doing so, we arrive at the integrable PDE

$$Z_\tau = \nu(Z^{-2}Z_\xi)_\xi, \quad (10.5.1.15)$$

which can be reduced to the linear heat equation (see Example 10.30 with $f \equiv 0$).

2. Following Polyanin & Zhurov (2012a), we consider the more general case $f(t,x) = f(t)x + g(t)$. We look for exact solutions in the special form

$$\zeta = Z(\xi,\tau), \qquad \xi = \varphi(t)x + \psi(t)u + \theta(t), \qquad \tau = \int \psi^2(t)\,dt,$$

where $\varphi = \varphi(t)$, $\psi = \psi(t)$, and $\theta = \theta(t)$ are determined by the linear ODE system

$$\varphi'_t + f\psi = 0, \qquad \psi'_t + \varphi = 0, \qquad \theta'_t + g\psi = 0.$$

As a result, we arrive at a linearizable equation of the form (10.5.1.15). ◀

Section 10.4.2 showed that the von Mises transformation (10.5.1.4) can successfully be used for constructing RF pairs and Bäcklund transformations for evolution equations of a fairly general form.

10.5.2. Crocco Hydrodynamic Type Transformation

Crocco transformations are used to reduce the order of the hydrodynamic boundary layer equations and some other nonlinear PDEs. In a Crocco transformation, a suitable first-order partial derivative is taken as a new independent variable, and a second-order derivative is taken as the new dependent one.

1°. Before discussing Crocco transformations for partial differential equations, we will first examine their simpler analogs for ordinary differential equations.

Let $u = u(x)$ be a function of one variable. On solving the autonomous nth-order ordinary differential equation of general form (10.5.1.1) for u, we rewrite the equation as

$$G(u'_x, u''_{xx}, \dots, u_x^{(n)}) = u. \tag{10.5.2.1}$$

Differentiating (10.5.2.2) with respect to x and then changing from the old variables x and u to

$$\eta = u'_x, \quad \Phi(\eta) = u''_{xx} \qquad \textit{(analogue of the Crocco transformation)}, \tag{10.5.2.2}$$

we arrive at an $(n-1)$st-order equation.

2°. Consider the nonlinear nth-order PDE with a mixed derivative (Polyanin & Zhurov, 2012b):

$$u_{tx} + [a(t)u + b(t)x]u_{xx} = F(t, u_x, u_{xx}, \dots, u_x^{(n)}). \tag{10.5.2.3}$$

Introduce the notation

$$\eta = u_x, \qquad \Phi = u_{xx}. \tag{10.5.2.4}$$

Dividing equation (10.5.2.3) by $u_{xx} = \Phi$, differentiating with respect to x, and taking into account (10.5.2.4), we obtain

$$\frac{\Phi_t}{\Phi} - \frac{u_{tx}\Phi_x}{\Phi^2} + a(t)\eta + b(t) = \frac{\partial}{\partial x}\frac{F\big(t, \eta, \Phi, \Phi_x, \dots, \Phi_x^{(n-2)}\big)}{\Phi}. \tag{10.5.2.5}$$

In (10.5.2.5), we change from the old variables to the Crocco variables (analogue of transformation (10.5.2.2)):

$$t,\ x,\ u = u(t,x) \quad \Longrightarrow \quad t,\ \eta,\ \Phi = \Phi(t,\eta), \tag{10.5.2.6}$$

where η and Φ are given by (10.5.2.4). Then the derivatives are transformed as follows:

$$\frac{\partial}{\partial x} = \eta_x\frac{\partial}{\partial \eta} = u_{xx}\frac{\partial}{\partial \eta} = \Phi\frac{\partial}{\partial \eta}, \qquad \frac{\partial}{\partial t} = \frac{\partial}{\partial t} + \eta_t\frac{\partial}{\partial \eta} = \frac{\partial}{\partial t} + u_{tx}\frac{\partial}{\partial \eta}.$$

As a result, equation (10.5.2.5) and, hence, the original equation (10.5.2.3) are reduced to the $(n-1)$st-order equation

$$\frac{a(t)\eta + b(t)}{\Phi} - \frac{\partial}{\partial t}\frac{1}{\Phi} = \frac{\partial}{\partial \eta}\left[\frac{1}{\Phi}F\big(t, \eta, \Phi, \Phi\Phi_\eta, \dots, \Phi_x^{(n-2)}\big)\right], \tag{10.5.2.7}$$

where the third- and higher-order derivatives are calculated by the formulas

$$u_x^{(k)} = \Phi_x^{(k-2)} = \Phi[\Phi_x^{(k-3)}]_\eta, \qquad \frac{\partial}{\partial x} = \Phi\frac{\partial}{\partial \eta}, \qquad k = 3, \dots, n.$$

Remark 10.26. *Equation (10.5.2.3) has the following characteristic property: if* $\widetilde{u}(t,x)$ *is a solution to the equation, then*

$$u = \widetilde{u}(t, x+\varphi(t)) + \frac{1}{a(t)}[b(t)\varphi(t) - \varphi_t'(t)], \qquad a(t) \not\equiv 0,$$

where $\varphi(t)$ *is an arbitrary function, is also a solution to the equation. For* $a(t) \equiv 0$*, the function* $u = \widetilde{u}(t,x) + \varphi(t)$ *is a solution to equation (10.5.2.3).*

Whenever a solution to the original equation (10.5.2.3) is known, formulas (10.5.2.4) define a solution to equation (10.5.2.7) in parametric form.

Suppose $\Phi = \Phi(t,\eta)$ is a solution to equation (10.5.2.7). Then the corresponding solution to the original equation (10.5.2.3) is determined by integrating relations (10.5.2.4); for $a(t) \not\equiv 0$ it can be represented in parametric form as

$$x = \int_{\eta_0}^{\eta} \frac{ds}{\Phi(t,s)} + \varphi(t), \quad u = \int_{\eta_0}^{\eta} \frac{s\,ds}{\Phi(t,s)} + \frac{1}{a(t)}[b(t)\varphi(t) - \varphi_t'(t)], \tag{10.5.2.8}$$

where $\varphi(t)$ is an arbitrary function and η_0 is an arbitrary constant for which the above integrals make sense.

► Example 10.40. Setting $a = -1$, $b = 0$, and $F = f(u_x)$ in (10.5.2.3), we have the *Calogero equation*:

$$u_{tx} = uu_x + f(u_x).$$

The Crocco transformation (10.5.2.6) with (10.5.2.4) leads this equation to the integrable first-order PDE

$$-\frac{\eta}{\Phi} - \frac{\partial}{\partial t}\frac{1}{\Phi} = \frac{\partial}{\partial \eta}\left[\frac{f(\eta)}{\Phi}\right],$$

which is linearized by substituting $\Phi = 1/\Psi$. ◄

► Example 10.41. Let us look at the nonlinear third-order equation (Aristov & Polyanin, 2009 and Aristov et al., 2009):

$$u_{tx} + uu_{xx} - mu_x^2 = \nu u_{xxx} + q(t)u_x + p(t), \tag{10.5.2.9}$$

which arises in fluid dynamics and is derived from the Navier–Stokes equations, with u being one of the fluid velocity components and ν being the kinematic viscosity. The Crocco transformation (10.5.2.6) reduces (10.5.2.9) to a second-order equation, which can be written in the form

$$\Phi_t + [m\eta^2 + q(t)\eta + p(t)]\Phi_\eta = [(2m-1)\eta + q(t)]\Phi + \nu\Phi^2\Phi_{\eta\eta}. \tag{10.5.2.10}$$

Note that in the degenerate case of inviscid fluid ($\nu = 0$), the original nonlinear equation (10.5.2.9) is reduced to the linear first-order equation (10.5.2.10). ◄

3°. Consider the nonlinear nth-order equation

$$c(t)u_{tx} + [a(t)u + b(t)x]u_{xx} + d(t)(u_x u_{tx} - u_t u_{xx}) = F(t, u_x, u_{xx}, \dots, u_x^{(n)}), \tag{10.5.2.11}$$

which becomes (10.5.2.3) if $c(t) = 1$ and $d(t) = 0$.

Dividing equation (10.5.2.11) by u_{xx}, differentiating with respect to x, and changing to the Crocco variables (10.5.2.6) with (10.5.2.4), we arrive at the $(n-1)$st-order equation

$$\frac{a(t)\eta + b(t)}{\Phi} - [d(t)\eta + c(t)]\frac{\partial}{\partial t}\frac{1}{\Phi} = \frac{\partial}{\partial \eta}\frac{\partial}{\partial \eta}\left[\frac{1}{\Phi}F\big(t, \eta, \Phi, \Phi\Phi_\eta, \dots, \Phi_x^{(n-2)}\big)\right].$$

► Example 10.42. The equation of a steady-state boundary layer of a non-Newtonian fluid on a flat plate has the form

$$u_y u_{xy} - u_x u_{yy} = [f(u_{yy})]_y, \tag{10.5.2.12}$$

where u is the stream function, x and y are the coordinates along and across the flow, and $f(w)$ is a function determined by the rheological properties of the fluid (for a power-law fluid, we have $f(w) = kw^m$). With the renaming $x \Rightarrow t$ and $y \Rightarrow x$ in equation (10.5.2.12), we have a special case of equation (10.5.2.11) with $n = 3$, $a(t) = b(t) = c(t) = 0$, $d(t) = 1$, and $F = [f(u_{xx})]_x$.

The Crocco transformation

$$x,\ y,\ u = u(x, y) \quad \Longrightarrow \quad x,\ \eta,\ \Phi = \Phi(x, \eta), \quad \text{where} \quad \eta = u_y, \quad \Phi = u_{yy},$$

decreases the order of equation (10.5.2.12) and brings it to the second-order equation

$$\eta\Phi_x = \Phi^2[f(\Phi)]_{\eta\eta}.$$

For $f(\Phi) = k/\Phi$ with $k = \text{const}$, this equation is linearized with the substitution $\Psi = 1/\Phi$. ◄

4°. Consider the evolution nth-order equation of a quite general form (Polyanin & Zhurov, 2012b):

$$u_t + [a(t)u + b(t)x]u_x = F(t, u_x, u_{xx}, \dots, u_x^{(n)}), \tag{10.5.2.13}$$

which includes as special cases the Burgers and Korteweg–de Vries equations as well as equations of the nonlinear theory of fluid flows through porous media and some other nonlinear equations of mathematical physics.

On differentiating (10.5.2.13) with respect to x, we obtain an $(n+1)$st-order equation with a mixed derivative of the form (10.5.2.3):

$$u_{tx} + [a(t)u + b(t)x]u_{xx} = -a(t)u_x^2 - b(t)u_x + \frac{\partial}{\partial x}F(t, u_x, u_{xx}, u_{xxx}, \dots, u_x^{(n)}). \tag{10.5.2.14}$$

In (10.5.2.14), by changing from t, x, and u to the Crocco variables (10.5.2.6), we arrive at the nth-order equation

$$\frac{3a(t)\eta + 2b(t)}{\Phi} - \frac{\partial}{\partial t}\frac{1}{\Phi} + [a(t)\eta^2 + b(t)\eta]\frac{\partial}{\partial \eta}\frac{1}{\Phi} = \frac{\partial^2}{\partial \eta^2} F(t, \eta, \Phi, \Phi\Phi_\eta, \ldots, \Phi_x^{(n-2)}). \tag{10.5.2.15}$$

Equations (10.5.2.13) and (10.5.2.15) are connected by a Bäcklund-type transformation:

$$u_t + [a(t)u + b(t)x]\eta = F(t, \eta, \Phi, \Phi_x, \ldots, \Phi_x^{(n-2)}), \quad u_x = \eta, \quad u_{xx} = \Phi. \tag{10.5.2.16}$$

► Example 10.43. Consider the nonlinear second-order equation

$$u_t + [a(t)u + b(t)x]u_x = \frac{f(t, u_x)}{u_{xx}} + g(t, u_x), \tag{10.5.2.17}$$

which is a special case of equation (10.5.2.13). The Bäcklund transformation (10.5.2.16) reduces (10.5.2.17) to the equation

$$\frac{3a(t)\eta + 2b(t)}{\Phi} - \frac{\partial}{\partial t}\frac{1}{\Phi} + [a(t)\eta^2 + b(t)\eta]\frac{\partial}{\partial \eta}\frac{1}{\Phi} = \frac{\partial^2}{\partial \eta^2}\left[\frac{f(t, \eta)}{\Phi} + g(t, \eta)\right],$$

which becomes linear after the substitution $\Phi = 1/\Psi$. ◄

Remark 10.27. *The sum $p(t)u + q(t)x$ can be added to the right-hand side of equation (10.5.2.13) and the first equation in transformation (10.5.2.16). Equation (10.5.2.15) will then contain additional terms.*

10.5.3. Transformations Based on Conservation Laws

1°. Consider a differential equation written as a conservation law,

$$[G(t, x, u, u_x, u_t, \ldots)]_t + [F(t, x, u, u_x, u_t, \ldots)]_x = 0. \tag{10.5.3.1}$$

In (10.5.3.1), let us change from the original variables to the new variables

$$t,\ x,\ u = u(t, x) \quad \Longrightarrow \quad t,\ z,\ u = u(t, z), \tag{10.5.3.2}$$

where the function $z = z(t, x)$ is determined by solving the system of first-order partial differential equations

$$z_t = F, \qquad z_x = -G. \tag{10.5.3.3}$$

Here and henceforth, the arguments of F and G are omitted for brevity. By virtue of equation (10.5.3.1), system (10.5.3.3) is consistent.

The transformation (10.5.3.2), (10.5.3.3) determines the passage from the variables t and x to the new independent variables t and z; the respective derivatives are calculated by the rule

$$\frac{\partial}{\partial x} = -G\frac{\partial}{\partial z}, \qquad \frac{\partial}{\partial t} = \frac{\partial}{\partial t} + F\frac{\partial}{\partial z}.$$

Here, F and G are the same as in (10.5.3.1).

Whenever a solution $u = u(t,x)$ of equation (10.5.3.1) is known, the corresponding function $z = z(t,x)$ can be found by integrating the relation

$$dz = F(t,x,u,u_x,u_t,\dots)\,dt - G(t,x,u,u_x,u_t,\dots)\,dx, \tag{10.5.3.4}$$

which is obtained from equations (10.5.3.3) taking into account the expression of the total differential $dz = z_t\,dt + z_x\,dx$. Relation (10.5.3.4) is convenient to treat as a short representation of the transformation (10.5.3.2), (10.5.3.3), which is sometimes referred to as a *reciprocal transformation.*

Remark 10.28. *The transformation (10.5.3.2), (10.5.3.3) preserves the order of the equation.*

Remark 10.29. *Quite often, one may encounter transformations (10.5.3.4) that are supplemented with a transformation of the unknown function in the form $w = \varphi(u)$.*

► Example 10.44. Consider the nonlinear heat equation

$$u_t = [f(u)u_x]_x, \tag{10.5.3.5}$$

which represents a special case of equation (10.5.3.1) for $G = -u$ and $F = f(u)u_x$.

In this case, transformation (10.5.3.4) has the form (Strampp, 1982 and Burgan et al., 1984):

$$dz = u\,dx + [f(u)u_x]\,dt \tag{10.5.3.6}$$

and determines a transformation from the variables x and t to the new independent variables z and t, when the respective derivatives are calculated by the rule

$$\frac{\partial}{\partial x} = u\frac{\partial}{\partial z}, \qquad \frac{\partial}{\partial t} = \frac{\partial}{\partial t} + [f(u)u_x]\frac{\partial}{\partial z}.$$

Applying transformation (10.5.3.6) to equation (10.5.3.5), we obtain

$$u_t = u^2[f(u)u_z]_z. \tag{10.5.3.7}$$

The substitution $u = 1/w$ reduces (10.5.3.7) to an equation of the form (10.5.3.5):

$$w_t = [w^{-2}f(1/w)w_z]_z. \tag{10.5.3.8}$$

Suppose $u = u(t,x)$ is a solution of equation (10.5.3.5). The corresponding function $z = z(t,x)$ is found by integrating relation (10.5.3.6):

$$z = \int_{x_0}^{x} u(t,y)\,dy + \int_{t_0}^{t} f(u(\tau,x_0))\,\big[u_x(\tau,x)\big]_{x=x_0}\,d\tau,$$

where x_0 and t_0 are arbitrary constants (for which the integrals are meaningful).

In the special case of $f(u) = a = \text{const}$, the linear heat equation (10.5.3.5) is reduced to the nonlinear equation $w_t = a(w^{-2}w_z)_z$ by the transformation (10.5.3.6) with $u = 1/w$.

On the other hand, when $f(u) = au^{-2}$, the nonlinear equation (10.5.3.5) is reduced to the linear equation $w_t = aw_{zz}$ by the transformation (10.5.3.6) with $u = 1/w$ (see Rosen, 1979 and Bluman & Kumei, 1980). ◄

► Example 10.45. Consider the third-order nonlinear equation

$$u_t = [f(u)u_x]_{xx}, \tag{10.5.3.9}$$

which represents a special case of equation (10.5.3.1) for $G = -u$ and $F = [f(u)u_x]_x$.

In this case, transformation (10.5.3.4) has the form

$$dz = u\,dx + [f(u)u_x]_x\,dt. \tag{10.5.3.10}$$

The respective derivatives are calculated as

$$\frac{\partial}{\partial x} = u\frac{\partial}{\partial z}, \qquad \frac{\partial}{\partial t} = \frac{\partial}{\partial t} + [f(u)u_x]_x\frac{\partial}{\partial z}.$$

Applying transformation (10.5.3.10) to equation (10.5.3.9), we obtain

$$u_t = u^2[uf(u)u_z]_{zz}. \tag{10.5.3.11}$$

The substitution $u = 1/w$ reduces (10.5.3.11) to an equation of the form (10.5.3.9):

$$w_t = [w^{-3}f(1/w)w_z]_{zz}.$$

In the special case of $f(u) = a = \text{const}$, the linear equation (10.5.3.9) is reduced to the nonlinear equation $w_t = a(w^{-3}w_z)_{zz}$ by the transformation (10.5.3.10) with $u = 1/w$.

On the other hand, when $f(u) = au^{-3}$, the nonlinear equation (10.5.3.9) is reduced to the linear equation $w_t = aw_{zzz}$ by the transformation (10.5.3.10) with $u = 1/w$. ◄

2°. Evolution equations of the form

$$u_t = F\bigl(t, x, u_x, u_{xx}, \dots, u_x^{(n)}\bigr)$$

are reduced, by the differentiation with respect to x followed by the substitution

$$v = u_x,$$

to the equation of the form (10.5.3.1):

$$-v_t + [F(t, x, v, v_x, \dots, v_x^{(n-1)})]_x = 0, \qquad \text{where } G = -v.$$

Problems and Exercises for Section 10.5

1. Reduce the second-order nonlinear PDE

$$u_x u_{tx} - u_t u_{xx} = f(t,u)u_{xx} + g(t,u)u_x^2 + h(t,u)$$

to a first-order linear PDE.

Hint. Apply the von Mises transformation (10.5.1.4) and then use the substitution $Z=\eta^2$.

2. Reduce the third-order nonlinear PDE

$$u_x u_{tx} - u_t u_{xx} = f(t,u)u_{xxx}/u_x$$

to a second-order linear PDE.

Hint. Apply the von Mises transformation (10.5.1.4), and then use the substitution $Z=\eta^2$.

3. Show that the third-order nonlinear PDE

$$u_x u_{tx} - u_t u_{xx} = a u_x^3 u_{xxx}$$

can be reduced, with the von Mises transformation (10.5.1.4) and substitution $Z = \eta^2$, to a second-order nonlinear PDE: $Z_t = aZ^2 Z_{uu}$. This equation is reducible to the linear heat equation; see Example 10.30 with $f(u) \equiv 0$.

4. Show that the nonlinear PDE of the third order

$$u_x u_{txx} - u_t u_{xxx} = f(t,u)u_x$$

can be reduced with the von Mises transformation (10.5.1.4) to a simple second-order PDE: $(\eta\eta_u)_t = f(t,u)$.

5. Show that the *generalized Calogero equation*

$$u_{tx} = [f(t,u_x) - a(t)u - b(t)x]u_{xx} + g(t,u_x),$$

can be reduced, using the Crocco variables (10.5.2.6) with (10.5.2.4), to a first-order PDE, which is linearized by substituting $\Phi = 1/\Psi$ (Polyanin & Zaitsev, 2012).

6. Show that the nonlinear PDE of the third order

$$u_t + a u u_x = \frac{f(t,u_x)}{u_{xx}^3} u_{xxx}$$

can be reduced with transformation (10.5.2.16), where $b(t) \equiv 0$ and $F = f(t,u_x)u_{xx}^{-3}u_{xxx} = f(t,\eta)\Phi^{-2}\Phi_\eta$, followed by the substitution $\Phi = 1/\Psi$ to a linear equation

$$\Psi_t - a\eta^2\Psi_\eta - 3a\eta\Psi = [f(t,\eta)\Psi]_{\eta\eta\eta}.$$

7. Reduce the Calogero equation

$$u_{xt} = u u_{xx} + f(u_x)$$

to an ordinary differential equation. Find the general solution of the original equation for $f(u_x) = a$.

Hint. Use a reciprocal transformation based on the conservation law

$$D_t[\Phi(u_x)] + D_x[-u\Phi(u_x)] = 0,$$

where $D_t = \dfrac{\partial}{\partial t}$, $D_x = \dfrac{\partial}{\partial x}$, and $\Phi(w) = \exp\Big[\displaystyle\int \frac{w\,dw}{f(w)}\Big]$.

8. Show that the nonlinear second-order equation

$$u_t = [f(u, u_x)]_x$$

can be reduced with the reciprocal transformation

$$dz = u\,dx + f(u, u_x)\,dt, \qquad u = 1/w,$$

to the following equation (Strampp, 1982):

$$w_t = -[wf(z^{-1}, -z^{-3}w_z]_z.$$

9. Show that the nonlinear hyperbolic heat equation

$$u_{tt} + u_t = a(u^{-2}u_x)_x$$

can be reduced with the reciprocal transformation (Rogers & Ruggeri, 1985)

$$\tau = t + \ln|u|, \quad dz = au^{-2}u_x dt + (u + u_t)\,dx, \quad w = 1/u \qquad (dz = z_t\,dt + z_x\,dx),$$

to the linear telegraph equation

$$w_{\tau\tau} + w_\tau = aw_{zz}.$$

10. Show that the nonlinear telegraph equation

$$au_{tt} + u_t = [f(u)u_x]_x$$

is connected with a similar equation of the form

$$aw_{\tau\tau} + w_\tau = [w^{-2}f(w^{-1})w_z]_z$$

by the reciprocal transformation (Ibragimov, 1994):

$$\tau = t + a\ln|u|, \quad dz = f(u)u_x dt + (u + au_t)\,dx, \quad w = 1/u \qquad (dz = z_t\,dt + z_x\,dx).$$

References for Chapter 10

Aksenov A.V. and Polyanin A.D. Symmetries, reductions and exact solutions of nonstationary Monge–Ampère type equations. *Mathematics*, 2025, Vol. 3, No. 7, 525.

Aristov S.N. and Polyanin A.D. Exact solutions of unsteady three-dimensional Navier–Stokes equations. *Doklady Physics*, 2009, Vol. 54, No. 7, pp. 316–321.

Aristov S.N., Knyazev D.V., and Polyanin A.D. Exact solutions of the Navier–Stokes equations with the linear dependence of velocity components of two space variables. *Theor. Foundations of Chemical Engineering*, 2009, Vol. 43, No. 5, pp. 642–662.

Barbashov B.M. and Chernikov N.A. Solution and quantization of a nonlinear two-dimensional model for a Born–Infeld type field. *Soviet Physics: JETP*, 1966, Vol. 23, No. 5, pp. 861–868.

Bluman G.W. and Kumei S. On the remarkable nonlinear diffusion equation $(\partial/\partial x)[a(u+b)^{-2}(\partial u/\partial x)] - (\partial u/\partial t) = 0$. *Math. Phys.*, 1980, Vol. 21, No. 5, pp. 1019–1023.

Burgan J.R., Munier A., Feix M.R., and Fyalkow E. Homology and the nonlinear heat diffusion equation. *SIAM J. Appl. Math.*, 1984, Vol. 44, No. 1, pp. 11–18.

Chakrabarty J. *Theory of Plasticity*. McGraw-Hill, New York, 1987.

Clarkson P.A., Fokas A.S., and Ablowitz M.J. Hodograph transformations on linearizable partial differential equations. *SIAM J. Appl. Math.*, 1989, Vol. 49, pp. 1188–1209.

Courant R. and Friedrichs R. *Supersonic Flow and Shock Waves.* Springer-Verlag, New York, 1985.

Courant R. and Hilbert D. *Methods of Mathematical Physics, Vol. 2.* Wiley–Interscience Publ., New York, 1989.

Dorodnitsyn V.A. and Svirshchevskii S.R. *On Lie–Bäcklund groups admitted by the heat equation with a source* (in Russian), Preprint 101. Keldysh Institute of Applied Mathematics, Academy of Sciences U.S.S.R., Moscow, 1983.

Ibragimov N.H. (ed.) *CRC Handbook of Lie Group Analysis of Differential Equations, Vol. 1, Symmetries, Exact Solutions and Conservation Laws.* CRC Press, Boca Raton, 1994.

Kamke E. *Differentialgleichungen: Lösungsmethoden und Lösungen, II, Partielle Differentialgleichungen Erster Ordnung für eine gesuchte Funktion.* Akad. Verlagsgesellschaft Geest & Portig, Leipzig, 1965.

Kumei S. and Bluman G.W. When nonlinear differential equations are equivalent to linear differential equations. *SIAM J. Appl. Math.*, 1982, Vol. 42, pp. 1157–1173.

Kuznetsova M.N. On nonlinear hyperbolic differential equations related to the Klein–Gordon equation by differential substitutions. *Ufimsk. Mat. Zh.*, 2012, Vol. 4, No. 3, pp. 86–103.

Kuznetsova M.N., Pekcan A., and Zhiber A.V. The Klein–Gordon equation and differential substitutions of the form $v = \varphi(u, u_x, u_y)$. *Symmetry, Integrability and Geometry: Methods and Applications* (SIGMA), 2012, Vol. 8, 090.

Lei H.-C. and Chang H.-W. A list of hodograph transformations and exactly linearizable systems. *Int. J. Non-Linear Mech.*, 1996, Vol. 31, No. 2, pp. 117–127.

Miura R.M. (ed.) *Bäcklund Transformations.* Springer-Verlag, Berlin, 1976a.

Miura R.M. *The Korteweg-de Vries equation: A survey of results. SIAM Review*, 1976b, Vol. 18, No. 3, pp. 412–459.

Nimmo J.J.C. and Crighton D.J. Bäcklund transformations for nonlinear parabolic equations: the general results. *Proc. Royal Soc. London A*, 1982, Vol. 384, pp. 381–401.

Polyanin A.D. and Zaitsev V.F. *Handbook of Nonlinear Partial Differential Equations, 2nd ed.* Chapman & Hall/CRC Press, Boca Raton–London, 2012.

Polyanin A.D. and Zaitsev V.F. *Handbook of Ordinary Differential Equations: Exact Solutions, Methods, and Problems.* CRC Press, Boca Raton–London, 2018.

Polyanin A.D. and Zhurov A.I. On RF-pairs, Bäcklund transformations and linearization of nonlinear equations. *Commun. Nonlinear Sci. Numer. Simulat.*, 2012a, Vol. 17, No. 2, pp. 536–544.

Polyanin A.D. and Zhurov A.I. On order reduction of non-linear equations of mechanics and mathematical physics, new integrable equations and exact solutions. *Int. J. Non-Linear Mech.*, 2012b, Vol. 47, No. 5, pp. 413–417.

Rogers C. and Ruggeri T. A reciprocal Backlund transformation: application to a nonlinear hyperbolic model in heat conduction, Lett. Nuovo Cimento, 1985, Vol. 44, pp. 289–296.

Rogers C. and Schief W.K. *Bäcklund and Darboux Transformations: Geometry and Modern Applications in Soliton Theory.* Cambridge Univ. Press, Cambridge, 2002.

Rogers C. and Shadwick W.F. *Bäcklund Transformations and Their Applications.* Academic Press, New York, 1982.

Rosen G. Nonlinear heat conduction in solid H_2. *Phys. Rev. B*, 1979, Vol. 19, No. 4, pp. 2398–2399.

Rozhdestvenskii B.L. and Yanenko N.N. *Systems of Quasilinear Equations and Their Applications to Gas Dynamics.* American Mathematical Society, Providence, RI, 1983.

Schlichting H. *Boundary Layer Theory.* McGraw-Hill, New York, 1981.

Siddiqui A.M., Kaloni P.N., and Chandna O.P. Hodograph transformation methods in non-Newtonian fluids. *J. Eng. Math.*, 1985, Vol. 19, pp. 203–216.

Strampp W. Bäcklund transformations for diffusion equations. *Physica D*, 1982, Vol. 6, No. 1, pp. 113–118.

Svirshchevskii S.R. Group properties of the hyperbolic system of heat transfer (in Russian), Preprint 20. Keldysh Institute of Applied Mathematics, Academy of Sciences U.S.S.R., Moscow, 1986.

Tikhonov A.N. and Samarskii A.A. *Equations of Mathematical Physics*. Dover Publ., New York, 1990.

Whitham G.B. *Linear and Nonlinear Waves*. Wiley, New York, 1974.

Zaitsev V.F. and Polyanin A.D. *Discrete-Group Methods for Integrating Equations of Nonlinear Mechanics*. CRC Press, Boca Raton, 1994.

Zaitsev V.F. and Polyanin A.D. Exact solutions and transformations of nonlinear heat and wave equations. *Doklady Mathematics*, 2001, Vol. 64, No. 3, pp. 416–420.

Zwillinger D. *Handbook of Differential Equations, 3rd ed.* Academic Press, San Diego, 1997.

11. Using Simple Solutions to Construct Complex Solutions

► **Preliminary remarks.** This chapter outlines several simple methods for constructing exact solutions to nonlinear partial differential equations. These methods do not require special training and involve minimal intermediate calculations. The central idea behind these methods is that *simple exact solutions of equations can be used to derive more complex solutions for the same equations.*

In particular, the chapter illustrates how, starting from simple solutions of equations, one can derive more complex exact solutions of the same equations using translation and scaling transformations. It describes a method for finding exact solutions to equations with multiple spatial variables, starting from solutions to related equations with only one spatial variable. The discussion includes instances where it is possible to create more intricate solutions by adding terms to simpler ones. Additionally, the chapter explores situations where similar simple solutions can be combined to generate a more complex composite solution through a nonlinear superposition of solutions. Furthermore, a method is presented for constructing complicated exact solutions of linear and nonlinear equations by incorporating a complex-valued parameter into simpler solutions.

The chapter illustrates the effectiveness of the described methods with numerous specific examples. Nonlinear heat equations, reaction-diffusion equations, nonlinear wave equations, Navier–Stokes equations, and others are considered.

11.1. Constructing Complex Solutions from Simple Solutions Using Translation and Scaling Transformations

11.1.1. Some Definitions. Simplest Transformations

We say that a partial differential equation

$$F(x, t, u, u_x, u_t, u_{xx}, u_{xt}, u_{tt}, \dots) = 0 \qquad (11.1.1.1)$$

is invariant under a one-parameter reversible transformation

$$x = X(\bar{x}, \bar{t}, \bar{u}, c), \quad t = T(\bar{x}, \bar{t}, \bar{u}, c), \quad u = U(\bar{x}, \bar{t}, \bar{u}, c), \qquad (11.1.1.2)$$

if substituting the expressions (11.1.1.2) into (11.1.1.1) yields an exactly the same equation

$$F(\bar{x}, \bar{t}, \bar{u}, \bar{u}_{\bar{x}}, \bar{u}_{\bar{t}}, \bar{u}_{\bar{x}\bar{x}}, \bar{u}_{\bar{x}\bar{t}}, \bar{u}_{\bar{t}\bar{t}}, \dots) = 0.$$

Notably, the free parameter c, appearing in transformation (11.1.1.2), can take values within a certain interval (c_1, c_2). However, this parameter is not contained in equation (11.1.1.1).

Transformations that preserve the form of equation (11.1.1.2) convert a solution of the equation into another solution of the same equation.

In what follows, we will consider only one-parameter transformations of the form

$$x = \bar{x} + a_1, \qquad t = \bar{t} + a_2, \qquad u = \bar{u} + a_3 \qquad (\textit{translation transformation});$$
$$x = b_1\bar{x}, \qquad t = b_2\bar{t}, \qquad u = b_3\bar{u} \qquad (\textit{scaling transformation}),$$

and composition of these transformations. Here, a_n and b_n $(n = 1,\ 2,\ 3)$ are constant quantities dependent on the free parameter c. We will refer to such transformations as the *simplest transformations*.

11.1.2. Constructing Complex Solutions from Simple Separable Solutions of a Special Form

Simple one-term solutions, expressed as a product of functions of different variables, can be most readily found using the method of separation of variables. The simplest solutions of this form, such as $u = Ax^\alpha t^\beta$, can be easily determined from the relevant equations using the method of undetermined coefficients. Below, we will describe a few methods for constructing more complex solutions based on these simpler ones.

Following Aksenov & Polyanin (2021, 2022), we will first examine simple separable solutions of the special form

$$u = t^k \varphi_1(x), \tag{11.1.2.1}$$

where k is some constant and $\varphi_1(x)$ is some function. Such solutions remain unchanged (invariant) under the scaling transformation

$$t = c\bar{t}, \qquad u = c^k \bar{u}. \tag{11.1.2.2}$$

The following statement is true, enabling us to derive more complex solutions based on solutions of the form (11.1.2.1).

Proposition 1. *Suppose an equation*

$$F(t, u, u_x, u_t, u_{xx}, u_{xt}, u_{tt}, \dots) = 0, \tag{11.1.2.3}$$

which is explicitly independent of the space variable x, has a simple solution of the form (11.1.2.1) *and remains unchanged under the scaling transformation* (11.1.2.2)*; that is, equation* (11.1.2.3) *has the same property as the original solution* (11.1.2.1)*. Then this equation also has a more complicated solution of the form*

$$u = t^k \varphi_2(z), \qquad z = x + m \ln t, \tag{11.1.2.4}$$

where m is an arbitrary constant.

Proposition 1 can be proved based on the invariance of PDE (11.1.2.3) under translation in the space variable x; see Aksenov & Polyanin (2021, 2022).

Remark 11.1. *In general, the form of the functions $\varphi_1(x)$ and $\varphi_2(z)$, appearing in the original solution (11.1.2.1) and more complex solution (11.1.2.4), can differ.*

We will now look at simple separable solutions of another form

$$u = x^n \psi_1(t), \tag{11.1.2.5}$$

where n is some constant, and $\psi_1(t)$ is some function. Such solutions remain unchanged (invariant) under the scaling transformation

$$x = c\bar{x}, \qquad u = c^n \bar{u}. \tag{11.1.2.6}$$

A more complicated solution than (11.1.2.5) can be obtained using Proposition 1, by appropriately renaming the constants and variables in (11.1.2.1)–(11.1.2.4). Additionally, below is another equivalent method for constructing a more complex solution, which may be more convenient to use in practice.

Proposition 2. *Suppose an equation*

$$F\left(x, u, u_x, u_t, u_{xx}, u_{xt}, u_{tt}, \ldots\right) = 0, \tag{11.1.2.7}$$

which is explicitly independent of time t, has a simple solution of the form (11.1.2.5) and remains unchanged under the scaling transformation (11.1.2.6); that is, equation (11.1.2.7) has the same property as the original solution (11.1.2.5). Then this equation also has a more complex solution of the form

$$u = e^{-npt} \psi_2(y), \qquad y = xe^{pt}, \tag{11.1.2.8}$$

where p is an arbitrary constant.

Proposition 2 can be proved based on the invariance of PDE (11.1.2.7) under translation in time t; see Aksenov & Polyanin (2021, 2022).

► Example 11.1. Let us consider the Boussinesq equation

$$u_t = a(uu_x)_x, \tag{11.1.2.9}$$

which describes an unsteady groundwater flow through a porous medium with a free surface.

Equation (11.1.2.9) has a simple exact solution

$$u = -\frac{x^2}{6at}, \tag{11.1.2.10}$$

which is a solution of the forms (11.1.2.1) and (11.1.2.5) simultaneously. Consider both possibilities for constructing more complex solutions based on solutions (11.1.2.10).

1°. Equation (11.1.2.9) is explicitly independent of the space variable x. Solution (11.1.2.10) and equation (11.1.2.9) remain unchanged under the scaling transformation $t = c\bar{t}$, $u = \bar{u}/c$. Therefore, by virtue of Proposition 1, equation (11.1.2.9) admits a more complex exact solution

$$u = \frac{\varphi(z)}{t}, \qquad z = x + m \ln t,$$

where m is an arbitrary constant, and the function $\varphi = \varphi(z)$ satisfies the second-order ODE

$$m\varphi'_z - \varphi = a(\varphi\varphi'_z)'_z. \tag{11.1.2.11}$$

It should be noted that equation (11.1.2.11) with $m = 0$, when $z = x$, admits a one-parameter family of solutions as a quadratic polynomial

$$\varphi = -\frac{x^2}{6a} + Cx - \frac{3aC^2}{2},$$

where C is an arbitrary constant. For $C = 0$, this solution leads to the original solution (11.1.2.10).

2°. Equation (11.1.2.9) is explicitly independent of time t. Solution (11.1.2.10) and equation (11.1.2.9) remain unchanged under the scaling transformation $x = c\bar{x}$, $u = c^2\bar{u}$. Therefore, by virtue of Proposition 2, equation (11.1.2.9) admits another exact solution

$$u = e^{-2pt}\psi(y), \qquad y = xe^{pt},$$

where $p \neq 0$ is an arbitrary constant, and the function $\psi = \psi(y)$ is described by the second-order ODE

$$py\psi'_y - 2p\psi = a(\psi\psi'_y)'_y.$$

◀

▶ **Example 11.2.** Now let us examine the Guderley equation

$$u_{xx} = au_y u_{yy}, \tag{11.1.2.12}$$

which is used for describing transonic gas flows (Guderley, 1962).

Equation (11.1.2.12) admits a simple exact solution

$$u = \frac{y^3}{3ax^2}, \tag{11.1.2.13}$$

which is a special case of the two forms of solution (11.1.2.1) and (11.1.2.5) simultaneously. Consider both possibilities for constructing more complex solutions based on solution (11.1.2.13).

1°. Equation (11.1.2.12) is explicitly independent of the variable y. Solution (11.1.2.13) and equation (11.1.2.12) remain unchanged under the scaling transformation $x = c\bar{x}$, $u = c^{-2}\bar{u}$. Therefore, by virtue of Proposition 1, equation (11.1.2.12) has a more complex exact solution of the form

$$u = x^{-2}\varphi(z), \qquad z = y + m \ln x,$$

where the function $\varphi = \varphi(z)$ is described by the second-order ODE

$$m^2\varphi''_{zz} - 5m\varphi'_z + 6\varphi = a\varphi'_z\varphi''_{zz}.$$

For $m = 0$, this equation admits a one-parameter family of solutions as a cubic polynomial

$$\varphi(z) = \frac{z^3}{3a} + Cz^2 + aC^2z + \frac{a^2C^3}{3},$$

where C is an arbitrary constant. For $C = 0$, this solution leads to the original solution (11.1.2.13).

2°. Equation (11.1.2.12) is explicitly independent of variable x. Solution (11.1.2.13) and equation (11.1.2.12) also remain unchanged under the scaling transformation $y = c\bar{y}$, $u = c^3\bar{u}$. Therefore, by virtue of Proposition 2, we can also obtain another, more complex exact solution:

$$u = e^{-3px}\psi(z), \qquad z = ye^{px},$$

where $p \neq 0$, and the function $\psi = \psi(z)$ is described by the second-order ODE

$$p^2z^2\psi''_{zz} - 5p^2z\psi'_z + 9p^2\psi = a\psi'_z\psi''_{zz}.$$

◄

► Example 11.3. The nonlinear wave equation

$$u_{tt} = a(u^b u_x)_x, \qquad b \neq 0, \tag{11.1.2.14}$$

arises in gas dynamics. It does not include the independent variables explicitly and admits the simple exact solution

$$u = a^{-1/b}x^{2/b}t^{-2/b}, \tag{11.1.2.15}$$

which belongs to both classes of solutions, (11.1.2.1) and (11.1.2.5). Therefore, based on solutions (11.1.2.15), we can construct two more complex solutions described below.

1°. Solution (11.1.2.15) and equation (11.1.2.14) are invariant under the scaling transformation $t = c\bar{t}$, $u = c^{-2/b}\bar{u}$. By virtue of Proposition 1, equation (11.1.2.14) admits a more complex solution of the form

$$u = t^{-2/b}\varphi(z), \qquad z = x + m\ln t,$$

where the function $\varphi = \varphi(z)$ is described by the second-order ODE

$$m^2\varphi''_{zz} - \frac{m(b+4)}{b}\varphi'_z + \frac{2(b+2)}{b^2}\varphi = a(\varphi^b\varphi'_z)'_z.$$

2°. Solution (11.1.2.15) and equation (11.1.2.14) are invariant under the scaling transformation $x = c\bar{x}$, $u = c^{2/b}\bar{u}$. Therefore, by virtue of Proposition 2, equation (11.1.2.14) admits another solution

$$u = e^{-2pt/b}\psi(y), \qquad y = xe^{pt},$$

where $p \neq 0$, and the function $\psi = \psi(y)$ satisfies the nonlinear second-order ODE

$$p^2y^2\psi''_{yy} + \frac{p^2(b-4)}{b}y\psi'_y + \frac{4p^2}{b^2}\psi = a(\psi^b\psi'_y)'_y.$$

◄

Below is another statement that enables the construction of exact solutions for nonlinear PDEs.

Proposition 3. *Suppose an equation*

$$F(u, u_x, u_t, u_{xx}, u_{xt}, u_{tt}, \dots) = 0 \qquad (11.1.2.16)$$

that is explicitly independent of the variables x and t and, hence, admits a traveling wave solution, remains unchanged under the scaling transformation of the unknown:

$$u = c\bar{u}.$$

Then this equation admits an exact solution, more complex than the traveling wave solution, which can be represented in two alternative forms:

$$\begin{aligned} u &= Ce^{kt}\varphi(px+qt) \quad \text{(first form)}, \\ u &= Ce^{mx}\psi(px+qt) \quad \text{(second form)}, \end{aligned} \qquad (11.1.2.17)$$

where C, k, p, and q are arbitrary constants ($pq \neq 0$), $m = kp/q$, and $\psi(z) = e^{-mz/p}\varphi(z)$. Also, there are simpler separable solutions of the forms $u = Ce^{kt}\varphi(x)$ and $u = Ce^{mx}\psi(t)$.

Proposition 3 can be proved based on the invariance of the nonlinear PDE (11.1.2.16) under translation in both independent variables, x and t; see Aksenov & Polyanin (2021, 2022).

► Example 11.4. Consider the nonlinear heat-type equation

$$u_t = au_{xx} + uf(u_x/u), \qquad (11.1.2.18)$$

where $f = f(\xi)$ is an arbitrary function.

Equation (11.1.2.18) is explicitly independent of the variables x and t and remains unchanged under the scaling transformation of the dependent variable $u = c\bar{u}$. Hence, by virtue of Proposition 3, this equation has solution of the form (11.1.2.17) (first form), where the function $\varphi = \varphi(z)$ satisfies the second-order nonlinear ODE:

$$k\varphi + q\varphi'_z = ap^2\varphi''_{zz} + \varphi f(p\varphi'_z/\varphi).$$ ◄

11.1.3. Generalization to Arbitrarily Many Spatial Variables

Propositions 1–3 (see above) can be generalized to the case of arbitrarily many spatial variables. We will illustrate this with specific examples.

► Example 11.5. Consider the parabolic Monge–Ampère equation with three independent variables

$$u_t = u_{xx}u_{yy} - u_{xy}^2, \qquad (11.1.3.1)$$

which arises in electron magnetohydrodynamics (see Smirnov & Chukbar, 2001 and Zaburdaev et al., 2014).

It is easy to verify that equation (11.1.3.1) has a simple solution as a product of the appropriate powers of the independent variables:

$$u = \frac{x^2y^2}{12t}. \qquad (11.1.3.2)$$

Below are several solutions that generalize solution (11.1.3.2) and can be obtained using Proposition 1 stated in Section 11.1.2.

1°. Equation (11.1.3.1) is explicitly independent of the space variable y. Solution (11.1.3.2) and equation (11.1.3.1) remain unchanged under the scaling transformation $t = c\bar{t}$, $u = c^{-1}\bar{u}$ $(x = \bar{x})$. Therefore, by virtue of Proposition 1 (to be precise, its obvious generalization), equation (11.1.3.1) also has a more complex solution of the form

$$u = \frac{x^2}{t} f(z), \qquad z = y + m \ln t, \tag{11.1.3.3}$$

where m is an arbitrary constant, and the function $f = f(z)$ is described by the ODE:

$$2ff''_{zz} - 4(f'_z)^2 = -f + mf'_z.$$

2°. Equation (11.1.3.1) is explicitly independent of the space variable y. Solution (11.1.3.2) and equation (11.1.3.1) remain unchanged under the scaling transformation $x = c\bar{x}$, $u = c^2\bar{u}$ $(t = \bar{t})$. Therefore, by virtue of Proposition 1, equation (11.1.3.1) also has a solution of the form

$$u = \frac{x^2}{t} g(\xi), \qquad \xi = y + \beta \ln x, \tag{11.1.3.4}$$

where β is an arbitrary constant, and the function $f = f(z)$ satisfies the ODE:

$$\beta g'_\xi g''_{\xi\xi} - 2gg''_{\xi\xi} + 4(g'_\xi)^2 = g.$$

3°. Equation (11.1.3.1) is explicitly independent of time t. Solution (11.1.3.2) and equation (11.1.3.1) remain unchanged under the scaling transformation $y = c\bar{y}$, $u = c^2\bar{u}$ $(x = \bar{x})$. Therefore, by virtue of Proposition 1, equation (11.1.3.1) also admits a solution of the form

$$u = x^2 y^2 \varphi(\eta), \qquad \eta = t + \gamma \ln y, \tag{11.1.3.5}$$

where γ is an arbitrary constant, and the function $\varphi = \varphi(\eta)$ satisfies the ODE:

$$2\gamma^2 \varphi\varphi''_{\eta\eta} - 4\gamma^2(\varphi'_\eta)^2 - 10\gamma\varphi\varphi'_\eta - 12\varphi^2 = \varphi'_\eta.$$

The space variables x and y in solutions (11.1.3.3)–(11.1.3.5) can be interchanged. ◄

Remark 11.2. *See the article by Polyanin & Aksenov (2024) for more complex solutions of the parabolic Monge–Ampère equation (11.1.3.1).*

► Example 11.6. Consider the nonlinear heat equation with n space variables

$$u_t = a \sum_{i=1}^{n} \frac{\partial}{\partial x_i}\left(u^k \frac{\partial u}{\partial x_i}\right), \qquad k \neq 0. \tag{11.1.3.6}$$

Equation (11.1.3.6) admits a simple separable solution

$$u = t^{-1/k}\varphi(x_1, \dots, x_n), \tag{11.1.3.7}$$

where the function $\varphi = \varphi(x_1, \dots, x_n)$ satisfies the stationary equation

$$-\frac{1}{k}\varphi = a\sum_{i=1}^{n}\frac{\partial}{\partial x_i}\left(\varphi^k\frac{\partial\varphi}{\partial x_i}\right).$$

Solution (11.1.3.7) and equation (11.1.3.6) remain unchanged under the scaling transformation $t = c\bar{t}$, $u = c^{-1/k}\bar{u}$. Therefore, by virtue of Proposition 1 (to be precise, its generalization), equation (11.1.3.6) also has a more complex solution of the form

$$u = t^{-1/k}\theta(z_1, \dots z_n), \qquad z_i = x_i + m_i \ln t,$$

where m_i are arbitrary constants, and the function $\theta = \theta(z_1, \dots z_n)$ satisfies the stationary equation

$$-\frac{1}{k}\theta + \sum_{i=1}^{n} m_i\frac{\partial\theta}{\partial z_i} = a\sum_{i=1}^{n}\frac{\partial}{\partial z_i}\left(\theta^k\frac{\partial\theta}{\partial z_i}\right).$$ ◄

11.1.4. Generalization to Systems of Equations of Mathematical Physics

Propositions 1–3 from Section 11.1.2 can also help find solutions to systems of PDEs.

► Example 11.7. Let us consider a system consisting of two nonlinear reaction-diffusion type equations (Aksenov & Polyanin, 2021):

$$\begin{aligned} u_t &= a(u^k u_x)_x + u^{k+1}f(u/v), \\ v_t &= b(v^k v_x)_x + v^{k+1}g(u/v), \end{aligned} \qquad (11.1.4.1)$$

where a, b, and k are some constants ($k \neq 0$), while $f(z)$ and $g(z)$ are arbitrary functions.

The PDE system (11.1.4.1) does not include the independent variables explicitly and has a simple solution of the form

$$u = t^{-1/k}\varphi(x), \qquad v = t^{-1/k}\psi(x), \qquad (11.1.4.2)$$

where the function $\varphi = \varphi(x)$ and $\psi = \psi(x)$ are described by the system of second-order ODEs

$$\begin{aligned} -\frac{\varphi}{k} &= a(\varphi^k\varphi'_x)'_x + \varphi^{k+1}f(\varphi/\psi), \\ -\frac{\psi}{k} &= b(\psi^k\psi'_x)'_x + \psi^{k+1}g(\varphi/\psi). \end{aligned}$$

Solution (11.1.4.2) and system of equations (11.1.4.1) remain unchanged under the scaling transformation $t = c\bar{t}$, $u = c^{-1/k}\bar{u}$, $v = c^{-1/k}\bar{v}$. By virtue of Proposition 1 (to be precise, its generalization), the system of equations (11.1.4.1) also has a more complex solution of the form

$$u = t^{-1/k}\Phi(z), \qquad v = t^{-1/k}\Psi(z), \qquad z = x + m\ln t,$$

where m is an arbitrary constant, and the functions $\Phi = \Phi(z)$ and $\Psi = \Psi(z)$ satisfy the ODE system

$$-\frac{\Phi}{k} + m\Phi'_z = a(\Phi^k \Phi'_z)'_z + \Phi^{k+1} f(\Phi/\Psi),$$
$$-\frac{\Psi}{k} + m\Psi'_z = b(\Psi^k \Psi'_z)'_z + \Psi^{k+1} g(\Phi/\Psi).$$

◀

► Example 11.8. Let us look at another PDE system comprised of two reaction-diffusion type equations (Aksenov & Polyanin, 2021):

$$u_t = a(u^k u_x)_x + uf(u/v),$$
$$v_t = b(v^k v_x)_x + vg(u/v), \qquad (11.1.4.3)$$

where a, b, and k are some constants ($k \neq 0$), while $f(z)$ and $g(z)$ are arbitrary functions.

The system of PDEs (11.1.4.3) does not include the independent variables explicitly and has a simple solution of the form

$$u = x^{2/k}\varphi(t), \qquad v = x^{2/k}\psi(t), \qquad (11.1.4.4)$$

where the functions $\varphi = \varphi(t)$ and $\psi = \psi(t)$ are described by the system of first-order ODEs

$$\varphi'_t = \frac{2a(k+2)}{k^2}\varphi^{k+1} + \varphi f(\varphi/\psi),$$
$$\psi'_t = \frac{2b(k+2)}{k^2}\psi^{k+1} + \psi g(\varphi/\psi).$$

Solution (11.1.4.4) and system (11.1.4.3) remain unchanged under the scaling transformation $x = c\bar{x}$, $u = c^{2/k}\bar{u}$, $v = c^{2/k}\bar{v}$. Therefore, by virtue of Proposition 2 (to be precise, its generalization), the system of equations (11.1.4.3) also has a more complex solution of the form

$$u = e^{-2mt/k}\Phi(z), \qquad v = e^{-2mt/k}\Psi(z), \qquad z = xe^{mt},$$

where the functions $\Phi = \Phi(z)$ and $\Psi = \Psi(z)$ are described by the ODE system:

$$mz\Phi'_z - \frac{2m}{k}\Phi = a(\Phi^k \Phi'_z)'_z + \Phi f(\Phi/\Psi),$$
$$mz\Psi'_z - \frac{2m}{k}\Psi = b(\Psi^k \Psi'_z)'_z + \Psi f(\Phi/\Psi),$$

where m is an arbitrary constant ($m \neq 0$). ◀

Problems and Exercises for Section 11.1

1. Consider the nonlinear heat equation

$$u_t = a(u^b u_x)_x.$$

(a) Find a simple exact solution of this equation in the form

$$u = Ax^\alpha t^\beta.$$

(b) Taking advantage of the solution from Item (a), construct two more complex solutions using scaling transformations and Propositions 1 and 2.

2. Consider the nonlinear equation

$$u_t = a u_x u_{xx}$$

(a) Find a simple exact solution of this equation in the form

$$u = A x^\alpha t^\beta.$$

(b) Taking advantage of the solution from Item (a), construct two more complex solutions using scaling transformations and Propositions 1 and 2.

3. Consider the equation describing a boundary layer on a flat plate for the stream function

$$u_y u_{xy} - u_x u_{yy} = \nu u_{yyy}.$$

(a) Show that this equation has a simple solution of the form

$$u = \frac{ax}{y}$$

and express the undetermined coefficient a in terms of ν.

(b) Taking advantage of the solution from Item (a), construct two more complex solutions using scaling transformations and Propositions 1 and 2.

4. The change in film thickness of a heavy viscous liquid flowing over a horizontal superhydrophobic surface, characterized by a variable surface tension coefficient, is described by a PDE with variable coefficients that can be written as follows (Aksenov et al., 2017):

$$u_t = [(au^3 + bx^{2/3}u^2)(u_x - c(x^2 u_{xx})_x)]_x, \qquad (11.1.4.5)$$

where a, b, and c are some constants.

(a) Show that this equation has a simple power-law solution

$$u = x^{2/3} f(t)$$

and derive an ODE for the function $f = f(t)$.

(b) Taking advantage of the solution from Item (a), construct a more complex solution using scaling transformation and Proposition 2.

5. Consider the following nonlinear system consisting of two wave-type equations (a system of Klein–Gordon type PDEs):

$$u_{tt} = a(u^b u_x)_x + u f(u/v),$$
$$v_{tt} = a(v^b v_x)_x + v g(u/v),$$

where a and b are some constants ($b \neq 0$), while $f(z)$ and $g(z)$ are arbitrary functions.

(a) Show that this system has a simple solution of the form

$$u = x^k \varphi(t), \quad v = x^k \psi(t).$$

Find the coefficient k and an ODE system for the functions $\varphi = \varphi(t)$ and $\psi = \psi(t)$.

(b) Taking advantage of the solution from Item (a), construct a more complex solution using scaling transformation and Proposition 2.

5. Consider the following nonlinear system consisting of two wave-type equations (a system of Klein–Gordon type PDEs):

$$u_{tt} = a(u^b u_x)_x + u^{b+1} f(u/v),$$
$$v_{tt} = a(v^b v_x)_x + v^{b+1} g(u/v),$$

where a and b are some constants ($b \neq 0$), while $f(z)$ and $g(z)$ are arbitrary functions.

(a) Show that this system has a simple solution of the form

$$u = t^k \varphi(x), \qquad v = t^k \psi(x).$$

Find the coefficient k and the ODE system for the functions $\varphi = \varphi(x)$ and $\psi = \psi(x)$.

(b) Taking advantage of the solution from Item (a), construct a more complex solution using scaling transformation and Proposition 1.

11.2. Constructing Complex Exact Solutions by Generalizing Simple Solutions

11.2.1. Constructing Complex Exact Solutions by Adding Terms to Simpler Solutions

In certain instances, simple solutions can be generalized by including one or more additional terms, resulting in more complex, generalized separable solutions (refer to Chapter 5). We will illustrate this reasoning through examples of the Boussinesq equation (11.1.2.9) and Guderley equation (11.1.2.12).

► Example 11.9. As noted earlier, the Boussinesq equation (11.1.2.9) has a simple separable solution (11.1.2.10) quadratic in the space variable x. We present it in the following form:

$$u = \varphi(t)x^2, \quad \varphi(t) = -1/(6at). \tag{11.2.1.1}$$

Let us try to seek a more complex solution to equation (11.1.2.9) in the additive form

$$u(x,t) = \varphi(t)x^2 + \psi(t)x^k, \quad k \neq 2, \tag{11.2.1.2}$$

where the first term coincides with solution (11.2.1.1). The second term in (11.2.1.2) includes a function $\psi(t)$ and a coefficient k, which need to be found.

On substituting (11.2.1.2) into (11.1.2.9) and after simple rearrangements, we obtain

$$(\varphi_t' - 6a\varphi^2)x^2 + [\psi_t' - a(k+1)(k+2)\varphi\psi]x^k - ak(2k-1)\psi^2 x^{2k-2} = 0. \tag{11.2.1.3}$$

Since this equation must hold identically for any x, the functional coefficients of like powers of x in (11.2.1.3) must be zero. This requirement results in two possible cases: $k = 0$ and $k = 1/2$ (both correspond to the vanishing of the coefficient of x^{2k-2}). These cases need to be analyzed separately.

1°. *First case.* On substituting $k=0$ into (11.2.1.3), we get the following system of ODEs for the functions $\varphi=\varphi(t)$ and $\psi=\psi(t)$:

$$\varphi'_t - 6a\varphi^2 = 0, \qquad \psi'_t - 2a\varphi\psi = 0.$$

Its general solution is expressed as

$$\varphi(t) = -\frac{1}{6a(t+C_1)}, \qquad \psi(t) = \frac{C_2}{|t+C_1|^{1/3}}, \tag{11.2.1.4}$$

where C_1 and C_2 are arbitrary constants.

2°. *Second case* (Berenblatt–Zeldovich dipole-type solution). On substituting $k=1/2$ into (11.2.1.3), we arrive at the following system of ODEs for $\varphi=\varphi(t)$ and $\psi=\psi(t)$:

$$\varphi'_t - 6a\varphi^2 = 0, \qquad \psi'_t - \tfrac{15}{4}a\varphi\psi = 0.$$

Its general solution is

$$\varphi(t) = -\frac{1}{6a(t+C_1)}, \qquad \psi(t) = \frac{C_2}{|t+C_1|^{5/8}}. \tag{11.2.1.5}$$

Considering formulas (11.2.1.2), (11.2.1.4), and (11.2.1.5), we ultimately obtain two generalized separable solutions with three arbitrary constants for equation (11.1.2.9):

$$u = -\frac{1}{6a(t+C_1)}(x+C_3)^2 + \frac{C_2}{|t+C_1|^{1/3}},$$
$$u = -\frac{1}{6a(t+C_1)}(x+C_3)^2 + \frac{C_2}{|t+C_1|^{5/8}}(x+C_3)^{1/2}.$$

For greater generality, an arbitrary translation in the space variable x has been added here. ◀

Remark 11.3. *The wave equation with a quadratic nonlinearity*

$$u_{tt} = a(uu_x)_x,$$

also admits solutions of the form (11.2.1.2) with $k=0$ and $k=1/2$.

▶ Example 11.10. Now let us return to the Guderley equation (11.1.2.12). It admits a simple exact solution (11.1.2.13), which can be written as

$$u = f(x)y^3, \qquad f(x) = 1/(3ax^2).$$

We will look for more complex (generalized separable) solutions for equation (11.1.2.12) in the form

$$u(x,y) = \varphi(x)y^k + \psi(x), \tag{11.2.1.6}$$

where the function $\varphi(x)$ and $\psi(x)$ and the constant $k\neq 0$ are to be determined; note that solution (11.1.2.13) is a special case of solution (11.2.1.6) with $k=3$ and $\psi=0$. Notably, such two-term solutions PDEs are quite common in practice and represent the simplest cases of generalized separable solutions.

On substituting (11.2.1.6) into (11.1.2.12) and rearranging, we obtain

$$\varphi''_{xx} y^k - ak^2(k-1)\varphi^2 y^{2k-3} + \psi''_{xx} = 0, \tag{11.2.1.7}$$

This equation include the power-law functions y^k and y^{2k-3} and must hold identically for any y.

We will consider two situations: $\psi''_{xx} = 0$ and $\psi''_{xx} \neq 0$.

1°. *First case.* For $\psi''_{xx} = 0$, we get a two-term separable equation, which can be satisfied if we set

$$k = 3, \quad \varphi''_{xx} - 18a\varphi^2 = 0. \tag{11.2.1.8}$$

The general solution of the autonomous ODE (11.2.1.8) can be represented in the implicit form

$$x = \pm \int (12a\varphi^3 + C_1)^{-1/2} d\varphi + C_2.$$

Also, this equation admits a power-law particular solution $\varphi = \frac{1}{3a}(x+C_1)^{-2}$, which leads to a three-parameter exact solution of equation (11.1.2.12):

$$u = \frac{1}{3a}(x + C_1)^{-2} y^3 + C_2 x + C_3, \tag{11.2.1.9}$$

where C_1, C_2, and C_3 are arbitrary constants.

2°. *Second case.* To balance the function $\psi''_{xx} \neq 0$ with the second term in equation (11.2.1.7), we must set $k = 3/2$. As a result, we obtain a two-term equation, which can be satisfied if we set

$$\varphi''_{xx} = 0, \quad \psi''_{xx} = \tfrac{9}{8} a\varphi^2.$$

These equations are easy to integrate and lead to a four-parameter exact solution of equation (11.1.2.12):

$$u = (C_1 x + C_2) y^{3/2} + \frac{3a}{32C_1^2}(C_1 x + C_2)^4 + C_3 x + C_4, \tag{11.2.1.10}$$

where $C_1, \ldots, C_4$ are arbitrary constants. ◀

► Example 11.11. Consider the equation of a hydrodynamic boundary layer on a flat plate

$$u_y u_{xy} - u_x u_{yy} = \nu u_{yyy}, \tag{11.2.1.11}$$

where u is the stream function and ν is kinematic viscosity. It is easy to verify that this equation admits a self-similar solution

$$u = F(\xi), \quad \xi = y/x, \tag{11.2.1.12}$$

where the function $F = F(\xi)$ satisfies the third-order ODE: $-(F'_z)^2 = \nu F'''_{zzz}$.

We will look for a more general solution of equation (11.2.1.11) by adding a function $\varphi(x)$ to (11.2.1.12):

$$u = F(\xi) + \varphi(x), \quad \xi = y/x.$$

Simple calculations show that $\varphi(x) = a \ln x$, where a is an arbitrary constant. As a result, we obtain a more complex (non-self-similar) solution of the boundary layer equation (11.2.1.11) of the form

$$u = F(\xi) + a \ln x, \qquad \xi = y/x,$$

where the function $F = F(\xi)$ satisfies the ODE: $-(F'_z)^2 - aF''_{zz} = \nu F'''_{zzz}$. ◄

11.2.2. Constructing Composite Solutions (Nonlinear Superposition of Solutions)

In certain situations, two similar yet distinct solutions to a nonlinear equation can be combined to create a more comprehensive composite solution. To illustrate this reasoning, we will use the examples of Guderley's equation and a nonlinear diffusion equation involving a second-order volume chemical reaction.

► Example 11.12. From the expressions (11.2.1.9) and (11.2.1.10) it follows that the Guderley equation (11.1.2.12) has two one-type solutions $u_1 = \varphi y^{3/2} + \psi$ and $u_2 = \varphi y^3 + \psi$, which differ from each other in the functions φ and ψ, as well as in the exponent of y. This circumstance suggests the idea of trying to construct a more general solution of equation (11.1.2.12), including both terms with different exponents at once. In other words, we are looking for a composite solution of the form

$$u(x, y) = \varphi_1(x) y^3 + \varphi_2(x) y^{3/2} + \psi(x). \tag{11.2.2.1}$$

Substituting it into equation (11.1.2.12) and collecting the terms with like powers of $y^{3n/2}$ ($n = 0,\ 1,\ 2$), we obtain

$$(\varphi_1'' - 18a\varphi_1^2)y^3 + (\varphi_2'' - \tfrac{45}{4}a\varphi_1\varphi_2)y^{3/2} + \psi'' - \tfrac{9}{8}a\varphi_2^2 = 0.$$

For this relation to hold for any y, we must equate the coefficients of $y^{3n/2}$ with zero. As a result, we arrive at the ODE system

$$\begin{aligned}
&\varphi_1'' - 18a\varphi_1^2 = 0,\\
&\varphi_2'' - \tfrac{45}{4}a\varphi_1\varphi_2 = 0,\\
&\psi'' - \tfrac{9}{8}a\varphi_2^2 = 0.
\end{aligned} \tag{11.2.2.2}$$

Thus, it is constructively proven that equation (11.1.2.12) has a solution of the form (11.2.2.1).

It can be shown that system (11.2.2.2) admits the exact solution

$$\begin{aligned}
\varphi_1 &= \frac{1}{3a}(x+C_1)^{-2},\\
\varphi_2 &= C_2(x+C_1)^{5/2} + C_3(x+C_1)^{-3/2},\\
\psi &= \frac{3a}{112}C_2^2(x+C_1)^7 + \frac{3}{8}aC_2C_3(x+C_1)^3 + \frac{9}{16}aC_3^2(x+C_1)^{-1} + C_4x + C_5,
\end{aligned}$$

where $C_1, \ldots, C_5$ are arbitrary constants. ◄

► Example 11.13. Now let us explore the nonlinear diffusion equation with a second-order volume chemical reaction

$$u_t = a(uu_x) - bu^2. \tag{11.2.2.3}$$

We will be constructing a composite solution for this equation in two steps: first, we will identify two relatively simple solutions, and then, using these solutions, we will create a composite solution.

1°. *Exponential solution in x.* We look for generalized separable solutions of equation (11.2.2.3) in the form

$$u(x,t) = \varphi(t)e^{\lambda x} + \psi(t), \tag{11.2.2.4}$$

where the functions $\varphi = \varphi(t)$ and $\psi = \psi(t)$ and the constant λ are subject to determination in the process of further analysis. On substituting (11.2.2.4) into (11.2.2.3) and collecting terms of like exponentials $e^{n\lambda x}$ ($n = 0,\ 1,\ 2$), we obtain

$$(b - 2a\lambda^2)\varphi^2 e^{2\lambda x} + [\varphi_t' + (2b - a\lambda^2)\varphi\psi]e^{\lambda x} + \psi_t' + b\psi^2 = 0.$$

Since this relation must hold identically for any x, the functional coefficients of $e^{n\lambda x}$ ($n = 0,\ 1,\ 2$) must be set equal to zero. As a result, we arrive at a simple differential-algebraic system:

$$b - 2a\lambda^2 = 0,$$
$$\varphi_t' + (2b - a\lambda^2)\varphi\psi = 0,$$
$$\psi_t' + b\psi^2 = 0.$$

It admits two solutions

$$\lambda = \pm\left(\frac{b}{2a}\right)^{1/2}, \quad \varphi = \frac{C_1}{|t + C_2|^{3/2}}, \quad \psi = \frac{1}{b(t + C_2)}, \tag{11.2.2.5}$$

where C_1 and C_2 are arbitrary constants.

2°. *Composite solution exponential in x.* It follows from (11.2.2.4) and (11.2.2.5) that equation (11.2.2.3) has two solutions: $u_1 = \varphi e^{-\lambda x} + \psi$ and $u_2 = \varphi e^{\lambda x} + \psi$. They only differ from each other in the sign of decay constant λ.

This circumstance suggests trying to construct a more general solution of equation (11.2.2.3) that includes both exponential terms at once. In other words, we look for a composite solution of the form

$$u(x,t) = \varphi_1(t)e^{-\lambda x} + \varphi_2(t)e^{\lambda x} + \psi(t), \quad \lambda = \left(\frac{b}{2a}\right)^{1/2}. \tag{11.2.2.6}$$

On substituting it into (11.2.2.3) and after simple rearrangements, we get

$$[(\varphi_1)_t' + \tfrac{3}{2}b\varphi_1\psi]e^{-\lambda x} + [(\varphi_2)_t' + \tfrac{3}{2}b\varphi_2\psi]e^{\lambda x} + \psi_t' + b(2\varphi_1\varphi_2 + \psi^2) = 0.$$

Equating the functional coefficients of $e^{n\lambda x}$ ($n = 0,\ \pm 1$) with zero, we arrive at the first-order ODE system

$$(\varphi_1)_t' + \tfrac{3}{2}b\varphi_1\psi = 0,$$
$$(\varphi_2)_t' + \tfrac{3}{2}b\varphi_2\psi = 0, \tag{11.2.2.7}$$
$$\psi_t' + b(2\varphi_1\varphi_2 + \psi^2) = 0.$$

This proves that the diffusion equation with a quadratic nonlinearity (11.2.2.3) admits a solution of the form (11.2.2.6).

On eliminating ψ from the first two equations in (11.2.2.7), we arrive at the relation $(\varphi_1)'_t/\varphi_1 = (\varphi_2)'_t/\varphi_2$. It follows that $\varphi_1 = A\varphi(t)$ and $\varphi_2 = B\varphi(t)$, where A and B are arbitrary constants, and $\varphi(t)$ is any function (at this stage). As a result, the generalized separable solution (11.2.2.6) becomes

$$u(x,t) = \varphi(t)(Ae^{-\lambda x} + Be^{\lambda x}) + \psi(t), \quad \lambda = \left(\frac{b}{2a}\right)^{1/2}, \tag{11.2.2.8}$$

where the functions $\varphi = \varphi(t)$ and $\psi = \psi(t)$ are described by the nonlinear ODE system consisting of two equations

$$\begin{aligned} &\varphi'_t + \tfrac{3}{2}b\varphi\psi = 0, \\ &\psi'_t + b(2AB\varphi^2 + \psi^2) = 0. \end{aligned} \tag{11.2.2.9}$$

This autonomous system reduces to a single ODE through the elimination of t. This ODE is homogeneous and, hence, can be integrated. It is noteworthy that for $AB > 0$, the system of equations (11.2.2.9) admits two simple particular solutions

$$\varphi = \pm\frac{1}{3b\sqrt{AB}\,(t+C)}, \quad \psi = \frac{2}{3b(t+C)},$$

which determine solution (11.2.2.8) as a product of functions with different arguments.

3°. *Trigonometric solution in x.* Formulas (11.2.2.6) and (11.2.2.8) implicitly assumed that $ab > 0$. If $ab < 0$, we get

$$\lambda = i\beta, \quad \beta = \left(-\frac{b}{2a}\right)^{1/2}, \quad i^2 = -1.$$

In this case, instead of exponential functions, solution (11.2.2.8) will include trigonometric functions. Hence, the solution can be written as

$$u(x,t) = \varphi(t)[A_1\cos(\beta x) + B_1\sin(\beta x)] + \psi(t), \quad \beta = \left(-\frac{b}{2a}\right)^{1/2}, \tag{11.2.2.10}$$

where A_1 and B_1 are arbitrary constants. On substituting (11.2.2.10) into the original equation (11.2.2.3) and performing similar calculations to those in Item 2°, we obtain the following nonlinear ODE system for determining the functions $\varphi = \varphi(t)$ and $\psi = \psi(t)$:

$$\begin{aligned} &\varphi'_t + \tfrac{3}{2}b\varphi\psi = 0, \\ &\psi'_t + b[\tfrac{1}{2}(A_1^2 + B_1^2)\varphi^2 + \psi^2] = 0. \end{aligned} \tag{11.2.2.11}$$

This system admits two simple particular solutions

$$\varphi = \pm\frac{2}{3b\sqrt{A_1^2 + B_1^2}\,(t+C)}, \quad \psi = \frac{2}{3b(t+C)},$$

which determine solution (11.2.2.10) as a product of functions with different arguments. ◄

11.3. Using Complex-Valued Parameters to Construct Exact Solutions for PDEs

11.3.1. Linear Partial Differential Equations

For linear partial differential equations, more complicated solutions can be constructed from simpler ones using the following statement.

Proposition 4. *Suppose a linear homogeneous partial differential equation in two independent variables, x and t, has a one-parameter solution of the form $u = \varphi(x,t,c)$, where c is an arbitrary parameter that does not appear in the original equation. Then this equation also has two two-parameter solutions*

$$u_1 = \operatorname{Re}\varphi(x,t,a+ib), \qquad u_2 = \operatorname{Im}\varphi(x,t,a+ib),$$

where a and b are arbitrary real constants, $\operatorname{Re} z$ and $\operatorname{Im} z$ are the real and imaginary parts of the complex number z, and $i^2 = -1$.

Two corollaries follow from Proposal 4.

Corollary 4.1. Let the left-hand side of the linear homogeneous partial differential equation be explicitly independent of time t (meaning that this equation is invariant under translation in t) and has a solution $u = \varphi(x,t)$. Then this equation also has two one-parameter families of solutions

$$u_1 = \operatorname{Re}\varphi(x,t+ia), \qquad u_2 = \operatorname{Im}\varphi(x,t+ia),$$

where a is an arbitrary real constant.

Corollary 4.2. Let the left-hand side of the linear homogeneous partial differential equation be explicitly independent of the space variable x (meaning that this equation is invariant under translation in x) and has a solution $u = \varphi(x,t)$. Then this equation also has two one-parameter families of solutions

$$u_1 = \operatorname{Re}\varphi(x+ia,t), \qquad u_2 = \operatorname{Im}\varphi(x+ia,t),$$

where a is an arbitrary real constant.

Remark 11.4. *The proofs of Proposition 4 and the two corollaries are detailed in the articles by Aksenov & Polyanin (2021, 2022).*

► Example 11.14. Consider the linear heat equation

$$u_t - u_{xx} = 0. \qquad (11.3.1.1)$$

It is easy to verify that this equation admits a simple exact solution of the exponential form

$$u = \exp(c^2 t + cx),$$

where c is an arbitrary parameter.

Using Proposition 4, we find two more complicated two-parameter families of exact solutions to equation (11.3.1.1):

$$u_1 = \operatorname{Re}\exp(c^2t + cx)|_{c=a+ib} = \exp[(a^2 - b^2)t + ax]\cos[b(2at + x)],$$
$$u_2 = \operatorname{Im}\exp(c^2t + cx)|_{c=a+ib} = \exp[(a^2 - b^2)t + ax]\sin[b(2at + x)].$$

◀

► Example 11.15. Consider the more complex linear heat equation

$$u_t = u_{xx} + \frac{1}{x}u_x. \tag{11.3.1.2}$$

It describes two-dimensional processes with axial symmetry, where x is a radial coordinate. It is easy to verify that equation (11.3.1.2) admits a translation transformation in t and has a particular solution

$$u = \frac{1}{t}\exp\Bigl(-\frac{x^2}{4t}\Bigr). \tag{11.3.1.3}$$

By performing a translation in solution (11.3.1.3) by an imaginary parameter along the variable t and using Corollary 4.1, we find two more complicated one-parameter families of solutions:

$$\begin{aligned}
u_1 &= \operatorname{Re}\frac{1}{t+ia}\exp\left(-\frac{x^2}{4(t+ia)}\right)\\
&= \frac{1}{t^2+a^2}\exp\left(-\frac{x^2t}{4(t^2+a^2)}\right)\left(t\cos\frac{ax^2}{4(t^2+a^2)} + a\sin\frac{ax^2}{4(t^2+a^2)}\right),\\
u_2 &= \operatorname{Im}\frac{1}{t+ia}\exp\left(-\frac{x^2}{4(t+ia)}\right)\\
&= \frac{1}{t^2+a^2}\exp\left(-\frac{x^2t}{4(t^2+a^2)}\right)\left(a\cos\frac{ax^2}{4(t^2+a^2)} - t\sin\frac{ax^2}{4(t^2+a^2)}\right).
\end{aligned}$$

◀

11.3.2. Nonlinear Partial Differential Equations

In certain instances, a solution to a nonlinear equation can be transformed into another solution by transitioning from real parameters to complex ones while ensuring that the transformed equation and solution remain real. We will illustrate this concept with specific examples.

► Example 11.16. Let us return to equation (11.2.2.3). It is easy to verify that the trigonometric solution (11.2.2.10) and the nonlinear ODE system (11.2.2.11) can be obtained from the exponential solution (see solution (11.2.2.8)):

$$u(x,t) = \varphi(t)(Ae^{-\lambda x} + Be^{\lambda x}) + \psi(t), \quad \lambda = \left(\frac{b}{2a}\right)^{1/2},$$

and the system of nonlinear ODEs (11.2.2.9) if we formally set

$$\begin{aligned} e^{\lambda x} &= e^{i\beta x} = \cos(\beta x) + i\sin(\beta x), \\ e^{-\lambda x} &= e^{-i\beta x} = \cos(\beta x) - i\sin(\beta x), \\ A &= \tfrac{1}{2}(A_1 + iB_1), \quad B = \tfrac{1}{2}(A_1 - iB), \\ A_1 &= A + B, \quad B_1 = i(B - A), \end{aligned} \qquad (11.3.2.1)$$

where $\beta = [-b/(2a)]^{1/2}$ and $i^2 = -1$. ◀

► Example 11.17. One can verify by a direct check that the parabolic Monge–Ampère equation with three independent variables

$$u_t = u_{xx}u_{yy} - u_{xy}^2 \qquad (11.3.2.2)$$

admits a simple radially symmetric exact solution

$$u = -\frac{(x^2 + y^2)^2}{48t}. \qquad (11.3.2.3)$$

Equation (11.3.2.2) will remain unchanged under the transformation $\bar{u} = -u$, $\bar{y} = iy$, where $i^2 = -1$. By making a similar transformation in (11.3.2.3), we obtain another exact solution of equation (11.3.2.2):

$$u = \frac{(x^2 - y^2)^2}{48t}.$$

The following statement presents a more general result.

Proposition 5. *Let* $u = f(t, x^2 + y^2)$ *be any radially symmetric solution of the parabolic Monge–Ampère equation* (11.3.2.2). *Then this equation also admits the solution* $u = -f(t, x^2 - y^2)$. ◀

► Example 11.18. Consider the nonlinear parabolic equation

$$u_t = au_{xx} + uf(u_x^2 - bu^2), \qquad (11.3.2.4)$$

where $f(z)$ is an arbitrary function.

It is easy to verify that equation (11.3.2.4) has a simple multiplicative separable solution exponential in x:

$$u = \psi(t)e^{\lambda x}, \qquad (11.3.2.5)$$

where the parameter λ and the function $\psi(t)$ are to be determined in the process of the further analysis. Substituting (11.3.2.5) into equation (11.3.2.4) yields two solutions of the form (11.3.2.5), where

$$\lambda = \pm\sqrt{b}, \qquad \psi_t' = [ab + f(0)]\psi.$$

The existence of two similar solutions corresponding to $\pm\lambda$ indicates the possibility of creating a linear combination to seek a more general composite solution of the form

$$u = \psi(t)(Ae^{\lambda x} + Be^{-\lambda x}), \qquad (11.3.2.6)$$

where A and B are some constants. A direct check confirms that function (11.3.2.6) for arbitrary A and B is indeed a solution to equation (11.3.2.4), and the function $\psi = \psi(t)$ satisfies the nonlinear separable first-order ODE

$$\psi_t' = ab\psi + \psi f(-4ABb\psi^2). \tag{11.3.2.7}$$

Substituting (11.3.2.1) into (11.3.2.6) and (11.3.2.7) yields a new solution that involves trigonometric functions of x:

$$u = \varphi(t)[A_1 \cos(\beta x) + B_1 \sin(\beta x)], \qquad \beta = \sqrt{-b},$$

where A_1 and B_1 are arbitrary constants, and the function $\psi = \psi(t)$ is described by nonlinear first-order ODE

$$\psi_t' = ab\psi + \psi f\big(-(A_1^2 + B_1^2)b\psi^2\big).$$

◄

Proposition 6. *Suppose a nonlinear PDE has an exact solution involving trigonometric functions*

$$u = F(x, t, A\cos(\beta x) + B\sin(\beta x), \beta^2), \tag{11.3.2.8}$$

where A, B, and β are free real parameters not involved in the equation. Then this equation also has an exact solution containing hyperbolic functions:

$$u = F(x, t, \bar{A}\cosh(\lambda x) + \bar{B}\sinh(\lambda x), -\lambda^2), \tag{11.3.2.9}$$

where $\bar{A}$, $\bar{B}$, and λ are free real parameters. The converse is also true: if an equation has a solution of the form (11.3.2.9), *then the equation also has an exact solution of the form* (11.3.2.8).

Solution (11.3.2.9) can be derived from (11.3.2.8) by renaming the parameters: $\beta = i\lambda$, $A = \bar{A}$, and $B = -i\bar{B}$, where $i^2 = -1$.

► Example 11.19. Consider the nonlinear fourth-order equation

$$u_y(\Delta u)_x - u_x(\Delta u)_y = \nu\Delta\Delta u, \qquad \Delta u = u_{xx} + u_{yy}, \tag{11.3.2.10}$$

which results from the two-dimensional steady-state Navier–Stokes equations in the planar case.

Equation (11.3.2.10) has an exact solution

$$u(x, y) = [\bar{A}\sinh(\lambda x) + \bar{B}\cosh(\lambda x)]e^{-\gamma y} + \frac{\nu}{\gamma}(\gamma^2 + \lambda^2)x.$$

Therefore, by virtue of Proposition 6, this equation also has the exact solution

$$u(x, y) = [A\sin(\beta x) + B\cos(\beta x)]e^{-\gamma y} + \frac{\nu}{\gamma}(\gamma^2 - \beta^2)x.$$

◄

Problems and Exercises for Section 11.3

1. The linear heat equation (11.3.1.1) has a particular solution

$$u = \frac{1}{\sqrt{t}} \exp\Big(-\frac{x^2}{4t}\Big).$$

Using this solution, derive more complicated solutions of this equation through translation with an imaginary parameter in the variable t (take advantage of Corollary 4.1).

2. Consider the linear wave equation

$$u_{tt} - u_{xx} = 0.$$

(a) Verify that this equation admits translation transformations in both independent variables and has a particular solution

$$u = \frac{x}{x^2 - t^2}.$$

(b) Using the exact solution from Item (a), derive more complicated solutions of this equation through translations with imaginary parameters in the variables t and x (take advantage of Corollaries 4.1 and 4.2).

3. Consider the linear wave-type PDE with delay

$$u_{tt} = w_{xx}, \quad w = u(x, t - \tau).$$

(a) Verify that this equation has an exponential particular solution

$$\tilde{u}(x,t) = \exp\big(\mu e^{\mu\tau/2} x + \mu t\big),$$

where μ is an arbitrary constant.

(b) By setting $\mu = i\alpha$ in the solution from Item (a) $\mu = i\alpha$ followed by extracting the real and imaginary parts, derive two more complicated solutions of the equation.

4. Consider the linear wave equation with variable coefficients

$$u_{tt} - (xu_x)_x = 0.$$

This equation is used to describe the propagation of localized disturbances in one-dimensional shallow water over a sloping bottom.

(a) Verify that this equation admits a translation transformation in time t and has a particular solution

$$u = \frac{Ct}{(4x - t^2)^{3/2}},$$

where C is an arbitrary constant.

(b) Using the exact solution from Item (a), derive more complicated solutions of this equation through translation with an imaginary parameter in t (take advantage of Corollary 4.1).

References for Chapter 11

Aksenov A.V. and Polyanin A.D. Methods for constructing complex solutions of nonlinear PDEs using simpler solutions. *Mathematics*, 2021, Vol. 9, No. 4, p. 345.

Aksenov A.V. and Polyanin A.D. Review of methods for constructing exact solutions of equations of mathematical physics based on simpler solutions. *Theor. Math. Phys.*, 2022, Vol. 211, No. 2, pp. 567–594.

Aksenov A.V., Sudarikova A.D., and Chicherin I.S. The surface tension effect on viscous liquid spreading along a superhydrophobic surface. *J. Phys. Conf. Ser.*, 2017, Vol. 788, 01200.

Guderley K.G. *The Theory of Transonic Flow*. Pergamon Press, Oxford–London, 1962.

Polyanin A.D., Aksenov A.V. Unsteady magnetohydrodynamics PDE of Monge–Ampère type: Symmetries, closed-form solutions, and reductions. *Mathematics*, 2024, Vol. 12, No. 13, 2127.

Smirnov V.V. and Chukbar K.V. "Phonons" in two-dimensional vortex lattices. *J. Exper. Theor. Physics*, 2001, Vol. 93, No. 1, pp. 126–135.

Zaburdaev V.Yu., Smirnov V.V., and Chukbar K.V. Nonlinear dynamics of electron vortex lattices. *Plasma Physics Rep.*, 2014, Vol. 30, No. 3, pp. 214–217.

12. Constructing Solutions of Complex Equations by Solutions of Simpler Equations

► **Preliminary remarks.** In practice, when studying complex nonlinear partial differential equations, there are instances where standard methods, or those that a researcher is well versed in, may either fail entirely or prove to be highly ineffective.* This chapter discusses non-standard constructive approaches and provides examples of nonrigorous reasoning that utilize heuristic structural analogies between the solutions of different equations. Not infrequently, these approaches enable the discovery of exact solutions to nonlinear PDEs in such challenging cases.

In particular, the chapter presents several simple yet quite effective methods for constructing exact solutions of nonlinear PDEs and delay PDEs, as well as functional PDEs. These methods typically require minimal intermediate calculations. The core idea behind these methods is that *exact solutions of simpler equations can serve as a basis for deriving solutions to more complex related equations or other classes of equations that have similar nonlinear terms.*

This chapter discusses how exact solutions for complex PDEs can be derived from exact solutions of simpler PDEs. It explores a method for finding exact solutions to equations involving multiple spatial variables based on solutions to similar equations with a single spatial variable. The problem is discussed in which cases exact solutions of non-autonomous nonlinear PDEs with coefficients explicitly dependent on time can be constructed using solutions of related simpler autonomous PDEs with coefficients that are independent of time. The chapter also demonstrates that, in some instances, it is possible to obtain exact solutions for nonlinear delay PDEs by using solutions of the corresponding simpler PDEs without delay. Additionally, the principle of structural analogy of solutions is presented, which enables the construction of exact solutions for functional PDEs with proportional arguments based on simpler PDEs that involve usual arguments.

The effectiveness of the methods described is illustrated by specific examples. Nonlinear equations of the reaction-diffusion type (with or without delay or with proportional arguments) and nonlinear wave-type equations are considered.

---

*The author of this book frequently faced such situations while analyzing nonlinear PDEs with constant or variable delay.

12.1. Constructing Solutions of Complex PDEs Using Solutions of Simpler PDEs

12.1.1. Using Solutions of Linear Equations to Construct Exact Solutions of Nonlinear Equations

In certain instances, exact solutions to complex nonlinear partial differential equations can be derived from solutions of simpler related equations.

The two examples below illustrate how, through non-rigorous reasoning, exact solutions of nonlinear hyperbolic equations can be obtained by utilizing solutions of their corresponding simpler linear equations.

► Example 12.1. Consider the model gas dynamics equation

$$u_t = a(u)u_x, \tag{12.1.1.1}$$

where $a(u)$ is an arbitrary function.

Instead of the nonlinear equation (12.1.1.1), we will explore an auxiliary simpler linear first-order equation:

$$u_t = au_x, \tag{12.1.1.2}$$

where a is an arbitrary constant. It is easy to verify that the general solution of equation (12.1.1.2) is

$$u = \Phi(x + at), \tag{12.1.1.3}$$

where $\Phi(z)$ is an arbitrary smooth function.

Assuming that the function $\Phi(z)$ is monotonic, we represent solution (12.1.1.3) in implicit form:

$$x + at = \varphi(u), \tag{12.1.1.4}$$

where $\varphi(u)$ is an arbitrary smooth function, the inverse of $\Phi(z)$.

By applying the principle of the analogy of solutions, which is a purely heuristic principle in this case, we replace the constant a in solution (12.1.1.4) with an arbitrary function $a(u)$. As a result, we obtain

$$x + a(u)t = \varphi(u). \tag{12.1.1.5}$$

We will now demonstrate that relation (12.1.1.5) represents an exact solution in implicit form for the nonlinear equation (12.1.1.2), which includes an arbitrary function $\varphi(u)$. By differentiating (12.1.1.5) with respect to x and t, we find that

$$u_x = \frac{1}{\varphi'_u - ta'_u}, \qquad u_t = \frac{a}{\varphi'_u - ta'_u}, \qquad a = a(u). \tag{12.1.1.6}$$

On substituting the derivatives (12.1.1.6) into the original equation (12.1.1.1), we obtain an identity.

Note that relation (12.1.1.5) with an arbitrary function $\varphi(u)$ is the general solution (in implicit form) of the original equation (12.1.1.1). ◄

► Example 12.2. Now let us find an exact solution of the nonlinear wave equation

$$u_{tt} = [a(u)u_x]_x, \tag{12.1.1.7}$$

where $a(u)$ is an arbitrary twice differentiable positive function.

To this end, we explore the simpler linear wave equation

$$u_{tt} = au_{xx}, \tag{12.1.1.8}$$

where a is an arbitrary positive constant. Equation (12.1.1.8) has the following traveling-wave particular solutions:

$$u = \Phi(x \pm \sqrt{a}\,t), \tag{12.1.1.9}$$

where $\Phi(z)$ is an arbitrary twice differentiable function.

Assuming that the function $\Phi(z)$ is monotonic, we represent solutions (12.1.1.9) in implicit form:

$$x \pm \sqrt{a}\,t = \varphi(u), \tag{12.1.1.10}$$

where $\varphi(u)$ is an arbitrary twice differentiable function, the inverse of $\Phi(z)$.

Applying the solution analogy principle and replacing the constant a in solution (12.1.1.10) with an arbitrary function $a(u)$, we obtain

$$\varphi(u) = x \pm t\sqrt{a(u)}. \tag{12.1.1.11}$$

Now we will demonstrate that relations (12.1.1.11), which include an arbitrary function $\varphi(u)$, represent exact solutions to the nonlinear wave equation (12.1.1.7). To be specific, we choose the upper sign in (12.1.1.11). Differentiating (12.1.1.11) with respect to x and t, we find that

$$u_x = \frac{1}{\varphi'_u - t(\sqrt{a})'_u}, \qquad u_t = \frac{\sqrt{a}}{\varphi'_u - t(\sqrt{a})'_u}, \qquad a = a(u). \tag{12.1.1.12}$$

It is evident that the partial derivatives are related by a differential constraint:

$$u_t = \sqrt{a}\,u_x. \tag{12.1.1.13}$$

Considering that, we find the second derivative with respect to t:

$$u_{tt} = (\sqrt{a}\,u_x)_t = (\sqrt{a}\,u_t)_x = (au_x)_x, \qquad a = a(u). \tag{12.1.1.14}$$

It follows that under condition (12.1.1.13), the left-hand side of the nonlinear equation (12.1.1.7) equals its right-hand side, suggesting that the implicit relation (12.1.1.11) represents an exact solution of this equation. ◄

12.1.2. Examples of Non-Rigorous Reasoning Using Structural Analogies Between Solutions of Different PDEs

In some situations, it is useful to exploit assumed structural analogies between solutions of related equations to construct exact solutions of nonlinear PDEs. Typically, more complex solutions can be obtained by suitably generalizing the structure of simpler solutions. To illustrate this logic of non-rigorous reasoning, we will give a few specific examples.

► Example 12.3. Consider the steady-state heat equation in an inhomogeneous anisotropic medium with a nonlinear source

$$a(x^k u_x)_x + b(y^n u_y)_y = f(u), \tag{12.1.2.1}$$

whose coefficients are power-law functions of the independent variables.

The steady-state heat equation in a homogeneous isotropic medium

$$u_{xx} + u_{yy} = f(u), \tag{12.1.2.2}$$

which is equation (12.1.2.1) with $a = b = 1$ and $k = n = 0$, is a much simpler version of the original PDE.

Equation (12.1.2.2) with arbitrary $f(u)$ admits the following two exact solutions:

$$\begin{aligned} &u = U_1(z_1), \quad z_1 = c_1 x + c_2 y \quad \text{(traveling wave solution);} \\ &u = U_2(z_2), \quad z_2 = x^2 + y^2 \quad \text{(radially symmetric solution),} \end{aligned} \tag{12.1.2.3}$$

where c_1 and c_2 are arbitrary constants.

Both solutions (12.1.2.3) can be combined by rewriting them in the more general form

$$u = U(z), \qquad z = \varphi(x) + \psi(y). \tag{12.1.2.4}$$

Assuming further that there is a structural analogy between the solutions of the auxiliary equation (12.1.2.2) and those of the more complex related equation (12.1.2.1), we will look for solutions to PDE (12.1.2.1) in the form (12.1.2.4), where the functions $\varphi(x)$, $\psi(y)$, and $U(z)$ are subject to determination in the course of the further analysis.

On substituting (12.1.2.4) into PDE (12.1.2.1) and after an appropriate analysis, we ultimately arrive at the function $\varphi(x)$ and $\psi(y)$ (see Example 6.15 in Chapter 6):

$$\varphi(x) = \frac{x^{2-k}}{a(2-k)^2}, \quad \psi(y) = \frac{y^{2-n}}{b(2-n)^2} \quad (k, n \neq 2).$$

We also obtain an ODE for $U(z)$:

$$zU''_{zz} + \frac{4-k-n}{(2-k)(2-n)} U'_z = f(U).$$

◄

Remark 12.1. *The coefficients of the original equation (12.1.2.1) are power-law functions of the variables and the exact solutions (12.1.2.3) to the auxiliary equation (12.1.2.2) also involve power-law functions. Considering that, instead of the general representation (12.1.2.4), we can seek solutions in a simpler form where the inner argument is represented as a combination of power-law functions:*

$$u = U(z), \qquad z = s_1 x^{m_1} + s_2 y^{m_2},$$

where m_1, m_2, s_1, and s_2 are undetermined constants that need to be identified. This is the solution structure for equation (12.1.2.1) that was used in the book by Zaitsev & Polyanin (1996).

► Example 12.4. Consider the parabolic Monge–Ampère type equation

$$u_t = u_{xx}u_{yy} - f(x)u_{xy}^2, \qquad (12.1.2.5)$$

where $f(x)$ is an arbitrary function.

To obtain a simpler analogue of PDE (12.1.2.5), we set $f(x) = a = \text{const}$, which yields

$$u_t = u_{xx}u_{yy} - au_{xy}^2. \qquad (12.1.2.6)$$

This equation has a simple solution as a product of the independent variables raised to appropriate powers:

$$u = A\frac{x^2y^2}{t}, \qquad A = \frac{1}{16a - 4}. \qquad (12.1.2.7)$$

At $a = 1$, this solution becomes solution (11.1.3.2) from Example 11.5.

Since the structure of equations (12.1.2.5) and (12.1.2.6) with respect to the independent variables y and t is the same, it is reasonable to assume that the structure of their solutions will also be the same in these variables. Therefore, when solving the more complex equation (12.1.2.5), we can expect that only the component dependent on x will differ. In short, we aim to look for solutions to equation (12.1.2.5) in the following form:

$$u = \frac{y^2}{t}\varphi(x). \qquad (12.1.2.8)$$

On substituting (12.1.2.8) into (12.1.2.5), we arrive at an ODE for $\varphi = \varphi(x)$:

$$2\varphi\varphi''_{xx} - 4f(x)(\varphi'_x)^2 = -\varphi.$$

◄

► Example 12.5. The direct method for constructing functional separable solutions typically relies on identifying exact solutions in explicit form:

$$u = U(z), \qquad z = \varphi(x) + \psi(t), \qquad (12.1.2.9)$$

where all three functions $U(z)$, $\varphi(x)$, and $\psi(t)$ are unknown and need to be determined (refer to Chapter 6 for details). After substituting the expression (12.1.2.9) into the partial differential equation under consideration, we arrive at a functional differential equation. In this equation, the highest order of the derivative of the function U with respect to z matches the highest order of the partial derivative present in the original equation.

In the analysis of several nonlinear partial differential equations below, we will demonstrate why the explicit representation of solutions in equation (12.1.2.9) is not entirely effective. To this end, we will consider the generalized Burgers equation

$$u_t + f(u)u_x = u_{xx}, \qquad (12.1.2.10)$$

which involves an arbitrary function $f(u)$. This equation is explicitly independent of x and t and has a traveling wave solution (which is a special case of solution (12.1.2.9)):

$$u = U(z), \qquad z = \kappa x + \lambda t, \qquad (12.1.2.11)$$

where κ and λ are arbitrary constants. Inserting (12.1.2.11) into (12.1.2.10), we obtain the ODE: $\lambda U'_z + \kappa f(U)U'_z = \kappa^2 U''_{zz}$. Integrating yields its solution in implicit form:

$$\kappa^2 \int \frac{dU}{\kappa F(U) + \lambda U + C_1} = \kappa x + \lambda t + C_2, \qquad F(U) = \int f(U)\,dU, \quad (12.1.2.12)$$

where C_1 and C_2 are arbitrary constants. The argument z on the right-hand side of solution (12.1.2.12) has been replaced with the original independent variables by using (12.1.2.11).

It is clear that even for simple functions such as $f(u) = e^u$, $f(u) = \ln u$, $f(u) = \sin u$, and $f(u) = \cos u$, the solution of equation (12.1.2.12) cannot be expressed in terms of elementary functions.

Similar conclusions can be drawn for the nonlinear wave equation

$$u_{tt} = [f(u)u_x]_x. \quad (12.1.2.13)$$

It has a traveling wave solution (12.1.2.11), which can be written in implicit form:

$$\int [\kappa^2 f(u) - \lambda^2]\,du = C_1(\kappa x + \lambda t) + C_2. \quad (12.1.2.14)$$

This solution cannot be expressed explicitly in terms of elementary functions at least for $f(u) = e^u$, $f = \ln u$, $f(u) = \sin u$, or $f(u) = \cos u$.

To summarize, we can conclude that nonlinear PDEs (12.1.2.10) and (12.1.2.13), which are frequently encountered in applications, admit traveling wave solutions that allow for representation in implicit form. Moreover, for the vast majority of elementary functions $f(u)$, these solutions cannot be written explicitly. Note that Section 6.5.1 lists two other nonlinear PDEs whose exact solutions have similar properties.

Since even the simplest traveling wave solutions of various PDEs usually cannot be represented explicitly, one would not expect that more complex solutions of the above and related nonlinear equations can be represented in explicit form.

Considering the aforesaid and assuming that there are structural analogies between solutions of related equations, we arrive at an implicit representation of exact solutions to nonlinear PDEs in the form

$$\int h(u)\,du = \varphi(x) + \psi(t), \quad (12.1.2.15)$$

where the function $h = h(u)$, $\varphi = \varphi(x)$, and $\psi = \psi(t)$ are subject to determination in the processes of the further analysis.

The representation of solutions in implicit form (12.1.2.15) is a natural generalization of the traveling wave solutions discussed earlier for four different equations, two PDEs in this section and two PDEs in Section 6.5.1. For example, solution (12.1.2.15) can be obtained by generalizing solution (12.1.2.14) of equation (12.1.2.13). This can be done as follows:

$$\kappa^2 f(u) - \lambda^2 \implies h(u), \qquad C_1 \kappa x \implies \varphi(x), \qquad C_1 \lambda t + C_2 \implies \psi(t).$$

It is noteworthy that seeking exact solutions to nonlinear PDEs in the implicit form (12.1.2.15) is technically easier than using explicit representation (12.1.2.9).

This is because, after calculating the partial derivatives $u_x = \varphi'_x/h$, $u_t = \psi'_t/h$, ... followed by substituting them into the PDE, the order of the highest derivative of h with respect to u will be one less than that of the highest partial derivative in the original equation.

The efficiency of implicit representations such as (12.1.2.15) was demonstrated in Section 6.5; for more details, see Polyanin (2019a, 2019b) and Polyanin & Zhurov (2022). ◀

12.1.3. Constructing Exact Solutions of Non-Autonomous Equations by Using Solutions of Autonomous Equations

We will give a general formulation of the important problem discussed below. Suppose we know an exact solution $u = u(x,t)$ of a nonlinear PDE dependent on free parameters $a_1, \dots, a_p$. The question arises: Can we derive any insights about the exact solutions of a more complex PDE, which is obtained by replacing these free parameters with arbitrary functions: $a_1(t), \dots, a_p(t)$?

In what follows, we will demonstrate that exact solutions to non-autonomous PDEs with coefficients explicitly dependent on time t, can often be constructed from solutions of related, simpler autonomous PDEs with coefficients independent of time. Following the study by Polyanin (2024), we will outline the reasoning process—although non-rigorous in this context—by exploring two specific examples. After analyzing these examples, we will formulate general conclusions as rigorous propositions.

► Example 12.6. Consider the nonlinear reaction-diffusion type equation

$$u_t = a_1 u u_{xx} + a_2, \tag{12.1.3.1}$$

where a_1 and a_2 are free parameters. Equation (12.1.3.1) admits a generalized separable solution quadratic in the space variable

$$u = \psi_1(t)x^2 + \psi_2(t), \tag{12.1.3.2}$$

where the functions $\psi_1 = \psi_1(t)$ and $\psi_2 = \psi_2(t)$ satisfy the nonlinear ODE system:

$$\begin{aligned} \psi_1' &= 2a_1\psi_1^2, \\ \psi_2' &= 2a_1\psi_1\psi_2 + a_2. \end{aligned} \tag{12.1.3.3}$$

This system is easy to integrate, but we will not need it further.

Now, by formally replacing the parameters a_1 and a_2 in equation (12.1.3.1) with arbitrary functions $a_1(t)$ and $a_2(t)$, we obtain

$$u_t = a_1(t) u u_{xx} + a_2(t). \tag{12.1.3.4}$$

It is easy to verify that equation (12.1.3.4) also admits an exact solution of the form (12.1.3.2), where the functions $\psi_1 = \psi_1(t)$ and $\psi_2 = \psi_2(t)$ satisfy the nonlinear ODE system (12.1.3.3), in which the parameters a_1 and a_2 should be replaced with arbitrary functions $a_1(t)$ and $a_2(t)$. ◀

This illustrative example allows us to make a generalization, formulated as follows.

Proposition 1. *Suppose an nonlinear autonomous partial differential equation*

$$u_t = F\big(x, u, u_x, \ldots, u_x^{(n)}; a_1, \ldots, a_p\big), \qquad (12.1.3.5)$$

dependent on free parameters a_1, ..., a_p, *has a generalized separable solution as a polynomial in the space variable* x:

$$u = \sum_{k=0}^{m} \psi_k(t) x^k. \qquad (12.1.3.6)$$

Then, the more complex nonautonomous partial differential equation

$$u_t = F\big(x, u, u_x, \ldots, u_x^{(n)}; a_1(t), \ldots, a_p(t)\big), \qquad (12.1.3.7)$$

obtained from (12.1.3.5) *by formally replacing the parameters* a_1, ..., a_p *with arbitrary functions* $a_1(t)$, ..., $a_p(t)$, *also admits an exact solution of the form* (12.1.3.6) *with other functions* $\psi_k(t)$.

This proposition is a special case of Proposition 2 formulated below.

► Example 12.7. Consider another nonlinear equation of the reaction-diffusion type

$$u_t = a_1 u_{xx} + u_x^2 + bu^2 + a_2 \qquad (b > 0), \qquad (12.1.3.8)$$

where a_1, a_2, and b are free parameters. Equation (12.1.3.8) admits a generalized separable solution involving a trigonometric function of x:

$$u = \psi_1(t) \cos(\sqrt{b}\, x) + \psi_2(t), \qquad (12.1.3.9)$$

where the functions $\psi_1 = \psi_1(t)$ and $\psi_2 = \psi_2(t)$ satisfy the nonlinear ODE system

$$\begin{aligned} \psi_1' &= 2b\psi_1\psi_2 - a_1 b\psi_1, \\ \psi_2' &= b(\psi_1^2 + \psi_2^2) + a_2. \end{aligned} \qquad (12.1.3.10)$$

Be formally replacing the parameters a_1 and a_2 in equation (12.1.3.8) with arbitrary functions $a_1(t)$ and $a_2(t)$, we get

$$u_t = a_1(t) u_{xx} + u_x^2 + bu^2 + a_2(t) \qquad (b > 0). \qquad (12.1.3.11)$$

It is easy to verify that equation (12.1.3.11) also admits an exact solution of the form (12.1.3.9), where the function $\psi_1 = \psi_1(t)$ and $\psi_2 = \psi_2(t)$ satisfy the nonlinear ODE system (12.1.3.10), in which a_1 and a_2 must be replaced with $a_1(t)$ and $a_2(t)$.

An attempt to replace the constant b in equation (12.1.3.8) with an arbitrary function $b(t)$ does not lead to success. This is because the function (12.1.3.9) with $b = b(t)$ is no longer an exact solution to the modified equation, which is a consequence of the fact that the coordinate function $\varphi(x) = \cos(\sqrt{b}\, x)$ in solution (12.1.3.9) explicitly depends on b. ◄

The above example allows us to make an straightforward generalization, formulated as follows.

Proposition 2. *Suppose an autonomous nonlinear partial differential equation* (12.1.3.5) *dependent on free parameters* $a_1, \ldots, a_p$ *has a generalized separable solution of the form*

$$u = \sum_{k=1}^{m} \varphi_k(x)\psi_k(t), \tag{12.1.3.12}$$

whose coordinate functions $\varphi_k(x)$ *are all independent of* $a_1, \ldots, a_p$, *and the function* $\psi_k = \psi_k(t)$ *are described by the autonomous ODE system*

$$\psi_k' = f_k(\psi_1, \ldots, \psi_m; a_1, \ldots, a_p), \qquad k = 1, \ldots, m. \tag{12.1.3.13}$$

Then, the more complex nonautonomous partial differential equation (12.1.3.7), *obtained from* (12.1.3.5) *by formally replacing the parameters* $a_1, \ldots, a_k$ *with arbitrary functions* $a_1(t), \ldots, a_p(t)$, *admits an exact solution of the form* (12.1.3.12), *where the coordinate functions* $\varphi_k(x)$ *remain the same, and the functions* $\psi_k = \psi_k(t)$ *are described by the nonautonomous ODE system*

$$\psi_k' = f_k\bigl(\psi_1, \ldots, \psi_m; a_1(t), \ldots, a_p(t)\bigr), \qquad k = 1, \ldots, m. \tag{12.1.3.14}$$

This proposition can be proven using Proposition 1 from Section 5.6.1. Indeed, we substitute (12.1.3.12) into the autonomous equation (12.1.3.5) and then eliminate the derivatives with respect to time using ODE (12.1.3.13). Further, replacing the functions $\psi_k(t)$ with arbitrary constants C_k, we obtain

$$F\left[\sum_{k=1}^{m} C_k\varphi_k(x)\right] = \sum_{k=1}^{m} f_k(C_1, \ldots, C_m; a_1, \ldots, a_p)\varphi_k(x), \tag{12.1.3.15}$$

where $F[u]$ denotes the right-hand side of equation (12.1.3.5). Relations (12.1.3.15) can be interpreted as follows. The finite-dimensional linear subspace

$$\mathscr{L}_m = \bigl\{\varphi_1(x), \ldots, \varphi_m(x)\bigr\},$$

whose elements are all possible linear combinations of the coordinate functions $\varphi_1(x), \ldots, \varphi_m(x)$ appearing in solution (12.1.3.12), is invariant under the nonlinear differential operator in the space variable x on the right-hand side of the autonomous equation (12.1.3.5). The nonlinear differential operator on the right-hand side of the more complex nonautonomous partial differential equation (12.1.3.7), obtained from (12.1.3.5) by formally replacing the parameters $a_1, \ldots, a_k$ with arbitrary functions $a_1(t), \ldots, a_p(t)$, depends on time t parametrically, meaning that it only includes derivatives with respect to x. Therefore, when the expression (12.1.3.12) is substituted into the right-hand side of equation (12.1.3.7), the functions $a_1(t), \ldots, a_p(t)$ will behave as constants, thus leading to an nonautonomous ODE system (12.1.3.14) for $\psi_k = \psi_k(t)$.

Remark 12.2. *The first derivative u_t on the left-hand sides of PDEs (12.1.3.5) and (12.1.3.7) can be replaced with any linear combination of derivatives with respect to t: $L[u] = \sum_{i=1}^{s} c_i(t)u_t^{(i)}$. In this case, the derivatives ψ_k' on the left hand sides of ODEs (12.1.3.13) and (12.1.3.14) in Proposition 2 must be replaced with $L[\psi_k]$.*

Proposition 2 can be rephrased in a shorter but less informative form, as stated below.

Proposition 2a. Suppose an autonomous partial differential equation (12.1.3.5) dependent on free parameters $a_1, \ldots, a_p$ has a solution (12.1.3.12), in which the linearly independent coordinate functions $\varphi_k(x)$ are independent of $a_1, \ldots, a_p$. Then, the more complex nonautonomous partial differential equation (12.1.3.7) obtained from (12.1.3.5) by formally replacing the parameters $a_1, \ldots, a_k$ with arbitrary functions $a_1(t), \ldots, a_p(t)$ will also admit an exact solution of the form (12.1.3.12) in which the coordinate functions $\varphi_k(x)$ remain unchanged, while the functions $\psi_k = \psi_k(t)$ satisfy an appropriate ODE system.

Propositions 1 and 2 can also be generalized to the case of nonlinear PDEs that admit more complex functional separable solutions. We will illustrate this with a specific example.

► Example 12.8. Consider the reaction-diffusion type equation with a logarithmic nonlinearity

$$u_t = a_1 u_{xx} + a_2 u \ln u, \tag{12.1.3.16}$$

where a_1 and a_2 are free parameters. Equation (12.1.3.16) admits a functional separable solution of the form (see Example 6.3):

$$u = \exp[\psi_1(t)x^2 + \psi_2(t)x + \psi_3(t)], \tag{12.1.3.17}$$

where the function $\psi_n = \psi_n(t)$ ($n = 1, 2, 3$) satisfy the nonlinear ODE system

$$\begin{aligned} \psi_1' &= 4a_1\psi_1^2 + a_2\psi_1, \\ \psi_2' &= 4a_1\psi_1\psi_2 + a_2\psi_2, \\ \psi_3' &= a_2\psi_3 + 2a_1\psi_1 + a_1\psi_2^2. \end{aligned} \tag{12.1.3.18}$$

Now, by formally replacing the parameters a_1 and a_2 in equation (12.1.3.16) with arbitrary functions $a_1(t)$ and $a_2(t)$, we get

$$u_t = a_1(t)u_{xx} + a_2(t)u \ln u, \tag{12.1.3.19}$$

It is easy to verify that equation (12.1.3.19) also admits an exact solution of the form (12.1.3.17), where the function $\psi_n = \psi_n(t)$ ($n = 1, 2, 3$) satisfy the nonlinear ODE system (12.1.3.18) in which the parameters a_1 and a_2 are replaced with arbitrary functions $a_1(t)$ and $a_2(t)$, respectively. ◄

This illustrative example allows us to make the following generalization.

Proposition 3. *Suppose an autonomous nonlinear partial differential equation* (12.1.3.5), *dependent of free parameters* $a_1, \dots, a_p$, *has a functional separable solution of the form*

$$u = U(z), \qquad z = \sum_{k=0}^{m} \psi_k(t)x^k, \tag{12.1.3.20}$$

where $U(z)$ *is some function independent of* $a_1, \dots, a_p$. *Then, the more complex nonautonomous partial differential equation* (12.1.3.7), *obtained from* (12.1.3.5) *by formally replacing the parameters* $a_1, \dots, a_p$ *with arbitrary functions* $a_1(t), \dots, a_p(t)$, *also admits an exact solution of the form* (12.1.3.20) *with the same function* $U(z)$ *but different functions* $\psi_k(t)$.

This proposition can be proven by changing to a new variable in PDE (12.1.3.5) with the substitution $u = U(z)$. As a result, the transformed equation will have a generalized separable solution to which Proposition 1 is applicable.

Proposition 3 can be generalized further as stated below.

Proposition 4. *Suppose an autonomous nonlinear partial differential equation* (12.1.3.5), *dependent of free parameters* $a_1, \dots, a_p$, *has a functional separable solution of the form*

$$u(x,t) = U(z), \quad \text{where} \quad z = \sum_{k=1}^{m} \varphi_k(x)\psi_k(t), \tag{12.1.3.21}$$

whose coordinate functions $\varphi_k(x)$ *and outer function* $U(z)$ *are all independent of* $a_1, \dots, a_p$, *while the functions* $\psi_k = \psi_k(t)$ *satisfy the autonomous ODE system* (12.1.3.13). *Then, the more complex nonautonomous partial differential equation* (12.1.3.7), *obtained from* (12.1.3.5) *by formally replacing the parameters* $a_1, \dots, a_p$ *with arbitrary functions* $a_1(t), \dots, a_p(t)$, *admits an exact solution* (12.1.3.21), *where the coordinate functions* $\varphi_k(x)$ *and outer function* $U(z)$ *remain unchanged, while the functions* $\psi_k = \psi_k(t)$ *satisfy the nonautonomous ODE system* (12.1.3.14).

Proposition 4 can be proven by changing to a new variable, z, in equations (12.1.3.5) and (12.1.3.7) with the substitution $u = U(z)$. As a result, the transformed equations will have generalized separable solutions to which Proposition 2 is applicable.

Proposition 4 admits can be rephrased in a shorted but less informative form, as stated below.

Proposition 4a. Suppose an autonomous partial differential equation (12.1.3.5), dependent on free parameters $a_1, \dots, a_p$, has a functional separable solution (12.1.3.21), in which the coordinate function $\varphi_k(x)$ and outer function $U(z)$ are independent of $a_1, \dots, a_p$. Then, the more complicated nonautonomous equation (12.1.3.7), obtained from (12.1.3.5) by formally replacing the parameters $a_1, \dots, a_p$ with arbitrary functions $a_1(t), \dots, a_p(t)$, also admits an exact solution of the form (12.1.3.21), in which the coordinate functions $\varphi_k(x)$ and function $U(z)$ remain unchanged, while the functions $\psi_k = \psi_k(t)$ satisfy an appropriate ODE system.

12.1.4. Using the Structural Analogy of Solutions for Different Types of Equations with the Same Nonlinearity

Employing structural analogies between solutions to different types of equations that have the same nonlinearities can often help in constructing exact solutions.

The following illustrative example, borrowed from Aksenov & Polyanin (2022), demonstrates how one could obtain exact solutions of nonlinear wave-type equations using solutions of reaction-diffusion equations.

► Example 12.9. Consider the reaction-diffusion equation with a quadratic nonlinearity

$$u_t = a(uu_x)_x + bu, \tag{12.1.4.1}$$

which admits two simple exact solutions presented below.

1°. Generalized separable solution:

$$u = \psi_1(t)x^2 + \psi_2(t), \tag{12.1.4.2}$$

where $\psi_1(t) = -be^{bt}(6ae^{bt} + C_1)^{-1}$ and $\psi_2(t) = C_2 e^{bt}(6ae^{bt} + C_1)^{-1/3}$; C_1 and C_2 are arbitrary constants.

2°. Generalized separable solution:

$$u = \psi_1(t)x^2 + \psi_2(t)\sqrt{x}, \tag{12.1.4.3}$$

where $\psi_1(t) = -be^{bt}(6ae^{bt} + C_1)^{-1}$ and $\psi_2(t) = C_2 e^{bt}(6ae^{bt} + C_1)^{-5/8}$; C_1 and C_2 are arbitrary constants.

Now, let us look as the nonlinear wave-type equation with a quadratic nonlinearity

$$u_{tt} = a(uu_x)_x + bu. \tag{12.1.4.4}$$

Equations (12.1.4.1) and (12.1.4.4) differ only in the order of the derivative with respect to variable t on the left-hand side. The right-hand sides of these equations, containing derivatives with respect to x and nonlinearities, are the same. Therefore, it can be assumed that the power-law components of their solutions with respect to x will also be the same. What will change are the t-dependent functional coefficients of the different powers of x.

In other words, we seek exact solutions to the wave-type PDE (12.1.4.4) in the same form as those to the reaction-diffusion PDE (12.1.4.1). As a result, we obtain the following two exact solutions of equation (12.1.4.4).

1°. A generalized separable solution of the form (12.1.4.2), where the functions $\psi_1 = \psi_1(t)$ and $\psi_2 = \psi_2(t)$ satisfy the nonlinear ODE system

$$\begin{aligned} \psi_1'' &= 6a\psi_1^2 + b\psi_1, \\ \psi_2'' &= 2a\psi_1\psi_2 + b\psi_2. \end{aligned}$$

2°. A generalized separable solution of the form (12.1.4.3), where the functions $\psi_1 = \psi_1(t)$ and $\psi_2 = \psi_2(t)$ satisfy the nonlinear ODE system

$$\begin{aligned} \psi_1'' &= 6a\psi_1^2 + b\psi_1, \\ \psi_2'' &= \tfrac{15}{4}a\psi_1\psi_2 + b\psi_2. \end{aligned}$$

◄

This example effectively illustrates a general fact that follows from the results discussed in Section 5.6.2 (also refer to Galaktionov & Svirshchevskii, 2007, and Polyanin & Zhurov, 2022). This fact can be formulated as follows.

Proposition 5. *Suppose a nonlinear partial differential equation*

$$u_t = F[u], \tag{12.1.4.5}$$

where $F[u] \equiv F(x, u, u_x, \ldots, u_x^{(n)})$, *has a generalized separable solution*

$$u = \sum_{k=1}^{m} \varphi_k(x)\psi_k(t). \tag{12.1.4.6}$$

Then the more complex partial differential equation

$$L_1[u] = L_2[w], \qquad w = F[u], \tag{12.1.4.7}$$

where L_1 *and* L_2 *are linear differential operators with respect to* t*:*

$$L_1[u] = \sum_{i=0}^{k} a_i(t)u_t^{(i)}, \quad L_2[w] = \sum_{j=0}^{m} b_j(t)w_t^{(j)},$$

also has a generalized separable solution of the form (12.1.4.6) *with the same functions* $\varphi_k(x)$ *but different functions* $\psi_k(t)$.

12.1.5. Using One-Dimensional Equations to Construct Solutions of Multidimensional Equations

In certain instances, solutions to nonlinear partial differential equations with a single space variable, x, can help construct solutions for related equations with multiple space variables, $\mathbf{x} = (x_1, \ldots, x_n)$. The specific examples given below illustrate how to do so in practice.

► Example 12.10. Consider the nonlinear reaction-diffusion equation in a single space variable

$$u_t = [f(u)u_x]_x + \frac{a}{f(u)} + b, \tag{12.1.5.1}$$

where $f(u)$ is an arbitrary function, and a and b are arbitrary constants. It can be verified that equation (12.1.5.1) admits the following exact solution in implicit form:

$$\int f(u)\,du = at + \theta(x), \tag{12.1.5.2}$$

where $\theta(x) = -\frac{1}{2}bx^2 + C_1x + C_2$, and C_1 and C_2 are arbitrary constants.

Equation (12.1.5.1) can be generalized to multiple dimensions:

$$u_t = \operatorname{div}\,[f(u)\nabla u] + \frac{a}{f(u)} + b.$$

We look for an exact solution to this equations in implicit form by replacing x in (12.1.5.2) with **x**:

$$\int f(u)\,du = at + \theta(\mathbf{x}).$$

It follows that $f(u)u_t = a$ and $f(u)\nabla u = \nabla\theta$. As a result, we obtain Poisson's equation $\Delta\theta + b = 0$ for the function $\theta = \theta(\mathbf{x})$. ◀

▶ Example 12.11. The nonlinear reaction-diffusion equation with a single space variable and constant delay

$$u_t = au_{xx} + uf(\bar{u}/u), \quad \bar{u} = u(x, t-\tau), \quad \tau > 0, \tag{12.1.5.3}$$

admits a multiplicative separable solution

$$u = \varphi(x)\psi(t). \tag{12.1.5.4}$$

The functions $\varphi = \varphi(x)$ and $\psi = \psi(t)$ are described by the following ODE and delay ODE:

$$\varphi''_{xx} = C\varphi, \tag{12.1.5.5}$$
$$\psi'_t = aC\psi + \psi f(\bar{\psi}/\psi), \quad \bar{\psi} = \psi(t-\tau), \tag{12.1.5.6}$$

where C is an arbitrary constant. Equation (12.1.5.5) is a linear second-order ODE with constant coefficients and is easily integrated. The first-order delay ODE (12.1.5.6) has an exponential exact solution $\psi = Ae^{\lambda t}$, where A is an arbitrary constant and λ is a root of the transcendental equation $\lambda = aC + f(e^{-\lambda\tau})$.

Now, consider the reaction-diffusion equation with delay

$$u_t = a\Delta u + uf(\bar{u}/u), \quad \bar{u} = u(\mathbf{x}, t-\tau), \tag{12.1.5.7}$$

which is a spatial generalization of equation (12.1.5.3). An exact solution to equation (12.1.5.7) can be obtained by replacing x in (12.1.5.4) with **x**:

$$u = \varphi(\mathbf{x})\psi(t). \tag{12.1.5.8}$$

Substituting (12.1.5.8) into (12.1.5.7) and separating the variables, we obtain the Helmholtz equation $\Delta\varphi = C\varphi$ for $\varphi = \varphi(\mathbf{x})$ and the first-order delay ODE (12.1.5.6) for $\psi = \psi(t)$. ◀

◈ *Problems and Exercises for Section 12.1*

1. Construct an exact solution for the nonlinear wave equation with two space variables

$$u_{tt} = [a(u)u_x]_x + [b(u)u_y]_y,$$

where $a(u)$ and $b(u)$ are arbitrary twice differentiable functions, by using the linear wave equation

$$u_{tt} = au_{xx} + bu_{yy}$$

and its traveling wave solution

$$u = u(z), \quad z = \lambda t + \beta x + \gamma y.$$

Hint. Take advantage of the reasoning used in Example 12.2. In the final stage, treat a, b, λ, β, and γ as functions of u.

2. Considering that the nonlinear equation

$$u_t = [f(u)u_x]_x + \frac{a}{f(u)} + b,$$

where $f(u)$ is an arbitrary function, and a and b are arbitrary constants, admits an exact solution in implicit form

$$\int f(u)\, du = at + \theta(x),$$

where $\theta(x)$ is some function, find an exact solution for the more complex equation

$$u_t = [f(u)u_x]_x + \frac{a(t)}{f(u)} + b(x),$$

where $a(t)$ and $b(x)$ are arbitrary functions.

Hint. Take advantage of the structural analogy of solutions for these equations.

3. Prove Proposition 4.

4. The heat equation with an exponential nonlinearity

$$u_t = a(e^{\lambda u}u_x)_x$$

has an additive separable solution of the form $u = \varphi(x) + \psi(t)$, where the functions $\varphi = \varphi(x)$ and $\psi = \psi(t)$ are described by the independent ordinary differential equations

$$(e^{\lambda\varphi}\varphi'_x)'_x = C, \quad \psi'_t = aCe^{\lambda\psi},$$

where C is an arbitrary constant, which are easy to integrate.

By analogy with the solution of the heat equation, find an exact solution of the exponentially nonlinear wave equation

$$u_{tt} = a(e^{\lambda u}u_x)_x.$$

5. The reaction-diffusion equation with a quadratic nonlinearity

$$u_t = a(uu_x)_x + bu^2$$

admits two families of different exact solutions, depending on the sign of the coefficient b:

$$u = \psi_1(t)\cos(\lambda x) + \psi_2\sin(\lambda x) + \psi_3(t), \qquad \text{if } b/a = 2\lambda^2 > 0;$$

$$u = \psi_1(t)\exp(-\beta x) + \psi_2\exp(\beta x) + \psi_3(t), \qquad \text{if } b/a = -2\beta^2 < 0.$$

By analogy with the solutions of the reaction-diffusion equation, find exact solutions of the quadratically nonlinear wave equation

$$u_{tt} = a(uu_x)_x + bu^2.$$

6. The reaction-diffusion equation with a single space variable and an exponential nonlinearity

$$u_t = a(e^{\lambda u}u_x)_x + be^{\lambda u} + c$$

admits a functional separable solution of the form

$$u = \frac{1}{\lambda}\ln[\varphi(t) + \psi(t)\theta(x)],$$

where the function $\theta = \theta(x)$ satisfies the linear second-order ODE with constant coefficients $a\theta''_{xx} + b\lambda\theta = 0$, while the functions $\varphi(t)$ and $\psi(t)$ satisfy first-order ODEs.

By analogy with the solutions of the one-dimensional reaction-diffusion equation, find an exact solution of the multidimensional reaction-diffusion equation with an exponential nonlinearity

$$u_t = a\,\mathrm{div}\,(e^{\lambda u}\nabla u) + be^{\lambda u} + c.$$

7. The classical nonlinear Schrödinger equation with a cubic nonlinearity

$$iu_t + u_{xx} + k|u|^2 u = 0, \quad i^2 = -1,$$

admits functional separable solutions:

$$\begin{aligned}
&(a) \quad u = r(x)\exp[Ct + \theta(x)],\\
&(b) \quad u = r(t)\exp[a(t)x^2 + b(t)x + c(t)],
\end{aligned}$$

where C is an arbitrary constant, and $r(x)$, $\theta(x)$, $r(t)$, $a(t)$, $b(t)$, and $c(t)$ are functions determined in the process of a subsequent analysis.

Show that the more complex nonlinear Schrödinger equation (Polyanin & Zaitsev, 2012)

$$iu_t + u_{xx} + f(|u|)u = 0, \quad i^2 = -1,$$

where $f(z)$ is an arbitrary function, also allows for solutions of the forms (a) and (b). Write the ODEs that the desired functions must satisfy.

8. Show that the nonlinear Schrödinger-type equation (Polyanin & Kudryashov, 2025a)

$$iu_t + [f(|u|)u]_{xx} + g(|u|)u = 0, \quad i^2 = -1,$$

where $f(z)$ and $g(z)$ are arbitrary functions, admits solutions of the forms (a) and (b) from Exercise 7.

12.2. Constructing Solutions of Delay PDEs Using Solutions of Simpler Equations

12.2.1. Constructing Exact Solutions of Nonlinear Delay PDEs Based on the Structure of Solutions of the Corresponding PDEs without Delay

In various fields such as biology, biophysics, biochemistry, medicine, control theory, ecology, and economics, there are systems whose dynamics depends not only on their current state at a specific moment in time but also on their state at previous moments. The differential equations that describe these processes include not only the unknown function $u = u(x,t)$ but also its delayed counterpart $w = u(x, t-\tau)$, where t represents time and $\tau > 0$ is a constant delay. In some situations, the delay may depend on time, $\tau = \tau(t)$.

The presence of a delay greatly complicates the analysis of these equations. While many nonlinear partial differential equations with constant delay allow for traveling wave solutions, $u = u(x + \lambda t)$, they do not possess self-similar solutions of the form $u = t^{\beta}\varphi(xt^{\lambda})$, which often have simpler partial differential equations without delay.

Using simple concrete examples, we will demonstrate how exact solutions to certain nonlinear delay PDEs can be derived using the solution structure of simpler PDEs without delay.

► Example 12.12. Consider the nonlinear reaction-diffusion equation with a constant delay

$$u_t = a(uu_x)_x + bw, \qquad w = u(x, t - \tau). \tag{12.2.1.1}$$

Equation (12.2.1.1) is more complex than the equation without delay (12.1.4.1), but they coincide at $\tau = 0$. The presence of delay in equation (12.2.1.1) does not affect the nonlinear term that contains derivatives with respect to x. Therefore, we can assume that the power-law structure of solutions in the variable x will remain the same in both equations, with the only changing components being the functional coefficients dependent on t.

In other words, we look for exact solutions to the delay PDE (12.2.1.1) in the same form as those to the simpler non-delay PDE (12.1.4.1). As a result, we get two exact solutions of the nonlinear delay PDE (12.2.1.1), which are listed below.

1°. A generalized separable solution of the form (12.1.4.2), that is,

$$u = \psi_1(t)x^2 + \psi_2(t),$$

with the functions $\psi_1 = \psi_1(t)$ and $\psi_2 = \psi_2(t)$ described by the nonlinear ODE system with delay:

$$\begin{aligned} \psi_1' &= 6a\psi_1^2 + b\bar{\psi}_1, \qquad \bar{\psi}_1 = \psi_1(t - \tau), \\ \psi_2' &= 2a\psi_1\psi_2 + b\bar{\psi}_2, \qquad \bar{\psi}_2 = \psi_2(t - \tau). \end{aligned}$$

2°. A generalized separable solution of the form (12.1.4.3), that is,

$$u = \psi_1(t)x^2 + \psi_2(t)\sqrt{x},$$

with the functions $\psi_1 = \psi_1(t)$ and $\psi_2 = \psi_2(t)$ described by the nonlinear ODE system with delay:

$$\begin{aligned} \psi_1' &= 6a\psi_1^2 + b\bar{\psi}_1, \qquad \bar{\psi}_1 = \psi_1(t - \tau), \\ \psi_2' &= \tfrac{15}{4}a\psi_1\psi_2 + b\bar{\psi}_2, \qquad \bar{\psi}_2 = \psi_2(t - \tau). \end{aligned}$$

◄

► Example 12.13. The reaction-diffusion equation with a logarithmic nonlinearity

$$u_t = au_{xx} + u(b \ln u + c), \tag{12.2.1.2}$$

has a functional separable solution

$$u = \exp[\psi_2(t)x^2 + \psi_1(t)x + \psi_0(t)], \tag{12.2.1.3}$$

with the functions $\psi_n = \psi_n(t)$ described by the nonlinear ODE system

$$\begin{aligned} \psi_2' &= 4a\psi_2^2 + b\psi_2, \\ \psi_1' &= 4a\psi_1\psi_2 + b\psi_1, \\ \psi_0' &= a(\psi_1^2 + 2\psi_2) + b\psi_0 + c. \end{aligned}$$

Now, let us explore the more complex nonlinear reaction-diffusion equation with a constant delay

$$u_t = au_{xx} + u(b \ln w + c), \qquad w = u(x, t - \tau). \tag{12.2.1.4}$$

In the special case $\tau = 0$, the delay PDE (12.2.1.4) coincides with the simpler PDE without delay (12.2.1.2). Just as for the non-delay PDE, we look for a solution of the delay PDE (12.2.1.4) in the form (12.2.1.3). As a result, we arrive at the following system of delay ODEs for the functional coefficients $\psi_n = \psi_n(t)$:

$$\begin{aligned}
\psi_2' &= 4a\psi_2^2 + b\bar{\psi}_2, \quad \bar{\psi}_2 = \psi_2(t-\tau),\\
\psi_1' &= 4a\psi_1\psi_2 + b\bar{\psi}_1, \quad \bar{\psi}_1 = \psi_1(t-\tau),\\
\psi_0' &= a(\psi_1^2 + 2\psi_2) + b\bar{\psi}_0 + c, \quad \bar{\psi}_0 = \psi_0(t-\tau).
\end{aligned}$$
◀

12.2.2. Constructing Exact Solutions of Nonlinear Delay PDEs Using Special Solutions of Simpler PDEs without Delay

We will outline a method for constructing exact solutions of nonlinear delay partial differential equations that leverages special solutions of simpler PDEs that do not involve delay.

Let us consider nonlinear PDEs without delay in two independent variables of the form

$$\Phi(x, u, u_x, u_t, u_{xx}, u_{xt}, u_{tt}, \dots; \beta_1, \dots, \beta_m) = 0, \tag{12.2.2.1}$$

where $u = u(x,t)$ is the unknown function, and $\beta_1, \dots, \beta_m$ are free parameters.

In certain instances, exact solutions of equation (12.2.2.1) can help in constructing exact solutions for more complex nonlinear equations with delay. The following statement holds true (Polyanin & Sorokin, 2021a and 2021b).

Proposition 6. *Suppose a PDE without delay* (12.2.2.1) *admits a generalized traveling wave solution representable in implicit form*

$$F(u) = kt + \theta(x), \tag{12.2.2.2}$$

where the constant k is determined by an algebraic (or transcendental) equation

$$P(k, \beta_1, \dots, \beta_m) = 0, \tag{12.2.2.3}$$

and the function $\theta = \theta(x)$ satisfies an ordinary differential equation

$$Q(x, \theta, \theta_x', \theta_{xx}'', \dots; \beta_1, \dots, \beta_m) = 0. \tag{12.2.2.4}$$

Then, the more complex nonlinear PDE with delay, obtained from (12.2.2.1) *by formally replacing the free parameters $\beta_1, \dots, \beta_m$ with delayed functions by the rule*

$$\beta_i \implies \varphi_i(F(u) - F(w)), \quad w = u(x, t-\tau), \quad i = 1, \dots, m,$$

where $\varphi_i(z)$ are given (sufficiently arbitrary) functions, also admits exact a solution of the form (12.2.2.2). *In this case, the constant k and function $\theta = \theta(x)$ are determined by equations* (12.2.2.3) *and* (12.2.2.4) *in which*

$$\beta_i = \varphi_i(k\tau), \quad i = 1, \dots, m.$$

Remark 12.3. *Along solutions of the form (12.2.2.2), we have* $F(w) = k(t-\tau)+\theta(x) = F(u) - k\tau$, *or*

$$F(u) - F(w) = k\tau = \text{const.} \tag{12.2.2.5}$$

Remark 12.4. *In simple cases, the function* θ *can be specified in explicit form as* $\theta = \theta(x, \beta_1, \dots, \beta_m)$ *rather than determined by ODE (12.2.2.4).*

► Example 12.14. To demonstrate the practical application of Proposition 6, we will consider the linear diffusion equation without delay

$$u_t = u_{xx} + \beta, \tag{12.2.2.6}$$

where β is a free parameter.

Equation (12.2.2.6) admits a simple separable solution that can be written in explicit form as

$$u = kt + \lambda x^2 + C_1 x + C_2, \tag{12.2.2.7}$$

where C_1, C_2, and λ are arbitrary constants, and the parameter k is expressed in terms of β and λ as follows:

$$k = 2\lambda + \beta. \tag{12.2.2.8}$$

Solution (12.2.2.7) is a special case of solutions (12.2.2.2) with $F(u) = u$ and $\theta(x) = \lambda x^2 + C_1 x + C_2$. We can see that $F(u) - F(w) = u - w = k\tau$. Following Proposition 6, we replace the parameter β in the linear equation (12.2.2.6) with $\varphi(u-w)$, where $\varphi(z)$ is an arbitrary function, to obtain the nonlinear PDE with delay

$$u_t = u_{xx} + \varphi(u - w).$$

It admits the exact solution (12.2.2.7) with the constant k determined by the algebraic (transcendental) equation

$$k = 2\lambda + \varphi(k\tau),$$

which is obtained from (12.2.2.8) by setting $\beta = \varphi(k\tau)$. ◄

► Example 12.15. Consider the nonlinear reaction-diffusion type equation without delay

$$u_t = [f(u)u_x]_x + \sigma + \frac{\beta}{f(u)}, \tag{12.2.2.9}$$

which includes an arbitrary function $f(u)$ and two free parameters, σ and β. It can be verified that PDE (12.2.2.9) admits a generalized traveling wave solution in implicit form

$$\int f(u)\,du = kt - \frac{1}{2}\sigma x^2 + C_1 x + C_2, \tag{12.2.2.10}$$

where C_1 and C_2 are arbitrary constants, and the constant k is connected with the parameter β by the linear relation

$$k = \beta. \tag{12.2.2.11}$$

Solution (12.2.2.10) has the form (12.2.2.2) with $F(u) = \int f(u)\,du$.

Following Proposition 6, we replace the parameters σ and β in equation (12.2.2.9) with $\varphi(F(u)-F(w))$ and $\psi(F(u)-F(w))$, respectively, where $\varphi(z)$ and $\psi(z)$ are arbitrary functions. As a result, we arrive at a more complex nonlinear reaction-diffusion type equation with delay

$$u_t = [f(u)u_x]_x + \varphi(F(u)-F(w)) + \frac{1}{f(u)}\psi(F(u)-F(w)),$$
$$F(u) = \int f(u)\,du.$$

It depends on three arbitrary functions and admits the exact solution

$$\int f(u)\,du = kt - \frac{1}{2}\varphi(k\tau)x^2 + C_1x + C_2,$$

where the constant k is determined by the algebraic (transcendental) equations

$$k = \psi(k\tau),$$

which is obtained from (12.2.2.11) by setting $\beta = \psi(k\tau)$). ◄

Proposition 7. *Suppose the PDE without delay* (12.2.2.1) *has a special functional separable solution*

$$F(u) = e^{kt}\theta(x), \qquad (12.2.2.12)$$

with the constant k determined from the algebraic (or transcendental) equation (12.2.2.3) *and the function $\theta = \theta(x)$ satisfying the ordinary differential equation* (12.2.2.4). *Then, the more complex nonlinear delay PDE obtained from* (12.2.2.1) *by formally replacing the free parameters $\beta_1, \dots, \beta_m$ with delayed functions by the rule*

$$\beta_i \implies \varphi_i\big(F(w)/F(u)\big), \quad w = u(x, t-\tau), \quad i = 1, \dots, m, \qquad (12.2.2.13)$$

where $\varphi_i(z)$ are given (sufficiently arbitrary) functions, also admits an exact solution of the form (12.2.2.12), *with the constant k and function $\theta = \theta(x)$ determined from equations* (12.2.2.3) *and* (12.2.2.4) *in which it is necessary to set*

$$\beta_i = \varphi_i(e^{-k\tau}), \quad i = 1, \dots, m. \qquad (12.2.2.14)$$

Remark 12.5. *Along solutions of the form (12.2.2.12), the following relations hold: $F(w) = e^{k(t-\tau)}\theta(x) = e^{-k\tau}F(u)$, or*

$$F(w)/F(u) = e^{-k\tau} = \text{const.}$$

Remark 12.6. *Proposition 7 can be reduced to Proposition 6. To this end, while assuming that $F(u) > 0$, we should take the logarithm of solution (12.2.2.12) and then rename the functions: $\ln F(u) \implies F(u)$ and $\ln\theta \implies \theta$. The case $F(u) < 0$ is treated likewise. In practice, however, solutions are often directly represented in the form (12.2.2.12), making it simpler and more convenient to use.*

► Example 12.16. Consider the linear diffusion equation

$$u_t = u_{xx} + \beta u, \qquad (12.2.2.15)$$

where β is a free parameter.

Equation (12.2.2.15) admits a separable solution of the form

$$u = e^{kt}\theta(x), \tag{12.2.2.16}$$

where k is an arbitrary constant, and the function $\theta = \theta(x)$ satisfies the linear second-order ODE with constant coefficients

$$\theta''_{xx} + (\beta - k)\theta = 0. \tag{12.2.2.17}$$

Solution (12.2.2.16) is a special case of solutions (12.2.2.12) with $F(u) = u$. We find that $F(w)/F(u) = w/u = e^{-k\tau}$. Following Proposition 7, we replace the parameter β in equation (12.2.2.15) with $\varphi(w/u)$, where $\varphi(z)$ is an arbitrary function, to arrive at the nonlinear PDE with delay

$$u_t = u_{xx} + u\varphi(w/u).$$

It admits the exact solution (12.2.2.16), where k is an arbitrary constant, and the function $\theta = \theta(x)$ satisfies the linear ODE with constant coefficients

$$\theta''_{xx} + [\varphi(e^{-k\tau}) - k]\theta = 0.$$

This equation is obtained by substituting the constant $\beta = \varphi(e^{-k\tau})$ into (12.2.2.17) and is easy to integrate. ◀

The studies by Polyanin & Sorokin (2021a and 2021b) constructed exact solutions to other reaction-diffusion and wave-type PDEs with delay by using solutions of simpler PDEs without delay.

12.2.3. Method of Functional Constraints for Nonlinear Delay PDEs

We consider the following class of nonlinear reaction-diffusion equations with a constant delay

$$\begin{gathered} u_t = au_{xx} + uf(z) + wg(z) + h(z), \\ w = u(x, t - \tau), \quad z = z(u, w), \end{gathered} \tag{12.2.3.1}$$

where $f(z)$, $g(z)$, and $h(z)$ are given functions, some of which may be arbitrary. Depending on the purpose of the study, the function $z = z(u, w)$ may either be unknown or prescribed. Additionally, throughout this presentation, we will occasionally consider more general equations in which the functions f, g, and h may also explicitly depend on one independent variable, x or t.

Remark 12.7. *Instead of having the first derivative u_t on the left-hand side of equation (12.2.3.1), there can be the second derivative u_{tt} or a linear combination of time derivatives: $L[u] = \sum_{m=1}^{n} c_m u_t^{(m)}$, where c_m are arbitrary constants.*

We will seek generalized separable solutions in the form

$$u = \sum_{n=1}^{N} \varphi_n(x)\psi_n(t), \qquad (12.2.3.2)$$

where the functions $\varphi_n(x)$ and $\psi_n(t)$ are subject to determination in the process of the further analysis.

Let us now present some simple considerations that served as a constructive basis for developing the functional constraints method. If we formally replace z in equation (12.2.3.1) with a function that depends on only one independent variable, either x or t, the equation will be significantly simplified, becoming linear. In this case, we can seek its exact solutions using the standard method of separation of variables.

Considering the above, the study by Polyanin & Zhurov (2014a) proposed and successfully tested the method of functional constraints. This method is based on seeking generalized separable solutions (12.2.3.2) for equations of the form (12.2.3.1) and also for more complex related equations by utilizing one of the following two additional *functional constraints*:

$$z(u,w) = p(x), \qquad w = u(x, t-\tau); \qquad (12.2.3.3)$$

$$z(u,w) = q(t), \qquad w = u(x, t-\tau). \qquad (12.2.3.4)$$

These constraints represent difference equations in the variable t, while the space variable x serves as a free parameter. The function $z = z(u,w)$ is the argument for the functions f, g, and h that appear in equation (12.2.3.1). The functions $p(x)$ and $q(t)$ depend on x and t implicitly, being expressed through the functions $\varphi_n(x)$ and $\psi_n(t)$, respectively. These functions are to be determined in the analysis of equation (12.2.3.3) or (12.2.3.4) while taking into account (12.2.3.2). Importantly, there is no need to find the general solution to equation (12.2.3.3) or (12.2.3.4); obtaining a particular solution would be sufficient.

A solution of the difference equation (12.2.3.3) (or (12.2.3.4)) together with formula (12.2.3.2) establishes the admissible structure for the desired exact solution. The final form of the solution is determined by substituting it back into equation (12.2.3.1). The constraints (12.2.3.3) and (12.2.3.4) will be referred to as a *functional constraint of the first kind* and a *functional constraint of the second kind*, respectively.

When either relation (12.2.3.3) or (12.2.3.4) is satisfied, equation (12.2.3.1) is 'linearized', which allows further use of the separation of variables.

Remark 12.8. *For definiteness, we will refer to the simplest (degenerate) functional constraint of the form $z(u,w) = \text{const}$ as a functional constraint of the first kind (12.2.3.3).*

Remark 12.9. *In some cases, the functional equations (12.2.3.3) and (12.2.3.4) can yield more complex solutions than (12.2.3.2), functional separable solutions. For examples of such solutions, see Polyanin & Zhurov (2014b) and Polyanin et al. (2024).*

Below are a few examples of using the functional constraints method to construct generalized separable solutions for various nonlinear delay PDEs of the form (12.2.3.1).

► Example 12.17. Following Polyanin & Zhurov (2014a), we consider a nonlinear reaction-diffusion equation with a constant delay that involves one arbitrary function dependent of the ratio w/u:

$$u_t = au_{xx} + uf(w/u), \quad w = u(x, t - \tau), \tag{12.2.3.5}$$

where $f(z)$ is an arbitrary function. This equation is a special case of equation (12.2.3.1) with $g = h = 0$ and $z = w/u$.

1°. In this case, the functional constraint of the second kind (12.2.3.4) is expressed as

$$w/u = q(t), \qquad w = u(x, t - \tau). \tag{12.2.3.6}$$

The difference equation (12.2.3.6) is satisfied identically if we use a simple multiplicative separable solution

$$u = \varphi(x)\psi(t), \tag{12.2.3.7}$$

which gives $q(t) = \psi(t - \tau)/\psi(t)$. On substituting (12.2.3.7) into (12.2.3.5) and separating the variables, we obtain equations for the functions $\varphi = \varphi(x)$ and $\psi = \psi(t)$:

$$\varphi''_{xx} = k\varphi, \tag{12.2.3.8}$$
$$\psi'(t) = ak\psi(t) + \psi(t)f(\psi(t - \tau)/\psi(t)), \tag{12.2.3.9}$$

where k is an arbitrary constant.

The general solution of ODE (12.2.3.8) is given by

$$\varphi(x) = \begin{cases} C_1 \cos(\sqrt{|k|}\,x) + C_2 \sin(\sqrt{|k|}\,x) & \text{if } k < 0; \\ C_1 \exp(-\sqrt{k}\,x) + C_2 \exp(\sqrt{k}\,x) & \text{if } k > 0; \\ C_1 x + C_2 & \text{if } k = 0, \end{cases} \tag{12.2.3.10}$$

where C_1 and C_2 are arbitrary constants. The delay ODE (12.2.3.9) admits an exponential particular solution

$$\psi(t) = C_3 e^{\lambda t},$$

where C_3 is an arbitrary constant, and λ is a root of the transcendental equation

$$\lambda = ak + f(e^{-\lambda\tau}).$$

2°. In the simplest case $p(x) = p_0 = \text{const}$, the functional constraint of the first kind (12.2.3.3) for equation (12.2.3.5) is expressed as

$$w/u = p_0, \qquad w = u(x, t - \tau). \tag{12.2.3.11}$$

We look for a solution to the difference equation (12.2.3.11) with $p_0 > 0$ in the form

$$u = e^{ct}v(x, t), \qquad v(x, t) = v(x, t - \tau), \tag{12.2.3.12}$$

where c is an arbitrary constant, and $v(x, t)$ is a τ-periodic function that needs to be determined. In this case, we have $w/u = p_0 = e^{-c\tau}$.

On substituting (12.2.3.12) into equation (12.2.3.5), we obtain a linear problem for determining the function v:

$$v_t = av_{xx} + bv, \qquad v(x,t) = v(x, t-\tau), \tag{12.2.3.13}$$

where $b = f(e^{-c\tau}) - c$.

For convenience, let $v = V_1(x,t;b)$ denote the general solution of problem (12.2.3.13). It is expressed as

$$\begin{aligned} V_1(x,t;b) &= \sum_{n=0}^{\infty} \exp(-\lambda_n x)\big[A_n \cos(\beta_n t - \gamma_n x) + B_n \sin(\beta_n t - \gamma_n x)\big] \\ &\quad + \sum_{n=1}^{\infty} \exp(\lambda_n x)\big[C_n \cos(\beta_n t + \gamma_n x) + D_n \sin(\beta_n t + \gamma_n x)\big], \end{aligned} \tag{12.2.3.14}$$

$$\beta_n = \frac{2\pi n}{\tau}, \quad \lambda_n = \left(\frac{\sqrt{b^2+\beta_n^2} - b}{2a}\right)^{1/2}, \quad \gamma_n = \left(\frac{\sqrt{b^2+\beta_n^2} + b}{2a}\right)^{1/2}, \tag{12.2.3.15}$$

where A_n, B_n, C_n, and D_n are arbitrary constants such that the series (12.2.3.14)–(12.2.3.15) and its derivatives $(V_1)_t$, $(V_1)_{xx}$ are convergent. For example, convergence can be ensured by setting $A_n = B_n = C_n = D_n = 0$ for $n > N$, where N is an arbitrary positive integer.

Let us single out the following special cases of solutions to problem (12.2.3.13):

(i) τ-periodic in time and decaying as $x \to \infty$; these are given by formulas (12.2.3.14)–(12.2.3.15) with $A_0 = B_0 = 0$ and $C_n = D_n = 0$, $n = 1, 2, \ldots$;

(ii) τ-periodic in time and bounded as $x \to \infty$; these are given by formulas (12.2.3.14)–(12.2.3.15) with $C_n = D_n = 0$, $n = 1, 2, \ldots$;

(iii) stationary solution; it is defined by formulas (12.2.3.14)–(12.2.3.15) with $A_n = B_n = C_n = D_n = 0$, $n = 1, 2, \ldots$

We have thus described the following exact solutions of equation (12.2.3.5):

$$u = e^{ct} V_1(x,t;b), \quad b = f(e^{-c\tau}) - c, \tag{12.2.3.16}$$

where c is an arbitrary constant, and the τ-periodic function $V_1(x,t;b)$ is defined by formulas (12.2.3.14)–(12.2.3.15).

3°. We look for a solution to the difference equation (12.2.3.11) with $p_0 < 0$ in the form

$$u = e^{ct} v(x,t), \quad v(x,t) = -v(x,t-\tau), \tag{12.2.3.17}$$

where c is an arbitrary constant, and $v(x,t)$ is a τ-*antiperiodic function*. In this case, $w/u = p_0 = -e^{-c\tau}$.

On substituting (12.2.3.17) into equation (12.2.3.5), we obtain a linear problem for the function v:

$$v_t = av_{xx} + bv, \quad v(x,t) = -v(x,t-\tau), \tag{12.2.3.18}$$

where $b = f(-e^{-c\tau}) - c$.

The general solution of problem (12.2.3.18), which we denote $v = V_2(x,t;b)$ for convenience, is given by

$$V_2(x,t;b) = \sum_{n=1}^{\infty} \exp(-\lambda_n x)\big[A_n \cos(\beta_n t - \gamma_n x) + B_n \sin(\beta_n t - \gamma_n x)\big]$$
$$+ \sum_{n=1}^{\infty} \exp(\lambda_n x)\big[C_n \cos(\beta_n t + \gamma_n x) + D_n \sin(\beta_n t + \gamma_n x)\big], \tag{12.2.3.19}$$

$$\beta_n = \frac{\pi(2n-1)}{\tau}, \quad \lambda_n = \left(\frac{\sqrt{b^2+\beta_n^2} - b}{2a}\right)^{1/2}, \quad \gamma_n = \left(\frac{\sqrt{b^2+\beta_n^2} + b}{2a}\right)^{1/2}, \tag{12.2.3.20}$$

where A_n, B_n, C_n, and D_n are arbitrary constants such that the series (12.2.3.19)–(12.2.3.20) and the respective derivatives $(V_2)_t$ and $(V_2)_{xx}$ are convergent. The τ-antiperiodic solutions of problem (12.2.3.18) decaying as $x \to \infty$ are given by (12.2.3.19)–(12.2.3.20) with $C_n = D_n = 0$, $n = 1, 2, \ldots$

We have thus identified the following exact solutions of equation (12.2.3.5):

$$u = e^{ct} V_2(x,t;b), \quad b = f(-e^{-c\tau}) - c, \tag{12.2.3.21}$$

where c is an arbitrary constant, and the τ-antiperiodic function $V_2(x,t;b)$ is defined by formulas (12.2.3.19)–(12.2.3.20).

Remark 12.10. *Solutions (12.2.3.14)–(12.2.3.15) and (12.2.3.19)–(12.2.3.20) are very similar in appearance. However, the first sum in the first solution starts with $n = 0$, while that in the second solution starts with $n = 1$; the values of β_n also differ.* ◀

► Example 12.18. Consider a nonlinear reaction-diffusion equation with a constant delay that includes one arbitrary function dependent on the difference $u - w$:

$$u_t = a u_{xx} + bu + f(u-w), \quad w = u(x, t-\tau), \tag{12.2.3.22}$$

where $f(z)$ is an arbitrary function. This equation is a special case of equation (12.2.3.1) with $f(z) = b$, $g = 0$, and $z = u - w$; for convenience, the function h has been renamed f.

1°. In this case, the functional constraint of the second kind (12.2.3.4) is expressed as

$$u - w = q(t), \qquad w = u(x, t-\tau). \tag{12.2.3.23}$$

The linear difference equation (12.2.3.23) can be satisfied if we use an additive separable solution

$$u = \varphi(x) + \psi(t), \tag{12.2.3.24}$$

which gives $q(t) = \psi(t) - \psi(t-\tau)$. On substituting (12.2.3.24) into (12.2.3.22) and separating the variables, we obtain equations for determining the functions $\varphi = \varphi(x)$ and $\psi = \psi(t)$:

$$a\varphi''_{xx} + b\varphi = k, \tag{12.2.3.25}$$
$$\psi'_t(t) = b\psi(t) + k + f(\psi(t) - \psi(t-\tau)), \tag{12.2.3.26}$$

where k is an arbitrary constant.

The general solution of equation (12.2.3.25) when $b \neq 0$ and $k = 0$ is given by

$$\varphi(x) = \begin{cases} C_1 \cos\big(\sqrt{b/a}\,x\big) + C_2 \sin\big(\sqrt{b/a}\,x\big) & \text{if } b > 0, \\ C_1 \exp\big(-\sqrt{|b/a|}x\big) + C_2 \exp\big(\sqrt{|b/a|}x\big) & \text{if } b < 0, \\ \dfrac{k}{2a}\,x^2 + C_1 x + C_2 & \text{if } b = 0, \end{cases} \tag{12.2.3.27}$$

where C_1 and C_2 are arbitrary constants. For $b > 0$, solution (12.2.3.24), (12.2.3.27) is periodic in the space variable x.

2°. The functional constraint of the first kind (12.2.3.3) for equation (12.2.3.22) is expressed as

$$u - w = p(x), \qquad w = u(x, t - \tau). \tag{12.2.3.28}$$

The difference equation (12.2.3.28) can be satisfied if we use, for example, the generalized separable solution

$$u = t\varphi(x) + \psi(x), \tag{12.2.3.29}$$

which gives $p(x) = \tau\varphi(x)$.

Substituting (12.2.3.29) into (12.2.3.22) yields ordinary differential equations for the functions $\varphi = \varphi(x)$ and $\psi = \psi(x)$:

$$a\varphi''_{xx} + b\varphi = 0, \tag{12.2.3.30}$$

$$a\psi''_{xx} + b\psi + f(\tau\varphi) - \varphi = 0. \tag{12.2.3.31}$$

Equation (12.2.3.30) coincides with (12.2.3.25) when $k = 0$, and its solution is given by (12.2.3.27). The linear nonhomogeneous ODE with constant coefficients (12.2.3.31) is easy to integrate.

We can obtain more exact solutions of equations (12.2.3.22), involving any number of arbitrary parameters, with the help of the above solutions (12.2.3.24) and (12.2.3.29) and the following proposition on nonlinear superposition of solutions.

Proposition 8. *Suppose $u_0(x, t)$ is some solution of the nonlinear delay equation (12.2.3.22), and the function $v = V_1(x, t; b)$ is any τ-periodic solution of the linear heat equation equations with a source (12.2.3.13). Then the sum*

$$u = u_0(x, t) + V_1(x, t; b) \tag{12.2.3.32}$$

is also a solution of equation (12.2.3.22). The general form of the function $V_1(x, t; b)$ is defined by formulas (12.2.3.14)–(12.2.3.15).

This proposition is proven by a direct substitution of formula (12.2.3.32) into the original delay equation (12.2.3.22) with the help of the equation for v (12.2.3.13).

Remark 12.11. *Examples of particular solutions $u_0(x, t)$ for the nonlinear equation (12.2.3.22) that can be utilized in formula (12.2.3.32) include a spatially homogeneous solution $u_0(t)$, a stationary solution $u_0(x)$, and a traveling wave solution $u_0 = u_0(\alpha x + \beta t)$.* ◀

Remark 12.12. *The studies by Polyanin & Zhurov (2014a, 2014b) (see also Polyanin et al., 2024) obtained many exact solutions for other nonlinear reaction-diffusion and wave-type delay PDEs using the functional constraints method.*

. Principle of Structural Analogy of Solutions for Nonlinear PDEs with Proportional Arguments

onsider nonlinear partial differential equations that involve proportional ar-. These equations include the unknown function $u = u(x,t)$ as well as a that stretches one or more independent variables, such as $u(px,t)$, $u(x,qt)$, $qt)$. The constants p and q serve as scaling parameters; for partial differen-tions with proportional delays, these parameters are bounded by the intervals 1 and $0 < q < 1$.

will outline a fairly general method for constructing exact solutions to non-DEs that involve proportional arguments, based on a principle proposed by & Sorokin (2021c); see also Polyanin et al. (2024).

ciple of structural analogy of solutions. The structure of exact solutions for ith proportional arguments

$$, u, u_x, u_t, u_{xx}, u_{xt}, u_{tt}, \dots, w, w_x, w_t, w_{xx}, w_{xt}, w_{tt}, \dots) = 0, \qquad (12.2.4.1)$$
$$(px, qt),$$

(but not always) determined by the structure of solutions to simpler partial ial equations with usual arguments:

$$t, u, u_x, u_t, u_{xx}, u_{xt}, u_{tt}, \dots, u, u_x, u_t, u_{xx}, u_{xt}, u_{tt}, \dots) = 0. \qquad (12.2.4.2)$$

tion (12.2.4.2) does not include the unknown function with proportional ts; it is obtained from (12.2.4.1) by formally replacing w with u.

application of the structural analogy principle for a second-order PDE with onal arguments, which is solved for u_t and explicitly independent of the x and t, is illustrated in Fig. 12.1.

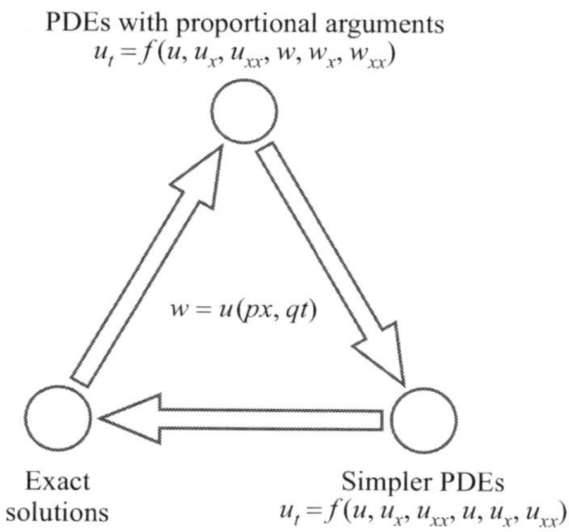

.1. A schematic of using the structural analogy principle to construct exact solutions ear PDEs with proportional arguments.

Let us illustrate the application of the structural analogy principle using examples of three PDEs with proportional arguments that have different types of solutions.

► Example 12.19. Consider the reaction-diffusion equation with proportional arguments (Aksenov & Polyanin, 2021):

$$u_t = au_{xx} + bw^k, \qquad w = u(px, qt), \tag{12.2.4.3}$$

which contains a power-law nonlinearity.

Following the structural analogy principle, we set $w = u$ in equation (12.2.4.3) to obtain a simpler nonlinear heat (diffusion) equation with usual arguments:

$$u_t = au_{xx} + bu^k. \tag{12.2.4.4}$$

This equation admits a self-similar solution (Dorodnitsyn, 1982):

$$u(x,t) = t^{\frac{1}{1-k}} U(z), \qquad z = xt^{-1/2}, \qquad k \neq 1. \tag{12.2.4.5}$$

Using the structural analogy principle, we look for a solution to the nonlinear equation with proportional arguments (12.2.4.3) in the same form (12.2.4.5). As a result, we obtain the following nonlinear ODE with proportional argument for the function $U = U(z)$:

$$\begin{aligned} &aU''_{zz} + \frac{1}{2} zU'_z - \frac{1}{1-k} U + bq^{\frac{k}{1-k}} W^k = 0, \\ &W = U(sz), \quad s = pq^{-1/2}. \end{aligned}$$

◄

► Example 12.20. Now, let us explore the reaction-diffusion equation with proportional arguments and exponential nonlinearity (Aksenov & Polyanin, 2021):

$$u_t = au_{xx} + be^{\lambda w}, \qquad w = u(px, qt). \tag{12.2.4.6}$$

Following the structural analogy principle, we set $w = u$ in equation (12.2.4.6) to obtain a simpler nonlinear heat (diffusion) equation with usual arguments and an exponential source:

$$u_t = au_{xx} + be^{\lambda u}.$$

This equation admits the following invariant solution (Dorodnitsyn, 1982):

$$u(x,t) = U(z) - \frac{1}{\lambda} \ln t, \qquad z = xt^{-1/2}, \qquad \lambda \neq 0. \tag{12.2.4.7}$$

Using the structural analogy principle, we look for a solution to the nonlinear PDE with proportional arguments (12.2.4.6) in the form (12.2.4.7). As a result, we obtain a nonlinear ODE with a proportional argument for the function $U = U(z)$:

$$\begin{aligned} &aU''_{zz} + \frac{1}{2} zU'_z + \frac{1}{\lambda} + \frac{b}{q} e^{\lambda W} = 0, \\ &W = U(sz), \quad s = pq^{-1/2}. \end{aligned}$$

◄

► Example 12.21. Consider the following PDE with proportional arguments and a logarithmic nonlinearity

$$u_t = au_{xx} + u(b \ln u + c \ln w + d), \qquad w = u(px, qt). \tag{12.2.4.8}$$

By setting $w = u$ in equation (12.2.4.8), we arrive at a simpler nonlinear equation with usual arguments

$$u_t = au_{xx} + u[(b + c) \ln u + d],$$

which admits a multiplicative separable solution

$$u(x, t) = \varphi(x)\psi(t). \tag{12.2.4.9}$$

Using the structural analogy principle, we look for a solution to the nonlinear PDE with proportional arguments (12.2.4.8) in the form (12.2.4.9). After separating the variables, we obtain two nonlinear ODEs with proportional arguments for the functions $\varphi = \varphi(x)$ and $\psi = \psi(t)$:

$$a\varphi''_{xx} + \varphi(b \ln \varphi + c \ln \bar{\varphi}) = K\varphi, \quad \bar{\varphi} = \varphi(px);$$
$$\psi'_t = \psi(b \ln \psi + c \ln \bar{\psi}) + (d + K)\psi, \quad \bar{\psi} = \psi(qt),$$

where K is an arbitrary constant. ◄

Remark 12.13. *The studies by Polyanin & Sorokin (2021a and 2023a) and Polyanin et al. (2024) obtained many exact solutions for other nonlinear reaction-diffusion and wave-type PDEs with proportional arguments and general PDEs with variable delay using the principle of structural analogy of solutions.*

12.2.5. Constructing Solutions to PDEs with Variable Time Delay Using Solutions of Simpler PDEs with Constant Delay

The general topic we are interested in can be formulated as follows. Let $u = u(x, t)$ denote an exact solution of a nonlinear partial differential equation that involves a constant delay τ. The question we want to explore is whether we can draw any conclusions about the exact solutions of a more complex PDE formed by substituting the constant delay τ with an arbitrary variable time delay $\tau(t)$.

It will be shown that, in many cases, constructing exact solutions of partial differential equations with an arbitrary variable time delay $\tau(t)$ can be achieved by utilizing solutions of related, simpler PDEs that have a constant delay τ. We will illustrate this through a sequence and logic of non-rigorous reasoning using two specific examples. After analyzing these examples, we will formulate general conclusions as rigorous propositions.

► Example 12.22. Consider the reaction-diffusion equation with a quadratic nonlinearity and a constant delay

$$u_t = a(uu_x)_x + b\bar{u}, \qquad \bar{u} = u(x, t - \tau), \tag{12.2.5.1}$$

which admits a generalized separable solution

$$u = \psi_1(t)x^2 + \psi_2(t). \tag{12.2.5.2}$$

The functions $\psi_1 = \psi_1(t)$ and $\psi_2 = \psi_2(t)$ satisfy the nonlinear ODE system with delay

$$\begin{aligned}\psi_1' &= 6a\psi_1^2 + b\bar{\psi}_1, \quad \bar{\psi}_1 = \psi_1(t-\tau),\\ \psi_2' &= 2a\psi_1\psi_2 + b\bar{\psi}_2, \quad \bar{\psi}_2 = \psi_2(t-\tau).\end{aligned} \tag{12.2.5.3}$$

Replacing the constant $\tau > 0$ in (12.2.5.1) with an arbitrary positive function of time $\tau(t)$, we obtain a more complex nonlinear reaction-diffusion type equation with a quadratic nonlinearity and variable delay:

$$u_t = a(uu_x)_x + b\bar{u}, \quad \bar{u} = u(x, t - \tau(t)). \tag{12.2.5.4}$$

The nonlinear terms involving derivatives with respect to x on the right-hand sides of equations (12.2.5.1) and (12.2.5.4) are identical. Therefore, it is logical to assume that the power-law structure of the solutions in the spatial variable x will also be the same for both equations. The only differences will lie in the t-dependent functional coefficients of the different powers of x.

In other words, we seek exact solutions to the reaction-diffusion PDE (12.2.5.4) in the same form (12.2.5.2) as those to the original PDE (12.2.5.1). As a result, we obtain an ODE system (12.2.5.3) for the functions $\psi_1 = \psi_1(t)$ and $\psi_2 = \psi_2(t)$, in which τ must be replaced with $\tau(t)$. ◀

The results presented in this example allow for the following generalization (Polyanin, 2024).

Proposition 9. *Suppose the nonlinear PDE with a constant delay*

$$u_t = F\big(x, u, u_x, \dots, u_x^{(n)}; \bar{u}, \bar{u}_x, \dots, \bar{u}_x^{(p)}\big), \quad \bar{u} = u(x, t-\tau), \tag{12.2.5.5}$$

where $n > p$, has a generalized separable solution

$$u = \sum_{k=1}^{m} \varphi_k(x)\psi_k(t), \tag{12.2.5.6}$$

whose coordinate functions $\varphi_k(x)$ are all independent of τ, and the function $\psi_k = \psi_k(t)$ are described by an ODE system with a constant delay

$$\psi_k' = f_k(\psi_1, \dots, \psi_m; \bar{\psi}_1, \dots, \bar{\psi}_m), \quad \bar{\psi}_k = \psi_k(t-\tau), \quad k = 1, \dots, m. \tag{12.2.5.7}$$

Then the more complex PDE with a variable delay

$$u_t = F\big(x, u, u_x, \dots, u_x^{(n)}; \bar{u}, \bar{u}_x, \dots, \bar{u}_x^{(p)}\big), \quad \bar{u} = u\big(x, t-\tau(t)\big), \tag{12.2.5.8}$$

obtained from equation (12.2.5.5) by formally replacing the constant τ with an arbitrary function of time $\tau(t)$, also admits an exact solution of the form (12.2.5.6), where the coordinate functions $\varphi_k(x)$ remain the same, while the functions $\psi_k = \psi_k(t)$ are described by the ODE system with a variable delay

$$\psi_k' = f_k(\psi_1, \dots, \psi_m; \bar{\psi}_1, \dots, \bar{\psi}_m), \quad \bar{\psi}_k = \psi_k\big(t-\tau(t)\big), \quad k = 1, \dots, m. \tag{12.2.5.9}$$

Remark 12.14. *Instead of having the first derivative u_t on the left-hand side of PDE (12.2.5.5), there can be any linear combination of time derivatives: $L[u]=\sum_{i=1}^{s} a_i u_t^{(i)}$, where the coefficients a_i may be arbitrary functions of time. In this case, the first derivatives ψ_k' on the left-hand sides of ODEs (12.2.5.7) and (12.2.5.9) in Proposition 9 must be replaced with $L[\psi_k]$.*

Proposition 9 can be generalized to nonlinear delay PDEs that allow for more complex functional separable solutions. We will illustrate this with a specific example.

► Example 12.23. Consider the reaction-diffusion type equation with a double logarithmic nonlinearity and a constant delay

$$u_t = u_{xx} + u(a \ln^2 u + b \ln \bar{u}), \qquad \bar{u} = u(x, t - \tau), \tag{12.2.5.10}$$

which admits, for $a > 0$, a functional separable solution of the form (Polyanin & Zhurov, 2013):

$$u = \exp\bigl[\psi_1(t)\varphi(x) + \psi_2(t)\bigr], \quad \varphi(x) = C_1 \cos\bigl(\sqrt{a}\,x\bigr) + C_2 \sin\bigl(\sqrt{a}\,x\bigr), \tag{12.2.5.11}$$

where C_1 and C_2 are arbitrary constants, and the functions $\psi_1 = \psi_1(t)$ and $\psi_2 = \psi_2(t)$ satisfy the nonlinear ODE system with a constant delay

$$\begin{aligned} \psi_1' &= 2a\psi_1\psi_2 - a\psi_1 + b\bar{\psi}_1, \quad \bar{\psi}_1 = \psi_1(t-\tau), \\ \psi_2' &= a(C_1^2 + C_2^2)\psi_1^2 + a\psi_2^2 + b\bar{\psi}_2, \quad \bar{\psi}_2 = \psi_2(t-\tau). \end{aligned} \tag{12.2.5.12}$$

Formally replacing the constant τ in equation (12.2.5.10) with an arbitrary function $\tau(t)$ yields the functional PDE:

$$u_t = u_{xx} + u(a \ln^2 u + b \ln \bar{u}), \qquad \bar{u} = u\bigl(x, t - \tau(t)\bigr). \tag{12.2.5.13}$$

It is easy to verify that equation (12.2.5.13) also admits an exact solution of the form (12.2.5.11), with the functions $\psi_1 = \psi_1(t)$ and $\psi_2 = \psi_2(t)$ satisfying the nonlinear ODE system (12.2.5.12) in which the constant τ must be replaced with the arbitrary function $\tau(t)$. ◄

The above example allows us to make a generalization, which we will express as a proposition (Polyanin, 2024).

Proposition 10. *Suppose the nonlinear PDE with a constant delay (12.2.5.5) has a functional separable solution of the form*

$$u(x,t) = U(z), \quad \text{where} \quad z = \sum_{k=1}^{m} \varphi_k(x)\psi_k(t),$$

whose coordinate functions $\varphi_k(x)$ and the outer function $U(z)$ are all independent of the parameter τ, and the functions $\psi_k = \psi_k(t)$ are described by the ODE system with a constant delay (12.2.5.7). Then the more complex PDE with a variable delay (12.2.5.8), obtained from equation (12.2.5.5) by formally replacing the constant τ with an arbitrary function of time $\tau(t)$, admits an exact solution of the form (12.2.5.6), where the coordinate functions $\varphi_k(x)$ and outer function $U(z)$ remain the same, while the functions $\psi_k = \psi_k(t)$ are described by the system of ODEs with a variable delay (12.2.5.9).

Table 12.1 presents several nonlinear reaction-diffusion equations with a variable delay that allow for separable exact solutions. The results are borrowed from Polyanin et al. (2024). It is assumed that the time-dependent delay $\tau = \tau(t)$ is an arbitrary positive continuously differentiable function. In particular, in cases involving proportional delay in the equations, one should set $\tau = (1-p)t$, which means that $t - \tau = pt$.

12.2.6. Constructing Solutions to PDEs with Anisotropic Time Delay Using Solutions of Simpler PDEs with Constant Delay

In this section, by saying a PDE with a spatially anisotropic time delay (briefly, a PDE with anisotropic time delay), we mean a functional differential equation that, in addition to the unknown function $u(x,t)$, also contains a delayed function of the form $u(x, t-\tau(x))$, where $\tau(x)$ is a given positive function, as well as partial derivatives of u with respect to x and t. PDEs with an anisotropic time delays can model delay systems in anisotropic and inhomogeneous media, in which the signal propagation speed depends on the chosen direction or varies at different points of the medium. For example, in medicine, this can be differences in the rate of transmission of a nerve impulse in "healthy" and "sick" tissues of the human body.

By exact solutions of a nonlinear PDE with anisotropic time delay, we understand solutions that can be expressed in terms of elementary functions, functions included in the PDE under consideration, indefinite integrals, or/and solutions of some ODEs without delay (systems of such ODEs).

Currently, there are no regular methods for constructing exact solutions to nonlinear PDEs with spatially anisotropic time delays or related functional PDEs. It should be noted that any PDEs with anisotropic time delays admit neither traveling-wave solutions nor self-similar solutions.

Further, to construct exact solutions to such equations, we use a simple but very important proposition, namely, exact solutions of certain PDEs with constant delay can serve as the basis for constructing solutions to more complex PDEs with variable anisotropic delay (a somewhat similar approach was used in Section 12.2.4 to construct solutions to nonlinear PDEs with constant or proportional delay using solutions of simpler non-delay PDEs and was referred to as the principle of structural analogy of solutions). This approach to constructing exact solutions, using the principle 'from simple to complex' was employed in the work of Polyanin & Sorokin (2023b) and is illustrated below with specific examples.

In all the equations discussed in this section, it is assumed that the anisotropic time delay $\tau(x)$ is an arbitrary positive continuous function.

► Example 12.24. Consider the nonlinear reaction-diffusion equation with delay

$$u_t = au_{xx} + f(u-w), \qquad w = u(x, t-\tau), \tag{12.2.6.1}$$

where $f(z)$ is an arbitrary function, and the time delay $\tau > 0$ is a constant.

Table 12.1. Reaction-diffusion type equations with a general variable delay, allowing for generalized or functional separable solutions. Notation: $\bar{u}=u(x,t-\tau(t))$, $\bar{\psi}=\psi(t-\tau(t))$, $f(z)$ and $g(z)$ are arbitrary functions, C_1, C_2, and C_3 are arbitrary constants.

Original equation	Form of exact solution	Determining equations
$u_t=au_{xx}+uf(\bar{u}/u)$	$u=[C_1\cos(\beta x)+C_2\sin(\beta x)]\psi(t)$; $u=[C_1\exp(-\beta x)+C_2\exp(\beta x)]\psi(t)$; $u=(C_1x+C_2)\psi(t)$	$\psi'_t=-a\beta^2\psi+\psi f(\bar{\psi}/\psi)$; $\psi'_t=a\beta^2\psi+\psi f(\bar{\psi}/\psi)$; $\psi'_t=\psi f(\bar{\psi}/\psi)$
$u_t=au_{xx}+bu\ln u$ $+uf(\bar{u}/u)$	$u=\varphi(x)\psi(t)$	$a\varphi''_{xx}=C_1\varphi-b\varphi\ln\varphi$, $\psi'_t=C_1\psi+b\psi\ln\psi$ $+\psi f(\bar{\psi}/\psi)$
$u_t=au_{xx}+f(u-\bar{u})$	$u=C_2x^2+C_1x+\psi(t)$	$\psi'_t=2C_2a+f(\psi-\bar{\psi})$
$u_t=au_{xx}+bu$ $+f(u-\bar{u})$	$u=C_1\cos(\lambda x)+C_2\sin(\lambda x)+\psi(t)$, where $\lambda=\sqrt{b/a}$ (for $ab>0$); $u=C_1\exp(-\lambda x)+C_2\exp(\lambda x)+\psi(t)$, where $\lambda=\sqrt{-b/a}$ (for $ab<0$)	$\psi'_t=b\psi+f(\psi-\bar{\psi})$; $\psi'_t=b\psi+f(\psi-\bar{\psi})$
$u_t=a(u^ku_x)_x$ $+uf(\bar{u}/u)$	$u=\varphi(x)\psi(t)$	$a(\varphi^k\varphi'_x)'_x=C_1\varphi$, $\psi'_t=C_1\psi^{k+1}+\psi f(\bar{\psi}/\psi)$
$u_t=a(u^ku_x)_x+bu^{k+1}$ $+uf(\bar{u}/u)$	$u=[C_1\cos(\beta x)+C_2\sin(\beta x)]^{\frac{1}{k+1}}\psi(t)$, where $\beta=\sqrt{b(k+1)/a}$, $b(k+1)>0$; $u=(C_1e^{-\beta x}+C_2e^{\beta x})^{\frac{1}{k+1}}\psi(t)$, where $\beta=\sqrt{-b(k+1)/a}$, $b(k+1)<0$; $u=C_1\exp\big(-\frac{b}{2a}x^2+C_2x\big)\psi(t)$ for $k=-1$; $u=\varphi(x)\psi(t)$ (generalizes previous solutions)	$\psi'_t=\psi f(\bar{\psi}/\psi)$; $\psi'_t=\psi f(\bar{\psi}/\psi)$; $\psi'_t=\psi f(\bar{\psi}/\psi)$; $a(\varphi^k\varphi'_x)'_x+b\varphi^{k+1}=C_1\varphi$, $\psi'_t=C_1\psi^{k+1}+\psi f(\bar{\psi}/\psi)$
$u_t=a(e^{\lambda u}u_x)_x$ $+f(u-\bar{u})$	$u=\dfrac{1}{\lambda}\ln(C_1\lambda x^2+C_2x+C_3)+\psi(t)$	$\psi'_t=2aC_1e^{\lambda\psi}+f(\psi-\bar{\psi})$
$u_t=a(e^{\lambda u}u_x)_x+be^{\lambda u}$ $+f(u-\bar{u})$	$u=\frac{1}{\lambda}\ln[C_1\cos(\beta x)+C_2\sin(\beta x)]+\psi(t)$, where $\beta=\sqrt{b\lambda/a}$, $b\lambda>0$; $u=\frac{1}{\lambda}\ln(C_1e^{-\beta x}+C_2e^{\beta x})+\psi(t)$, where $\beta=\sqrt{-b\lambda/a}$, $b\lambda<0$; $u=\varphi(x)+\psi(t)$ (generalizes previous solutions)	$\psi'_t=f(\psi-\bar{\psi})$; $\psi'_t=f(\psi-\bar{\psi})$; $a(e^{\lambda\varphi}\varphi'_x)'_x+be^{\lambda\varphi}=C_1$, $\psi'_t=C_1e^{\lambda\psi}+f(\psi-\bar{\psi})$
$u_t=[(a\ln u+b)u_x]_x$ $-cu\ln u+uf(\bar{u}/u)$	$u=\exp(\pm\lambda x)\psi(t)$, $\lambda=\sqrt{c/a}$	$\psi'_t=\lambda^2(a+b)\psi+\psi f(\bar{\psi}/\psi)$
$u_t=a[f'(u)u_x]_x+b$ $+\frac{1}{f'(u)}g(f(u)-f(\bar{u}))$	$f(u)=\psi(t)-\dfrac{b}{2a}x^2+C_1x+C_2$	$\psi'_t=g(\psi-\bar{\psi})$
$u_t=a[uf'(u)u_x]_x$ $+\frac{1}{f'(u)}[bf(u)+cf(\bar{u})]$	$f(u)=\varphi(t)x+\psi(t)$	$\varphi'_t=b\varphi+c\bar{\varphi}$, $\psi'_t=b\psi+c\bar{\psi}+a\varphi^2$

It is easy to verify that PDE (12.2.6.1) admits an additive separable solution of the form

$$u=bt+\psi(x), \tag{12.2.6.2}$$

where b is an arbitrary constant and the function $\psi = \psi(x)$ satisfies the second-order linear ODE:

$$a\psi''_{xx} + f(b\tau) - b = 0.$$

Its general solution is written as

$$\psi = \frac{1}{2a}[b - f(b\tau)]x^2 + C_1 x + C_2,$$

where C_1 and C_2 are arbitrary constants.

A more complex reaction-diffusion equation with anisotropic delay obtained from (12.2.6.1) by formally replacing the constant τ with an arbitrary function $\tau(x)$ (using the principle "from simple to complex"), namely:

$$u_t = au_{xx} + f(u - w), \quad w = u\big(x, t - \tau(x)\big), \tag{12.2.6.3}$$

admits an exact solution of the same form (12.2.6.2) with the function $\psi = \psi(x)$ satisfying the second-order linear ODE:

$$a\psi''_{xx} + f(b\tau(x))^k - b = 0,$$

which is easily integrated.

Thus, in this case, we can say that a solution of the simpler PDE (12.2.6.1) generates a solution for the more complex PDE (12.2.6.3). ◄

For clarity, the process for constructing exact solutions to PDEs with anisotropic delay based on exact solutions from simpler PDEs with constant delay, which was used in Example 12.24 and will be used further, is schematically depicted in Figure 12.2.

In certain cases, functional PDEs with anisotropic time delays and their exact solutions can be derived by analogy from simpler PDEs with constant delays, which allow for solutions of the form

$$F(u) = \varphi(x)t + \psi(x) \qquad \text{or} \qquad F(u) = e^{\varphi(x)t}\psi(x), \tag{12.2.6.4}$$

and satisfy the following two conditions:

(i) The original simpler PDEs with a constant delay (such equations are further called *generating equations*) are reduced to ODEs dependent only on the spatial argument x;

(ii) The parameters and functional coefficients of the original PDEs with constant delay do not explicitly depend on τ.

In the simplest cases, we have $F(u) = u$ and $\varphi(x) = \text{const}$ in solutions (12.2.6.4).

► Example 12.25. It is easy to verify that the PDE with a constant delay (12.2.6.1) has a more complex generalized separable solution than (12.2.6.2). This solution is a special case of the first solution in (12.2.6.4) for $F(u) = u$:

$$u = \varphi(x)t + \psi(x), \tag{12.2.6.5}$$

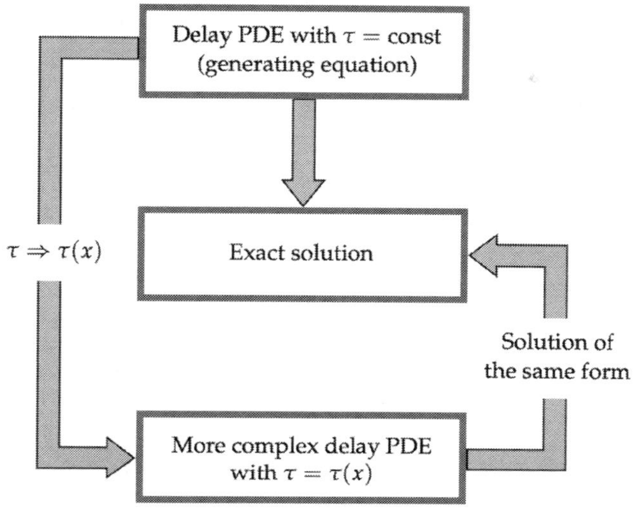

Figure 12.2. A schematic of using exact solutions for PDEs with a constant delay to construct exact solutions to PDEs with anisotropic delay.

where the functions $\varphi = \varphi(x)$ and $\psi = \psi(x)$ are described by the linear system of ODEs:

$$\begin{gathered} \varphi''_{xx} = 0, \\ a\psi''_{xx} + f(\tau\varphi) - \varphi = 0. \end{gathered} \tag{12.2.6.6}$$

Integrating equations (12.2.6.6) in order, we get

$$\begin{aligned} \varphi(x) &= C_1 x + C_2, \\ \psi(x) &= \frac{1}{a}\int_0^x (x-\xi)[\varphi(\xi) - f(\tau\varphi(\xi))]\,d\xi + C_3 x + C_4, \end{aligned} \tag{12.2.6.7}$$

where C_1, C_2, C_3, and C_4 are arbitrary constants.

A more complex reaction-diffusion equation with anisotropic delay (12.2.6.3) obtained from (12.2.6.1) by formally replacing the constant τ with an arbitrary function $\tau(x)$ admits an exact solution of the same form (12.2.6.5) with the functions $\varphi = \varphi(x)$ and $\psi = \psi(x)$ satisfying the second-order linear ODEs (12.2.6.6) and defined by formulas (12.2.6.7) with $\tau = \tau(x)$. ◄

► Example 12.26. Let us now consider the generating reaction-diffusion equation with constant delay

$$u_t = au_{xx} + uf(w/u), \qquad w = u(x, t-\tau), \tag{12.2.6.8}$$

where $f(z)$ is an arbitrary function.

Equation (12.2.6.8) has a multiplicative separable solution that represents a special case of the second solution in (12.2.6.4) for $F(u) = u$ and $\varphi(x) = \text{const}$:

$$u = e^{\lambda t}\psi(x), \tag{12.2.6.9}$$

where λ is an arbitrary constant and the function $\psi = \psi(x)$ satisfies the second-order constant-coefficient linear ODE:

$$a\psi''_{xx} + [f(e^{-\lambda\tau}) - \lambda]\psi = 0. \tag{12.2.6.10}$$

The more complex reaction-diffusion PDE with anisotropic time delay

$$u_t = au_{xx} + uf(w/u), \qquad w = u(x, t - \tau(x)),$$

which is obtained from (12.2.6.8) by formally replacing the constant τ with an arbitrary function $\tau(x)$, admits an exact solution of the same form (12.2.6.9), where the function $\psi = \psi(x)$ satisfies the second-order linear ODEs (12.2.6.10) with $\tau = \tau(x)$. ◄

The study by Polyanin & Sorokin (2023b) obtained exact solutions of other reaction-diffusion and wave-type PDEs with anisotropic time delays by using solutions to simpler PDEs with constant delay.

12.2.7. Generating Equations Method

Here, we describe a method to construct exact solutions for some nonlinear systems of delay PDEs using simpler solutions from individual (isolated) delay PDEs.

Consider two different independent nonlinear PDEs with a constant delay $\tau > 0$:

$$F\big(u, \bar{u}, u_x, u_t, u_{xx}, u_{xt}, u_{tt}, f(z_1)\big) = 0, \qquad \bar{u} = u(x, t - \tau), \qquad z_1 = z_1(u, \bar{u}); \tag{12.2.7.1}$$

$$G\big(v, \bar{v}, v_x, v_t, v_{xx}, v_{xt}, v_{tt}, g(z_2)\big) = 0, \qquad \bar{v} = v(x, t - \tau), \qquad z_2 = z_2(v, \bar{v}), \tag{12.2.7.2}$$

which include predefined functions F, G, z_1, and z_2 and arbitrary functions of a single argument $f(z_1)$ and $g(z_2)$.

We assume that equations (12.2.7.1) and (12.2.7.2) have generalized separable solutions of the form

$$u = \sum_{n=1}^{N_1} \varphi_{1n}(x)\psi_{1n}(t), \quad v = \sum_{n=1}^{N_2} \varphi_{2n}(x)\psi_{2n}(t), \tag{12.2.7.3}$$

both satisfying similar functional constraints. In other words, we assume that either of the following two situations is possible:

$$\begin{aligned} &z_1(u, \bar{u}) = p_1(x), \quad z_2(v, \bar{v}) = p_2(x) \quad \text{(functional constraints of the first kind)};\\ &z_1(u, \bar{u}) = q_1(t), \quad z_2(v, \bar{v}) = q_2(t) \quad \text{(functional constraints of the second kind)}. \end{aligned} \tag{12.2.7.4}$$

The method for generating equations is based on the following principle (Polyanin & Zhurov, 2015):

The principle of constructing delay PDE systems and their exact solutions. Suppose the isolated nonlinear delay PDEs (12.2.7.1) and (12.2.7.2) admit generalized separable solutions (12.2.7.3), each satisfying two functional constraints of the same form (12.2.7.4). Then, the more complex nonlinear system consisting of two coupled nonlinear delay PDEs

$$F\big(u, \bar{u}, u_x, u_t, u_{xx}, u_{xt}, u_{tt}, f(z_1, z_2)\big) = 0, \qquad \bar{u} = u(x, t-\tau), \qquad z_1 = z_1(u, \bar{u}); \tag{12.2.7.5}$$

$$G\big(v, \bar{v}, v_x, v_t, v_{xx}, v_{xt}, v_{tt}, g(z_1, z_2)\big) = 0, \qquad \bar{v} = v(x, t-\tau), \qquad z_2 = z_2(v, \bar{v}), \tag{12.2.7.6}$$

where $f(z_1, z_2)$ and $g(z_1, z_2)$ are arbitrary functions of two arguments, admits exact solution of the form (12.2.7.3).

In what follows, we will refer to the independent equations (12.2.7.1) and (12.2.7.2) as *generating equations*.

Remark 12.15. *Equations (12.2.7.1) and (12.2.7.2) and system (12.2.7.5)–(12.2.7.6) can additionally depend on x and t explicitly and include higher derivatives of the unknowns with respect to these variables.*

Remark 12.16. *In simple cases, the generating PDEs (12.2.7.1) and (12.2.7.2) can be the same up to elementary redesignations of the determining parameters and arbitrary functions.*

► Example 12.27. We will use the reaction-diffusion type equation with a constant delay

$$u_t = a(u^k u_x)_x + u f(\bar{u}/u), \qquad \bar{u} = u(x, t-\tau), \tag{12.2.7.7}$$

where $z = \bar{u}/u$, as a single generating equation. It admits a multiplicative separable solution $u = \varphi(x)\psi(t)$, which satisfies a functional constraint of the second kind (12.2.7.4).

Following Remark 12.16, we will get another generating equation by renaming the determining parameters and arbitrary function in (12.2.7.7):

$$v_t = b(v^m v_x)_x + v g(\bar{v}/v), \qquad \bar{v} = v(x, t-\tau). \tag{12.2.7.8}$$

By construction, this equation also has a multiplicative separable solution satisfying a functional constraint of the second kind (12.2.7.4).

By applying the generating equations method and using PDEs (12.2.7.7) and (12.2.7.8), we arrive at the following nonlinear system of coupled reaction-diffusion equations with delay

$$\begin{aligned} u_t &= a(u^k u_x)_x + u f(\bar{u}/u, \bar{v}/v), \quad && \bar{u} = u(x, t-\tau), \\ v_t &= b(v^m v_x)_x + v g(\bar{u}/u, \bar{v}/v), \quad && \bar{v} = v(x, t-\tau). \end{aligned} \tag{12.2.7.9}$$

The PDE system (12.2.7.9) inherits the solution structure of the generating equations; that is, it admits multiplicative separable solutions

$$u = \varphi_1(x)\psi_1(t), \qquad v = \varphi_2(x)\psi_2(t). \tag{12.2.7.10}$$

On substituting (12.2.7.10) into (12.2.7.9) and performing simple rearrangements, we obtain the following system of two independent ODEs for the functions $\varphi_1 = \varphi_1(t)$ and $\varphi_2 = \varphi_2(t)$, and two coupled delay ODEs for the functions $\psi_1 = \psi_1(t)$ and $\psi_2 = \psi_2(t)$:

$$\begin{aligned} &(\varphi_1^k \varphi_1')' = C_1 \varphi_1, \\ &(\varphi_2^m \varphi_2')' = C_2 \varphi_2, \\ &\psi_1' = aC_1 \psi_1^{k+1} + \psi_1 f(\bar{\psi}_1/\psi_1, \bar{\psi}_2/\psi_2), \quad \bar{\psi}_1 = \psi_1(t-\tau), \\ &\psi_2' = bC_2 \psi_2^{m+1} + \psi_2 g(\bar{\psi}_1/\psi_1, \bar{\psi}_2/\psi_2), \quad \bar{\psi}_2 = \psi_2(t-\tau), \end{aligned} \tag{12.2.7.11}$$

where C_1 and C_2 are arbitrary constants.

The general solutions to the first two autonomous ODEs in (12.2.7.11) can be written in implicit form. For $k, m \neq 0$ and $k, m \neq -2$, these equations have the following particular solutions:

$$\varphi_1 = \left[\frac{C_1 k^2 x^2}{2(k+2)}\right]^{1/k}, \qquad \varphi_2 = \left[\frac{C_2 m^2 x^2}{2(m+2)}\right]^{1/m}.$$

For $k = m = 0$, the first two ODEs in (12.2.7.11) are linear, and their general solutions are expressed explicitly in terms of exponential or trigonometric functions. For zero values of k and m, the system of coupled ODEs with delay that consists of the two last equations in (12.2.7.11) admits the exponential particular solution

$$\psi_1 = Ae^{\beta t}, \qquad \psi_2 = Be^{\gamma t},$$

where A and B are arbitrary constants, and the constants β and γ are determined by the system of algebraic (transcendental) equations

$$\begin{aligned} \beta &= aC_1 + f(e^{-\tau\beta}, e^{-\tau\gamma}), \\ \gamma &= bC_2 + g(e^{-\tau\beta}, e^{-\tau\gamma}). \end{aligned}$$

◄

► Example 12.28. Just as before, we take the reaction-diffusion equation with a constant delay (12.2.7.7) as the first generating PDE; it has a multiplicative separable solution $u = \varphi_1(x)\psi_1(t)$. As the second generating PDE, we use another reaction-diffusion equation with a constant delay

$$v_t = b(e^{\lambda v} v_x)_x + g(v - \bar{v}), \qquad \bar{v} = v(x, t-\tau), \tag{12.2.7.12}$$

which differs significantly from the first PDE and admits an additive separable solution of the form $v = \varphi_2(x) + \psi_2(t)$.

We can use the generating equations methods because both equations (12.2.7.7) and (12.2.7.12) have separable solutions that satisfy two different functional constraints of the second kind (12.2.7.4). As a result, we obtain the following nonlinear system of coupled reaction-diffusion equations with delay

$$\begin{aligned} u_t &= a(u^k u_x)_x + uf(\bar{u}/u, v - \bar{v}), \quad \bar{u} = u(x, t-\tau), \\ v_t &= b(e^{\lambda v} v_x)_x + g(\bar{u}/u, v - \bar{v}), \quad \bar{v} = v(x, t-\tau). \end{aligned} \tag{12.2.7.13}$$

It admits a mixed multiplicative/additive separable solution

$$u = \varphi_1(x)\psi_1(t), \quad v = \varphi_2(x) + \psi_2(t). \tag{12.2.7.14}$$

Substituting (12.2.7.14) into (12.2.7.13) and performing simple rearrangements, we arrive at the following system consisting of two independent ODEs for the functions $\varphi_1 = \varphi_1(x)$ and $\varphi_2 = \varphi_2(x)$ and two coupled delay ODEs for the functions $\psi_1 = \psi_1(t)$ and $\psi_2 = \psi_2(t)$:

$$\begin{aligned}
&(\varphi_1^k \varphi_1')' = C_1\varphi_1,\\
&(e^{\lambda\varphi_2}\varphi_2')' = C_2,\\
&\psi_1' = aC_1\psi_1^{k+1} + \psi_1 f(\bar\psi_1/\psi_1, \psi_2 - \bar\psi_2), \quad \bar\psi_1 = \psi_1(t-\tau),\\
&\psi_2' = bC_2 e^{\lambda\psi_2} + g(\bar\psi_1/\psi_1, \psi_2 - \bar\psi_2), \qquad \bar\psi_2 = \psi_2(t-\tau),
\end{aligned} \tag{12.2.7.15}$$

where C_1 and C_2 are arbitrary constants.

For solutions of the first ODE in (12.2.7.15), see Example 12.27. The general solution of the second ODE in (12.2.7.15) is written as

$$\varphi_2(x) = \begin{cases} \frac{1}{\lambda}\ln\bigl(\frac{1}{2}C_2\lambda x^2 + C_3 x + C_4\bigr) & \text{for } \lambda \neq 0,\\ \frac{1}{2}C_2 x^2 + C_3 x + C_4 & \text{for } \lambda = 0,\end{cases}$$

where C_3 and C_4 are arbitrary constants.

For $k = \lambda = 0$, the system of coupled delay ODEs that consists of the last two equations in (12.2.7.15) admits a particular solution of the form

$$\psi_1 = Ae^{\beta t}, \quad \psi_2 = \gamma t + B,$$

where A and B are arbitrary constants, and the constants β and γ are determined from the system of algebraic (transcendental) equations

$$\begin{aligned}
\beta &= aC_1 + f(e^{-\tau\beta}, \tau\gamma),\\
\gamma &= bC_2 + g(e^{-\tau\beta}, \tau\gamma).
\end{aligned}$$

◄

For other examples of applying the method of generating equations, refer to Polyanin & Zhurov (2015) and Polyanin et al. (2024).

➾ *Problems and Exercises for Section 12.2*

1. The quadratically nonlinear wave equation

$$u_{tt} = a(uu_x)_x + bu$$

coincides with equation (12.1.4.4) and has two generalized separable solutions, (12.1.4.2) and (12.1.4.3).

Find exact solutions to the more complex nonlinear wave equation with a constant delay

$$u_{tt} = a(uu_x)_x + bw, \quad w = u(x, t-\tau)$$

taking advantage of the solution structure (by analogy) of the above simpler wave PDE without delay.

2. Consider the nonlinear reaction-diffusion type equation without delay

$$u_t = [a(x)f(u)u_x]_x + \sigma + \frac{\beta}{f(u)}, \tag{12.2.7.16}$$

which generalizes PDE (12.2.2.9) and involves two arbitrary functions, $a(x)$ and $f(u)$, and two free parameters, σ and β. It can be verified that equation (12.2.7.16) admits a generalized traveling wave solution in implicit form

$$\int f(u)\,du = kt - \sigma \int \frac{x\,dx}{a(x)} + C_1 \int \frac{dx}{a(x)} + C_2, \tag{12.2.7.17}$$

where C_1 and C_2 are arbitrary constants, and the constant k is linearly related to the parameter β: $k = \beta$.

Using Proposition 6 and relation (12.2.7.17), which is a special case of solution (12.2.2.2), find an exact solution for the nonlinear reaction-diffusion type equation with delay

$$u_t = [a(x)f(u)u_x]_x + \varphi(F(u) - F(w)) + \frac{1}{f(u)}\psi(F(u) - F(w)),$$
$$F(u) = \int f(u)\,du,$$

where $\varphi(z)$ and $\psi(z)$ are arbitrary functions.

3. Consider the reaction-diffusion equation with a quadratic nonlinearity

$$u_t = (uu_x)_x + au + bu^2, \tag{12.2.7.18}$$

where a and b are free parameters.

For $b > 0$, equation (12.2.7.18) admits a separable solution in explicit form

$$u = e^{kt}\sqrt{|C_1 \cos(\sigma x) + C_2 \sin(\sigma x)|}, \qquad \sigma = \sqrt{2b}, \tag{12.2.7.19}$$

which is a special case of solution (12.2.2.12); the parameter k satisfies the linear relation $k = a$.

Using Proposition 7, with $\beta_1 = a$ and $\beta_2 = b$, and solution (12.2.7.19), find an exact solution for the nonlinear PDE with delay

$$u_t = (uu_x)_x + u\varphi_1(w/u) + u^2\varphi_2(w/u), \qquad w = u(x, t - \tau),$$

where $\varphi_1(z)$ and $\varphi_2(z)$ are arbitrary functions.

4. The classical nonlinear Schrödinger equation with a cubic nonlinearity

$$iu_t + u_{xx} + k|u|^2 u = 0, \qquad i^2 = -1,$$

admits functional separable solutions of the form

$$\text{(a)} \quad u = r(x)\exp[Ct + \theta(x)],$$
$$\text{(b)} \quad u = r(t)\exp[a(t)x^2 + b(t)x + c(t)],$$

where C is an arbitrary constant, and the functions $r(x)$, $\theta(x)$, $r(t)$, $a(t)$, $b(t)$, and $c(t)$ are determined in the course of a further analysis.

Show that the more complex nonlinear Schrödinger equation with delay (Polyanin & Kudryashov, 2024):

$$iu_t + u_{xx} + f(|w|)u = 0, \qquad w = u(x, t - \tau),$$

where $f(z)$ is an arbitrary function, also allows for solutions of the form (a) and (b). Write the ODEs and delay ODEs that the desired functions satisfy.

Prove that exact solutions with the same structure are admitted by a more general delay PDE of the form (Polyanin & Kudryashov, 2025b):

$$iu_t + u_{xx} + F(u, |w|)u = 0, \qquad w = u(x, t - \tau),$$

where $F(z_1, z_2)$ is an arbitrary function of two arguments.

5. Taking advantage of the principle of structural analogy of solutions, construct an exact solution to the nonlinear reaction-diffusion equation with proportional arguments

$$u_t = au_{xx} + bu^m w^k, \qquad w = u(px, qt).$$

Take into account that the simpler PDE with $p = q = 1$ admits a self-similar solution of the form

$$u(x, t) = t^{\frac{1}{1-m-k}} U(z), \qquad z = xt^{-1/2}.$$

6. The reaction-diffusion equation with a logarithmic nonlinearity

$$u_t = au_{xx} + u(b \ln u + c), \tag{12.2.7.20}$$

admits an exact solution of the form

$$u(x, t) = \exp[\psi_2(t)x^2 + \psi_1(t)x + \psi_0(t)].$$

Taking advantage of the principle of structural analogy of solutions, construct an exact solution to the more complex reaction-diffusion equation with a proportional argument

$$u_t = au_{xx} + u(b_1 \ln u + b_2 \ln w + c), \qquad w = u(x, qt).$$

7. The Klein–Gordon type equation with a power-law nonlinearity

$$u_{tt} = au_{xx} + bu^k$$

admits, for $k \neq 1$, a self-similar solution of the form

$$u(x, t) = t^{\frac{2}{1-k}} U(z), \qquad z = x/t.$$

Taking advantage of the principle of structural analogy of solutions, construct an exact solution to the more complex PDE with proportional arguments (Polyanin & Sorokin, 2023a):

$$u_{tt} = au_{xx} + bw^k, \qquad w = u(px, qt).$$

8. The Klein–Gordon type equation with an exponential nonlinearity

$$u_{tt} = au_{xx} + be^{\mu u}$$

admits an exact solution of the form

$$u(x, t) = U(z) - \frac{2}{\mu} \ln t, \qquad z = x/t.$$

Taking advantage of the principle of structural analogy of solutions, construct an exact solution to the more complex PDE with proportional arguments (Polyanin & Sorokin, 2023a):

$$u_{tt} = au_{xx} + be^{\mu u+\lambda w}, \qquad w = u(px, qt).$$

9. Find an exact solution to the nonlinear reaction-diffusion PDE with anisotropic time delay

$$u_t = a(e^{\lambda u}u_x)_x + b + e^{\lambda u}f(u - w), \qquad w = u(x, t - \tau(x)),$$

where $f(z)$ and $\tau(x)$ are arbitrary functions (the solution is given in Polyanin & Sorokin, 2023b).

10. Find an exact solution to the nonlinear wave-type PDE with anisotropic time delay

$$u_{tt} = au_{xx} + uf(w/u), \qquad w = u(x, t - \tau(x)),$$

where $f(z)$ and $\tau(x)$ are arbitrary functions (the solution is given in Polyanin & Sorokin, 2023b).

11. Consider two different reaction-diffusion type equations with a constant delay:

$$\begin{aligned} u_t &= a_1u_{xx} + bu + f(u - \bar{u}), && \bar{u} = u(x, t - \tau); \\ v_t &= a_2v_{xx} + vg(\bar{v}/v), && \bar{v} = v(x, t - \tau), \end{aligned} \qquad (12.2.7.21)$$

where $f(z_1)$ and $g(z_2)$ are arbitrary functions. The first equation has an additive separable solution, and the second equation admits a multiplicative separable solution.

By applying the generating equations method to equation (12.2.7.21), find a solution to the system of reaction-diffusion type equations with a constant delay

$$\begin{aligned} u_t &= a_1u_{xx} + bu + f(u - \bar{u}, \bar{v}/v), \\ v_t &= a_2v_{xx} + vg(u - \bar{u}, \bar{v}/v), \end{aligned}$$

where $f(z_1, z_2)$ and $g(z_1, z_2)$ are arbitrary functions of two arguments.

References for Chapter 12

Aksenov A.V. and Polyanin A.D. Methods for constructing complex solutions of nonlinear PDEs using simpler solutions. *Mathematics*, 2021, Vol. 9, No. 4, p. 345.

Aksenov A.V. and Polyanin A.D. Review of methods for constructing exact solutions of equations of mathematical physics based on simpler solutions. *Theor. Math. Phys.*, 2022, Vol. 211, No. 2, pp. 567–594.

Dorodnitsyn V.A. On invariant solutions of the equation of non-linear heat conduction with a source. *USSR Comput. Math. Math. Phys.*, 1982, Vol. 22, pp. 115–122.

Galaktionov V.A. and Svirshchevskii S.R. *Exact Solutions and Invariant Subspaces of Nonlinear Partial Differential Equations in Mechanics and Physics*. Chapman & Hall/CRC Press, Boca Raton, 2007.

Polyanin A.D. Construction of exact solutions in implicit form for PDEs: New functional separable solutions of non-linear reaction-diffusion equations with variable coefficients. *Int. J. Non-Linear Mech.*, 2019a, Vol. 111, pp. 95–105.

Polyanin A.D. Construction of functional separable solutions in implicit form for non-linear Klein–Gordon type equations with variable coefficients. *Int. J. Non-Linear Mech.*, 2019b, Vol. 114, pp. 29–40.

Polyanin A.D. Principle of structural analogy of solutions and its applications to nonlinear PDEs and delay PDEs. *J. Math. Sci.*, 2024; doi:10.1007/s10958-024-07234-6.

Polyanin A.D. and Kudryashov N.A. Nonlinear Schrödinger equations with delay: Closed-form and generalized separable solutions. *Contemporary Mathematics*, 2024, Vol. 5, No. 4, p. 5783.

Polyanin A.D. and Kudryashov N.A. Closed-form solutions of the nonlinear Schrödinger equation with arbitrary dispersion and potential. *Chaos Solitons and Fractals*, 2025a, Vol. 191, p. 115822.

Polyanin A.D. and Kudryashov N.A. Exact solutions and reductions of nonlinear Schrödinger equations with delay. *J. Comp. Appl. Math.*, 2025b, Vol. 462, p. 116477.

Polyanin A.D. and Sorokin V.G. A method for constructing exact solutions of nonlinear delay PDEs. *J. Math. Analysis & Applications*, 2021a, Vol. 494, p. 124619.

Polyanin A.D. and Sorokin V.G. Construction of exact solutions to nonlinear PDEs with delay using solutions of simpler PDEs without delay. *Commun. Nonlinear Science & Numer. Simulation*, 2021b, Vol. 95, p. 105634.

Polyanin A.D. and Sorokin V.G. Nonlinear pantograph-type diffusion PDEs: Exact solutions and the principle of analogy. *Mathematics*, 2021c, Vol. 9, No. 5, p. 511.

Polyanin A.D. and Sorokin V.G. Reductions and exact solutions of nonlinear wave-type PDEs with proportional and more complex delays. *Mathematics*, 2023a, Vol. 11, No. 3, p. 516.

Polyanin A.D. and Sorokin V.G. Exact solutions of reaction-diffusion PDEs with anisotropic time delay. *Mathematics*, 2023b, Vol. 11, No. 14, p. 3111.

Polyanin A.D., Sorokin V.G., and Zhurov A.I. *Delay Ordinary and Partial Differential Equations*. CRC Press, Boca Raton–London, 2024.

Polyanin A.D. and Zhurov A.I. Exact solutions of linear and non-linear differential-difference heat and diffusion equations with finite relaxation time. *Int. J. Non-Linear Mech.*, 2013, Vol. 54, pp. 115–126.

Polyanin A.D. and Zhurov A.I. Functional constraints method for constructing exact solutions to delay reaction-diffusion equations and more complex nonlinear equations. *Comm. Nonlinear Sci. Numer. Simul.*, 2014a, Vol. 19, No. 3, pp. 417–430.

Polyanin A.D. and Zhurov A.I. The functional constraints method: Application to non-linear delay reaction-diffusion equations with varying transfer coefficients. *Int. J. Non-Linear Mech.*, 2014b, Vol. 67, pp. 267–277.

Polyanin A.D. and Zhurov A.I. The generating equations method: Constructing exact solutions to delay reaction-diffusion systems and other non-linear coupled delay PDEs. *Int. J. Non-Linear Mechanics*, 2015, Vol. 71, pp. 104–115.

Polyanin A.D. and Zhurov A.I. *Separation of Variables and Exact Solutions to Nonlinear PDEs*. CRC Press, Boca Raton–London, 2022.

Zaitsev V.F. and Polyanin A.D. *Handbook of Partial Differential Equations: Exact Solutions* (in Russian). Mezhdunarodnaya Programma Obrazovaniya, Moscow, 1996.

Index

D

F

G

H

I

J

K

L

M

N

S

T

U

V

W